Springer Texts in Statistics

Series Editors:
G. Casella
S. Fienberg
I. Olkin

For further volumes:
http://www.springer.com/series/417

Anirban DasGupta

Probability for Statistics and Machine Learning

Fundamentals and Advanced Topics

Anirban DasGupta
Department of Statistics
Purdue University
150 N. University Street
West Lafayette, IN 47907, USA
dasgupta@stat.purdue.edu

Mathematica® is a registered trademark of Wolfram Research, Inc.

ISBN 978-1-4419-9633-6   e-ISBN 978-1-4419-9634-3
DOI 10.1007/978-1-4419-9634-3
Springer New York Dordrecht Heidelberg London

Library of Congress Control Number: 2011924777

© Springer Science+Business Media, LLC 2011
All rights reserved. This work may not be translated or copied in whole or in part without the written permission of the publisher (Springer Science+Business Media, LLC, 233 Spring Street, New York, NY 10013, USA), except for brief excerpts in connection with reviews or scholarly analysis. Use in connection with any form of information storage and retrieval, electronic adaptation, computer software, or by similar or dissimilar methodology now known or hereafter developed is forbidden.
The use in this publication of trade names, trademarks, service marks, and similar terms, even if they are not identified as such, is not to be taken as an expression of opinion as to whether or not they are subject to proprietary rights.

Printed on acid-free paper

Springer is part of Springer Science+Business Media (www.springer.com)

*To Persi Diaconis, Peter Hall, Ashok Maitra,
and my mother, with affection*

Preface

This is the companion second volume to my undergraduate text *Fundamentals of Probability: A First Course*. The purpose of my writing this book is to give graduate students, instructors, and researchers in statistics, mathematics, and computer science a lucidly written unique text at the confluence of probability, advanced stochastic processes, statistics, and key tools for machine learning. Numerous topics in probability and stochastic processes of current importance in statistics and machine learning that are widely scattered in the literature in many different specialized books are all brought together under one fold in this book. This is done with an extensive bibliography for each topic, and numerous worked-out examples and exercises. Probability, with all its models, techniques, and its poignant beauty, is an incredibly powerful tool for anyone who deals with data or randomness. The content and the style of this book reflect that philosophy; I emphasize lucidity, a wide background, and the far-reaching applicability of probability in science.

The book starts with a self-contained and fairly complete review of basic probability, and then traverses its way through the classics, to advanced modern topics and tools, including a substantial amount of statistics itself. Because of its nearly encyclopaedic coverage, it can serve as a graduate text for a year-long probability sequence, or for focused short courses on selected topics, for self-study, and as a nearly unique reference for research in statistics, probability, and computer science. It provides an extensive treatment of most of the standard topics in a graduate probability sequence, and integrates them with the basic theory and many examples of several core statistical topics, as well as with some tools of major importance in machine learning. This is done with unusually detailed bibliographies for the reader who wants to dig deeper into a particular topic, and with a huge repertoire of worked-out examples and exercises. The total number of worked-out examples in this book is 423, and the total number of exercises is 808. An instructor can rotate the exercises between semesters, and use them for setting exams, and a student can use them for additional exam preparation and self-study. I believe that the book is unique in its range, unification, bibliographic detail, and its collection of problems and examples.

Topics in core probability, such as distribution theory, asymptotics, Markov chains, martingales, Poisson processes, random walks, and Brownian motion are covered in the first 14 chapters. In these chapters, a reader will also find basic coverage of such core statistical topics as confidence intervals, likelihood functions, maximum likelihood estimates, posterior densities, sufficiency, hypothesis testing, variance stabilizing transformations, and extreme value theory, all illustrated with many examples. In Chapters 15, 16, and 17, I treat three major topics of great application potential, empirical processes and VC theory, probability metrics, and large deviations. Chapters 18, 19, and 20 are specifically directed to the statistics and machine-learning community, and cover simulation, Markov chain Monte Carlo, the exponential family, bootstrap, the EM algorithm, and kernels.

The book does not make formal use of measure theory. I do not intend to minimize the role of measure theory in a rigorous study of probability. However, I believe that a large amount of probability can be taught, understood, enjoyed, and applied without needing formal use of measure theory. We do it around the world every day. At the same time, some theorems cannot be proved without at least a mention of some measure theory terminology. Even some definitions require a mention of some measure theory notions. I include some unavoidable mention of measure-theoretic terms and results, such as the strong law of large numbers and its proof, the dominated convergence theorem, monotone convergence, Lebesgue measure, and a few others, but only in the advanced chapters in the book.

Following the table of contents, I have suggested some possible courses with different themes using this book. I have also marked the nonroutine and harder exercises in each chapter with an asterisk. Likewise, some specialized sections with reference value have also been marked with an asterisk. Generally, the exercises and the examples come with a caption, so that the reader will immediately know the content of an exercise or an example. The end of the proof of a theorem has been marked by a □ sign.

My deepest gratitude and appreciation are due to Peter Hall. I am lucky that the style and substance of this book are significantly molded by Peter's influence. Out of habit, I sent him the drafts of nearly every chapter as I was finishing them. It didn't matter where exactly he was, I always received his input and gentle suggestions for improvement. I have found Peter to be a concerned and warm friend, teacher, mentor, and guardian, and for this, I am extremely grateful.

Mouli Banerjee, Rabi Bhattacharya, Burgess Davis, Stewart Ethier, Arthur Frazho, Evarist Giné, T. Krishnan, S. N. Lahiri, Wei-Liem Loh, Hyun-Sook Oh, B. V. Rao, Yosi Rinott, Wen-Chi Tsai, Frederi Viens, and Larry Wasserman graciously went over various parts of this book. I am deeply indebted to each of them. Larry Wasserman, in particular, suggested the chapters on empirical processes, VC theory, concentration inequalities, the exponential family, and Markov chain Monte Carlo. The Springer series editors, Peter Bickel, George Casella, Steve Fienberg, and Ingram Olkin have consistently supported my efforts, and I am so very thankful to them. Springer's incoming executive editor Marc Strauss saw through the final production of this book extremely efficiently, and I have much enjoyed working with him. I appreciated Marc's gentility and his thoroughly professional handling of the transition of the production of this book to his oversight. Valerie Greco did an astonishing job of copyediting the book. The presentation, display, and the grammar of the book are substantially better because of the incredible care and thoughtfulness that she put into correcting my numerous errors. The staff at SPi Technologies, Chennai, India did an astounding and marvelous job of producing this book. Six anonymous reviewers gave extremely gracious and constructive comments, and their input has helped me in various dimensions to make this a better book. Doug Crabill is the greatest computer systems administrator, and with an infectious pleasantness has bailed me out of my stupidity far too many times. I also want to mention my fond memories and deep-rooted feelings for the Indian Statistical Institute, where I had all of my college education. It was just a wonderful place for research, education, and friendships. Nearly everything that I know is due to my years at the Indian Statistical Institute, and for this I am thankful.

This is the third time that I have written a book in contract with John Kimmel. John is much more than a nearly unique person in the publishing world. To me, John epitomizes sensitivity and professionalism, a singular combination. I have now known John for almost six years, and it is very very difficult not to appreciate and admire him a whole lot for his warmth, style, and passion for the subjects of statistics and probability. Ironically, the day that this book entered production, the news came that John was leaving Springer. I will remember John's contribution to my professional growth with enormous respect and appreciation.

West Lafayette, Indiana   Anirban DasGupta

Contents

Suggested Courses with Different Themes xix

1 Review of Univariate Probability 1
    1.1   Experiments and Sample Spaces 1
    1.2   Conditional Probability and Independence.......................... 5
    1.3   Integer-Valued and Discrete Random Variables 8
        1.3.1   CDF and Independence....................................... 9
        1.3.2   Expectation and Moments.................................. 13
    1.4   Inequalities 19
    1.5   Generating and Moment-Generating Functions 22
    1.6   * Applications of Generating Functions to a Pattern Problem 26
    1.7   Standard Discrete Distributions 28
    1.8   Poisson Approximation to Binomial 34
    1.9   Continuous Random Variables.. 36
    1.10 Functions of a Continuous Random Variable 42
        1.10.1 Expectation and Moments.................................. 45
        1.10.2 Moments and the Tail of a CDF 49
    1.11 Moment-Generating Function and Fundamental Inequalities........ 51
        1.11.1 * Inversion of an MGF and Post's Formula 53
    1.12 Some Special Continuous Distributions.............................. 54
    1.13 Normal Distribution and Confidence Interval for a Mean 61
    1.14 Stein's Lemma.. 66
    1.15 * Chernoff's Variance Inequality 68
    1.16 * Various Characterizations of Normal Distributions 69
    1.17 Normal Approximations and Central Limit Theorem 71
        1.17.1 Binomial Confidence Interval 74
        1.17.2 Error of the CLT 76
    1.18 Normal Approximation to Poisson and Gamma 79
        1.18.1 Confidence Intervals.. 80
    1.19 * Convergence of Densities and Edgeworth Expansions............. 82
    References... 92

2 Multivariate Discrete Distributions ... 95
- 2.1 Bivariate Joint Distributions and Expectations of Functions ... 95
- 2.2 Conditional Distributions and Conditional Expectations ... 100
  - 2.2.1 Examples on Conditional Distributions and Expectations ... 101
- 2.3 Using Conditioning to Evaluate Mean and Variance ... 104
- 2.4 Covariance and Correlation ... 107
- 2.5 Multivariate Case ... 111
  - 2.5.1 Joint MGF ... 112
  - 2.5.2 Multinomial Distribution ... 114
- 2.6 * The Poissonization Technique ... 116

3 Multidimensional Densities ... 123
- 3.1 Joint Density Function and Its Role ... 123
- 3.2 Expectation of Functions ... 132
- 3.3 Bivariate Normal ... 136
- 3.4 Conditional Densities and Expectations ... 140
  - 3.4.1 Examples on Conditional Densities and Expectations ... 142
- 3.5 Posterior Densities, Likelihood Functions, and Bayes Estimates ... 147
- 3.6 Maximum Likelihood Estimates ... 152
- 3.7 Bivariate Normal Conditional Distributions ... 154
- 3.8 * Useful Formulas and Characterizations for Bivariate Normal ... 155
  - 3.8.1 Computing Bivariate Normal Probabilities ... 157
- 3.9 * Conditional Expectation Given a Set and Borel's Paradox ... 158
- References ... 165

4 Advanced Distribution Theory ... 167
- 4.1 Convolutions and Examples ... 167
- 4.2 Products and Quotients and the t- and F-Distribution ... 172
- 4.3 Transformations ... 177
- 4.4 Applications of Jacobian Formula ... 178
- 4.5 Polar Coordinates in Two Dimensions ... 180
- 4.6 * n-Dimensional Polar and Helmert's Transformation ... 182
  - 4.6.1 Efficient Spherical Calculations with Polar Coordinates ... 182
  - 4.6.2 Independence of Mean and Variance in Normal Case ... 185
  - 4.6.3 The t Confidence Interval ... 187
- 4.7 The Dirichlet Distribution ... 188
  - 4.7.1 * Picking a Point from the Surface of a Sphere ... 191
  - 4.7.2 * Poincaré's Lemma ... 191
- 4.8 * Ten Important High-Dimensional Formulas for Easy Reference ... 191
- References ... 197

Contents xiii

5 Multivariate Normal and Related Distributions 199
   5.1 Definition and Some Basic Properties 199
   5.2 Conditional Distributions 202
   5.3 Exchangeable Normal Variables 205
   5.4 Sampling Distributions Useful in Statistics 207
       5.4.1 ∗ Wishart Expectation Identities 208
       5.4.2 ∗ Hotelling's T^2 and Distribution of Quadratic Forms 209
       5.4.3 ∗ Distribution of Correlation Coefficient 212
   5.5 Noncentral Distributions 213
   5.6 Some Important Inequalities for Easy Reference 214
   References 218

6 Finite Sample Theory of Order Statistics and Extremes 221
   6.1 Basic Distribution Theory 221
   6.2 More Advanced Distribution Theory 225
   6.3 Quantile Transformation and Existence of Moments 229
   6.4 Spacings 233
       6.4.1 Exponential Spacings and Réyni's Representation 233
       6.4.2 Uniform Spacings 234
   6.5 Conditional Distributions and Markov Property 235
   6.6 Some Applications 238
       6.6.1 ∗ Records 238
       6.6.2 The Empirical CDF 241
   6.7 ∗ Distribution of the Multinomial Maximum 243
   References 247

7 Essential Asymptotics and Applications 249
   7.1 Some Basic Notation and Convergence Concepts 250
   7.2 Laws of Large Numbers 254
   7.3 Convergence Preservation 259
   7.4 Convergence in Distribution 262
   7.5 Preservation of Convergence and Statistical Applications 267
       7.5.1 Slutsky's Theorem 268
       7.5.2 Delta Theorem 269
       7.5.3 Variance Stabilizing Transformations 272
   7.6 Convergence of Moments 274
       7.6.1 Uniform Integrability 275
       7.6.2 The Moment Problem and Convergence in Distribution 277
       7.6.3 Approximation of Moments 278
   7.7 Convergence of Densities and Scheffé's Theorem 282
   References 292

8 Characteristic Functions and Applications 293
- 8.1 Characteristic Functions of Standard Distributions 294
- 8.2 Inversion and Uniqueness 298
- 8.3 Taylor Expansions, Differentiability, and Moments 302
- 8.4 Continuity Theorems 303
- 8.5 Proof of the CLT and the WLLN 305
- 8.6 ∗ Producing Characteristic Functions 306
- 8.7 Error of the Central Limit Theorem 308
- 8.8 Lindeberg–Feller Theorem for General Independent Case 311
- 8.9 ∗ Infinite Divisibility and Stable Laws 315
- 8.10 ∗ Some Useful Inequalities 317
- References 322

9 Asymptotics of Extremes and Order Statistics 323
- 9.1 Central-Order Statistics 323
  - 9.1.1 Single-Order Statistic 323
  - 9.1.2 Two Statistical Applications 325
  - 9.1.3 Several Order Statistics 326
- 9.2 Extremes 328
  - 9.2.1 Easily Applicable Limit Theorems 328
  - 9.2.2 The Convergence of Types Theorem 332
- 9.3 ∗ Fisher–Tippett Family and Putting it Together 333
- References 338

10 Markov Chains and Applications 339
- 10.1 Notation and Basic Definitions 340
- 10.2 Examples and Various Applications as a Model 340
- 10.3 Chapman–Kolmogorov Equation 345
- 10.4 Communicating Classes 349
- 10.5 Gambler's Ruin 352
- 10.6 First Passage, Recurrence, and Transience 354
- 10.7 Long Run Evolution and Stationary Distributions 359
- References 374

11 Random Walks 375
- 11.1 Random Walk on the Cubic Lattice 375
  - 11.1.1 Some Distribution Theory 378
  - 11.1.2 Recurrence and Transience 379
  - 11.1.3 ∗ Pólya's Formula for the Return Probability 382
- 11.2 First Passage Time and Arc Sine Law 383
- 11.3 The Local Time 387
- 11.4 Practically Useful Generalizations 389
- 11.5 Wald's Identity 390
- 11.6 Fate of a Random Walk 392

	11.7	Chung–Fuchs Theorem 394
	11.8	Six Important Inequalities 396
	References..400

12 Brownian Motion and Gaussian Processes 401
- 12.1 Preview of Connections to the Random Walk 402
- 12.2 Basic Definitions 403
  - 12.2.1 Condition for a Gaussian Process to be Markov 406
  - 12.2.2 * Explicit Construction of Brownian Motion 407
- 12.3 Basic Distributional Properties 408
  - 12.3.1 Reflection Principle and Extremes..........................410
  - 12.3.2 Path Properties and Behavior Near Zero and Infinity.......412
  - 12.3.3 * Fractal Nature of Level Sets 415
- 12.4 The Dirichlet Problem and Boundary Crossing Probabilities 416
  - 12.4.1 Recurrence and Transience...................................418
- 12.5 The Local Time of Brownian Motion 419
- 12.6 Invariance Principle and Statistical Applications 421
- 12.7 Strong Invariance Principle and the KMT Theorem 425
- 12.8 Brownian Motion with Drift and Ornstein–Uhlenbeck Process......427
  - 12.8.1 Negative Drift and Density of Maximum....................427
  - 12.8.2 * Transition Density and the Heat Equation 428
  - 12.8.3 * The Ornstein–Uhlenbeck Process 429
- References..435

13 Poisson Processes and Applications...437
- 13.1 Notation 438
- 13.2 Defining a Homogeneous Poisson Process............................439
- 13.3 Important Properties and Uses as a Statistical Model 440
- 13.4 * Linear Poisson Process and Brownian Motion: A Connection 448
- 13.5 Higher-Dimensional Poisson Point Processes 450
  - 13.5.1 The Mapping Theorem 452
- 13.6 One-Dimensional Nonhomogeneous Processes 453
- 13.7 * Campbell's Theorem and Shot Noise 456
  - 13.7.1 Poisson Process and Stable Laws 458
- References..462

14 Discrete Time Martingales and Concentration Inequalities 463
- 14.1 Illustrative Examples and Applications in Statistics..................463
- 14.2 Stopping Times and Optional Stopping 468
  - 14.2.1 Stopping Times 469
  - 14.2.2 Optional Stopping 470
  - 14.2.3 Sufficient Conditions for Optional Stopping Theorem 472
  - 14.2.4 Applications of Optional Stopping 474

	14.3	Martingale and Concentration Inequalities................................477
		14.3.1 Maximal Inequality 477
		14.3.2 * Inequalities of Burkholder, Davis, and Gundy 480
		14.3.3 Inequalities of Hoeffding and Azuma 483
		14.3.4 * Inequalities of McDiarmid and Devroye 485
		14.3.5 The Upcrossing Inequality 488
	14.4	Convergence of Martingales 490
		14.4.1 The Basic Convergence Theorem 490
		14.4.2 Convergence in L_1 and L_2 493
	14.5	* Reverse Martingales and Proof of SLLN 494
	14.6	Martingale Central Limit Theorem 497
	References..503

15 Probability Metrics 505
- 15.1 Standard Probability Metrics Useful in Statistics.....................505
- 15.2 Basic Properties of the Metrics 508
- 15.3 Metric Inequalities 515
- 15.4 Differential Metrics for Parametric Families........................519
  - 15.4.1 * Fisher Information and Differential Metrics 520
  - 15.4.2 * Rao's Geodesic Distances on Distributions 522
- References..525

16 Empirical Processes and VC Theory 527
- 16.1 Basic Notation and Definitions 527
- 16.2 Classic Asymptotic Properties of the Empirical Process 529
  - 16.2.1 Invariance Principle and Statistical Applications 531
  - 16.2.2 * Weighted Empirical Process 534
  - 16.2.3 The Quantile Process 536
  - 16.2.4 Strong Approximations of the Empirical Process..........537
- 16.3 Vapnik–Chervonenkis Theory 538
  - 16.3.1 Basic Theory 538
  - 16.3.2 Concrete Examples 540
- 16.4 CLTs for Empirical Measures and Applications 543
  - 16.4.1 Notation and Formulation 543
  - 16.4.2 Entropy Bounds and Specific CLTs........................544
  - 16.4.3 Concrete Examples 547
- 16.5 Maximal Inequalities and Symmetrization.......................547
- 16.6 * Connection to the Poisson Process 551
- References..557

17 Large Deviations 559
- 17.1 Large Deviations for Sample Means 560
  - 17.1.1 The Cramér–Chernoff Theorem in $\mathcal{R}$......................560
  - 17.1.2 Properties of the Rate Function 564
  - 17.1.3 Cramér's Theorem for General Sets 566

Contents   xvii

    17.2 The Gärtner–Ellis Theorem and Markov Chain Large Deviations 567
    17.3 The t-Statistic 570
    17.4 Lipschitz Functions and Talagrand's Inequality 572
    17.5 Large Deviations in Continuous Time................................. 574
        17.5.1 ∗ Continuity of a Gaussian Process......................... 576
        17.5.2 ∗ Metric Entropy of T and Tail of the Supremum 577
    References.. 582

18 The Exponential Family and Statistical Applications 583
    18.1 One-Parameter Exponential Family 583
        18.1.1 Definition and First Examples 584
    18.2 The Canonical Form and Basic Properties 589
        18.2.1 Convexity Properties 590
        18.2.2 Moments and Moment Generating Function 591
        18.2.3 Closure Properties 594
    18.3 Multiparameter Exponential Family.................................. 596
    18.4 Sufficiency and Completeness 600
        18.4.1 ∗ Neyman–Fisher Factorization and Basu's Theorem 602
        18.4.2 ∗ Applications of Basu's Theorem to Probability........... 604
    18.5 Curved Exponential Family 607
    References.. 612

19 Simulation and Markov Chain Monte Carlo 613
    19.1 The Ordinary Monte Carlo... 615
        19.1.1 Basic Theory and Examples................................. 615
        19.1.2 Monte Carlo P-Values 622
        19.1.3 Rao–Blackwellization 623
    19.2 Textbook Simulation Techniques...................................... 624
        19.2.1 Quantile Transformation and Accept–Reject............. 624
        19.2.2 Importance Sampling and Its Asymptotic Properties 629
        19.2.3 Optimal Importance Sampling Distribution 633
        19.2.4 Algorithms for Simulating from Common Distributions 634
    19.3 Markov Chain Monte Carlo.. 637
        19.3.1 Reversible Markov Chains 639
        19.3.2 Metropolis Algorithms 642
    19.4 The Gibbs Sampler... 645
    19.5 Convergence of MCMC and Bounds on Errors 651
        19.5.1 Spectral Bounds 653
        19.5.2 ∗ Dobrushin's Inequality and Diaconis–Fill–Stroock Bound 657
        19.5.3 ∗ Drift and Minorization Methods 659

	19.6	MCMC on General Spaces 662

- 19.6 MCMC on General Spaces 662
  - 19.6.1 General Theory and Metropolis Schemes 662
  - 19.6.2 Convergence 666
  - 19.6.3 Convergence of the Gibbs Sampler 670
- 19.7 Practical Convergence Diagnostics 673
- References 686

20 Useful Tools for Statistics and Machine Learning 689
- 20.1 The Bootstrap 689
  - 20.1.1 Consistency of the Bootstrap 692
  - 20.1.2 Further Examples 696
  - 20.1.3 ∗ Higher-Order Accuracy of the Bootstrap 699
  - 20.1.4 Bootstrap for Dependent Data 701
- 20.2 The EM Algorithm 704
  - 20.2.1 The Algorithm and Examples 706
  - 20.2.2 Monotone Ascent and Convergence of EM 711
  - 20.2.3 ∗ Modifications of EM 714
- 20.3 Kernels and Classification 715
  - 20.3.1 Smoothing by Kernels 715
  - 20.3.2 Some Common Kernels in Use 717
  - 20.3.3 Kernel Density Estimation 719
  - 20.3.4 Kernels for Statistical Classification 724
  - 20.3.5 Mercer's Theorem and Feature Maps 732
- References 744

A Symbols, Useful Formulas, and Normal Table 747
- A.1 Glossary of Symbols 747
- A.2 Moments and MGFs of Common Distributions 750
- A.3 Normal Table 755

Author Index 757

Subject Index 763

Suggested Courses with Different Themes

Duration	Theme	Chapters
15 weeks	Beginning graduate	2–7, 9
15 weeks	Advanced graduate	7, 8, 10, 11, 12, 13, 14
15 weeks	Special topics for statistics students	9, 10, 15, 16, 17, 18, 20
15 weeks	Special topics for computer science students	4, 11, 14, 16, 17, 18, 19
8 weeks	Summer course for statistics students	11, 12, 14, 20
8 weeks	Summer course for computer science students	14, 16, 18, 20
8 weeks	Summer course on modeling and simulation	4, 10, 13, 19

Chapter 1
Review of Univariate Probability

Probability is a universally accepted tool for expressing degrees of confidence or doubt about some proposition in the presence of incomplete information or uncertainty. By convention, probabilities are calibrated on a scale of 0 to 1; assigning something a zero probability amounts to expressing the belief that we consider it impossible, whereas assigning a probability of one amounts to considering it a certainty. Most propositions fall somewhere in between. Probability statements that we make can be based on our past experience, or on our personal judgments. Whether our probability statements are based on past experience or subjective personal judgments, they obey a common set of rules, which we can use to treat probabilities in a mathematical framework, and also for making decisions on predictions, for understanding complex systems, or as intellectual experiments and for entertainment. Probability theory is one of the most applicable branches of mathematics. It is used as the primary tool for analyzing statistical methodologies; it is used routinely in nearly every branch of science, such as biology, astronomy and physics, medicine, economics, chemistry, sociology, ecology, finance, and many others. A background in the theory, models, and applications of probability is almost a part of basic education. That is how important it is.

For a classic and lively introduction to the subject of probability, we recommend Feller (1968,1971). Among numerous other expositions of the theory of probability, a variety of examples on various topics can be seen in Ross (1984), Stirzaker (1994), Pitman (1992), Bhattacharya and Waymire (2009), and DasGupta (2010). Ash (1972), Chung (1974), Breiman (1992), Billingsley (1995), and Dudley (2002) are masterly accounts of measure-theoretic probability.

1.1 Experiments and Sample Spaces

Treatment of probability theory starts with the consideration of a *sample space*. The sample space is the set of all possible outcomes in some physical experiment. For example, if a coin is tossed twice and after each toss the face that shows is recorded, then the possible outcomes of this particular coin-tossing experiment, say

ξ are HH, HT, TH, TT, with H denoting the occurrence of heads and T denoting the occurrence of tails. We call

$$\Omega = \{HH, HT, TH, TT\}$$

the sample space of the experiment ξ.

In general, a sample space is a general set Ω, finite or infinite. An easy example where the sample space Ω is infinite is to toss a coin until the first time heads show up and record the number of the trial at which the first head appeared. In this case, the sample space Ω is the *countably infinite* set

$$\Omega = \{1, 2, 3, \ldots\}.$$

Sample spaces can also be *uncountably infinite*; for example, consider the experiment of choosing a number at random from the interval $[0, 1]$. The sample space of this experiment is $\Omega = [0, 1]$. In this case, Ω is an uncountably infinite set. In all cases, individual elements of a sample space are denoted as ω. The first task is to define *events* and to explain the meaning of the probability of an event.

Definition 1.1. Let Ω be the sample space of an experiment ξ. Then any subset A of Ω, including the empty set ϕ and the entire sample space Ω is called an *event*.

Events may contain even one single sample point ω, in which case the event is a *singleton set* $\{\omega\}$. We want to assign probabilities to events. But we want to assign probabilities in a way that they are logically consistent. In fact, this cannot be done in general if we insist on assigning probabilities to arbitrary collections of sample points, that is, arbitrary subsets of the sample space Ω. We can only define probabilities for such subsets of Ω that are tied together like a family, the exact concept being that of a σ-field. In most applications, including those cases where the sample space Ω is infinite, events that we would want to normally think about will be members of such an appropriate σ-field. So we do not mention the need for consideration of σ-fields any further, and get along with thinking of events as subsets of the sample space Ω, including in particular the empty set ϕ and the entire sample space Ω itself.

Here is a definition of what counts as a legitimate probability on events.

Definition 1.2. Given a sample space Ω, a probability or a *probability measure* on Ω is a function P on subsets of Ω such that

(a) $P(A) \geq 0$ for any $A \subseteq \Omega$;
(b) $P(\Omega) = 1$;
(c) Given disjoint subsets $A_1, A_2, \ldots$ of Ω, $P(\cup_{i=1}^{\infty} A_i) = \sum_{i=1}^{\infty} P(A_i)$.

Property (c) is known as *countable additivity*. Note that it is not something that can be proved, but it is like an assumption or an *axiom*. In our experience, we have seen that operating as if the assumption is correct leads to useful and credible answers in many problems, and so we accept it as a reasonable assumption. Not all

1.1 Experiments and Sample Spaces

probabilists agree that countable additivity is natural; but we do not get into that debate in this book. One important point is that finite additivity is subsumed in countable additivity; that is if there are some finite number m of disjoint subsets $A_1, A_2, \ldots, A_m$ of Ω, then $P(\cup_{i=1}^m A_i) = \sum_{i=1}^m P(A_i)$. Also, it is useful to note that the last two conditions in the definition of a probability measure imply that $P(\phi)$, the probability of the empty set or the *null event*, is zero.

One notational convention is that strictly speaking, for an event that is just a singleton set $\{\omega\}$, we should write $P(\{\omega\})$ to denote its probability. But to reduce clutter, we simply use the more convenient notation $P(\omega)$.

One pleasant consequence of the axiom of countable additivity is the following basic result. We do not prove it here as it is a simple result; see DasGupta (2010) for a proof.

Theorem 1.1. *Let $A_1 \supset A_2 \supset A_3 \supset \cdots$ be an infinite family of subsets of a sample space Ω such that $A_n \downarrow A$. Then, $P(A_n) \to P(A)$ as $n \to \infty$.*

Next, the concept of equally likely sample points is a very fundamental one.

Definition 1.3. Let Ω be a finite sample space consisting of N sample points. We say that the sample points are *equally likely* if $P(\omega) = \frac{1}{N}$ for each sample point ω.

An immediate consequence, due to the addivity axiom, is the following useful formula.

Proposition. *Let Ω be a finite sample space consisting of N equally likely sample points. Let A be any event and suppose A contains n distinct sample points. Then*

$$P(A) = \frac{n}{N} = \frac{\textit{Number of sample points favorable to A}}{\textit{Total number of sample points}}.$$

Let us see some examples.

Example 1.1 (The Shoe Problem). Suppose there are five pairs of shoes in a closet and four shoes are taken out at random. What is the probability that among the four that are taken out, there is at least one complete pair?

The total number of sample points is $\binom{10}{4} = 210$. Because selection was done completely at random, we assume that all sample points are equally likely. At least one complete pair would mean two complete pairs, or exactly one complete pair and two other nonconforming shoes. Two complete pairs can be chosen in $\binom{5}{2} = 10$ ways. Exactly one complete pair can be chosen in $\binom{5}{1}\binom{4}{2} \times 2 \times 2 = 120$ ways. The $\binom{5}{1}$ term is for choosing the pair that is complete; the $\binom{4}{2}$ term is for choosing two incomplete pairs, and then from each incomplete pair, one chooses the left or the right shoe. Thus, the probability that there will be at least one complete pair among the four shoes chosen is $(10 + 120)/210 = 13/21 = .62$.

Example 1.2 (Five-Card Poker). In five-card poker, a player is given 5 cards from a full deck of 52 cards at random. Various named hands of varying degrees of rarity exist. In particular, we want to calculate the probabilities of $A = $ *two pairs* and

$B =$ a *flush*. Two pairs is a hand with 2 cards each of 2 different denominations and the fifth card of some other denomination; a flush is a hand with 5 cards of the same suit, but the cards cannot be of denominations in a sequence.

Then, $P(A) = \binom{13}{2}[\binom{4}{2}]^2 \binom{44}{1}/\binom{52}{5} = .04754$.

To find $P(B)$, note that there are 10 ways to select 5 cards from a suit such that the cards are in a sequence, namely, $\{A, 2, 3, 4, 5\}, \{2, 3, 4, 5, 6\}, \ldots, \{10, J, Q, K, A\}$, and so,

$$P(B) = \binom{4}{1}\left(\binom{13}{5} - 10\right) \bigg/ \binom{52}{5} = .00197.$$

These are basic examples of counting arguments that are useful whenever there is a finite sample space and we assume that all sample points are equally likely.

A major result in combinatorial probability is the *inclusion–exclusion formula*, which says the following.

Theorem 1.2. *Let* $A_1, A_2, \ldots, A_n$ *be n general events. Let*

$$S_1 = \sum_{i=1}^{n} P(A_i); \; S_2 = \sum_{1 \leq i < j \leq n} P(A_i \cap A_j); \; S_3 = \sum_{1 \leq i < j < k \leq n} P(A_i \cap A_j \cap A_k); \cdots$$

Then,

$$P(\cup_{i=1}^{n} A_i) = \sum_{i=1}^{n} P(A_i) - \sum_{1 \leq i < j \leq n} P(A_i \cap A_j) + \sum_{1 \leq i < j < k \leq n} P(A_i \cap A_j \cap A_k)$$
$$- \cdots + (-1)^{n+1} P(A_1 \cap A_2 \cap \cdots \cap A_n)$$
$$= S_1 - S_2 + S_3 - \cdots + (-1)^{n+1} S_n.$$

Example 1.3 (Missing Suits in a Bridge Hand). Consider a specific player, say North, in a Bridge game. We want to calculate the probability that North's hand is void in at least one suit. Towards this, denote the suits as 1, 2, 3, 4 and let $A_i =$ North's hand is void in suit i.

Then, by the inclusion exclusion formula,

$$P(\text{North's hand is void in at least one suit})$$
$$= P(A_1 \cup A_2 \cup A_3 \cup A_4)$$

$= 4\binom{39}{13}/\binom{52}{13} - 6\binom{26}{13}/\binom{52}{13} + 4\binom{13}{13}/\binom{52}{13} = .051$, which is small, but not very small.

1.2 Conditional Probability and Independence

The inclusion–exclusion formula can be hard to apply exactly, because the quantities S_j for large indices j can be difficult to calculate. However, fortunately, the inclusion–exclusion formula leads to bounds in both directions for the probability of the union of n general events. We have the following series of bounds.

Theorem 1.3 (Bonferroni Bounds). *Given n events $A_1, A_2, \ldots, A_n$, let $p_n = P(\cup_{i=1}^{n} A_i)$. Then,*

$$p_n \leq S_1; \; p_n \geq S_1 - S_2; \; p_n \leq S_1 - S_2 + S_3; \; \ldots.$$

In addition,

$$P(\cap_{i=1}^{n} A_i) \geq 1 - \sum_{i=1}^{n} P(A_i^c).$$

1.2 Conditional Probability and Independence

Both conditional probability and independence are fundamental concepts for probabilists and statisticians alike. Conditional probabilities correspond to updating one's beliefs when new information becomes available. Independence corresponds to irrelevance of a piece of new information, even when it is made available. In addition, the assumption of independence can and does significantly simplify development, mathematical analysis, and justification of tools and procedures.

Definition 1.4. Let A, B be general events with respect to some sample space Ω, and suppose $P(A) > 0$. The *conditional probability* of B given A is defined as

$$P(B|A) = \frac{P(A \cap B)}{P(A)}.$$

Some immediate consequences of the definition of a conditional probability are the following.

Theorem 1.4. *(a)* **(Multiplicative Formula).** *For any two events A, B such that $P(A) > 0$, one has $P(A \cap B) = P(A)P(B|A)$;*
(b) For any two events A, B such that $0 < P(A) < 1$, one has $P(B) = P(B|A)P(A) + P(B|A^c)P(A^c)$;
(c) **(Total Probability Formula).** *If $A_1, A_2, \ldots, A_k$ form a partition of the sample space Ω, (i.e., $A_i \cap A_j = \phi$ for all $i \neq j$, and $\cup_{i=1}^{k} A_i = \Omega$), and if $0 < P(A_i) < 1$ for all i, then*

$$P(B) = \sum_{i=1}^{k} P(B|A_i) P(A_i).$$

(d) **(Hierarchical Multiplicative Formula).** Let $A_1, A_2, \ldots, A_k$ be k general events in a sample space Ω. Then

$$P(A_1 \cap A_2 \cap \cdots \cap A_k) = P(A_1) P(A_2|A_1) P(A_3|A_1 \cap A_2) \cdots$$
$$\times P(A_k|A_1 \cap A_2 \cap \cdots \cap A_{k-1}).$$

Example 1.4. One of two urns has a red and b black balls, and the other has c red and d black balls. One ball is chosen at random from each urn, and then one of these two balls is chosen at random. What is the probability that this ball is red?

If each ball selected from the two urns is red, then the final ball is definitely red. If one of those two balls is red, then the final ball is red with probability 1/2. If none of those two balls is red, then the final ball cannot be red.

Thus,

$$P(\text{The final ball is red}) = a/(a+b) \times c/(c+d) + 1/2$$
$$\times [a/(a+b) \times d/(c+d) + b/(a+b) \times c/(c+d)]$$
$$= \frac{2ac + ad + bc}{2(a+b)(c+d)}.$$

As an example, suppose $a = 99, b = 1, c = 1, d = 1$.

$$\text{Then} \quad \frac{2ac + ad + bc}{2(a+b)(c+d)} = .745.$$

Although the total percentage of red balls in the two urns is more than 98%, the chance that the final ball selected would be red is just about 75%.

Example 1.5 (A Clever Conditioning Argument). Coin A gives heads with probability s and coin B gives heads with probability t. They are tossed alternately, starting off with coin A. We want to find the probability that the first head is obtained on coin A.

We find this probability by conditioning on the outcomes of the first two tosses; more precisely, define

$$A_1 = \{H\} = \text{First toss gives H}; \quad A_2 = \{TH\}; \quad A_3 = \{TT\}.$$

Let also,

$$A = \text{The first head is obtained on coin A}.$$

One of the three events A_1, A_2, A_3 must happen, and they are also mutually exclusive. Therefore, by the total probability formula,

$$P(A) = \sum_{i=1}^{3} P(A_i) P(A|A_i) = s \times 1 + (1-s)t \times 0 + (1-s)(1-t)P(A)$$
$$\Rightarrow P(A) = s/[1 - (1-s)(1-t)] = s/(s + t - st).$$

1.2 Conditional Probability and Independence

As an example, let $s = .4, t = .5$. Note that coin A is biased against heads. Even then, $s/(s + t - st) = .57 > .5$. We see that *there is an advantage in starting first.*

Definition 1.5. A collection of events $A_1, A_2, \ldots, A_n$ is said to be *mutually independent* (or just independent) if for each $k, 1 \leq k \leq n$, and any k of the events, $A_{i_1}, \ldots, A_{i_k}$, $P(A_{i_1} \cap \cdots A_{i_k}) = P(A_{i_1}) \cdots P(A_{i_k})$. They are called *pairwise independent* if this property holds for $k = 2$.

Example 1.6 (Lotteries). Although many people buy lottery tickets out of an expectation of good luck, probabilistically speaking, buying lottery tickets is usually a waste of money. Here is an example. Suppose in a weekly state lottery, five of the numbers $00, 01, \ldots, 49$ are selected without replacement at random, and someone holding exactly those numbers wins the lottery. Then, the probability that someone holding one ticket will be the winner in a given week is

$$\frac{1}{\binom{50}{5}} = 4.72 \times 10^{-7}.$$

Suppose this person buys a ticket every week for 40 years. Then, the probability that he will win the lottery on at least one week is $1 - (1 - 4.72 \times 10^{-7})^{52 \times 40} = .00098 < .001$, still a very small probability. We assumed in this calculation that the weekly lotteries are all mutually independent, a reasonable assumption. The calculation would fall apart if we did not make this independence assumption.

It is not uncommon to see the conditional probabilities $P(A|B)$ and $P(B|A)$ confused with each other. Suppose in some group of lung cancer patients, we see a large percentage of smokers. If we define B to be the event that a person is a smoker, and A to be the event that a person has lung cancer, then all we can conclude is that in our group of people $P(B|A)$ is large. But we cannot conclude from just this information that smoking increases the chance of lung cancer, that is, that $P(A|B)$ is large. In order to calculate a conditional probability $P(A|B)$ when we know the other conditional probability $P(B|A)$, a simple formula known as *Bayes' theorem* is useful. Here is a statement of a general version of Bayes' theorem.

Theorem 1.5. *Let $\{A_1, A_2, \ldots, A_m\}$ be a partition of a sample space Ω. Let B be some fixed event. Then*

$$P(A_j|B) = \frac{P(B|A_j)P(A_j)}{\sum_{i=1}^{m} P(B|A_i)P(A_i)}.$$

Example 1.7 (Multiple Choice Exams). Suppose that the questions in a multiple choice exam have five alternatives each, of which a student has to pick one as the correct alternative. A student either knows the truly correct alternative with probability .7, or she randomly picks one of the five alternatives as her choice. Suppose a particular problem was answered correctly. We want to know what the probability is that the student really knew the correct answer.

Define

$$A = \text{The student knew the correct answer,}$$
$$B = \text{The student answered the question correctly.}$$

We want to compute $P(A|B)$. By Bayes' theorem,

$$P(A|B) = \frac{P(B|A)P(A)}{P(B|A)P(A) + P(B|A^c)P(A^c)} = \frac{1 \times .7}{1 \times .7 + .2 \times .3} = .921.$$

Before the student answered the question, our probability that she would know the correct answer to the question was .7; but once she answered it correctly, the posterior probability that she knew the correct answer increases to .921. This is exactly what Bayes' theorem does; it updates our *prior* belief to the *posterior* belief, when new evidence becomes available.

1.3 Integer-Valued and Discrete Random Variables

In some sense, the entire subject of probability and statistics is about distributions of random variables. Random variables, as the very name suggests, are quantities that vary, over time, or from individual to individual, and the reason for the variability is some underlying random process. Depending on exactly how an underlying experiment ξ ends, the random variable takes different values. In other words, the value of the random variable is determined by the sample point ω that prevails, when the underlying experiment ξ is actually conducted. We cannot know a priori the value of the random variable, because we do not know a priori which sample point ω will prevail when the experiment ξ is conducted. We try to understand the behavior of a random variable by analyzing the probability structure of that underlying random experiment.

Random variables, like probabilities, originated in gambling. Therefore, the random variables that come to us more naturally, are integer-valued random variables; for examples, the sum of the two rolls when a die is rolled twice. Integer-valued random variables are special cases of what are known as discrete random variables. Discrete or not, a common mathematical definition of all random variables is the following.

Definition 1.6. Let Ω be a sample space corresponding to some experiment ξ and let $X : \Omega \to \mathcal{R}$ be a function from the sample space to the real line. Then X is called a *random variable*.

Discrete random variables are those that take a finite or a countably infinite number of possible values. In particular, all integer-valued random variables are discrete. From the point of view of understanding the behavior of a random variable, the important thing is to know the probabilities with which X takes its different possible values.

1.3 Integer-Valued and Discrete Random Variables

Definition 1.7. Let $X : \Omega \to \mathcal{R}$ be a discrete random variable taking a finite or countably infinite number of values $x_1, x_2, x_3, \ldots$. The probability distribution or the *probability mass function* (pmf) of X is the function $p(x) = P(X = x), x = x_1, x_2, x_3, \ldots$, and $p(x) = 0$, otherwise.

It is common to not explicitly mention the phrase "$p(x) = 0$ otherwise," and we generally follow this convention. Some authors use the phrase *mass function* instead of *probability mass function*.

For any pmf, one must have $p(x) \geq 0$ for any x, and $\sum_i p(x_i) = 1$. Any function satisfying these two properties for some set of numbers $x_1, x_2, x_3, \ldots$ is a valid pmf.

1.3.1 CDF and Independence

A second important definition is that of a *cumulative distribution function* (CDF). The CDF gives the probability that a random variable X is less than or equal to any given number x. It is important to understand that the notion of a CDF is universal to all random variables; it is not limited to only the discrete ones.

Definition 1.8. The *cumulative distribution function* of a random variable X is the function $F(x) = P(X \leq x), x \in \mathcal{R}$.

Definition 1.9. Let X have the CDF $F(x)$. Any number m such that $P(X \leq m) \geq .5$, and also $P(X \geq m) \geq .5$ is called a median of F, or equivalently, a median of X.

Remark. The median of a random variable need not be unique. A simple way to characterize all the medians of a distribution is available.

Proposition. *Let X be a random variable with the CDF $F(x)$. Let m_0 be the first x such that $F(x) \geq .5$, and let m_1 be the last x such that $P(X \geq x) \geq .5$. Then, a number m is a median of X if and only if $m \in [m_0, m_1]$.*

The CDF of any random variable satisfies a set of properties. Conversely, any function satisfying these properties is a valid CDF; that is, it will be the CDF of some appropriately chosen random variable. These properties are given in the next result.

Theorem 1.6. *A function $F(x)$ is the CDF of some real-valued random variable X if and only if it satisfies all of the following properties.*

(a) $0 \leq F(x) \leq 1 \; \forall x \in \mathcal{R}$.
(b) $F(x) \to 0$ as $x \to -\infty$, and $F(x) \to 1$ as $x \to \infty$.
(c) Given any real number a, $F(x) \downarrow F(a)$ as $x \downarrow a$.
(d) Given any two real numbers $x, y, x < y, F(x) \leq F(y)$.

Property (c) is called continuity from the right, *or simply* right continuity. *It is clear that a CDF need not be continuous from the left; indeed, for discrete random variables, the CDF has a jump at the values of the random variable, and at the jump points, the CDF is not left continuous. More precisely, one has the following result.*

Proposition. *Let $F(x)$ be the CDF of some random variable X. Then, for any x,*

(a) $P(X = x) = F(x) - \lim_{y \uparrow x} F(y) = F(x) - F(x-)$, *including those points x for which $P(X = x) = 0$.*
(b) $P(X \geq x) = P(X > x) + P(X = x) = (1 - F(x)) + (F(x) - F(x-)) = 1 - F(x-)$.

Example 1.8 (Bridge). Consider the random variable

$$X = \text{Number of aces in North's hand in a Bridge game.}$$

Clearly, X can take any of the values $x = 0, 1, 2, 3, 4$. If $X = x$, then the other $13 - x$ cards in North's hand must be non-ace cards. Thus, the pmf of X is

$$P(X = x) = \frac{\binom{4}{x}\binom{48}{13-x}}{\binom{52}{13}}, \quad x = 0, 1, 2, 3, 4.$$

In decimals, the pmf of X is:

x	0	1	2	3	4
p(x)	.304	.439	.213	.041	.003

The CDF of X is a jump function, taking jumps at the values $0, 1, 2, 3, 4$, namely the possible values of X. The CDF is

$$\begin{aligned} F(x) &= 0 && \text{if } x < 0; \\ &= .304 && \text{if } 0 \leq x < 1; \\ &= .743 && \text{if } 1 \leq x < 2; \\ &= .956 && \text{if } 2 \leq x < 3; \\ &= .997 && \text{if } 3 \leq x < 4; \\ &= 1 && \text{if } x \geq 4. \end{aligned}$$

Example 1.9 (Indicator Variables). Consider the experiment of rolling a fair die twice and now define a random variable Y as follows.

$Y = 1$ if the sum of the two rolls X is an even number;
$Y = 0$ if the sum of the two rolls X is an odd number.

If we let A be the event that X is an even number, then $Y = 1$ if A happens, and $Y = 0$ if A does not happen. Such random variables are called *indicator random variables* and are immensely useful in mathematical calculations in many complex situations.

1.3 Integer-Valued and Discrete Random Variables

Definition 1.10. Let A be any event in a sample space Ω. The *indicator random variable* for A is defined as

$$I_A = 1 \text{ if A happens.}$$
$$I_A = 0 \text{ if A does not happen.}$$

Thus, the distribution of an indicator variable is simply $P(I_A = 1) = P(A)$; $P(I_A = 0) = 1 - P(A)$.

An indicator variable is also called a *Bernoulli variable* with parameter p, where p is just $P(A)$. We later show examples of uses of indicator variables in calculation of *expectations*.

In applications, we are sometimes interested in the distribution of a function, say $g(X)$, of a basic random variable X. In the discrete case, the distribution of a function is found in the obvious way.

Proposition (Function of a Random Variable). *Let X be a discrete random variable and $Y = g(X)$ a real-valued function of X. Then, $P(Y = y) = \sum_{x: g(x)=y} p(x)$.*

Example 1.10. Suppose X has the pmf

$$p(x) = \frac{c}{1+x^2}, \quad x = 0, \pm 1, \pm 2, \pm 3.$$

Suppose we want to find the distribution of two functions of X:

$$Y = g(X) = X^3; \quad Z = h(X) = \sin\left(\frac{\pi}{2}X\right).$$

First, the constant c must be explicitly evaluated. By directly summing the values,

$$\sum_x p(x) = \frac{13c}{5} \Rightarrow c = \frac{5}{13}.$$

Note that $g(X)$ is a one-to-one function of X, but $h(X)$ is not one-to-one. The values of Y are $0, \pm 1, \pm 8, \pm 27$. For example, $P(Y = 0) = P(X = 0) = c = 5/13$; $P(Y = 1) = P(X = 1) = c/2 = 5/26$, and so on. In general, for $y = 0, \pm 1, \pm 8, \pm 27$, $P(Y = y) = P(X = y^{1/3}) = \frac{c}{1+y^{2/3}}$, with $c = 5/13$.

However, $Z = h(X)$ is not a one-to-one function of X. The possible values of Z are as follows.

x	$h(x)$
−3	1
−2	0
−1	−1
0	0
1	1
2	0
3	−1

So, for example, $P(Z = 0) = P(X = -2) + P(X = 0) + P(X = 2) = \frac{7}{5}c = 7/13$. The pmf of $Z = h(X)$ is:

z	-1	0	1
$P(Z = z)$	3/13	7/13	3/13

A key concept in probability is that of independence of a collection of random variables. The collection could be finite or infinite. In the infinite case, we want each finite subcollection of the random variables to be independent. The definition of independence of a finite collection is as follows.

Definition 1.11. Let $X_1, X_2, \ldots, X_k$ be $k \geq 2$ discrete random variables defined on the same sample space Ω. We say that $X_1, X_2, \ldots, X_k$ are *independent* if $P(X_1 = x_1, X_2 = x_2, \ldots, X_k = x_k) = P(X_1 = x_1)P(X_2 = x_2) \cdots P(X_k = x_k)$, $\forall\, x_1, x_2, \ldots, x_k$.

It follows from the definition of independence of random variables that if X_1, X_2 are independent, then any function of X_1 and any function of X_2 are also independent. In fact, we have a more general result.

Theorem 1.7. *Let $X_1, X_2, \ldots, X_k$ be $k \geq 2$ discrete random variables, and suppose they are independent. Let $U = f(X_1, X_2, \ldots, X_i)$ be some function of $X_1, X_2, \ldots, X_i$, and $V = g(X_{i+1}, \ldots, X_k)$ be some function of $X_{i+1}, \ldots, X_k$. Then, U and V are independent.*

This result is true of any types of random variables $X_1, X_2, \cdots, X_k$, not just discrete ones.

A common notation of wide use in probability and statistics is now introduced.

If $X_1, X_2, \ldots, X_k$ are independent, and moreover have the same CDF, say F, then we say that $X_1, X_2, \ldots, X_k$ are iid (or IID) and write $X_1, X_2, \ldots, X_k \stackrel{iid}{\sim} F$. The abbreviation iid (IID) means independent and identically distributed.

Example 1.11 (Two Simple Illustrations). Consider the experiment of tossing a fair coin (or any coin) four times. Suppose X_1 is the number of heads in the first two tosses, and X_2 is the number of heads in the last two tosses. Then, it is intuitively clear that X_1, X_2 are independent, because the last two tosses carry no information regarding the first two tosses. The independence can be easily mathematically verified by using the definition of independence.

Next, consider the experiment of drawing 13 cards at random from a deck of 52 cards. Suppose X_1 is the number of aces and X_2 is the number of clubs among the 13 cards. Then, X_1, X_2 are not independent. For example, $P(X_1 = 4, X_2 = 0) = 0$, but $P(X_1 = 4)$, and $P(X_2 = 0)$ are both > 0, and so $P(X_1 = 4)P(X_2 = 0) > 0$. So, X_1, X_2 cannot be independent.

1.3.2 Expectation and Moments

By definition, a random variable takes different values on different occasions. It is natural to want to know what value it takes on average. Averaging is a very primitive concept. A simple average of just the possible values of the random variable will be misleading, because some values may have so little probability that they are relatively inconsequential. The average or the mean value, also called the expected value of a random variable is a weighted average of the different values of X, weighted according to how important the value is. Here is the definition.

Definition 1.12. Let X be a discrete random variable. We say that the *expected value* of X exists if $\sum_i |x_i| p(x_i) < \infty$, in which case the expected value is defined as

$$\mu = E(X) = \sum_i x_i p(x_i).$$

For notational convenience, we simply write $\sum_x x p(x)$ instead of $\sum_i x_i p(x_i)$. The expected value is also known as *the expectation or the mean* of X.

If the set of possible values of X is infinite, then the infinite sum $\sum_x x p(x)$ can take different values on rearranging the terms of the infinite series unless $\sum_x |x| p(x) < \infty$. So, as a matter of definition, we have to include the qualification that $\sum_x |x| p(x) < \infty$.

If the sample space Ω of the underlying experiment is finite or countably infinite, then we can also calculate the expectation by averaging directly over the sample space.

Proposition (Change of Variable Formula). *Suppose the sample space Ω is finite or countably infinite and X is a discrete random variable with expectation μ. Then,*

$$\mu = \sum_x x p(x) = \sum_\omega X(\omega) P(\omega),$$

where $P(\omega)$ is the probability of the sample point ω.

Important Point. Although it is not the focus of this chapter, in applications we are often interested in more than one variable at the same time. To be specific, consider two discrete random variables X, Y defined on a common sample space Ω. Then we could construct new random variables out of X and Y, for example, $XY, X + Y, X^2 + Y^2$, and so on. We can then talk of their expectations as well. Here is a general definition of expectation of a function of more than one random variable.

Definition 1.13. Let $X_1, X_2, \ldots, X_n$ be n discrete random variables, all defined on a common sample space Ω, with a finite or a countably infinite number of sample points. We say that the expectation of a function $g(X_1, X_2, \ldots, X_n)$ exists if $\sum_\omega |g(X_1(\omega), X_2(\omega), \ldots, X_n(\omega))| P(\omega) < \infty$, in which case, the expected value of $g(X_1, X_2, \ldots, X_n)$ is defined as

$$E[g(X_1, X_2, \ldots, X_n)] = \sum_\omega g(X_1(\omega), X_2(\omega), \ldots, X_n(\omega)) P(\omega).$$

The next few results summarize the most fundamental properties of expectations.

Proposition. *(a) If there exists a finite constant c such that $P(X = c) = 1$, then $E(X) = c$.*
(b) If X, Y are random variables defined on the same sample space Ω with finite expectations, and if $P(X \leq Y) = 1$, then $E(X) \leq E(Y)$.
(c) If X has a finite expectation, and if $P(X \geq c) = 1$, then $E(X) \geq c$. If $P(X \leq c) = 1$, then $E(X) \leq c$.

Proposition (Linearity of Expectations). *Let $X_1, X_2, \ldots, X_n$ be random variables defined on the same sample space Ω, and $c_1, c_2, \ldots, c_n$ any real-valued constants. Then, provided $E(X_i)$ exists for every X_i,*

$$E\left(\sum_{i=1}^n c_i X_i\right) = \sum_{i=1}^n c_i E(X_i).$$

in particular, $E(cX) = cE(X)$ and $E(X_1 + X_2) = E(X_1) + E(X_2)$, whenever the expectations exist.

The following fact also follows easily from the definition of the pmf of a function of a random variable. The result says that the expectation of a function of a random variable X can be calculated directly using the pmf of X itself, without having to calculate the pmf of the function.

Proposition (Expectation of a Function). *Let X be a discrete random variable on a sample space Ω with a finite or countable number of sample points, and $(Y = g(X))$ a function of X. Then,*

$$E(Y) = \sum_\omega Y(\omega) P(w) = \sum_x g(x) p(x),$$

provided $E(Y)$ exists.

Caution. If $g(X)$ is a linear function of X, then, of course, $E(g(X)) = g(E(X))$. But, in general, the two things are not equal. For example, $E(X^2)$ is not the same as $(E(X))^2$; indeed, $E(X^2) > (E(X))^2$ for any random variable X that is not a constant.

A very important property of independent random variables is the following factorization result on expectations.

Theorem 1.8. *Suppose $X_1, X_2, \ldots, X_n$ are independent random variables. Then, provided each expectation exists,*

$$E(X_1 X_2 \cdots X_n) = E(X_1) E(X_2) \cdots E(X_n).$$

Let us now show some more illustrative examples.

1.3 Integer-Valued and Discrete Random Variables

Example 1.12. Let X be the number of heads obtained in two tosses of a fair coin. The pmf of X is $p(0) = p(2) = 1/4, p(1) = 1/2$. Therefore, $E(X) = 0 \times 1/4 + 1 \times 1/2 + 2 \times 1/4 = 1$. Because the coin is fair, we expect it to show heads 50% of the number of times it is tossed, which is 50% of 2, that is, 1.

Example 1.13 (Dice Sum). Let X be the sum of the two rolls when a fair die is rolled twice. The pmf of X is $p(2) = p(12) = 1/36; p(3) = p(11) = 2/36; p(4) = p(10) = 3/36; p(5) = p(9) = 4/36; p(6) = p(8) = 5/36; p(7) = 6/36$. Therefore, $E(X) = 2 \times 1/36 + 3 \times 2/36 + 4 \times 3/36 + \cdots + 12 \times 1/36 = 7$. This can also be seen by letting $X_1 = $ the face obtained on the first roll, $X_2 = $ the face obtained on the second roll, and by using $E(X) = E(X_1 + X_2) = E(X_1) + E(X_2) = 3.5 + 3.5 = 7$.

Let us now make this problem harder. Suppose that a fair die is rolled 10 times and X is the sum of all 10 rolls. The pmf of X is no longer so simple; it will be cumbersome to write it down. But, if we let $X_i = $ the face obtained on the ith roll, it is still true by the linearity of expectations that $E(X) = E(X_1 + X_2 + \cdots + X_{10}) = E(X_1) + E(X_2) + \cdots + E(X_{10}) = 3.5 \times 10 = 35$. We can easily compute the expectation, although the pmf would be difficult to write down.

Example 1.14 (A Random Variable Without a Finite Expectation). Let X take the positive integers $1, 2, 3, \ldots$ as its values with the pmf

$$p(x) = P(X = x) = \frac{1}{x(x+1)}, \quad x = 1, 2, 3, \ldots.$$

This is a valid pmf, because obviously $\frac{1}{x(x+1)} > 0$ for any $x = 1, 2, 3, \ldots$, and also the infinite series $\sum_{x=1}^{\infty} \frac{1}{x(x+1)}$ sums to 1, a fact from calculus. Now,

$$E(X) = \sum_{x=1}^{\infty} xp(x) = \sum_{x=1}^{\infty} x \frac{1}{x(x+1)} = \sum_{x=1}^{\infty} \frac{1}{x+1} = \sum_{x=2}^{\infty} \frac{1}{x} = \infty,$$

also a fact from calculus.

This example shows that not all random variables have a finite expectation. Here, the reason for the infiniteness of $E(X)$ is that X takes large integer values x with probabilities $p(x)$ that are not adequately small. The large values are realized sufficiently often that on average X becomes larger than any given finite number.

The zero–one nature of indicator random variables is extremely useful for calculating expectations of certain integer-valued random variables whose distributions are sometimes so complicated that it would be difficult to find their expectations directly from definition. We describe the technique and some illustrations of it below.

Proposition. *Let X be an integer-valued random variable such that it can be represented as $X = \sum_{i=1}^{m} c_i I_{A_i}$ for some m, constants $c_1, c_2, \ldots, c_m$, and suitable events $A_1, A_2, \ldots, A_m$. Then, $E(X) = \sum_{i=1}^{m} c_i P(A_i)$.*

Example 1.15 (Coin Tosses). Suppose a coin that has probability p of showing heads in any single toss is tossed n times, and let X denote the number of times in the n tosses that a head is obtained. Then, $X = \sum_{i=1}^{n} I_{A_i}$, where A_i is the event that a head is obtained in the ith toss. Therefore, $E(X) = \sum_{i=1}^{n} P(A_i) = \sum_{i=1}^{n} p = np$.

A direct calculation of the expectation would involve finding the pmf of X and obtaining the sum $\sum_{x=0}^{n} xP(X = x)$; it can also be done that way, but that is a much longer calculation.

The random variable X of this example is a *binomial random variable* with parameters n and p. Its pmf is given by the formula $P(X = x) = \binom{n}{x} p^x (1-p)^{n-x}, x = 0, 1, 2, \ldots, n$.

Example 1.16 (Consecutive Heads in Coin Tosses). Suppose a coin with probability p for heads in a single toss is tossed n times. How many times can we expect to see a head followed by at least one more head? For example, if $n = 5$, and we see the outcomes HTHHH, then we see a head followed by at least one more head twice.

Define A_i = The ith and the $(i+1)$th toss both result in heads. Then

$$X = \text{number of times a head is followed by at least one more head} = \sum_{i=1}^{n-1} I_{A_i},$$

and so $E(X) = \sum_{i=1}^{n-1} P(A_i) = \sum_{i=1}^{n-1} p^2 = (n-1)p^2$. For example, if a fair coin is tossed 20 times, we can expect to see a head followed by another head about five times ($19 \times .5^2 = 4.75$).

Another useful technique for calculating expectations of nonnegative integer-valued random variables is based on the CDF of the random variable, rather than directly on the pmf. This method is useful when calculating probabilities of the form $P(X > x)$ is logically more straightforward than directly calculating $P(X = x)$. Here is the expectation formula based on the tail CDF.

Theorem 1.9 (Tailsum Formula). *Let X take values $0, 1, 2, \ldots$. Then*

$$E(X) = \sum_{n=0}^{\infty} P(X > n).$$

Example 1.17 (Family Planning). Suppose a couple will have children until they have at least one child of each sex. How many children can they expect to have? Let X denote the childbirth at which they have a child of each sex for the first time. Suppose the probability that any particular childbirth will be a boy is p, and that all births are independent. Then,

$$P(X > n) = P(\text{the first } n \text{ children are all boys or all girls}) = p^n + (1-p)^n.$$

Therefore, $E(X) = 2 + \sum_{n=2}^{\infty} [p^n + (1-p)^n] = 2 + p^2/(1-p) + (1-p)^2/p = \frac{1}{p(1-p)} - 1$. If boys and girls are equally likely on any childbirth, then this says that a couple waiting to have a child of each sex can expect to have three children.

1.3 Integer-Valued and Discrete Random Variables

The expected value is calculated with the intention of understanding what a typical value is of a random variable. But two very different distributions can have exactly the same expected value. A common example is that of a return on an investment in a stock. Two stocks may have the same average return, but one may be much riskier than the other, in the sense that the variability in the return is much higher for that stock. In that case, most risk-averse individuals would prefer to invest in the stock with less variability. Measures of risk or variability are of course not unique. Some natural measures that come to mind are $E(|X-\mu|)$, known as the *mean absolute deviation*, or $P(|X-\mu| > k)$ for some suitable k. However, neither of these two is the most common measure of variability. The most common measure is the *standard deviation* of a random variable.

Definition 1.14. Let a random variable X have a finite mean μ. The *variance* of X is defined as

$$\sigma^2 = E[(X-\mu)^2],$$

and the *standard deviation* of X is defined as $\sigma = \sqrt{\sigma^2}$.

It is easy to prove that $\sigma^2 < \infty$ if and only if $E(X^2)$, the *second moment* of X, is finite. It is not uncommon to mistake the standard deviation for the mean absolute deviation, but they are not the same. In fact, an inequality always holds.

Proposition. $\sigma \geq E(|X-\mu|)$, and σ is strictly greater unless X is a constant random variable, namely, $P(X=\mu) = 1$.

We list some basic properties of the variance of a random variable.

Proposition.

(a) $Var(cX) = c^2 Var(X)$ for any real c.
(b) $Var(X+k) = Var(X)$ for any real k.
(c) $Var(X) \geq 0$ for any random variable X, and equals zero only if $P(X=c)=1$ for some real constant c.
(d) $Var(X) = E(X^2) - \mu^2$.

The quantity $E(X^2)$ is called the second moment *of X. The definition of a general moment is as follows.

Definition 1.15. Let X be a random variable, and $k \geq 1$ a positive integer. Then $E(X^k)$ is called the *kth moment of X*, and $E(X^{-k})$ is called the *kth inverse moment of X*, provided they exist.

We therefore have the following relationships involving moments and the variance.

$$\text{Variance} = \text{Second Moment} - (\text{First Moment})^2.$$

$$\text{Second Moment} = \text{Variance} + (\text{First Moment})^2.$$

Statisticians often use the third moment around the mean as a measure of lack of symmetry in the distribution of a random variable. The point is that if a random variable X has a symmetric distribution, and has a finite mean μ, then all odd moments around the mean, namely, $E[(X-\mu)^{2k+1}]$ will be zero, if the moment exists.

In particular, $E[(X-\mu)^3]$ will be zero. Likewise, statisticians also use the fourth moment around the mean as a measure of how spiky the distribution is around the mean. To make these indices independent of the choice of unit of measurement (e.g., inches or centimeters), they use certain scaled measures of asymmetry and peakedness. Here are the definitions.

Definition 1.16. (a) Let X be a random variable with $E[|X|^3] < \infty$. The *skewness* of X is defined as

$$\beta = \frac{E[(X-\mu)^3]}{\sigma^3}.$$

(b) Suppose X is a random variable with $E[X^4] < \infty$. The *kurtosis* of X is defined as

$$\gamma = \frac{E[(X-\mu)^4]}{\sigma^4} - 3.$$

The skewness β is zero for symmetric distributions, but the converse need not be true. The kurtosis γ is necessarily ≥ -2, but can be arbitrarily large, with spikier distributions generally having a larger kurtosis. But a very good interpretation of γ is not really available. We later show that $\gamma = 0$ for all *normal distributions*; hence the motivation for subtracting 3 in the definition of γ.

Example 1.18 (Variance of Number of Heads). Consider the experiment of two tosses of a fair coin and let X be the number of heads obtained. Then, we have seen that $p(0) = p(2) = 1/4$, and $p(1) = 1/2$. Thus, $E(X^2) = 0 \times 1/4 + 1 \times 1/2 + 4 \times 1/4 = 3/2$, and $E(X) = 1$. Therefore, $\text{Var}(X) = E(X^2) - \mu^2 = 3/2 - 1 = \frac{1}{2}$, and the standard deviation is $\sigma = \sqrt{.5} = .707$.

Example 1.19 (A Random Variable with an Infinite Variance). If a random variable has a finite variance, then it can be shown that it must have a finite mean. This example shows that the converse need not be true.

Let X be a discrete random variable with the pmf

$$P(X = x) = \frac{c}{x(x+1)(x+2)}, \quad x = 1, 2, 3, \ldots,$$

where the normalizing constant $c = 4$. The expected value of X is

$$E(X) = \sum_{x=1}^{\infty} x \times \frac{4}{x(x+1)(x+2)} = 4 \sum_{x=1}^{\infty} \frac{1}{(x+1)(x+2)} = 4 \times 1/2 = 2.$$

Therefore, by direct verification, X has a finite expectation. Let us now examine the second moment of X.

$$E(X^2) = \sum_{x=1}^{\infty} x^2 \times \frac{4}{x(x+1)(x+2)} = 4 \sum_{x=1}^{\infty} x \times \frac{1}{(x+1)(x+2)} = \infty,$$

because the series

$$\sum_{x=1}^{\infty} x \times \frac{1}{(x+1)(x+2)}$$

is not finitely summable, a fact from calculus. Because $E(X^2)$ is infinite, but $E(X)$ is finite, $\sigma^2 = E(X^2) - [E(X)]^2$ must also be infinite.

If a collection of random variables is independent, then just like the expectation, the variance also adds up. Precisely, one has the following very useful fact.

Theorem 1.10. *Let $X_1, X_2, \ldots, X_n$ be n independent random variables. Then,*

$$Var(X_1 + X_2 + \cdots + X_n) = Var(X_1) + Var(X_2) + \cdots + Var(X_n).$$

An important corollary of this result is the following variance formula for the mean, $\bar{X}$, of n independent and identically distributed random variables.

Corollary 1.1. *Let $X_1, X_2, \ldots, X_n$ be independent random variables with a common variance $\sigma^2 < \infty$. Let $\bar{X} = \frac{X_1 + \cdots + X_n}{n}$. Then $Var(\bar{X}) = \frac{\sigma^2}{n}$.*

1.4 Inequalities

The mean and the variance, together, have earned the status of being the two most common summaries of a distribution. A relevant question is whether μ, σ are useful summaries of the distribution of a random variable. The answer is a qualified yes. The inequalities below suggest that knowing just the values of μ, σ, it is in fact possible to say something useful about the full distribution.

Theorem 1.11. *(a)* **(Chebyshev's Inequality).** *Suppose $E(X) = \mu$ and $Var(X) = \sigma^2$, assumed to be finite. Let k be any positive number. Then*

$$P(|X - \mu| \geq k\sigma) \leq \frac{1}{k^2}.$$

(b) **(Markov's Inequality).** *Suppose X takes only nonnegative values, and suppose $E(X) = \mu$, assumed to be finite. Let c be any postive number. Then,*

$$P(X \geq c) \leq \frac{\mu}{c}.$$

The virtue of these two inequalities is that they make no restrictive assumptions on the random variable X. Whenever μ, σ are finite, Chebyshev's inequality is applicable, and whenever μ, is finite, Markov's inequality applies, provided the random variable is nonnegative. However, the universal nature of these inequalities also makes them typically quite conservative.

Although Chebyshev's inequality usually gives conservative estimates for tail probabilities, it does imply a major result in probability theory in a special case.

Theorem 1.12 (Weak Law of Large Numbers). *Let $X_1, X_2, \ldots$ be iid random variables, with $E(X_i) = \mu$, $Var(X_i) = \sigma^2 < \infty$. Then, for any $\epsilon > 0$, $P(|\bar{X} - \mu| > \epsilon) \to 0$, as $n \to \infty$.*

There is a stronger version of the weak law of large numbers, which says that in fact, with certainty, $\bar{X}$ will converge to μ as $n \to \infty$. The precise mathematical statement is that

$$P\left(\lim_{n \to \infty} \bar{X} = \mu\right) = 1.$$

The only condition needed is that $E(|X_i|)$ should be finite. This is called the strong law of large numbers. *It is impossible to prove it without using much more sophisticated concepts and techniques than we are using here. The strong law of large numbers is treated later in the book. Inequalities better than Chebyshev's or Markov's inequality are available under additional restrictions on the distribution of the underlying random variable X. We state three other inequalities that can sometimes give bounds better than what Chebyshev's or Markov's inequality can give.*

Theorem 1.13. *(a)* **(Cantelli's Inequality).** *Suppose $E(X) = \mu$, $Var(X) = \sigma^2$, assumed to be finite. Then,*

$$P(X - \mu \geq k\sigma) \leq \frac{1}{k^2 + 1}.$$

$$P(X - \mu \leq -k\sigma) \leq \frac{1}{k^2 + 1}.$$

(b) **(Paley–Zygmund Inequality).** *Suppose X takes only nonnegative values, with $E(X) = \mu$, $Var(X) = \sigma^2$, assumed to be finite. Then, for $0 < c < 1$,*

$$P(X > c\mu) \geq (1 - c)^2 \frac{\mu^2}{\mu^2 + \sigma^2}.$$

(c) **(Alon–Spencer Inequality).** *Suppose X takes only nonnegative integer values, with $E(X) = \mu$, $Var(X) = \sigma^2$, assumed to be finite. Then,*

$$P(X = 0) \leq \frac{\sigma^2}{\mu^2 + \sigma^2}.$$

These inequalities may be seen in Rao (1973), Paley and Zygmund (1932), and Alon and Spencer (2000, p. 58), respectively.

The area of probability inequalities is an extremely rich and diverse area. The reason for it is that inequalities are tremendously useful in giving approximate answers when the exact answer to a problem, or a calculation, is very hard or perhaps even impossible to obtain. We periodically present and illustrate inequalities over the rest of the book. Some really basic inequalities based on moments are presented in the next theorem.

1.4 Inequalities

Theorem 1.14. *(a)* **(Cauchy–Schwarz Inequality).** *Let X, Y be two random variables such that $E(X^2)$ and $E(Y^2)$ are finite. Then,*

$$E(|XY|) \le \sqrt{E(X^2)}\sqrt{E(Y^2)}.$$

(b) **(Hölder's Inequality).** *Let X, Y be two random variables, and $1 < p < \infty$ a real number such that $E(|X|^p) < \infty$. Let $q = \frac{p}{p-1}$, and suppose $E(|Y|^q) < \infty$. Then,*

$$E(|XY|) \le [E(|X|^p)]^{\frac{1}{p}}[E(|Y|^q)]^{\frac{1}{q}}.$$

(c) **(Minkowski's Inequality).** *Let X, Y be two random variables, and $p \ge 1$ a real number such that $E(|X|^p), E(|Y|^p) < \infty$. Then,*

$$[E(|X + Y|^p)]^{\frac{1}{p}} \le [E(|X|^p)]^{\frac{1}{p}} + [E(|Y|^p)]^{\frac{1}{p}},$$

and, in particular, if $E(|X|), E(|Y|)$ are both finite, then,

$$E(|X + Y|) \le E(|X|) + E(|Y|),$$

known as the **triangular inequality**.

(d) **(Lyapounov Inequality).** *Let X be a random variable, and $0 < \alpha < \beta$ such that $E(|X|^\beta) < \infty$. Then,*

$$[E(|X|^\alpha)]^{\frac{1}{\alpha}} \le [E(|X|^\beta)]^{\frac{1}{\beta}}.$$

Example 1.20 (Application of Cauchy–Schwarz Inequality). The most useful applications of Holder's inequality and the Cauchy–Schwarz inequality are to continuous random variables, which we have not discussed yet. We give a simple application of the Cauchy–Schwarz inequality to a dice problem.

Suppose X, Y are the maximum and the minimum of two rolls of a fair die. Also let X_1 be the first roll and X_2 be the second roll. Note that $XY = X_1 X_2$. Therefore,

$$\begin{aligned}
E\left(\sqrt{X}\sqrt{Y}\right) &= E\left(\sqrt{XY}\right) = E\left(\sqrt{X_1 X_2}\right) \\
&= E\left(\sqrt{X_1}\sqrt{X_2}\right) = E\left(\sqrt{X_1}\right) E\left(\sqrt{X_2}\right) \\
&= \left[E\left(\sqrt{X_1}\right)\right]^2 = \frac{1}{36}\left(\sqrt{1} + \cdots + \sqrt{6}\right)^2 \\
&= \frac{1}{36} \times (10.83)^2 = 3.26.
\end{aligned}$$

Therefore, by the Cauchy–Schwarz inequality,

$$\sqrt{E(X)}\sqrt{E(Y)} \ge 3.26$$

$$\Rightarrow \sqrt{E(X)}\sqrt{7 - E(X)} \ge 3.26$$

(because, $E(X) + E(Y) = E(X_1) + E(X_2) = 7$)

$$\Rightarrow \sqrt{m(7-m)} \geq 3.26$$

(writing m for $E(X)$)

$$\Rightarrow m(7-m) \geq 10.63$$
$$\Rightarrow m \leq 4.77,$$

because the quadratic $m(7-m) - 10.63 = 0$ has the two roots $m = 2.23, 4.77$.

It is interesting that this bound is reasonably accurate, as the exact value of $m = E(X)$ is $\frac{161}{36} = 4.47$.

1.5 Generating and Moment-Generating Functions

Studying distributions of random variables and their basic quantitative properties, such as expressions for moments, occupies a central role in both statistics and probability. It turns out that a function called the probability-generating function is often a very useful mathematical tool in studying distributions of random variables. The moment-generating function, which is related to the probability-generating function, is also extremely useful as a mathematical tool in numerous problems.

Definition 1.17. The *probability generating function* (pgf), also called simply the *generating function*, of a nonnegative integer-valued random variable X is defined as $G(s) = G_X(s) = E(s^X) = \sum_{x=0}^{\infty} s^x P(X = x)$, provided the expectation is finite.

In this definition, 0^0 is to be understood as being equal to 1. Note that $G(s)$ is always finite for $|s| \leq 1$, but it could be finite over a larger interval, depending on the specific random variable X.

Two basic properties of the generating function are the following.

Theorem 1.15. *(a) Suppose $G(s)$ is finite in some open interval containing the origin. Then, $G(s)$ is infinitely differentiable in that open interval, and $P(X = k) = \frac{G^{(k)}(0)}{k!}, k \geq 0$, where $G^{(0)}(0)$ means $G(0)$.*
(b) If $\lim_{s \uparrow 1} G^{(k)}(s)$ is finite, then $E[X(X-1)\cdots(X-k+1)]$ exists and is finite, and $G^{(k)}(1) = \lim_{s \uparrow 1} G^{(k)}(s) = E[X(X-1)\cdots(X-k+1)]$.

Definition 1.18. $E[X(X-1)\cdots(X-k+1)]$ is called the kth *factorial moment* of X.

Remark. The kth factorial moment of X exists if and only if the kth moment $E(X^k)$ exists.

One of the most important properties of generating functions is the following.

1.5 Generating and Moment-Generating Functions

Theorem 1.16. *Let $X_1, X_2, \ldots, X_n$ be independent random variables, with generating functions $G_1(s), G_2(s), \ldots, G_n(s)$. Then the generating function of $X_1 + X_2 + \cdots + X_n$ equals*

$$G_{X_1+X_2+\cdots+X_n}(s) = \prod_{i=1}^{n} G_i(s).$$

One reason that the generating function is useful as a tool is its distribution determining property, *in the following sense.*

Theorem 1.17. *Let $G(s)$ and $H(s)$ be the generating functions of two random variables X, Y. If $G(s) = H(s)$ in any nonempty open interval, then X, Y have the same distribution.*

Summarizing, then, one can find from the generating function of a nonnegative integer-valued random variable X, the pmf of X, and every moment of X, including the moments that are infinite.

Example 1.21 (Discrete Uniform Distribution). Suppose X has the discrete uniform distribution on $\{1, 2, \ldots, n\}$. Then, its generating function is

$$G(s) = E[s^X] = \sum_{x=1}^{n} s^x P(X = x) = \frac{1}{n} \sum_{x=1}^{n} s^x$$
$$= \frac{s(s^n - 1)}{n(s - 1)},$$

by summing the geometric series $\sum_{x=1}^{n} s^x$. As a check, if we differentiate $G(s)$ once, we get

$$G'(s) = \frac{1 + s^n[n(s-1) - 1]}{n(s-1)^2}.$$

On applying L'Hospital's rule, we get that $G'(1) = \frac{n+1}{2}$ which, therefore, is the mean of X.

Example 1.22 (The Poisson Distribution). Consider a nonnegative integer-valued random variable X with the pmf $p(x) = e^{-1}\frac{1}{x!}, x = 0, 1, 2, \ldots$. This is indeed a valid pmf. First, it is clear that $p(x) \geq 0$ for any x. Also,

$$\sum_{x=0}^{\infty} p(x) = \sum_{x=0}^{\infty} e^{-1}\frac{1}{x!} = e^{-1} \sum_{x=0}^{\infty} \frac{1}{x!} = e^{-1} e = 1.$$

We find the generating function of this distribution. The generating function is

$$G(s) = E[s^X] = \sum_{x=0}^{\infty} s^x e^{-1} \frac{1}{x!}$$
$$= e^{-1} \sum_{x=0}^{\infty} \frac{s^x}{x!} = e^{-1} e^s = e^{s-1}.$$

The first derivative of $G(s)$ is $G'(s) = e^{s-1}$, and therefore $G'(1) = e^0 = 1$. From our theorem above, we conclude that $E(X) = 1$. Indeed, the pmf that we have in this example is the pmf of the so-called Poisson distribution with mean one. The pmf of the Poisson distribution with a general mean λ is $p(x) = \frac{e^{-\lambda}\lambda^x}{x!}, x = 0, 1, 2, \ldots$. The Poisson distribution is an extremely important distribution in probability theory and is studied in more detail below.

We have defined the probability-generating function only for nonnegative integer-valued random variables. The moment-generating function is usually discussed in the context of general random variables, not necessarily integer-valued, or discrete. The two functions are connected. Here is the formal definition.

Definition 1.19. Let X be a real-valued random variable. The *moment-generating function (mgf)* of X is defined as

$$\psi_X(t) = \psi(t) = E[e^{tX}],$$

whenever the expectation is finite.

Note that the mgf $\psi(t)$ of a random variable X always exists and is finite if $t = 0$, and $\psi(0) = 1$. It may or may not exist when $t \neq 0$. If it does exist for t in a nonempty open interval containing zero, then many properties of X can be derived by using the mgf $\psi(t)$; it is an extremely useful tool. If X is a nonnegative integer-valued random variable, then writing s^X as $e^{X \log s}$, it follows that the (probability) generating function $G(s)$ is equal to $\psi(\log s)$, whenever $G(s) < \infty$. Thus, the two generating functions, namely the probability-generating function, and the moment-generating function are connected.

The following theorem explains the name of a moment-generating function.

Theorem 1.18. *(a) Suppose the mgf $\psi(t)$ of a random variable X is finite in some open interval containing zero. Then, $\psi(t)$ is infinitely differentiable in that open interval, and for any $k \geq 1$,*

$$E(X^k) = \psi^{(k)}(0).$$

(b) **(Distribution-Determining Property).** *If $\psi_1(t), \psi_2(t)$ are the mgfs of two random variables X, Y, and if $\psi_1(t) = \psi_2(t)$ in some nonempty open interval containing zero, then X, Y have the same distribution.*

(c) If $X_1, X_2, \ldots, X_n$ are independent random variables, and if each X_i has a mgf $\psi_i(t)$, existing in some open interval around zero, then $X_1 + X_2 + \cdots + X_n$ also has a mgf in that open interval, and

$$\psi_{X_1+X_2+\cdots+X_n}(t) = \prod_{i=1}^{n} \psi_i(t).$$

1.5 Generating and Moment-Generating Functions

Example 1.23 (Discrete Uniform Distribution). Let X have the pmf $P(X = x) = \frac{1}{n}, x = 1, 2, \ldots, n$. Then, its mgf is

$$\psi(t) = E[e^{tX}] = \frac{1}{n}\sum_{k=1}^{n} e^{tk} = \frac{e^t(e^{nt} - 1)}{n(e^t - 1)}.$$

By direct differentiation,

$$\psi'(t) = \frac{e^t(1 + ne^{(n+1)t} - (n+1)e^{nt})}{n(e^t - 1)^2}.$$

On applying L'Hospital's rule twice, we get the previously derived fact that $E(X) = \frac{n+1}{2}$.

Example 1.24. Suppose X takes only two values 0 and 1, with $P(X = 1) = p$, $P(X = 0) = 1 - p, 0 < p < 1$. Thus, X is a Bernoulli variable with parameter p. Then, the mgf of X is

$$\psi(t) = E[e^{tX}] = pe^t + (1 - p).$$

If we differentiate this, we get $\psi'(t) = pe^t, \psi''(t) = pe^t$. Therefore, $\psi'(0) = pe^0 = p$, and also $\psi''(0) = p$. From the general properties of mgfs, it then follows that $E(X) = \psi'(0) = p$, and $E(X^2) = \psi''(0) = p$. Now go back to the pmf of X that we started with in this example, and note that indeed, by direct calculation, $E(X) = E(X^2) = p$.

Closely related to the moments of a random variable are certain quantities known as *cumulants*. Cumulants arise in accurate approximation of the distribution of sums of independent random variables. They are also used for statistical modeling purposes in some applied sciences. The name *cumulant* was coined by Sir Ronald Fisher (Fisher (1929)), although they were discussed in the literature by others prior to Fisher's coining the cumulant term. We define and describe some basic facts about cumulants below; this material is primarily for reference purposes, and may be omitted at first reading.

We first need to define *central moments* of a random variable because cumulants are related to them.

Definition 1.20. Let a random variable X have a finite jth moment for some specified $j \geq 1$. The jth *central moment* of X is defined as $\mu_j = E[(X - \mu)^j]$, where $\mu = E(X)$.

Remark. Note that $\mu_1 = E(X - \mu) = 0$ and $\mu_2 = E(X - \mu)^2 = \sigma^2$, the variance of X. If X has a distribution *symmetric about zero*, then every odd-order central moment $E[(X - \mu)^{2k+1}]$ is easily proved to be zero, provided it exists.

Definition 1.21. Let X have a finite mgf $\psi(t)$ is some neighborhood of zero, and let $K(t) = \log \psi(t)$, when it exists. The rth *cumulant* of X is defined as

$\kappa_r = \frac{d^r}{dt^r} K(t)|_{t=0}$. Equivalently, the cumulants of X are the coefficients in the power series expansion $K(t) = \sum_{n=1}^{\infty} \kappa_n \frac{t^n}{n!}$, within its radius of convergence.

Note that $K(t) = \log \psi(t)$ implies that $e^{K(t)} = \psi(t)$. By equating coefficients in the power series expansion of $e^{K(t)}$ with those in the power series expansion of $\psi(t)$, it is easy to express the first few moments (and therefore, the first few central moments) in terms of the cumulants. Indeed, denoting $c_i = E(X^i), \mu = E(X) = c_1, \mu_i = E(X - \mu)^i, \sigma^2 = \mu_2$, one obtains the expressions

$$c_1 = \kappa_1; \quad c_2 = \kappa_2 + \kappa_1^2; \quad c_3 = \kappa_3 + 3\kappa_1\kappa_2 + \kappa_1^3.$$
$$c_4 = \kappa_4 + 4\kappa_1\kappa_3 + 3\kappa_2^2 + 6\kappa_1^2\kappa_2 + \kappa_1^4.$$

The corresponding expressions for the central moments are much simpler:

$$\sigma^2 = \kappa_2; \quad \mu_3 = \kappa_3; \quad \mu_4 = \kappa_4 + 3\kappa_2^2.$$

In general, the cumulants satisfy the recursion relations

$$\kappa_n = c_n - \sum_{k=1}^{n-1} \binom{n-1}{k-1} c_{n-k} \kappa_k.$$

These result in the specific expressions

$$\kappa_2 = \mu_2; \quad \kappa_3 = \mu_3; \quad \kappa_4 = \mu_4 - 3\mu_2^2.$$

High-order cumulants have quite complex expressions in terms of the central moments μ_j; the corresponding expressions in terms of the c_j are even more complex.

The derivations of these expressions stated above involve straight differentiation. We do not present the algebra. It is useful to know these expressions for some problems in statistics.

1.6 * Applications of Generating Functions to a Pattern Problem

We now describe some problems in discrete probability that are generally known as problems of patterns. Generating functions turn out to be crucially useful in analyzing many of these problems. Suppose a coin with probability p for heads is tossed repeatedly. How long does it take before we see three heads in succession for the first time? Questions such as this which pertain to waiting times for seeing one or more specified patterns are particularly amenable to use of the generating function.

1.6 Applications of Generating Functions to a Pattern Problem

Theorem 1.19. *Suppose a coin with probability p for heads is tossed repeatedly, and $N = N_r$ is the first toss at which a head run of length r is obtained. Then*

$$E(N) = \sum_{k=1}^{r} \left(\frac{1}{p}\right)^k;$$

$$Var(N) = \frac{1 - p^{1+2r} - qp^r(1+2r)}{q^2 p^{2r}}.$$

Proof. Let $p_k = P(N = k)$. The trick is to write a *recursion relation* for the sequence p_k and then convert it to a generating function problem. This technique has been found to be successful in solving numerous hard combinatorial problems.

Clearly, $p_1 = p_2 = \cdots = p_{r-1} = 0$. Also, $p_r = p^r$. The first head run of length r occurs at the $(r+1)$th trial if and only if the first trial is a tail and the last r trials are all heads; therefore, $p_{r+1} = qp^r$, where $q = 1 - p$. Similarly, $p_{r+2} = q^2 p^r + pq \times p^r = qp_{r+1} + pqp_r$. For a general $k \geq r+1$, we have the recursion relation

$$p_k = qp_{k-1} + pqp_{k-2} + \cdots p^{r-1}qp_{k-r}.$$

Multiplying by s^k and summing over k, we get

$$G(s) = p^r s^r + qsG(s) + pqs^2 G(s) + \cdots + p^{r-1}qs^r G(s)$$

$$\Rightarrow G(s) = \frac{p^r s^r}{1 - qs(1 + ps + \cdots + (ps)^{r-1})} = \frac{p^r s^r (1-ps)}{1 - s + qp^r s^{r+1}},$$

on summing the geometric series $1 + ps + \cdots + (ps)^{r-1}$, and on using the fact that $p + q = 1$. Thus, by using a very clever recursion, the generating function of N has been obtained; it is:

$$G(s) = \frac{p^r s^r (1-ps)}{1 - s + qp^r s^{r+1}}.$$

Inasmuch as we have a closed-form formula for $G(s)$, we can determine p_k for any specified k by simply repeated differentiation; we can also obtain the expected value of N, as $E(N) = G'(1)$.

By using the fact that $G''(1) = E[N(N-1)]$, we can obtain the second moment, and from there the variance of N. □

Example 1.25 (Run of Heads). If the coin is a fair coin, we get the result that the expected number of tosses necessary to get the first head run of length r is $2 + 2^2 + \cdots + 2^r$; for example, it takes on average 14 tosses of a fair coin to obtain a run of three heads for the first time.

On computing using the variance formula in the theorem, one can see that the variance of N is very large for $r > 3$; sometimes one has to wait a very long time to see a run of four or more consecutive heads.

1.7 Standard Discrete Distributions

A few special discrete distributions arise very frquently in applications. Either the underlying probability mechanism of a problem is such that one of these distributions is truly the correct distribution for that problem, or the problem may be such that one of these distributions is a very good choice to model that problem. The special distributions we present are the Binomial, the geometric, the negative binomial, the hypergeometric, and the Poisson.

The Binomial Distribution. The binomial distribution represents a sequence of independent coin tossing experiments. Suppose a coin with probability $p, 0 < p < 1$ for heads in a single trial is tossed independently a prespecified number of times, say n times, $n \geq 1$. Let X be the number of times in the n tosses that a head is obtained. Then the pmf of X is:

$$P(X = x) = \binom{n}{x} p^x (1-p)^{n-x}, \quad x = 0, 1, \ldots, n,$$

the $\binom{n}{x}$ term giving the choice of the x tosses out of the n tosses in which the heads occur.

Coin tossing, of course, is just an artifact. Suppose a trial can result in only one of two outcomes, called a *success*(S) or a *failure*(F), the probability of obtaining a success being p in any trial. Such a trial is called a *Bernoulli trial*. Suppose a Bernoulli trial is repeated independently a prespecified number of times, say n times, Let X be the number of times in the n trials that a success is obtained. Then X has the pmf given above, and we say that X has a *binomial distribution with parameters n and p*, and write $X \sim \text{Bin}(n, p)$.

The Geometric Distribution. Suppose a coin with probability $p, 0 < p < 1$, for heads in a single trial is repeatedly tossed until a head is obtained for the first time. Assume that the tosses are independent. Let X be the number of the toss at which the very first head is obtained. Then the pmf of X is:

$$P(X = x) = p(1-p)^{x-1}, \quad x = 1, 2, 3, \ldots.$$

We say that X has a *geometric distribution with parameter p*, and we write $X \sim \text{Geo}(p)$. A geometric distribution measures a waiting time for the first success in a sequence of independent Bernoulli trials, each with the same success probability p; that is, the coin cannot change from one toss to another.

The Negative Binomial Distribution. The negative binomial distribution is a generalization of a geometric distribution, when we repeatedly toss a coin with probability p for heads, independently, until a total number of r heads has been obtained, where r is some fixed integer ≥ 1. The case $r = 1$ corresponds to the geometric distribution. Let X be the number of the first toss at which the rth success is obtained. Then the pmf of X is:

1.7 Standard Discrete Distributions

$$P(X = x) = \binom{x-1}{r-1} p^r (1-p)^{x-r}, \quad x = r, r+1, \ldots,$$

the term $\binom{x-1}{r-1}$ simply giving the choice of the $r-1$ tosses among the first $x-1$ tosses where the first $r-1$ heads were obtained. We say that X has a *negative binomial distribution with parameters r and p*, and we write $X \sim NB(r, p)$.

The Hypergeometric Distribution. The hypergeometric distribution also represents the number of successes in a prespecified number of Bernoulli trials, but the trials happen to be dependent. A typical example is that of a finite population in which there are in all N objects, of which some D are of type I, and the other $N - D$ are of type II. A *without replacement sample* of size n, $1 \leq n < N$ is chosen at random from the population. Thus, the selected sampling units are necessarily different. Let X be the number of units or individuals of type I among the n units chosen. Then the pmf of X is:

$$P(X = x) = \frac{\binom{D}{x}\binom{N-D}{n-x}}{\binom{N}{n}},$$

$n - N + D \leq x \leq D$. We say that such an X has a *hypergeometric distribution with parameters n, D, N*, and we write $X \sim \text{Hypergeo}(n, D, N)$.

The Poisson Distribution. The Poisson distribution is perhaps the most used and useful distribution for modeling nonnegative integer-valued random variables.

The pmf of a *Poisson distribution with parameter λ* is:

$$P(X = x), \frac{e^{-\lambda} \lambda^x}{x!}, \quad x = 0, 1, 2, \ldots;$$

by using the power series expansion of $e^\lambda = \sum_{x=0}^{\infty} \frac{\lambda^x}{x!}$, it follows that this is indeed a valid pmf.

Three specific situations where a Poisson distribution is almost routinely adopted as a model are the following.

(A) The number of times a specific event happens in a specified period of time, for example, the number of phone calls received by someone over a 24-hour period.
(B) The number of times a specific event or phenomenon is observed in a specified amount of area or volume, for example, the number of bacteria of a certain kind in one liter of a sample of water, or the number of misprints per page of a book, and so on.
(C) The number of times a success is obtained when a Bernoulli trial with success probability p is repeated independently n times, with p being small and n being large, such that the product np has a *moderate value*, say between .5 and 10. Thus, although the *true* distribution is a binomial, a Poisson distribution is used as an effective and convenient *approximation*.

We now present the most important properties of these special discrete distributions.

Theorem 1.20. *Let $X \sim \text{Bin}(n, p)$. Then,*

(a) $\mu = E(X) = np; \quad \sigma^2 = \text{Var}(X) = np(1-p);$
(b) *The mgf of X equals $\psi(t) = (pe^t + 1 - p)^n$ at any t;*
(c) $E[(X - \mu)^3] = np(1 - 3p + 2p^2).$
(d) $E[(X - \mu)^4] = np(1-p)[1 + 3(n-2)p(1-p)].$

Theorem 1.21 (∗Mean Absolute Deviation and Mode). *Let $X \sim \text{Bin}(n, p)$. Let v denote the smallest integer $> np$ and let $m = \lfloor np + p \rfloor$. Then,*

(a) $E|X - np| = 2v(1-p)P(X = v).$
(b) *The mode of X equals m. In particular, if np is an integer, then the mode is exactly np; if np is not an integer, then the mode is one of the two integers just below and just above np.*

Theorem 1.22. *(a) Let $X \sim \text{Geo}(p)$. Let $q = 1 - p$. Then,*

$$E(X) = \frac{1}{p}; \quad \text{Var}(X) = \frac{q}{p^2}.$$

(b) Let $X \sim NB(r, p), r \geq 1$. Then,

$$E(X) = \frac{r}{p}; \quad \text{Var}(X) = \frac{rq}{p^2}.$$

Furthermore, the mgf and the (probability) generating function of X equal

$$\psi(t) = \left(\frac{pe^t}{1 - qe^t}\right)^r, \quad t < \log\left(\frac{1}{q}\right).$$

$$G(s) = \left(\frac{ps}{1 - qs}\right)^r, \quad s < \frac{1}{q}.$$

Theorem 1.23. *Let $X \sim \text{Hypergeo}(n, D, N)$, and let $p = \frac{D}{N}$. Then,*

$$E(X) = np; \quad \text{Var}(X) = np(1-p)\left(\frac{N-n}{N-1}\right).$$

Problems that should truly be modeled as hypergeometric distribution problems are often analyzed as if they were binomial distribution problems. That is, the fact that samples have been taken without replacement is ignored, and one pretends the successive draws are independent. When does it not matter that the dependence between the trials is ignored? Intuitively, we would think that if the population size N were large, and neither D nor $N - D$ were small, the trials would act as if they were independent trials. The following theorem justifies this intuition.

1.7 Standard Discrete Distributions

Theorem 1.24 (Convergence of Hypergeometric to Binomial). Let $X = X_N \sim$ Hypergeo(n, D, N), where $D = D_N$ and N are such that $N \to \infty$, $\frac{D}{N} \to p$, $0 < p < 1$. Then, for any fixed n, and for any fixed x,

$$P(X = x) = \frac{\binom{D}{x}\binom{N-D}{n-x}}{\binom{N}{n}} \to \binom{n}{x} p^x (1-p)^{n-x},$$

as $N \to \infty$.

Theorem 1.25. Let $X \sim$ Poi(λ). Then,

(a) $E(X) = Var(X) = \lambda$.
(b) $E(X - \lambda)^3 = \lambda$; $E(X - \lambda)^4 = 3\lambda^2 + \lambda$;
(c) The mgf of X equals

$$\psi(t) = e^{\lambda(e^t - 1)}.$$

(d) The integer part of λ is always a mode of X. If λ is itself an integer, then λ and $\lambda - 1$ are both modes of X.

Let us now see some illustrative examples.

Example 1.26 (Guessing on a Multiple Choice Exam). A multiple choice test with 20 questions has five possible answers for each question. A completely unprepared student picks the answer for each question at random and independently. Suppose X is the number of questions that the student answers correctly.

We identify each question with a Bernoulli trial and a correct answer as a success. Because there are 20 questions and the student picks the answer at random from five choices, $X \sim$ Bin(n, p), with $n = 20, p = \frac{1}{5} = .2$. We can now answer any question we want about X.
For example,

$$P(\text{the student gets every answer wrong}) = P(X = 0) = .8^{20} = .0115,$$

whereas,

$$P(\text{the student gets every answer right}) = P(X = 20) = .2^{20} = 1.05 \times 10^{-14},$$

a near impossibility. Suppose the instructor has decided that it will take at least 13 correct answers to pass this test. Then,

$$P(\text{the student will pass}) = \sum_{x=13}^{20} \binom{20}{x} .2^x .8^{20-x} = .000015,$$

still a very very small probability.

Example 1.27 (Meeting Someone with the Same Birthday). Suppose you were born on October 15. How many different people do you have to meet before you find someone who was also born on October 15? Under the usual conditions of equally likely birthdays, and independence of the birthdays of all people that you will meet, the number of people X you have to meet to find the first person with the same birthday as yours is geometric: $X \sim \text{Geo}(p)$ with $p = \frac{1}{365}$. The pmf of X is $P(X = x) = p(1-p)^{x-1}$. Thus, for any given k,

$$P(X > k) = \sum_{x=k+1}^{\infty} p(1-p)^{x-1} = p \sum_{x=k}^{\infty} (1-p)^x = (1-p)^k.$$

For example, the chance that you will have to meet more than 1000 people to find someone with the same birthday as yours is $(364/365)^{1000} = .064$.

Example 1.28 (Lack of Memory of Geometric Distribution). Let $X \sim \text{Geo}(p)$, and suppose m, n are given positive integers. Then, X has the interesting property

$$P(X > m+n | X > n) = P(X > m).$$

That is, suppose you are waiting for some event to happen for the first time. You have tried, say, 20 times, and you still have not succeeded. You may feel that it is due anytime now. But the chance that it will take another ten tries is the same as if you just started, and forget that you have been patient for long time and have already tried very hard for success.

The proof is simple. Indeed,

$$P(X > m+n | X > n) = \frac{P(X > m+n)}{P(X > n)} = \frac{\sum_{x>m+n} p(1-p)^{x-1}}{\sum_{x>n} p(1-p)^{x-1}}$$

$$= \frac{(1-p)^{m+n}}{(1-p)^n} = (1-p)^m = P(X > m).$$

Example 1.29 (A Classic Example: Capture–Recapture). An ingenious use of the hypergeometric distribution in estimating the size of a finite population is the *capture–recapture* method. It was originally used for estimating the total number of fish in a body of water, such as a pond. Let N be the number of fish in the pond. In this method, a certain number of fish, say D of them are initially captured and tagged with a safe mark or identification device, and are returned to the water. Then, a second sample of n fish is recaptured from the water. Assuming that the fish population has not changed in any way in the intervening time, and that the initially captured fish remixed with the fish population homogeneously, the number of fish in the second sample, say X, that bear the mark is a hypergeometric random variable, namely, $X \sim \text{Hypergeo}(n, D, N)$. We know that the expected value of a hypergeometric random variable is $n\frac{D}{N}$. If we set, as a formalism, $X = n\frac{D}{N}$. and solve for N, we get $N = \frac{nD}{X}$. This is an estimate of the total number of fish in the pond. Although the idea is extremely original, this estimate can run into various kinds of

1.7 Standard Discrete Distributions

difficulties if, for example, the first catch of fish cluster around after being returned, or hide, or if the fish population has changed between the two catches due to death or birth, and of course if X turns out to be zero. Modifications of this estimate (known as the *Petersen estimate*) are widely used in wildlife estimation, census, and by the government for estimating tax frauds and number of people afflicted with some infection.

Example 1.30 (Events over Time). April receives three phone calls at her home on the average per day. On what percentage of days does she receive no phone calls; more than five phone calls?

Because the number of calls received in a 24-hour period counts the occurrences of an event in a fixed time period, we model $X =$ number of calls received by April on one day as a Poisson random variable with mean 3. Then,

$$P(X = 0) = e^{-3} = .0498; \quad P(X > 5) = 1 - P(X \leq 5) = 1 - \sum_{x=0}^{5} e^{-3} 3^x / x!$$

$$= 1 - .9161 = .0839.$$

Thus, she receives no calls on 4.98% of the days and she receives more than five calls on 8.39% of the days. It is important to understand that X has only been modeled as a Poisson random variable, and other models could also be reasonable.

Example 1.31 (A Hierarchical Model with a Poisson Base). Suppose a chick lays a Poi(λ) number of eggs in some specified period of time, say a month. Each egg has a probability p of actually developing. We want to find the distribution of the number of eggs that actually develop during that period of time.

Let $X \sim$ Poi(λ) denote the number of eggs the chick lays, and Y the number of eggs that develop. For example,

$$P(Y = 0) = \sum_{x=0}^{\infty} P(Y = 0 | X = x) P(X = x) = \sum_{x=0}^{\infty} (1-p)^x \frac{e^{-\lambda} \lambda^x}{x!}$$

$$= e^{-\lambda} \sum_{x=0}^{\infty} \frac{(\lambda(1-p))^x}{x!} = e^{-\lambda} e^{\lambda(1-p)} = e^{-p\lambda}.$$

In general,

$$P(Y = y) = \sum_{x=y}^{\infty} \binom{x}{y} p^y (1-p)^{x-y} \frac{e^{-\lambda} \lambda^x}{x!}$$

$$= \frac{(p/(1-p))^y}{y!} e^{-\lambda} \sum_{x=y}^{\infty} \frac{1}{(x-y)!} (1-p)^x \lambda^x$$

$$= \frac{(p/(1-p))^y}{y!} e^{-\lambda} (\lambda(1-p))^y \sum_{n=0}^{\infty} \frac{(\lambda(1-p))^n}{n!},$$

on writing $n = x - y$ in the summation,

$$= \frac{(\lambda p)^y}{y!} e^{-\lambda} e^{\lambda(1-p)} = \frac{e^{-\lambda p}(\lambda p)^y}{y!},$$

and so, we recognize that $Y \sim \text{Poi}(\lambda p)$. What is interesting here is that the distribution still remains Poisson, under assumptions that seem to be very realistic physically.

1.8 Poisson Approximation to Binomial

A binomial random variable is the sum of n indicator variables. When the expectation of these indicator variables, namely p is small, and the number of summands n is large, the Poisson distribution provides a good approximation to the binomial. The Poisson distribution can also sometimes serve as a good approximation when the indicators are independent, but have different expectations p_i, or when the indicator variables have some weak dependence. We start with the Poisson approximation to the binomial when n is large, and p is small.

Theorem 1.26. *Let $X_n \sim \text{Bin}(n, p_n), n \geq 1$. Suppose $np_n \to \lambda, 0 < \lambda < \infty$, as $n \to \infty$, Let $Y \sim \text{Poi}(\lambda)$. Then, for any given $k, 0 \leq k < \infty$,*

$$P(X_n = k) \to P(Y = k),$$

as $n \to \infty$.

In fact, the convergence is not just pointwise for each fixed k, but it is uniform in k. This follows from the next theorem.

Theorem 1.27 (Le Cam, Barbour and Hall, Steele). *Let $X_n = B_1 + B_2 + \cdots + B_n$, where B_i are independent Bernoulli variables with parameters $p_i = p_{i,n}$. Let $Y_n \sim \text{Poi}(\lambda)$, where $\lambda = \lambda_n = \sum_{i=1}^{n} p_i$. Then,*

$$\sum_{k=0}^{\infty} |P(X_n = k) - P(Y_n = k)| \leq 2 \frac{1 - e^{-\lambda}}{\lambda} \sum_{i=1}^{n} p_i^2.$$

Here is an application of this Poisson approximation result.

Example 1.32 (Lotteries). Consider a weekly lottery in which 3 numbers from 25 numbers are selected at random, and a person holding exactly those 3 numbers is the winner of the lottery. Suppose the person plays for n weeks, for large n. What is the probability that he will win the lottery at least once (at least twice)?

Let X be the number of weeks that the player wins. Then, assuming the weekly lotteries to be independent, $X \sim \text{Bin}(n, p)$, where $p = 1/\binom{25}{3} = \frac{1}{2300} = .00043$.

1.8 Poisson Approximation to Binomial

Because p is small, and n is supposed to be large, $X \stackrel{\text{approx.}}{\sim} \text{Poi}(\lambda)$, $\lambda = np = .00043n$. Therefore,

$$P(X \geq 1) = 1 - P(X = 0) \approx 1 - e^{-.00043n},$$

and,

$$P(X \geq 2) = 1 - P(X = 0) - P(X = 1) \approx 1 - e^{-.00043n} - .00043n e^{-.00043n}$$
$$= 1 - (1 + .00043n)e^{-.00043n}.$$

We can compute these for various n. If he plays for five years,

$$1 - e^{-.00043n} = 1 - e^{-.00043 \times 5 \times 52} = .106,$$

and,

$$1 - (1 + .00043n)e^{-.00043n} = .006.$$

If he plays for ten years,

$$1 - e^{-.00043n} = 1 - e^{-.00043 \times 10 \times 52} = .200,$$

and,

$$1 - (1 + .00043n)e^{-.00043n} = .022.$$

We can see that the chances of any luck are at best moderate even after prolonged tries.

Sums of random variables arise very naturally in practical applications. For example, the revenue over a year is the sum of the monthly revenues; the time taken to finish a test with ten problems is the sum of the times taken to finish the individual problems, and so on. Sometimes we can reasonably assume that the various random variables being added are independent. Thus, the following general question is an important one.

Suppose $X_1, X_2, \ldots, X_k$ are k independent random variables, and suppose we know the distributions of the individual X_i. What is the distribution of the sum $X_1 + X_2 + \cdots + X_k$?

In general, this is a very difficult question. Interestingly, if the individual X_i have one of the distinguished distributions we have discussed in this chapter, then their sum is also often a distribution of that same type.

Theorem 1.28. *(a) Suppose $X_1, X_2, \ldots, X_k$ are k independent binomial random variables, with $X_i \sim \text{Bin}(n_i, p)$. Then $X_1 + X_2 + \cdots + X_k \sim \text{Bin}(n_1 + n_2 + \cdots + n_k, p)$;*
(b) Suppose $X_1, X_2, \ldots, X_k$ are k independent negative binomial random variables, with $X_i \sim \text{NB}(r_i, p)$. Then $X_1 + X_2 + \cdots + X_k \sim \text{NB}(r_1 + r_2 + \cdots + r_k, p)$;
(c) Suppose $X_1, X_2, \ldots, X_k$ are k independent Poisson random variables, with $X_i \sim \text{Poi}(\lambda_i)$. Then $X_1 + X_2 + \cdots + X_k \sim \text{Poi}(\lambda_1 + \lambda_2 + \cdots \lambda_k)$.

1.9 Continuous Random Variables

Discrete random variables serve as good examples to develop probabilistic intuition, but they do not account for all the random variables that one studies in theory and in applications. We now introduce the so called *continuous random variables*, which typically take all values in some nonempty interval, such as the unit interval, the entire real line, or the like, The right probabilistic paradigm for continuous variables cannot be pmfs. Instead of pmfs, we operate with a density function for the variable. The density function fully describes the distribution.

Definition 1.22. Let X be a real-valued random variable taking values in $\mathcal{R}$, the real line. A function $f(x)$ is called the *density function* or the *probability density function* (pdf) of X if

$$\text{For all } a, b, \quad -\infty < a \leq b < \infty, \quad P(a \leq X \leq b) = \int_a^b f(x) dx;$$

in particular, for a function $f(x)$ to be a density function of some random variable, it must satisfy:

$$f(x) \geq 0 \, \forall x \in \mathcal{R}; \quad \int_{-\infty}^{\infty} f(x) dx = 1.$$

The statement that $P(a \leq X \leq b) = \int_a^b f(x) dx$ is the same as saying that if we plot the density function $f(x)$, then the area under the graph between a and b will give the probability that X is between a and b, while the statement that $\int_{-\infty}^{\infty} f(x) dx = 1$ is the same as saying that the area under the entire graph must be one. This is a visually helpful way to think of probabilities for continuous random variables; larger areas under the graph of the density function correspond to larger probabilities.

The density function $f(x)$ can in principle be used to calculate the probability that the random variable X belongs to a general set A, not just an interval. Indeed, $P(X \in A) = \int_A f(x) dx$.

Caution. Integrals over completely general sets A in the real line are not defined. To make this completely rigorous, one has to use measure theory and concepts of a *Lebesgue integral*. However, generally we only want to calculate $P(X \in A)$ for sets A that are countable union of intervals. For such sets, defining the integral $\int_A f(x) dx$ would not be a problem and we can proceed as if we were just calculating ordinary integrals.

The definition of the cumulative distribution function remains the same as before.

Definition 1.23. Let X be a continuous random variable with a pdf $f(x)$. Then the CDF of X is defined as

$$F(x) = P(X \leq x) = P(X < x) = \int_{-\infty}^{x} f(t) dt.$$

1.9 Continuous Random Variables

Remark. At any point x_0 at which $f(x)$ is continuous, the CDF $F(x)$ is differentiable, and $F'(x_0) = f(x_0)$. In particular, if $f(x)$ is continuous everywhere, then $F'(x) = f(x)$ at all x.

Again, to be strictly rigorous, one really needs to say in the above sentence that $F'(x) = f(x)$ at almost all x, a concept in measure theory.

Example 1.33 (Using the Density to Calculate a Probability). Suppose X has the uniform density on $[0, 1]$ defined by $f(x) = 1, 0 \leq x \leq 1$, and $f(x) = 0$ otherwise. We write $X \sim U[0, 1]$. Consider the events

$$A = \{X \text{ is between } .4 \text{ and } .6\};$$
$$B = \{X(1 - X) \leq .21\};$$
$$C = \left\{\sin\left(\frac{\pi}{2}X\right) \geq \frac{1}{\sqrt{2}}\right\};$$
$$D = \{X \text{ is a rational number}\}.$$

We calculate each of $P(A)$, $P(B)$, $P(C)$, and $P(D)$. Recall that the probability of any event, say E, is calculated as $P(E) = \int_E f(x)dx$, where $f(x)$ is the density function: here $f(x) = 1$ on $[0, 1]$. Then,

$$P(A) = \int_{.4}^{.6} dx = .2.$$

Next, note that $x(1 - x) = .21$ has two roots in $[0,1]$, namely $x = .3, .7$, and $x(1 - x) \leq .21$ if $x \leq .3$ or $\geq .7$. Therefore,

$$P(B) = P(X \leq .3) + P(X \geq .7) = \int_0^{.3} dx + \int_{.7}^1 dx = .3 + .3 = .6.$$

For the event C, $\sin(\frac{\pi}{2}X) \geq \frac{1}{\sqrt{2}}$ if (and only if)

$$\frac{\pi}{2}X \geq \frac{\pi}{4} \Rightarrow X \geq \frac{1}{2}.$$

Thus,

$$P(C) = P\left(X \geq \frac{1}{2}\right) = \int_{\frac{1}{2}}^1 dx = \frac{1}{2}.$$

Finally, the set of rationals in $[0,1]$ is a countable set. Therefore,

$$P(D) = \sum_{x;x \text{ is rational}} P(X = x) = \sum_{x;x \text{ is rational}} 0 = 0.$$

Example 1.34 (From CDF to PDF and Median). Consider the function $F(x) = 0$, if $x < 0; = 1 - e^{-x}$ if $0 \leq x < \infty$. This is a nonnegative nondecreasing function, that goes to one as $x \to \infty$, is continuous at any real number x, and is also differentiable at any x except $x = 0$. Thus, it is the CDF of a continuous random variable, and the PDF can be obtained by the relation $f(x) = F'(x) = e^{-x}, 0 < x < \infty$, and $f(x) = F'(x) = 0, x < 0$. At $x = 0$, $F(x)$ is not differentiable. But we can define the PDF in any manner we like at one specific point; so to be specific, we write our PDF as

$$f(x) = e^{-x} \quad \text{if } 0 \leq x < \infty;$$
$$= 0, \quad \text{if } x < 0.$$

This density is called *the standard exponential density* and is enormously important in practical applications.

From the formula for the CDF, we see that $F(m) = .5 \Rightarrow 1 - e^{-m} = .5 \Rightarrow e^{-m} = .5 \Rightarrow m = \log 2 = .693$. Thus, we have established that the standard exponential density has median $\log 2 = .693$.

In general, given a number p, there can be infinitely many values x such that $F(x) = p$. Any such value splits the distribution into two parts, $100p\%$ of the probability below it, and $100(1-p)\%$ above. Such a value is called the pth quantile or percentile of F. However, in order to give a prescription for choosing a unique value when there is more than one x at which $F(x) = p$, the following definition is adopted.

Definition 1.24. Let X have the CDF $F(x)$. Let $0 < p < 1$. The pth *quantile* or the pth *percentile* of X is defined to be the first x such that $F(x) \geq p$:

$$F^{-1}(p) = \inf\{x : F(x) \geq p\}.$$

The function $F^{-1}(p)$ is also sometimes denoted as $Q(p)$ and is called the *quantile function of F or X*.

Remark. Statisticians call $Q(.25)$ and $Q(.75)$ the first and the third *quartile* of F or X.

The distribution of a continuous random variable is completely described if we describe either its density function, or its CDF. For flexible modeling, it is useful to know how to create new densities or new CDFs out of densities or CDFs that we have already thought of. This is similar to generating new functions out of old functions in calculus. The following theorem describes some standard methods to make new densities or CDFs out of already available ones.

Theorem 1.29. *(a) Let $f(x)$ be any density function. Then, for any real number μ and any $\sigma > 0$,*

$$g(x) = g_{\mu,\sigma}(x) = \frac{1}{\sigma} f\left(\frac{x-\mu}{\sigma}\right)$$

is also a valid density function.

1.9 Continuous Random Variables

(b) Let $f_1, f_2, \ldots, f_k$ be k densities for some $k, 2 \leq k < \infty$, and let $p_1, p_2, \ldots, p_k$ be k constants such that each $p_i \geq 0$, and $\sum_{i=1}^{k} p_i = 1$. Then,

$$f(x) = \sum_{i=1}^{k} p_i f_i(x)$$

is also a valid density function.

Densities of the form in part (a) are called location scale parameter densities, *with μ as a location and σ as a scale parameter. Densities of the form in part (b) are known as* finite mixtures. *Both types are enormously useful in statistics.*

Two other very familiar concepts in probability and statistics are those of symmetry *and* unimodality. *Symmetry of a density function means that around some point, the density has two halves that are exact mirror images of each other. Unimodality means that the density has just one peak point at some value. We give the formal definitions.*

Definition 1.25. A density function $f(x)$ is called *symmetric* around a number M if $f(M+u) = f(M-u) \ \forall \, u > 0$. In particular, $f(x)$ is symmetric around zero if $f(u) = f(-u) \ \forall \, u > 0$.

Definition 1.26. A density function $f(x)$ is called *strictly unimodal* at (or around) a number M if $f(x)$ is increasing for $x < M$, and decreasing for $x > M$.

Example 1.35 (The Triangular Density). Consider the density function

$$f(x) = cx, \quad 0 \leq x \leq \frac{1}{2};$$
$$= c(1-x), \quad \frac{1}{2} \leq x \leq 1,$$

where c is a normalizing constant. It is easily verified that $c = 4$. This density consists of two different linear segments on $[0, \frac{1}{2}]$ and $[\frac{1}{2}, 1]$. A plot of this density see Fig. 1.1 looks like a triangle, and it is called the *triangular density* on $[0, 1]$. Note that it is symmetric and strictly unimodal.

Example 1.36 (The Double Exponential Density). We have previously seen the standard exponential density on $[0, \infty)$ defined as $e^{-x}, x \geq 0$. We can extend this to the negative real numbers by writing $-x$ for x in the above formula; that is, simply define the density to be e^x for $x \leq 0$. Then, we have an overall function that equals

$$e^{-x} \text{ for } x \geq 0.$$
$$e^{x} \text{ for } x \leq 0.$$

This function integrates to

$$\int_0^\infty e^{-x} dx + \int_{-\infty}^0 e^x dx = 1 + 1 = 2.$$

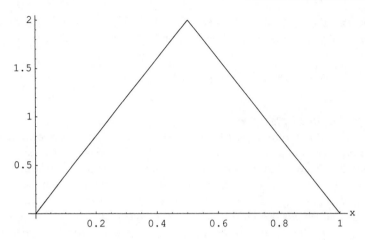

Fig. 1.1 Triangular density on [0,1]

So, if we use a *normalizing constant* of $\frac{1}{2}$, then we get a valid density on the entire real line:

$$f(x) = \frac{1}{2}e^{-x} \quad \text{for } x \geq 0.$$

$$f(x) = \frac{1}{2}e^{x} \quad \text{for } x \leq 0.$$

The two lines can be combined into one formula as

$$f(x) = \frac{1}{2}e^{-|x|}, \quad -\infty < x < \infty.$$

This is the *standard double exponential density*, and is symmetric, unimodal, and has a *cusp* at $x = 0$; see Fig. 1.2.

Example 1.37 (The Normal Density). The double exponential density tapers off to zero at the linear exponential rate at both tails (i.e., as $x \to \pm\infty$). If we force the density to taper off at a quadratic exponential rate, then we will get a function like e^{-ax^2}, for some chosen $a > 0$. Although this is obviously nonnegative, and also has a finite integral over the whole real line, it does not integrate to one. So we need a normalizing constant to make it a valid density function. Densities of this form are called *normal densities*, and occupy the central place among all distributions in the theory and practice of probability and statistics. Gauss, while using the method of least squares for analyzing astronomical data, used the normal distribution to justify least squares methods; the normal distribution is also often called the *Gaussian distribution*, although de Moivre and Laplace both worked with it before

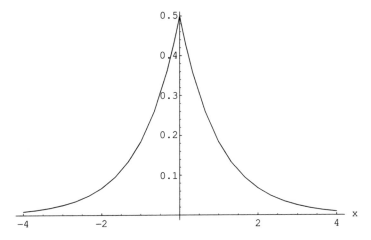

Fig. 1.2 Standard double exponential density

Gauss. Physical data on many types of variables approximately fit a normal distribution. The theory of statistical methods is often best understood when the underlying distribution is normal. The normal distributions have many unique properties not shared by any other distribution. Because of all these reasons, the normal density, also called *the bell curve*, is the most used, important, and well-studied distribution.

Let
$$f(x) = f(x|\mu,\sigma) = ce^{-\frac{(x-\mu)^2}{2\sigma^2}}, \quad -\infty < x < \infty,$$

where c is a normalizing constant. The normalizing constant can be proved to be equal to $\frac{1}{\sigma\sqrt{2\pi}}$. Thus, a normal density with parameters μ and σ is given by

$$f(x|\mu,\sigma) = \frac{1}{\sigma\sqrt{2\pi}} e^{-\frac{(x-\mu)^2}{2\sigma^2}}, \quad -\infty < x < \infty.$$

We write $X \sim N(\mu,\sigma^2)$; we show later that the two parameters μ and σ^2 are the mean and the variance of this distribution. Note that the $N(\mu,\sigma^2)$ density is a location-scale parameter density.

If $\mu = 0$ and $\sigma = 1$, this simplifies to the formula

$$\frac{1}{\sqrt{2\pi}} e^{-\frac{x^2}{2}}, \quad -\infty < x < \infty,$$

and is universally denoted by the notation $\phi(x)$. It is called the *standard normal density*. The standard normal density, then, is:

$$\phi(x) = \frac{1}{\sqrt{2\pi}} e^{-\frac{x^2}{2}}, \quad -\infty < x < \infty.$$

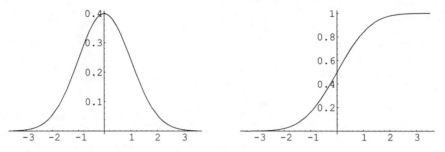

Fig. 1.3 The standard normal density and the CDF

Consequently, the CDF of the standard normal density is the function $\int_{-\infty}^{x} \phi(t)dt$. It is not possible to express the CDF in terms of the elementary functions. It is standard practice to denote it by using the notation $\Phi(x)$, and compute it using widely available tables or software for a given x needed in a specific application.

A plot of the standard normal density and its CDF are given in Fig. 1.3. Note the bell shape of the density function $\phi(x)$.

1.10 Functions of a Continuous Random Variable

As for discrete random variables, we are often interested in the distribution of some function $g(X)$ of a continuous random variable X. For example, X could measure the input into some production process, and $g(X)$ could be a function that describes the output. For one-to-one functions $g(X)$, one has the following important formula.

Theorem 1.30 (The Jacobian Formula). *Let X have a continuous pdf $f(x)$ and a CDF $F(x)$, and suppose $Y = g(X)$ is a strictly monotone function of X with a nonzero derivative. Then Y has the pdf*

$$f_Y(y) = \frac{f(g^{-1}(y))}{|g'(g^{-1}(y))|},$$

where y belongs to the range of g.

Example 1.38 (Simple Linear Transformations). Suppose X is any continuous random variable with a pdf $f(x)$ and let $Y = g(X)$ be the linear function (a location and scale change on X) $g(X) = a + bX, b \neq 0$. This is obviously a strictly monotone function, as $b \neq 0$. Take $b > 0$. Then the inverse function of g is $g^{-1}(y) = \frac{y-a}{b}$, and of course $g'(x) \equiv b$. Putting it all together, from the theorem above,

$$f_Y(y) = \frac{f(g^{-1}(y))}{|g'(g^{-1}(y))|} = \frac{1}{b} f\left(\frac{y-a}{b}\right);$$

1.10 Functions of a Continuous Random Variable

in general, whether b is positive or negative, the formula is:

$$f_Y(y) = \frac{1}{|b|} f\left(\frac{y-a}{b}\right).$$

Example 1.39 (From Exponential to Uniform). Suppose X has the standard exponential density $f(x) = e^{-x}, x \geq 0$. Let $Y = g(X) = e^{-X}$. Again, $g(X)$ is a strictly monotone function, and the inverse function is found as follows.

$$g(x) = e^{-x} = y \Rightarrow x = -\log y = g^{-1}(y).$$

Also, $g'(x) = -e^{-x}$,

$$\Rightarrow f_Y(y) = \frac{f(g^{-1}(y))}{g'(g^{-1}(y))} = \frac{e^{-(-\log y)}}{|e^{-(-\log y)}|}.$$

$$= \frac{y}{y} = 1, \quad 0 \leq y \leq 1.$$

We have thus proved that if X has a standard exponential density, then $Y = e^{-X}$ is *uniformly distributed* on $[0, 1]$.

There is actually nothing special about choosing X to be the standard exponential; the following important result says that what we saw in the above example is completely general for all continuous random variables.

Theorem 1.31. *Let X have a continuous CDF $F(x)$. Consider the new random variables $Y = 1 - F(X)$ and $Z = F(X)$. Then both Y, Z are distributed as $U[0, 1]$.*

It is useful to remember this result in informal notation:

$$F(X) = U, \text{ and } F^{-1}(U) = X.$$

The implication is a truly useful one. Suppose for purposes of computer experiments, we want to have computer-simulated values of some *random variable X that has some CDF F and the quantile function $Q = F^{-1}$. Then, all we need to do is to have the computer generate $U[0, 1]$ values, say $u_1, u_2, \ldots, u_n$, and use $x_1 = F^{-1}(u_1), x_2 = F^{-1}(u_2), \ldots, x_n = F^{-1}(u_n)$ as the set of simulated values for our random variable of interest, namely X. Thus, the problem can be reduced to simulation of uniform values, a simple task. The technique has so many uses that there is a name for this particular function $Z = F^{-1}(U)$ of a uniform random variable U.*

Definition 1.27 (Quantile Transformation). Let U be a $U[0, 1]$ random variable and let $F(x)$ be a continuous CDF. Then the function of U defined as $X = F^{-1}(U)$ is called the *quantile transformation of U*, and it has exactly the CDF F.

What we have shown here is that we can simply start with a $U[0, 1]$ random variable and convert it to any other continuous random variable X we want by simply using a transformation of U, and that transformation is the quantile transformation.

Example 1.40 (The Cauchy Distribution). The Cauchy density, like the normal and the double exponential, is also symmetric and unimodal, but the properties are very different. It is such an atypical density that we often think of the Cauchy density first when we look for a counterexample to a conjecture. There is a very interesting way to obtain a Cauchy density from a uniform density by using the quantile transformation. We describe that derivation in this example.

Suppose a person holds a flashlight in her hand, and standing one foot away from an infinitely long wall, points the beam of light in a *random direction*. Here, by random direction, we mean that the point of landing of the light ray makes an angle X with the individual (considered to be a straight line one foot long), and this angle $X \sim U[-\pi/2, \pi/2]$. Let Y be the horizontal distance of the point at which the light lands from the person, with Y being considered negative if the light lands on the person's left, and it being considered positive if it lands on the person's right.

Then, by elementary trigonometry,

$$\tan(X) = \frac{Y}{1} \Rightarrow Y = \tan(X).$$

Now $g(X) = \tan X$ is a strictly monotone function of X, and the inverse function is $g^{-1}(y) = \arctan(y), -\infty < y < \infty$. Also, $g'(x) = 1 + \tan^2 x$. Putting it all together,

$$f_Y(y) = \frac{\frac{1}{\pi}}{1 + [\tan(\arctan y)]^2} = \frac{1}{\pi(1 + y^2)}, \quad -\infty < y < \infty.$$

This is the *standard Cauchy density*.

The Cauchy density is particularly notorious for its heavy tail.

Example 1.41 (An Interesting Function that Is Not Strictly Monotone). Suppose X has the standard normal density $f(x) = \frac{1}{\sqrt{2\pi}} e^{-x^2/2}$ on $(-\infty, \infty)$. We want to find the density of $Y = g(X) = X^2$. However, we immediately realize that X^2 is not a strictly monotone function on the whole real line (its graph is a parabola). Thus, the general formula given above for densities of strictly monotone functions cannot be applied in this problem. We attack the problem directly. Thus,

1.10 Functions of a Continuous Random Variable

$$P(Y \leq y) = P(X^2 \leq y) = P(X^2 \leq y, X > 0) + P(X^2 \leq y, X < 0)$$
$$= P(0 < X \leq \sqrt{y}) + P(-\sqrt{y} \leq X < 0)$$
$$= F(\sqrt{y}) - F(0) + [F(0) - F(-\sqrt{y})] = F(\sqrt{y}) - F(-\sqrt{y}),$$

where F is the CDF of X, that is, the standard normal CDF.

Inasmuch as we have obtained the CDF of Y, we now differentiate to get the pdf of Y:

$$f_Y(y) = \frac{d}{dy}[F(\sqrt{y}) - F(-\sqrt{y})] = \frac{f(\sqrt{y})}{2\sqrt{y}} - \frac{f(-\sqrt{y})}{-2\sqrt{y}}$$

(by use of the chain rule)

$$= \frac{f(\sqrt{y})}{2\sqrt{y}} + \frac{f(\sqrt{y})}{2\sqrt{y}}$$

(because f is symmetric around zero, i.e., $f(-u) = f(u)$ for any u)

$$= \frac{2f(\sqrt{y})}{2\sqrt{y}} = \frac{f(\sqrt{y})}{\sqrt{y}} = \frac{e^{-y/2}}{\sqrt{2\pi y}},$$

$y > 0$. This is a very special density in probability and statistics, and is called the *chi-square density with one degree of freedom*. We have thus proved that the square of a standard normal random variable has a chi-square distribution with one degree of freedom.

There is an analogous Jacobian formula for transformations $g(X)$ that are not one-to-one. Basically, we need to break the problem up into disjoint intervals, on each of which the function g is one-to-one, apply the usual Jacobian technique on each such subinterval, and then piece them together. Here is the formula.

Theorem 1.32 (Density of a Nonmonotone Transformation). *Let X have a continuous pdf $f(x)$ and let $Y = g(X)$ be a transformation of X such that for a given y, the equation $g(x) = y$ has at most countably many roots, say $x_1, x_2, \ldots$, where the x_i depend on the given y. Assume also that g has a nonzero derivative at each x_i. Then, Y has the pdf*

$$f_Y(y) = \sum_i \frac{f(x_i)}{|g'(x_i)|}.$$

1.10.1 Expectation and Moments

For discrete random variables, expectation was seen to be equal to $\sum_x x P(X = x)$. Of course, for continuous random variables, the analogous sum $\sum_x x f(x)$ is not defined. The correct definition of expectation for continuous random variables replaces sums by integrals.

Definition 1.28. Let X be a continuous random variable with a pdf $f(x)$. We say that the expectation of X exists if $\int_{-\infty}^{\infty} |x| f(x) dx < \infty$, in which case the expectation, or the expected value, or the mean of X is defined as

$$E(X) = \mu = \int_{-\infty}^{\infty} x f(x) dx.$$

Suppose X is a continuous random variable with a pdf $f(x)$ and $Y = g(X)$ is a function of X. If Y has a density, say $f_Y(y)$, then we can compute the expectation as $\int y f_Y(y) dy$, or as $\int g(x) f(x) dx$. Because Y need not always be a continuous random variable just because X is, it may not in general have a density $f_Y(y)$; but the second expression is always applicable and correct.

Theorem 1.33. *Let X be a continuous random variable with pdf $f(x)$. Let $g(X)$ be a function of X. The expectation of $g(X)$ exists if and only if $\int_{-\infty}^{\infty} |g(x)| f(x) dx < \infty$, in which case the expectation of $g(X)$ is*

$$E[g(X)] = \int_{-\infty}^{\infty} g(x) f(x) dx.$$

The definitions of moments and the variance remain the same as in the discrete case.

Definition 1.29. Let X be a continuous random variable with pdf $f(x)$. Then the kth moment of X is defined to be $E(X^k), k \geq 1$. We say that the kth moment does not exist if $E(|X|^k) = \infty$.

Corollary. *Suppose X is a continuous random variable with pdf $f(x)$. Then its variance, provided it exists, is equal to*

$$\sigma^2 = \int_{-\infty}^{\infty} (x - \mu)^2 f(x) dx = \int_{-\infty}^{\infty} x^2 f(x) dx - \mu^2.$$

One simple observation that saves calculations, but is sometimes overlooked, is the following fact; the proof of it merely uses the integration result that the integral of the product of an odd function and an even function on a symmetric interval is zero, if the integral exists.

Proposition. *Suppose X has a distribution symmetric around some number a; that is, $X - a$ and $a - X$ have the same distribution. Then, $E[(X - a)^{2k+1}] = 0$, for any $k \geq 0$ for which the expectation $E[(X - a)^{2k+1}]$ exists.*

For example, if X has a distribution symmetric about zero, then any odd moment (e.g., $E(X), E(X^3)$, etc.), provided it exists, must be zero. There is no need to calculate it; it is automatically zero.

Example 1.42 (Area of a Random Triangle). Suppose an equilateral triangle is constructed by choosing the common side length X to be uniformly distributed on $[0, 1]$. We want to find the mean and the variance of the area of the triangle.

1.10 Functions of a Continuous Random Variable

For a general triangle with sides a, b, c, the area equals

$$\text{Area} = \sqrt{s(s-a)(s-b)(s-c)},$$

where $s = \frac{a+b+c}{2}$. When all the side lengths are equal, say, to a, this reduces to $\frac{\sqrt{3}}{4}a^2$. Therefore, in this example, we want the mean and variance of $Y = \frac{\sqrt{3}}{4}X^2$. The mean is

$$E(Y) = \frac{\sqrt{3}}{4}E(X^2) = \frac{\sqrt{3}}{4}\frac{1}{3} = \frac{1}{4\sqrt{3}}.$$

The variance equals

$$\text{var}(Y) = E(Y^2) - [E(Y)]^2 = \frac{3}{16}E(X^4) - \frac{1}{48} = \frac{3}{16}\frac{1}{5} - \frac{1}{48}$$
$$= \frac{3}{80} - \frac{1}{48} = \frac{1}{60}.$$

For the next example, we need the definition of the *Gamma function*. It repeatedly necessary for us to work with the Gamma function in this text.

Definition 1.30. The Gamma function is defined as

$$\Gamma(\alpha) = \int_0^\infty e^{-x} x^{\alpha-1} dx, \quad \alpha > 0.$$

In particular,

$$\Gamma(n) = (n-1)!, \text{ for any positive integer } n.$$
$$\Gamma(\alpha+1) = \alpha\Gamma(\alpha) \ \forall \alpha > 0.$$
$$\Gamma\left(\frac{1}{2}\right) = \sqrt{\pi}.$$

Example 1.43 (Moments of Exponential). Let X have the standard exponential density. Then, all its moments exist, and indeed,

$$E(X^n) = \int_0^\infty x^n e^{-x} dx = \Gamma(n+1) = n!.$$

In particular,
$$E(X) = 1; \quad E(X^2) = 2,$$

and therefore, $\text{Var}(X) = E(X^2) - [E(X)]^2 = 2 - 1 = 1$. Thus, the standard exponential density has the same mean and variance.

Example 1.44 (Absolute Value of a Standard Normal). This is often required in calculations in statistical theory. Let X have the standard normal distribution; we want to find $E(|X|)$. By definition,

$$E(|X|) = \int_{-\infty}^{\infty} |x| f(x) dx = \frac{1}{\sqrt{2\pi}} \int_{-\infty}^{\infty} |x| e^{-x^2/2} dx = \frac{2}{\sqrt{2\pi}} \int_{0}^{\infty} x e^{-x^2/2} dx$$

(because $|x|e^{-x^2/2}$ is an even function of x on $(-\infty, \infty)$)

$$= \frac{2}{\sqrt{2\pi}} \int_0^\infty \left[\frac{d}{dx}\left(-e^{-x^2/2}\right)\right] dx = \frac{2}{\sqrt{2\pi}} \left(-e^{-x^2/2}\right)\Big|_0^\infty$$
$$= \frac{2}{\sqrt{2\pi}} = \sqrt{\frac{2}{\pi}}.$$

Example 1.45 (A Random Variable Whose Expectation Does Not Exist). Consider the standard Cauchy random variable with the density $f(x) = \frac{1}{\pi(1+x^2)}$, $-\infty < x < \infty$. Recall that for $E(X)$ to exist, we must have $\int_{-\infty}^{\infty} |x| f(x) dx < \infty$. But,

$$\int_{-\infty}^{\infty} |x| f(x) dx = \frac{1}{\pi} \int_{-\infty}^{\infty} \frac{|x|}{1+x^2} dx \geq \frac{1}{\pi} \int_0^\infty \frac{x}{1+x^2} dx$$
$$\geq \frac{1}{\pi} \int_0^M \frac{x}{1+x^2} dx$$

(for any $M < \infty$)

$$= \frac{1}{2\pi} \log(1 + M^2),$$

and on letting $M \to \infty$, we see that

$$\int_{-\infty}^{\infty} |x| f(x) dx = \infty.$$

Therefore the expectation of a standard Cauchy random variable, or synonymously, the expectation of a standard Cauchy distribution does not exist.

Example 1.46 (Moments of the Standard Normal). In contrast to the standard Cauchy, every moment of a standard normal variable exists. The basic reason is that the tail of the standard normal density is too thin. A formal proof is as follows.

Fix $k \geq 1$. Then,

$$|x|^k e^{-x^2/2} = |x|^k e^{-x^2/4} e^{-x^2/4} \leq C e^{-x^2/4},$$

where C is a finite constant such that $|x|^k e^{-x^2/4} \leq C$ for any real number x (such a constant C does exist). Therefore,

$$\int_{-\infty}^{\infty} |x|^k e^{-x^2/2} dx \leq C \int_{-\infty}^{\infty} e^{-x^2/4} dx < \infty.$$

Hence, by definition, for any $k \geq 1$, $E(X^k)$ exists.

1.10 Functions of a Continuous Random Variable

Now, take k to be an odd integer, say $k = 2n + 1, n \geq 0$. Then,

$$E(X^k) = \frac{1}{\sqrt{2\pi}} \int_{-\infty}^{\infty} x^{2n+1} e^{-x^2/2} dx = 0,$$

because x^{2n+1} is an *odd function* and $e^{-x^2/2}$ is an *even function*. Thus, every odd moment of the standard normal distribution is zero.

Next, take k to be an even integer, say $k = 2n, n \geq 1$. Then,

$$E(X^k) = \frac{1}{\sqrt{2\pi}} \int_{-\infty}^{\infty} x^{2n} e^{-x^2/2} dx = \frac{2}{\sqrt{2\pi}} \int_{0}^{\infty} x^{2n} e^{-x^2/2} dx$$

$$= \frac{2}{\sqrt{2\pi}} \int_{0}^{\infty} z^n e^{-z/2} \frac{1}{2\sqrt{z}} dz = \frac{1}{\sqrt{2\pi}} \int_{0}^{\infty} z^{n-1/2} e^{-z/2} dz,$$

on making the substitution $z = x^2$.

Now make a further substitution $u = \frac{z}{2}$. Then we get,

$$E(X^{2n}) = \frac{1}{\sqrt{2\pi}} \int_{0}^{\infty} (2u)^{n-1/2} e^{-u} 2du = \frac{2^n}{\sqrt{\pi}} \int_{0}^{\infty} u^{n-1/2} e^{-u} du.$$

Now, we recognize $\int_0^\infty u^{n-1/2} e^{-u} du$ to be $\Gamma(n + \frac{1}{2})$, and so, we get the formula

$$E(X^{2n}) = \frac{2^n \Gamma(n + \frac{1}{2})}{\sqrt{\pi}}, \quad n \geq 1.$$

By using the *Gamma duplication formula*

$$\Gamma\left(n + \frac{1}{2}\right) = \sqrt{\pi} 2^{1-2n} \frac{(2n-1)!}{(n-1)!},$$

this reduces to

$$E(X^{2n}) = \frac{(2n)!}{2^n n!}, \quad n \geq 1.$$

1.10.2 Moments and the Tail of a CDF

We now describe methods to calculate moments of a random variable from its survival function, namely $\bar{F}(x) = 1 - F(x)$. There are important relationships between the existence of moments and the rapidity with which the survival function goes to zero as $|x| \to \infty$.

(a) Let X be a nonnegative random variable and suppose $E(X)$ exists. Then
$$x\overline{F}(x) = x[(1 - F(x))] \to 0, \text{ as } x \to \infty.$$

(b) Let X be a nonnegative random variable and suppose $E(X)$ exists. Then $E(X) = \int_0^\infty \overline{F}(x)dx$.

(c) Let X be a nonnegative random variable and suppose $E(X^k)$ exists, where $k \geq 1$ is a given positive integer. Then
$$x^k\overline{F}(x) = x^k[1 - F(x)] \to 0, \text{ as } x \to \infty.$$

(d) Let X be a nonnegative random variable and suppose $E(X^k)$ exists. Then
$$E(X^k) = \int_0^\infty (kx^{k-1})[1 - F(x)]dx.$$

(e) Let X be a general real-valued random variable and suppose $E(X)$ exists. Then
$$x[1 - F(x) + F(-x)] \to 0, \text{ as } x \to \infty.$$

(f) Let X be a general real-valued random variable and suppose $E(X)$ exists. Then
$$E(X) = \int_0^\infty [1 - F(x)]dx - \int_{-\infty}^0 F(x)dx.$$

Example 1.47 (Expected Value of the Minimum of Several Uniform Variables). Suppose $X_1, X_2, \ldots, X_n$ are independent $U[0, 1]$ random variables, and let $m_n = \min\{X_1, X_2, \ldots, X_n\}$ be their minimum. By virtue of the independence of $X_1, X_2, \ldots, X_n$,

$$P(m_n > x) = P(X_1 > x, X_2 > x, \ldots, X_n > x)$$
$$= \prod_{i=1}^n P(X_i > x) = (1 - x)^n, \quad 0 < x < 1,$$

and $P(m_n > x) = 0$ if $x \geq 1$. Therefore, by the above theorem,

$$E(m_n) = \int_0^\infty P(m_n > x)dx = \int_0^1 P(m_n > x)dx = \int_0^1 (1 - x)^n dx$$
$$= \int_0^1 x^n dx = \frac{1}{n+1}.$$

1.11 Moment-Generating Function and Fundamental Inequalities

The definition previously given of the moment-generating function of a random variable is completely general. We work out a few examples for some continuous random variables.

Example 1.48 (Moment-Generating Function of Standard Exponential). Let X have the standard exponential density. Then,

$$E(e^{tX}) = \int_0^\infty e^{tx} e^{-x} dx = \int_0^\infty e^{-(1-t)x} dx = \frac{1}{1-t},$$

if $t < 1$, and it equals $+\infty$ if $t \geq 1$. Thus, the mgf of the standard exponential distribution is finite if and only if $t < 1$. So, the moments can be found by differentiating the mgf, namely, $E(X^n) = \psi^{(n)}(0)$. Now, at any $t < 1$, by direct differentiation, $\psi^{(n)}(t) = \frac{n!}{(1-t)^{n+1}} \Rightarrow E(X^n) = \psi^{(n)}(0) = n!$, a result we have derived before directly.

Example 1.49 (Moment-Generating Function of Standard Normal). Let X have the standard normal density. Then,

$$E(e^{tX}) = \frac{1}{\sqrt{2\pi}} \int_{-\infty}^\infty e^{tx} e^{-x^2/2} dx = \frac{1}{\sqrt{2\pi}} \int_{-\infty}^\infty e^{-(x-t)^2/2} dx \times e^{t^2/2}$$

$$= \frac{1}{\sqrt{2\pi}} \int_{-\infty}^\infty e^{-z^2/2} dz \times e^{t^2/2} = 1 \times e^{t^2/2} = e^{t^2/2},$$

because $\frac{1}{\sqrt{2\pi}} \int_{-\infty}^\infty e^{-z^2/2} dz$ is the integral of the standard normal density, and so must be equal to one.

We have therefore proved that the mgf of the standard normal distribution exists at any real t and equals $\psi(t) = e^{t^2/2}$.

The mgf is useful in deriving inequalities on probabilities of tail values of a random variable that have proved to be extremely useful in many problems in statistics and probability. In particular, these inequalities typically give much sharper bounds on the probability that a random variable would be far from its mean value than Chebyshev's inequality can give. Such probabilities are called *large deviation probabilities*. We treat large deviations in detail in Chapter 17. We present a particular large deviation inequality below and then present some neat applications.

Theorem 1.34 (Chernoff–Bernstein Inequality). *Let X have the mgf $\psi(t)$, and assume that $\psi(t) < \infty$ for $t < t_0$ for some $t_0, 0 < t_0 \leq \infty$. Let $\kappa(t) = \log \psi(t)$, and for a real number x, define*

$$I(x) = \sup_{0 < t < t_0} [tx - \kappa(t)].$$

Then,
$$P(X \geq x) \leq e^{-I(x)}.$$

See Bernstein (1927) and Chernoff (1952) for this inequality and other refinements of it.

To apply the Chernoff–Bernstein inequality, it is necessary to be able to find the mgf $\psi(t)$ and then be able to find the function $I(x)$, which is called the rate function of X.

We now show an example.

Example 1.50 (Testing the Bound in the Standard Normal Case). Suppose X is a standard normal variable. Then the exact value of the probability $P(X > x) = 1 - P(X \leq x) = 1 - \Phi(x)$ is easily computable, although no formula can be written for it. The Chebyshev inequality will give for $x > 0$,

$$P(X > x) = \frac{1}{2}P(|X| > x) \leq \frac{1}{2x^2}.$$

To apply the Chernoff–Bernstein bound, use the formula $\psi(t) = e^{t^2/2} \Rightarrow \kappa(t) = t^2/2 \Rightarrow I(x) = \sup_{t>0}[tx - t^2/2] = x^2/2$. Therefore,

$$P(X > x) \leq e^{-I(x)} = e^{-x^2/2}.$$

Obviously, the Chernoff–Bernstein bound is much smaller than the Chebyshev bound for large x. There arc numerous other moment inequalities on positive and general real-valued random variables. They have a variety of uses in theoretical calculations. We present a few fundamental moment inequalities that are special.

Theorem 1.35 (Jensen's Inequality). *Let X be a random variable with a finite mean, and $g(x) : \mathcal{R} \to \mathcal{R}$ a convex function. Then $g(E(X)) \leq E(g(X))$.*

Example 1.51. Let X be any random variable with a finite mean μ. Consider the function $g(x) = e^{ax}$, where a is a real number. Then, by the second derivative test, g is a convex function on the entire real line, and therefore, by Jensen's inequality

$$E(e^{aX}) \geq e^{a\mu}.$$

Here are two other important moment inequalities.

Theorem 1.36. *(a)* **(Lyapounov Inequality).** *Given a nonnegative random variable X, and $0 < \alpha < \beta$,*

$$(EX^\alpha)^{\frac{1}{\alpha}} \leq (EX^\beta)^{\frac{1}{\beta}}.$$

1.11 Moment-Generating Function and Fundamental Inequalities

(b) **(Log Convexity Inequality of Lyapounov).** *Given a nonnegative random variable X, and $0 \leq \alpha_1 < \alpha_2 \leq \frac{\beta}{2}$,*

$$EX^{\alpha_1} EX^{\beta-\alpha_1} \geq EX^{\alpha_2} EX^{\beta-\alpha_2}.$$

We finish with an example of a paradox of expectations.

Example 1.52 (An Expectation Paradox). Suppose X, Y are two positive nonconstant independent random variables, with the same distribution; for example, X, Y could be independent variables with a uniform distribution on $[5, 10]$. We need the assumption that the common distribution of X and Y is such that $E(\frac{1}{X}) = E(\frac{1}{Y}) < \infty$.

Let $R = \frac{X}{Y}$. Then, by Jensen's inequality,

$$E(R) = E\left(\frac{X}{Y}\right) = E(X)E\left(\frac{1}{Y}\right) > E(X)\frac{1}{E(Y)} = 1.$$

So, we have proved that $E(\frac{X}{Y}) > 1$. But we can repeat exactly the same argument to conclude that $E(\frac{Y}{X}) > 1$. So, we seem to have the paradoxical conclusion that we expect X to be somewhat larger than Y, and we also expect Y to be somewhat larger than X.

There are many other such examples of paradoxes of expectations.

1.11.1 * Inversion of an MGF and Post's Formula

The moment-generating function uniquely determines the distribution. Therefore, in principle, if we knew the mgf of a distribution, we should be able to find the distribution to which the mgf corresponds. In practice, this *inversion* is difficult, and often we find the distribution corresponding to an mgf by inspection. There are theoretical formulas for inverting an mgf. One of these formulas uses complex variable methods, and the other uses only real variable methods. The latter, called *Post's inversion formula*, requires that we can calculate all derivatives (at least of large orders) of the given mgf. This may be impractical. However, with the use of symbolic software, and efficient numerical means, the Post inversion formula may be usable in some cases. It is given below; see Widder (1989, p. 463) for a proof.

Proposition. *Let X be a nonnegative continuous random variable with density $f(x)$. Let $\psi(t), t \geq 0$ be the one-sided mgf $\psi(t) = E[e^{-tX}]$. Suppose that f is everywhere continuous. Then,*

$$f(x) = \lim_{k \to \infty} \frac{(-1)^k}{k!} \left(\frac{k}{x}\right)^{k+1} \psi^{(k)}\left(\frac{k}{x}\right), \quad x > 0.$$

1.12 Some Special Continuous Distributions

A number of densities, by virtue of their popularity in modeling, or because of their special theoretical properties, are considered to be special. We discuss, when suitable, their moments, the form of the CDF, the mgf, shape, and modal properties, and interesting inequalities. Classic references to standard continuous distributions are Johnson et al. (1994), and Kendall and Stuart (1976); Everitt (1998) contains many unusual facts.

Definition 1.31. Let X have the pdf

$$f(x) = \frac{1}{b-a}, \quad a \leq x \leq b,$$
$$= 0 \quad \text{otherwise,}$$

where $-\infty < a < b < \infty$ are given real numbers.

Then we say that X has the uniform distribution on $[a, b]$ and write $X \sim U[a, b]$. The basic properties of a uniform density are given next.

Theorem 1.37. *(a) If $X \sim U[0, 1]$, then $a + (b-a)X \sim U[a, b]$, and if $X \sim U[a, b]$, then $\frac{X-a}{b-a} \sim U[0, 1]$.*

(b) The CDF of the $U[a, b]$ distribution equals:

$$F(x) = 0, \quad x < a.$$
$$= \frac{x-a}{b-a}, \quad a \leq x \leq b.$$
$$= 1, \quad x > b.$$

(c) The mgf of the $U[a, b]$ distribution equals $\psi(t) = \frac{e^{tb} - e^{ta}}{(b-a)t}$.

(d) The nth moment of the $U[a, b]$ distribution equals

$$E(X^n) = \frac{b^{n+1} - a^{n+1}}{(b-a)(n+1)}.$$

(e) The mean and the variance of the $U[a, b]$ distribution equal

$$\mu = \frac{a+b}{2}; \quad \sigma^2 = \frac{(b-a)^2}{12}.$$

Example 1.53. A point is selected at random on the unit interval, dividing it into two pieces with total length 1. Find the probability that the ratio of the length of the shorter piece to the length of the longer piece is less than $1/4$.

Let $X \sim U[0, 1]$; we want $P(\frac{\min\{X, 1-X\}}{\max\{X, 1-X\}} < 1/4)$. This happens only if $X < 1/5$ or $> 4/5$. Therefore, the required probability is $P(X < 1/5) + P(X > 4/5) = 1/5 + 1/5 = 2/5$.

1.12 Some Special Continuous Distributions

We defined the standard exponential density in the previous section. We now introduce the general exponential density. Exponential densities are used to model waiting times (e.g., waiting times for an elevator or at a supermarket checkout) or failure times (e.g., the time till the first failure of some equipment) or renewal times (e.g., time elapsed between successive earthquakes at a location), and so on. The exponential density also has some very interesting theoretical properties.

Definition 1.32. A nonnegative random variable X has the exponential distribution with parameter $\lambda > 0$ if it has the pdf $f(x) = \frac{1}{\lambda}e^{-x/\lambda}, x > 0$. We write $X \sim \text{Exp}(\lambda)$.

Here are the basic properties of an exponential density.

Theorem 1.38. Let $X \sim \text{Exp}(\lambda)$. Then,

(a) $\frac{X}{\lambda} \sim \text{Exp}(1)$.
(b) The CDF $F(x) = 1 - e^{-x/\lambda}$, $x > 0$, (and zero for $x \leq 0$.)
(c) $E(X^n) = \lambda^n n!$, $n \geq 1$.
(d) The mgf $\psi(t) = \frac{1}{1-\lambda t}$, $t < 1/\lambda$.

Example 1.54 (Mean Is Larger Than Median for Exponential). Suppose $X \sim \text{Exp}(4)$. What is the probability that $X > 4$?

Since $X/4 \sim \text{Exp}(1)$,

$$P(X > 4) = P(X/4 > 1) = \int_1^\infty e^{-x} dx = e^{-1} = .3679,$$

quite a bit smaller than 50%. This implies that the median of the distribution has to be smaller than 4, where 4 is the mean. Indeed, the median is a number m such that $F(m) = \frac{1}{2}$ (the median is unique in this example) $\Rightarrow 1 - e^{-m/4} = \frac{1}{2} \Rightarrow m = 4 \log 2 = 2.77$.

This phenomenon that the mean is larger than the median is quite typical of distributions that have a long right tail, as does the exponential.

In general, if $X \sim \text{Exp}(\lambda)$, the median of X is $\lambda \log 2$.

Example 1.55 (Lack of Memory of the Exponential Distribution). The exponential densities have a lack of memory property similar to the one we established for the geometric distribution. Let $X \sim \text{Exp}(\lambda)$, and let s, t be positive numbers. The lack of memory property is that $P(X > s+t | X > s) = P(X > t)$. So, suppose that X is the waiting time for an elevator, and suppose that you have already waited $s = 3$ minutes. Then the probability that you have to wait another two minutes is the same as the probability that you would have to wait two minutes if you just arrived. This is not true if the waiting time distribution is something other than an exponential.

The proof of the property is simple:

$$P(X > s+t | X > s) = \frac{P(X > s+t)}{P(X > s)} = \frac{e^{-(s+t)/\lambda}}{e^{-s/\lambda}}$$

$$= e^{-t/\lambda} = P(X > t).$$

Example 1.56 (The Weibull Distribution). Suppose $X \sim \text{Exp}(1)$, and let $Y = X^\alpha$, where $\alpha > 0$ is a constant. This is a strictly monotone function with the inverse function $y^{1/\alpha}$, thus the density of Y is

$$f_Y(y) = \frac{f(y^{1/\alpha})}{|g'(y^{1/\alpha})|} = e^{-y^{1/\alpha}} \times \frac{1}{\alpha y^{(\alpha-1)/\alpha}}$$

$$= \frac{1}{\alpha} y^{(1-\alpha)/\alpha} e^{-y^{1/\alpha}}, \quad y > 0.$$

This final answer can be made to look a little simpler by writing $\beta = \frac{1}{\alpha}$. If we do so, the density becomes

$$\beta y^{\beta-1} e^{-y^\beta}, \quad y > 0.$$

We can introduce an extra scale parameter akin to what we do for the exponential case itself. In that case, we have the general two-parameter Weibull density

$$f(y|\beta, \lambda) = \frac{\beta}{\lambda} \left(\frac{x}{\lambda}\right)^{\beta-1} e^{-(\frac{x}{\lambda})^\beta}, \quad y > 0.$$

This is the *Weibull density* with parameters β, λ.

The exponential density is decreasing on $[0, \infty)$. A generalization of the exponential density with a mode usually at some strictly positive number m is the Gamma distribution. It includes the exponential as a special case, and can be very skewed, or even almost a bell-shaped density. We later show that it also arises naturally, as the density of the sum of a number of independent exponential random variables.

Definition 1.33. A positive random variable X is said to have a Gamma distribution with shape parameter α and scale parameter λ if it has the pdf

$$f(x|\alpha, \lambda) = \frac{e^{-x/\lambda} x^{\alpha-1}}{\lambda^\alpha \Gamma(\alpha)}, \quad x > 0, \alpha, \lambda > 0;$$

we write $X \sim G(\alpha, \lambda)$. The Gamma density reduces to the exponential density with mean λ when $\alpha = 1$; for $\alpha < 1$, the Gamma density is decreasing and unbounded, whereas for large α, it becomes nearly a bell-shaped curve. A plot of some Gamma densities in Fig. 1.4 reveals these features.

The basic facts about a Gamma distribution are given in the following theorem.

Theorem 1.39. *(a) The CDF of the $G(\alpha, \lambda)$ density is the* normalized incomplete Gamma function

$$F(x) = \frac{\gamma(\alpha, x/\lambda)}{\Gamma(\alpha)},$$

where $\gamma(\alpha, x) = \int_0^x e^{-t} t^{\alpha-1} dt$.

1.12 Some Special Continuous Distributions

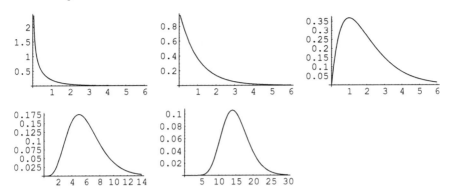

Fig. 1.4 Plot of Gamma density with lambda = 1, alpha = .5, 1, 2, 6, 15

(b) The nth moment equals

$$E(X^n) = \lambda^n \frac{\Gamma(\alpha + n)}{\Gamma(\alpha)}, \quad n \geq 1.$$

(c) The mgf equals

$$\psi(t) = (1 - \lambda t)^{-\alpha}, \quad t < \frac{1}{\lambda}.$$

(d) The mean and the variance equal

$$\mu = \alpha\lambda; \quad \sigma^2 = \alpha\lambda^2.$$

An important consequence of the mgf formula is the following result.

Corollary. *Suppose $X_1, X_2, \ldots, X_n$ are independent* $\text{Exp}(\lambda)$ *variables. Then $X_1 + X_2 + \cdots + X_n \sim G(n, \lambda)$.*

Proof. $X_1, X_2, \ldots, X_n$ are independent, thus for $t < 1/\lambda$,

$$\begin{aligned}
E(e^{t(X_1+X_2+\cdots+X_n)}) &= E(e^{tX_1}e^{tX_2}\cdots e^{tX_n}) \\
&= E(e^{tX_1})E(e^{tX_2})\cdots E(e^{tX_n}) \\
&= (1-\lambda t)^{-1}(1-\lambda t)^{-1}\cdots(1-\lambda t)^{-1} \\
&= (1-\lambda t)^{-n},
\end{aligned}$$

which agrees with the mgf of a $G(n, \lambda)$ distribution, and therefore, by the distribution determining property of mgfs, it follows that $X_1 + X_2 + \cdots + X_n \sim G(n, \lambda)$. □

Example 1.57 (The General Chi-Square Distribution). We saw in the previous section that the distribution of the square of a standard normal variable is the chi-square distribution with one degree of freedom. A natural question is what is the distribution of the sum of squares of several independent standard normal variables.

Although we do not yet have the technical tools necessary to derive this distribution, it turns out that this distribution is in fact a Gamma distribution. Precisely, if $X_1, X_2, \ldots, X_m$ are m independent standard normal variables, then $T = \sum_{i=1}^{m} X_i^2$ has a $G(\frac{m}{2}, 2)$ distribution, and therefore has the density

$$f_m(t) = \frac{e^{-t/2} t^{m/2-1}}{2^{m/2} \Gamma(\frac{m}{2})}, \quad t > 0.$$

This is called the *chi-square density with m degrees of freedom*, and arises in numerous contexts in statistics and probability. We write $T \sim \chi_m^2$. From the general formulas for the mean and variance of a Gamma distribution, we get that

$$\text{Mean of a } \chi_m^2 \text{ distribution} = m;$$
$$\text{Variance of a } \chi_m^2 \text{ distribution} = 2m.$$

The chi-square density is rather skewed for small m, but becomes approximately bell-shaped when m gets large; we have seen this for general Gamma densities.

One especially important context in which the chi-square distribution arises is in consideration of the distribution of the *sample variance* for iid normal observations. The sample variance of a set of n random variables $X_1, X_2, \ldots, X_n$ is defined as $s^2 = \frac{1}{n-1} \sum_{i=1}^{n} (X_i - \bar{X})^2$, where $\bar{X} = \frac{X_1 + \cdots + X_n}{n}$ is the mean of $X_1, \ldots, X_n$. The name sample variance derives from the following property.

Theorem 1.40. *Suppose $X_1, \ldots, X_n$ are independent with a common distribution F having a finite variance σ^2. Then, for any n, $E(s^2) = \sigma^2$.*

Proof. First note the algebraic identity

$$\sum_{i=1}^{n}(X_i - \bar{X})^2 = \sum_{i=1}^{n}(X_i^2 - 2X_i\bar{X} + \bar{X}^2) = \sum_{i=1}^{n} X_i^2 - 2n\bar{X}^2 + n\bar{X}^2 = \sum_{i=1}^{n} X_i^2 - n\bar{X}^2.$$

Therefore,

$$E(s^2) = \frac{1}{n-1} E\left[\sum_{i=1}^{n} X_i^2 - n\bar{X}^2\right] = \frac{1}{n-1}\left[n(\sigma^2 + \mu^2) - n\left(\frac{\sigma^2}{n} + \mu^2\right)\right] = \sigma^2.$$

If, in particular, $X_1, \ldots, X_n$ are iid $N(\mu, \sigma^2)$, then $\frac{X_i - \bar{X}}{\sigma}$ are also normally distributed, each with mean zero. However, they are no longer independent. If we sum their squares, then the sum of the squares will still be distributed as a chi square, but there will be a loss of one degree of freedom, due to the fact that $X_i - \bar{X}$ are not independent, even though the X_i are independent.

We state this important fact formally. □

Theorem 1.41. *Suppose $X_1, \ldots, X_n$ are iid $N(\mu, \sigma^2)$. Then $\frac{\sum_{i=1}^{n}(X_i - \bar{X})^2}{\sigma^2} \sim \chi_{n-1}^2$.*

Example 1.58 *(Inverse Gamma Distribution).* Suppose $X \sim G(\alpha, \lambda)$. The distribution of $\frac{1}{X}$ is called the inverse Gamma distribution. We derive its density.

1.12 Some Special Continuous Distributions

Because $Y = g(X) = \frac{1}{X}$ is a strictly monotone function with the inverse function $g^{-1}(y) = \frac{1}{y}$, and because the derivative of g is $g'(x) = -\frac{1}{x^2}$, the density of Y is

$$f_Y(y) = \frac{f\left(\frac{1}{y}\right)}{\left|g'\left(\frac{1}{y}\right)\right|} = \frac{e^{-1/(\lambda y)} y^{1-\alpha}}{\lambda^\alpha \Gamma(\alpha)} \cdot \frac{1}{y^2}$$

$$= \frac{e^{-1/(\lambda y)} y^{-1-\alpha}}{\lambda^\alpha \Gamma(\alpha)}, \quad y > 0.$$

The inverse Gamma density is extremely skewed for small values of α; furthermore, the right tail is so heavy for small α, that the mean does not exist if $\alpha \leq 1$. Inverse Gamma distributions are quite popular in studies of economic inequality, reliability problems, and as prior distributions in Bayesian statistics.

For continuous random variables that take values between 0 and 1, the most standard family of densities is the family of Beta densities. Their popularity is due to their analytic tractability, and due to the large variety of shapes that Beta densities can take when the parameter values change. It is a generalization of the $U[0, 1]$ density.

Definition 1.34. X is said to have a Beta density with parameters α and β if it has the density

$$f(x) = \frac{x^{\alpha-1}(1-x)^{\beta-1}}{B(\alpha, \beta)}, \quad 0 \leq x \leq 1, \alpha, \beta > 0,$$

where $B(\alpha, \beta) = \frac{\Gamma(\alpha)\Gamma(\beta)}{\Gamma(\alpha+\beta)}$. We write $X \sim Be(\alpha, \beta)$. An important point is that by its very notation, $\frac{1}{B(\alpha,\beta)}$ must be the normalizing constant of the function $x^{\alpha-1}(1-x)^{\beta-1}$; thus, another way to think of $B(\alpha, \beta)$ is that for any $\alpha, \beta > 0$,

$$B(\alpha, \beta) = \int_0^1 x^{\alpha-1}(1-x)^{\beta-1} dx.$$

This fact is repeatedly useful in the following.

Theorem 1.42. *Let* $X \sim Be(\alpha, \beta)$.

(a) *The CDF equals*

$$F(x) = \frac{B_x(\alpha, \beta)}{B(\alpha, \beta)},$$

where $B_x(\alpha, \beta)$ *is the incomplete Beta function* $\int_0^x t^{\alpha-1}(1-t)^{\beta-1} dt$.

(b) *The nth moment equals*

$$E(X^n) = \frac{\Gamma(\alpha+n)\Gamma(\alpha+\beta)}{\Gamma(\alpha+\beta+n)\Gamma(\alpha)}.$$

(c) The mean and the variance equal

$$\mu = \frac{\alpha}{\alpha + \beta}; \quad \sigma^2 = \frac{\alpha\beta}{(\alpha + \beta)^2(\alpha + \beta + 1)}.$$

(d) The mgf equals

$$\psi(t) = {}_1F_1(\alpha, \alpha + \beta, t),$$

where ${}_1F_1(a, b, z)$ denotes the confluent hypergeometric function.

Example 1.59 *(Square of a Beta).* Suppose X has a Beta density. Then X^2 also takes values in $[0, 1]$, but it does not have a Beta density. To have a specific example, suppose $X \sim Be(7,7)$. Then the density of $Y = X^2$ is

$$f_Y(y) = \frac{f(\sqrt{y})}{2\sqrt{y}} = \frac{y^3(1-\sqrt{y})^6}{B(7,7)2\sqrt{y}}$$

$$= 6006 y^{5/2}(1-\sqrt{y})^6, \quad 0 \le y \le 1.$$

Clearly, this is not a Beta density.

In practical applications, certain types of random variables consistently exhibit a long right tail, in the sense that a lot of small values are mixed with a few large or excessively large values in the distributions of these random variables. Economic variables such as wealth typically manifest such heavy tail phenomena. Other examples include sizes of oil fields, insurance claims, stock market returns, and river height in a flood among others. The tails are sometimes so heavy that the random variable may not even have a finite mean. *Extreme value distributions* are common and increasingly useful models for such applications. A brief introduction to two specific extreme value distributions is provided next. These two distributions are the *Pareto* distribution and the *Gumbel* distribution. One peculiarity of semantics is that the Gumbel distribution is often called the *Gumbel law*.

A random variable X is said to have the Pareto density with parameters θ and α if it has the density

$$f(x) = \frac{\alpha\theta^\alpha}{x^{\alpha+1}}, \quad x \ge \theta > 0, \alpha > 0.$$

We write $X \sim Pa(\alpha, \theta)$. The density is monotone decreasing. It may or may not have a finite expectation, depending on the value of α. It never has a finite mgf in any nonempty interval containing zero. The basic facts about a Pareto density are given in the next result.

Theorem 1.43. *Let $X \sim Pa(\alpha, \theta)$.*

(a) The CDF of X equals

$$F(x) = 1 - \left(\frac{\theta}{x}\right)^\alpha, \quad x \ge \theta,$$

and zero for $x < \theta$.

(b) The nth moment exists if and only if $n < \alpha$, in which case

$$E(X^n) = \frac{\alpha \theta^n}{\alpha - n}.$$

(c) For $\alpha > 1$, the mean exists; for $\alpha > 2$, the variance exists. Furthermore, they equal

$$E(X) = \frac{\alpha \theta}{\alpha - 1}; \quad Var(X) = \frac{\alpha \theta^2}{(\alpha - 1)^2 (\alpha - 2)}.$$

We next define the Gumbel law. A random variable X is said to have the Gumbel density with parameters μ, σ if it has the density

$$f(x) = \frac{1}{\sigma} e^{-e^{-\frac{x-\mu}{\sigma}}} e^{-\frac{x-\mu}{\sigma}}, \quad -\infty < x < \infty, -\infty < \mu < \infty, \sigma > 0.$$

If $\mu = 0$ and $\sigma = 1$, the density is called the standard Gumbel density. Thus, the standard Gumbel density has the formula $f(x) = e^{-e^{-x}} e^{-x}, -\infty < x < \infty$. The density converges extremely fast (at a superexponential rate) at the left tail, but at only a regular exponential rate at the right tail. Its relation to the density of the maximum of a large number of independent normal variables makes it a special density in statistics and probability; see Chapter 7. The basic facts about a Gumbel density are collected together in the result below. All Gumbel distributions have a finite mgf $\psi(t)$ at any t. But no simple formula for it is possible.

Theorem 1.44. *Let X have the Gumbel density with parameters μ, σ. Then,*

(a) *The CDF equals*

$$F(x) = e^{-e^{-\frac{x-\mu}{\sigma}}}, \quad -\infty < x < \infty.$$

(b) $E(X) = \mu - \gamma \sigma$, *where* $\gamma \approx .577216$ *is the* Euler constant.
(c) $Var(X) = \frac{\pi^2}{6} \sigma^2$.
(d) *The mgf of X exists everywhere.*

1.13 Normal Distribution and Confidence Interval for a Mean

Empirical data on many types of variables across disciplines tend to exhibit unimodality and only a small amount of skewness. It is quite common to use a normal distribution as a model for such data. The normal distribution occupies the central place among all distributions in probability and statistics. There is also the *central limit theorem*, which says that the sum of many small independent quantities approximately follows a normal distribution. By a combination of reputation, convenience, mathematical justification, empirical experience, and habit, the normal distribution has become the most ubiquitous of all distributions. Detailed algebraic properties can be seen in Rao (1973), Kendall and Stuart (1976), and Feller (1971). Petrov (1975) is a masterly account of the role of the normal distribution in the limit theorems of probability.

We have actually already defined a normal density. But let us recall the definition here.

Definition 1.35. A random variable X is said to have a *normal distribution* with parameters μ and σ^2 if it has the density

$$f(x) = \frac{1}{\sigma\sqrt{2\pi}} e^{-\frac{(x-\mu)^2}{2\sigma^2}}, \quad -\infty < x < \infty,$$

where μ can be any real number, and $\sigma > 0$. We write $X \sim N(\mu, \sigma^2)$. If $X \sim N(0, 1)$, we call it a *standard normal variable*.

The *density* of a standard normal variable is denoted as $\phi(x)$, and equals the function

$$\phi(x) = \frac{1}{\sqrt{2\pi}} e^{-\frac{x^2}{2}}, \quad -\infty < x < \infty,$$

and the CDF is denoted as $\Phi(x)$. Note that the *standard normal density* is symmetric and unimodal about zero. The general $N(\mu, \sigma^2)$ density is symmetric and unimodal about μ.

By definition of a CDF,

$$\Phi(x) = \int_{-\infty}^{x} \phi(z) dz.$$

The CDF $\Phi(x)$ cannot be written in terms of the elementary functions, but can be computed at a given value x, and tables of the values of $\Phi(x)$ are widely available. For example, here are some selected values.

Example 1.60 (Standard Normal CDF at Selected Values).

x	$\Phi(x)$
-4	.00003
-3	.00135
-2	.02275
-1	.15866
0	.5
1	.84134
2	.97725
3	.99865
4	.99997

Here are the most basic properties of a normal distribution.

1.13 Normal Distribution and Confidence Interval for a Mean

Theorem 1.45. (a) If $X \sim N(\mu, \sigma^2)$, then $Z = \frac{X-\mu}{\sigma} \sim N(0, 1)$, and if $Z \sim N(0, 1)$, then $X = \mu + \sigma Z \sim N(\mu, \sigma^2)$.

In words, if X is any normal random variable, then its standardized version is always a standard normal variable.

(b) If $X \sim N(\mu, \sigma^2)$, then

$$P(X \leq x) = \Phi\left(\frac{x-\mu}{\sigma}\right) \quad \forall x.$$

In particular, $P(X \leq \mu) = P(Z \leq 0) = .5$; that is, the median of X is μ.

(c) Every moment of any normal distribution exists, and the odd central moments $E[(X - \mu)^{2k+1}]$ are all zero.

(d) If $Z \sim N(0, 1)$, then

$$E(Z^{2k}) = \frac{(2k)!}{2^k k!}, \quad k \geq 1.$$

(e) The mgf of the $N(\mu, \sigma^2)$ distribution exists at all real t, and equals

$$\psi(t) = e^{t\mu + \frac{t^2\sigma^2}{2}}.$$

(f) If $X \sim N(\mu, \sigma^2)$,

$$E(X) = \mu; \quad \text{var}(X) = \sigma^2; \quad E(X^3) = \mu^3 + 3\mu\sigma^2.$$
$$E(X^4) = \mu^4 + 6\mu^2\sigma^2 + 3\sigma^4.$$

(g) If $X \sim N(\mu, \sigma^2)$, then $\kappa_1 = \mu, \kappa_2 = \sigma^2$, and $\kappa_r = 0 \ \forall r > 2$, where κ_j is the jth cumulant of X.

An important consequence of part (b) of this theorem is the following result.

Corollary. Let $X \sim N(\mu, \sigma^2)$, and let $0 < \alpha < 1$. Let $Z \sim N(0, 1)$. Suppose x_α is the $(1-\alpha)$th quantile (also called percentile) of X, and z_α is the $(1-\alpha)$th quantile of Z. Then

$$x_\alpha = \mu + \sigma z_\alpha.$$

Example 1.61 (Setting a Thermostat). Suppose that when the thermostat is set at d degrees Celsius, the actual temperature of a certain room is a normal random variable with parameters $\mu = d$ and $\sigma = .5$.

If the thermostat is set at 75°C, what is the probability that the actual temperature of the room will be below 74°C?

By standardizing to an $N(0, 1)$ random variable,

$$P(X < 74) = P(Z < (74 - 75)/.5) = P(Z < -2) = .02275.$$

Next, what is the lowest setting of the thermostat that will maintain a temperature of at least 72°C with a probability of .99?

We want to find the value of d that makes $P(X \geq 72) = .99 \Rightarrow P(X < 72) = .01$. Now, from a standard normal table, $P(Z < -2.326) = .01$. Therefore, we want to find d that makes $d + \sigma \times (-2.326) = 72 \Rightarrow d - .5 \times 2.326 = 72 \Rightarrow d = 72 + .5 \times 2.326 = 73.16°C$.

Example 1.62 (Rounding a Normal Variable). Suppose $X \sim N(0, \sigma^2)$ and suppose the absolute value of X is rounded to the nearest integer. We have seen that the expected value of $|X|$ itself is $\sigma\sqrt{2/\pi}$. How does rounding affect the expected value?

Denote the rounded value of $|X|$ by Y. Then, $Y = 0 \Leftrightarrow |X| < .5; Y = 1 \Leftrightarrow .5 < |X| < 1.5; \cdots$, and so on. Therefore,

$$E(Y) = \sum_{i=1}^{\infty} iP(i - 1/2 < |X| < i + 1/2) = \sum_{i=1}^{\infty} iP(i - 1/2 < X < i + 1/2)$$

$$+ \sum_{i=1}^{\infty} iP(-i - 1/2 < X < -i + 1/2)$$

$$= 2\sum_{i=1}^{\infty} i[\Phi((i + 1/2)/\sigma) - \Phi((i - 1/2)/\sigma)] = 2\sum_{i=1}^{\infty}[1 - \Phi((i + 1/2)/\sigma)],$$

on some manipulation.

For example, if $\sigma = 1$, then this equals $2\sum_{i=1}^{\infty}[1 - \Phi(i + 1/2)] = .76358$, whereas the unrounded $|X|$ has the expectation $\sqrt{2/\pi} = .79789$. The effect of rounding is not serious when $\sigma = 1$.

A plot of the expected value of Y and the expected value of $|X|$ is shown in Fig. 1.5 to study the effect of rounding.

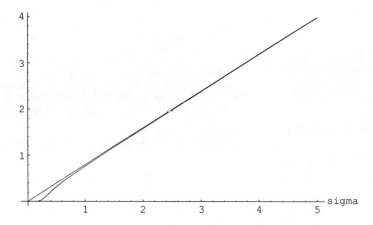

Fig. 1.5 Expected value of rounded and unrounded $|X|$ when X is N(0, sigma^2)

1.13 Normal Distribution and Confidence Interval for a Mean

We can see that the effect of rounding is uniformly small. There is classic literature on corrections needed in computing means, variances, and higher moments when data are rounded. These are known as *Sheppard's corrections*. Kendall and Stuart (1976) give a thorough treatment of these needed corrections.

Example 1.63 (Lognormal Distribution). Lognormal distributions are common models in studies of economic variables, such as income and wealth, because they can adequately describe the skewness that one sees in data on such variables. If $X \sim N(\mu, \sigma^2)$, then the distribution of $Y = e^X$ is called a *lognormal distribution with parameters* μ, σ^2. Note that the lognormal name can be confusing; a lognormal variable is not the logarithm of a normal variable. A better way to remember its meaning is *log is normal*.

$Y = e^X$ is a strictly monotone function of X, therefore by the usual formula for the density of a monotone function, Y has the pdf

$$f_Y(y) = \frac{1}{y\sigma\sqrt{2\pi}} e^{-\frac{(\log y - \mu)^2}{2\sigma^2}}, \quad y > 0;$$

this is called the lognormal density with parameters μ, σ^2. A lognormal variable is defined as e^X for a normal variable X, thus its mean and variance are easily found from the mgf of a normal variable. A simple calculation shows that

$$E(Y) = e^{\mu + \frac{\sigma^2}{2}}; \quad \text{Var}(Y) = (e^{\sigma^2} - 1)e^{2\mu + \sigma^2}.$$

One of the main reasons for the popularity of the lognormal distribution is its skewness; the lognormal density is extremely skewed for large values of σ. The coefficient of skewness has the formula

$$\beta = (2 + e^{\sigma^2})\sqrt{e^{\sigma^2} - 1} \to \infty, \quad \text{as } \sigma \to \infty.$$

Note that the lognormal densities do not have a finite mgf at any $t > 0$, although all its moments are finite. It is also the only standard continuous distribution that is not determined by its moments. That is, there exist other distributions besides the lognormal all of whose moments exactly coincide with the moments of a given lognormal distribution. This is not true of any other distribution with a name that we have come across in this chapter. For example, the normal and the Poisson distributions are all determined by their moments.

We had remarked in the above that sums of many independent variables tend to be approximately normally distributed. A precise version of this is the central limit theorem, which we study in the next section. What is interesting is that sums of any number of independent normal variables are exactly normally distributed. Here is the result.

Theorem 1.46. *Let* $X_1, X_2, \ldots, X_n, n \geq 2$ *be independent random variables, with* $X_i \sim N(\mu_i, \sigma_i^2)$. *Let* $S_n = \sum_{i=1}^{n} X_i$. *Then,*

$$S_n \sim N\left(\sum_{i=1}^n \mu_i, \sum_{i=1}^n \sigma_i^2\right).$$

An important consequence is the following result.

Corollary. *Suppose $X_i, 1 \leq i \leq n$ are independent, and each distributed as $N(\mu, \sigma^2)$. Then $\overline{X} = \frac{S_n}{n} \sim N(\mu, \frac{\sigma^2}{n})$.*

The theorem above implies that any linear function of independent normal variables is also normal,

$$\sum_{i=1}^n a_i X_i \sim N\left(\sum_{i=1}^n a_i \mu_i, \sum_{i=1}^n a_i^2 \sigma_i^2\right).$$

Example 1.64 (Confidence Interval and Margin of Error). Suppose some random variable $X \sim N(\mu, \sigma^2)$, and we have n independent observations $X_1, X_2, \ldots, X_n$ on this variable X; another way to put it is that $X_1, X_2, \ldots, X_n$ are iid $N(\mu, \sigma^2)$. Therefore, $\overline{X} \sim N(\mu, \sigma^2/n)$, and we have

$$P(\overline{X} - 1.96\sigma/\sqrt{n} \leq \mu \leq \overline{X} + 1.96\sigma/\sqrt{n})$$
$$= P(-1.96\sigma/\sqrt{n} \leq \overline{X} - \mu \leq 1.96\sigma/\sqrt{n})$$
$$= P\left(-1.96 \leq \frac{\overline{X} - \mu}{\sigma/\sqrt{n}} \leq 1.96\right) = \Phi(1.96) - \Phi(-1.96) = .95,$$

from a standard normal table.

Thus, with a 95% probability, for any n, μ is between $\overline{X} \pm 1.96\sigma/\sqrt{n}$. Statisticians call the interval of values $\overline{X} \pm 1.96\sigma/\sqrt{n}$ a 95% confidence interval for μ, with a margin of error $1.96\sigma/\sqrt{n}$.

A tight confidence interval will correspond to a small margin of error. For example, if we want a margin of error $\leq .1$, then we need $1.96\sigma/\sqrt{n} \leq .1 \Leftrightarrow \sqrt{n} \geq 19.6\sigma \Leftrightarrow n \geq 384.16\sigma^2$. Statisticians call such a calculation a *sample size calculation*.

1.14 Stein's Lemma

In 1981, Charles Stein gave a simple lemma for a normal distribution, and extended it to the case of a finite number of independent normal variables, which seems innocuous on its face, but has proved to be a really powerful tool in numerous areas of statistics. It has also had its technical influence on the area of Poisson approximations, which we briefly discussed in this chapter. We present the basic lemma, its extension to the case of several independent variables, and show some applications. It would not be possible to give more than just a small glimpse of the applications of Stein's lemma here; the applications are too varied. Regrettably, no comprehensive book or review of the various applications of Stein's lemma is available at this time.

1.14 Stein's Lemma

The original article is Stein (1981); Wasserman (2006) and Diaconis and Zabell (1991) are two of the best sources to learn more about Stein's lemma.

Theorem 1.47. *(a) Let $X \sim N(\mu, \sigma^2)$, and suppose $g : \mathcal{R} \to \mathcal{R}$ is such that g is differentiable at all but at most a finite number of points, and*

(i) *For some $\lambda < 1$, $g(x)e^{-\frac{\lambda x^2}{2\sigma^2}} \to 0$, as $x \to \pm\infty$.*
(ii) *$E[|g'(X)|] < \infty$.*

Then,
$$E[(X - \mu)g(X)] = \sigma^2 E[g'(X)].$$

(b) Let $X_1, X_2, \ldots, X_k$ be independent $N(\mu_i, \sigma^2)$ variables, and suppose $g : \mathcal{R}^k \to \mathcal{R}$ is such that g has a partial derivative with respect to each x_i at all but at most a finite number of points. Then,

$$E[(X_i - \mu_i)g(X_1, X_2, \ldots, X_k)] = \sigma^2 E\left[\frac{\partial}{\partial X_i} g(X_1, X_2, \ldots, X_k)\right].$$

Proof. We prove part (a). By definition of expectation, and by using integration by parts,

$$\begin{aligned}
E[(X-\mu)g(X)] &= \frac{1}{\sigma\sqrt{2\pi}} \int_{-\infty}^{\infty} (x-\mu)g(x)e^{-(x-\mu)^2/(2\sigma^2)} dx \\
&= -\sigma^2 \frac{1}{\sigma\sqrt{2\pi}} \int_{-\infty}^{\infty} g(x)\left[\frac{d}{dx} e^{-(x-\mu)^2/(2\sigma^2)}\right] dx \\
&= -\sigma^2 \frac{1}{\sigma\sqrt{2\pi}} g(x) e^{-(x-\mu)^2/(2\sigma^2)} \Big|_{-\infty}^{\infty} \\
&\quad + \sigma^2 \frac{1}{\sigma\sqrt{2\pi}} \int_{-\infty}^{\infty} g'(x) e^{-(x-\mu)^2/(2\sigma^2)} dx \\
&= -\sigma \frac{1}{\sqrt{2\pi}} g(x) e^{-\frac{\lambda x^2}{2\sigma^2}} e^{-\frac{(1-\lambda)x^2}{2\sigma^2} + \frac{\mu x}{\sigma^2} - \frac{\mu^2}{2\sigma^2}} \Big|_{-\infty}^{\infty} \\
&\quad + \sigma^2 \frac{1}{\sigma\sqrt{2\pi}} \int_{-\infty}^{\infty} g'(x) e^{-(x-\mu)^2/(2\sigma^2)} dx \\
&= 0 + \sigma^2 E[g'(X)] = \sigma^2 E[g'(X)],
\end{aligned}$$

because by assumption

$$g(x)e^{-\frac{\lambda x^2}{2\sigma^2}} \to 0, \text{ as } x \to \pm\infty,$$

and,

$$e^{-\frac{(1-\lambda)x^2}{2\sigma^2} + \frac{\mu x}{\sigma^2}} \text{ is uniformly bounded in } x.$$

The principal applications of Stein's lemma are in statistical theory. Here, we show a simple application. □

Example 1.65. Suppose $X \sim N(\mu, \sigma^2)$, and $g(x)$ is a differentiable and uniformly bounded function with a bounded derivative; that is, $|g(x)| \leq M < \infty\ \forall x$., and $|g'(x)| \leq C < \infty\ \forall x$., By Stein's lemma,

$$\begin{aligned} E(X + g(X) - \mu)^2 &= E[(X - \mu)^2 + g^2(X) + 2(X - \mu)g(X)] \\ &= E[(X - \mu)^2] + E[g^2(X)] + 2E[(X - \mu)g(X)] \\ &= \sigma^2 + E[g^2(X)] + 2\sigma^2 E[g'(X)] \leq \sigma^2 + M^2 + 2C\sigma^2 \end{aligned}$$

because $g^2(x) \leq M^2\ \forall x$, and $g'(x) \leq C\ \forall x$.

Therefore, by applying Markov's inequality,

$$P(|X + g(X) - \mu| \geq k) \leq \frac{E[(X + g(X) - \mu)^2]}{k^2} \leq \frac{\sigma^2(1 + 2C) + M^2}{k^2}.$$

An example of such a function $g(x)$ would be $g(x) = \frac{cx}{1+x^2}$, where c is any real number. Its maximum value is $M = c/2$, and its derivative is bounded by $C = |c|$. Plugging into the inequality above,

$$P\left(\left|\left(1 + \frac{c}{1 + X^2}\right)X - \mu\right| \geq k\right) \leq \frac{(1 + 2|c|)\sigma^2 + c^2/4}{k^2}.$$

Functions of this general form $(1 + \frac{c}{1+X^2})X$ are of interest in statistical theory.

1.15 * Chernoff's Variance Inequality

In 1981, Herman Chernoff gave a proof of an inequality for the normal distribution that essentially says that a smoothing operation, such as integration, is going to reduce the variance of a function. The inequality has since been extensively generalized; see Chernoff (1981) for this inequality. We present this inequality, but present a different proof, which works more generally.

Theorem 1.48. *Let $X \sim N(0, 1)$, and let $g : \mathcal{R} \to \mathcal{R}$ be a function such that g is once continuously differentiable, and $E[(g'(X))^2] < \infty$. Then,*

$$\mathrm{Var}(g(X)) \leq E[(g'(X))^2],$$

with equality holding if and only if $g(X)$ is a linear function, $g(X) = a + bX$ for some a, b.

Proof. We need to use the fact that for any random variable Y, $[E(Y)]^2 \leq E(Y^2)$; we choose the variable Y suitably in the proof below. By the fundamental theorem of calculus,

$$(g(x) - g(0))^2 = \left(\int_0^x g'(t)dt\right)^2 = \left(x \frac{1}{x} \int_0^x g'(t)dt\right)^2$$

$$= x^2 \left(\frac{1}{x} \int_0^x g'(t)dt\right)^2 \leq x^2 \left(\frac{1}{x} \int_0^x [g'(t)]^2 dt\right).$$

$$= x \int_0^x [g'(t)]^2 dt,$$

by identifying Y here with a uniform random variable on $[0, x]$,

$$\text{Var} g(X) = E[g(X) - E(g(X))]^2 \leq E[g(X) - g(0)]^2$$

$$\leq \int_{-\infty}^{\infty} \left[\int_0^x (g'(t))^2 dt\right] x\phi(x) dx$$

$$= \int_{-\infty}^{\infty} \left[\int_0^x (g'(t))^2 dt\right] (-\phi'(x))dx = \int_{-\infty}^{\infty} [g'(x)]^2 \phi(x) dx$$

(by integration by parts)

$$= E[(g'(X))^2]. \qquad \square$$

Example 1.66. Let $X \sim N(0, 1)$. As a simple example, let $g(X) = (X-a)^2$, where a is a general constant. With some algebra, the exact variance of $g(X)$ can be found; we use Chernoff's inequality to find an upper bound on the variance.

Clearly, $g'(x) = 2(x - a)$, and so, by Chernoff's inequality, $\text{Var}[(X - a)^2] \leq E[4(X - a)^2] = 4(1 + a^2)$.

Example 1.67. Consider a general cubic polynomial $g(X) = a + bX + cX^2 + dX^3$, and suppose that $X \sim N(0, 1)$. The derivative of g is $g'(x) = 3dx^2 + 2cx + b \Rightarrow (g'(x))^2 = 9d^2x^4 + 12cdx^3 + (4c^2 + 6bd)x^2 + 4bcx + b^2$. Because X is standard normal, $E(X) = E(X^3) = 0, E(X^2) = 1, E(X^4) = 3$. Thus, by Chernoff's inequality, for a general cubic polynomial,

$$\text{Var}(a + bX + cX^2 + dX^3) \leq 27d^2 + 4c^2 + b^2 + 6bd.$$

1.16 * Various Characterizations of Normal Distributions

Normal distributions possess a huge number of characterization properties, that is, properties that are not satisfied by any other distribution in large classes of distributions, and sometimes in the class of all distributions. It is not possible to document many of these characterizing properties here; we present a subjective selection of some elegant characterizations of normal distributions. The proofs of most of these characterizing properties are entirely nontrivial, and we do not give the proofs here. This section primarily has reference value. The most comprehensive reference on characterizations of the normal distribution is Kagan, Linnik, and Rao (1973).

Theorem 1.49 (Cramér–Lévy Theorem). *Suppose for some given $n \geq 2$, and independent random variables $X_1, X_2, \cdots, X_n$, $S_n = X_1 + X_2 + \cdots + X_n$ has a normal distribution with some mean and some variance. Then each X_i is necessarily normally distributed.*

Theorem 1.50 (Chi-Square Distribution of Sample Variance). *Suppose X_1, $X_2, \ldots, X_n$ are independent $N(\mu, \sigma^2)$ variables. Then*

$$\frac{\sum_{i=1}^{n}(X_i - \bar{X})^2}{\sigma^2} \sim \chi_{n-1}^2.$$

Conversely, for given $n \geq 2$, if $X_1, X_2, \ldots, X_n$ are independent variables with some common distribution, symmetric about the mean μ, and having a finite variance σ^2, and if

$$\frac{\sum_{i=1}^{n}(X_i - \bar{X})^2}{\sigma^2} \sim \chi_{n-1}^2,$$

then the common distribution of the X_i must be $N(\mu, \sigma^2)$ for some μ.

Theorem 1.51 (Independence of Sample Mean and Sample Variance). *Suppose $X_1, X_2, \ldots, X_n$ are independent $N(\mu, \sigma^2)$ variables. Then, for any $n \geq 2$, $\bar{X}$ and $\sum_{i=1}^{n}(X_i - \bar{X})^2$ are independent. Conversely, for given $n \geq 2$, if $X_1, X_2, \ldots, X_n$ are independent variables with some common distribution, and if $\bar{X}$ and $\sum_{i=1}^{n}(X_i - \bar{X})^2$ are independent, then the common distribution of the X_i must be $N(\mu, \sigma^2)$ for some μ, σ^2.*

Theorem 1.52 (Independence and Spherical Symmetry). *Suppose for given $n \geq 2$, $X_1, X_2, \ldots, X_n$ are independent variables with a common density $f(x)$. If the product $f(x_1)f(x_2)\cdots f(x_n)$ is a spherically symmetric function, that is, if $f(x_1)f(x_2)\cdots f(x_n) = g(x_1^2 + x_2^2 + \cdots + x_n^2)$ for some function g, then $f(x)$ must be the density of $N(0, \sigma^2)$ for some σ^2.*

Theorem 1.53 (Independence of Sum and Difference). *Suppose X, Y are independent random variables, each with a finite variance. Then $X + Y$ and $X - Y$ are independent if and only if X, Y are normally distributed with an equal variance.*

Theorem 1.54 (Independence of Two Linear Functions). *Suppose $X_1, X_2, \ldots, X_n$ are independent variables, and $L_1 = \sum_{i=1}^{n} a_i X_i$, $L_2 = \sum_{i=1}^{n} b_i X_i$ are two different linear functions. If L_1, L_2 are independent, then for each i for which the product $a_i b_i \neq 0$, X_i must be normally distributed.*

Theorem 1.55 (Characterization Through Stein's Lemma). *Let X have a finite mean μ and a finite variance σ^2. If $E[(X - \mu)g(X)] = \sigma^2 E[g'(X)]$ for every differentiable function g with $E[|g'(X)|] < \infty$, then X must be distributed as $N(\mu, \sigma^2)$.*

Theorem 1.56 (Characterization Through Chernoff's Inequality). *Suppose X is a continuous random variable with a finite variance σ^2. Let*

$$\mathcal{G} = \{g : g \text{ is continuously differentiable}, \ E[(g'(X))^2] < \infty\}.$$

Let
$$B(g) = \frac{\mathrm{Var}(g(X))}{\sigma^2 E[(g'(X))^2]}.$$

If $\sup_{g \in \mathcal{G}} B(g) = 1$, then X must be distributed as $N(\mu, \sigma^2)$ for some μ.

1.17 Normal Approximations and Central Limit Theorem

Many of the special discrete and special continuous distributions that we have discussed can be well approximated by a normal distribution, for suitable configurations of their underlying parameters. Typically, the normal approximation works well when the parameter values are such that the skewness of the distribution is small. For example, binomial distributions are well approximated by a normal when n is large and p is not too small or too large. Gamma distributions are well approximated by a normal when the shape parameter α is large. There is a unifying mathematical result here. The unifying mathematical result is one of the most important results in all of mathematics, and is called the *central limit theorem*. The subject of central limit theorems is incredibly diverse. In this section, we present the basic or the *canonical central limit theorem*, and present its applications to certain problems with which we are already familiar. Among numerous excellent references on central limit theorems, we recommend Feller (1968, 1971) for lucid exposition and examples. The subject of central limit theorems also has a really interesting history; we recommend Le Cam (1986) and Stigler (1986) for reading some history of the central limit theorem. Careful and comprehensive mathematical treatment is available in Hall (1992) and Bhattacharya and Rao (1986). For a diverse selection of examples, see DasGupta (2008).

Theorem 1.57 (Central Limit Theorem). *For $n \geq 1$, let $X_1, X_2, \ldots, X_n$ be n independent random variables, each having the same distribution, and suppose this common distribution, say F, has a finite mean μ, and a finite variance σ^2. Let $S_n = X_1 + X_2 + \cdots + X_n, \bar{X} = \bar{X}_n = \frac{X_1 + X_2 + \cdots + X_n}{n}$. Then, as $n \to \infty$,*

(a) $P\left(\frac{S_n - n\mu}{\sqrt{n\sigma^2}} \leq x\right) \to \Phi(x) \; \forall \, x \in \mathcal{R};$
(b) $P\left(\frac{\sqrt{n}(\bar{X} - \mu)}{\sigma} \leq x\right) \to \Phi(x) \; \forall \, x \in \mathcal{R}.$

In words, for large n,
$$S_n \approx N\left(n\mu, n\sigma^2\right).$$
$$\bar{X} \approx N\left(\mu, \frac{\sigma^2}{n}\right).$$

A very important case in which the general central limit theorem applies is the binomial distribution. The CLT allows us to approximate clumsy binomial probabilities involving large factorials by simple and accurate normal approximations. We first give the exact result on normal approximation of the binomial.

Theorem 1.58 (de Moivre–Laplace Central Limit Theorem). *Let $X = X_n \sim \text{Bin}(n, p)$. Then, for any fixed p and $x \in \mathcal{R}$,*

$$P\left(\frac{X - np}{\sqrt{np(1-p)}} \leq x\right) \to \Phi(x),$$

as $n \to \infty$.

The de Moivre–Laplace CLT tells us that if $X \sim \text{Bin}(n, p)$, then we can approximate the $\leq$ type probability $P(X \leq k)$ as

$$P(X \leq k) = P\left(\frac{X - np}{\sqrt{np(1-p)}} \leq \frac{k - np}{\sqrt{np(1-p)}}\right)$$
$$\approx \Phi\left(\frac{k - np}{\sqrt{np(1-p)}}\right).$$

Note that, in applying the normal approximation in the binomial case, we are using a *continuous* distribution to approximate a discrete distribution taking only integer values. The quality of the approximation improves, sometimes dramatically, if we fill up the gaps between the successive integers. That is, pretend *that an event of the form $X = x$ really corresponds to $x - \frac{1}{2} \leq X \leq x + \frac{1}{2}$. In that case, in order to approximate $P(X \leq k)$, we in fact expand the domain of the event to $k + \frac{1}{2}$, and approximate $P(X \leq k)$ as*

$$P(X \leq k) \approx \Phi\left(\frac{k + \frac{1}{2} - np}{\sqrt{np(1-p)}}\right).$$

This adjusted normal approximation is called normal approximation with a continuity correction. *Continuity correction should always be done while computing a normal approximation to a binomial probability. Here are the continuity-corrected normal approxomation formulas for easy reference:*

$$P(X \leq k) \approx \Phi\left(\frac{k + \frac{1}{2} - np}{\sqrt{np(1-p)}}\right).$$

$$P(m \leq X \leq k) \approx \Phi\left(\frac{k + \frac{1}{2} - np}{\sqrt{np(1-p)}}\right) - \Phi\left(\frac{m - \frac{1}{2} - np}{\sqrt{np(1-p)}}\right).$$

Example 1.68 (Coin Tossing). This is the simplest example of a normal approximation of binomial probabilities. We solve a number of problems by applying the normal approximation method.

1.17 Normal Approximations and Central Limit Theorem

First, suppose a fair coin is tossed 100 times. What is the probability that we obtain between 45 and 55 heads? Denoting X as the number of heads obtained in 100 tosses, $X \sim \text{Bin}(n, p)$, with $n = 100, p = .5$. Therefore, by using the continuity corrected normal approximation,

$$P(45 \leq X \leq 55) \approx \Phi\left(\frac{55.5 - 50}{\sqrt{12.5}}\right) - \Phi\left(\frac{44.5 - -50}{\sqrt{12.5}}\right)$$
$$= \Phi(1.56) - \Phi(-1.56) = .9406 - .0594 = .8812.$$

So, the probability that the percentage of heads is between 45% and 55% is high, but not really high if we toss the coin 100 times. Here is the next question. How many times do we need to toss a fair coin to be 99% sure that the percentage of heads will be between 45 and 55%? The percentage of heads is between 45 and 55% if and only if the number of heads is between $.45n$ and $.55n$. Using the continuity corrected normal approximation, again, we want

$$.99 = \Phi\left(\frac{.55n + .5 - .5n}{\sqrt{.25n}}\right) - \Phi\left(\frac{.45n - .5 - .5n}{\sqrt{.25n}}\right)$$

$$\Rightarrow .99 = 2\Phi\left(\frac{.55n + .5 - .5n}{\sqrt{.25n}}\right) - 1$$

(because, for any real number x, $\Phi(x) - \Phi(-x) = 2\Phi(x) - 1$)

$$\Rightarrow \Phi\left(\frac{.55n + .5 - .5n}{\sqrt{.25n}}\right) = 995$$

$$\Rightarrow \Phi\left(\frac{.05n + .5}{\sqrt{.25n}}\right) = .995.$$

Now, from a standard normal table, we find that $\Phi(2.575) = .995$. Therefore, we equate

$$\frac{.05n + .5}{\sqrt{.25n}} = 2.575$$

$$\Rightarrow .05n + .5 = 2.575 \times .5\sqrt{n} = 1.2875\sqrt{n}.$$

Writing $\sqrt{n} = x$, we have here a quadratic equation $.05x^2 - 1.2875x + .5 = 0$ to solve. The root we want is $x = 25.71$, and squaring it gives $n \geq (25.71)^2 = 661.04$. Thus, an *approximate value of n* such that in n tosses of a fair coin, the percentage of heads will be between 45 and 55% with a 99% probability is $n = 662$. Most people find that the value of n needed is higher than what they would have guessed.

Example 1.69 (Random Walk). The theory of random walks is one of the most beautiful areas of probability. Here, we give an introductory example that makes use of the normal approximation to a binomial.

Suppose a drunkard is standing at time zero (say 11:00 PM) at some point, and every second he either moves one step to the right, or one step to the left, with equal probability, of where he is at that time. What is the probability that after two minutes, he will be ten or more steps away from where he started? Note that the drunkard will take 120 steps in 2 minutes.

Let the drunkard's movement at the ith step be denoted as X_i; then, $P(X_i = \pm 1) = .5$. So, we can think of X_i as $X_i = 2Y_i - 1$, where $Y_i \sim \text{Ber}(.5), 1 \leq i \leq n = 120$. If we assume that the drunkard's successive movements $X_1, X_2, \ldots$ are independent, then $Y_1, Y_2, \ldots$ are also independent, and so, $S_n = Y_1 + Y_2 + \cdots Y_n \sim \text{Bin}(n, .5)$. Furthermore,

$$|X_1 + X_2 + \cdots + X_n| \geq 10 \Leftrightarrow |2(Y_1 + Y_2 + \cdots + Y_n) - n| \geq 10.$$

So, we want to find

$$P(|2(Y_1 + Y_2 + \cdots + Y_n) - n| \geq 10)$$

$$= P\left(S_n - \frac{n}{2} \geq 5\right) + P\left(S_n - \frac{n}{2} \leq -5\right)$$

$$= P\left(\frac{S_n - \frac{n}{2}}{\sqrt{.25n}} \geq \frac{5}{\sqrt{.25n}}\right) + P\left(\frac{S_n - \frac{n}{2}}{\sqrt{.25n}} \leq -\frac{5}{\sqrt{.25n}}\right).$$

Using the normal approximation, this is approximately equal to $2[1 - \Phi(\frac{5}{\sqrt{.25n}})] = 2[1 - \Phi(.91)] = 2(1 - .8186) = .3628$.

We present four simulated walks of this drunkard in Fig. 1.6 over a two-minute interval consisting of 120 steps. The different simulations show that the drunkard's random walk could evolve in different ways.

1.17.1 Binomial Confidence Interval

The normal approximation to the binomial distribution forms the basis for most of the confidence intervals for the parameter p in common use. We describe two of these in this section, the *Wald confidence interval* and the *score confidence interval* for p. The Wald interval used to be the textbook interval, but the score interval is gaining in popularity due to recent research establishing unacceptably poor properties of the Wald interval. The derivation of each interval is sketched below.

Let $X \sim \text{Bin}(n, p)$. By the normal approximation to the Bin(n, p) distribution, for large n, $X \approx N(np, np(1-p))$, and therefore, the standardized binomial variable $\frac{X - np}{\sqrt{np(1-p)}} \approx N(0, 1)$. This implies

1.17 Normal Approximations and Central Limit Theorem

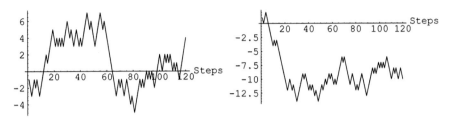

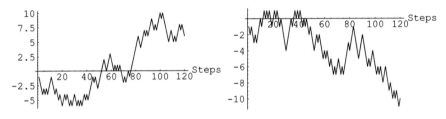

Fig. 1.6 Four simulated random walks

$$P\left(-z_{\frac{\alpha}{2}} \leq \frac{X - np}{\sqrt{np(1-p)}} \leq z_{\frac{\alpha}{2}}\right) \approx 1 - \alpha$$

$$\Rightarrow P\left(-z_{\frac{\alpha}{2}}\sqrt{np(1-p)} \leq X - np \leq z_{\frac{\alpha}{2}}\sqrt{np(1-p)}\right) \approx 1 - \alpha$$

$$\Rightarrow P\left(-z_{\frac{\alpha}{2}}\sqrt{\frac{p(1-p)}{n}} \leq \frac{X}{n} - p \leq z_{\frac{\alpha}{2}}\sqrt{\frac{p(1-p)}{n}}\right) \approx 1 - \alpha$$

$$\Rightarrow P\left(\frac{X}{n} - z_{\frac{\alpha}{2}}\sqrt{\frac{p(1-p)}{n}} \leq p \leq \frac{X}{n} + z_{\frac{\alpha}{2}}\sqrt{\frac{p(1-p)}{n}}\right) \approx 1 - \alpha.$$

This last probability statement almost looks like a confidence statement on the parameter p, but not quite, because $\sqrt{\frac{p(1-p)}{n}}$ is not computable. So, we cannot use $\frac{X}{n} \pm z_{\frac{\alpha}{2}}\sqrt{\frac{p(1-p)}{n}}$ as a confidence interval for p. We remedy this by substituting $\hat{p} = \frac{X}{n}$ in $\sqrt{\frac{p(1-p)}{n}}$, to finally result in the confidence interval

$$\hat{p} \pm z_{\frac{\alpha}{2}}\sqrt{\frac{\hat{p}(1-\hat{p})}{n}}.$$

This is *the Wald confidence interval* for p.

An alternative and much better confidence interval for p can be constructed by manipulating the normal approximation of the binomial in a different way. The steps proceed as follows. Writing once again $\hat{p}$ for $\frac{X}{n}$,

$$P\left(-z_{\frac{\alpha}{2}} \leq \frac{\hat{p} - p}{\sqrt{\frac{p(1-p)}{n}}} \leq z_{\frac{\alpha}{2}}\right) \approx 1 - \alpha$$

$$\Rightarrow P\left((\hat{p} - p)^2 \leq z_{\frac{\alpha}{2}}^2 \frac{p(1-p)}{n}\right) \approx 1 - \alpha$$

$$\Rightarrow P\left(p^2\left(1 + \frac{z_{\frac{\alpha}{2}}^2}{n}\right) - p\left(2\hat{p} + \frac{z_{\frac{\alpha}{2}}^2}{n}\right) + \hat{p}^2 \leq 0\right) \approx 1 - \alpha.$$

Now the quadratic equation

$$p^2\left(1 + \frac{z_{\frac{\alpha}{2}}^2}{n}\right) - p\left(2\hat{p} + \frac{z_{\frac{\alpha}{2}}^2}{n}\right) + \hat{p}^2 = 0$$

has the two real roots

$$p = p_\pm = \frac{\hat{p} + \frac{z_{\frac{\alpha}{2}}^2}{2n}}{1 + \frac{z_{\frac{\alpha}{2}}^2}{n}} \pm \frac{z_{\frac{\alpha}{2}} \sqrt{n}}{n + z_{\frac{\alpha}{2}}^2} \sqrt{\hat{p}(1 - \hat{p}) + \frac{z_{\frac{\alpha}{2}}^2}{4n}}.$$

This is the *score confidence interval* for p. It is established theoretically and empirically in Brown, Cai, and DasGupta (2001, 2002) that the score confidence interval performs much better than the Wald interval, even for very large n.

1.17.2 Error of the CLT

A famous theorem in probability places an upper bound on the error of the normal approximation in the central limit theorem. If we make this upper bound itself small, then we can be confident that the normal approximation will be accurate. This upper bound on the error of the normal approximation is known as the *Berry–Esseen bound*. Specialized to the binomial case, it says the following; a proof can be seen in Bhattacharya and Rao (1986) or in Feller (1968). The general Berry–Esseen bound is treated in this text in Chapter 8.

Theorem 1.59 (Berry–Esseen Bound for Normal Approximation). *Let* $X \sim \text{Bin}(n, p)$, *and let* $Y \sim N(np, np(1-p))$. *Then for any real number* x,

$$|P(X \leq x) - P(Y \leq x)| \leq \frac{4}{5} \frac{1 - 2p(1-p)}{\sqrt{np(1-p)}}.$$

1.17 Normal Approximations and Central Limit Theorem

It should be noted that the Berry–Esseen bound is rather conservative. Thus, accurate normal approximations are produced even when the upper bound, a conservative one, is .1 or so. We do not recommend use of the Berry–Esseen bound to decide when a normal approximation to the binomial can be accurately done. The bound is simply too conservative. However, it is good to know this bound due to its classic nature.

We finish with two more examples on the application of the general central limit theorem.

Example 1.70 (Sum of Uniforms). We can approximate the distribution of the sum of n independent uniforms on a general interval $[a, b]$ by a suitable normal distribution. However, it is interesting to ask what is the exact density of the sum of n independent uniforms on a general interval $[a, b]$. Because a uniform random variable on a general interval $[a, b]$ can be transformed to a uniform on the unit interval $[-1, 1]$ by a linear transformation and vice versa, we ask what is the exact density of the sum of n independent uniforms on $[-1, 1]$. We want to compare this exact density to a normal approximation for various values of n.

When $n = 2$, the density of the sum is a triangular density on $[-2, 2]$, which is a piecewise linear polynomial. In general, the density of the sum of n independent uniforms on $[-1, 1]$ is a piecewise polynomial of degree $n-1$, there being n different arcs in the graph of the density. The exact formula is:

$$f_n(x) = \frac{1}{2^n(n-1)!} \sum_{k=0}^{\lfloor \frac{n+x}{2} \rfloor} (-1)^k \binom{n}{k} (n + x - 2k)^{n-1}, \quad \text{if } |x| \leq n;$$

see Feller (1971, p. 27).

On the other hand, the CLT approximates the density of the sum by the $N(0, \frac{n}{3})$ density. It would be interesting to compare plots of the exact and the approximating normal density for various n. We see from Figs. 1.7, 1.8, and 1.9 that the normal approximation is already nearly exact when $n = 8$.

Example 1.71 (Unwise Use of the CLT). Suppose the checkout time at a supermarket has a mean of 4 minutes and a standard deviation of 1 minute. You have just joined the queue in a lane where there are eight people ahead of you. From just this information, can you say anything useful about the chances that you can be finished checking out within half an hour?

With the information provided being only on the mean and the variance of an individual checkout time, but otherwise nothing about the distribution, a possibility is to use the CLT, although here n is only 9, which is not large. Let $X_i, 1 \leq i \leq 8$, be the checkout times taken by the eight customers ahead of you, and X_9 the time taken by you yourself. If we use the CLT, then we have

$$S_n = \sum_{i=1}^{9} X_i \approx N(36, 9).$$

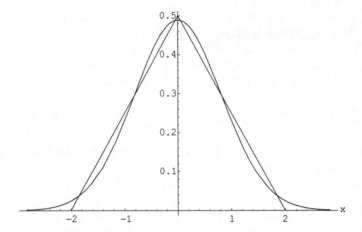

Fig. 1.7 Exact and approximating normal density for sum of uniforms; n = 2

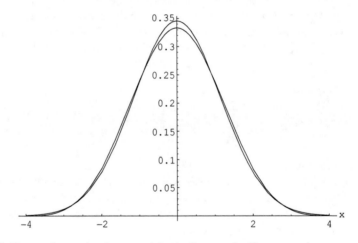

Fig. 1.8 Exact and approximating normal density for sum of uniforms; n = 4

Therefore,

$$P(S_n \leq 30) \approx \Phi\left(\frac{30-36}{3}\right) = \Phi(-2) = .0228.$$

In situations such as this where the information available is extremely limited, we do sometimes use the CLT, but it is not very wise because the value of n is so small. It may be better to model the distribution of checkout times, and answer the question under that chosen model.

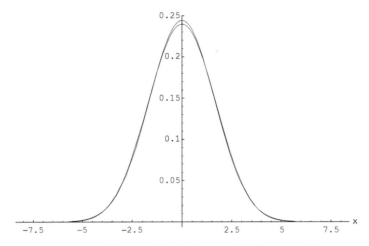

Fig. 1.9 Exact and approximating normal density for sum of uniforms; n = 8

1.18 Normal Approximation to Poisson and Gamma

A Poisson variable with an integer parameter $\lambda = n$ can be thought of as the sum of n independent Poisson variables, each with mean 1. Likewise, a Gamma variable with parameters $\alpha = n$ and λ can be thought of as the sum of n independent exponential variables, each with mean λ. So, in these two cases the CLT already implies that a normal approximation to the Poisson and the Gamma holds, when n is large. However, even if the Poisson parameter λ is not an integer, and even if the Gamma parameter α is not an integer, if λ is large, or if α is large, a normal approximation still holds. These results can be proved directly, by using the mgf technique. Here are the normal approximation results for general Poisson and Gamma distributions.

Theorem 1.60. *Let* $X \sim \text{Poi}(\lambda)$. *Then*

$$P\left(\frac{X - \lambda}{\sqrt{\lambda}} \leq x\right) \to \Phi(x), \quad \text{as } \lambda \to \infty,$$

for any real number x.
  Notationally, for large λ,

$$X \approx N(\lambda, \lambda).$$

Theorem 1.61. *Let* $X \sim G(\alpha, \lambda)$. *Then, for every fixed* λ,

$$P\left(\frac{X - \alpha\lambda}{\lambda\sqrt{\alpha}} \leq x\right) \to \Phi(x), \quad \text{as } \alpha \to \infty,$$

for any real number x.

Notationally, for large α,

$$X \approx N(\alpha\lambda, \alpha\lambda^2).$$

Example 1.72 (Nuclear Accidents). Suppose the probability of having any nuclear accidents in any single nuclear plant during a given year is .0005, and that a country has 100 such nuclear plants. What is the probability that there will be at least six nuclear accidents in the country during the next 250 years?

Let X_{ij} be the number of accidents in the ith year in the jth plant. We assume that each X_{ij} has a common Poisson distribution. The parameter, say θ of this common Poisson distribution, is determined from the equation $e^{-\theta} = 1 - .0005 = .9995 \Rightarrow \theta = -\log(.9995) = .0005$. Assuming that these X_{ij} are all independent, the number of accidents T in the country during 250 years has a Poi(λ) distribution, where $\lambda = \theta \times 100 \times 250 = .0005 \times 100 \times 250 = 12.5$. If we now do a normal approximation with continuity correction,

$$P(T \geq 6) \approx 1 - \Phi\left(\frac{5.5 - 12.5}{\sqrt{12.5}}\right)$$
$$= 1 - \Phi(-1.98) = .9761.$$

So we see that although the chances of having any accidents in a particular plant in any particular year are small, collectively, and in the long run the chances are high that there will be quite a few such accidents.

1.18.1 Confidence Intervals

In Example 1.64, we described a *confidence interval* for a normal mean. A major use of the various normal approximations described above is the construction of confidence intervals for an unknown parameter of interest. The parameter can be essentially anything. Thus, suppose that based on a sample of observations $X_1, X_2, \ldots, X_n$, a parameter θ is estimated by $\hat{\theta} = \hat{\theta}(X_1, \ldots, X_n)$. What we need is a theorem that allows us to approximate the distribution of $\hat{\theta}$ by a suitable normal distribution. To fix notation, suppose that for large n, $\hat{\theta}$ is approximately distributed as $N(\theta, \sigma^2(\theta))$ for some explicit and computable function $\sigma(\theta)$. Then, we can construct a confidence interval for θ, namely an interval $L \leq \theta \leq U$, and make probability statements of the form $P(L \leq \theta \leq U) \approx p$ for some suitable p. Here is an important illustration.

Example 1.73 (Confidence Interval for a Poisson Mean). The normal approximation to the Poisson distribution can be used to find a confidence interval for the mean of a Poisson distribution. We have already seen an example of a confidence interval for a normal mean in this chapter. We now work out the Poisson case, using the normal approximation to Poisson.

1.18 Normal Approximation to Poisson and Gamma

Suppose $X \sim \text{Poi}(\lambda)$. By the normal approximation theorem, if λ is large, then $\frac{X-\lambda}{\sqrt{\lambda}} \approx N(0,1)$. Now, a standard normal random variable Z has the property $P(-1.96 \leq Z \leq 1.96) = .95$. Because $\frac{X-\lambda}{\sqrt{\lambda}} \approx N(0,1)$, we have

$$P\left(-1.96 \leq \frac{X-\lambda}{\sqrt{\lambda}} \leq 1.96\right) \approx .95$$

$$\Leftrightarrow P\left(\frac{(X-\lambda)^2}{\lambda} \leq 1.96^2\right) \approx .95$$

$$\Leftrightarrow P((X-\lambda)^2 - 1.96^2 \lambda \leq 0) \approx .95$$

$$\Leftrightarrow P(\lambda^2 - \lambda(2X + 1.96^2) + X^2 \leq 0) \approx .95. \quad (*)$$

Now the quadratic equation

$$\lambda^2 - \lambda(2X + 1.96^2) + X^2 = 0$$

has the roots

$$\lambda = \lambda_\pm = \frac{(2X + 1.96^2) \pm \sqrt{(2X + 1.96^2)^2 - 4X^2}}{2}$$

$$= \frac{(2X + 1.96^2) \pm \sqrt{14.76 + 15.37X}}{2}$$

$$= (X + 1.92) \pm \sqrt{3.69 + 3.84X}.$$

The quadratic $\lambda^2 - \lambda(2X + 1.96^2) + X^2$ is ≤ 0 when λ is between these two values $\lambda_\pm$. So we can rewrite $(*)$ as

$$P\left((X + 1.92) - \sqrt{3.69 + 3.84X} \leq \lambda \leq (X + 1.92) + \sqrt{3.69 + 3.84X}\right) \approx .95 \quad (**).$$

In statistics, one often treats the parameter λ as unknown, and uses the data value X to estimate the unknown λ. The statement $(**)$ is interpreted as saying that with approximately 95% probability, λ will fall inside the interval of values

$$(X + 1.92) - \sqrt{3.69 + 3.84X} \leq \lambda \leq (X + 1.92) + \sqrt{3.69 + 3.84X},$$

and so the interval

$$\left[(X + 1.92) - \sqrt{3.69 + 3.84X}, (X + 1.92) + \sqrt{3.69 + 3.84X}\right]$$

is called an *approximate 95% confidence interval for* λ. We see that it is derived from the normal approximation to a Poisson distribution.

1.19 ∗ Convergence of Densities and Edgeworth Expansions

If in the central limit theorem, each individual X_i is a continuous random variable with a density $f(x)$, then the sum $S_n = \sum_{i=1}^{n} X_i$ also has a density for each n, and hence, the standardized sum $\frac{S_n - n\mu}{\sigma\sqrt{n}}$ also has a density for each n. It is natural to ask if the densitity of $\frac{S_n - n\mu}{\sigma\sqrt{n}}$ converges to the standard normal density when $n \to \infty$. This is true, under suitable conditions on the basic density $f(x)$. We present a result in this direction. It is useful to have a general result which ensures that under suitable conditions, in the central limit theorem the density of $Z_n = \frac{S_n - n\mu}{\sigma\sqrt{n}}$ converges to the $N(0, 1)$ density. The result below is not the best available result in this direction, but it often applies and is easy to state; a proof can be seen Bhattacharya and Rao (1986).

Theorem 1.62 (Gnedenko's Local Limit Theorem). *Suppose $X_1, X_2, \ldots$ are independent random variables with a density $f(x)$, mean μ, and variance σ^2. If $f(x)$ is uniformly bounded, then the density function of $Z_n = \frac{S_n - n\mu}{\sigma\sqrt{n}}$ converges uniformly on the real line $\mathcal{R}$ to the standard normal density $\phi(x) = \frac{1}{\sqrt{2\pi}} e^{-\frac{x^2}{2}}$.*

One criticism of the normal approximation in the various cases we have described is that any normal distribution is symmetric about its mean, and so, by employing a normal approximation we necessarily ignore any skewness that may be present in the true distribution that we are approximating. For instance, if the individual X_i have an exponential density, then the true density of the sum S_n is a Gamma density, which always has a skewness. But a normal approximation ignores that, and as a result, the quality of the approximation can be poor, unless n is quite large. Refined approximations that address this criticism are available.

We present refined density approximations that adjust the normal approximation for skewness, and one which also adjusts for kurtosis. They are collectively known as **Edgeworth expansions**; similar higher-order approximations are known for the CDF.

Suppose $X_1, X_2, \ldots, X_n$ are continuous random variables with a density $f(x)$. Suppose each individual X_i has four finite moments. Let $\mu, \sigma^2, \beta, \gamma$ denote the mean, variance, coefficient of skewness, and coefficient of kurtosis of the common distribution of the X_i. Let

$$Z_n = \frac{S_n - n\mu}{\sigma\sqrt{n}} = \frac{\sqrt{n}(\bar{X} - \mu)}{\sigma}.$$

Define the following three successively more refined density approximations for the density of Z_n:

$$\hat{f}_{n,0}(x) = \phi(x).$$

$$\hat{f}_{n,1}(x) = \left(1 + \frac{\beta(x^3 - 3x)}{6\sqrt{n}}\right)\phi(x).$$

$$\hat{f}_{n,2}(x) = \left(1 + \frac{\beta(x^3 - 3x)}{6\sqrt{n}} + \left[\gamma \frac{x^4 - 6x^2 + 3}{24}\right.\right.$$
$$\left.\left. + \beta^2 \frac{x^6 - 15x^4 + 45x^2 - 15}{72}\right]\frac{1}{n}\right)\phi(x).$$

The functions $\hat{f}_{n,0}(x)$, $\hat{f}_{n,1}(x)$, and $\hat{f}_{n,2}(x)$ are called the CLT approximation, the first-order Edgeworth expansion, and the second-order Edgeworth expansion for the density of the mean.

The approximations are of the form

$$\phi(x) + \frac{p_1(x)}{\sqrt{n}}\phi(x) + \frac{p_2(x)}{n}\phi(x) + \cdots.$$

The relevant polynomials $p_1(x)$, $p_2(x)$ are related to some very special polynomials, known as Hermite polynomials. Hermite polynomials are obtained from successive differentiation of the standard normal density $\phi(x)$. Precisely, the jth Hermite polynomial $H_j(x)$ is defined by the relation

$$\frac{d}{dx^j}\phi(x) = (-1)^j H_j(x)\phi(x).$$

In particular,

$H_1(x) = x;$ $H_2(x) = x^2 - 1;$ $H_3(x) = x^3 - 3x;$ $H_4(x) = x^4 - 6x^2 + 3;$
$H_5(x) = x^5 - 10x^3 + 15x;$ $H_6(x) = x^6 - 15x^4 + 45x^2 - 15.$

By comparing the formulas for the refined density approximations to the formulas for the Hermite polynomials, the connection becomes obvious. They arise in the density approximation formulas as a matter of fact; there is no intuition for it.

Exercises

Exercise 1.1. The population of Danville is 20,000. Can it be said with certainty that there must be two or more people in Danville with exactly the same three initials?

Exercise 1.2. The letters in the word FULL are rearranged at random. What is the probability that it still spells FULL?

Exercise 1.3 (Skills Exercise). Let E, F, and G be three events. Find expressions for the following events:

(a) Only E occurs
(b) Both E and G occur, but not F

(c) All three occur
(d) At least one of the events occurs
(e) At most two of them occur

Exercise 1.4. An urn contains 5 red, 5 black, and 5 white balls. If 3 balls are chosen without replacement at random, what is the probability that they are of exactly 2 different colors?

Exercise 1.5 (Matching Problem). Four men throw their watches into the sea, and the sea brings each man one watch back at random. What is the probability that at least one man gets his own watch back?

Exercise 1.6. Which is more likely:

(a) Obtaining at least one six in six rolls of a fair die,
or,
(b) Obtaining at least one double six in six rolls of a pair of fair dice.

Exercise 1.7 * **(The General Shoes Problem).** There are n pairs of shoes of n distinct colors in a closet and $2m$ are pulled out at random from the $2n$ shoes. What is the probability that there is at least one complete pair among the shoes pulled?

Exercise 1.8. *There are n people are lined up at random for a photograph. What is the probability that a specified set of r people happen to be next to each other?

Exercise 1.9. Calculate the probability that in Bridge, the hand of at least one player is void in a particular suit.

Exercise 1.10 * **(The Rumor Problem).** In a town with n residents, someone starts a rumor by saying it to one of the other $n-1$ residents. Thereafter, each recipient passes the rumor on to one of the other residents, chosen at random. What is the probability that by the kth time that the rumor has been told, it has not come back to someone who has already heard it?

Exercise 1.11. Jen will call Cathy on Saturday with a 60% probability. She will call Cathy on Sunday with a 80% probability. The probability that she will call on neither of the two days is 10%. What is the probability that she will call on Sunday if she calls on Saturday?

Exercise 1.12. Two distinct cards are drawn, one at a time, from a deck of 52 cards. The first chosen card is the ace of spades. What is the probability that the second card is neither an ace nor a spade?

Exercise 1.13. Suppose $P(A) = P(B) = .9$. Give a useful lower bound on $P(B|A)$.

Exercise 1.14. * The probability that a coin will show all heads or all tails when tossed four times is .25. What is the probability that it will show two heads and two tails?

Exercises

Exercise 1.15 *(Conditional Independence).** Events A, B are called *conditionally independent* given C, if $P(A \cap B|C) = P(A|C)P(B|C)$.

(a) Give an example of events A, B, C such that A, B are not independent, but they are conditionally independent given C.
(b) Give an example of events A, B, C such that A, B are independent, but are not conditionally independent given C.

Exercise 1.16 (Polygraphs). Polygraphs are routinely administered to job applicants for sensitive government positions. Suppose someone actually lying fails the polygraph 90% of the time. But someone telling the truth also fails the polygraph 15% of the time. If a polygraph indicates that an applicant is lying, what is the probability that he is in fact telling the truth? Assume a general prior probability p that the person is telling the truth.

Exercise 1.17 *(Random Matrix).** The diagonal elements a, c of a 2×2 symmetric matrix are chosen independently at random from $1, 2, \ldots, 5$, and the off-diagonal element is chosen at random from $1, \ldots, \min(a, c)$. Find the probability that the matrix is nonsingular.

Exercise 1.18 (The Parking Problem). At a parking lot, there are 12 places arranged in a row. A man observed that there were 8 cars parked, and that the four empty places were adjacent to each other. Given that there are 4 empty places, is this arrangement surprising?

Exercise 1.19. Suppose a fair die is rolled twice and suppose X is the absolute value of the difference of the two rolls. Find the pmf and the CDF of X and plot the CDF. Find a median of X; is the median unique?

Exercise 1.20 *(A Two-Stage Experiment).** Suppose a fair die is rolled once and the number observed is N. Then a fair coin is tossed N times. Let X be the number of heads obtained. Find the pmf, the CDF, and the expected value of X. Does the expected value make sense intuitively?

Exercise 1.21. *Find a discrete random variable X such that $E(X) = E(X^3) = 0$, $E(X^2) = E(X^4) = 1$.

Exercise 1.22 *(Waiting Time).** An urn contains four red and four green balls that are taken out without replacement, one at a time, at random. Let X be the first draw at which a green ball is taken out. Find the pmf and the expected value of X.

Exercise 1.23 *(Runs).** Suppose a fair die is rolled n times. By using the indicator variable method, find the expected number of times that a six is followed by at least two other sixes. Now compute the value when $n = 100$.

Exercise 1.24. *Suppose a couple will have children until they have at least two children of each sex. By using the tail sum formula, find the expected value of the number of children the couple will have.

Exercise 1.25. Suppose X has pmf $P(X = \frac{1}{n}) = \frac{1}{2^n}, n \geq 1$. Find the mean of X.

Exercise 1.26 (A Calculus Calculation). The best quadratic predictor of some random variable Y is $a + bX + cX^2$, where a, b, and c are chosen to minimize $E[(Y - (a + bX + cX^2))^2]$. Determine a, b, and c.

Exercise 1.27 * **(Tail Sum Formula for the Second Moment).** Let X be a nonnegative integer-valued random variable. Show that $E(X^2) - E(X) = 2\sum_{n=1}^{\infty} nP(X > n)$.

Exercise 1.28 * **(Obtaining Equality in Chebyshev's Inequality).** Consider a discrete random variable X with the pmf $P(X = \pm k) = p, P(X = 0) = 1 - 2p$, where k is a fixed positive number, and $0 < p < \frac{1}{2}$.

(a) Find the mean and variance of X.
(b) Find $P(|X - \mu| \geq k\sigma)$.
(c) Can you now choose p in such a way that $P(|X - \mu| \geq k\sigma)$ becomes equal to $\frac{1}{k^2}$?

Exercise 1.29 * **(Variance of a Product).** Suppose X_1, X_2 are independent random variables. Give a sufficient condition for it to be true that $\text{Var}(X_1 X_2) = \text{Var}(X_1)\text{Var}(X_2)$.

Exercise 1.30 (Existence of Some Moments, but Not All). Give an example of a random variable X taking the values $1, 2, 3, \ldots$ such that $E(X^k) < \infty$ for any $k < p$ (p is specified), but $E(X^p) = \infty$.

Exercise 1.31. Find the generating function and the mgf of the random variable X with the pmf $P(X = n) = \frac{1}{2^n}, n = 1, 2, 3, \ldots$.

Exercise 1.32 (MGF of a Linear Function). Suppose X has the mgf $\psi(t)$. Find an expression for the mgf of $aX + b$, where a, b are real constants.

Exercise 1.33 (Convexity of the MGF). Suppose X has the mgf $\psi(t)$, finite in some open interval. Show that $\psi(t)$ is convex in that open interval.

Exercise 1.34. Suppose $G(s), H(s)$ are both generating functions. Show that $pG(s) + (1-p)H(s)$ is also a valid generating function for any p in $(0, 1)$. What is an interesting interpretation of the distribution that has $pG(s) + (1-p)H(s)$ as its generating function?

Exercise 1.35. * Give an example of a random variable X such that X has a finite mgf at any t, but X^2 does not have a finite mgf at any $t > 0$.

Exercise 1.36. Suppose a fair coin is tossed n times. Find the probability that exactly half of the tosses result in heads, when $n = 10, 30, 50$; where does the probability seem to converge as n becomes large?

Exercises

Exercise 1.37. Suppose one coin with probability .4 for heads, one with probability .6 for heads, and one that is a fair coin are each tossed once. Find the pmf of the total number of heads obtained; is it a binomial distribution?

Exercise 1.38. In repeated rolling of a fair die, find the minimum number of rolls necessary in order for the probability of at least one six to be:

$$(a) \geq .5; \quad (b) \geq .9.$$

Exercise 1.39 * **(Distribution of Maximum).** Suppose n numbers are drawn at random from $\{1, 2, \ldots, N\}$. What is the probability that the largest number drawn is a specified number k if sampling is (a) with replacement; (b) without replacement?

Exercise 1.40 * **(Poisson Approximation).** One hundred people will each toss a fair coin 200 times. Approximate the probability that at least 10 of the 100 people would each have obtained exactly 100 heads and 100 tails.

Exercise 1.41. Suppose a fair coin is tossed repeatedly. Find the probability that 3 heads will be obtained before 4 tails.
Generalize to r heads and s tails.

Exercise 1.42 * **(A Pretty Question).** Suppose X is a Poisson distributed random variable. Can three different values of X have an equal probability?

Exercise 1.43 * **(Poisson Approximation).** There are 20 couples seated at a rectangular table, husbands on one side and the wives on the other, in a random order. Using a Poisson approximation, find the probability that exactly two husbands are seated directly across from their wives; at least three are; at most three are.

Exercise 1.44 **(Poisson Approximation).** There are 5 coins on a desk, with probabilities .05, .1, .05, .01, and .04 for heads. By using a Poisson approximation, find the probability of obtaining at least one head when the five coins are each tossed once.
Is the number of heads obtained binomially distributed in this problem?

Exercise 1.45. Let $X \sim \text{Bin}(n, p)$. Prove that $P(X \text{ is even}) = \frac{1}{2} + \frac{(1-2p)^n}{2}$.
Hence, show that $P(X \text{ is even})$ is larger than $\frac{1}{2}$ for any n if $p < \frac{1}{2}$, but that it is larger than $\frac{1}{2}$ for only even values of n, if $p > \frac{1}{2}$.

Exercise 1.46. Let $f(x) = c|x|(1 + x)(1 - x), -1 \leq x \leq 1$.

(a) Find the normalizing constant c that makes $f(x)$ a density function.
(b) Find the CDF corresponding to this density function. Plot it.
(c) Use the CDF to find:

$$P(X < -.5); \quad P(X > .5); \quad P(-.5 < X < .5).$$

Exercise 1.47. Show that for every $p, 0 \leq p \leq 1$, the function $f(x) = p \sin x + (1-p) \cos x, 0 \leq x \leq \pi/2$ (and $f(x) = 0$ otherwise), is a density function. Find its CDF and use it to find all the medians.

Exercise 1.48. * Give an example of a density function on $[0, 1]$ by giving a formula such that the density is finite at zero, unbounded at one, has a unique minimum in the open interval $(0, 1)$ and such that the median is .5.

Exercise 1.49 * **(A Mixed Distribution).** Suppose the damage claims on a particular type of insurance policy are uniformly distributed on $[0, 5]$ (in thousands of dollars), but the maximum payout by the insurance company is 2500 dollars. Find the CDF and the expected value of the payout, and plot the CDF. What is unusual about this CDF?

Exercise 1.50 * **(Random Division).** Jen's dog broke her six-inch long pencil off at a random point on the pencil. Find the density function and the expected value of the ratio of the lengths of the shorter piece and the longer piece of the pencil.

Exercise 1.51 (Square of a PDF Need Not Be a PDF). Give an example of a density function $f(x)$ on $[0, 1]$ such that $cf^2(x)$ cannot be a density function for any c.

Exercise 1.52 (Percentiles of the Standard Cauchy). Find the pth percentile of the standard Cauchy density for a general p, and compute it for $p = .75$.

Exercise 1.53 * **(Functional Similarity).** Suppose X has the standard Cauchy density. Show that $X - \frac{1}{X}$ also has a Cauchy density.
Can you find another function with this property on your own?
Hint: Think of simple rational functions.

Exercise 1.54 * **(An Intriguing Identity).** Suppose X has the standard Cauchy density. Give a rigorous proof that $P(X > 1) = P(X > 2) + P(X > 3)$.

Exercise 1.55 * **(Integer Part).** Suppose X has a uniform distribution on $[0, 10.5]$. Find the expected value of the integer part of X.

Exercise 1.56 (The Density Function of the Density Function). Suppose X has a density function $f(x)$. Find the density function of $f(X)$ when $f(x)$ is the standard normal density.

Exercise 1.57 (Minimum of Exponentials). Let $X_1, X_2, \ldots, X_n$ be n independent standard exponential random variables. Find an expression for $E[\min\{X_1, \ldots, X_n\}]$.

Exercise 1.58. Suppose X is a positive random variable with mean one. Show that $E(\log X) \leq 0$.

Exercise 1.59. Suppose X is a positive random variable with four finite moments. Show that $E(X)E(X^3) \geq [E(X^2)]^2$.

Exercises

Exercise 1.60 (Rate Function for Exponential). Derive the rate function $I(x)$ of the Chernoff–Bernstein inequality for the standard exponential density, and hence derive a bound for $P(X > x)$.

Exercise 1.61 *(Rate Function for the Double Exponential). Derive the rate function $I(x)$ of the Chernoff–Bernstein inequality for the double exponential density and hence derive a bound for $P(X > x)$.

Exercise 1.62. X is uniformly distributed on some interval $[a, b]$. If its mean is 2, and variance is 3, what are the values of a, b?

Exercise 1.63. Let $X \sim U[0, 1]$. Find the density of each of the following:

(a) $X^3 - 3X$;
(b) $\left(X - \frac{1}{2}\right)^2$;
(c) $\left(\sin\left(\frac{\pi}{2} X\right)\right)^4$.

Exercise 1.64 *(Mode of a Beta Density). Show that if a Beta density has a mode in the open interval $(0, 1)$, then we must have $\alpha > 1, \alpha + \beta > 2$, in which case, the mode is unique and equals $\frac{\alpha-1}{\alpha+\beta-2}$.

Exercise 1.65. An exponential random variable with mean 4 is known to be larger than 6. What is the probability that it is larger than 8?

Exercise 1.66 *(Sum of Gammas). Suppose X, Y are independent random variables, and $X \sim G(\alpha, \lambda), Y \sim G(\beta, \lambda)$. Find the distribution of $X + Y$ by using moment-generating functions.

Exercise 1.67 (Inverse Gamma Moments). Suppose $X \sim G(\alpha, \lambda)$. Find a formula for $E[(\frac{1}{X})^n]$, when this expectation exists.

Exercise 1.68 (Product of Chi Squares). Suppose $X_1, X_2, \ldots, X_n$ are independent chi square variables, with $X_i \sim \chi^2_{m_i}$. Find the mean and variance of $\prod_{i=1}^n X_i$.

Exercise 1.69 *(Chi-Square Skewness). Let $X \sim \chi^2_m$. Find the coefficient of skewness of X and prove that it converges to zero as $m \to \infty$.

Exercise 1.70 *(A Relation Between Poisson and Gamma). Suppose $X \sim \text{Poi}(\lambda)$. Prove by repeated integration by parts that

$$P(X \leq n) = P(G(n + 1, 1) > \lambda),$$

where $G(n + 1, 1)$ means a Gamma random variable with parameters $n + 1$ and 1.

Exercise 1.71 *(A Relation Between Binomial and Beta). Suppose $X \sim \text{Bin}(n, p)$. Prove that
$$P(X \leq k - 1) = P(B(k, n - k + 1) > p),$$
where $B(k, n - k + 1)$ means a Beta random variable with parameters $k, n - k + 1$.

Exercise 1.72. Suppose X has the standard Gumbel density. Find the density of e^{-X}.

Exercise 1.73. Suppose X is uniformly distributed on $[0, 1]$. Find the density of $\log\log\frac{1}{X}$.

Exercise 1.74. Let $Z \sim N(0, 1)$. Find

$$P\left(.5 < \left|Z - \frac{1}{2}\right| < 1.5\right); \quad P(1 + Z + Z^2 > 0); \quad P\left(\frac{e^Z}{1 + e^Z} > \frac{3}{4}\right);$$
$$P(\Phi(Z) < .5).$$

Exercise 1.75. Let $Z \sim N(0, 1)$. Find the density of $\frac{1}{Z}$. Is the density bounded?

Exercise 1.76. The 25th and the 75th percentile of a normally distributed random variable are -1 and $+1$. What is the probability that the random variable is between -2 and $+2$?

Exercise 1.77 (Standard Normal CDF in Terms of the Error Function). In some places, instead of the standard normal CDF, one sees use of the *error function* $\text{erf}(x) = (2/\sqrt{\pi})\int_0^x e^{-t^2}dt$.
Express $\Phi(x)$ in terms of $\text{erf}(x)$.

Exercise 1.78 * (An Interesting Calculation). Suppose $X \sim N(\mu, \sigma^2)$. Prove that

$$E[\Phi(X)] = \Phi\left(\mu/\sqrt{1 + \sigma^2}\right).$$

Exercise 1.79 * (Useful Normal Distribution Formulas). Prove the following primitive (indefinite integral) formulas.

(a) $\int x^2\phi(x)dx = \Phi(x) - x\phi(x)$.
(b) $\int [\phi(x)]^2 dx = 1/(2\sqrt{\pi})\Phi(x\sqrt{2})$.
(c) $\int \phi(x)\phi(a + bx)dx = (1/t)\phi(a/t)\Phi(tx + a/t)$, where $t = \sqrt{1 + b^2}$.
(d) $\int x\phi(x)\Phi(bx)dx = b/(\sqrt{2\pi}t)\Phi(tx) - \phi(x)\Phi(bx)$.

Exercise 1.80 * (Useful Normal Distribution Formulas). Prove the following definite integral formulas, with t as in the previous exercise:

(a) $\int_0^\infty x\phi(x)\Phi(bx)dx = 1/(2\sqrt{2\pi})[1 + b/t]$.
(b) $\int_{-\infty}^\infty x\phi(x)\Phi(bx)dx = b/(\sqrt{2\pi}t)$.
(c) $\int_{-\infty}^\infty \phi(x)\Phi(a + bx)dx = \Phi(a/t)$.
(d) $\int_0^\infty \phi(x)[\Phi(bx)]^2 dx = 1/(2\pi)[\arctan b + \arctan\sqrt{1 + 2b^2}]$.
(e) $\int_{-\infty}^\infty \phi(x)[\Phi(bx)]^2 dx = 1/\pi \arctan\sqrt{1 + 2b^2}$.

Exercises

Exercise 1.81 (Median and Mode of Lognormal). Show that a general lognormal density is unimodal, and find its mode and median.

Hint: For the median, remember that a lognormal variable is e^X, where X is a normal variable.

Exercise 1.82 (Margin of Error of a Confidence Interval). Suppose $X_1, X_2, \ldots, X_n$ are independent $N(\mu, 10)$ variables. What is the smallest n such that the margin of error of a 95% confidence interval for μ is at most .05?

Exercise 1.83. Suppose $X \sim N(0, 1)$, $Y \sim N(0, 9)$, and X, Y are independent. Find the value of $P((X - Y)^2 > 5)$.

Exercise 1.84. A fair die is rolled 25 times. Let X be the number of times a six is obtained. Find the exact value of $P(X = 6)$, and compare it to a normal approximation of $P(X = 6)$.

Exercise 1.85. A basketball player has a history of converting 80% of his free throws. Find a normal approximation with a continuity correction of the probability that he will make between 18 and 22 throws out of 25 free throws.

Exercise 1.86 (Airline Overbooking). An airline knows from past experience that 10% of fliers with a confirmed reservation do not show up for the flight. Suppose a flight has 250 seats. How many reservations over 250 can the airline permit, if they want to be 95% sure that no more than two passengers with a confirmed reservation would have to be bumped?

Exercise 1.87. * Suppose $X_1, X_2, \ldots, X_n$ are independent $N(0, 1)$ variables. Find an approximation to the probability that $\sum_{i=1}^n X_i$ is larger than $\sum_{i=1}^n X_i^2$, when $n = 10, 20, 30$.

Exercise 1.88 (A Product Problem). Suppose $X_1, X_2, \ldots, X_{30}$ are 30 independent variables, each distributed as $U[0, 1]$. Find an approximation to the probability that their *geometric mean* exceeds .4; exceeds .5.

Exercise 1.89 (Comparing a Poisson Approximation and a Normal Approximation). Suppose 1.5% of residents of a town never read a newspaper. Compute the exact value, a Poisson approximation, and a normal approximation of the probability that at least one resident in a sample of 50 residents never reads a newspaper.

Exercise 1.90 (Anything That Can Happen Will Eventually Happen). If you predict in advance the outcomes of 10 tosses of a fair coin, the probability that you get them all correct is $(.5)^{10}$, which is very small. Show that if 2000 people each try to predict the 10 outcomes correctly, the chance that at least one of them succeeds is better than 85%.

Exercise 1.91 *(Random Walk).** Consider the drunkard's random walk example. Find the probability that the drunkard will be at least 10 steps over on the right from his starting point after 200 steps. Compute a normal approximation.

Exercise 1.92 (Test Your Intuition). Suppose a fair coin is tossed 100 times. Which is more likely: you will get exactly 50 heads, or you will get more than 60 heads?

Exercise 1.93 *(Density of Uniform Sums).** Give a direct proof that the density of $\frac{S_n}{\sqrt{n/3}}$ at zero converges to $\phi(0)$, where S_n is the sum of n independent $U[-1, 1]$ variables.

Exercise 1.94 (Confidence Interval for Poisson mean). Derive a formula for an approximate 99% confidence interval for a Poisson mean, by using the normal approximation to a Poisson distribution. Compare your formula to the formula for an approximate 95% confidence interval that was worked out in text. Compute the 95% and the 99% confidence interval if $X = 5, 8, 12$.

References

Alon, N. and Spencer, J. (2000). *The Probabilistic Method*, Wiley, New York.
Ash, R. (1972). *Real Analysis and Probability*, Academic Press, New York.
Barbour, A. and Hall, P. (1984). On the rate of Poisson convergence, *Math. Proc. Camb. Phil. Soc.*, 95, 473–480.
Bernstein, S. (1927). *Theory of Probability*, Nauka, Moscow.
Bhattacharya, R.N. and Rao, R.R. (1986). *Normal Approximation and Asymptotic Expansions*, Robert E. Krieger, Melbourne, FL.
Bhattacharya, R.N. and Waymire, E. (2009). *A Basic Course in Probability Theory*, Springer, New York.
Billingsley, P.(1995). *Probability and Measure*, Third Edition, John Wiley, New York.
Breiman, L. (1992). *Probability*, Addison-Wesley, New York.
Brown, L., Cai, T., and DasGupta, A. (2001). Interval estimation for a binomial proportion, *Statist. Sci.*, 16, 101–133.
Brown, L., Cai, T., and DasGupta, A. (2002). Confidence intervals for a binomial proportion and asymptotic expansions, *Ann. Statist.*, 30, 160–201.
Chernoff, H. (1952). A measure of asymptotic efficiency for tests of a hypothesis based on the sum of observations, *Ann. Math. Statist.*, 23, 493–507.
Chernoff, H. (1981). A note on an inequality involving the normal distribution, *Ann. Prob.*, 9, 533–535.
Chung, K. L. (1974). *A Course in Probability*, Academic Press, New York.
DasGupta, A. (2008). *Asymptotic Theory of Statistics and Probability*, Springer, New York.
DasGupta, A. (2010). *Fundamentals of Probability: A First Course*, Springer, New York.
Diaconis, P. and Zabell, S. (1991). Closed form summation formulae for classical distributions, *Statist. Sci.*, 6, 284–302.
Dudley, R. (2002). *Real Analysis and Probability*, Cambridge University Press, Cambridge, UK.
Everitt, B. (1998). *Cambridge Dictionary of Statistics*, Cambridge University Press, New York.
Feller, W. (1968). *Introduction to Probability Theory and its Applications*, Vol. I, Wiley, New York.
Feller, W. (1971). *Introduction to Probability Theory and Its Applications*, Vol. II, Wiley, New York.
Fisher, R.A. (1929). Moments and product moments of sampling distributions, *Proc. London Math. Soc.*, 2, 199–238.
Hall, P. (1992). *The Bootstrap and Edgeworth Expansion*, Springer-Verlag, New York.

References

Johnson, N., Kotz, S., and Balakrishnan, N. (1994). *Continuous Univariate Distributions*, Vol. I, Wiley, New York.

Kagan, A., Linnik, Y., and Rao, C.R. (1973). *Characterization Problems in Mathematical Statistics*, Wiley, New York.

Kendall, M.G. and Stuart, A. (1976). *Advanced Theory of Statistics*, Vol. I, Wiley, New York.

Le Cam, L. (1960). An approximation theorem for the Poisson binomial distribution, *Pacific J. Math.*, 10, 1181–1197.

Le Cam, L. (1986). The central limit theorem around 1935, Statist. Sci., 1, 78–96.

Paley, R.E. and Zygmund, A. (1932). A note on analytic functions in the unit circle, *Proc. Camb. Philos. Soc.*, 28, 266–272.

Petrov, V. (1975). *Limit Theorems of Probability Theory*, Oxford University Press, Oxford, UK.

Pitman, J. (1992). *Probability*, Springer-Verlag, New York.

Rao, C.R. (1973), *Linear Statistical Inference and Applications*, Wiley, New York.

Ross, S. (1984). *A First Course in Probability*, Macmillan, New York.

Steele, J.M. (1994). Le Cam's inequality and Poisson approximations, *Amer. Math Month.*, 101, 48–54.

Stein, C. (1981). Estimation of the mean of a multivariate normal distribution, *Ann. Stat.*, 9, 1135–1151.

Stigler, S. (1986). *The History of Statistics*, Belknap Press, Cambridge, MA.

Stirzaker, D. (1994). *Elementary Probability*, Cambridge University Press, London.

Wasserman, L. (2006). *All of Nonparametric Statistics*, Springer, New York.

Widder, D. (1989). *Advanced Calculus*, Dover, New York.

Chapter 2
Multivariate Discrete Distributions

We have provided a detailed overview of distributions of one discrete or one continuous random variable in the previous chapter. But often in applications, we are just naturally interested in two or more random variables simultaneously. We may be interested in them simultaneously because they provide information about each other, or because they arise simultaneously as part of the data in some scientific experiment. For instance, on a doctor's visit, the physician may check someone's blood pressure, pulse rate, blood cholesterol level, and blood sugar level, because together they give information about the general health of the patient. In such cases, it becomes essential to know how to operate with many random variables simultaneously. This is done by using *joint distributions*. Joint distributions naturally lead to considerations of *marginal* and *conditional distributions*. We study joint, marginal, and conditional distributions for discrete random variables in this chapter. The concepts of these various distributions for continuous random variables are not different; but the techniques are mathematically more sophisticated. The continuous case is treated in the next chapter.

2.1 Bivariate Joint Distributions and Expectations of Functions

We present the fundamentals of joint distributions of two variables in this section. The concepts in the multivariate case are the same, although the technicalities are somewhat more involved. We treat the multivariate case in a later section. The idea is that there is still an underlying experiment ξ, with an associated sample space Ω. But now we have two or more random variables on the sample space Ω. Random variables being functions on the sample space Ω, we now have multiple functions, say $X(\omega), Y(\omega), \ldots$, and so on Ω. We want to study their *joint behavior*.

Example 2.1 (Coin Tossing). Consider the experiment ξ of tossing a fair coin three times. Let X be the number of heads among the first two tosses, and Y the number of heads among the last two tosses. If we consider X and Y individually, we realize immediately that they are each Bin(2, .5) random variables. But the individual distributions hide part of the full story. For example, if we knew that X was 2, then that would imply that Y must be at least 1. Thus, their joint behavior cannot be fully understood from their individual distributions; we must study their *joint distribution*.

Here is what we mean by their joint distribution. The sample space Ω of this experiment is

$$\Omega = \{HHH, HHT, HTH, HTT, THH, THT, TTH, TTT\}.$$

Each sample point has an equal probability $\frac{1}{8}$. Denoting the sample points as $\omega_1, \omega_2, \ldots, \omega_8$, we see that if ω_1 prevails, then $X(\omega_1) = Y(\omega_1) = 2$. but if ω_2 prevails, then $X(\omega_2) = 2, Y(\omega_2) = 1$. The combinations of all possible values of (X, Y) are

$$(0,0), (0,1), (0,2), (1,0), (1,1), (1,2), (2,0), (2,1), (2,2).$$

The joint distribution of (X, Y) provides the probability $p(x, y) = P(X=x, Y=y)$ for each such combination of possible values (x, y). Indeed, by direct counting using the eight equally likely sample points, we see that

$$p(0,0) = \frac{1}{8}, \quad p(0,1) = \frac{1}{8}, \quad p(0,2) = 0, \quad p(1,0) = \frac{1}{8}, \quad p(1,1) = \frac{1}{4};$$
$$p(1,2) = \frac{1}{8}, \quad p(2,0) = 0, \quad p(2,1) = \frac{1}{8}, \quad p(2,2) = \frac{1}{8}.$$

For example, why is $p(0,1)\frac{1}{8}$? This is because the combination $(X = 0, Y = 1)$ is favored by only one sample point, namely TTH. It is convenient to present these nine different probabilities in the form of a table as follows.

	Y		
X	0	1	2
0	$\frac{1}{8}$	$\frac{1}{8}$	0
1	$\frac{1}{8}$	$\frac{1}{4}$	$\frac{1}{8}$
2	0	$\frac{1}{8}$	$\frac{1}{8}$

Such a layout is a convenient way to present the joint distribution of two discrete random variables with a small number of values. The distribution itself is called *the joint pmf*; here is a formal definition.

Definition 2.1. Let X, Y be two discrete random variables with respective sets of values $x_1, x_2, \ldots,$ and $y_1, y_2, \ldots,$ defined on a common sample space Ω. The *joint pmf* of X, Y is defined to be the function $p(x_i, y_j) = P(X=x_i, Y=y_j), i, j \geq 1$, and $p(x, y) = 0$ at any other point (x, y) in $\mathcal{R}^2$.

2.1 Bivariate Joint Distributions and Expectations of Functions

The requirements of a joint pmf are that

(i) $p(x, y) \geq 0 \ \forall (x, y)$;

(ii) $\sum_i \sum_j p(x_i, y_j) = 1$.

Thus, if we write the joint pmf in the form of a table, then all entries should be nonnegative, and the sum of all the entries in the table should be one.

As in the case of a single variable, we can define a CDF for more than one variable also. For the case of two variables, here is the definition of a CDF.

Definition 2.2. Let X, Y be two discrete random variables, defined on a common sample space Ω. The joint CDF, or simply the CDF, of (X, Y) is a function $F : \mathcal{R}^2 \to [0, 1]$ defined as $F(x, y) = P(X \leq x, Y \leq y), x, y \in \mathcal{R}$.

Like the joint pmf, the CDF also characterizes the joint distribution of two discrete random variables. But it is not very convenient or even interesting to work with the CDF in the case of discrete random variables. It is much preferable to work with the pmf when dealing with discrete random variables.

Example 2.2 (Maximum and Minimum in Dice Rolls). Suppose a fair die is rolled twice, and let X, Y be the larger and the smaller of the two rolls (note that X can be equal to Y). Each of X, Y takes the individual values $1, 2, \ldots, 6$, but we have necessarily $X \geq Y$. The sample space of this experiment is

$$\{11, 12, 13, \ldots, 64, 65, 66\}.$$

By direct counting, for example, $p(2, 1) = \frac{2}{36}$. Indeed, $p(x, y) = \frac{2}{36}$ for each $x, y = 1, 2, \ldots, 6, x > y$, and $p(x, y) = \frac{1}{36}$ for $x = y = 1, 2, \ldots, 6$. Here is what the joint pmf looks like in the form of a table:

X \ Y	1	2	3	4	5	6
1	$\frac{1}{36}$	0	0	0	0	0
2	$\frac{1}{18}$	$\frac{1}{36}$	0	0	0	0
3	$\frac{1}{18}$	$\frac{1}{18}$	$\frac{1}{36}$	0	0	0
4	$\frac{1}{18}$	$\frac{1}{18}$	$\frac{1}{18}$	$\frac{1}{36}$	0	0
5	$\frac{1}{18}$	$\frac{1}{18}$	$\frac{1}{18}$	$\frac{1}{18}$	$\frac{1}{36}$	0
6	$\frac{1}{18}$	$\frac{1}{18}$	$\frac{1}{18}$	$\frac{1}{18}$	$\frac{1}{18}$	$\frac{1}{36}$

The individual pmfs of X, Y are easily recovered from the joint distribution. For example,

$$P(X = 1) = \sum_{y=1}^{6} P(X = 1, Y = y) = \frac{1}{36}, \quad \text{and}$$

$$P(X = 2) = \sum_{y=1}^{6} P(X = 2, Y = y) = \frac{1}{18} + \frac{1}{36} = \frac{1}{12},$$

and see on. The individual pmfs are obtained by summing the joint probabilities over all values of the other variable. They are:

x	1	2	3	4	5	6
$p_X(x)$	$\frac{1}{36}$	$\frac{3}{36}$	$\frac{5}{36}$	$\frac{7}{36}$	$\frac{9}{36}$	$\frac{11}{36}$
y	1	2	3	4	5	6
$p_Y(y)$	$\frac{11}{36}$	$\frac{9}{36}$	$\frac{7}{36}$	$\frac{5}{36}$	$\frac{3}{36}$	$\frac{1}{36}$

From the individual pmf of X, we can find the expectation of X. Indeed,

$$E(X) = 1 \times \frac{1}{36} + 2 \times \frac{3}{36} + \cdots + 6 \times \frac{11}{36} = \frac{161}{36}.$$

Similarly, $E(Y) = \frac{91}{36}$. The individual pmfs are called *marginal pmfs*, and here is the formal definition.

Definition 2.3. Let $p(x, y)$ be the joint pmf of (X, Y). The *marginal pmf* of a function $Z = g(X, Y)$ is defined as $p_Z(z) = \sum_{(x,y):g(x,y)=z} p(x, y)$. In particular,

$$p_X(x) = \sum_y p(x, y); \quad p_Y(y) = \sum_x p(x, y),$$

and for any event A,

$$P(A) = \sum_{(x,y) \in A} p(x, y).$$

Example 2.3. Consider a joint pmf given by the formula

$$p(x, y) = c(x + y), \quad 1 \leq x, y \leq n,$$

where c is a normalizing constant.

First of all, we need to evaluate c by equating

$$\sum_{x=1}^{n} \sum_{y=1}^{n} p(x, y) = 1$$

$$\Leftrightarrow c \sum_{x=1}^{n} \sum_{y=1}^{n} (x + y) = 1$$

$$\Leftrightarrow c \sum_{x=1}^{n} \left[nx + \frac{n(n+1)}{2} \right] = 1$$

$$\Leftrightarrow c \left[\frac{n^2(n+1)}{2} + \frac{n^2(n+1)}{2} \right] = 1$$

$$\Leftrightarrow cn^2(n+1) = 1$$

$$\Leftrightarrow c = \frac{1}{n^2(n+1)}.$$

2.1 Bivariate Joint Distributions and Expectations of Functions

The joint pmf is symmetric between x and y (because $x + y = y + x$), and so, X, Y have the same marginal pmf. For example, X has the pmf

$$p_X(x) = \sum_{y=1}^{n} p(x, y) = \frac{1}{n^2(n+1)} \sum_{y=1}^{n} (x+y)$$

$$= \frac{1}{n^2(n+1)} \left[nx + \frac{n(n+1)}{2} \right]$$

$$= \frac{x}{n(n+1)} + \frac{1}{2n}, \quad 1 \leq x \leq n.$$

Suppose now we want to compute $P(X > Y)$. This can be found by summing $p(x, y)$ over all combinations for which $x > y$. But this longer calculation can be avoided by using a *symmetry argument* that is often very useful. Note that because the joint pmf is symmetric between x and y, we must have $P(X > Y) = P(Y > X) = p$ (say). But, also,

$$P(X > Y) + P(Y > X) + P(X = Y) = 1 \Rightarrow 2p + P(X = Y) = 1$$

$$\Rightarrow p = \frac{1 - P(X = Y)}{2}.$$

Now,

$$P(X = Y) = \sum_{x=1}^{n} p(x, x) = c \times \sum_{x=1}^{n} 2x$$

$$= \frac{1}{n^2(n+1)} n(n+1) = \frac{1}{n}.$$

Therefore, $P(X > Y) = p = \frac{n-1}{2n} \approx \frac{1}{2}$, for large n.

Example 2.4 (Dice Rolls Revisited). Consider again the example of two rolls of a fair die, and suppose X, Y are the larger and the smaller of the two rolls. We have worked out the joint distribution of (X, Y) in Example 2.2. Suppose we want to find the distribution of the difference, $X - Y$. The possible values of $X - Y$ are $0, 1, \ldots, 5$, and we find $P(X - Y = k)$ by using the joint distribution of (X, Y):

$$P(X - Y = 0) = p(1, 1) + p(2, 2) + \cdots + p(6, 6) = \frac{1}{6};$$

$$P(X - Y = 1) = p(2, 1) + p(3, 2) + \cdots + p(6, 5) = \frac{5}{18};$$

$$P(X - Y = 2) = p(3, 1) + p(4, 2) + p(5, 3) + p(6, 4) = \frac{2}{9};$$

$$P(X - Y = 3) = p(4, 1) + p(5, 2) + p(6, 3) = \frac{1}{6};$$

$$P(X - Y = 4) = p(5, 1) + p(6, 2) = \frac{1}{9};$$

$$P(X - Y = 5) = p(6, 1) = \frac{1}{18}.$$

There is no way to find the distribution of $X - Y$ except by using the joint distribution of (X, Y).

Suppose now we also want to know the expected value of $X - Y$. Now that we have the distribution of $X - Y$ worked out, we can find the expectation by directly using the definition of expectation:

$$E(X - Y) = \sum_{k=0}^{5} k P(X - Y = k)$$
$$= \frac{5}{18} + \frac{4}{9} + \frac{1}{2} + \frac{4}{9} + \frac{5}{18} = \frac{35}{18}.$$

But, we can also use linearity of expectations and find $E(X - Y)$ as

$$E(X - Y) = E(X) - E(Y) = \frac{161}{36} - \frac{91}{36} = \frac{35}{18}$$

(see Example 2.2 for $E(X), E(Y)$).

A third possible way to compute $E(X - Y)$ is to treat $X - Y$ as a function of (X, Y) and use the joint pmf of (X, Y) to find $E(X - Y)$ as $\sum_x \sum_y (x - y) p(x, y)$. In this particular example, this is an unncessarily laborious calculation, because luckily we can find $E(X - Y)$ by other quicker means in this example, as we just saw. But in general, one has to resort to the joint pmf to calculate the expectation of a function of (X, Y). Here is the formal formula.

Theorem 2.1 (Expectation of a Function). *Let (X, Y) have the joint pmf $p(x, y)$, and let $g(X, Y)$ be a function of (X, Y). We say that the expectation of $g(X, Y)$ exists if $\sum_x \sum_y |g(x, y)| p(x, y) < \infty$, in which case,*

$$E[g(X, Y)] = \sum_x \sum_y g(x, y) p(x, y).$$

2.2 Conditional Distributions and Conditional Expectations

Sometimes we want to know what the expected value is of one of the variables, say X, if we knew the value of the other variable Y. For example, in the die tossing experiment above, what should we expect the larger of the two rolls to be if the smaller roll is known to be 2?

To answer this question, we have to find the probabilities of the various values of X, conditional on knowing that Y equals some given y, and then average by using these conditional probabilities. Here are the formal definitions.

Definition 2.4 (Conditional Distribution). Let (X, Y) have the joint pmf $p(x, y)$. The *conditional distribution of X given $Y = y$* is defined to be

$$p(x|y) = P(X = x|Y = y) = \frac{p(x, y)}{p_Y(y)},$$

2.2 Conditional Distributions and Conditional Expectations

and the *conditional expectation of X given Y = y* is defined to be

$$E(X|Y = y) = \sum_x xp(x|y) = \frac{\sum_x xp(x, y)}{p_Y(y)} = \frac{\sum_x xp(x, y)}{\sum_x p(x, y)}.$$

The conditional distribution of Y given $X = x$ and the conditional expectation of Y given $X = x$ are defined analogously, by switching the roles of X and Y in the above definitions.

We often casually write $E(X|y)$ to mean $E(X|Y = y)$.

Two easy facts that are nevertheless often useful are the following.

Proposition. *Let X, Y be random variables defined on a common sample space Ω. Then,*

(a) $E(g(Y)|Y = y) = g(y), \forall y,$ *and for any function g;*
(b) $E(Xg(Y)|Y = y) = g(y)E(X|Y = y) \ \forall y,$ *and for any function g.*

Recall that in Chapter 1, we defined two random variables to be independent if $P(X \le x, Y \le y) = P(X \le x)P(Y \le y) \ \forall \ x, y \in \mathcal{R}$. This is of course a correct definition; but in the case of discrete random variables, it is more convenient to think of independence in terms of the pmf. The definition below puts together some equivalent definitions of independence of two discrete random variables.

Definition 2.5 (Independence). Let (X, Y) have the joint pmf $p(x, y)$. Then X, Y are said to be *independent* if

$$p(x|y) = p_X(x), \ \forall \ x, y \text{ such that } p_Y(y) > 0;$$
$$\Leftrightarrow p(y|x) = p_Y(y). \ \forall \ x, y \text{ such that } p_X(x) > 0;$$
$$\Leftrightarrow p(x, y) = p_X(x)p_Y(y), \quad \forall \ x, y;$$
$$\Leftrightarrow P(X \le x, Y \le y) = P(X \le x)P(Y \le y) \quad \forall \ x, y.$$

The third equivalent condition in the above list is usually the most convenient one to verify and use.

One more frequently useful fact about conditional expectations is the following.

Proposition. *Suppose X, Y are independent random variables. Then, for any function $g(X)$ such that the expectations below exist, and for any y,*

$$E[g(X)|Y = y] = E[g(X)].$$

2.2.1 Examples on Conditional Distributions and Expectations

Example 2.5 (Maximum and Minimum in Dice Rolls). In the experiment of two rolls of a fair die, we have worked out the joint distribution of X, Y, where X is the larger and Y the smaller of the two rolls. Using this joint distribution, we can now find the conditional distributions. For instance,

$$P(Y = 1|X = 1) = 1; \quad P(Y = y|X = 1) = 0, \text{ if } y > 1;$$
$$P(Y = 1|X = 2) = \frac{1/18}{1/18 + 1/36} = \frac{2}{3};$$
$$P(Y = 2|X = 2) = \frac{1/36}{1/18 + 1/36} = \frac{1}{3};$$
$$P(Y = y|X = 2) = 0, \quad \text{if } y > 2;$$
$$P(Y = y|X = 6) = \frac{1/18}{5/18 + 1/36} = \frac{2}{11}, \quad \text{if } 1 \le y \le 5;$$
$$P(Y = 6|X = 6) = \frac{1/36}{5/18 + 1/36} = \frac{1}{11}.$$

Example 2.6 (Conditional Expectation in a 2×2 Table). Suppose X, Y are binary variables, each taking only the values 0, 1 with the following joint distribution.

	Y	
X	0	1
0	s	t
1	u	v

We want to evaluate the conditional expectation of X given $Y = 0, 1$, respectively. By using the definition of conditional expectation,

$$E(X|Y = 0) = \frac{0 \times p(0,0) + 1 \times p(1,0)}{p(0,0) + p(1,0)} = \frac{u}{s + u};$$
$$E(X|Y = 1) = \frac{0 \times p(0,1) + 1 \times p(1,1)}{p(0,1) + p(1,1)} = \frac{v}{t + v}.$$

Therefore,

$$E(X|Y = 1) - E(X|Y = 0) = \frac{v}{t + v} - \frac{u}{s + u} = \frac{vs - ut}{(t + v)(s + u)}.$$

It follows that we can now have the single formula

$$E(X|Y = y) = \frac{u}{s + u} + \frac{vs - ut}{(t + v)(s + u)} y,$$

$y = 0, 1$. We now realize that the conditional expectation of X given $Y = y$ is a linear function of y in this example. This is the case whenever both X, Y are binary variables, as they were in this example.

2.2 Conditional Distributions and Conditional Expectations

Example 2.7 (Conditional Expectation in Dice Experiment). Consider again the example of the joint distribution of the maximum and the minimum of two rolls of a fair die. Let X denote the maximum, and Y the minimum. We find $E(X|Y = y)$ for various values of y.

By using the definition of $E(X|Y = y)$, we have, for example,

$$E(X|Y = 1) = \frac{1 \times \frac{1}{36} + \frac{1}{18}[2 + \cdots + 6]}{\frac{1}{36} + \frac{5}{18}} = \frac{41}{11} = 3.73;$$

as another example,

$$E(X|Y = 3) = \frac{3 \times \frac{1}{36} + \frac{1}{18} \times 15}{\frac{1}{36} + \frac{3}{18}} = \frac{33}{7} = 4.71;$$

and,

$$E(X|Y = 5) = \frac{5 \times \frac{1}{36} + 6 \times \frac{1}{18}}{\frac{1}{36} + \frac{1}{18}} = \frac{17}{3} = 5.77.$$

We notice that $E(X|Y = 5) > E(X|Y = 3) > E(X|Y = 1)$; in fact, it is true that $E(X|Y = y)$ is increasing in y in this example. This does make intuitive sense.

Just as in the case of a distribution of a single variable, we often also want a measure of variability in addition to a measure of average for conditional distributions. This motivates defining a *conditional variance*.

Definition 2.6 (Conditional Variance). Let (X, Y) have the joint pmf $p(x, y)$. Let $\mu_X(y) = E(X|Y = y)$. The *conditional variance* of X given $Y = y$ is defined to be

$$\text{Var}(X|Y = y) = E[(X - \mu_X(y))^2|Y = y] = \sum_x (x - \mu_X(y))^2 p(x|y).$$

We often write casually $\text{Var}(X|y)$ to mean $\text{Var}(X|Y = y)$.

Example 2.8 (Conditional Variance in Dice Experiment). We work out the conditional variance of the maximum of two rolls of a die given the minimum. That is, suppose a fair die is rolled twice, and X, Y are the larger and the smaller of the two rolls; we want to compute $\text{Var}(X|y)$.

For example, if $y = 3$, then $\mu_X(y) = E(X|Y = y) = E(X|Y = 3) = 4.71$ (see the previous example). Therefore,

$$\text{Var}(X|y) = \sum_x (x - 4.71)^2 p(x|3)$$

$$= \frac{(3 - 4.71)^2 \times \frac{1}{36} + (4 - 4.71)^2 \times \frac{1}{18} + (5 - 4.71)^2 \times \frac{1}{18} + (6 - -4.71)^2 \times \frac{1}{18}}{\frac{1}{36} + \frac{1}{18} + \frac{1}{18} + \frac{1}{18}}$$

$$= 1.06.$$

To summarize, given that the minimum of two rolls of a fair die is 3, the expected value of the maximum is 4.71 and the variance of the maximum is 1.06.

These two values, $E(X|y)$ and $\text{Var}(X|y)$, change as we change the given value y. Thus $E(X|y)$ and $\text{Var}(X|y)$ are functions of y, and for each separate y, a new calculation is needed. If X, Y happen to be independent, then of course whatever y is, $E(X|y) = E(X)$, and $\text{Var}(X|y) = \text{Var}(X)$.

The next result is an important one in many applications.

Theorem 2.2 (Poisson Conditional Distribution). *Let X, Y be independent Poisson random variables, with means λ, μ. Then the conditional distribution of X given $X + Y = t$ is $\text{Bin}(t, p)$, where $p = \frac{\lambda}{\lambda+\mu}$.*

Proof. Clearly, $P(X = x | X + Y = t) = 0 \; \forall x > t$. For $x \leq t$,

$$P(X = x | X + Y = t) = \frac{P(X = x, X + Y = t)}{P(X + Y = t)}$$

$$= \frac{P(X = x, Y = t - x)}{P(X + Y = t)}$$

$$= \frac{e^{-\lambda}\lambda^x}{x!} \frac{e^{-\mu}\mu^{t-x}}{(t-x)!} \frac{t!}{e^{-(\lambda+\mu)}(\lambda+\mu)^t}$$

(on using the fact that $X + Y \sim \text{Poi}(\lambda + \mu)$; see Chapter 1)

$$= \frac{t!}{x!(t-x)!} \frac{\lambda^x \mu^{t-x}}{(\lambda+\mu)^t}$$

$$= \binom{t}{x} \left(\frac{\lambda}{\lambda+\mu}\right)^x \left(\frac{\mu}{\lambda+\mu}\right)^{t-x},$$

which is the pmf of the $\text{Bin}(t, \frac{\lambda}{\lambda+\mu})$ distribution. □

2.3 Using Conditioning to Evaluate Mean and Variance

Conditioning is often an extremely effective tool to calculate probabilities, means, and variances of random variables with a complex or clumsy joint distribution. Thus, in order to calculate the mean of a random variable X, it is sometimes greatly convenient to follow an *iterative process*, whereby we first evaluate the mean of X after conditioning on the value y of some suitable random variable Y, and then average over y. The random variable Y has to be chosen judiciously, but is often clear from the context of the specific problem. Here are the precise results on how this technique works; it is important to note that the next two results hold for any kind of random variables, not just discrete ones.

2.3 Using Conditioning to Evaluate Mean and Variance

Theorem 2.3 (Iterated Expectation Formula). *Let X, Y be random variables defined on the same probability space Ω. Suppose $E(X)$ and $E(X|Y = y)$ exist for each y. Then,*

$$E(X) = E_Y[E(X|Y = y)];$$

thus, in the discrete case,

$$E(X) = \sum_y \mu_X(y) p_Y(y),$$

where $\mu_X(y) = E(X|Y = y)$.

Proof. We prove this for the discrete case. By definition of conditional expectation,

$$\mu_X(y) = \frac{\sum_x x p(x, y)}{p_Y(y)}$$

$$\Rightarrow \sum_y \mu_X(y) p_Y(y) = \sum_y \sum_x x p(x, y) = \sum_x \sum_y x p(x, y)$$

$$= \sum_x x \sum_y p(x, y) = \sum_x x p_X(x) = E(X).$$

The corresponding variance calculation formula is the following. The proof of this uses the iterated mean formula above, and applies it to $(X - \mu_X)^2$. $\square$

Theorem 2.4 (Iterated Variance Formula). *Let X, Y be random variables defined on the same probability space Ω. Suppose $\mathrm{Var}(X)$, $\mathrm{Var}(X|Y = y)$ exist for each y. Then,*

$$\mathrm{Var}(X) = E_Y[\mathrm{Var}(X|Y = y)] + \mathrm{Var}_Y[E(X|Y = y)].$$

Remark. These two formulas for iterated expectation and iterated variance are valid for all types of variables, not just the discrete ones. Thus, these same formulas still hold when we discuss joint distributions for continuous random variables in the next chapter.

Some operational formulas that one should be familiar with are summarized below.

Conditional Expectation and Variance Rules.

$E(g(X)|X = x) = g(x); \quad E(g(X)h(Y)|Y = y) = h(y)E(g(X)|Y = y);$

$E(g(X)|Y = y) = E(g(X))$ if X, Y are independent;

$\mathrm{Var}(g(X)|X = x) = 0; \quad \mathrm{Var}(g(X)h(Y)|Y = y) = h^2(y)\mathrm{Var}(g(X)|Y = y);$

$\mathrm{Var}(g(X)|Y = y) = \mathrm{Var}(g(X))$ if X, Y are independent.

Let us see some applications of the two iterated expectation and iterated variance formulas.

Example 2.9 (A Two-Stage Experiment). Suppose n fair dice are rolled. Those that show a six are rolled again. What are the mean and the variance of the number of sixes obtained in the second round of this experiment?

Define Y to be the number of dice in the first round that show a six, and X the number of dice in the second round that show a six. Given $Y = y$, $X \sim \text{Bin}(y, \frac{1}{6})$, and Y itself is distributed as $\text{Bin}(n, \frac{1}{6})$. Therefore,

$$E(X) = E[E(X|Y = y)] = E_Y\left[\frac{y}{6}\right] = \frac{n}{36}.$$

Also,

$$\text{Var}(X) = E_Y[\text{Var}(X|Y = y)] + \text{Var}_Y[E(X|Y = y)]$$
$$= E_Y\left[y\frac{1}{6}\frac{5}{6}\right] + \text{Var}_Y\left[\frac{y}{6}\right]$$
$$= \frac{5}{36}\frac{n}{6} + \frac{1}{36}n\frac{1}{6}\frac{5}{6}$$
$$= \frac{5n}{216} + \frac{5n}{1296} = \frac{35n}{1296}.$$

Example 2.10. Suppose a chicken lays a Poisson number of eggs per week with mean λ. Each egg, independently of the others, has a probability p of fertilizing. We want to find the mean and the variance of the number of eggs fertilized in a week.

Let N denote the number of eggs hatched and X the number of eggs fertilized. Then, $N \sim \text{Poi}(\lambda)$, and given $N = n$, $X \sim \text{Bin}(n, p)$. Therefore,

$$E(X) = E_N[E(X|N = n)] = E_N[np] = p\lambda,$$

and,

$$\text{Var}(X) = E_N[\text{Var}(X|N = n)] + \text{Var}_N(E(X|N = n))$$
$$= E_N[np(1 - p)] + \text{Var}_N(np) = \lambda p(1 - p) + p^2\lambda = p\lambda.$$

Interestingly, the number of eggs actually fertilized has the same mean and variance $p\lambda$, (Can you see why?)

Remark. In all of these examples, it was important to choose the variable Y wisely on which one should condition. The efficiency of the technique depends on this very crucially.

Sometimes a formal generalization of the iterated expectation formula when a third variable Z is present is useful. It is particularly useful in hierarchical statistical modeling of distributions, where an ultimate marginal distribution for some X is constructed by first conditioning on a number of auxiliary variables, and then gradually unconditioning them. We state the more general iterated expectation formula; its proof is exactly similar to that of the usual iterated expectation formula.

Theorem 2.5 (Higher-Order Iterated Expectation). *Let X, Y, Z be random variables defined on the same sample space Ω. Assume that each conditional expectation below and the marginal expectation $E(X)$ exist. Then,*

$$E(X) = E_Y[E_{Z|Y}\{E(X|Y = y, Z = z)\}].$$

2.4 Covariance and Correlation

We know that variance is additive for independent random variables; that is, if $X_1, X_2, \ldots, X_n$ are independent random variables, then $\text{Var}(X_1 + X_2 + \cdots + X_n) = \text{Var}(X_1) + \cdots + \text{Var}(X_n)$. In particular, for two independent random variables X, Y, $\text{Var}(X+Y) = \text{Var}(X) + \text{Var}(Y)$. However, in general, variance is not additive. Let us do the general calculation for $\text{Var}(X + Y)$.

$$\begin{aligned}\text{Var}(X + Y) &= E(X + Y)^2 - [E(X + Y)]^2 \\ &= E(X^2 + Y^2 + 2XY) - [E(X) + E(Y)]^2 \\ &= E(X^2) + E(Y^2) + 2E(XY) - [E(X)]^2 - [E(Y)]^2 - 2E(X)E(Y) \\ &= E(X^2) - [E(X)]^2 + E(Y^2) - [E(Y)]^2 + 2[E(XY) - E(X)E(Y)] \\ &= \text{Var}(X) + \text{Var}(Y) + 2[E(XY) - E(X)E(Y)].\end{aligned}$$

We thus have the extra term $2[E(XY) - E(X)E(Y)]$ in the expression for $\text{Var}(X + Y)$; of course, when X, Y are independent, $E(XY) = E(X)E(Y)$, and so the extra term drops out. But, in general, one has to keep the extra term. The quantity $E(XY) - E(X)E(Y)$ is called the *covariance* of X and Y.

Definition 2.7 (Covariance). Let X, Y be two random variables defined on a common sample space Ω, such that $E(XY), E(X), E(Y)$ all exist. The *covariance* of X and Y is defined as

$$\text{Cov}(X, Y) = E(XY) - E(X)E(Y) = E[(X - E(X))(Y - E(Y))].$$

Remark. Covariance is a measure of whether two random variables X, Y tend to increase or decrease together. If a larger value of X generally causes an increment in the value of Y, then often (but not always) they have a positive covariance. For example, taller people tend to weigh more than shorter people, and height and weight usually have a positive covariance.

Unfortunately, however, covariance can take arbitrary positive and arbitrary negative values. Therefore, by looking at its value in a particular problem, we cannot judge whether it is a large value. We cannot compare a covariance with a standard to judge if it is large or small. A renormalization of the covariance cures this problem, and calibrates it to a scale of -1 to $+1$. We can judge such a quantity as large, small, or moderate; for example, .95 would be large positive, .5 moderate, and .1 small. The renormalized quantity is the *correlation coefficient* or simply the *correlation* between X and Y.

Definition 2.8 (Correlation). Let X, Y be two random variables defined on a common sample space Ω, such that $\text{Var}(X), \text{Var}(Y)$ are both finite. The *correlation* between X, Y is defined to be

$$\rho_{X,Y} = \frac{\text{Cov}(X,Y)}{\sqrt{\text{Var}(X)}\sqrt{\text{Var}(Y)}}.$$

Some important properties of covariance and correlation are put together in the next theorem.

Theorem 2.6 (Properties of Covariance and Correlation). *Provided that the required variances and the covariances exist,*

(a) $\text{Cov}(X,c) = 0$ for any X and any constant c;
(b) $\text{Cov}(X,X) = \text{var}(X)$ for any X;
(c) $\text{Cov}\left(\sum_{i=1}^{n} a_i X_i, \sum_{j=1}^{m} b_j Y_j\right) = \sum_{i=1}^{n} \sum_{j=1}^{m} a_i b_j \text{Cov}(X_i, Y_j)$,

and in particular,

$$\text{Var}(aX+bY) = \text{Cov}(aX+bY, aX+bY)$$
$$= a^2 \text{Var}(X) + b^2 \text{Var}(Y) + 2ab \text{Cov}(X,Y),$$

and,

$$\text{Var}\left(\sum_{i=1}^{n} X_i\right) = \sum_{i=1}^{n} \text{Var}(X_i) + 2 \sum_{i<j=1}^{n} \text{Cov}(X_i, X_j);$$

(d) For any two independent random variables X, Y, $\text{Cov}(X,Y) = \rho_{X,Y} = 0$;
(e) $\rho_{a+bX, c+dY} = \text{sgn}(bd) \rho_{X,Y}$, where $\text{sgn}(bd) = 1$ if $bd > 0$, and $= -1$ if $bd < 0$.
(f) Whenever $\rho_{X,Y}$ is defined, $-1 \leq \rho_{X,Y} \leq 1$.
(g) $\rho_{X,Y} = 1$ if and only if for some a, some $b > 0$, $P(Y = a+bX) = 1$; $\rho_{X,Y} = -1$ if and only if for some a, some $b < 0$, $P(Y = a+bX) = 1$.

Proof. For part (a), $\text{Cov}(X,c) = E(cX) - E(c)E(X) = cE(X) - cE(X) = 0$. For part (b), $\text{Cov}(X,X) = E(X^2) - [E(X)]^2 = \text{var}(X)$. For part (c),

$$\text{Cov}\left(\sum_{i=1}^{n} a_i X_i, \sum_{j=1}^{m} b_j Y_j\right)$$

$$= E\left[\sum_{i=1}^{n} a_i X_i \times \sum_{j=1}^{m} b_j Y_j\right] - E\left(\sum_{i=1}^{n} a_i X_i\right) E\left(\sum_{j=1}^{m} b_j Y_j\right)$$

$$= E\left(\sum_{i=1}^{n}\sum_{j=1}^{m} a_i b_j X_i Y_j\right) - \left[\sum_{i=1}^{n} a_i E(X_i)\right] \times \left[\sum_{j=1}^{m} b_j E(Y_j)\right]$$

2.4 Covariance and Correlation

$$= \sum_{i=1}^{n}\sum_{j=1}^{m} a_i b_j E(X_i, Y_j) - \sum_{i=1}^{n} a_i \sum_{j=1}^{m} b_j E(X_i)E(Y_j)$$

$$= \sum_{i=1}^{n}\sum_{j=1}^{m} a_i b_j [E(X_i, Y_j) - E(X_i)E(Y_j)]$$

$$= \sum_{i=1}^{n}\sum_{j=1}^{m} a_i b_j \; \text{Cov}(X_i, Y_j).$$

Part (d) follows on noting that $E(XY) = E(X)E(Y)$ if X, Y are independent. For part (e), first note that $\text{Cov}(a + bX, c + dY) = bd\,\text{Cov}(X, Y)$ by using part (a) and part (c). Also, $\text{Var}(a + bX) = b^2\text{Var}(X)$, $\text{Var}(c + dY) = d^2\text{Var}(Y)$

$$\Rightarrow \rho_{a+bX, c+dY} = \frac{bd\,\text{Cov}(X, Y)}{\sqrt{b^2\text{Var}(X)}\sqrt{d^2\text{Var}(Y)}}$$

$$= \frac{bd\,\text{Cov}(X, Y)}{|b|\sqrt{\text{Var}(X)}|d|\sqrt{\text{Var}(Y)}}$$

$$= \frac{bd}{|bd|}\rho_{X,Y} = \text{sgn}(bd)\rho_{X,Y}.$$

The proof of part (f) uses the Cauchy–Schwarz inequality (see Chapter 1) that for any two random variables U, V, $[E(UV)]^2 \leq E(U^2)E(V^2)$. Let

$$U = \frac{X - E(X)}{\sqrt{\text{Var}(X)}}, \quad V = \frac{Y - E(Y)}{\sqrt{\text{Var}(Y)}}.$$

Then, $E(U^2) = E(V^2) = 1$, and

$$\rho_{X,Y} = E(UV) \leq E(U^2)E(V^2) = 1.$$

The lower bound $\rho_{X,Y} \geq -1$ follows similarly.

Part (g) uses the condition for equality in the Cauchy–Schwarz inequality: in order that $\rho_{X,Y} = \pm 1$, one must have $[E(UV)]^2 = E(U^2)E(V^2)$ in the argument above, which implies the statement in part (g). □

Example 2.11 (Correlation Between Minimum and Maximum in Dice Rolls). Consider again the experiment of rolling a fair die twice, and let X, Y be the maximum and the minimum of the two rolls. We want to find the correlation between X, Y.

The joint distribution of (X, Y) was worked out in Example 2.2. From the joint distribution,

$$E(XY) = \frac{1}{36} + \frac{2}{18} + \frac{4}{36} + \frac{3}{18} + \frac{6}{18} + \frac{9}{36} + \cdots + \frac{30}{18} + \frac{36}{36} = \frac{49}{4}.$$

The marginal pmfs of X, Y were also worked out in Example 2.2. From the marginal pmfs, by direct calculation, $E(X) = 161/36$, $E(Y) = 91/36$, $\text{Var}(X) = \text{Var}(Y) = 2555/1296$. Therefore,

$$\rho_{X,Y} = \frac{E(XY) - E(X)E(Y)}{\sqrt{\text{Var}(X)}\sqrt{\text{Var}(Y)}}$$
$$= \frac{49/4 - 161/36 \times 91/36}{2555/1296} = \frac{35}{73} = .48.$$

The correlation between the maximum and the minimum is in fact positive for any number of rolls of a die, although the correlation will converge to zero when the number of rolls converges to ∞.

Example 2.12 (Correlation in the Chicken–Eggs Example). Consider again the example of a chicken laying a Poisson number of eggs N with mean λ, and each egg fertilizing, independently of others, with probability p. If X is the number of eggs actually fertilized, we want to find the correlation between the number of eggs laid and the number fertilized, that is, the correlation between X and N.

First,

$$E(XN) = E_N[E(XN|N=n)] = E_N[nE(X|N=n)]$$
$$= E_N[n^2 p] = p(\lambda + \lambda^2).$$

Next, from our previous calculations, $E(X) = p\lambda$, $E(N) = \lambda$, $\text{Var}(X) = p\lambda$, $\text{Var}(N) = \lambda$. Therefore,

$$\rho_{X,N} = \frac{E(XN) - E(X)E(N)}{\sqrt{\text{Var}(X)}\sqrt{\text{Var}(N)}}$$
$$= \frac{p(\lambda + \lambda^2) - p\lambda^2}{\sqrt{p\lambda}\sqrt{\lambda}} = \sqrt{p}.$$

Thus, the correlation goes up with the fertility rate of the eggs.

Example 2.13 (Best Linear Predictor). Suppose X and Y are two jointly distributed random variables, and either by necessity, or by omission, the variable Y was not observed. But X was observed, and there may be some information in the X value about Y. The problem is to predict Y by using X. Linear predictors, because of their functional simplicity, are appealing. The mathematical problem is to choose the *best linear predictor* $a + bX$ of Y, where best is defined as the predictor that minimizes the mean squared error $E[Y - (a+bX)]^2$. We show that the answer has something to do with the covariance between X and Y.

By breaking the square, $R(a,b)$

$$= E[Y - (a+bX)]^2 = a^2 + b^2 E(X^2) + 2ab E(X) - 2a E(Y) - 2b E(XY) + E(Y^2).$$

2.5 Multivariate Case

To minimize this with respect to a, b, we partially differentiate $R(a, b)$ with respect to a, b, and set the derivatives equal to zero:

$$\frac{\partial}{\partial a} R(a, b) = 2a + 2bE(X) - 2E(Y) = 0$$
$$\Leftrightarrow a + bE(X) = E(Y);$$

$$\frac{\partial}{\partial b} R(a, b) = 2bE(X^2) + 2aE(X) - 2E(XY) = 0$$
$$\Leftrightarrow aE(X) + bE(X^2) = E(XY).$$

Simultaneously solving these two equations, we get

$$b = \frac{E(XY) - E(X)E(Y)}{\text{Var}(X)}, \quad a = E(Y) - \frac{E(XY) - E(X)E(Y)}{\text{Var}(X)} E(X).$$

These values do minimize $R(a, b)$ by an easy application of the second derivative test. So, the best linear predictor of Y based on X is

$$\text{best linear predictor of } Y = E(Y) - \frac{\text{Cov}(X, Y)}{\text{Var}(X)} E(X) + \frac{\text{Cov}(X, Y)}{\text{Var}(X)} X$$
$$= E(Y) + \frac{\text{Cov}(X, Y)}{\text{Var}(X)} [X - E(X)].$$

The best linear predictor is also known as the *regression line of Y on X*. It is of widespread use in statistics.

Example 2.14 (Zero Correlation Does Not Mean Independence). If X, Y are independent, then necessarily $\text{Cov}(X, Y) = 0$, and hence the correlation is also zero. The converse is not true. Take a three-valued random variable X with the pmf $P(X = \pm 1) = p, P(X = 0) = 1 - 2p, 0 < p < \frac{1}{2}$. Let the other variable Y be $Y = X^2$. Then, $E(XY) = E(X^3) = 0$, and $E(X)E(Y) = 0$, because $E(X) = 0$. Therefore, $\text{Cov}(X, Y) = 0$. But X, Y are certainly not independent; for example, $P(Y = 0 | X = 0) = 1$, but $P(Y = 0) = 1 - 2p \neq 0$.

Indeed, if X has a distribution symmetric around zero, and if X has three finite moments, then X and X^2 always have a zero correlation, although they are not independent.

2.5 Multivariate Case

The extension of the concepts for the bivariate discrete case to the multivariate discrete case is straightforward. We give the appropriate definitions and an important example, namely that of the *multinomial distribution*, an extension of the binomial distribution.

Definition 2.9. Let $X_1, X_2, \ldots, X_n$ be discrete random variables defined on a common sample space Ω, with X_i taking values in some countable set $\mathcal{X}_i$. The *joint pmf* of $(X_1, X_2, \ldots, X_n)$ is defined as $p(x_1, x_2, \ldots, x_n) = P(X_1 = x_1, \ldots, X_n = x_n), x_i \in \mathcal{X}_i$, and zero otherwise..

Definition 2.10. Let $X_1, X_2, \ldots, X_n$ be random variables defined on a common sample space Ω. The *joint CDF* of $X_1, X_2, \ldots, X_n$ is defined as $F(x_1, x_2, \ldots, x_n) = P(X_1 \leq x_1, X_2 \leq x_2, \ldots, X_n \leq x_n), x_1, x_2, \ldots, x_n \in \mathcal{R}$.

The requirements of a joint pmf are the usual:

(i) $p(x_1, x_2, \ldots, x_n) \geq 0 \ \forall \ x_1, x_2, \ldots, x_n \in \mathcal{R}$;
(ii) $\sum_{x_1 \in \mathcal{X}_1, \ldots, x_n \in \mathcal{X}_n} p(x_1, x_2, \ldots, x_n) = 1$.

The requirements of a joint CDF are somewhat more complicated.

The requirements of a CDF are that

(i) $0 \leq F \leq 1 \ \forall (x_1, \ldots, x_n)$.
(ii) F is nondecreasing in each coordiante.
(iii) F equals zero if one or more of the $x_i = -\infty$.
(iv) F equals one if all the $x_i = +\infty$.
(v) F assigns a nonnegative probability to every n dimensional rectangle

$$[a_1, b_1] \times [a_2, b_2] \times \cdots \times [a_n, b_n].$$

This last condition, (v), is a notationally clumsy condition to write down. If $n = 2$, it reduces to the simple inequality that

$$F(b_1, b_2) - F(a_1, b_2) - F(b_1, a_2) + F(a_1, a_2) \geq 0 \ \forall a_1 \leq b_1, a_2 \leq b_2.$$

Once again, we mention that it is not convenient or interesting to work with the CDF for discrete random variables; for discrete variables, it is preferable to work with the pmf.

2.5.1 Joint MGF

Analogous to the case of one random variable, we can define the joint mgf for several random variables. The definition is the same for all types of random variables, discrete or continuous, or other mixed types. As in the one-dimensional case, the joint mgf of several random variables is also a very useful tool. First, we repeat the definition of expectation of a function of several random random variables; see Chapter 1, where it was first introduced and defined. The definition below is equivalent to what was given in Chapter 1.

Definition 2.11. Let $X_1, X_2, \ldots, X_n$ be discrete random variables defined on a common sample space Ω, with X_i taking values in some countable set $\mathcal{X}_i$.

2.5 Multivariate Case

Let the joint pmf of $X_1, X_2, \ldots, X_n$ be $p(x_1, \ldots, x_n)$. Let $g(x_1, \ldots, x_n)$ be a real-valued function of n variables. We say that $E[g(X_1, X_2, \ldots, X_n)]$ exists if $\sum_{x_1 \in \mathcal{X}_1, \ldots, x_n \in \mathcal{X}_n} |g(x_1, \ldots, x_n)| p(x_1, \ldots, x_n) < \infty$, in which case, the expectation is defined as

$$E[g(X_1, X_2, \ldots, X_n)] = \sum_{x_1 \in \mathcal{X}_1, \ldots, x_n \in \mathcal{X}_n} g(x_1, \ldots, x_n) p(x_1, \ldots, x_n).$$

A corresponding definition when $X_1, X_2, \ldots, X_n$ are all continuous random variables is given in the next chapter.

Definition 2.12. Let $X_1, X_2, \ldots, X_n$ be n random variables defined on a common sample space Ω. The *joint moment-generating function* of $X_1, X_2, \ldots, X_n$ is defined to be

$$\psi(t_1, t_2, \ldots, t_n) = E[e^{t_1 X_1 + t_2 X_2 + \cdots + t_n X_n}] = E[e^{\mathbf{t}'\mathbf{X}}],$$

provided the expectation exists, and where $\mathbf{t}'\mathbf{X}$ denotes the inner product of the vectors $\mathbf{t} = (t_1, \ldots, t_n), \mathbf{X} = (X_1, \ldots, X_n)$.

Note that the joint moment-generating function (mgf) always exists at the origin, namely, $\mathbf{t} = (0, \ldots, 0)$, and equals one at that point. It may or may not exist at other points $\mathbf{t}$. If it does exist in a nonempty rectangle containing the origin, then many important characteristics of the joint distribution of $X_1, X_2, \ldots, X_n$ can be derived by using the joint mgf. As in the one-dimensional case, it is a very useful tool. Here is the moment-generation property of a joint mgf.

Theorem 2.7. *Suppose $\psi(t_1, t_2, \ldots, t_n)$ exists in a nonempty open rectangle containing the origin $\mathbf{t} = \mathbf{0}$. Then a partial derivative of $\psi(t_1, t_2, \ldots, t_n)$ of every order with respect to each t_i exists in that open rectangle, and furthermore,*

$$E\left(X_1^{k_1} X_2^{k_2} \cdots X_n^{k_n}\right) = \frac{\partial^{k_1 + k_2 + \cdots + k_n}}{\partial t_1^{k_1} \cdots \partial t_n^{k_n}} \psi(t_1, t_2, \ldots, t_n)|t_1 = 0, t_2 = 0, \ldots, t_n = 0.$$

A corollary of this result is sometimes useful in determining the covariance between two random variables.

Corollary. *Let X, Y have a joint mgf in some open rectangle around the origin $(0, 0)$. Then,*

$$\mathrm{Cov}(X, Y) = \frac{\partial^2}{\partial t_1 \partial t_2} \psi(t_1, t_2)|_{0,0} - \left(\frac{\partial}{\partial t_1} \psi(t_1, t_2)|_{0,0}\right)\left(\frac{\partial}{\partial t_2} \psi(t_1, t_2)|_{0,0}\right).$$

We also have the distribution-determining property, as in the one-dimensional case.

Theorem 2.8. *Suppose $(X_1, X_2, \ldots, X_n)$ and $(Y_1, Y_2, \ldots, Y_n)$ are two sets of jointly distributed random variables, such that their mgfs $\psi_{\mathbf{X}}(t_1, t_2, \ldots, t_n)$ and $\psi_{\mathbf{Y}}(t_1, t_2, \ldots, t_n)$ exist and coincide in some nonempty open rectangle containing the origin. Then $(X_1, X_2, \ldots, X_n)$ and $(Y_1, Y_2, \ldots, Y_n)$ have the same joint distribution.*

Remark. It is important to note that the last two theorems are not limited to discrete random variables; they are valid for general random variables. The proofs of these two theorems follow the same arguments as in the one-dimensional case, namely that when an mgf exists in a nonempty open rectangle, it can be differentiated infinitely often with respect to each variable t_i inside the expectation; that is, the order of the derivative and the expectation can be interchanged.

2.5.2 Multinomial Distribution

One of the most important multivariate discrete distributions is the multinomial distribution. The multinomial distribution corresponds to n balls being distributed to k cells, independently, with each ball having the probability p_i of being dropped into the ith cell. The random variables under consideration are $X_1, X_2, \ldots, X_k$, where X_i is the number of balls that get dropped into the ith cell. Then their joint pmf is the *multinomial pmf* defined below.

Definition 2.13. A multivariate random vector $(X_1, X_2, \ldots, X_k)$ is said to have a multinomial distribution with parameters $n, p_1, p_2, \ldots, p_k$ if it has the pmf

$$P(X_1 = x_1, X_2 = x_2, \ldots, X_k = x_k) = \frac{n!}{x_1! x_2! \cdots x_k!} p_1^{x_1} p_2^{x_2} \cdots p_k^{x_k},$$

$$x_i \geq 0, \sum_{i=1}^{k} x_i = n,$$

$p_i \geq 0, \sum_{i=1}^{k} p_i = 1$.

We write $(X_1, X_2, \ldots, X_k) \sim \text{Mult}(n, p_1, \ldots, p_k)$ to denote a random vector with a multinomial distribution.

Example 2.15 (Dice Rolls). Suppose a fair die is rolled 30 times. We want to find the probabilities that

(i) Each face is obtained exactly five times.
(ii) The number of sixes is at least five.

If we denote the number of times face number i is obtained as X_i, then $(X_1, X_2, \ldots, X_6) \sim \text{Mult}(n, p_1, \ldots, p_6)$, where $n = 30$ and each $p_i = \frac{1}{6}$. Therefore,

$$P(X_1 = 5, X_2 = 5, \ldots, X_6 = 5)$$

$$= \frac{30!}{(5!)^6} \left(\frac{1}{6}\right)^5 \cdots \left(\frac{1}{6}\right)^5$$

$$= \frac{30!}{(5!)^6} \left(\frac{1}{6}\right)^{30}$$

$$= .0004.$$

2.5 Multivariate Case

Next, each of the 30 rolls will either be a 6 or not, independently of the other rolls, with probability $\frac{1}{6}$, and so, $X_6 \sim \text{Bin}(30, \frac{1}{6})$. Therefore,

$$P(X_6 \geq 5) = 1 - P(X_6 \leq 4) = 1 - \sum_{x=0}^{4} \binom{30}{x} \left(\frac{1}{6}\right)^x \left(\frac{5}{6}\right)^{30-x}$$
$$= .5757.$$

Example 2.16 (Bridge). Consider a Bridge game with four players, North, South, East, and West. We want to find the probability that North and South together have two or more aces. Let X_i denote the number of aces in the hands of player $i, i = 1, 2, 3, 4$; we let $i = 1, 2$ mean North and South. Then, we want to find $P(X_1 + X_2 \geq 2)$.

The joint distribution of (X_1, X_2, X_3, X_4) is $\text{Mult}(4, \frac{1}{4}, \frac{1}{4}, \frac{1}{4}, \frac{1}{4})$ (think of each ace as a ball, and the four players as cells). Then, $(X_1 + X_2, X_3 + X_4) \sim \text{Mult}(4, \frac{1}{2}, \frac{1}{2})$. Therefore,

$$P(X_1 + X_2 \geq 2) = \frac{4!}{2!2!}\left(\frac{1}{2}\right)^4 + \frac{4!}{3!1!}\left(\frac{1}{2}\right)^4 + \frac{4!}{4!0!}\left(\frac{1}{2}\right)^4$$
$$= \frac{11}{16}.$$

Important formulas and facts about the multinomial distribution are given in the next theorem.

Theorem 2.9. *Let $(X_1, X_2, \ldots, X_k) \sim \text{Mult}(n, p_1, p_2, \ldots, p_k)$. Then,*

(a) $E(X_i) = np_i$; $\text{Var}(X_i) = np_i(1 - p_i)$;
(b) $\forall i, X_i \sim \text{Bin}(n, p_i)$;
(c) $\text{Cov}(X_i, X_j) = -np_i p_j$, $\forall i \neq j$;
(d) $\rho_{X_i, X_j} = -\sqrt{\frac{p_i p_j}{(1-p_i)(1-p_j)}}$, $\forall i \neq j$;
(e) $\forall m, 1 \leq m < k, (X_1, X_2, \ldots, X_m)|(X_{m+1} + X_{m+2} + \ldots + X_k) = s \sim \text{Mult}(n - s, \theta_1, \theta_2, \ldots, \theta_m)$,

where $\theta_i = \frac{p_i}{p_1 + p_2 + \ldots + p_m}$.

Proof. Define W_{ir} as the indicator of the event that the rth ball lands in the ith cell. Note that for a given i, the variables W_{ir} are independent. Then,

$$X_i = \sum_{r=1}^{n} W_{ir},$$

and therefore, $E(X_i) = \sum_{r=1}^{n} E[W_{ir}] = np_i$, and $\text{Var}(X_i) = \sum_{r=1}^{n} \text{Var}(W_{ir}) = np_i(1 - p_i)$. Part (b) follows from the definition of a multinomial experiment (the trials are identical and independent, and each ball either lands or not in the ith cell). For part (c),

$$\text{Cov}(X_i, X_j) = \text{Cov}\left(\sum_{r=1}^{n} W_{ir}, \sum_{s=1}^{n} W_{js}\right)$$

$$= \sum_{r=1}^{n}\sum_{s=1}^{n} \text{Cov}(W_{ir}, W_{js})$$

$$= \sum_{r=1}^{n} \text{Cov}(W_{ir}, W_{jr})$$

(because $\text{Cov}(W_{ir}, W_{js})$ would be zero when $s \neq r$)

$$= \sum_{r=1}^{n}[E(W_{ir} W_{jr}) - E(W_{ir})E(W_{jr})]$$

$$= \sum_{r=1}^{n}[0 - p_i p_j] = -n p_i p_j.$$

Part (d) follows immediately from part (c) and part (a). Part (e) is a calculation, and is omitted. $\square$

Example 2.17 (MGF of the Multinomial Distribution). Let $(X_1, X_2, \ldots, X_k) \sim$ Mult$(n, p_1, p_2, \ldots, p_k)$. Then the mgf $\psi(t_1, t_2, \ldots, t_k)$ exists at all **t**, and a formula follows easily. Indeed,

$$E[e^{t_1 X_1 + \cdots + t_k X_k}] = \sum_{x_i \geq 0, \sum_{i=1}^{k} x_i = n} \frac{n!}{x_1! \cdots x_k!} e^{t_1 x_1} e^{t_2 x_2} \cdots e^{t_k x_k} p_1^{x_1} p_2^{x_2} cdots p_k^{x_k}$$

$$= \sum_{x_i \geq 0, \sum_{i=1}^{k} x_i = n} \frac{n!}{x_1! \cdots x_k!} (p_1 e^{t_1})^{x_1} (p_2 e^{t_2})^{x_2} \cdots (p_k e^{t_k})^{x_k}$$

$$= (p_1 e^{t_1} + p_2 e^{t_2} + \cdots + p_k e^{t_k})^n,$$

by the *multinomial expansion identity*

$$(a_1 + a_2 + \cdots + a_k)^n = \sum_{x_i \geq 0, \sum_{i=1}^{k} x_i = n} \frac{n!}{x_1! \cdots x_k!} a_1^{x_1} a_2^{x_2} \cdots a_k^{x_k}.$$

2.6 * The Poissonization Technique

Calculation of complex multinomial probabilities often gets technically simplified by taking the number of balls to be a random variable, specifically, a Poisson random variable. We give the Poissonization theorem and some examples in this section.

2.6 The Poissonization Technique

Theorem 2.10. *Let $N \sim \text{Poi}(\lambda)$, and suppose given $N = n$, $(X_1, X_2, \ldots, X_k) \sim \text{Mult}(n, p_1, p_2, \ldots, p_k)$. Then, marginally, $X_1, X_2, \ldots, X_k$ are independent Poisson, with $X_i \sim \text{Poi}(\lambda p_i)$.*

Proof. By the total probability formula,

$$P(X_1 = x_1, X_2 = x_2, \ldots, X_k = x_k)$$

$$= \sum_{n=0}^{\infty} P(X_1 = x_1, X_2 = x_2, \ldots, X_k = x_k | N = n) \frac{e^{-\lambda} \lambda^n}{n!}$$

$$= \sum_{n=0}^{\infty} \frac{(x_1 + x_2 + \cdots + x_k)!}{x_1! x_2! \cdots x_k!} p_1^{x_1} p_2^{x_2} \cdots p_k^{x_k} \frac{e^{-\lambda} \lambda^n}{n!} I_{n = x_1 + x_2 + \cdots + x_k}$$

$$= e^{-\lambda} \lambda^{x_1} \lambda^{x_2} \cdots \lambda^{x_k} p_1^{x_1} p_2^{x_2} \cdots p_k^{x_k} \frac{1}{x_1! x_2! \cdots x_k!}$$

$$= e^{-\lambda} (\lambda p_1)^{x_1} (\lambda p_2)^{x_2} \cdots (\lambda p_k)^{x_k} \frac{1}{x_1! x_2! \cdots x_k!}$$

$$= \prod_{i=1}^{k} \frac{e^{-\lambda p_i} (\lambda p_i)^{x_i}}{x_i!},$$

which establishes that the joint marginal pmf of $(X_1, X_2, \ldots, X_k)$ is the product of k Poisson pmfs, and so $X_1, X_2, \ldots, X_k$ must be marginally independent, with $X_i \sim \text{Poi}(\lambda p_i)$. □

Corollary. *Let A be a set in the k-dimensional Euclidean space $\mathcal{R}^k$. Let $(Y_1, Y_2, \ldots, Y_k) \sim \text{Mult}(n, p_1, p_2, \ldots, p_k)$. Then, $P((Y_1, Y_2, \ldots, Y_k) \in A)$ equals $n! c(n)$, where $c(n)$ is the coefficient of λ^n in the power series expansion of $e^{\lambda} P((X_1, X_2, \ldots, X_k) \in A)$. Here $X_1, X_2, \ldots, X_k$ are as above: they are independent Poisson variables, with $X_i \sim \text{Poi}(\lambda p_i)$.*

The corollary is simply a restatement of the identity

$$P((X_1, X_2, \ldots, X_k) \in A) = e^{-\lambda} \sum_{n=0}^{\infty} \frac{\lambda^n}{n!} P((Y_1, Y_2, \ldots, Y_k) \in A).$$

Example 2.18 (No Empty Cells). Suppose n balls are distributed independently and at random into k cells. We want to find a formula for the probability that no cell remains empty.

We use the Poissonization technique to solve this problem. We want a formula for $P(Y_1 \neq 0, Y_2 \neq 0, \ldots, Y_k \neq 0)$.

Marginally, each $X_i \sim \text{Poi}(\frac{\lambda}{k})$, and therefore,

$$P(X_1 > 0, X_2 > 0, \ldots, X_k > 0) = (1 - e^{-\lambda/k})^k$$
$$\Rightarrow e^{\lambda} P(X_1 > 0, X_2 > 0, \ldots, X_k > 0) = e^{\lambda} (1 - e^{-\lambda/k})^k$$

$$= \sum_{x=0}^{k}(-1)^x \binom{k}{x} e^{\lambda(1-x/k)}$$

$$= \sum_{x=0}^{k}(-1)^x \binom{k}{x} \sum_{n=0}^{\infty} \frac{(\lambda(1-x/k))^n}{n!}$$

$$= \sum_{n=0}^{\infty} \frac{\lambda^n}{n!}[\sum_{x=0}^{k}(-1)^x \binom{k}{x}(1-x/k)^n].$$

Therefore, by the above corollary,

$$P(Y_1 \neq 0, Y_2 \neq 0, \ldots, Y_k \neq 0) = \sum_{x=0}^{k}(-1)^x \binom{k}{x}(1-x/k)^n.$$

Exercises

Exercise 2.1. Consider the experiment of picking one word at random from the sentence
  ALL IS WELL IN THE NEWELL FAMILY
Let X be the length of the word selected and Y the number of Ls in it. Find in a tabular form the joint pmf of X and Y, their marginal pmfs, means, and variances, and the correlation between X and Y.

Exercise 2.2. A fair coin is tossed four times. Let X be the number of heads, Z the number of tails, and $Y = |X - Z|$. Find the joint pmf of (X, Y), and $E(Y)$.

Exercise 2.3. Consider the joint pmf $p(x, y) = cxy, 1 \leq x \leq 3, 1 \leq y \leq 3$.

(a) Find the normalizing constant c.
(b) Are X, Y independent? Prove your claim.
(c) Find the expectations of X, Y, XY.

Exercise 2.4. Consider the joint pmf $p(x, y) = cxy, 1 \leq x \leq y \leq 3$.

(a) Find the normalizing constant c.
(b) Are X, Y independent? Prove your claim.
(c) Find the expectations of X, Y, XY.

Exercise 2.5. A fair die is rolled twice. Let X be the maximum and Y the minimum of the two rolls. By using the joint pmf of (X, Y) worked out in text, find the pmf of $\frac{X}{Y}$, and hence the mean of $\frac{X}{Y}$.

Exercise 2.6. A hat contains four slips of paper, numbered 1, 2, 3, and 4. Two slips are drawn at random, without replacement. X is the number on the first slip and Y the sum of the two numbers drawn. Write in a tabular form the joint pmf of (X, Y). Hence find the marginal pmfs. Are X, Y independent?

Exercises

Exercise 2.7 * (**Conditional Expectation in Bridge**). Let X be the number of clubs in the hand of North and Y the number of clubs in the hand of South in a Bridge game. Write a general formula for $E(X|Y = y)$, and compute $E(X|Y = 3)$. How about $E(Y|X = 3)$?

Exercise 2.8. A fair die is rolled four times. Find the probabilities that:

(a) At least 1 six is obtained;
(b) Exactly 1 six and exactly one two is obtained,
(c) Exactly 1 six, 1 two, and 2 fours are obtained.

Exercise 2.9 (**Iterated Expectation**). A household has a Poisson number of cars with mean 1. Each car that a household possesses has, independently of the other cars, a 20% chance of being an SUV. Find the mean number of SUVs a household possesses.

Exercise 2.10 (**Iterated Variance**). Suppose $N \sim \text{Poi}(\lambda)$, and given $N = n$, X is distributed as a uniform on $\{0, 1, \ldots, n\}$. Find the variance of the marginal distribution of X.

Exercise 2.11. Suppose X and Y are independent $\text{Geo}(p)$ random variables. Find $P(X \geq Y); P(X > Y)$.

Exercise 2.12. * Suppose X and Y are independent $\text{Poi}(\lambda)$ random variables. Find $P(X \geq Y); P(X > Y)$.

Hint: This involves a Bessel function of a suitable kind.

Exercise 2.13. Suppose X and Y are independent and take the values 1, 2, 3, 4 with probabilities .2, .3, .3, .2. Find the pmf of $X + Y$.

Exercise 2.14. Two random variables have the joint pmf $p(x, x + 1) = \frac{1}{n+1}, x = 0, 1, \ldots, n$. Answer the following questions with as little calculation as possible.

(a) Are X, Y independent?
(b) What is the variance of $Y - X$?
(c) What is $\text{Var}(Y|X = 1)$?

Exercise 2.15 (**Binomial Conditional Distribution**). Suppose X, Y are independent random variables, and that $X \sim \text{Bin}(m, p), Y \sim \text{Bin}(n, p)$. Show that the conditional distribution of X given $X + Y = t$ is a hypergeometric distribution; identify the parameters of this hypergeometric distribution.

Exercise 2.16 * (**Poly-Hypergeometric Distribution**). A box has D_1 red, D_2 green, and D_3 blue balls. Suppose n balls are picked at random without replacement from the box. Let X, Y, Z be the number of red, green, and blue balls selected. Find the joint pmf of (X, Y, Z).

Exercise 2.17 (Bivariate Poisson). Suppose U, V, W are independent Poisson random variables, with means λ, μ, η. Let $X = U + W$; $Y = V + W$.

(a) Find the marginal pmfs of X, Y.
(b) Find the joint pmf of (X, Y).

Exercise 2.18. Suppose a fair die is rolled twice. Let X, Y be the two rolls. Find the following with as little calculation as possible:

(a) $E(X + Y | Y = y)$.
(b) $E(XY | Y = y)$.
(c) $\text{Var}(X^2 Y | Y = y)$.
(d) $\rho_{X+Y, X-Y}$.

Exercise 2.19 (A Waiting Time Problem). In repeated throws of a fair die, let X be the throw in which the first six is obtained, and Y the throw in which the second six is obtained.

(a) Find the joint pmf of (X, Y).
(b) Find the expectation of $Y - X$.
(c) Find $E(Y - X | X = 8)$.
(d) Find $\text{Var}(Y - X | X = 8)$.

Exercise 2.20 * (Family Planning). A couple want to have a child of each sex, but they will have at most four children. Let X be the total number of children they will have and Y the number of girls at the second childbirth. Find the joint pmf of (X, Y), and the conditional expectation of X given $Y = y, y = 0, 2$.

Exercise 2.21 (A Standard Deviation Inequality). Let X, Y be two random variables. Show that $\sigma_{X+Y} \leq \sigma_X + \sigma_Y$.

Exercise 2.22 * (A Covariance Fact). Let X, Y be two random variables. Suppose $E(X|Y = y)$ is nondecreasing in y. Show that $\rho_{X,Y} \geq 0$, assuming the correlation exists.

Exercise 2.23 (Another Covariance Fact). Let X, Y be two random variables. Suppose $E(X|Y = y)$ is a finite constant c. Show that $\text{Cov}(X, Y) = 0$.

Exercise 2.24 (Two-Valued Random Variables). Suppose X, Y are both two-valued random variables. Prove that X and Y are independent if and only if they have a zero correlation.

Exercise 2.25 * (A Correlation Inequality). Suppose X, Y each have mean 0 and variance 1, and a correlation ρ. Show that $E(\max\{X^2, Y^2\}) \leq 1 + \sqrt{1 - \rho^2}$.

Exercise 2.26 (A Covariance Inequality). Let X be any random variable, and $g(X), h(X)$ two functions such that they are both nondecreasing or both nonincreasing. Show that $\text{Cov}(g(X), h(X)) \geq 0$.

Exercises

Exercise 2.27 (**Joint MGF**). Suppose a fair die is rolled four times. Let X be the number of ones and Y the number of sixes. Find the joint mgf of X and Y, and hence, the covariance between X, Y.

Exercise 2.28 (**MGF of Bivariate Poisson**). Suppose U, V, W are independent Poisson random variables, with means λ, μ, η. Let $X = U + W$; $Y = V + W$. Find the joint mgf of X, Y, and hence $E(XY)$.

Exercise 2.29 (**Joint MGF**). In repeated throws of a fair die, let X be the throw in which the first six is obtained, and Y the throw in which the second six is obtained. Find the joint mgf of X, Y, and hence the covariance between X and Y.

Exercise 2.30 * (**Poissonization**). A fair die is rolled 30 times. By using the Poissonization theorem, find the probability that the maximum number of times any face appears is 9 or more.

Exercise 2.31 * (**Poissonization**). Individuals can be of one of three genotypes in a population. Each genotype has the same percentage of individuals. A sample of n individuals from the population will be taken. What is the smallest n for which with probability $\geq .9$, there are at least five individuals of each genotype in the sample?

Chapter 3
Multidimensional Densities

Similar to several discrete random variables, we are frequently interested in applications in studying several continuous random variables simultaneously. And similar to the case of one continuous random variable, again we do not speak of pmfs of several continuous variables, but of a pdf, jointly for all the continuous random variables. The joint density function completely characterizes the joint distribution of the full set of continuous random variables. We refer to the entire set of random variables as a random vector. Both the calculation aspects, as well as the application aspects of multidimensional density functions are generally sophisticated. As such, the ability to use and operate with multidimensional densities is among the most important skills one needs to have in probability and also in statistics. The general concepts and calculations are discussed in this chapter. Some special multidimensional densities are introduced separately in later chapters.

3.1 Joint Density Function and Its Role

Exactly as in the one-dimensional case, it is important to note the following points.

(a) The joint density function of all the variables does not equal the probability of a specific point in the multidimensional space; the probability of any specific point is still zero.
(b) The joint density function reflects the relative importance of a particular point. Thus, the probability that the variables together belong to a small set around a specific point, say $\mathbf{x} = (x_1, x_2, \ldots, x_n)$ is roughly equal to the volume of that set multiplied by the density function at the specific point $\mathbf{x}$. This volume interpretation for probabilities is useful for intuitive understanding of distributions of multidimensional continuous random variables.
(c) For a general set A in the multidimensional space, the probability that the random vector $\mathbf{X}$ belongs to A is obtained by integrating the joint density function over the set A.

These are all just the most natural extensions of the corresponding one-dimensional facts to the present multidimensional case. We now formally define a joint density function.

Definition 3.1. Let $\mathbf{X} = (X_1, X_2, \ldots, X_n)$ be an n-dimensional random vector, taking values in $\mathcal{R}^n$, for some $n, 1 < n < \infty$. We say that $f(x_1, x_2, \ldots, x_n)$ is the *joint density* or simply the density of $\mathbf{X}$ if for all $a_1, a_2, \ldots, a_n, b_1, b_2, \ldots, b_n, -\infty < a_i \leq b_i < \infty$,

$$P(a_1 \leq X_1 \leq b_1, a_2 \leq X_2 \leq b_2, \ldots, a_n \leq X_n \leq b_n)$$
$$= \int_{a_n}^{b_n} \cdots \int_{a_2}^{b_2} \int_{a_1}^{b_1} f(x_1, x_2, \ldots, x_n) \, dx_1 dx_2 \cdots dx_n.$$

In order that a function $f : \mathcal{R}^n \to \mathcal{R}$ be a density function of some n-dimensional random vector, it is necessary and sufficient that

(i) $f(x_1, x_2, \ldots, x_n) \geq 0 \ \forall \ (x_1, x_2, \ldots, x_n) \in \mathcal{R}^n$;

(ii) $\int_{\mathcal{R}^n} f(x_1, x_2, \ldots, x_n) \, dx_1 dx_2 \cdots dx_n = 1.$

The definition of the joint CDF is the same as that given in the discrete case. But now the joint CDF is an integral of the density rather than a sum. Here is the precise definition.

Definition 3.2. Let $\mathbf{X}$ be an n-dimensional random vector with the density function $f(x_1, x_2, \ldots, x_n)$. The *joint CDF* or simply the CDF of $\mathbf{X}$ is defined as

$$F(x_1, x_2, \ldots, x_n) = \int_{-\infty}^{x_n} \cdots \int_{-\infty}^{x_1} f(t_1, \ldots, t_n) \, dt_1 \cdots dt_n.$$

As in the one-dimensional case, both the CDF and the density completely specify the distribution of a continuous random vector and one can be obtained from the other. We know how to obtain the CDF from the density; the reverse relation is that (for almost all $(x_1, x_2, \ldots, x_n)$),

$$f(x_1, x_2, \ldots, x_n) = \frac{\partial^n}{\partial x_1 \cdots \partial x_n} F(x_1, x_2, \ldots, x_n).$$

Again, the qualification *almost all* is necessary for a rigorous description of the interrelation between the CDF and the density, but we operate as though the identity above holds for all $(x_1, x_2, \ldots, x_n)$,

Analogous to the case of several discrete variables, the marginal densities are obtained by integrating out (instead of summing) all the other variables. In fact, all lower-dimensional marginals are obtained that way. The precise statement is the following.

3.1 Joint Density Function and Its Role

Proposition. Let $\mathbf{X} = (X_1, X_2, \ldots, X_n)$ be a continuous random vector with a joint density $f(x_1, x_2, \ldots, x_n)$. Let $1 \leq p < n$. Then the marginal joint density of $(X_1, X_2, \ldots, X_p)$ is given by

$$f_{1,2,\ldots,p}(x_1, x_2, \ldots, x_p) = \int_{-\infty}^{\infty} \cdots \int_{-\infty}^{\infty} f(x_1, x_2, \ldots, x_n)\, dx_{p+1} \cdots dx_n.$$

At this stage, it is useful to give a characterization of independence of a set of n continuous random variables by using the density function.

Proposition. Let $\mathbf{X} = (X_1, X_2, \ldots, X_n)$ be a continuous random vector with a joint density $f(x_1, x_2, \ldots, x_n)$. Then, $X_1, X_2, \ldots, X_n$ are independent if and only if the joint density factorizes as

$$f(x_1, x_2, \ldots, x_n) = \prod_{i=1}^{n} f_i(x_i),$$

where $f_i(x_i)$ is the marginal density function of X_i.

Proof. If the joint density factorizes as above, then on integrating both sides of this factorization identity, one gets $F(x_1, x_2, \ldots, x_n) = \prod_{i=1}^{n} F_i(x_i) \; \forall \; (x_1, x_2, \ldots, x_n)$, which is the definition of independence.

Conversely, if they are independent, then take the identity

$$F(x_1, x_2, \ldots, x_n) = \prod_{i=1}^{n} F_i(x_i),$$

and partially differentiate both sides successively with respect to $x_1, x_2, \ldots, x_n$, and it follows that the joint density factorizes as $f(x_1, x_2, \ldots, x_n) = \prod_{i=1}^{n} f_i(x_i)$. □

Let us see some initial examples.

Example 3.1 (Bivariate Uniform). Consider the function

$$f(x, y) = 1 \quad \text{if } 0 \leq x \leq 1, 0 \leq y \leq 1;$$
$$= 0 \quad \text{otherwise.}$$

Clearly, f is always nonnegative, and

$$\int_{-\infty}^{\infty} \int_{-\infty}^{\infty} f(x, y) dx dy = \int_{0}^{1} \int_{0}^{1} f(x, y) dx dy$$
$$= \int_{0}^{1} \int_{0}^{1} dx dy = 1.$$

Therefore, f is a valid bivariate density function. The marginal density of X is

$$f_1(x) = \int_{-\infty}^{\infty} f(x, y) dy$$
$$= \int_0^1 f(x, y) dy = \int_0^1 dy = 1,$$

if $0 \leq x \leq 1$, and zero otherwise. Thus, marginally, $X \sim U[0, 1]$, and similarly, marginally, $Y \sim U[0, 1]$. Furthermore, clearly, for all x, y the joint density $f(x, y)$ factorizes as $f(x, y) = f_1(x) f_2(y)$, and so X, Y are independent too. The joint density $f(x, y)$ of this example is called the *bivariate uniform density*. It gives the constant density of one to all points (x, y) in the unit square $[0, 1] \times [0, 1]$ and zero density outside the unit square. The bivariate uniform, therefore, is the same as putting two independent $U[0, 1]$ variables together as a bivariate vector.

Example 3.2 (Uniform in a Triangle). Consider the function

$$f(x, y) = c, \quad \text{if } x, y \geq 0, x + y \leq 1,$$
$$= 0 \text{ otherwise.}$$

The set of points $x, y \geq 0, x + y \leq 1$ forms a triangle in the plane with vertices at $(0, 0), (1, 0)$, and $(0, 1)$; thus, it is just half the unit square see Fig. 3.1. The normalizing constant c is easily evaluated:

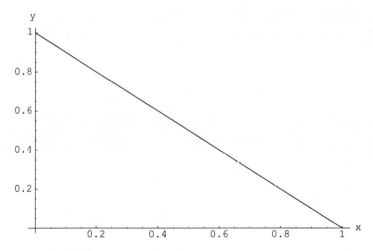

Fig. 3.1 Uniform density on a triangle equals $c = 2$ on this set

3.1 Joint Density Function and Its Role

$$1 = \int_{x,y:x,y\geq 0, x+y\leq 1} cdxdy$$

$$= \int_0^1 \int_0^{1-y} cdxdy$$

$$= c \int_0^1 (1-y)dy$$

$$= \frac{c}{2}$$

$$\Rightarrow c = 2.$$

The marginal density of X is

$$f_1(x) = \int_0^{1-x} 2dy = 2(1-x), \quad 0 \leq x \leq 1.$$

Similarly, the marginal density of Y is

$$f_2(y) = 2(1-y), \quad 0 \leq y \leq 1.$$

Contrary to the previous example, X, Y are not independent now. There are many ways to see this. For example,

$$P\left(X > \frac{1}{2} \middle| Y > \frac{1}{2}\right) = 0.$$

But, $P\left(X > \frac{1}{2}\right) = \int_{\frac{1}{2}}^1 2(1-x)dx = \frac{1}{4} \neq 0$. So, X, Y cannot be independent. We can also see that the joint density $f(x, y)$ does not factorize as the product of the marginal densities, and so X, Y cannot be independent.

Example 3.3. Consider the function $f(x, y) = xe^{-x(1+y)}, x, y \geq 0$. First, let us verify that it is a valid density function.

It is obviously nonnegative. Furthermore,

$$\int_{-\infty}^{\infty} \int_{-\infty}^{\infty} f(x, y)dxdy = \int_0^{\infty} \int_0^{\infty} xe^{-x(1+y)}dxdy$$

$$= \int_0^{\infty} \frac{1}{(1+y)^2}dy$$

$$= \int_1^{\infty} \frac{1}{y^2}dy = 1.$$

Hence, $f(x, y)$ is a valid joint density. It is plotted in Fig. 3.2. Next, let us find the marginal densities:

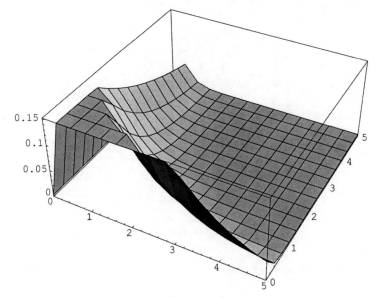

Fig. 3.2 The density f(x, y) = x Exp (−x(1+y))

$$f_1(x) = \int_0^\infty xe^{-x(1+y)}dy = x\int_0^\infty e^{-x(1+y)}dy$$
$$= x\int_1^\infty e^{-xy}dy = x\frac{e^{-x}}{x} = e^{-x}, \quad x \geq 0.$$

Therefore, marginally, X is a standard exponential. Next,

$$f_2(y) = \int_0^\infty xe^{-x(1+y)}dx = \frac{1}{(1+y)^2}, \quad y \geq 0.$$

Clearly, we do not have the factorization identity $f(x, y) = f_1(x)f_2(y) \; \forall \, x, y$; thus, X, Y are not independent.

Example 3.4 (Nonuniform Joint Density with Uniform Marginals). Let (X, Y) have the joint density function $f(x, y) = c - 2(c - 1)(x + y - 2xy), x, y \in [0, 1], 0 < c < 2$. This is nonnegative in the unit square, as can be seen by considering the cases $c < 1, c = 1, c > 1$ separately. Also,

$$\int_0^1 \int_0^1 f(x, y)dxdy$$
$$= c - 2(c - 1)\int_0^1 \int_0^1 (x + y - 2xy)dxdy$$
$$= c - 2(c - 1)\int_0^1 \left(\frac{1}{2} + y - y\right)dy = c - (c - 1) = 1.$$

3.1 Joint Density Function and Its Role

Now, the marginal density of X is

$$f_1(x) = \int_0^1 f(x, y) dy$$
$$= c - 2(c-1)\left[x + \frac{1}{2} - x\right] = 1.$$

Similarly, the marginal density of Y is also the constant function 1. So each marginal is uniform, although the joint density is not uniform if $c \neq 1$.

Example 3.5 (Using the Density to Calculate a Probability). Suppose (X, Y) has the joint density $f(x, y) = 6xy^2, x, y \geq 0, x + y \leq 1$. Thus, this is yet another density on the triangle with vertices at $(0,0), (1,0),$ and $(0,1)$. We want to find $P(X + Y < \frac{1}{2})$. By definition,

$$P\left(X + Y < \frac{1}{2}\right) = \int_{(x,y); x, y \geq 0, x+y < \frac{1}{2}} 6xy^2 dx dy$$
$$= 6 \int_0^{\frac{1}{2}} \int_0^{\frac{1}{2}-y} xy^2 dx dy$$
$$= 6 \int_0^{\frac{1}{2}} y^2 \frac{(\frac{1}{2}-y)^2}{2} dy$$
$$= 3 \int_0^{\frac{1}{2}} y^2 \left(\frac{1}{2} - y\right)^2 dy$$
$$= 3 \times \frac{1}{960} = \frac{1}{320}.$$

This example gives an elementary illustration of the need to work out the limits of the iterated integrals carefully while using a joint density to calculate the probability of some event. In fact, properly finding the limits of the iterated integrals is the part that requires the greatest care when working with joint densities.

Example 3.6 (Uniform Distribution in a Circle). Suppose C denotes the unit circle in the plane:

$$C = \{(x, y) : x^2 + y^2 \leq 1\}.$$

We pick a point (X, Y) at random from C; what that means is that (X, Y) has the density

$$f(x, y) = c, \quad \text{if } (x, y) \in C,$$

and is zero otherwise. Because

$$\int_C f(x, y) dx dy = c \int_C dx dy = c \times \text{Area of } C = c\pi = 1,$$

we have that the normalizing constant $c = \frac{1}{\pi}$. Let us find the marginal densities. First,

$$f_1(x) = \int_{y:x^2+y^2 \leq 1} \frac{1}{\pi} dy = \frac{1}{\pi} \int_{-\sqrt{1-x^2}}^{\sqrt{1-x^2}} dy$$

$$= \frac{2\sqrt{1-x^2}}{\pi}, \quad -1 \leq x \leq 1.$$

The joint density $f(x, y)$ is symmetric between x, y (i.e., $f(x, y) = f(y, x)$) thus Y has the same marginal density as X; that is,

$$f_2(y) = \frac{2\sqrt{1-y^2}}{\pi}, \quad -1 \leq y \leq 1.$$

Because $f(x, y) \neq f_1(x) f_2(y)$, X, Y are not independent. Note that if X, Y has a joint uniform density in the unit square, we find them to be independent; but now, when they have a uniform density in the unit circle, we find them to be not independent. In fact, the following general rule holds.

Suppose a joint density $f(x, y)$ can be written in a form $g(x)h(y), (x, y) \in S$, and $f(x, y)$ zero otherwise. Then, X, Y are independent if and only if S is a rectangle (including squares).

Example 3.7 (An Interesting Property of Exponential Variables). Suppose X, Y are independent $\text{Exp}(\lambda), \text{Exp}(\mu)$ variables. We want to find $P(X \leq Y)$. A possible application is the following. Suppose you have two televisions at your home, a plasma unit with a mean lifetime of five years, and an ordinary unit with a mean lifetime of ten years. What is the probability that the plasma tv will fail before the ordinary one?

From our general definition of probabilities of events, we need to calculate $\int_{x,y>0, x \leq y} f(x, y) dxdy$. In general, there need not be an interesting answer for this integral. But, here in the independent exponential case, there is.

Since X, Y are independent, the joint density is $f(x, y) = \frac{1}{\lambda \mu} e^{-x/\lambda - y/\mu}$, $x, y > 0$. Therefore,

$$P(X \leq Y) = \int_{x,y>0, x \leq y} \frac{1}{\lambda \mu} e^{-x/\lambda - y/\mu} dxdy$$

$$= \frac{1}{\lambda \mu} \int_0^\infty \int_0^y e^{-x/\lambda - y/\mu} dxdy$$

$$= \frac{1}{\mu} \int_0^\infty e^{-y/\mu} \int_0^{y/\lambda} e^{-x} dxdy$$

$$= \frac{1}{\mu} \int_0^\infty e^{-y/\mu} \left(1 - e^{-y/\lambda}\right) dy$$

$$= 1 - \frac{1}{\mu} \int_0^\infty e^{-y(1/\mu + 1/\lambda)} dy$$

3.1 Joint Density Function and Its Role

$$= 1 - \frac{\frac{1}{\mu}}{\frac{1}{\mu} + \frac{1}{\lambda}} = 1 - \frac{\lambda}{\lambda + \mu}$$

$$= \frac{\mu}{\lambda + \mu} = \frac{1}{1 + \frac{\lambda}{\mu}}.$$

Thus, the probability that X is less than Y depends in a very simple way on just the quantity $\frac{E(X)}{E(Y)}$.

Example 3.8 (Curse of Dimensionality). A phenomenon that complicates the work of a probabilist in high dimensions (i.e., when dealing with a large number of random variables simultaneously) is that the major portion of the probability in the joint distribution lies away from the central region of the variable space. As a consequence, sample observations taken from the high-dimensional distribution tend to leave the central region sparsely populated. Therefore, it becomes difficult to learn about what the distribution is doing in the central region. This phenomenon has been called *the curse of dimensionality*.

As an example, consider n independent $U[-1, 1]$ random variables, $X_1, X_2, \ldots, X_n$, and suppose we ask what the probability is that $\mathbf{X} = (X_1, X_2, \ldots, X_n)$ lies in the inscribed sphere

$$B_n = \{(x_1, x_2, \ldots, x_n) : x_1^2 + x_2^2 + \cdots + x_n^2 \leq 1\}.$$

By definition, the joint density of $X_1, X_2, \ldots, X_n$ is

$$f(x_1, x_2, \ldots, x_n) = c, \quad -1 \leq x_i \leq 1, 1 \leq i \leq n,$$

where $c = \frac{1}{2^n}$. Also, by definition of probability,

$$P(\mathbf{X} \in B_n) = \int_{B_n} c \, dx_1 dx_2 \cdots dx_n$$

$$= \frac{\text{Vol}(B_n)}{2^n},$$

where Vol (B_n) is the volume of the n-dimensional unit sphere B_n, and equals

$$\text{Vol}(B_n) = \frac{\pi^{\frac{n}{2}}}{\Gamma\left(\frac{n}{2} + 1\right)}.$$

Thus, finally,

$$P(\mathbf{X} \in B_n) = \frac{\pi^{\frac{n}{2}}}{2^n \Gamma\left(\frac{n}{2} + 1\right)}.$$

This is a very pretty formula. Let us evaluate this probability for various values of n, and examine the effect of increasing the number of dimensions on this probability. Here is a table.

n	$P(\mathbf{X} \in B_n)$
2	.785
3	.524
4	.308
5	.164
6	.081
10	.002
12	.0003
15	.00001
18	3.13×10^{-7}

We see that in ten dimensions, there is a 1 in 500 chance that a uniform random vector will fall in the central inscribed sphere, and in 18 dimensions, the chance is much less than one in a million. Therefore, when you are dealing with a large number of random variables at the same time, you will need a huge amount of sample data to learn about the behavior of their joint distribution in the central region; most of the data will come from the corners! You must have a huge amount of data to have at least some data points in your central region. As stated above this phenomenon has been termed *the curse of dimensionality*.

3.2 Expectation of Functions

Expectations for multidimensional densities are defined analogously to the one-dimensional case. Here is the definition.

Definition 3.3. Let $(X_1, X_2, \ldots, X_n)$ have a joint density function $f(x_1, x_2, \ldots, x_n)$, and let $g(x_1, x_2, \ldots, x_n)$ be a real-valued function of $x_1, x_2, \ldots, x_n$. We say that the expectation of $g(X_1, X_2, \ldots, X_n)$ exists if

$$\int_{\mathcal{R}^n} |g(x_1, x_2, \ldots, x_n)| f(x_1, x_2, \ldots, x_n) \, dx_1 dx_2 \cdots dx_n < \infty,$$

in which case the expected value of $g(X_1, X_2, \ldots, X_n)$ is defined as

$$E[g(X_1, X_2, \ldots, X_n)] = \int_{\mathcal{R}^n} g(x_1, x_2, \ldots, x_n) f(x_1, x_2, \ldots, x_n) \, dx_1 dx_2 \cdots dx_n.$$

Remark. It is clear from the definition that the expectation of each individual X_i can be evaluated by either interpreting X_i as a function of the full vector $(X_1, X_2, \ldots, X_n)$, or by simply using the marginal density $f_i(x)$ of X_i; that is,

3.2 Expectation of Functions

$$E(X_i) = \int_{\mathcal{R}^n} x_i f(x_1, x_2, \ldots, x_n) \, dx_1 dx_2 \cdots dx_n$$
$$= \int_{-\infty}^{\infty} x f_i(x) dx.$$

A similar comment applies to any function $h(X_i)$ of just X_i alone. All the properties of expectations that we have previously established, for example, linearity of expectations, continue to hold in the multidimensional case. Thus,

$$E[ag(X_1, X_2, \ldots, X_n) + bh(X_1, X_2, \ldots, X_n)]$$
$$= aE[g(X_1, X_2, \ldots, X_n)] + bE[h(X_1, X_2, \ldots, X_n)].$$

We work out some examples now.

Example 3.9 (Bivariate Uniform). Two numbers X, Y are picked independently at random from $[0, 1]$. What is the expected distance between them?

Thus, if X, Y are independent $U[0, 1]$, we want to compute $E(|X - Y|)$, which is

$$E(|X - Y|) = \int_0^1 \int_0^1 |x - y| dx dy$$
$$= \int_0^1 \left[\int_0^y (y - x) dx + \int_y^1 (x - y) dx \right] dy$$
$$= \int_0^1 \left[\left(y^2 - \frac{y^2}{2} \right) + \left(\frac{1 - y^2}{2} - y(1 - y) \right) \right] dy$$
$$= \int_0^1 \left[\frac{1}{2} - y + y^2 \right] dy$$
$$= \frac{1}{2} - \frac{1}{2} + \frac{1}{3} = \frac{1}{3}.$$

Example 3.10 (Independent Exponentials). Suppose X, Y are independently distributed as $\text{Exp}(\lambda)$, $\text{Exp}(\mu)$, respectively. We want to find the expectation of the minimum of X and Y. The calculation below requires patience, but is not otherwise difficult.

Denote $W = \min\{X, Y\}$. Then,

$$E(W) = \int_0^\infty \int_0^\infty \min\{x, y\} \frac{1}{\lambda \mu} e^{-x/\lambda} e^{-y/\mu} dx dy$$
$$= \int_0^\infty \int_0^y x \frac{1}{\lambda \mu} e^{-x/\lambda} e^{-y/\mu} dx dy + \int_0^\infty \int_y^\infty y \frac{1}{\lambda \mu} e^{-x/\lambda} e^{-y/\mu} dx dy$$

$$= \int_0^\infty \frac{1}{\mu} e^{-y/\mu} \left[\int_0^y x \frac{1}{\lambda} e^{-x/\lambda} dx \right] dy$$

$$+ \int_0^\infty \frac{1}{\mu} e^{-y/\mu} \left[\int_y^\infty y \frac{1}{\lambda} e^{-x/\lambda} dx \right] dy$$

$$= \int_0^\infty \frac{1}{\mu} e^{-y/\mu} \left[\lambda - \lambda e^{-y/\lambda} - y e^{-y/\lambda} \right] dy$$

$$+ \int_0^\infty \frac{1}{\mu} e^{-y/\mu} y e^{-y/\lambda} dy$$

(on integrating the x integral in the first term by parts)

$$= \frac{\lambda \mu^2}{(\lambda + \mu)^2} + \frac{\mu \lambda^2}{(\lambda + \mu)^2}$$

(once again, by integration by parts)

$$= \frac{\lambda \mu}{\lambda + \mu} = \frac{1}{\frac{1}{\lambda} + \frac{1}{\mu}},$$

a very pretty result.

Example 3.11 (Use of Polar Coordinates). Suppose a point (x, y) is picked at random from inside the unit circle. We want to find its expected distance from the center of the circle.

Thus, let (X, Y) have the joint density

$$f(x, y) = \frac{1}{\pi}, \quad x^2 + y^2 \leq 1,$$

and zero otherwise.

We find $E[\sqrt{X^2 + Y^2}]$. By definition,

$$E[\sqrt{X^2 + Y^2}]$$

$$= \frac{1}{\pi} \int_{(x,y):x^2+y^2 \leq 1} \sqrt{x^2 + y^2} \, dx \, dy.$$

It is now very useful to make a transformation by using the polar coordinates

$$x = r \cos \theta, \quad y = r \sin \theta,$$

3.2 Expectation of Functions

with $dxdy = r\,dr\,d\theta$. Therefore,

$$E\left[\sqrt{X^2+Y^2}\right] = \frac{1}{\pi}\int_{(x,y):x^2+y^2\leq 1}\sqrt{x^2+y^2}\,dxdy$$

$$= \frac{1}{\pi}\int_0^1\int_{-\pi}^{\pi}r^2\,d\theta\,dr$$

$$= 2\int_0^1 r^2\,dr = \frac{2}{3}.$$

We later show various calculations finding distributions of functions of many continuous variables where transformation to polar and spherical coordinates often simplifies the integrations involved.

Example 3.12 (A Spherically Symmetric Density). Suppose (X, Y) has a joint density function

$$f(x,y) = \frac{c}{(1+x^2+y^2)^{\frac{3}{2}}}, \quad x, y \geq 0,$$

where c is a positive normalizing constant. We prove below that this is a valid joint density and evaluate the normalizing constant c. Note that $f(x, y)$ depends on x, y only through $x^2 + y^2$; such a density function is called *spherically symmetric*, because the density $f(x, y)$ takes the same value at all points on the perimeter of a circle given by $x^2 + y^2 = k$.

To prove that f is a valid density, first note that it is obviously nonnegative. Next, by making a transformation to polar coordinates, $x = r\cos\theta$, $y = r\sin\theta$,

$$\int_{x>0,y>0} f(x,y)\,dxdy = c\int_0^\infty\int_0^{\frac{\pi}{2}} \frac{r}{(1+r^2)^{\frac{3}{2}}}\,d\theta\,dr$$

(here, $0 \leq \theta \leq \frac{\pi}{2}$, as x, y are both positive)

$$= c\frac{\pi}{2}\int_0^\infty \frac{r}{(1+r^2)^{\frac{3}{2}}}\,dr = c\frac{\pi}{2}\times 1 = c\frac{\pi}{2}$$

$$\Rightarrow c = \frac{2}{\pi}.$$

We show that $E(X)$ does not exist. Note that it then follows that $E(Y)$ also does not exist, because $f(x, y) = f(y, x)$ in this example. The expected value of X is, again, by transforming to polar coordinates,

$$E(X) = \frac{2}{\pi}\int_0^\infty\int_0^{\frac{\pi}{2}} \frac{r^2}{(1+r^2)^{\frac{3}{2}}}\cos\theta\,d\theta\,dr$$

$$= \frac{2}{\pi}\int_0^\infty \frac{r^2}{(1+r^2)^{\frac{3}{2}}}\,dr$$

$$= \infty,$$

because the final integrand $\frac{r^2}{(1+r^2)^{\frac{3}{2}}}$ behaves like the function $\frac{1}{r}$ for large r, and $\int_k^\infty \frac{1}{r}\,dr$ diverges for any positive k.

3.3 Bivariate Normal

The bivariate normal density is one of the most important densities for two jointly distributed continuous random variables, just as the univariate normal density is for one continuous variable. Many correlated random variables across applied and social sciences are approximately distributed as a bivariate normal. A typical example is the joint distribution of two size variables, such as height and weight.

Definition 3.4. The function $f(x, y) = \frac{1}{2\pi} e^{-\frac{x^2+y^2}{2}}$, $-\infty < x, y < \infty$ is called the *bivariate standard normal density*.

Clearly, we see that $f(x, y) = \phi(x)\phi(y) \ \forall \ x, y$. Therefore, the bivariate standard normal distribution corresponds to a pair of independent standard normal variables X, Y. If we make a linear transformation

$$U = \mu_1 + \sigma_1 X$$
$$V = \mu_2 + \sigma_2 \left[\rho X + \sqrt{1-\rho^2} Y\right],$$

then we get the general *five-parameter bivariate normal density*, with means μ_1, μ_2, standard deviations σ_1, σ_2, and correlation $\rho_{U,V} = \rho$; here, $-1 < \rho < 1$.

Definition 3.5. The density of the five-parameter bivariate normal distribution is

$$f(x, y) = \frac{1}{2\pi\sigma_1\sigma_2\sqrt{1-\rho^2}} e^{-\frac{1}{2(1-\rho^2)}[\frac{(x-\mu_1)^2}{\sigma_1^2} + \frac{(y-\mu_2)^2}{\sigma_2^2} - \frac{2\rho(x-\mu_1)(y-\mu_2)}{\sigma_1\sigma_2}]},$$

$-\infty < x, y < \infty$.

If $\mu_1 = \mu_2 = 0, \sigma_1 = \sigma_2 = 1$, then the bivariate normal density has just the parameter ρ, and it is denoted as $SBVN(\rho)$.

If we sample observations from a general bivariate normal distribution, and plot the data points as points in the plane, then they would roughly plot out to an elliptical shape. The reason for this approximate elliptical shape is that the exponent in the formula for the density function is a quadratic form in the variables. Figure 3.3 is a simulation of 1000 values from a bivariate normal distribution. The roughly elliptical shape is clear. It is also seen in the plot that the center of the point cloud is quite close to the true means of the variables, which were chosen to be $\mu_1 = 4.5, \mu_2 = 4$.

From the representation we have given above of the general bivariate normal vector (U, V) in terms of independent standard normals X, Y, it follows that

$$E(UV) = \rho\sigma_1\sigma_2 + \mu_1\mu_2$$
$$\Rightarrow \text{Cov}(U, V) = \rho\sigma_1\sigma_2.$$

The symmetric matrix with the variances as diagonal entries and the covariance as the off-diagonal entry is called the *variance–covariance matrix*, or the *dispersion*

3.3 Bivariate Normal

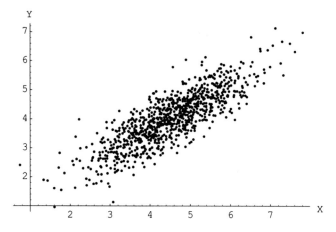

Fig. 3.3 Simulation of a bivariate normal with means 4.5, 4; variances 1; correlation .75

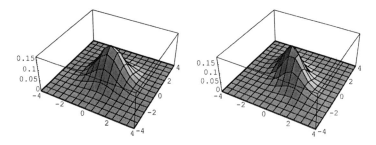

Fig. 3.4 Bivariate normal densities with zero means, unit variances, and rho = 0, .5

matrix, or sometimes simply the *covariance matrix* of (U, V). Thus, the covariance matrix of (U, V) is

$$\Sigma = \begin{pmatrix} \sigma_1^2 & \rho\sigma_1\sigma_2 \\ \rho\sigma_1\sigma_2 & \sigma_2^2 \end{pmatrix}.$$

A plot of the $SBVN(\rho)$ density is provided in Fig. 3.4 for $\rho = 0, .5$; the zero correlation case corresponds to independence. We see from the plots that the bivariate density has a unique peak at the mean point $(0, 0)$ and falls off from that point like a mound. The higher the correlation, the more the density concentrates near a plane. In the limiting case, when $\rho = \pm 1$, the density becomes fully concentrated on a plane, and we call it a *singular bivariate normal*.

When $\rho = 0$, the bivariate normal density does factorize into the product of the two marginal densities. Therefore, if $\rho = 0$, then U, V are actually independent, and so, in that case, $P(U > \mu_1, V > \mu_2) = P(\text{Each variable is larger than its mean value}) = \frac{1}{2}\frac{1}{2} = \frac{1}{4}$. When the parameters are general, one has the following classic formula.

Theorem 3.1 (A Classic Bivariate Normal Formula). *Let (U, V) have the five-parameter bivariate normal density with parameters $\mu_1, \mu_2, \sigma_1, \sigma_2, \rho$. Then,*

$$P(U > \mu_1, V > \mu_2) = P(U < \mu_1, V < \mu_2)$$
$$= \frac{1}{4} + \frac{\arcsin \rho}{2\pi}$$

A derivation of this formula can be seen in Tong (1990).

Example 3.13. Suppose a bivariate normal vector (U, V) has correlation ρ. Then, by applying the formula above, whatever σ_1, σ_2 are,

$$P(U > \mu_1, V > \mu_2) = 1/4 + 1/(2\pi)\arcsin\left[\frac{1}{2}\right] = \frac{1}{3},$$

when $\rho = \frac{1}{2}$. When $\rho = .75$, the probability increases to .385. In the limit, when $\rho \to 1$, the probability tends to .5. That is, when $\rho \to 1$, all the probability becomes confined to the first and the third quadrants $\{U > \mu_1, V > \mu_2\}$, and $\{U < \mu_1, V < \mu_2\}$, with the probability of each of these two quadrants approaching .5.

Another important property of a bivariate normal distribution is the following result.

Theorem 3.2. *Let (U, V) have a general five-parameter bivariate normal distribution. Then, any linear function $aU + bV$ of (U, V) is normally distributed:*

$$aU + bV \sim N\left(a\mu_1 + b\mu_2, a^2\sigma_1^2 + b^2\sigma_2^2 + 2ab\rho\sigma_1\sigma_2\right).$$

In particular, each of U, V is marginally normally distributed:

$$U \sim N\left(\mu_1, \sigma_1^2\right), \quad V \sim N\left(\mu_2, \sigma_2^2\right).$$

If $\rho = 0$, then U, V are independent with $N\left(\mu_1, \sigma_1^2\right), N\left(\mu_2, \sigma_2^2\right)$ marginal distributions.

Proof. First note that $E(aU + bV) = a\mu_1 + b\mu_2$ by linearity of expectations, and $\text{Var}(aU + bV) = a^2\text{Var}(U) + b^2\text{Var}(V) + 2ab\,\text{Cov}(U, V)$ by the general formula for the variance of a linear combination of two jointly distributed random variables (see Chapter 2). But $\text{Var}(U) = \sigma_1^2, \text{Var}(V) = \sigma_2^2$, and $\text{Cov}(U, V) = \rho\sigma_1\sigma_2$. Therefore, $\text{Var}(aU + bV) = a^2\sigma_1^2 + b^2\sigma_2^2 + 2ab\rho\sigma_1\sigma_2$.

Therefore, we only have to prove that $aU + bV$ is normally distributed. For this, we use our representation of U, V in terms of a pair of independent standard normal variables X, Y:

$$U = \mu_1 + \sigma_1 X$$
$$V = \mu_2 + \sigma_2\left[\rho X + \sqrt{1 - \rho^2}Y\right].$$

3.3 Bivariate Normal

Multiplying the equations by a, b and adding, we get the representation

$$aU + bV = a\mu_1 + b\mu_2 + \left[a\sigma_1 X + b\sigma_2 \rho X + b\sigma_2 \sqrt{1-\rho^2} Y\right]$$
$$= a\mu_1 + b\mu_2 + \left[(a\sigma_1 + b\sigma_2 \rho)X + b\sigma_2 \sqrt{1-\rho^2} Y\right].$$

That is, $aU + bV$ can be represented as a linear function $cX + dY + k$ of two independent standard normal variables X, Y, and so $aU + bV$ is necessarily normally distributed (see Chapter 1). □

In fact, a result stronger than the previous theorem holds. What is true is that any two linear functions of U, V will again be distributed as a bivariate normal. Here is the stronger result.

Theorem 3.3. *Let (U, V) have a general five-parameter bivariate normal distribution. Let $Z = aU + bV, W = cU + dV$ be two linear functions, such that $ad - bc \neq 0$. Then, (Z, W) also has a bivariate normal distribution, with parameters*

$$E(Z) = a\mu_1 + b\mu_2, E(W) = c\mu_1 + d\mu_2;$$
$$\text{Var}(Z) = a^2\sigma_1^2 + b^2\sigma_2^2 + 2ab\rho\sigma_1\sigma_2;$$
$$\text{Var}(W) = c^2\sigma_1^2 + d^2\sigma_2^2 + 2cd\rho\sigma_1\sigma_2;$$
$$\rho_{Z,W} = \frac{ac\sigma_1^2 + bd\sigma_2^2 + (ad+bc)\rho\sigma_1\sigma_2}{\sqrt{\text{Var}(Z)\text{Var}(W)}}.$$

The proof of this theorem is similar to the proof of the previous theorem, and the details are omitted.

Example 3.14 (Independence of Mean and Variance). Suppose X_1, X_2 are two iid $N(\mu, \sigma^2)$ variables. Then, of course, they are also jointly bivariate normal. Define now two linear functions

$$Z = X_1 + X_2, \quad W = X_1 - X_2.$$

Because (X_1, X_2) has a bivariate normal distribution, so does (Z, W). However, plainly,

$$\text{Cov}(Z, W) = \text{Cov}(X_1 + X_2, X_1 - X_2) = \text{Var}(X_1) - \text{Var}(X_2) = 0.$$

Therefore, Z, W must actually be independent. As a consequence, Z and W^2 are also independent.

Now note that the sample variance of X_1, X_2 is

$$s^2 = \left(X_1 - \frac{X_1+X_2}{2}\right)^2 + \left(X_2 - \frac{X_1+X_2}{2}\right)^2 = \frac{(X_1-X_2)^2}{2} = \frac{W^2}{2}.$$

And, of course, $\bar{X} = \frac{X_1+X_2}{2} = \frac{Z}{2}$. Therefore, it follows that $\bar{X}$ and s^2 are independent.

This is true not just for two observations, but for any number of iid observations from a normal distribution. This is proved after we introduce *multivariate normal distributions*, and it is also proved in Chapter 18 by using Basu's theorem.

Example 3.15 (Normal Marginals Do Not Guarantee Joint Normal). Although joint bivariate normality of two random variables implies that each variable must be marginally a univariate normal, the converse is in general not true.

Let $Z \sim N(0,1)$, and let U be a two-valued random variable with the pmf $P(U = \pm 1) = \frac{1}{2}$. Take U and Z to be independent. Define now $X = U|Z|$, and $Y = Z$.

Then, each of X, Y has a standard normal distribution. That X has a standard normal distribution is easily seen in many ways, for example, by just evaluating its CDF. Take $x > 0$; then,

$$P(X \leq x) = P(X \leq x | U = -1) \times \frac{1}{2} + P(X \leq x | U = 1) \times \frac{1}{2}$$

$$= 1 \times \frac{1}{2} + P(|Z| \leq x) \times \frac{1}{2}$$

$$= \frac{1}{2} + \frac{1}{2} \times [2\Phi(x) - 1] = \Phi(x).$$

Similarly, also for $x \leq 0$, $P(X \leq x) = \Phi(x)$.

But, jointly, X, Y cannot be bivariate normal, because $X^2 = U^2 Z^2 = Z^2 = Y^2$ with probability one. That is, the joint distribution of (X, Y) lives on just the two lines $y = \pm x$, and so is certainly not bivariate normal.

3.4 Conditional Densities and Expectations

The conditional distribution for continuous random variables is defined analogously to the discrete case, with pmfs replaced by densities. The formal definitions are as follows.

Definition 3.6 (Conditional Density). Let (X, Y) have a joint density $f(x, y)$. The *conditional density* of X given $Y = y$ is defined as

$$f(x|y) = f(x|Y = y) = \frac{f(x, y)}{f_Y(y)}, \quad \forall y \text{ such that } f_Y(y) > 0.$$

3.4 Conditional Densities and Expectations

The *conditional expectation* of X given $Y = y$ is defined as

$$E(X|y) = E(X|Y = y) = \int_{-\infty}^{\infty} x f(x|y) dx$$
$$= \frac{\int_{-\infty}^{\infty} x f(x, y) dx}{\int_{-\infty}^{\infty} f(x, y) dx},$$

$\forall y$ such that $f_Y(y) > 0$.

For fixed x, the conditional expectation $E(X|y) = \mu_X(y)$ is a number. As we vary y, we can think of $E(X|y)$ as a function of y. The corresponding function of Y is written as $E(X|Y)$ and is a random variable. It is very important to keep this notational distinction in mind.

The conditional density of Y given $X = x$ and the conditional expectation of Y given $X = x$ are defined analogously. That is, for instance,

$$f(y|x) = \frac{f(x, y)}{f_X(x)}, \quad \forall x \text{ such that } f_X(x) > 0.$$

An important relationship connecting the two conditional densities is the following result.

Theorem 3.4 (Bayes Theorem for Conditional Densities). *Let (X, Y) have a joint density $f(x, y)$. Then, $\forall x, y$, such that $f_X(x) > 0$, $f_Y(y) > 0$,*

$$f(y|x) = \frac{f(x|y) f_Y(y)}{f_X(x)}.$$

Proof.

$$\frac{f(x|y) f_Y(y)}{f_X(x)} = \frac{\frac{f(x,y)}{f_Y(y)} f_Y(y)}{f_X(x)}$$
$$= \frac{f(x, y)}{f_X(x)} = f(y|x).$$

Thus, we can convert one conditional density to the other one by using Bayes' theorem; note the similarity to Bayes' theorem discussed in Chapter 1. □

Definition 3.7 (Conditional Variance). Let (X, Y) have a joint density $f(x, y)$. The *conditional variance* of X given $Y = y$ is defined as

$$\text{Var}(X|y) = \text{Var}(X|Y = y) = \frac{\int_{-\infty}^{\infty} (x - \mu_X(y))^2 f(x, y) dx}{\int_{-\infty}^{\infty} f(x, y) dx},$$

$\forall y$ such that $f_Y(y) > 0$, where $\mu_X(y)$ denotes $E(X|y)$.

Remark. All the facts and properties about conditional pmfs and conditional expectations that were presented in the previous chapter for discrete random variables continue to hold verbatim in the continuous case, with densities replacing the pmfs in their statements. In particular, the iterated expectation and variance formula, and all the rules about conditional expectations and variance in Section 2.3 hold in the continuous case.

An important optimizing property of the conditional expectation is that the best predictor of Y based on X among all possible predictors is the conditional expectation of Y given X. Here is the exact result.

Proposition (Best Predictor). *Let (X, Y) be jointly distributed random variables (of any kind). Suppose $E(Y^2) < \infty$. Then $E_{X,Y}[(Y - E(Y|X))^2] \leq E_{X,Y}[(Y - g(X))^2]$, for any function $g(X)$. Here, the notation $E_{X,Y}$ stands for expectation with respect to the joint distribution of X, Y.*

Proof. Denote $\mu_Y(x) = E(Y|X = x)$. Then, by the property of the mean of any random variable U that $E(U - E(U))^2 \leq E(U - a)^2$ for any a, we get that here,

$$E[(Y - \mu_Y(x))^2 | X = x] \leq E[(Y - g(x))^2 | X = x],$$

for any x.

Inasmuch as this inequality holds for any x, it also holds on taking an expectation:

$$E_X\left[E\left[(Y - \mu_Y(x))^2 | X = x\right]\right] \leq E_X\left[E\left[(Y - g(x))^2 | X = x\right]\right]$$
$$\Rightarrow E_{X,Y}\left[(Y - \mu_Y(X))^2\right] \leq E_{X,Y}\left[(Y - g(X))^2\right],$$

where the final line is a consequence of the iterated expectation formula (see Chapter 2). □

We now show a number of examples.

3.4.1 Examples on Conditional Densities and Expectations

Example 3.16 (Uniform in a Triangle). Consider the joint density

$$f(x, y) = 2, \quad \text{if } x, y \geq 0, x + y \leq 1.$$

By using the results derived in Example 3.2,

$$f(x|y) = \frac{f(x, y)}{f_Y(y)} = \frac{1}{1 - y},$$

3.4 Conditional Densities and Expectations

if $0 \leq x \leq 1 - y$, and is zero otherwise. Thus, we have the interesting conclusion that given $Y = y$, X is distributed uniformly in $[0, 1-y]$. Consequently,

$$E(X|y) = \frac{1-y}{2}, \quad \forall y, 0 < y < 1.$$

Also, the conditional variance of X given $Y = y$ is, by the general variance formula for uniform distributions,

$$\text{Var}(X|y) = \frac{(1-y)^2}{12}.$$

Example 3.17 (Uniform Distribution in a Circle). Let (X, Y) have a uniform density in the unit circle, $f(x, y) = \frac{1}{\pi}$, $x^2 + y^2 \leq 1$, We find the conditional expectation of X given $Y = y$. First, the conditional density is

$$f(x|y) = \frac{f(x,y)}{f_Y(y)} = \frac{\frac{1}{\pi}}{\frac{2\sqrt{1-y^2}}{\pi}} = \frac{1}{2\sqrt{1-y^2}},$$

$-\sqrt{1-y^2} \leq x \leq \sqrt{1-y^2}$.

Thus, we have the interesting result that the conditional density of X given $Y = y$ is uniform on $[-\sqrt{1-y^2}, \sqrt{1-y^2}]$. It being an interval symmetric about zero, we have in addition the result that for any y, $E(X|Y = y) = 0$.

Let us now find the conditional variance. The conditional distribution of X given $Y = y$ is uniform on $[-\sqrt{1-y^2}, \sqrt{1-y^2}]$, therefore by the general variance formula for uniform distributions,

$$\text{Var}(X|y) = \frac{\left(2\sqrt{1-y^2}\right)^2}{12} = \frac{1-y^2}{3}.$$

Thus, the conditional variance decreases as y moves away from zero, which makes sense intuitively, because as y moves away from zero, the line segment in which x varies becomes smaller.

Example 3.18 (A Two-Stage Experiment). Suppose X is a positive random variable with density $f(x)$, and given $X = x$, a number Y is chosen at random between 0 and x. Suppose, however, that you are only told the value of Y, and the x value is kept hidden from you. What is your guess for x?

The formulation of the problem is:

$$X \sim f(x); \quad Y|X = x \sim U[0, x]; \quad \text{we want to find } E(X|Y = y).$$

To find $E(X|Y = y)$, our first task would be to find $f(x|y)$, the conditional density of X given $Y = y$. This is, by its definition,

$$f(x|y) = \frac{f(x, y)}{f_Y(y)} = \frac{f(y|x)f(x)}{f_Y(y)}$$

$$= \frac{\frac{1}{x}I_{\{x \geq y\}}f(x)}{\int_y^\infty \frac{1}{x}f(x)dx}.$$

Therefore,

$$E(X|Y = y) = \int_y^\infty xf(x|y)dx$$

$$= \frac{\int_y^\infty x\frac{1}{x}f(x)dx}{\int_y^\infty \frac{1}{x}f(x)dx} = \frac{1 - F(y)}{\int_y^\infty \frac{1}{x}f(x)dx},$$

where F denotes the CDF of X.

Suppose now, in particular, that $f(x)$ is the $U[0, 1]$ density. Then, by plugging into this general formula,

$$E(X|Y = y) = \frac{1 - F(y)}{\int_y^\infty \frac{1}{x}f(x)dx} = \frac{1 - y}{-\log y}, \quad 0 < y < 1.$$

The important thing to note is that although X has marginally a uniform density and expectation $\frac{1}{2}$, given $Y = y$, X is not uniformly distributed, and $E(X|Y = y)$ is not $\frac{1}{2}$. Indeed, as Fig. 3.5 shows, $E(X|Y = y)$ is an increasing function of y, increasing from zero at $y = 0$ to one at $y = 1$.

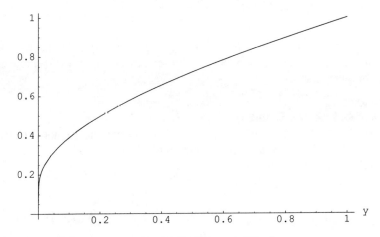

Fig. 3.5 Plot of E(X|Y = y) when X is U[0,1], Y|X = x is U[0, x]

3.4 Conditional Densities and Expectations

Example 3.19 ($E(X|Y = y)$ Exists for Any y, but $E(X)$ Does Not). Consider the setup of the preceding example once again (i.e., $X \sim f(x)$) and given $X = x$, $Y \sim U[0, x]$. Suppose $f(x) = \frac{1}{x^2}, x \geq 1$. Then the marginal expectation $E(X)$ does not exist, because $\int_1^\infty x \frac{1}{x^2} dx = \int_1^\infty \frac{1}{x} dx$ diverges.

However, from the general formula in the preceding example,

$$E(X|Y = y) = \frac{1 - F(y)}{\int_y^\infty \frac{f(x)}{x} dx} = \frac{\frac{1}{y}}{\frac{1}{2y^2}} = 2y,$$

and thus, $E(X|Y = y)$ exists for every y.

Example 3.20 (Using Conditioning to Evaluate Probabilities). We described the iterated expectation technique in the last chapter to calculate expectations. It turns out that it is in fact also a really useful way to calculate probabilities. Because the probability of any event A is also the expectation of $X = I_A$, by the iterated expectation technique, we can calculate $P(A)$ as

$$P(A) = E(I_A) = E(X) = E_Y[E(X|Y = y)] = E_Y[P(A|Y = y)],$$

by using a conditioning variable Y judiciously. The choice of the conditioning variable Y is usually clear from the particular context. Here is an example.

Let X, Y be independent $U[0, 1]$ random variables. Then $Z = XY$ also takes values in $[0, 1]$, and suppose we want to find an expression for $P(Z \leq z)$. We can do this by using the iterated expectation technique:

$$P(XY \leq z) = E[I_{XY \leq z}] = E_Y[E(I_{XY \leq z}|Y = y)]$$
$$= E_Y[E(I_{Xy \leq z}|Y = y)] = E_Y\left[E\left(I_{X \leq \frac{z}{y}}|Y = y\right)\right]$$
$$= E_Y[E(I_{X \leq \frac{z}{y}})]$$

(because X and Y are independent)

$$= E_Y\left[P\left(X \leq \frac{z}{y}\right)\right].$$

Now, note that $P(X \leq \frac{z}{y})$ is $\frac{z}{y}$ if $\frac{z}{y} \leq 1 \Leftrightarrow y \geq z$, and $P(X \leq \frac{z}{y}) = 1$ if $y < z$. Therefore,

$$E_Y\left[P\left(X \leq \frac{z}{y}\right)\right] = \int_0^z 1 dy + \int_z^1 \frac{z}{y} dy$$
$$= z - z \log z,$$

$0 < z \leq 1$. So, the final answer to our problem is $P(XY \leq z) = z - z \log z, 0 < z \leq 1$.

Example 3.21 (Power of the Iterated Expectation Formula). Let X, Y, Z be three independent $U[0, 1]$ random variables. We find the probability that $X^2 \geq YZ$ by once again using the iterated expectation formula.

Towards this,

$$\begin{aligned}P(X^2 \geq YZ) &= 1 - P(X^2 < YZ) \\ &= 1 - E[I_{X^2 < YZ}] = 1 - E_{Y,Z}[E(I_{X^2 < YZ}|Y = y, Z = z)] \\ &= 1 - E_{Y,Z}[E(I_{X^2 < yz}|Y = y, Z = z)] \\ &= 1 - E_{Y,Z}[E(I_{X^2 < yz})]\end{aligned}$$

(because X, Y, Z are independent)

$$\begin{aligned}&= 1 - E_{Y,Z}\big[P(X^2 < yz)\big] = 1 - E_{Y,Z}\big[\sqrt{yz}\big] \\ &= 1 - E_Y\big[\sqrt{Y}\big]E_Z\big[\sqrt{Z}\big] \\ &= 1 - \left(\frac{2}{3}\right)^2 = \frac{5}{9}.\end{aligned}$$

Once again, we see the power of identifying probabilities as expectations of indicator variables and the power of using the iterated expectation formula.

Example 3.22 (Conditional Density Given the Sum). Suppose X, Y are two independent Exp(1) variables. What is the conditional density of X given that $X + Y = t$? Denote $X + Y = T$. Then, we know from Chapter 1 that $T \sim G(2, 1)$. Also, by definition of probabilities for jointly continuous random variables, by denoting the joint density of (X, Y) as $f(x, y)$,

$$\begin{aligned}P(X \leq x, T \leq t) &= \int_{u \leq x, u+v \leq t} f(u, v) du\,dv \\ &= \int_{0 < u \leq x, 0 < u+v \leq t} e^{-u-v} du\,dv \\ &= \int_0^x e^{-u} \left[\int_0^{t-u} e^{-v} dv\right] du \\ &= \int_0^x e^{-u}(1 - e^{u-t}) du \\ &= \int_0^x e^{-u} du - \int_0^x e^{-t} du \\ &= 1 - e^{-x} - xe^{-t},\end{aligned}$$

for $x > 0, t > x$.

Therefore, the joint density of X and T is

$$\begin{aligned}f_{X,T}(x, t) &= \frac{\partial^2}{\partial x \partial t}[1 - e^{-x} - xe^{-t}] \\ &= e^{-t}, \quad 0 < x < t < \infty.\end{aligned}$$

Now, therefore, from the definition of conditional densities,

$$f(x|t) = \frac{f_{X,T}(x,t)}{f_T(t)}$$
$$= \frac{e^{-t}}{te^{-t}} = \frac{1}{t},$$

$0 < x < t$.

That is, given that the sum $X + Y = t$, X is distributed uniformly on $[0, t]$. In particular,

$$E(X|X+Y=t) = \frac{t}{2}, \quad \mathrm{Var}(X|X+Y=t) = \frac{t^2}{12}.$$

To complete the example, we mention a quick trick to compute the conditional expectation. Note that by symmetry,

$$E(X|X+Y=t) = E(Y|X+Y=t)$$
$$\Rightarrow t = E(X+Y|X+Y=t) = 2E(X|X+Y=t)$$
$$\Rightarrow E(X|X+Y=t) = \frac{t}{2}.$$

So, if we wanted just the conditional expectation, then the conditional density calculation would not be necessary in this case. This sort of symmetry argument is often very useful in reducing algebraic calculations. But one needs to be absolutely sure that the symmetry argument will be valid in a given problem.

3.5 Posterior Densities, Likelihood Functions, and Bayes Estimates

In Bayesian statistics, parameters of distributions, being unknown, are formally assigned a probability distribution, called a *prior distribution*. For example, if $X \sim \mathrm{Bin}(n, p)$, then the binomial distribution is interpreted to be the conditional distribution of X given that another random variable taking values in $[0, 1]$ is equal to p. This other variable Y is assigned a density $g(p)$, which reflects the statistician's a priori belief about the value of that success probability. One then uses Bayes' theorem to find a conditional density for the parameter given the observed value, x, of X. This density is called *the posterior density of the parameter*. The posterior density combines the a priori information with the information coming from the data value x to form a final density for the parameter. One then uses this posterior density to make statements about the parameter, for example, *the posterior probability that the parameter is $> .6$ is $< .25$*. One can use the mean of the posterior density as an estimate for the true value of the unknown parameter, and so on. Some examples of this Bayesian approach are worked out below. But, first we formally define a posterior density.

Definition 3.8. Suppose for some fixed $n \geq 1$, $(X_1, \ldots, X_n)$ have the joint density, or the joint pmf, $f(x_1, \ldots, x_n | \theta)$, where θ is a real-valued parameter, taking values in an interval (a, b), where a, b may be $\pm \infty$. Formally, consider θ itself to be a random variable, and suppose θ has a density $g(\theta)$ on (a, b). Then, the conditional density of θ given $X_1 = x_1, \ldots, X_n = x_n$ is called the *posterior density* of θ, and is given by

$$f(\theta | x_1, \ldots, x_n) = \frac{f(x_1, \ldots, x_n | \theta) g(\theta)}{\int_a^b f(x_1, \ldots, x_n | \theta) g(\theta) d\theta}.$$

The function $l(\theta) = f(x_1, \ldots, x_n | \theta)$ is called the *likelihood function*, the function $g(\theta)$ is called the *prior density*, and the conditional expectation of θ given $X_1 = x_1, \ldots, X_n = x_n$, $E(\theta | X_1 = x_1, \ldots, X_n = x_n)$, if it exists, is called the *posterior mean* of θ.

Remark. Note that in the expression for the posterior density, only the numerator depends on θ. The denominator depends only on $x_1, \ldots, x_n$, because in the denominator, θ is being completely integrated out. So, we should think of the denominator in the expression for the posterior density to be merely a normalizing constant.

Note also that if (a, b) is a bounded interval, and we take g to be the uniform density on (a, b), then, apart from the normalizing constant in the denominator, the posterior density of θ is exactly the same as the likelihood function.

Example 3.23 (Posterior Density for Exponential Mean). Suppose we have a single observation $X \sim \text{Exp}(\lambda)$, and that λ has the marginal density $g(\lambda) = 2\lambda, 0 < \lambda < 1$. Then, by Bayes' theorem,

$$f(\lambda | x) = \frac{f(x|\lambda)g(\lambda)}{f_X(x)}$$

$$= \frac{f(x|\lambda)g(\lambda)}{\int_0^1 f(x, \lambda) d\lambda} = \frac{f(x|\lambda)g(\lambda)}{\int_0^1 f(x|\lambda)g(\lambda) d\lambda}$$

$$= \frac{\frac{1}{\lambda} e^{-\frac{x}{\lambda}} \times 2\lambda}{\int_0^1 \frac{1}{\lambda} e^{-\frac{x}{\lambda}} \times 2\lambda d\lambda} = \frac{2e^{-\frac{x}{\lambda}}}{\int_0^1 2e^{-\frac{x}{\lambda}} d\lambda}$$

$$= \frac{e^{-\frac{x}{\lambda}}}{k(x)},$$

where $k(x)$ denotes the integral $\int_0^1 e^{-\frac{x}{\lambda}} d\lambda$, which exists, but does not have a simple final formula. Thus, finally, the posterior density of λ given that the data value $X = x$, is

$$f(\lambda | x) = \frac{e^{-\frac{x}{\lambda}}}{k(x)}, \quad 0 < \lambda < 1.$$

3.5 Posterior Densities, Likelihood Functions, and Bayes Estimates

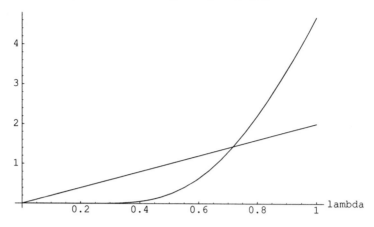

Fig. 3.6 Prior and posterior density in exponential example

We give a plot of the prior density for λ, along with the posterior density for λ in Fig. 3.6. A comparison of the two density plots explains the effect of the data value $X = x$ on updating the prior density to the posterior density. We see from the plots that the data value ($x = 3$) makes larger λ values more likely under the posterior than they were under the prior.

Example 3.24 (Posterior Mean for Binomial p). Suppose $X \sim \text{Bin}(n, p)$, where the probability of success p is treated as an unknown parameter. For example, you may take a sample of n people independently from a population and count how many are vegetarians. Then p will correspond to the fraction of vegetarians in the entire population, and it seems likely that you cannot really know what that proportion is in the entire population.

In the Bayesian approach, you have to assign the parameter p a distribution. For simplicity of calculations, suppose we give p the $U[0, 1]$ prior. So, the Bayes model is:
$$p \sim U[0, 1], \ X|p \sim \text{Bin}(n, p).$$
The posterior density, by definition, is the conditional density of p given $X = x$, x being the actual observed value of X. Then, from Bayes' theorem for conditional densities,

$$f(p|x) = \frac{f(x|p)g(p)}{\int_0^1 f(x|p)g(p)dp} = \frac{\binom{n}{x}p^x(1-p)^{n-x}}{\int_0^1 \binom{n}{x}p^x(1-p)^{n-x}dp}$$

$$= \frac{p^x(1-p)^{n-x}}{\int_0^1 p^x(1-p)^{n-x}dp} = \frac{p^x(1-p)^{n-x}}{\frac{\Gamma(x+1)\Gamma(n-x+1)}{\Gamma(n+2)}}.$$

Here, in the last line, the denominator is obtained by actually doing the integration. But, if we did not bother to do the integration in the denominator, and just looked at the numerator, we would have realized that apart from an as yet unevaluated constant term that the denominator will contribute, the posterior density is a Beta density with parameters $x + 1$ and $n - x + 1$, respectively. From the formula for the mean of a Beta density(see Chapter 1), we get the additional formula that the conditional expectation.

$$E(p|X = x) = \frac{x + 1}{x + 1 + n - x + 1} = \frac{x + 1}{n + 2}.$$

This is called the *posterior mean*. So, if you believe in your $U[0, 1]$ prior for p, then as a Bayesian you may wish to estimate p by $\frac{x+1}{n+2}$, the posterior mean. This is different from $\frac{x}{n}$, the more common estimate of p. The slight alteration is caused by treating p as a random variable, and by adopting the Bayesian approach.

More generally, if p has the general Beta prior density, $p \sim \text{Be}(\alpha, \beta)$, then the same calculation as above shows that the posterior density is another Beta, and it is the $\text{Be}(x + \alpha, n - x + \beta)$ density.

Example 3.25 (Posterior Density for a Poisson Mean). Suppose we have n iid sample values $X_1, X_2, \ldots, X_n$ from a Poisson distribution with mean λ. The parameter λ is considered unknown. Therefore, in the Bayesian approach, we have to choose a prior distribution for it. Suppose we choose a standard exponential prior for λ: the Bayes model is:

$$f(x_1, x_2, \ldots, x_n \mid \lambda) = \prod_{i=1}^{n} \frac{e^{-\lambda} \lambda^{x_i}}{x_i!} = \frac{e^{-n\lambda} \lambda^{\sum_{i=1}^{n} x_i}}{\prod_{i=1}^{n} x_i!};$$

and $g(\lambda) = e^{-\lambda}, \lambda > 0$.

Therefore, by definition of a posterior density,

$$f(\lambda \mid x_1, x_2, \ldots, x_n) = \frac{f(x_1, x_2, \ldots, x_n \mid \lambda) g(\lambda)}{\int_0^\infty f(x_1, x_2, \ldots, x_n \mid \lambda) g(\lambda) d\lambda}$$

$$= \frac{e^{-(n+1)\lambda} \lambda^{\sum_{i=1}^{n} x_i}}{\int_0^\infty e^{-(n+1)\lambda} \lambda^{\sum_{i=1}^{n} x_i} d\lambda} = \frac{e^{-(n+1)\lambda} \lambda^{\sum_{i=1}^{n} x_i}}{\frac{\Gamma(\sum_{i=1}^{n} x_i + 1)}{(n+1)^{1+\sum_{i=1}^{n} x_i}}}$$

$$= \frac{(n+1)^{1+\sum_{i=1}^{n} x_i} e^{-(n+1)\lambda} \lambda^{\sum_{i=1}^{n} x_i}}{\Gamma(\sum_{i=1}^{n} x_i + 1)}.$$

Once again, the integration in the denominator did not really need to be done. By simply looking at the numerator, we would recognize this to be a Gamma density with shape parameter $\sum_{i=1}^{n} x_i + 1$ and scale parameter $\frac{1}{n+1}$. From the general formula for the mean of a Gamma density(see Chapter 1), we get the posterior mean formula $E(\lambda \mid x_1, x_2, \ldots, x_n) = \frac{\sum_{i=1}^{n} x_i + 1}{n+1}$. Once again, it is a slight alteration of the estimate one may have thought of intuitively, namely the estimate $\frac{\sum_{i=1}^{n} x_i}{n}$.

3.5 Posterior Densities, Likelihood Functions, and Bayes Estimates

Example 3.26 (Posterior Density of a Normal Mean). Suppose $X_1, X_2, \ldots, X_n \sim N(\mu, 1)$, and suppose μ is assigned a standard normal prior density. Then, by definition of the posterior density,

$$f(\mu|x_1,\ldots,x_n) = \frac{e^{-\frac{1}{2}\sum_{i=1}^{n}(x_i-\mu)^2}e^{-\frac{\mu^2}{2}}}{\int_{-\infty}^{\infty} e^{-\frac{1}{2}\sum_{i=1}^{n}(x_i-\mu)^2}e^{-\frac{\mu^2}{2}}d\mu}$$

$$= \frac{e^{-\frac{n+1}{2}\mu^2+\mu(\sum_{i=1}^{n}x_i)}}{\int_{-\infty}^{\infty} e^{-\frac{n+1}{2}\mu^2+\mu(\sum_{i=1}^{n}x_i)}d\mu}$$

$$= \frac{e^{-\frac{n+1}{2}\left[\mu-\frac{\sum_{i=1}^{n}x_i}{n+1}\right]^2}}{\int_{-\infty}^{\infty} e^{-\frac{n+1}{2}\left[\mu-\frac{\sum_{i=1}^{n}x_i}{n+1}\right]^2}d\mu}.$$

Once again, it is not necessary to work out the integral in the denominator, although it is certainly not difficult to do so. All we need to recognize is that the numerator, after doing all the algebra that we did, has reduced to yet another normal density on μ, namely, a normal density with mean $\frac{\sum_{i=1}^{n}x_i}{n+1}$ and variance $\frac{1}{n+1}$. If we did go through the chores of actually performing the integration in the denominator, we would surely find it to be just the normalizing constant of this normal density. The conclusion is that if $X_1, X_2, \ldots, X_n \sim N(\mu, 1)$, and μ has a standard normal prior, then the posterior density of μ is the $N\left(\frac{\sum_{i=1}^{n}x_i}{n+1}, \frac{1}{n+1}\right)$ density. In particular, the posterior mean is $E(\mu|x_1,\ldots,x_n) = \frac{\sum_{i=1}^{n}x_i}{n+1}$, and the variance of the posterior distribution is $\frac{1}{n+1}$. Note that the posterior variance does not depend on $x_1,\ldots,x_n$! This rather remarkable fact is entirely specific to the choice of a normal prior density; any normal prior will result in this constant posterior variance property.

Remark. In each of the last three examples, we obtained something interesting. In the binomial example, the prior density used was $U[0, 1]$, which is a special case of a Beta, and we found the posterior density to be another Beta. In the Poisson example, the prior density used was the standard exponential, which is a special case of a Gamma, and we found the posterior density to be another Gamma. And, in the normal example, the prior density used was the standard normal, and we found the posterior density to be another normal. In other words, the functional form of the prior and the posterior density are the same in each of these three examples. In updating the prior to the posterior, we only updated the parameters of the prior, but not the functional form. This happens more generally, but only for special types of priors, and the special type depends on the specific problem. This is considered to be of such great convenience, that priors which satisfy this neat updating property have been given a name; they are called conjugate priors. Here is a formal definition.

Definition 3.9. Let $X \sim f(x|\theta)$, and suppose $\theta \sim g(\theta)$, where g belongs to some family of densities $\mathcal{G}$. The family $\mathcal{G}$ is called a *family of conjugate priors* for the model f if the posterior density $f(\theta|x)$ is also a member of $\mathcal{G}$ for any x.

The Beta family is a conjugate family for the binomial case. The Gamma family is conjugate for the Poisson case. Normal distributions on μ form a conjugate family for the mean μ in the normal case. Conjugate families are not unique, and in each new problem, one has to find a convenient one by inspection.

3.6 Maximum Likelihood Estimates

The posterior density combines the likelihood function $l(\theta)$ and the prior density $g(\theta)$ with suitable normalization. Fisher's idea was to use just the likelihood function itself as the yardstick for assessing the credibility of each θ for being the true value of the parameter θ. If the likelihood function $l(\theta)$ is large at some θ, that θ value is consistent with the data that where obtained; on the other hand, if the likelihood function $l(\theta)$ is small at some θ, that θ value is inconsistent with the data that were obtained. Fisher suggested maximizing the likelihood function over all possible values of θ, and using the maxima as an estimate of θ. This is the celebrated *maximum likelihood estimate*.

Many think that maximum likelihood is the greatest conceptual invention in the history of statistics. Although in some high or infinite-dimensional problems computation and performance of maximum likelihood estimates are less than desirable, or even poor, in a vast majority of models in practical use, MLEs are about the best that one can do. They have many asymptotic optimality properties that translate into fine performance in finite samples. We give a few illustrative examples, after defining an MLE.

Definition 3.10. Suppose given a parameter θ, $X^{(n)} = (X_1, \ldots, X_n)$ have a joint pdf or joint pmf $f(x_1, \ldots, x_n | \theta), \theta \in \Theta$. Any value $\hat{\theta} = \hat{\theta}(X_1, \ldots, X_n)$ at which the likelihood function $l(\theta) = f(x_1, \ldots, x_n | \theta)$ is maximized is called a *maximum likelihood estimate (MLE)* of θ, provided $\hat{\theta} \in \Theta$, and $l(\hat{\theta}) < \infty$.

It is important to understand that an MLE need not exist, or be unique. But, in many examples, it exists and is unique for any dataset $X^{(n)}$. In maximizing the likelihood function over θ, any pure constant terms not involving θ may be ignored. Also, in many standard models, it is more convenient to maximize $L(\theta) = \log l(\theta)$; it simplifies the algebra without affecting the correctness of the final answer.

Example 3.27 (MLE of Binomial p). Let $X_1, \ldots, X_n \stackrel{iid}{\sim} \text{Ber}(p), 0 < p < 1$. Then, writing $X = \sum_{i=1}^{n} X_i$ (the total number of successes in these n trials),

$$l(p) = p^X (1-p)^{(n-X)} \Rightarrow L(p) = \log l(p) = X \log p + (n - X) \log(1 - p).$$

For $0 < X < n$, $L(p)$ has a unique stationary point, namely a point at which the first derivative $L'(p) = 0$. This point is $p = \frac{X}{n}$. Furthermore, it is easily verified that

3.6 Maximum Likelihood Estimates

$L''(p) < 0$ for all $p \in (0,1)$; that is, $L(p)$ is strictly concave. So, for $0 < X < n$, there is a unique MLE of p, and it is just the common sense estimate $\frac{X}{n}$, the sample proportion of successes. If $X = 0$ or n, the likelihood function is maximized at a boundary value $p = 0$ or 1. In those two cases, an MLE of p does not exist.

Example 3.28 (Mean of an Exponential). Let $X_1, \ldots, X_n \stackrel{iid}{\sim} \text{Exp}(\lambda)$, $\lambda > 0$. Then,

$$l(\lambda) = \frac{e^{-\frac{1}{\lambda}\sum_{i=1}^{n} X_i}}{\lambda^n} \Rightarrow L(\lambda) = -\frac{1}{\lambda}\sum_{i=1}^{n} X_i - n \log \lambda.$$

$L(\lambda)$ has a unique stationary point, it being $\lambda = \frac{\sum_{i=1}^{n} X_i}{n} = \overline{X}$. Furthermore, $L''(\lambda) < 0$ at $\lambda = \overline{X}$. Also note that $l(\lambda) \to 0$ as $\lambda \to 0$ or ∞. These three facts together imply that for all possible datasets $X_1, \ldots, X_n$, there is a unique MLE of λ, and it is the sample mean $\overline{X}$.

Example 3.29 (MLE of Normal Mean and Variance). Let $X_1, \ldots, X_n \stackrel{iid}{\sim} N(\mu, \sigma^2)$, $-\infty < \mu < \infty, \sigma^2 > 0$. This is a two-parameter example. The likelihood function is

$$l(\mu, \sigma^2) = \frac{e^{-\frac{1}{2\sigma^2}\sum_{i=1}^{n}(X_i - \mu)^2}}{(\sigma^2)^{\frac{n}{2}}}.$$

Maximizing a function of two variables by calculus methods has to be done carefully, because the second derivative tests are subtle and must be carefully applied. We instead obtain the MLEs of μ and σ^2 directly, as follows. The argument uses a sequence of simple inequalities. As usual, let $\overline{X} = \frac{\sum_{i=1}^{n} X_i}{n}$, and also let $s_0^2 = \frac{1}{n}\sum_{i=1}^{n}(X_i - \overline{X})^2$; note that this is different from the sample variance $s^2 = \frac{1}{n-1}\sum_{i=1}^{n}(X_i - \overline{X})^2$. The argument below shows that the unique MLEs of μ, σ^2 are $\overline{X}$ and s_0^2. This follows from the straightforward inequalities

$$l(\mu, \sigma^2) \leq l(\overline{X}, \sigma^2) \leq l(\overline{X}, s_0^2) = \frac{e^{-\frac{n}{2}}}{s_0^n} < \infty,$$

and therefore, $(\overline{X}, s_0^2)$ is the unique global maxima of $l(\mu, \sigma^2)$.

Example 3.30 (Endpoint of a Uniform). This is an example where the MLE cannot be obtained by finding stationary points of the likelihood function, and is found by examining the shape of the likelihood function. Let $X_1, \ldots, X_n \stackrel{iid}{\sim} U[0, \theta]$, $\theta > 0$. Then, the individual densities are $f(x_i \mid \theta) = \frac{1}{\theta} I_{0 \leq x_i \leq \theta}$. Therefore,

$$l(\theta) = \prod_{i=1}^{n} \frac{1}{\theta} I_{0 \leq x_i \leq \theta} = \frac{1}{\theta^n} I_{\theta \geq \max(x_1, \ldots, x_n)} I_{\min(x_1, \ldots, x_n) \geq 0}$$

$$= \frac{1}{\theta^n} I_{\theta \geq \max(x_1, \ldots, x_n)},$$

because under the model, with probability one under any $\theta > 0$, $\min(X_1, \ldots, X_n)$ is greater than zero.

The likelihood function is therefore zero on $(0, \max(x_1,\ldots,x_n))$, and on $[\max(x_1,\ldots,x_n), \infty)$ it is strictly decreasing, with a finite value at the jump point $\theta = \max(x_1,\ldots,x_n)$. Therefore, $X_{(n)} = \max(X_1,\ldots,X_n)$ is the unique MLE of θ for all data values $X_1,\ldots,X_n$.

3.7 Bivariate Normal Conditional Distributions

Suppose (X, Y) have a joint bivariate normal distribution. A very important property of the bivariate normal is that *each conditional distribution*, the distribution of Y given $X = x$, and that of X given $Y = y$ is a univariate normal, for any x and any y. This really helps in easily computing conditional probabilities involving one variable, when the other variable is held fixed at some specific value.

Theorem 3.5. *Let (X, Y) have a bivariate normal distribution with parameters $\mu_1, \mu_2, \sigma_1, \sigma_2, \rho$. Then,*

(a) $X|Y = y \sim N\left(\mu_1 + \rho\frac{\sigma_1}{\sigma_2}(y - \mu_2), \sigma_1^2(1 - \rho^2)\right)$;

(b) $Y|X = x \sim N\left(\mu_2 + \rho\frac{\sigma_2}{\sigma_1}(x - \mu_1), \sigma_2^2(1 - \rho^2)\right)$.

In particular, the conditional expectations of X given $Y = y$ and of Y given $X = x$ are linear functions of y and x, respectively:

$$E(X|Y = y) = \mu_1 + \rho\frac{\sigma_1}{\sigma_2}(y - \mu_2);$$

$$E(Y|X = x) = \mu_2 + \rho\frac{\sigma_2}{\sigma_1}(x - \mu_1),$$

and the variance of each conditional distribution is a constant, and does not depend on the conditioning values x or y.

The proof of this theorem involves some tedious integration manipulations, and we omit it; the details of the proof are available in Tong (1990).

Remark. We see here that the conditional expectation is linear in the bivariate normal case. Specifically, take $E(Y|X = x) = \mu_2 + \rho\frac{\sigma_2}{\sigma_1}(x - \mu_1)$. Previously, we have seen in Chapter 2 that the conditional expectation $E(Y|X)$ is, in general, the best predictor of Y based on X. Now we see that the conditional expectation is a linear predictor in the bivariate normal case, and it is the best predictor and therefore, also the best linear predictor. In Chapter 2, we called the best linear predictor the regression line of Y on X. Putting it all together, we have the very special result that in the bivariate normal case, the regression line of Y on X and the best overall predictor are the same:

For bivariate normal distributions, the conditional expectation of one variable given the other coincides with the regression line of that variable on the other variable.

Example 3.31. Suppose incomes of husbands and wives in a population are bivariate normal with means 75 and 60 (in thousands of dollars), standard deviations 20 each, and a correlation of .75. We want to know in what percentage of those families where the wife earns 80,000 dollars, the family income exceeds 175,000 dollars.

Denote the income of the husband and the wife by X and Y. Then, we want to find $P(X + Y > 175 | Y = 80)$. By the above theorem. $X|Y = 80 \sim N(75 + .75(80 - 60), 400(1 - .75^2)) = N(90, 175)$. Therefore,

$$P(X + Y > 175 | Y = 80) = P(X > 95 | Y = 80)$$
$$= P\left(Z > \frac{95 - 90}{\sqrt{175}}\right) = P(Z > .38)$$
$$= .3520,$$

where Z denotes a standard normal variable.

Example 3.32 (Galton's Observation: Regression to the Mean). This example is similar to the previous example, but makes an interesting different point. It is often found that students who get a very good grade on the first midterm, do not do as well on the second midterm. We can try to explain it by doing a bivariate normal calculation.

Denote the grade on the first midterm by X, that on the second midterm by Y, and suppose X, Y are jointly bivariate normal with means 70, standard deviations 10, and a correlation .7. Suppose a student scored 90 on the first midterm. What are the chances that she will get a lower grade on the second midterm?

This is

$$P(Y < X | X = 90) = P(Y < 90 | X = 90)$$
$$= P\left(Z < \frac{90 - 84}{\sqrt{51}}\right) = P(Z < .84)$$
$$= .7995,$$

where Z is a standard normal variable, and we have used the fact that $Y|X = 90 \sim N(70 + .7(90 - 70), 100(1 - .7^2)) = N(84, 51)$.

Thus, with a fairly high probability, the student will not be able to match her first midterm grade on the second midterm. The phenomenon of *regression to mediocrity* was popularized by Galton, who noticed that the offspring of very tall parents tended to be much closer to being of just about average height, and the extreme tallness in the parents was not commonly passed on to the children.

3.8 * Useful Formulas and Characterizations for Bivariate Normal

A number of extremely elegant characterizations and also some very neat formulas for useful quantities are available for a general bivariate normal distribution. Another practical issue is the numerical computation of bivariate normal probabilities.

Although tables are widely available for the univariate standard normal, for the bivariate normal corresponding tables are found only in specialized sources, and are sketchy. Thus, a simple and reasonably accurate approximation formula is practically useful. We deal with these issues in this section.

First, we need some notation. For jointly distributed random variables X, Y with means μ_1, μ_2, standard deviations σ_1, σ_2, and positive integers r, s, we denote

$$\lambda_{r,s} = \frac{E[(X-\mu_1)^r(Y-\mu_2)^s]}{\sigma_1^r \sigma_2^s}; \quad \nu_{r,s} = \frac{E[|X-\mu_1|^r |Y-\mu_2|^s]}{\sigma_1^r \sigma_2^s}.$$

We then have the following useful formulas for a general bivariate normal distribution.

Theorem 3.6 (Bivariate Normal Formulas).

(a) *The joint mgf of the bivariate normal distribution is given by*

$$\psi(t_1, t_2) = e^{t_1\mu_1 + t_2\mu_2 + \frac{1}{2}[t_1^2 \sigma_1^2 + t_2^2 \sigma_2^2 + 2\rho t_1 t_2 \sigma_1 \sigma_2]};$$

(b) $\lambda_{r,s} = \lambda_{s,r}; \quad \nu_{r,s} = \nu_{s,r};$
(c) $\lambda_{r,s} = 0$ *if* $r+s$ *is odd;*
(d) $\lambda_{1,1} = \rho; \lambda_{1,3} = 3\rho;$

$$\lambda_{2,2} = 1 + 2\rho^2; \lambda_{2,4} = 3(1 + 4\rho^2);$$

$$\lambda_{3,3} = 3\rho(3 + 2\rho^2); \lambda_{4,4} = 3(3 + 24\rho^2 + 8\rho^4);$$

(e) $\nu_{1,1} = \frac{2}{\pi}\left[\sqrt{1-\rho^2} + \rho \arcsin \rho\right]; \nu_{3,3} = \frac{2}{\pi}\left[(4+11\rho^2)\sqrt{1-\rho^2} + 3\rho(3+2\rho^2)\arcsin \rho\right].$

(f) $E(\max\{X, Y\}) = p\mu_1 + (1-p)\mu_2 + \delta\tau,$

where

$$p = \Phi\left(\frac{\Delta}{\tau}\right), \quad \delta = \phi\left(\frac{\Delta}{\tau}\right),$$

and

$$\Delta = \mu_1 - \mu_2; \quad \tau^2 = \sigma_1^2 + \sigma_2^2 - 2\rho\sigma_1\sigma_2.$$

These formulas are proved in Kamat (1953).

Among the numerous characterizations of the bivariate normal distribution, a few stand out in their clarity and in being useful. We state these characterizations below.

3.8 Useful Formulas and Characterizations for Bivariate Normal

Theorem 3.7 (Characterizations). *Let X, Y be jointly distributed random variables.*

(a) *X, Y are jointly bivariate normal if and only if every linear combination $aX + bY$ has a univariate normal distribution.*
(b) *X, Y are jointly bivariate normal if and only if X is univariate normal, and $\forall x, \, Y|X = x \sim N(a + bx, c^2)$, for some a, b, c.*
(c) *X, Y are jointly bivariate normal if and only if $\forall x, \, \forall y, \, Y|X = x$, and $X|Y = y$ are univariate normals, and in addition, either one of the marginals is a univariate normal, or one of the conditional variance functions is a constant function.*

See Kagan et al. (1973), and Patel and Read (1996) for these and other characterizations of the bivariate normal distribution.

3.8.1 Computing Bivariate Normal Probabilities

There are many approximations to the CDF of a general bivariate normal distribution. The most accurate ones are too complex for quick use. The relatively simple approximations are not computationally accurate for all configurations of the arguments and the parameters. Keeping a balance between simplicity and accuracy, we present here two approximations.

Mee–Owen Approximation. Let (X, Y) have the general five-parameter bivariate normal distribution. Then,

$$P(X \leq \mu_1 + h\sigma_1, Y \leq \mu_2 + k\sigma_2) \approx \Phi(h)\Phi\left(\frac{k - c}{\tau}\right),$$

where $c = -\rho \frac{\phi(h)}{\Phi(h)}, \tau^2 = 1 + \rho h c - c^2$.

Cox–Wermuth Approximation.

$$P(X \geq \mu_1 + h\sigma_1, Y \geq \mu_2 + k\sigma_2) \approx \Phi(-h)\Phi\left(\frac{\rho\mu(h) - k}{\sqrt{1 - \rho^2}}\right),$$

where $\mu(h) = \frac{\phi(h)}{1-\Phi(h)}$.

See Mee and Owen (1983) and Cox and Wermuth (1991) for the motivation behind these approximations. See Plackett (1954) for reducing the dimension of the integral for computing multivariate normal probabilities. Genz (1993) provides some comparison of the different algorithms and approximations for computing multivariate normal probabilities.

Example 3.33 (Testing the Approximations). We compute the Mee–Owen approximation, the Cox–Wermuth approximation, and the corresponding exact probabilities

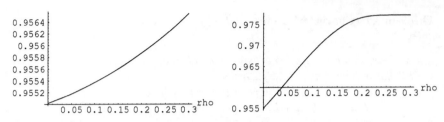

Fig. 3.7 P(X < 2, Y < 2) and the Mee–Owen approximation in the standardized case

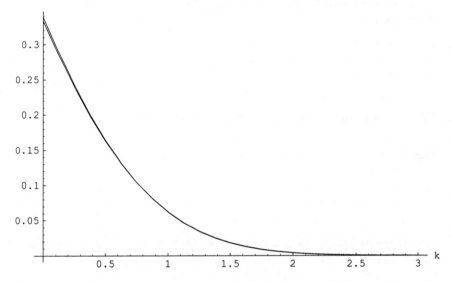

Fig. 3.8 P(X > k, Y > k) and Cox–Wermuth approximation in the standardized case, rho = .5

in two trial cases. We can see from Fig. 3.8 that the Cox–Wermuth approximation is nearly exact in the trial case. The Mee–Owen approximation in Fig. 3.7 is reasonable, but not very accurate. It should also be noted that the quantity τ^2 in the Mee–Owen approximation can be negative, in which case the approximation is not usable. Generally, the Mee–Owen approximation is inaccurate or unusable if h, k, ρ are large. The Cox–Wermuth approximation should not be used when ρ is large (>.75 or so).

3.9 * Conditional Expectation Given a Set and Borel's Paradox

In applications, one is very often interested in finding the expectation of one random variable given that another random variable belongs to some set, rather than given that it is exactly equal to some value. For instance, we may want to know what the average income of husbands is among those families where the wife earns more than $100,000.

3.9 Conditional Expectation Given a Set and Borel's Paradox

The mathematical formulation of the problem is to find $E(X|Y \in A)$, for some given set A. It is not possible to talk rigorously about this without using measure theory. In fact, even defining $E(X|Y \in A)$ can be a problem. We limit ourselves to special types of sets A.

Definition 3.11. Let (X, Y) have a joint density $f(x, y)$, with marginal densities $f_X(x)$, $f_Y(y)$. Let A be a subset of the real line such that $P(Y \in A) > 0$. Then

$$E(X|Y \in A) = \frac{\int_{-\infty}^{\infty} \int_A x f(x,y) dy dx}{\int_{-\infty}^{\infty} \int_A f(x,y) dy dx}.$$

Remark. When the conditioning event A has probability zero, we can get into paradoxical situations when we try to compute $E(X|Y \in A)$. What happens is that it may be possible to rewrite the conditioning event $Y \in A$ as an equivalent event $V \in B$ for some carefully chosen function $V = V(X, Y)$. Yet, when we compute $E(X|Y \in A)$ and $E(X|V \in B)$, we arrive at different answers! The paradox arises because of subtleties of *measure zero* sets. It is not possible to describe how one avoids such a paradox without the knowledge of abstract measure theory. We do however, give an example illustrating this paradox, popularly known as *Borel's paradox*.

Example 3.34 (Borel's Paradox). Let (X, Y) have the joint density

$$f(x, y) = 1, \quad \text{if } 0 \leq x \leq 1, -x \leq y \leq 1 - x,$$

and zero otherwise. Then, the marginal density of Y is

$$f_Y(y) = \int_{-y}^{1} dx = 1 + y, \quad \text{if } -1 \leq y \leq 0, \quad \text{and}$$

$$f_Y(y) = \int_{0}^{1-y} = 1 - y, \quad \text{if } 0 \leq y \leq 1.$$

Therefore, the conditional density of X given $Y = 0$ is

$$f(x|Y = 0) = 1, \quad 0 \leq x \leq 1.$$

This is just the uniform density on $[0, 1]$, and so we get $E(X|Y = 0) = .5$.

Now transform (X, Y) by the one-to-one transformation $(X, Y) \to (U, V)$, where $U = X$, $V = \frac{X+Y}{X}$. The Jacobian of the transformation is $J = u$, and hence the joint density of (U, V) is

$$f_{U,V}(u, v) = u, \quad 0 < u < 1, 0 < v < \frac{1}{u}.$$

In the transformed variables, $X = U$ and $Y = U(V - 1)$. So, $Y = 0 \Leftrightarrow V = 1$. Yet, when, we compute the conditional density of U given $V = 1$, we get after a little algebra

$$f(u|V = 1) = 2u, \ 0 \leq u \leq 1.$$

So, $E(U|V = 1) = \frac{2}{3}$, which, paradoxically, is different from .5!

Example 3.35 (Mean Residual Life). In survival analysis and medicine, a quantity of great interest is the mean residual life. Suppose that a person afflicted with some disease has survived five years. How much longer can the patient be expected to survive? Thus, suppose X is a continuous random variable with density $f(x)$. We want to find $E(X - c \mid X \geq c)$. Assuming that $P(X \geq c) > 0$,

$$E(X - c \mid X \geq c) = E(X \mid X \geq c) - c$$
$$= \frac{\int_c^\infty x f(x) dx}{1 - F(c)} - c.$$

As one specific example, let $X \sim \text{Exp}(\lambda)$. Then, from this formula,

$$E(X - c \mid X \geq c)$$
$$= \frac{\int_c^\infty x \frac{1}{\lambda} e^{-x/\lambda} dx}{e^{-c/\lambda}} - c$$
$$= \frac{(c + \lambda) e^{-c/\lambda}}{e^{-c/\lambda}} - c$$
$$= \lambda,$$

which is independent of c. We recognize that this is just the lack of memory property of an exponential distribution manifesting itself in the mean residual life calculation.

In contrast, suppose $X \sim N(0, 1)$ (of course, in reality a survival time X cannot have mean zero!). Then,

$$E(X - c \mid X \geq c) = E(X \mid X \geq c) - c$$
$$= \frac{\int_c^\infty x \phi(x) dx}{1 - \Phi(c)} - c$$
$$= \frac{\phi(c)}{1 - \Phi(c)} - c = \frac{1}{R(c)} - c,$$

where $R(c) = \frac{1 - \Phi(c)}{\phi(c)}$ is the *Mills ratio*. The calculation shows that the Mills ratio arises very naturally in a calculation of interest in survival analysis.

Note that now the mean residual life is no longer independent of c. Take c to be positive. From Laplace's expansion for the Mills ratio (see Chapter 1),

$$\frac{1}{R(c)} - c \approx \frac{c^3}{c^2 - 1} - c = \frac{c}{c^2 - 1} \approx \frac{1}{c}.$$

That is, the mean residual life is approximately equal to $\frac{1}{c}$, which is a decreasing function of c. So, unlike in the exponential case, if survival time is normal, then a patient who has survived a long time is increasingly more unlikely to survive too much longer.

Exercises

Exercise 3.1. Suppose (X, Y) have the joint density $f(x, y) = cxy, x, y \in [0, 1]$.

(a) Find the normalizing constant c.
(b) Are X, Y independent?
(c) Find the marginal densities and expectations of X, Y.
(d) Find the expectation of XY.

Exercise 3.2. Suppose (X, Y) have the joint density $f(x, y) = cxy, x, y \geq 0; x + y \leq 1$.

(a) Find the normalizing constant c.
(b) Are X, Y independent?
(c) Find the marginal densities and expectations of X, Y.
(d) Find the expectation of XY.

Exercise 3.3. Suppose (X, Y) have the joint density $f(x, y) = ce^{-y}, 0 \leq x \leq y < \infty$.

(a) Find the normalizing constant c.
(b) Are X, Y independent?
(c) Find the marginal densities and expectations of X, Y.
(d) Find the conditional expectation of X given $Y = y$.
(e) Find the conditional expectation of Y given $X = x$.
(f) Find the correlation between X and Y.

Exercise 3.4 (Uniform in a Triangle). Suppose X, Y are uniformly distributed in the triangle bounded by $-1 \leq x \leq 1, y \geq 0$, and the two lines $y = 1 + x$ and $y = 1 - x$.

(a) Find $P(X \geq -.5)$.
(b) Find $P(Y \geq .5)$.
(c) Find the marginal densities and expectations of X, Y.

Exercise 3.5 (Uniform Distribution in a Sphere). Suppose (X, Y, Z) has the density $f(x, y, z) = c$, if $x^2 + y^2 + z^2 \leq 1$.

(a) Find the constant c.
(b) Are any of X, Y, or Y, Z or X, Z pairwise independent?
(c) Find the marginal densities and expectations of X, Y, Z.

(d) Find the conditional expectation of X given $Y = y$, and the conditional expectation of X given $Z = z$.
(e) Find the conditional expectation of X given (both) $Y = y, Z = z$.
(f) Find the correlation between any pair, say X and Y.

Exercise 3.6. Suppose X, Y are independent $U[0, 1]$ variables. Find the conditional expectation $E(|X - Y| \, | \, Y = y)$.

Exercise 3.7 (Uniform in a Triangle). Suppose X, Y are uniformly distributed in the triangle $x, y \geq 0, x + y \leq 1$. Find the conditional expectation $E(|X - Y| \, | \, Y = y)$.

Exercise 3.8. Suppose X, Y, Z are independent $U[0, 1]$ variables. Find $P(|X - Y| > |Y - Z|)$.

Exercise 3.9 * (Iterated Expectation). Suppose X, Y are independent standard exponential variables. Find $E(X\sqrt{X + Y})$.

Exercise 3.10 (Expectation of a Quotient). Suppose X, Y are independent, and $X \sim Be(2, 2), Y \sim Be(3, 3)$. Find $E(\frac{X^2}{Y^2})$.

Exercise 3.11. Suppose X, Y, Z are three independent standard exponential variables. Find $P(X < 2Y < 3Z)$.

Exercise 3.12 * (Conceptual). Suppose $X \sim U[0, 1]$, and $Y = 2X$. What is the joint distribution of (X, Y)? Does the joint distribution have a density?

Exercise 3.13 * (Breaking a Stick). Suppose $X \sim U[0, 1]$, and given that $X = x, Y \sim U[0, x]$. Let $U = 1 - X, V = Y, W = X - Y$. Find the expectation of the maximum of U, V, W.
This amounts to breaking a stick, and then breaking the left piece again.

Exercise 3.14 (Iterated Expectation). Suppose $X_1 \sim U[0, 1]$, and for $n \geq 2, X_n$ given that $X_{n-1} = x$ is distributed as $U[0, x]$. What is $E(X_n)$, and its limit as $n \to \infty$?

Exercise 3.15 (An MLE). Let $X_1, \ldots, X_n \stackrel{iid}{\sim} \text{Poi}(\lambda), \lambda > 0$. Show that a unique MLE of λ exists unless each X_i is equal to zero, in which case an MLE does not exist.

Exercise 3.16 * (Nonunique MLE). Let $X_1, \ldots, X_n \stackrel{iid}{\sim} U[\theta - 1, \theta + 1], -\infty < \theta < \infty$. Show that the MLE of θ is not unique. Find all the MLEs.

Exercise 3.17 * (A Difficult to Find MLE). Let $X_1, \ldots, X_n \stackrel{iid}{\sim} C(\mu, 1), n \geq 2$. Show that the likelihood function is, in general, multimodal. Consider the following data values: $-10, 0, 2, 5, 14$. Plot the likelihood function and find the MLE of μ.

Exercises

Exercise 3.18. Let $X_1, \ldots, X_n \overset{iid}{\sim} N(\theta, \theta), \theta > 0$. Show that there is a unique MLE of θ, and find it.

Exercise 3.19 (MLE in a Genetics Problem). According to Mendel's law, the genotypes $aa, Aa,$ and AA in a population with genetic equilibrium with respect to a single gene having two alleles have proportions $f^2, 2f(1-f)$, and $(1-f)^2$ in the population. Suppose n individuals are sampled from the population and the number of observed individuals of each genotype are n_1, n_2, n_3, respectively. Find the MLE of f.

Exercise 3.20 *(MLE in a Discrete Parameter Problem).** Two independent proofreaders A and B are asked to read a manuscript containing N errors; $N \geq 0$ is unknown. n_1 errors are found by A alone, n_2 by B alone, and n_{12} by both. What is the MLE of N? State your assumptions.

Exercise 3.21 *(MLE for Double Exponential Case).** Let $X_1, \ldots, X_n \overset{iid}{\sim}$ DoubleExp$(\mu, 1)$. Show that the sample median is one MLE of μ; is it the only MLE?

Exercise 3.22 *(MLE of Common Mean).** Suppose $X_1, X_2, \ldots, X_m$ are iid $N(\mu, \sigma_1^2)$ and $Y_1, Y_2, \ldots, Y_n$ are iid $N(\mu, \sigma_2^2)$, and all $m+n$ observations are independent. Find the MLE of μ.

Exercise 3.23 *(MLE Under a Constraint).** Let $X_1, \ldots, X_n \overset{iid}{\sim} N(\mu, 1)$, where we know that $\mu \geq 0$. Show that there is a unique MLE of μ, and find it.

Hint: Think intuitively.

Exercise 3.24 *(MLE in the Gamma Case).** Let $X_1, \ldots, X_n \overset{iid}{\sim} G(\alpha, \lambda), \alpha > 0, \lambda > 0$. Show that there is a unique MLE of (α, λ), which is the only stationary point of the logarithm of the likelihood function. Compute it for the following simple dataset $(n = 8) : .5, 1, 1.4, 2, 1, 2.5, 1.5, 2$.

Exercise 3.25 (Bivariate Normal Probability). Suppose X, Y are jointly bivariate normal with zero means, unit standard deviations, and correlation ρ. Find all values of ρ for which $\frac{1}{4} \leq P(X > 0, Y > 0) \leq \frac{5}{12}$.

Exercise 3.26. Suppose X, Y are jointly bivariate normal with zero means, unit standard deviations, and correlation $\rho = .75$. Find $P(Y > 2 | X = 1)$.

Exercise 3.27. Suppose X, Y are jointly bivariate normal with general parameters. Characterize all constants a, b such that $X + Y$ and $aX + bY$ are independent.

Exercise 3.28 *(Probability of a Diamond).** Suppose X, Y, Z are independent $U[-1, 1]$ variables. Find the probability that $|X| + |Y| + |Z| \leq 1$.

Exercise 3.29 (Missing the Bus). A bus arrives at a random time between 9:00 AM and 9:15 AM at a stop. Tim will arrive at that stop at a random time between 9:00 AM and 9:15 AM, independently of the bus, and will wait for (at most) five minutes at the stop. Find the probability that Tim will meet the bus.

Exercise 3.30. Cathy and Jen plan to meet at a cafe and each will arrive at the cafe at a random time between 11:00 AM and 11:30 AM, independently of each other. Find the probability that the first to arrive has to wait between 5 and 10 minutes for the other to arrive.

Exercise 3.31 (Bivariate Normal Probability). Suppose the amounts of oil (in barrels) lifted on a given day from two wells are jointly bivariate normal, with means 150 and 200, and variances 100 and 25, and correlation .5. What is the probability that the total amount lifted is larger than 400 barrels on one given day? The probability that the amounts lifted from the two wells on one day differ by more than 50 barrels?

Exercise 3.32 *(Conceptual).** Suppose (X, Y) have a bivariate normal distribution with zero means, unit standard deviations, and correlation $\rho, -1 < \rho < 1$.
  What is the joint distribution of $(X + Y, X - Y, Y)$? Does this joint distribution have a density?

Exercise 3.33. Suppose $X \sim N(\mu, \sigma^2)$. Find the correlation between X and Y, where $Y = X^2$. Find all values of (μ, σ) for which the correlation is zero.

Exercise 3.34 *(Maximum Correlation).** Suppose (X, Y) has a general bivariate normal distribution with a positive correlation ρ. Show that among all functions $g(X), h(Y)$ with finite variances, the correlation between $g(X)$ and $h(Y)$ is maximized when $g(X) = X, h(Y) = Y$.

Exercise 3.35 (Bivariate Normal Calculation). Suppose $X \sim N(0, 1)$, and given $X = x, Y \sim N(x + 1, 1)$.

(a) What is the marginal distribution of Y?
(b) What is the correlation between X and Y?
(c) What is the conditional distribution of X given $Y = y$?

Exercise 3.36 *(Uniform Distribution in a Sphere).** Suppose X, Y, Z are uniformly distributed in the unit sphere. Find the mean and the variance of the distance of the point (X, Y, Z) from the origin.

Exercise 3.37 *(Uniform Distribution in a Sphere).** Suppose X, Y, Z are uniformly distributed in the unit sphere.

(a) Find the marginal density of (X, Y).
(b) Find the marginal density of X.

Exercise 3.38. Suppose X, Y, Z are independent exponentials with means $\lambda, 2\lambda, 3\lambda$. Find $P(X < Y < Z)$.

Exercise 3.39 *(Mean Residual Life).** Suppose $X \sim N(\mu, \sigma^2)$. Derive a formula for the mean residual life and investigate its monotonicity behavior with respect to each of σ, c, μ, each time holding the other two fixed.

Exercise 3.40 * **(Bivariate Normal Conditional Calculation).** Suppose the systolic blood pressure X and fasting blood sugar, Y are jointly distributed as a bivariate normal in some population with means 120, 105, standard deviations 10, 20, and correlation 0.7. Find the average fasting blood sugar of those with a systolic blood pressure greater than 140.

Exercise 3.41 * **(Buffon's Needle).** Suppose the plane is gridded by a series of parallel lines, drawn h units apart. A needle of length l is dropped at random on the plane. Let $p(l, h)$ be the probability that the needle intersects one of the parallel lines. Show that

(a) $p(l, h) = \frac{2l}{h\pi}$, if $l \leq h$;

(b) $p(l, h) = \frac{2l}{h\pi} - \frac{2}{h\pi}[\sqrt{l^2 - h^2} + h \arcsin(\frac{h}{l})] + 1$, if $l > h$.

References

Cox, D. and Wermuth, N. (1991). A simple approximation for bivariate and trivariate normal integrals, *Internat. Statist. Rev.*, 59, 263–269.

Genz, A. (1993). Comparison of methods for the computation of multivariate norwal probabilities, Computing Sciences and Statistics, 25, 400–405.

Kagan, A., Linnik, Y., and Rao, C. R. (1973). *Characterization Problems in Mathematical Statistics*, Wiley, New York.

Kamat, A. (1953). Incomplete and absolute moments of the multivariate normal distribution, with applications, *Biometrika*, 40, 20–34.

Mee, R. and Owen. D. (1983). A simple approximation for bivariate normal probability, *J. Qual. Tech.*, 15, 72–75.

Patel, J. and Read, C. (1996). *Handbook of the Normal Distribution*, Marcel Dekker, New York.

Plackett, R. (1954). A reduction formula for multivariate normal probabilities, *Biometrika*, 41, 351–360.

Tong, Y. (1990). *Multivariate Normal Distribution*, Springer-Verlag, New York.

Chapter 4
Advanced Distribution Theory

Studying distributions of functions of several random variables is of primary interest in probability and statistics. For example, the original variables $X_1, X_2, \ldots, X_n$ could be the inputs into some process or system, and we may be interested in the output, which is some suitable function of these input variables. Sums, products, and quotients are special functions that arise quite naturally in applications. These are discussed with a special emphasis in this chapter, although the general theory is also presented. Specifically, we present the classic theory of polar transformations and the Helmert transformation in arbitrary dimensions, and the development of the Dirichlet, t- and the F-distribution. The t- and the F-distribution arise in numerous problems in statistics, and the Dirichlet distribution has acquired an extremely special role in modeling and also in Bayesian statistics. In addition, these techniques and results are among the most sophisticated parts of distribution theory.

4.1 Convolutions and Examples

Definition 4.1. Let X, Y be independent random variables. The distribution of their sum $X + Y$ is called the *convolution* of X and Y.

Remark. Usually, we study convolutions of two continuous or two discrete random variables. But, in principle, one could be continuous and the other discrete.

Example 4.1. Suppose X, Y have a joint density function $f(x, y)$, and suppose we want to find the density of their sum, namely $X + Y$. Denote the conditional density of X given $Y = y$ by $f_{X|y}(x|y)$ and the conditional CDF, namely, $P(X \leq u|Y = y)$ by $F_{X|Y}(u)$. Then, by the iterated expectation formula,

$$P(X + Y \leq z) = E[I_{X+Y \leq z}]$$
$$= E_Y[E(I_{X+Y \leq z}|Y = y)] = E_Y[E(I_{X+y \leq z}|Y = y)]$$
$$= E_Y[P(X \leq z - y|Y = y)]$$
$$= E_Y[F_{X|Y}(z - y)] = \int_{-\infty}^{\infty} F_{X|Y}(z - y) f_Y(y) dy.$$

In particular, if X and Y are independent, then the conditional CDF $F_{X|Y}(u)$ will be the same as the marginal CDF $F_X(u)$ of X. In this case, the expression above simplifies to

$$P(X+Y \le z) = \int_{-\infty}^{\infty} F_X(z-y) f_Y(y) dy.$$

The density of $X+Y$ can be obtained by differentiating the CDF of $X+Y$:

$$\begin{aligned} f_{X+Y}(z) &= \frac{d}{dz} P(X+Y \le z) \\ &= \frac{d}{dz} \int_{-\infty}^{\infty} F_X(z-y) f_Y(y) dy \\ &= \int_{-\infty}^{\infty} \left[\frac{d}{dz} F_X(z-y) f_Y(y) \right] dy \end{aligned}$$

(assuming that the derivative can be carried inside the integral)

$$= \int_{-\infty}^{\infty} f_X(z-y) f_Y(y) dy.$$

Indeed, this is the general formula for the density of the sum of two real-valued independent continuous random variables.

Theorem 4.1. *Let X, Y be independent real-valued random variables with densities $f_X(x)$, $f_Y(y)$, respectively. Let $Z = X + Y$ be the sum of X and Y. Then, the density of the convolution is*

$$f_Z(z) = \int_{-\infty}^{\infty} f_X(z-y) f_Y(y) dy.$$

More generally, if X, Y are not necessarily independent, and have joint density $f(x, y)$, then $Z = X + Y$ has the density

$$f_Z(z) = \int_{-\infty}^{\infty} f_{X|Y}(z-y) f_Y(y) dy.$$

Definition 4.2. If X, Y are independent continuous random variables with a common density $f(x)$, then the density of the convolution is denoted as $f * f$. In general, if $X_1, X_2, \ldots, X_n$ are n independent continuous random variables with a common density $f(x)$, then the density of their sum $X_1 + X_2 + \cdots + X_n$ is called the *n-fold convolution of f* and is denoted by $f^{*(n)}$.

Example 4.2 (Sum of Exponentials). Suppose X, Y are independent $\text{Exp}(\lambda)$ variables, and we want to find the density of $Z = X + Y$. By the convolution formula, for $z > 0$,

4.1 Convolutions and Examples

$$f_Z(z) = \int_{-\infty}^{\infty} \frac{1}{\lambda} e^{-\frac{z-y}{\lambda}} I_{y<z} \frac{1}{\lambda} e^{-\frac{y}{\lambda}} I_{y>0} dy$$

$$= \frac{1}{\lambda^2} \int_0^z e^{-\frac{z}{\lambda}} dy$$

$$= \frac{z e^{-\frac{z}{\lambda}}}{\lambda^2},$$

which is the density of a Gamma distribution with parameters 2 and λ. Recall that we had proved this earlier in Chapter 1 by using mgfs.

Example 4.3 (Difference of Exponentials). Let U, V be independent standard exponentials. We want to find the density of $Z = U - V$. Writing $X = U$, and $Y = -V$, we notice that $Z = X + Y$, and X, Y are still independent. However, now Y is a *negative exponential*, and so has density $f_Y(y) = e^y I y < 0$. It is also important to note that Z can now take any real value, positive or negative. Substituting into the formula for the convolution density,

$$f_Z(z) = \int_{-\infty}^{\infty} e^{-(z-y)} (Iy < z) e^y (Iy < 0) dy.$$

Now, first consider $z > 0$. Then this last expression becomes

$$f_Z(z) = \int_{-\infty}^{0} e^{-(z-y)} e^y dy = e^{-z} \int_{-\infty}^{0} e^{2y} dy$$

$$= \frac{1}{2} e^{-z}.$$

On the other hand, for $z < 0$, the convolution formula becomes

$$f_Z(z) = \int_{-\infty}^{z} e^{-(z-y)} e^y dy = e^{-z} \int_{-\infty}^{z} e^{2y} dy$$

$$= e^{-z} \frac{1}{2} e^{2z} = \frac{1}{2} e^z.$$

Combining the two cases, we can write the single formula

$$f_Z(z) = \frac{1}{2} e^{-|z|}, \quad -\infty < z < \infty;$$

that is, if X, Y are independent standard exponentials, then the difference $X - Y$ has a standard double exponential density. This representation of the double exponential is often useful. Also note that although the standard exponential distribution is obviously not symmetric, the distribution of the difference of two independent exponentials is symmetric. This is a useful technique for symmetrizing a random variable.

Definition 4.3 (Symmetrization of a Random Variable). Let X_1, X_2 be independent random variables with a common distribution F. Then $X_s = X_1 - X_2$ is called the *symmetrization of F* or *symmetrization of X_1*.

If X_1 is a continuous random variable with density $f(x)$, then its symmetrization has the density

$$f_s(z) = \int_{-\infty}^{\infty} f(z+y) f(y) dy.$$

Example 4.4 (A Neat General Formula). Suppose X, Y are positive random variables with a joint density of the form $f(x, y) = g(x + y)$. What is the density of the convolution?

Note that now X, Y are in general not independent, because a joint density of the form $g(x+y)$ does not in general factorize into the product form necessary for independence. First, the conditional density

$$f_{X|Y}(x) = \frac{g(x+y)}{\int_0^\infty g(x+y)dx}$$

$$= \frac{g(x+y)}{\int_y^\infty g(x)dx} = \frac{g(x+y)}{\bar{G}(y)},$$

writing $\bar{G}(y)$ for $\int_y^\infty g(x)dx$. Also, the marginal density of Y is

$$f_Y(y) = \int_0^\infty g(x+y)dx = \int_y^\infty g(x)dx = \bar{G}(y).$$

Substituting into the general case formula for the density of a sum,

$$f_Z(z) = \int_0^z \frac{g(z)}{\bar{G}(y)} \bar{G}(y) dy = z g(z),$$

a very neat formula.

As an application, consider the example of (X, Y) being uniformly distributed in a triangle with the joint density $f(x, y) = 2, x, y \geq 0, x + y \leq 1$. Identifying the function g as $g(z) = 2 I_{0 \leq z \leq 1}$, we have, from our general formula above, that in this case $Z = X + Y$ has the density $f_Z(z) = 2z, 0 \leq z \leq 1$.

Example 4.5 (Sums of Cauchy Variables). Let X, Y be independent standard Cauchy random variables with the common density function $f(x) = \frac{1}{\pi(1+x^2)}$, $-\infty < x < \infty$. Then, the density of the convolution is

$$f_Z(z) = \int_{-\infty}^{\infty} f_X(z-y) f_Y(y) dy = \int_{-\infty}^{\infty} f(z-y) f(y) dy$$

$$= \frac{1}{\pi^2} \int_{-\infty}^{\infty} \frac{1}{(1+(z-y)^2)(1+y^2)} dy$$

$$= \frac{1}{\pi^2} \frac{2\pi}{4+z^2}$$

4.1 Convolutions and Examples

on a partial fraction expansion of

$$\frac{1}{(1+(z-y)^2)(1+y^2)} = \frac{2}{\pi(4+z^2)}.$$

Therefore, the density of $W = \frac{Z}{2} = \frac{X+Y}{2}$ would be $\frac{1}{\pi(1+w^2)}$, which is, remarkably, the same standard Cauchy density with which we had started.

By using characteristic functions, which we discuss in Chapter 8, it can be shown that if $X_1, X_2, \ldots, X_n$ are independent standard Cauchy variables, then for any $n \geq 2$, their average $\bar{X} = \frac{X_1+X_2+\cdots+X_n}{n}$ also has the standard Cauchy distribution.

Example 4.6 (Normal–Poisson Convolution). Here is an example of the convolution of one continuous and one discrete random variable. Let $X \sim N(0, 1)$ and $Y \sim \text{Poi}(\lambda)$. Then their sum $Z = X + Y$ is still continuous, and has the density

$$f_Z(z) = \sum_{y=0}^{\infty} \phi(z-y) \frac{e^{-\lambda} \lambda^y}{y!}.$$

More generally, if $X \sim N(0, \sigma^2)$, and $Y \sim \text{Poi}(\lambda)$. Then the density of the sum is

$$f_Z(z) = \frac{1}{\sigma} \sum_{y=0}^{\infty} \phi\left(\frac{z-y}{\sigma}\right) \frac{e^{-\lambda} \lambda^y}{y!}.$$

This is not expressible in terms of the elementary functions. However, it is interesting to plot the density. Figure 4.1 shows an unconventional density for the convolution with multiple local maxima and shoulders.

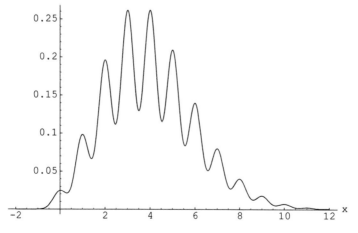

Fig. 4.1 Convolution of $N(0, .09)$ and $\text{Poi}(4)$

For purposes of summary and easy reference, we list some convolutions of common types below.

Distribution of Summands	Distribution of Sum
$X_i \sim \text{Bin}(n_i, p)$	$\text{Bin}(\sum n_i, p)$
$X_i \sim \text{Poi}(\lambda_i)$	$\text{Poi}(\sum \lambda_i)$
$X_i \sim \text{NB}(r_i, p)$	$\text{NB}(\sum r_i, p)$
$X_i \sim \text{Exp}(\lambda)$	$\text{Gamma}(n, \lambda)$
$X_i \sim N(\mu_i, \sigma_i^2)$	$N(\sum \mu_i, \sum \sigma_i^2)$
$X_i \sim C(\mu_i, \sigma_i^2)$	$C(\sum \mu_i, (\sum \sigma_i)^2)$
$X_i \sim U[a, b]$	See Chapter 1

4.2 Products and Quotients and the t- and F-Distribution

Suppose X, Y are two random variables. Then two other functions that arise naturally in many applications are the product XY, and the quotient $\frac{X}{Y}$. Following exactly the same technique as for convolutions, one can find the density of each of XY and $\frac{X}{Y}$. More precisely, one first finds the CDF by using the iterated expectation technique, exactly as we did for convolutions, and then differentiates the CDF to obtain the density. Here are the density formulas; they are extremely important and useful. They are proved in exactly the same way that the formula for the density of the convolution was obtained above; you would condition, and then take an iterated expectation. Therefore, the formal detail is omitted.

Theorem 4.2. *Let X, Y be continuous random variables with a joint density $f(x, y)$. Let $U = XY, V = \frac{X}{Y}$. Then the densities of U, V are given by*

$$f_U(u) = \int_{-\infty}^{\infty} \frac{1}{|x|} f\left(x, \frac{u}{x}\right) dx;$$

$$f_V(v) = \int_{-\infty}^{\infty} |y| f(vy, y) dy.$$

Example 4.7 (Product and Quotient of Uniforms). Suppose X, Y are independent $U[0, 1]$ random variables. Then, by the above theorem, the density of the product $U = XY$ is

$$f_U(u) = \int_{-\infty}^{\infty} \frac{1}{|x|} f\left(x, \frac{u}{x}\right) dx$$

$$= \int_u^1 \frac{1}{x} \times 1 dx = -\log u, \quad 0 < u < 1.$$

4.2 Products and Quotients and the t- and F-Distribution

Next, again by the above theorem, the density of the quotient $V = \frac{X}{Y}$ is

$$f_V(v) = \int_{-\infty}^{\infty} |y| f(vy, y) dy$$

$$= \int_0^{\min\{\frac{1}{v},1\}} y \, dy$$

$$= \frac{(\min\{\frac{1}{v}, 1\})^2}{2}, \quad 0 < v < \infty;$$

thus, the density of the quotient V is

$$f_V(v) = \frac{1}{2}, \quad \text{if } 0 < v \le 1;$$

$$= \frac{1}{2v^2}, \quad \text{if } v > 1.$$

The density of the quotient is plotted in Fig. 4.2; we see that it is continuous, but not differentiable at $v = 1$.

Example 4.8 (Ratio of Standard Normals). The distribution of the ratio of two independent standard normal variables is an interesting one; we show now that it is in fact a standard Cauchy distribution. Indeed, by applying the general formula, the density of the quotient $V = \frac{X}{Y}$ is

$$f_V(v) = \int_{-\infty}^{\infty} |y| f(vy, y) dy$$

$$= \frac{1}{2\pi} \int_{-\infty}^{\infty} |y| e^{-\frac{y^2}{2}(1+v^2)} dy$$

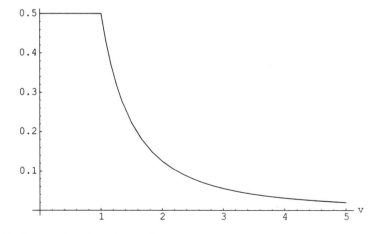

Fig. 4.2 Density of quotient of two uniforms

$$= \frac{1}{\pi} \int_0^\infty y e^{-\frac{y^2}{2}(1+v^2)} dy$$

$$= \frac{1}{\pi(1+v^2)}, \quad -\infty < v < \infty,$$

by making the substitution $t = \sqrt{1+v^2}\, y$ in the integral on the last line. This proves that the quotient has a standard Cauchy distribution.

It is important to note that zero means for the normal variables are essential for this result. If either X or Y has a nonzero mean, the quotient has a complicated distribution, and is definitely not Cauchy. The distribution is worked out in Hinkley (1969). It is also highly interesting that there are many other distributions F such that if X, Y are independent with the common distribution F, then the quotient $\frac{X}{Y}$ is distributed as a standard Cauchy. One example of such a distribution F is a continuous distribution with the density

$$f(x) = \frac{\sqrt{2}}{\pi} \frac{1}{1+x^4}, -\infty < x < \infty.$$

Example 4.9 (The F-Distribution). Let $X \sim G(\alpha, 1), Y \sim G(\beta, 1)$, and suppose X, Y are independent. The distribution of the ratio $R = \frac{X/\alpha}{Y/\beta}$ arises in statistics in many contexts and is called an *F-distribution*. We derive the explicit form of the density here.

First, we find the density of $\frac{X}{Y}$, from which the density of $R = \frac{X/\alpha}{Y/\beta}$ follows easily. Again, by applying the general formula for the density of a quotient, the density of the quotient $V = \frac{X}{Y}$ is

$$f_V(v) = \int_{-\infty}^\infty |y| f(vy, y) dy$$

$$= \frac{1}{\Gamma(\alpha)\Gamma(\beta)} \int_0^\infty y e^{-y(1+v)} (vy)^{\alpha-1} y^{\beta-1} dy$$

$$= \frac{1}{\Gamma(\alpha)\Gamma(\beta)} v^{\alpha-1} \int_0^\infty e^{-y(1+v)} y^{\alpha+\beta-1} dy$$

$$= \frac{1}{\Gamma(\alpha)\Gamma(\beta)} v^{\alpha-1} \frac{\Gamma(\alpha+\beta)}{(1+v)^{\alpha+\beta}}$$

$$= \frac{\Gamma(\alpha+\beta)}{\Gamma(\alpha)\Gamma(\beta)} \frac{v^{\alpha-1}}{(1+v)^{\alpha+\beta}}, \quad 0 < v < \infty.$$

To complete the example, notice now that $R = cV$, where $c = \frac{\beta}{\alpha}$. Therefore, the density of R is immediately obtained from the density of V. Indeed,

$$f_R(r) = \frac{1}{c} f_V\left(\frac{r}{c}\right),$$

4.2 Products and Quotients and the t- and F-Distribution

where f_V is the function we just derived above. If we simplify $f_R(r)$, we get the final expression

$$f_R(r) = \frac{\left(\frac{\beta}{\alpha}\right)^\beta r^{\alpha-1}}{B(\alpha,\beta)\left(r + \frac{\beta}{\alpha}\right)^{\alpha+\beta}}, \quad r > 0.$$

This is the F-density with parameters α, β; it is common in statistics to refer to 2α and 2β as the degrees of freedom of the distribution.

Example 4.10 (The Student t-Distribution). Once again, the t-distribution is one that arises frequently in statistics. Suppose, $X \sim N(0,1), Z \sim \chi_m^2$, and suppose X, Z are independent. Let $Y = \sqrt{\frac{Z}{m}}$. Then the distribution of the quotient $V = \frac{X}{Y}$ is called t-*distribution with m degrees of freedom.* We derive its density in this example.

Recall that Z has the density

$$\frac{e^{-z/2}z^{m/2-1}}{2^{m/2}\Gamma(\frac{m}{2})}, \quad z > 0.$$

Therefore, $Y = \sqrt{\frac{Z}{m}}$ has the density

$$f_Y(y) = \frac{m^{m/2}e^{-my^2/2}y^{m-1}}{2^{m/2-1}\Gamma(\frac{m}{2})}, \quad y > 0.$$

Because, by hypothesis, X and Z are independent, it follows that X and Y are also independent, and so their joint density $f(x, y)$ is just the product of the marginal densities of X and Y.

Once again, by applying our general formula for the density of a quotient,

$$f_V(v) = \int_{-\infty}^{\infty} |y| f(vy, y) dy$$

$$= \frac{m^{m/2}}{\sqrt{2\pi}\, 2^{m/2-1}\Gamma(\frac{m}{2})} \int_0^\infty y e^{-v^2 y^2/2} e^{-my^2/2} y^{m-1} dy$$

$$= \frac{m^{m/2}}{\sqrt{2\pi}\, 2^{m/2-1}\Gamma(\frac{m}{2})} \int_0^\infty e^{-(v^2+m)y^2/2} y^m dy$$

$$= \frac{m^{m/2}}{\sqrt{2\pi}\, 2^{m/2-1}\Gamma(\frac{m}{2})} \times \frac{\Gamma(\frac{m+1}{2}) 2^{(m-1)/2}}{(m+v^2)^{(m+1)/2}}$$

$$= \frac{m^{m/2}\Gamma(\frac{m+1}{2})}{\sqrt{\pi}\Gamma(\frac{m}{2})} \frac{1}{(m+v^2)^{(m+1)/2}}$$

$$= \frac{\Gamma(\frac{m+1}{2})}{\sqrt{m\pi}\Gamma(\frac{m}{2})(1+\frac{v^2}{m})^{(m+1)/2}}, \quad -\infty < v < \infty.$$

This is the density of the *Student t-distribution with m degrees of freedom*.

Note that when the degree of freedom $m = 1$, this becomes just the standard Cauchy density. The t-distribution was first derived in 1908 by William Gossett under the pseudonym *Student*. The distribution was later named the *Student t-distribution* by Ronald Fisher.

The t density, just like the standard normal, is symmetric and unimodal around zero, although with tails much heavier than those of the standard normal for small values of m. However, as $m \to \infty$, the density converges pointwise to the standard normal density, and then the t and the standard normal density look almost the same. We give a plot of the t density for a few degrees of freedom in Fig. 4.3, and of the standard normal density for a visual comparison.

Example 4.11 (An Interesting Gaussian Factorization). We exhibit independent random variables X, Y in this example such that XY has a standard normal distribution. Note that if we allow Y to be a constant random variable, then we can always write such a factorization. After all, we can take X to be standard normal, and Y to be 1! So, we will exhibit *nonconstant* X, Y such that they are independent, and XY has a standard normal distribution.

For this, let X have the density $xe^{-x^2/2}, x > 0$, and let Y have the density $\frac{1}{\pi\sqrt{1-y^2}}, -1 < y < 1$. Then, by our general formula for the density of a product, the product $U = XY$ has the density

$$f_U(u) = \int_{-\infty}^{\infty} \frac{1}{|x|} f\left(x, \frac{u}{x}\right) dx$$

$$= \frac{1}{\pi} \int_{|u|}^{\infty} \frac{1}{x} xe^{-x^2/2} \frac{1}{\sqrt{1-\frac{u^2}{x^2}}} dx$$

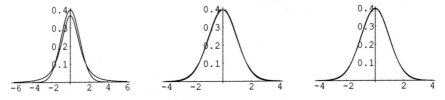

Fig. 4.3 t Density for m = 3, 20, 30 degrees of freedom with N(0,1) density superimposed

$$= \frac{1}{\pi} \int_{|u|}^{\infty} \frac{xe^{-x^2/2}}{\sqrt{x^2 - u^2}} dx$$

$$= \frac{1}{\pi} \sqrt{\frac{\pi}{2}} e^{-u^2/2} = \frac{1}{\sqrt{2\pi}} e^{-u^2/2},$$

where the final integral is obtained by the substitution $x^2 = u^2 + z^2$.

4.3 Transformations

The simple technique that we used in the previous section to derive the density of a sum or a product does not extend to functions of a more complex nature. Consider the simple case of just two continuous variables X, Y with some joint density $f(x, y)$, and suppose we want to find the density of some function $U = g(X, Y)$. Then, the general technique is to pair up U with another function $V = h(X, Y)$, and first obtain the joint CDF of (U, V) from the joint CDF of (X, Y). The pairing up has to be done carefully: only some judicious choices of V will work in a given example. Having found the joint CDF of (U, V), by differentiation one finds the joint density of (U, V), and then finally integrates v out to obtain the density of just U. Fortunately, this agenda does work out, because the transformation from (X, Y) to (U, V) can be treated as just a change of variable in manipulation with double integrals, and calculus tells us how to find double integrals by making suitable changes of variables (i.e., substitutions). Indeed, the method works out for any number of jointly distributed variables, $X_1, X_2, \ldots, X_n$, and a function $U = g(X_1, X_2, \ldots, X_n)$, and the reason it works out is that the whole method is just a change of variables in manipulating a multivariate integral.

Here is the theorem on density of a multivariate transformation, a major theorem in multivariate distribution theory. It is really nothing but the change of variable theorem of multivariate calculus. After all, probabilities in the continuous case are integrals, and an integral can be evaluated by changing variables to a new set of coordinates. If we do that, then we have to put in the Jacobian term coming from making the change of variable. Translated into densities, the theorem is the following.

Theorem 4.3 (Multivariate Jacobian Formula). *Let* $\mathbf{X} = (X_1, X_2, \ldots, X_n)$ *have the joint density function* $f(x_1, x_2, \ldots, x_n)$, *such that there is an open set* $S \subseteq \mathcal{R}^n$ *with* $P(\mathbf{X} \in S) = 1$. *Suppose* $u_i = g_i(x_1, x_2, \ldots, x_n), 1 \leq i \leq n$ *are* n *real-valued functions of* $x_1, x_2, \ldots, x_n$ *such that*

(a) $(x_1, x_2, \ldots, x_n) \to (g_1(x_1, x_2, \ldots, x_n), \ldots, g_n(x_1, x_2, \ldots, x_n))$ *is a one-to-one function of* $(x_1, x_2, \ldots, x_n)$ *on S with range space T.*
(b) *The inverse functions* $x_i = h_i(u_1, u_2, \ldots, u_n), 1 \leq i \leq n$, *are continuously differentiable on T with respect to each u_j.*

(c) The Jacobian determinant

$$J = \begin{vmatrix} \frac{\partial x_1}{\partial u_1} & \frac{\partial x_1}{\partial u_2} & \cdots & \frac{\partial x_1}{\partial u_n} \\ \frac{\partial x_2}{\partial u_1} & \frac{\partial x_2}{\partial u_2} & \cdots & \frac{\partial x_2}{\partial u_n} \\ \vdots & \vdots & \vdots & \vdots \\ \frac{\partial x_n}{\partial u_1} & \frac{\partial x_n}{\partial u_2} & \cdots & \frac{\partial x_n}{\partial u_n} \end{vmatrix}$$

is nonzero.

Then the joint density of $(U_1, U_2, \ldots, U_n)$ is given by

$$f_{U_1,U_2,\ldots,U_n}(u_1, u_2, \ldots, u_n) = f(h_1(u_1, u_2, \ldots, u_n), \ldots, h_n(u_1, u_2, \ldots, u_n)) \times |J|,$$

where $|J|$ denotes the absolute value of the Jacobian determinant J, and the notation f on the right-hand side means the original joint density of $(X_1, X_2, \ldots, X_n)$.

4.4 Applications of Jacobian Formula

We now show a number of examples of its applications to finding the density of interesting transformations. We emphasize that quite often only one of the functions $u_i = g_i(x_1, x_2, \ldots, x_n)$ is provided, whose density function we want to find. But unless that function is a really simple one, its density cannot be found directly without invoking the Jacobian theorem given here. It is necessary to make up the remaining $(n-1)$ functions, and then obtain their joint density by using this Jacobian theorem. Finally, one would integrate out all these other coordinates to get the density function of just u_i. The other $(n-1)$ functions need to be found judiciously.

Example 4.12 (A Relation Between Exponential and Uniform). Let X, Y be independent standard exponentials, and define $U = \frac{X}{X+Y}$. We want to find the density of U. We have to pair it up with another function V in order to use the Jacobian theorem. We choose $V = X+Y$. We have here a one-to-one function for $x > 0, y > 0$. Indeed, the inverse functions are

$$x = x(u, v) = uv; \quad y = y(u, v) = v - uv = v(1-u).$$

The partial derivatives of the inverse functions are

$$\frac{\partial x}{\partial u} = v; \quad \frac{\partial x}{\partial v} = u; \quad \frac{\partial y}{\partial u} = -v; \quad \frac{\partial y}{\partial v} = 1 - u.$$

4.4 Applications of Jacobian Formula

Thus, the Jacobian determinant equals $J = v(1 - u) + uv = v$. By invoking the Jacobian theorem, the joint density of U, V is

$$f_{U,V}(u, v) = e^{-uv}e^{-v(1-u)}|v| = ve^{-v},$$

$0 < u < 1, v > 0$.

Thus, the joint density of U, V has factorized into a product form on a rectangle; the marginals are

$$f_U(u) = 1, \quad 0 < u < 1; \quad f_V(v) = ve^{-v}, \quad v > 0,$$

and the rectangle being $(0, 1) \times (0, \infty)$. Therefore, we have proved that if X, Y are independent standard exponentials, then $\frac{X}{X+Y}$ and $X + Y$ are independent, and they are, respectively, uniform and a Gamma. Of course, we already knew that $X + Y \sim G(2, 1)$ from our mgf proof in Chapter 1. In Chapter 18 we show that this result can also be proved by using Basu's theorem.

Example 4.13 (A Relation Between Gamma and Beta). The previous example generalizes in a nice way. Let X, Y be independent variables, distributed respectively as $G(\alpha, 1), G(\beta, 1)$. Let again, $U = \frac{X}{X+Y}, V = X + Y$. Then, from our previous example, the Jacobian determinant is still $J = v$. Therefore, the joint density of U, V is

$$f_{U,V}(u, v) = \frac{1}{\Gamma(\alpha)\Gamma(\beta)} e^{-v}(uv)^{\alpha-1}(v(1-u))^{\beta-1}v$$

$$= \frac{1}{\Gamma(\alpha)\Gamma(\beta)} u^{\alpha-1}(1-u)^{\beta-1} e^{-v} v^{\alpha+\beta-1}, \quad 0 < u < 1, v > 0.$$

Once again, we have factorized the joint density of U and V as the product of the marginal densities, with (U, V) varying in the rectangle $(0, 1) \times (0, \infty)$, the marginal densities being

$$f_V(v) = \frac{e^{-v}v^{\alpha+\beta-1}}{\Gamma(\alpha+\beta)}, \quad v > 0,$$

$$f_U(u) = \frac{\Gamma(\alpha+\beta)}{\Gamma(\alpha)\Gamma(\beta)} u^{\alpha-1}(1-u)^{\beta-1}, \quad 0 < u < 1.$$

That is, if X, Y are independent Gamma variables, then $\frac{X}{X+Y}$ and $X + Y$ are independent, and they are respectively distributed as a Beta and a Gamma. Of course, we already knew from an mgf argument that $X + Y$ is a Gamma.

This relationship between the Gamma and the Beta distribution is useful in simulating values from a Beta distribution.

4.5 Polar Coordinates in Two Dimensions

Example 4.14 (Transformation to Polar Coordinates). We have already worked out a few examples where we transformed two variables to their polar coordinates, in order to calculate expectations of suitable functions, when the variables have a spherically symmetric density. We now use a transformation to polar coordinates to do distributional calculations. In any spherically symmetric situation, transformation to polar coordinates is a technically useful device, and gets the answers out quickly for many problems.

Let (X, Y) have a spherically symmetric joint density given by $f(x, y) = g(\sqrt{x^2 + y^2})$. Consider the polar transformation $r = \sqrt{X^2 + Y^2}, \theta = \arctan \frac{Y}{X}$. This is a one-to-one transformation, with the inverse functions given by

$$x = r \cos \theta, \quad y = r \sin \theta.$$

The partial derivatives of the inverse functions are

$$\frac{\partial x}{\partial r} = \cos \theta, \quad \frac{\partial x}{\partial \theta} = -r \sin \theta, \quad \frac{\partial y}{\partial r} = \sin \theta, \quad \frac{\partial y}{\partial \theta} = r \cos \theta.$$

Therefore, the Jacobian determinant is $J = r \cos^2 \theta + r \sin^2 \theta = r$. By the Jacobian theorem, the density of (r, θ) is

$$f_{r,\theta}(r, \theta) = r g(r),$$

with r, θ belonging to a suitable rectangle, which depends on the exact set of values (x, y) on which the original joint density $f(x, y)$ is strictly positive. But, in any case, we have established that the joint density of (r, θ) factorizes into the product form on a rectangle, and so in any spherically symmetric situation, the polar coordinates r and θ are independent, a very convenient fact. Always, in a spherically symmetric case, r will have the density $crg(r)$ on some interval and for some suitable normalizing constant c, and θ will have a uniform density on some interval.

Now consider three specific choices of the original density function. First consider the uniform case:

$$f(x, y) = \frac{1}{\pi}, \quad 0 < x^2 + y^2 < 1.$$

Then $g(r) = \frac{1}{\pi}, 0 < r < 1$. So in this case, r has the density $2r, 0 < r < 1$, and θ has the uniform density $\frac{1}{2\pi}, -\pi < \theta < \pi$.

Next consider the case of two independent standard normals. Indeed, in this case, the joint density is spherically symmetric, namely,

$$f(x, y) = \frac{1}{2\pi} e^{-(x^2+y^2)/2}, \quad -\infty < x, y < \infty.$$

4.5 Polar Coordinates in Two Dimensions

Thus, $g(r) = \frac{1}{2\pi}e^{-r^2/2}, r > 0$. Therefore, in this case r has the *Weibull density* $re^{-r^2/2}, r > 0$, and θ, again is uniform on $(-\pi, \pi)$.

Finally, consider the case of two independent *folded standard normals*, that is, each of X, Y has the density $\sqrt{\frac{2}{\pi}}e^{-x^2/2}, x > 0$. In this case, r varies on $(0, \infty)$, but θ varies on $(0, \frac{\pi}{2})$. Thus, r and θ are still independent, but this time, θ is uniform on $(0, \frac{\pi}{2})$, whereas r still has the same Weibull density $re^{-r^2/2}, r > 0$.

Example 4.15 (Usefulness of the Polar Transformation). Suppose (X, Y) are jointly uniform in the unit circle. We use the joint density of (r, θ) to find the answers to a number of questions.

First, by using the polar transformation,

$$E(X + Y) = E[r(\cos\theta + \sin\theta)]$$
$$= E(r)E(\cos\theta + \sin\theta).$$

Now, $E(r) < \infty$, and

$$E(\cos\theta + \sin\theta) = \frac{1}{2\pi}\int_{-\pi}^{\pi}(\cos\theta + \sin\theta)$$
$$= \frac{1}{2\pi}(0 + 0) = 0.$$

Therefore, $E(X + Y) = 0$. It should be noted that actually each of X, Y has mean zero in this case, and so we could have proved that $E(X + Y) = 0$ directly too.

Next, suppose we want to find the probability that (X, Y) lies in the intersection of the spherical shell $\frac{1}{4} \leq \sqrt{X^2 + Y^2} \leq \frac{3}{4}$ and the cone $X, Y > 0, \frac{1}{\sqrt{3}} \leq \frac{X}{Y} \leq \sqrt{3}$. This looks like a hard problem! But polar coordinates will save us. Transforming to the polar coordinates, this probability is

$$P\left(\frac{1}{4} \leq r \leq \frac{3}{4}, \frac{\pi}{6} \leq \theta \leq \frac{\pi}{3}\right)$$
$$= P\left(\frac{1}{4} \leq r \leq \frac{3}{4}\right)P\left(\frac{\pi}{6} \leq \theta \leq \frac{\pi}{3}\right)$$
$$= \int_{\frac{1}{4}}^{\frac{3}{4}} 2r\,dr \times \frac{1}{12} = \frac{1}{24}.$$

It would have been a much more tedious calculation to do this using the original rectangular coordinates.

Example 4.16 (Product of n Uniforms). Let $X_1, X_2, \ldots, X_n$ be independent $U[0, 1]$ variables, and suppose we want to find the density of the product $U = U_n = \prod_{i=1}^{n} X_i$. According to our general discussion, we have to choose $n - 1$ other functions, and then apply the Jacobian theorem. Define

$$u_1 = x_1, \quad u_2 = x_1x_2, \quad u_3 = x_1x_2x_3, \ldots, \quad u_n = x_1x_2,\ldots,x_n.$$

This is a one-to-one transformation, and the inverse functions are $x_i = \frac{u_i}{u_{i-1}}, 2 \leq i \leq n; x_1 = u_1$. Thus, the Jacobian matrix of the partial derivatives is lower triangular, and therefore the Jacobian determinant equals the product of the diagonal elements

$$J = \prod_{i=1}^{n} \frac{\partial x_i}{\partial u_i} = \frac{1}{\prod_{i=1}^{n-1} u_i}.$$

Now applying the Jacobian density theorem, the joint density of $U_1, U_2, \ldots, U_n$ is

$$f_{U_1, U_2, \ldots, U_n}(u_1, u_2, \ldots, u_n) = \frac{1}{\prod_{i=1}^{n-1} u_i},$$

$0 < u_n < u_{n-1} < \cdots < u_1 < 1$.

On integrating out $u_1, u_2, \ldots, u_{n-1}$, we get the density of U_n:

$$f_{U_n}(u) = \int_u^1 \int_{u_{n-1}}^1 \cdots \int_{u_2}^1 \frac{1}{u_1 u_2 \cdots u_{n-1}} du_1 du_2 \cdots du_{n-2} du_{n-1}$$

$$= \frac{|(\log u)^{n-1}|}{(n-1)!},$$

$0 < u < 1$. This example illustrates that applying the Jacobian theorem needs careful manipulation with multiple integrals, and skills in using the Jacobian technique are very important in deriving distributions of functions of many variables.

4.6 * n-Dimensional Polar and Helmert's Transformation

We saw evidence of practical advantages of transforming to polar coordinates in two dimensions in the previous section. As a matter of fact, as was remarked in Example 4.14, in any spherically symmetric situation in any number of dimensions, transformation to the *n-dimensional polar coordinates* is a standard technical device. The transformation from rectangular to the polar coordinates often greatly simplifies the algebraic complexity of the calculations. We first present the *n*-dimensional polar transformation in this section.

4.6.1 *Efficient Spherical Calculations with Polar Coordinates*

Definition 4.4. For $n \geq 3$, the *n*-dimensional polar transformation is a one-to-one mapping from $\mathcal{R}^n \to [0, \infty) \times \prod_{i=1}^{n-2} \Theta_i \times \Theta_{n-1}$, where $\Theta_{n-1} = [0, 2\pi]$, and for $i \leq n-2, \Theta_i = [0, \pi]$, with the mapping defined by

$$x_1 = \rho \cos \theta_1,$$
$$x_2 = \rho \sin \theta_1 \cos \theta_2,$$

4.6 n-Dimensional Polar and Helmert's Transformation

$$x_3 = \rho \sin\theta_1 \sin\theta_2 \cos\theta_3,$$
$$\vdots$$
$$x_{n-1} = \rho \sin\theta_1 \sin\theta_2 \cdots \sin\theta_{n-2} \cos\theta_{n-1},$$
$$x_n = \rho \sin\theta_1 \sin\theta_2 \cdots \sin\theta_{n-2} \sin\theta_{n-1}.$$

The transformation has the useful property that $x_1^2 + x_2^2 + \cdots + x_n^2 = \rho^2 \ \forall\ (x_1, x_2, \ldots, x_n) \in \mathcal{R}^n$, that is, ρ is the length of the vector $\mathbf{x} = (x_1, x_2, \ldots, x_n)$. The Jacobian determinant of the transformation equals

$$J = \rho^{n-1} \sin^{n-2}\theta_1 \sin^{n-3}\theta_2 \cdots \sin\theta_{n-2}.$$

Consequently, by the Jacobian density theorem, we have the following result.

Theorem 4.4 (Joint Density of Polar Coordinates). *Let $X_1, X_2, \ldots, X_n$ be n continuous random variables with a joint density $f(x_1, x_2, \ldots, x_n)$. Then the joint density of $(\rho, \theta_1, \theta_2, \ldots, \theta_{n-1})$ is given by*

$$p(\rho, \theta_1, \theta_2, \ldots, \theta_{n-1}) = \rho^{n-1} f(x_1, x_2, \ldots, x_n) |\sin^{n-2}\theta_1 \sin^{n-3}\theta_2 \cdots \sin\theta_{n-2}|,$$

where in the right side, one writes for $x_1, x_2, \ldots, x_n$, their defining expressions in terms of $\rho, \theta_1, \theta_2, \ldots, \theta_{n-1}$, as provided above.

In particular, if $X_1, X_2, \ldots, X_n$ have a spherically symmetric joint density

$$f(x_1, x_2, \ldots, x_n) = g\left(\sqrt{x_1^2 + +x_2^2 + \cdots + x_n^2}\right)$$

for some function g, then the joint density of $(\rho, \theta_1, \theta_2, \ldots, \theta_{n-1})$ equals

$$p(\rho, \theta_1, \theta_2, \ldots, \theta_{n-1}) = \rho^{n-1} g(\rho) |\sin^{n-2}\theta_1 \sin^{n-3}\theta_2 \cdots \sin\theta_{n-2}|,$$

and ρ is distributed independently of the angles $(\theta_1, \theta_2, \ldots, \theta_{n-1})$.

Example 4.17 (Rederiving the Chi-Square Distribution). Suppose that $X_1, X_2, \ldots, X_n$ are independent standard normals, so that their joint density is

$$f(x_1, x_2, \ldots, x_n) = \frac{1}{(2\pi)^{n/2}} e^{-\frac{1}{2}\sum_{i=1}^{n} x_i^2}, \quad -\infty < x_i < \infty, i = 1, 2, \ldots, n.$$

Thus, the joint density is spherically symmetric with

$$f(x_1, x_2, \ldots, x_n) = g\left(\sqrt{x_1^2 + +x_2^2 + \cdots + x_n^2}\right),$$

where

$$g(\rho) = \frac{1}{(2\pi)^{n/2}} e^{-\frac{\rho^2}{2}}.$$

Therefore, from our general theorem above, $\rho = \sqrt{\sum_{i=1}^{n} X_i^2}$ has the density $c\rho^{n-1} e^{-\frac{\rho^2}{2}}$ for some normalizing constant c. Making the transformation $W = \rho^2$, we get from the general formula for the density of a monotone transformation in one dimension (see Chapter 1) that W has the density

$$f_W(w) = k e^{-w/2} w^{n/2 - 1}, \quad w > 0.$$

It follows that $W = \rho^2 = \sum_{i=1}^{n} X_i^2$ has a χ_n^2 density, because the constant k must necessarily be $\frac{1}{2^{n/2} \Gamma \frac{n}{2}}$, which makes $f_W(w)$ exactly equal to the χ_n^2 density. Recall that this was previously proved by using the mgf technique. Now we have a polar transformation proof of it.

Example 4.18 (Curse of Dimensionality). Suppose $X_1, X_2, \ldots, X_n$ are jointly uniform in the n-dimensional unit sphere B_n, with the density $f(x_1, x_2, \ldots, x_n) = \frac{1}{\text{Vol}(B_n)}$, where

$$\text{Vol}(B_n) = \frac{\pi^{\frac{n}{2}}}{\Gamma(\frac{n}{2} + 1)}$$

is the volume of B_n. Therefore, $f(x_1, x_2, \ldots, x_n)$ is spherically symmetric with $f(x_1, x_2, \ldots, x_n) = g\left(\sqrt{x_1^2 + +x_2^2 + \cdots + x_n^2}\right)$, where

$$g(\rho) = \frac{\Gamma(\frac{n}{2} + 1)}{\pi^{\frac{n}{2}}}, \quad 0 < \rho < 1.$$

Hence, by our general theorem above, ρ has the density $c\rho^{n-1}$ for some normalizing constant c. The normalizing constant is easily evaluated:

$$1 = \int_0^1 c\rho^{n-1} d\rho = \frac{c}{n}$$
$$\Rightarrow c = n.$$

Thus, the length of an n-dimensional vector picked at random from the unit sphere has the density $n\rho^{n-1}, 0 < \rho < 1$. As a consequence, the expected length of an n-dimensional vector picked at random from the unit sphere equals

$$E(\rho) = n \int_0^1 \rho^n d\rho = \frac{n}{n+1},$$

which is very close to one for large n. So, one can expect that a point chosen at random from a high-dimensional sphere would be very close to the boundary, rather than the center. Once again, we see the curse of dimensionality in action.

Transformation to polar coordinates also results in some striking formulas and properties for general spherically symmetric distributions. They are collected together in the following theorem. We do not prove this theorem in the text, as each part only requires a transformation to the n-dimensional polar coordinates, and then straightforward calculations.

Theorem 4.5 (**General Spherically Symmetric Facts**). *Let* $\mathbf{X} = (X_1, X_2, \ldots, X_n)$ *have a spherically symmetric joint density* $f(x_1, x_2, \ldots, x_n) = g\left(\sqrt{x_1^2 + \cdots + x_n^2}\right)$. *Then,*

(a) *For any* $m < n$, *the distribution of* $(X_1, X_2, \ldots, X_m)$ *is also spherically symmetric.*

(b) $\rho = \|\mathbf{X}\|$ *has the density* $c\rho^{n-1}g(\rho)$, *where* c *is the normalizing constant* $\frac{1}{\int_0^\infty \rho^{n-1}g(\rho)d\rho}$.

(c) *Let* $\mathbf{U} = \frac{\mathbf{X}}{\|\mathbf{X}\|}$, *and* $\rho = \|\mathbf{X}\|$. *Then* $\mathbf{U}$ *and* ρ *are independent, and* $\mathbf{U}$ *is distributed uniformly on the boundary of the n-dimensional unit sphere.*

   Conversely, if an n-dimensional random vector $\mathbf{Z}$ *can be represented as* $\mathbf{Z} = \mathbf{U}\rho$, *where* $\mathbf{U}$ *is distributed uniformly on the boundary of the n-dimensional unit sphere,* ρ *is some nonnegative random variable, and* $\rho, \mathbf{U}$ *are independent, then* $\mathbf{Z}$ *has a spherically symmetric distribution.*

(d) *For any unit vector* $\mathbf{c}$, $c_1 X_1 + \cdots + c_n X_n$ *has the same distribution as* X_1.

(e) *If* $E(|X_i|) < \infty$, *then for any vector* $\mathbf{c}$, $E(X_i \mid \sum_{i=1}^n c_i X_i = t) = \frac{c_i}{\sum_{i=1}^n c_i^2} t$.

(f) *If* $\mathbf{X}$ *is uniformly distributed on the boundary of the n-dimensional unit sphere, then a lower-dimensional projection* $(X_1, X_2, \ldots, X_k)(k < n)$ *has the density*

$$f_k(x_1, x_2, \ldots, x_k) = c\left(1 - \sum_{i=1}^k x_i^2\right)^{(n-k)/2 - 1}, \quad \sum_{i=1}^k x_i^2 < 1,$$

where the normalizing constant c *equals*

$$c = \frac{\Gamma\left(\frac{n}{2}\right)}{\Gamma\left(\frac{k}{2}\right)\Gamma\left(\frac{n-k}{2}\right)};$$

in particular, if $n = 3$, *then each* $|X_i| \sim U[0, 1]$, *and each* $X_i \sim U[-1, 1]$.

4.6.2 Independence of Mean and Variance in Normal Case

Another transformation of technical use in spherically symmetric problems is *Helmert's orthogonal transformation*. It transforms an n-dimensional vector to another n-dimensional vector by making an orthogonal transformation. In a spherically symmetric situation, the orthogonal transformation will not affect the joint distribution. At the same time, the transformation may unveil structures in the problem that were not initially apparent. Specifically, when $X_1, X_2, \ldots, X_n$ are independent standard normals, the joint density of $X_1, X_2, \ldots, X_n$ is spherically symmetric. And Helmert's transformation leads to a number of very important results in this case. Although quicker proofs of some of these results for the normal case are now available, nevertheless the utility of Helmert's transformation in spherically symmetric situations makes it an important tool. We first need to recall a few definitions and facts from linear algebra.

Definition 4.5. An $n \times n$ real matrix P is called orthogonal if $PP' = P'P = I_n$, where I_n is the $n \times n$ identity matrix.

Any orthogonal matrix P has the property that it leaves lengths unchanged, that is, if $\mathbf{x}$ is an n-dimensional vector, then $\mathbf{x}$ and $P\mathbf{x}$ have the same length: $||\mathbf{x}|| = ||P\mathbf{x}||$. Furthermore, the determinant of any orthogonal matrix must be $\pm 1 (|P| = \pm 1)$.

Definition 4.6 (Helmert's Transformation). Let $\mathbf{X} = (X_1, X_2, \ldots, X_n)$ be an n-dimensional random vector. The Helmert transformation of $\mathbf{X}$ is the orthogonal transformation $\mathbf{Y} = P\mathbf{X}$, where P is the $n \times n$ orthogonal matrix

$$P = \begin{pmatrix} \frac{1}{\sqrt{n}} & \frac{1}{\sqrt{n}} & \frac{1}{\sqrt{n}} & \cdots & \frac{1}{\sqrt{n}} \\ -\frac{1}{\sqrt{2}} & \frac{1}{\sqrt{2}} & 0 & \cdots & 0 \\ \frac{1}{\sqrt{6}} & \frac{1}{\sqrt{6}} & -\frac{2}{\sqrt{6}} & \cdots & 0 \\ & & \vdots & & \\ \frac{1}{\sqrt{n(n-1)}} & \frac{1}{\sqrt{n(n-1)}} & \frac{1}{\sqrt{n(n-1)}} & \cdots & -\frac{n-1}{\sqrt{n(n-1)}} \end{pmatrix}$$

Verbally, in the first row, every element is $\frac{1}{\sqrt{n}}$, and in the subsequent rows, say the ith row, every element after the diagonal element in that row is zero.

Two important properties of the Helmert transformation are the following.

Proposition. *For all $\mathbf{X}$ and $\mathbf{Y} = P\mathbf{X}$,*

$$\sum_{i=1}^{n} Y_i^2 = \sum_{i=1}^{n} X_i^2; \sum_{i=2}^{n} Y_i^2 = \sum_{i=1}^{n} (X_i - \bar{X})^2,$$

where $\bar{X} = \frac{\sum_{i=1}^{n} X_i}{n}$

Proof. P is an orthogonal matrix, thus $\sum_{i=1}^{n} Y_i^2 = \sum_{i=1}^{n} X_i^2$. Also, $\sum_{i=2}^{n} Y_i^2 = \sum_{i=1}^{n} Y_i^2 - Y_1^2 = \sum_{i=1}^{n} X_i^2 - n\bar{X}^2$, by definition of Y_1, because the first row of P has all entries equal to $\frac{1}{\sqrt{n}}$.

These two properties lead to the following two important results. □

Theorem 4.6 (Independence of Mean and Variance in Normal Case). *Suppose $X_1, X_2, \ldots, X_n$ are independent $N(\mu, \sigma^2)$ variables. Then $\bar{X}$ and $\sum_{i=1}^{n}(X_i - \bar{X})^2$ are independently distributed.*

Proof. First consider the case $\mu = 0, \sigma = 1$. The Jacobian determinant of the transformation $\mathbf{x} \to \mathbf{y} = P\mathbf{x}$ is $|P| = \pm 1$. Therefore, by the Jacobian density theorem, the joint density of $Y_1, \ldots, Y_n$ is

$$f_{Y_1,\ldots,Y_n}(y_1,\ldots,y_n) = f_{X_1,\ldots,X_n}(x_1,\ldots,x_n)|J|$$
$$= \frac{1}{(2\pi)^{n/2}} e^{-\frac{\sum_{i=1}^n x_i^2}{2}} = \frac{1}{(2\pi)^{n/2}} e^{-\frac{\sum_{i=1}^n y_i^2}{2}},$$

which proves that $Y_1,\ldots,Y_n$ too are independent standard normal variables.

As a consequence, $(Y_2,\ldots,Y_n)$ is independent of Y_1, and hence $\sum_{i=2}^n Y_i^2$ is independent of Y_1, which means that $\sum_{i=1}^n (X_i - \bar{X})^2$ is independent of $\bar{X}$, due to our earlier observation that $\sum_{i=2}^n Y_i^2 = \sum_{i=1}^n (X_i - \bar{X})^2$, and that $Y_1 = \sqrt{n}\bar{X}$.

Consider now the case of general μ, σ. Because $(X_1, X_2, \ldots, X_n)$ has the representation $(X_1, X_2, \ldots, X_n) = (\mu, \mu, \ldots, \mu) + \sigma(Z_1, Z_2, \ldots, Z_n)$, where $Z_1, Z_2, \ldots, Z_n$ are independent standard normals, one has $\bar{X} = \mu + \sigma\bar{Z}$, and $\sum_{i=1}^n (X_i - \bar{X})^2 = \sigma^2 \sum_{i=1}^n (Z_i - \bar{Z})^2$. Therefore, the independence of $\bar{X}$ and $\sum_{i=1}^n (X_i - \bar{X})^2$ follows from their independence in the special standard normal case. □

Remark. The independence of $\bar{X}$ and $\sum_{i=1}^n (X_i - \bar{X})^2$ is a signature property of the normal distribution; we observed it earlier in Chapter 1. A very important consequence of their independence is the following result, of enormous importance in statistics.

Theorem 4.7. *Let $X_1, X_2, \ldots, X_n$ be independent $N(\mu, \sigma^2)$ variables. Then,*

(a) $\frac{\sum_{i=1}^n (X_i - \bar{X})^2}{\sigma^2} \sim \chi_{n-1}^2$;

(b) *The t statistic* $t = \frac{\sqrt{n}(\bar{X}-\mu)}{s} \sim t(n-1)$, *a t distribution with $n-1$ degrees of freedom, where s is defined as $(n-1)s^2 = \sum_{i=1}^n (X_i - \bar{X})^2$.*

Proof. It is enough to prove this theorem when $\mu = 0, \sigma = 1$, by the same argument made in the preceding theorem. So we assume that $\mu = 0, \sigma = 1$. Because $\sum_{i=1}^n (X_i - \bar{X})^2 = \sum_{i=2}^n Y_i^2$, and $Y_2, \ldots, Y_n$ are independent $N(0,1)$ variables, it follows that $\sum_{i=1}^n (X_i - \bar{X})^2$ has a χ_{n-1}^2 distribution. For part (b), note that $\sqrt{n}\bar{X} \sim N(0,1)$, and write s as $s = \sqrt{\frac{(n-1)s^2}{n-1}}$, and so s is the square root of a χ_{n-1}^2 random variable divided by $n-1$, its degrees of freedom. It therefore follows that $\frac{\sqrt{n}\bar{X}}{s}$ has a t-distribution, from the definition of a t-distribution. □

4.6.3 The t Confidence Interval

The result in Theorem 4.7 leads to one of the mainstays of statistical methodology, namely the *t confidence interval* for the mean of a normal distribution, when the variance σ^2 is unknown. In Section 1.13, we described how to construct a confidence interval for μ when we know the value of σ^2. The interval derived there is $\bar{X} \pm z_{\frac{\alpha}{2}} \frac{s}{\sqrt{n}}$, where $z_{\frac{\alpha}{2}} = \Phi^{-1}(1-\frac{\alpha}{2})$. Obviously, this interval cannot be used if we do not know the value of σ. However, we can easily remedy this slight problem by simply using part (b) of Theorem 4.7, which says that if $X_1, \ldots, X_n$ are iid $N(\mu, \sigma^2)$, then

$\frac{\sqrt{n}(\bar{X}-\mu)}{s} \sim t(n-1)$. This result implies, with $t_{\frac{\alpha}{2},n-1}$ denoting the $1-\frac{\alpha}{2}$ quantile of the $t(n-1)$ distribution,

$$P\left(-t_{\frac{\alpha}{2},n-1} \leq \frac{\sqrt{n}(\bar{X}-\mu)}{s} \leq t_{\frac{\alpha}{2},n-1}\right) = 1-\alpha$$

$$\Leftrightarrow P\left(-t_{\frac{\alpha}{2},n-1}\frac{s}{\sqrt{n}} \leq \bar{X}-\mu \leq t_{\frac{\alpha}{2},n-1}\frac{s}{\sqrt{n}}\right) = 1-\alpha$$

$$\Leftrightarrow P\left(\bar{X}-t_{\frac{\alpha}{2},n-1}\frac{s}{\sqrt{n}} \leq \mu \leq \bar{X}+t_{\frac{\alpha}{2},n-1}\frac{s}{\sqrt{n}}\right) = 1-\alpha.$$

The interval $\bar{X} \pm t_{\frac{\alpha}{2},n-1}(\frac{s}{\sqrt{n}})$ is called the $100(1-\alpha)\%\,t$ confidence interval. It is based on the assumption of $X_1, \ldots, X_n$ being iid $N(\mu, \sigma^2)$ for some μ and some σ^2. In practice, it is often used for very nonnormal or even correlated data. This is unjustified and in fact wrong.

4.7 The Dirichlet Distribution

The Jacobian density formula, when suitably applied to a set of independent Gamma random variables, results in a hugely useful and important density for random variables in a *simplex*. In the plane, the standard simplex is the triangle with vertices at $(0,0), (0,1)$, and $(1,0)$. In the general n dimensions, the standard simplex is the set of all n-dimensional vectors $\mathbf{x} = (x_1, \ldots, x_n)$ such that each $x_i \geq 0$, and $\sum_{i=1}^{n} x_i \leq 1$. If we define an additional x_{n+1} as $x_{n+1} = 1 - \sum_{i=1}^{n} x_i$, then $(x_1, \ldots, x_{n+1})$ forms a vector of proportions adding to one. Thus, the Dirichlet distribution can be used in any situation where an entity has to necessarily fall into one of $n+1$ mutually exclusive subclasses, and we want to study the proportion of individuals belonging to the different subclasses. Indeed, when statisticians want to model an ensemble of fractional variables adding to one, they often first look at the Dirichlet distribution as their model. See Aitchison (1986). Dirichlet distributions are also immensely important in Bayesian statistics. Fundamental work on the use of Dirichlet distributions in Bayesian modeling and on calculations using the Dirichlet distribution has been done in Ferguson (1973), Blackwell (1973), and Basu and Tiwari (1982).

Let $X_1, X_2, \ldots, X_{n+1}$ be independent Gamma random variables, with $X_i \sim G(\alpha_i, 1)$. Define

$$p_i = \frac{X_i}{\sum_{j=1}^{n+1} X_j}, \quad 1 \leq i \leq n,$$

and denote $p_{n+1} = 1 - \sum_{i=1}^{n} p_i$. Then, we have the following theorem.

Theorem 4.8. $\mathbf{p} = (p_1, p_2, \ldots, p_n)$ *has the joint density*

$$f(p_1, p_2, \ldots, p_n) = \frac{\Gamma\left(\sum_{i=1}^{n+1} \alpha_i\right)}{\prod_{i=1}^{n+1} \Gamma(\alpha_i)} \prod_{i=1}^{n+1} p_i^{\alpha_i - 1}.$$

4.7 The Dirichlet Distribution

Proof. This is proved by using the Jacobian density theorem. The transformation

$$(x_1, x_2, \ldots, x_{n+1}) \to \left(p_1, p_2, \ldots, p_n, \sum_{j=1}^{n+1} X_j\right)$$

is a one-to-one transformation with the Jacobian determinant $J = (\sum_{j=1}^{n+1} X_j)^n$. Inasmuch as $X_1, X_2, \ldots, X_{n+1}$ are independent Gamma random variables, applying the Jacobian density theorem, we get the joint density of $(p_1, p_2, \ldots, p_n, \sum_{j=1}^{n+1} X_j)$ as

$$f_{p_1, p_2, \ldots, p_n, s}(p_1, p_2, \ldots, p_n, s) = \frac{1}{\prod_{i=1}^{n+1} \Gamma(\alpha_i)} e^{-s} s^{m + \sum_{i=1}^{n+1} \alpha_i - 1} \times \prod_{i=1}^{n+1} p_i^{\alpha_i - 1}.$$

If we now integrate s out (on $(0, \infty)$), we get the joint density of $p_1, p_2, \ldots, p_n$, as stated in this theorem. □

Definition 4.7 (Dirichlet Density). An n-dimensional vector $\mathbf{p} = (p_1, p_2, \ldots, p_n)$ is said to have *the Dirichlet distribution with parameter vector* $\alpha = (\alpha_1, \ldots, \alpha_{n+1})$, $\alpha_i > 0$, if it has the joint density

$$f(p_1, p_2, \ldots, p_n) = \frac{\Gamma(\sum_{i=1}^{n+1} \alpha_i)}{\prod_{i=1}^{n+1} \Gamma(\alpha_i)} \prod_{i=1}^{n+1} p_i^{\alpha_i - 1},$$

$p_i \geq 0$, $\sum_{i=1}^{n} p_i \leq 1$.
We write $\mathbf{p} \sim \mathcal{D}_n(\alpha)$.

Remark. When $n = 1$, the Dirichlet density reduces to a Beta density with parameters α_1, α_2. Simple integrations give the following moment formulas.

Proposition. *Let* $\mathbf{p} \sim \mathcal{D}_n(\alpha)$. *Then,*

$$E(p_i) = \frac{\alpha_i}{t}, \quad \text{Var}(p_i) = \frac{\alpha_i(t - \alpha_i)}{t^2(t+1)}, \quad \text{Cov}(p_i, p_j) = -\frac{\alpha_i \alpha_j}{t^2(t+1)}, \quad i \neq j,$$

where $t = \sum_{i=1}^{n+1} \alpha_i$.

Thus, notice that the covariances (and hence the correlations) are always negative.

A convenient fact about the Dirichlet density is that lower-dimensional marginals are also Dirichlet distributions. So are the conditional distributions of suitably renormalized subvectors given the rest.

Theorem 4.9 (Marginal and Conditional Distributions).

(a) *Let* $\mathbf{p} \sim \mathcal{D}_n(\alpha)$. *Fix* $m < n$, *and let* $\mathbf{p}_m = (p_1, \ldots, p_m)$, *and* $\alpha_m = (\alpha_1, \ldots, \alpha_m, t - \sum_{i=1}^{m} \alpha_i)$. *Then* $\mathbf{p}_m \sim \mathcal{D}_m(\alpha_m)$. *In particular, each* $p_i \sim Be(\alpha_i, t - \alpha_i)$.

(b) Let $\mathbf{p} \sim \mathcal{D}(\alpha)$. Fix $m < n$, and let $q_i = \frac{p_i}{1-\sum_{i=1}^m p_i}, i = m+1,\ldots,n$. Let $\boldsymbol{\beta}_m = (\alpha_{m+1},\ldots,\alpha_{n+1})$. Then,

$$(q_{m+1},\ldots,q_n)|(p_1,\ldots,p_m) \sim \mathcal{D}_{n-m}(\boldsymbol{\beta}_m).$$

These two results follow in a straightforward manner from the definition of conditional densities, and the functional form of a Dirichlet density.

It also follows from the representation of a Dirichlet random vector in terms of independent Gamma variables that sums of a subset of the coordinates must have Beta distributions.

Theorem 4.10 (Sums of Subvectors).

(a) Let $\mathbf{p} \sim \mathcal{D}_n(\alpha)$. Fix $m < n$, and let $S_m = S_{m,n} = \sum_{i=1}^m p_i$. Then $S_m \sim Be(\sum_{i=1}^m \alpha_i, t - \sum_{i=1}^m \alpha_i)$.

(b) More generally, suppose the entire Dirichlet vector $\mathbf{p}$ is partitioned into k subvectors,

$$(p_1,\ldots,p_{m_1});(p_{m_1+1},\ldots,p_{m_2});\ldots;(p_{m_{k-1}+1},\ldots,p_n).$$

Let $S_1, S_2,\ldots,S_k$ be the sums of the coordinates of these k subvectors. Then

$$(S_1, S_2,\ldots,S_k) \sim \mathcal{D}_k\left(\sum_{i=1}^{m_1}\alpha_i, \sum_{i=m_1+1}^{m_2}\alpha_i,\ldots,\sum_{i=m_{k-1}+1}^{n}\alpha_i\right).$$

The Dirichlet density is obtained by definition as the density of functions of independent Gamma variables, thus it also has connections to the normal distribution, by virtue of the fact that sums of squares of independent standard normal variables are chi square, which are, after all, Gamma random variables. Here is the connection to the normal distribution.

Theorem 4.11 (Dirichlet and Normal). Let $X_1, X_2,\ldots,X_n$ be independent standard normal variables. Fix $k, 1 \leq k < n$. Let $||X|| = \sqrt{\sum_{i=1}^n X_i^2}$, the length of $\mathbf{X}$. Let $Y_i = \frac{X_i}{||X||}, i = 1,2,\ldots,k$. Then,

(a) $(Y_1^2, Y_2^2,\ldots,Y_k^2)$ has the density $\mathcal{D}_k(\alpha)$, where $\alpha = (1/2, 1/2,\ldots,1/2, (n-k)/2)$;

(b) $(Y_1, Y_2,\ldots,Y_k)$ has the density $\frac{\Gamma(\frac{n}{2})}{\pi^{k/2}\Gamma(\frac{n-k}{2})}(1 - \sum_{i=1}^k y_i^2)^{(n-k)/2-1}$, $\sum_{i=1}^k y_i^2 < 1$.

Proof. Part (a) is a consequence of the definition of a Dirichlet distribution in terms of independent Gamma variables, and the fact that marginals of a Dirichlet distribution are also Dirichlet. For part (b), first make the transformation from Y_i^2 to $|Y_i|, 1 \leq i \leq k$, and use the Jacobian density theorem. Then, observe that the joint density of $(Y_1, Y_2,\ldots,Y_k)$ is symmetric, and so the joint density of $(Y_1, Y_2,\ldots,Y_k)$ and that of $(|Y_1|, |Y_2|,\ldots,|Y_k|)$ are given by essentially the same function, apart from a normalizing constant. □

4.7.1 ∗ Picking a Point from the Surface of a Sphere

Actually, part (b) of this last theorem brings out a very interesting connection between the normal distribution and the problem of picking a point at random from the boundary of a high-dimensional sphere. If $\mathbf{U_n} = (U_{n1}, U_{n2}, \ldots, U_{nn})$ is uniformly distributed on the boundary of the n-dimensional unit sphere, and if we take $k < n$, hold k fixed and let $n \to \infty$, then part (b) leads to the very useful fact that the joint distribution of $(U_{n1}, U_{n2}, \ldots, U_{nk})$ is approximately the same as the joint distribution of k independent $N(0, \frac{1}{n})$-variables. That is, if a point was picked at random from the surface of a high-dimensional sphere, and if we then looked at a low-dimensional projection of that point, the projection would act as a set of nearly independent normal variables with zero means and variances $\frac{1}{n}$. This is known as *Poincaré's lemma*. Compare this with Theorem 15.5, where the exact density of a lower-dimensional projection was worked out; Poincaré's lemma can be derived from there.

4.7.2 ∗ Poincaré's Lemma

Theorem 4.12 (Poincaré's Lemma). *Let $\mathbf{U_n} = (U_{n1}, U_{n2}, \ldots, U_{nn})$ be uniformly distributed on the boundary of the n-dimensional unit sphere. Let $k \geq 1$ be a fixed positive integer. Then,*

$$P\left(\sqrt{n} U_{n1} \leq x_1, \sqrt{n} U_{n2} \leq x_2, \ldots, \sqrt{n} U_{nk} \leq x_k\right) \to \prod_{i=1}^{k} \Phi(x_i),$$

$\forall \, (x_1, x_2, \ldots, x_k) \in \mathcal{R}^k.$

4.8 ∗ Ten Important High-Dimensional Formulas for Easy Reference

Suppose an n-dimensional random vector has a spherically symmetric joint density $g(\sqrt{\sum_{i=1}^{n} x_i^2})$. Then, by making the n-dimensional polar transformation, we can reduce the expectation of a function $h(\sum_{i=1}^{n} X_i^2)$ to just a one-dimensional integral, although in principle, an expectation requires integration on the n-dimensional space. The structure of spherical symmetry enables us to make this drastic reduction to just a one-dimensional integral. Similar other reductions follow in many other problems, and essentially all of them are consequences of making a transformation just right for that problem, and then working out the Jacobian. We state a number of such frequently useful formulas in this section.

Theorem 4.13.

(a) Volume of n-dimensional unit sphere $= \frac{\pi^{\frac{n}{2}}}{\Gamma(\frac{n}{2}+1)}$.

(b) Surface area of n-dimensional unit sphere $= \frac{n\pi^{\frac{n}{2}}}{\Gamma(\frac{n}{2}+1)}$.

(c) Volume of n-dimensional simplex $= \int_{x_i \geq 0, x_1+\cdots+x_n \leq 1} dx_1 \cdots dx_n = \frac{1}{n!}$.

(d) $\int_{x_i \geq 0, \sum_{i=1}^{n}(\frac{x_i}{c_i})^{\alpha_i} \leq 1} f\left(\sum_{i=1}^{n}\left(\frac{x_i}{c_i}\right)^{\alpha_i}\right) \prod_{i=1}^{n} x_i^{p_i-1} dx_1 \cdots dx_n$

$$= \frac{\prod_{i=1}^{n} c_i^{p_i}}{\prod_{i=1}^{n} \alpha_i} \frac{\prod_{i=1}^{n} \Gamma(\frac{p_i}{\alpha_i})}{\Gamma(\sum_{i=1}^{n} \frac{p_i}{\alpha_i})} \int_0^1 f(t) t^{\sum_{i=1}^{n} \frac{p_i}{\alpha_i}-1} dt, \quad (c_i, \alpha_i, p_i > 0).$$

(e) $\int_{x_i \geq 0, \sum_{i=1}^{n} x_i^{\alpha_i} \leq 1} \frac{\prod_{i=1}^{n} x_i^{p_i-1}}{\left(\sum_{i=1}^{n} x_i^{\alpha_i}\right)^{\mu}} dx_1 \cdots dx_n$

$$= \frac{\prod_{i=1}^{n} \Gamma\left(\frac{p_i}{\alpha_i}\right)}{\prod_{i=1}^{n} \alpha_i \times \left(\sum_{i=1}^{n} \frac{p_i}{\alpha_i} - \mu\right) \times \Gamma(\sum_{i=1}^{n} \frac{p_i}{\alpha_i})}, \quad \left(\sum_{i=1}^{n} \frac{p_i}{\alpha_i} > \mu\right).$$

(f) $\int_{\sum_{i=1}^{n} x_i^2 \leq 1} \left(\sum_{i=1}^{n} p_i x_i\right)^{2m+1} dx_1 \cdots dx_n = 0 \; \forall p_1, \ldots, p_n, \; \forall m \geq 1$.

(g) $\int_{\sum_{i=1}^{n} x_i^2 \leq 1} \left(\sum_{i=1}^{n} p_i x_i\right)^{2m} dx_1 \cdots dx_n$

$$= \frac{(2m-1)!}{2^{2m-1}(m-1)!} \frac{\pi^{\frac{n}{2}}}{\Gamma(\frac{n}{2}+m+1)} \times \left(\sum_{i=1}^{n} p_i^2\right)^m, \quad \forall p_1, \ldots, p_n, \; \forall m \geq 1.$$

(h) $\int_{\sum_{i=1}^{2n} x_i^2 \leq 1} e^{\sum_{i=1}^{2n} c_i x_i} dx_1 \cdots dx_n = \frac{(2\pi)^n I_n\left(\sqrt{\sum_{i=1}^{2n} c_i^2}\right)}{\left(\sum_{i=1}^{2n} c_i^2\right)^{\frac{n}{2}}}$,

where $I_n(z)$ denotes the Bessel function defined by that notation.

(i) $\int_{\sum_{i=1}^{n} x_i^2 \leq 1} e^{\sum_{i=1}^{n} c_i x_i} dx_1 \cdots dx_n = \pi^{\frac{n}{2}} \sum_{k=0}^{\infty} \frac{\left(\sum_{i=1}^{n} c_i^2\right)^k}{4^k k! \Gamma\left(\frac{n}{2}+k+1\right)}$.

(j) $\int_{\sum_{i=1}^{n} x_i^2 \leq r^2} f\left(\sqrt{\sum_{i=1}^{n} x_i^2}\right) dx_1 \cdots dx_n = \frac{2\pi^{\frac{n}{2}}}{\Gamma(\frac{n}{2})} \int_0^r t^{n-1} f(t) dt$.

Exercises

Exercise 4.1. Suppose $X \sim U[0,1]$, and Y has the density $2y, 0 < y < 1$, and that X, Y are independent. Find the density of XY and of $\frac{X}{Y}$.

Exercises

Exercise 4.2. Suppose $X \sim U[0, 1]$, and Y has the density $2y, 0 < y < 1$, and that X, Y are independent. Find the density of $X + Y, X - Y, |X - Y|$.

Exercise 4.3. Suppose (X, Y) have the joint pdf $f(x, y) = c(x+y)e^{-x-y}, x, y > 0$.

(a) Are X, Y independent?
(b) Find the normalizing constant c.
(c) Find the density of $X + Y$.

Exercise 4.4. Suppose X, Y have the joint density $cxy, 0 < x < y < 1$.

(a) Are X, Y independent?
(b) Find the normalizing constant c.
(c) Find the density of XY.

Exercise 4.5 * **(A Conditioning Argument).** Suppose a fair coin is tossed twice and the number of heads obtained is N. Let X, Y be independent $U[0, 1]$ variables, and independent of N. Find the density of NXY.

Exercise 4.6. Suppose $X \sim U[0, a], Y \sim U[0, b], Z \sim U[0, c], 0 < a < b < c$, and that X, Y, Z are independent. Let $m = \min\{X, Y, Z\}$. Find expressions for $P(m = X), P(m = Y), P(m = Z)$.

Exercise 4.7. Suppose X, Y are independent standard exponential random variables. Find the density of XY, and of $\frac{XY}{(X+Y)^2}$.

Hint: Use $\frac{Y}{X+Y} = 1 - \frac{X}{X+Y}$, and see the examples in text.

Exercise 4.8 * **(Uniform in a Circle).** Suppose (X, Y) are jointly uniform in the unit circle. By transforming to polar coordinates, find the expectations of $\frac{XY}{X^2+Y^2}$, and of $\frac{XY}{\sqrt{X^2+Y^2}}$.

Exercise 4.9 * **(Length of Bivariate Uniform).** Suppose X, Y are independent $U[0, 1]$ variables.

(a) Find the density of $X^2 + Y^2$, and $P(X^2 + Y^2 \leq 1)$.
(b) Show that $E(\sqrt{X^2 + Y^2}) \approx .765$.

Hint: It is best to do this directly, and not try polar coordinates.

Exercise 4.10. Suppose (X, Y) have the joint CDF $F(x, y) = x^3 y^2, 0 \leq x, y \leq 1$. Find the density of XY and of $\frac{X}{Y}$.

Exercise 4.11 * **(Distance Between Two Random Points).** Suppose $P = (X, Y)$ and $Q = (Z, W)$ are two independently picked points from the unit circle, each according to a uniform distribution in the circle. What is the average distance between P and Q?

Exercise 4.12 * **(Distance from the Boundary).** A point is picked uniformly from the unit square. What is the expected value of the distance of the point from the boundary of the unit square?

Exercise 4.13. Suppose X, Y are independent standard normal variables. Find the values of $P(\frac{X}{Y} < 1)$, and of $P(X < Y)$. Why are they not the same?

Exercise 4.14 (A Normal Calculation). A marksman is going to take two shots at a bull's eye. The distance of the first and the second shot from the bull's eye are distributed as that of $(|X|, |Y|)$, where $X \sim N(0, \sigma^2)$, $Y \sim N(0, \tau^2)$, and X, Y are independent. Find a formula for the probability that the second shot is closer to the target.

Exercise 4.15 * **(Quotient in Bivariate Normal).** Suppose (X, Y) have a bivariate normal distribution with zero means, unit standard deviations, and a correlation ρ. Show that $\frac{X}{Y}$ still has a Cauchy distribution.

Exercise 4.16 * **(Product of Beta).** Suppose X, Y are independent $Be(\alpha, \beta)$, $Be(\gamma, \delta)$ random variables. Find the density of XY. Do you recognize the form?

Exercise 4.17 * **(Product of Normals).** Suppose X, Y are independent standard normal variables. Find the density of XY.

Hint: The answer involves a Bessel function K_0.

Exercise 4.18 * **(Product of Cauchy).** Suppose X, Y are independent standard Cauchy variables. Derive a formula for the density of XY.

Exercise 4.19. Prove that the square of a t random variable has an F-distribution.

Exercise 4.20 * **(Box–Mueller Transformation).** Suppose X, Y are independent $U[0, 1]$ variables. Let $U = \sqrt{-2 \log X} \cos(2\pi Y)$, $V = \sqrt{-2 \log X} \sin(2\pi Y)$. Show that U, V are independent and that each is standard normal.

Remark. This is a convenient method to generate standard normal values, by using only uniform random numbers.

Exercise 4.21 * **(Deriving a General Formula).** Suppose (X, Y, Z) have a joint density of the form $f(x, y, z) = g(x + y + z)$, $x, y, z > 0$. Find a formula for the density of $X + Y + Z$.

Exercise 4.22. Suppose (X, Y, Z) have a joint density $f(x, y, z) = \frac{6}{(1+x+y+z)^4}$, $x, y, z > 0$. Find the density of $X + Y + Z$.

Exercise 4.23 (Deriving a General Formula). Suppose $X \sim U[0, 1]$ and Y is an arbitrary continuous random variable. Derive a general formula for the density of $X + Y$.

Exercise 4.24 (Convolution of Uniform and Exponential). Let $X \sim U[0, 1]$, $Y \sim \text{Exp}(\lambda)$, and X, Y are independent. Find the density of $X + Y$.

Exercise 4.25 (Convolution of Uniform and Normal). Let $X \sim U[0, 1]$, $Y \sim N(\mu, \sigma^2)$, and X, Y are independent. Find the density of $X + Y$.

Exercise 4.26 (Convolution of Uniform and Cauchy). Let $X \sim U[0, 1]$, $Y \sim C(0, 1)$, and X, Y are independent. Find the density of $X + Y$.

Exercise 4.27 (Convolution of Uniform and Poisson). Let $X \sim U[0, 1]$, $Y \sim \text{Poi}(\lambda)$, and let X, Y be independent. Find the density of $X + Y$.

Exercise 4.28 * **(Bivariate Cauchy).** Suppose (X, Y) has the joint pdf $f(x, y) = \frac{c}{(1+x^2+y^2)^{3/2}}$, $-\infty < x, y < \infty$.

(a) Find the normalizing constant c.
(b) Are X, Y independent?
(c) Find the densities of the polar coordinates r, θ.
(d) Find $P(X^2 + Y^2 \leq 1)$.

Exercise 4.29. * Suppose X, Y, Z are independent standard exponentials. Find the joint density of
$$\frac{X}{X+Y+Z}, \frac{X+Y}{X+Y+Z}, X+Y+Z.$$

Exercise 4.30 (Correlation). Suppose X, Y are independent $U[0, 1]$ variables. Find the correlation between $X + Y$ and $\sqrt{XY}$.

Exercise 4.31 (Correlation). Suppose X, Y are jointly uniform in the unit circle. Find the correlation between XY and $X^2 + Y^2$.

Exercise 4.32 * **(Sum and Difference of General Exponentials).** Suppose $X \sim \text{Exp}(\lambda)$, $Y \sim \text{Exp}(\mu)$, and that X, Y are independent. Find the density of $X + Y$ and of $X - Y$.

Exercise 4.33 * **(Double Exponential Convolution).** Suppose X, Y are independent standard double exponentials, each with the density $\frac{1}{2}e^{-|x|}$, $-\infty < x < \infty$. Find the density of $X + Y$.

Exercise 4.34. * Let X, Y, Z be independent standard normals. Show that $\frac{X+YZ}{\sqrt{1+Z^2}}$ has a normal distribution.

Exercise 4.35 * **(Decimal Expansion of a Uniform).** Let $X \sim U[0, 1]$ and suppose $X = .n_1 n_2 n_3 \cdots$ is the decimal expansion of X. Find the marginal and joint distribution of $n_1, n_2, \ldots, n_k$, for $k \geq 1$.

Exercise 4.36 * **(Integer Part and Fractional Part).** Let X be a standard exponential variable. Find the joint distribution of the integer part and the fractional part of X. Note that they do not have a joint density.

Exercise 4.37 * **(Factorization of Chi Square).** Suppose X has a chi square distribution with one degree of freedom. Find nonconstant independent random variables Y, Z such that YZ has the same distribution as X.

Hint: Look at text.

Exercise 4.38 * (**Multivariate Cauchy**). Suppose $X_1, X_2, \ldots, X_n$ have the joint density

$$f(x_1, \ldots, x_n) = \frac{c}{(1 + x_1^2 + \cdots + x_n^2)^{\frac{3}{2}}},$$

where c is a normalizing constant.
Find the density of $X_1^2 + X_2^2 + \cdots + X_n^2$.

Exercise 4.39 (**Ratio of Independent Chi Squares**). Suppose $X_1, X_2, \ldots, X_m$ are independent $N(\mu, \sigma^2)$ variables, and $Y_1, Y_2, \ldots, Y_n$ are independent $N(\theta, \tau^2)$ variables. Assume also that all $m + n$ variables are independent. Show that

$$\frac{(n-1)\tau^2 \sum_{i=1}^{m}(X_i - \bar{X})^2}{(m-1)\sigma^2 \sum_{i=1}^{n}(Y_i - \bar{Y})^2}$$

has an F distribution.

Exercise 4.40 * (**An Example due to Larry Shepp**). Let X, Y be independent standard normals. Show that $\frac{XY}{\sqrt{X^2+Y^2}}$ has a normal distribution.

Exercise 4.41 (**Dirichlet Calculation**). Suppose $(p_1, p_2, p_3) \sim \mathcal{D}_3(1, 1, 1, 1)$.

(a) What is the marginal density of each of p_1, p_2, p_3?
(b) What is the marginal density of $p_1 + p_2$?
(c) Find the conditional probability $P(p_3 > \frac{1}{4} \mid p_1 + p_2 = \frac{1}{2})$.

Exercise 4.42 * (**Dirichlet Calculation**). Suppose (p_1, p_2, p_3, p_4) has a Dirichlet distribution with a general parameter vector α. Find each of the following.

(a) $\text{Var}(p_1 + p_2 - p_3 - p_4)$.
(b) $P(p_1 + p_2 > p_3 + p_4)$.
(c) $E(p_3 + p_4 \mid p_1 + p_2 = c)$.
(d) $\text{Var}(p_3 + p_4 \mid p_1 + p_2 = c)$.

Exercise 4.43 * (**Dirichlet Cross Moments**). Suppose $\mathbf{p} \sim \mathcal{D}_n(\alpha)$. Let $r, s \geq 1$ be integers. Show that

$$E(p_i^r p_j^s) = \frac{(\alpha_i + r - 1)(\alpha_j + s - 1)}{t^2(t+1)},$$

where $t = \sum_{i=1}^{n+1} \alpha_i$.

References

Aitchison, J. (1986). *Statistical Analysis of Compositional Data*, Chapman and Hall, New York.

Basu, D. and Tiwari, R. (1982). *A Note on Dirichlet Processes, in Statistics and Probability, Essays in Honor of C. R. Rao*, 89-103, J. K. Ghosh and G. Kallianpur, Eds., North-Holland, Amsterdam.

Blackwell, D. (1973). Discreteness of Ferguson selections, *Ann. Stat.*, 1, 2, 356–358.

Ferguson, T. (1973). A Bayesian analysis of some nonparametric problems, *Ann. Stat.*, 1, 209–230.

Hinkley, D. (1969). On the ratio of two correlated normal random variables, *Biometrika*, 56, 3, 635–639.

Chapter 5
Multivariate Normal and Related Distributions

Multivariate normal distribution is the natural extension of the bivariate normal to the case of several jointly distributed random variables. Dating back to the works of Galton, Karl Pearson, Edgeworth, and later Ronald Fisher, the multivariate normal distribution has occupied the central place in modeling jointly distributed continuous random variables. There are several reasons for its special status. Its mathematical properties show a remarkable amount of intrinsic structure; the properties are extremely well studied; statistical methodologies in common use often have their best or optimal performance when the variables are distributed as multivariate normal; and, there is the *multidimensional central limit theorem* and its various consequences which imply that many kinds of functions of independent random variables are approximately normally distributed, in some suitable sense. We present some of the multivariate normal theory and facts with examples in this chapter.

5.1 Definition and Some Basic Properties

As in the bivariate case, a general multivariate normal distribution is defined as the distribution of a linear function of a standard normal vector. Here is the definition.

Definition 5.1. Let $n \geq 1$. A random vector $\mathbf{Z} = (Z_1, Z_2, \ldots, Z_n)$ is said to have an n-dimensional standard normal distribution if the Z_i are independent univariate standard normal variables, in which case their joint density is

$$f(z_1, z_2, \ldots, z_n) = \frac{1}{(2\pi)^{\frac{n}{2}}} e^{-\frac{\sum_{i=1}^n z_i^2}{2}} \quad -\infty < z_i < \infty, i = 1, 2, \ldots, n.$$

Definition 5.2. Let $n \geq 1$, and let B be an $n \times n$ real matrix of rank $k \leq n$. Suppose $\mathbf{Z}$ has an n-dimensional standard normal distribution. Let μ be any n-dimensional vector of real constants. Then $\mathbf{X} = \mu + B\mathbf{Z}$ is said to have a *multivariate normal distribution with parameters* μ *and* Σ, where $\Sigma = BB'$ is an $n \times n$ real symmetric nonnegative definite (nnd) matrix. If $k < n$, the distribution is called a *singular multivariate normal*. If $k = n$, then Σ is positive definite and the distribution of

X is called a *nonsingular multivariate normal* or often, just a multivariate normal. We use the notation $\mathbf{X} \sim N_n(\mu, \Sigma)$, or sometimes $\mathbf{X} \sim MVN(\mu, \Sigma)$.

We treat only nonsingular multivariate normals in this chapter.

Theorem 5.1 (Density of Multivariate Normal). *Suppose B is full rank. Then, the joint density of $X_1, X_2, \ldots, X_n$ is*

$$f(x_1, x_2, \ldots, x_n) = \frac{1}{(2\pi)^{\frac{n}{2}} |\Sigma|^{\frac{1}{2}}} e^{-\frac{1}{2}(\mathbf{x}-\mu)'\Sigma^{-1}(\mathbf{x}-\mu)},$$

where $\mathbf{x} = (x_1, x_2, \ldots, x_n) \in \mathcal{R}^n$.

Proof. By definition of the multivariate normal distribution, and the definition of matrix product, $X_i = \mu_i + \sum_{j=1}^n b_{ij} Z_j, 1 \leq i \leq n$. This is a one-to-one transformation because B is full rank, and the partial derivatives are $\frac{\partial x_i}{\partial z_j} = b_{ij}$. Hence the determinant of the matrix of the partial derivatives $\frac{\partial x_i}{\partial z_j}$ is $|B|$. The Jacobian determinant is the reciprocal of this determinant, and hence, is $|B|^{-1} = |\Sigma|^{-\frac{1}{2}}$, because $|\Sigma| = |BB'| = |B||B'| = |B|^2$. Furthermore, $\sum_{i=1}^n z_i^2 = \mathbf{z}'\mathbf{z} = (B^{-1}(\mathbf{x}-\mu))'(B^{-1}(\mathbf{x}-\mu)) = (\mathbf{x}-\mu)'(B')^{-1} B^{-1}(\mathbf{x}-\mu) = (\mathbf{x}-\mu)'\Sigma^{-1}(\mathbf{x}-\mu)$. Now the theorem follows from the Jacobian density theorem. □

It follows from the linearity of expectations, and the linearity of covariance, that $E(X_i) = \mu_i$, and $\text{Cov}(X_i, X_j) = \sigma_{ij}$, the (i, j)th element of Σ. The vector of expectations is usually called the *mean vector*, and the matrix of pairwise covariances is called the *covariance matrix*. Thus, we have the following facts about the physical meanings of μ, Σ:

Proposition. *The mean vector of $\mathbf{X}$ equals $E(\mathbf{X}) = \mu$; The covariance matrix of $\mathbf{X}$ equals Σ.*

Example 5.1. Figure 5.1 of a simulation of 1000 values from a bivariate normal distribution shows the elliptical shape of the point cloud, as would be expected from the fact that the formula for the density function is a quadratic form in the variables. It is also seen in the plot that the center of the point cloud is quite close to the true means of the variables, namely $\mu_1 = 4.5, \mu_2 = 4$, which were used for the purpose of the simulation.

An important property of a multivariate normal distribution is the property of *closure under linear transformations*; that is, any number of linear functions of $\mathbf{X}$ will also have a multivariate normal distribution. The precise closure property is as follows.

Theorem 5.2 (Density of Linear Transformations). *Let $\mathbf{X} \sim N_n(\mu, \Sigma)$, and let $A_{k \times n}$ be a matrix of rank $k, k \leq n$. Then $\mathbf{Y} = A\mathbf{X} \sim N_k(A\mu, A\Sigma A')$. In particular, marginally, $X_i \sim N(\mu_i, \sigma_i^2)$, where $\sigma_i^2 = \sigma_{ii}, 1 \leq i \leq n$, and all lower-dimensional marginals are also multivariate normal in the corresponding dimension.*

5.1 Definition and Some Basic Properties

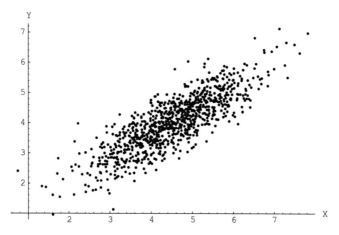

Fig. 5.1 Simulation of a bivariate normal with means 4.5, 4; variances 1; correlation .75

Proof. Let B be such that $\Sigma = BB'$. Then, writing $\mathbf{X} = \mu + B\mathbf{Z}$, we get $\mathbf{Y} = A(\mu + B\mathbf{Z}) = A\mu + AB\mathbf{Z}$, and therefore, using the same representation of a multivariate normal vector once again, $\mathbf{Y} \sim N_k(A\mu, (AB)(AB)') = N_k(A\mu, ABB'A') = N_k(A\mu, A\Sigma A')$. □

Corollary (MGF of Multivariate Normal). *Let* $\mathbf{X} \sim N_n(\mu, \Sigma)$. *Then the mgf of* $\mathbf{X}$ *exists at all points in* $\mathcal{R}^n$ *and is given by*

$$\psi_{\mu,\Sigma}(\mathbf{t}) = e^{t'\mu + \frac{t'\Sigma t}{2}}.$$

This follows on simply observing that the theorem above implies that $\mathbf{t}'\mathbf{X} \sim N(t'\mu, t'\Sigma t)$, *and by then using the formula for the mgf of a univariate normal distribution.* □

As we just observed, any linear combination $\mathbf{t}'\mathbf{X}$ of a multivariate normal vector is univariate normal. A remarkable fact is that the converse is also true.

Theorem 5.3 (Characterization of Multivariate Normal). *Let* $\mathbf{X}$ *be an n-dimensional random vector, and suppose each linear combination* $\mathbf{t}'\mathbf{X}$ *has a univariate normal distribution. Then* $\mathbf{X}$ *has a multivariate normal distribution.*

See Tong (1990, p. 29) for a proof of this result.

Example 5.2. Suppose $(X_1, X_2, X_3) \sim N_3(\mu, \Sigma)$, where

$$\mu = \begin{pmatrix} 0 \\ 1 \\ 2 \end{pmatrix}$$

$$\Sigma = \begin{pmatrix} 1 & 0 & 1 \\ 0 & 2 & 2 \\ 1 & 2 & 3 \end{pmatrix}.$$

We want to find the joint distribution of $(X_1 - X_2, X_1 + X_2 + X_3)$. We recognize this to be a linear function $A\mathbf{X}$, where

$$A = \begin{pmatrix} 1 & -1 & 0 \\ 1 & 1 & 1 \end{pmatrix}.$$

Therefore, by the theorem above,

$$(X_1 - X_2, X_1 + X_2 + X_3) \sim N_2(A\mu, A\Sigma A'),$$

and by direct matrix multiplication,

$$A\mu = \begin{pmatrix} -1 \\ 3 \end{pmatrix}$$

and

$$A\Sigma A' = \begin{pmatrix} 3 & -2 \\ -2 & 12 \end{pmatrix}.$$

In particular, $X_1 - X_2 \sim N(-1, 3)$, $X_1 + X_2 + X_3 \sim N(3, 12)$, and $\mathrm{Cov}(X_1 - X_2, X_1 + X_2 + X_3) = -2$. Therefore, the correlation between $X_1 - X_2$ and $X_1 + X_2 + X_3$ equals $\frac{-2}{\sqrt{3}\sqrt{12}} = -\frac{1}{3}$.

5.2 Conditional Distributions

As in the bivariate normal case, zero correlation between a particular pair of variables implies that the particular pair must be independent. And, as in the bivariate normal case, all lower-dimensional conditional distributions are also multivariate normal.

Theorem 5.4. *Suppose* $\mathbf{X} \sim N_n(\mu, \Sigma)$. *Then,* X_i, X_j *are independent if and only if* $\sigma_{ij} = 0$. *More generally, if* $\mathbf{X}$ *is partitioned as* $\mathbf{X} = (\mathbf{X}_1, \mathbf{X}_2)$, *where* $\mathbf{X}_1$ *is* k-*dimensional, and* $\mathbf{X}_2$ *is* $n - k$ *dimensional, and if* Σ *is accordingly partitioned as*

$$\Sigma = \begin{pmatrix} \Sigma_{11} & \Sigma_{12} \\ \Sigma_{21} & \Sigma_{22} \end{pmatrix},$$

where Σ_{11} *is* $k \times k$, Σ_{22} *is* $(n - k) \times (n - k)$, *then* $\mathbf{X}_1$ *and* $\mathbf{X}_2$ *are independent if and only if* Σ_{12} *is the null matrix.*

Proof. The first statement follows immediately because we have proved that all lower-dimensional marginals are also multivariate normal. Therefore, any pair $(X_i, X_j), i \neq j$, is bivariate normal. Now use the bivariate normal property that a zero covariance implies independence.

5.2 Conditional Distributions

The second part uses the argument that if Σ_{12} is the null matrix, then the joint density of (X_1, X_2) factorizes in a product form, on some calculation, and therefore, X_1 and X_2 must be independent. Alternatively, the second part also follows from the next theorem given immediately below. □

Theorem 5.5 (Conditional Distributions). *Let $n \geq 2$, and $1 \leq k < n$. Let $X = (X_1, X_2)$, where X_1 is k-dimensional, and X_2 is $n - k$ dimensional. Suppose $X \sim N_n(\mu, \Sigma)$. Partition μ, Σ as*

$$\mu = \begin{pmatrix} \mu_1 \\ \mu_2 \end{pmatrix}$$

$$\Sigma = \begin{pmatrix} \Sigma_{11} & \Sigma_{12} \\ \Sigma_{21} & \Sigma_{22} \end{pmatrix},$$

where μ_1 is k-dimensional, μ_2 is $n - k$ dimensional, and $\Sigma_{11}, \Sigma_{12}, \Sigma_{22}$ are, respectively, $k \times k, k \times n - k, n - k \times n - k$. Finally, let $\Sigma_{1.2} = \Sigma_{11} - \Sigma_{12} \Sigma_{22}^{-1} \Sigma_{21}$. Then,

$$X_1 \mid X_2 = x_2 \sim N_k(\mu_1 + \Sigma_{12} \Sigma_{22}^{-1}(x_2 - \mu_2), \Sigma_{1.2}).$$

This involves some tedious manipulations with matrices, and we omit the proof. See Tong (1990, pp. 33–35) for the details. An important result that follows from the conditional covariance matrix formula in the above theorem is the following; it says that once you take out the effect of X_2 on X_1, X_2 and the residual will actually become independent.

Corollary. *Let $X = (X_1, X_2) \sim N_n(\mu, \Sigma)$. Then, $X_1 - E(X_1 \mid X_2)$ and X_2 are independent.*

Example 5.3. Let $(X_1, X_2, X_3) \sim N(\mu, \Sigma)$, where $\mu_1 = \mu_2 = \mu_3 = 0$, $\sigma_{ii} = 1, \sigma_{12} = \frac{1}{2}, \sigma_{13} = \sigma_{23} = \frac{3}{4}$. It is easily verified that Σ is positive definite. We want to find the conditional distribution of (X_1, X_2) given $X_3 = x_3$.

By the above theorem, the conditional mean is

$$(0, 0) + (\sigma_{13}, \sigma_{23}) \frac{1}{\sigma_{33}}(x_3 - 0) = \left(\frac{3}{4}, \frac{3}{4}\right) x_3 = \left(\frac{3x_3}{4}, \frac{3x_3}{4}\right).$$

And the conditional covariance matrix is found from the formula in the above theorem as

$$\begin{pmatrix} \frac{7}{16} & -\frac{1}{16} \\ -\frac{1}{16} & \frac{7}{16} \end{pmatrix}.$$

In particular, given $X_3 = x_3$, $X_1 \sim N(\frac{3x_3}{4}, \frac{7}{16})$; the distribution of X_2 given $X_3 = x_3$ is the same normal distribution. Finally, given $X_3 = x_3$, the correlation between X_1 and X_2 is

$$-\frac{\frac{1}{16}}{\sqrt{\frac{7}{16}}\sqrt{\frac{7}{16}}} = -\frac{1}{7} < 0,$$

although the unconditional correlation between X_1 and X_2 is positive, because $\sigma_{12} = \frac{1}{2} > 0$. The correlation between X_1 and X_2 given $X_3 = x_3$ is called the *partial correlation between X_1 and X_2 given $X_3 = x_3$*.

Example 5.4. Suppose $(X_1, X_2, X_3) \sim N_3(\mu, \Sigma)$, where

$$\Sigma = \begin{pmatrix} 4 & 1 & 0 \\ 1 & 2 & 1 \\ 0 & 1 & 3 \end{pmatrix}.$$

Suppose we want to find all a, b such that $X_3 - aX_1 - bX_2$ is independent of (X_1, X_2). The answer to this question depends only on Σ; so, we leave the mean vector μ unspecified.

To answer this question, we first have to note that the three variables $X_3 - aX_1 - bX_2, X_1, X_2$ together have a multivariate normal distribution, by the general theorem on multivariate normality of linear transformations. Because of this fact, $X_3 - aX_1 - bX_2$ is independent of (X_1, X_2) if and only if each of Cov$(X_3 - aX_1 - bX_2, X_1)$, Cov$(X_3 - aX_1 - bX_2, X_2)$ is zero. These are equivalent to

$$\sigma_{31} - a\sigma_{11} - b\sigma_{21} = 0; \quad \text{and} \quad \sigma_{32} - a\sigma_{12} - b\sigma_{22} = 0$$
$$\Leftrightarrow 4a + b = 0; \qquad\qquad a + 2b = 1$$
$$\Rightarrow a = -\tfrac{1}{7}, \qquad\qquad b = \tfrac{4}{7}.$$

We conclude this section with a formula for a quadrant probability in the trivariate normal case; no such clean formula is possible in general in dimensions four and more. The formula given below is proved in Tong (1990).

Proposition (Quadrant Probability). *Let (X_1, X_2, X_3) have a multivariate normal distribution with means $\mu_i, i = 1, 2, 3$, and pairwise correlations ρ_{ij}. Then,*

$$P(X_1 > \mu_1, X_2 > \mu_2, X_3 > \mu_3) = \frac{1}{8} + \frac{1}{4\pi}[\arcsin \rho_{12} + \arcsin \rho_{23} + \arcsin \rho_{13}].$$

Example 5.5 (A Sample Size Problem). Suppose X_1, X_2, X_3 are jointly trivariate normal, each with mean zero, and suppose that $\rho_{X_i, X_j} = \rho$ for each pair $X_i, X_j, i \neq j$. Then, by the above proposition, the probability of the first quadrant is $\frac{1}{8} + \frac{3}{4\pi} \arcsin(\rho) = p$ (say). Now suppose that n points (X_{i1}, X_{i2}, X_{i3}) have been simulated from this trivariate normal distribution. Let T be the number of such points that fall into the first quadrant. We then have that $T \sim \text{Bin}(n, p)$.

Suppose now we need to have at least 100 points to fall in the first quadrant, with a large probability, say probability .9. How many points do we need to simulate to ensure this?

By the central limit theorem,

$$P(T \geq 100) \approx 1 - \Phi\left(\frac{99.5 - np}{\sqrt{np(1-p)}}\right) = .9$$

$$\Rightarrow \Phi\left(\frac{99.5 - np}{\sqrt{np(1-p)}}\right) = .1$$

$$\Rightarrow \frac{99.5 - np}{\sqrt{np(1-p)}} = -1.28$$

$$\Rightarrow \sqrt{n} = \frac{1.28\sqrt{p(1-p)} + \sqrt{1.64p(1-p) + 398p}}{2p}$$

(on solving the quadratic equation in $\sqrt{n}$ from the line before). For instance, if $\rho = .5$, then $p = .25$, and by plugging into the last formula, we get $\sqrt{n} = 21.1 \Rightarrow n \approx 445$.

5.3 Exchangeable Normal Variables

In some applications, a set of random variables is thought to have identical marginal distributions, but is not independent. If we also impose the assumption that any pair has the same bivariate joint distribution, then in the special multivariate normal case there is a simple description of the joint distribution of the entire set of variables. This is related to the concept of *an exchangeable sequence*, which we first define.

Definition 5.3. A sequence of random variables $\{X_1, X_2, \ldots, X_n\}$ is called a finitely exchangeable sequence if the joint distribution of $(X_1, X_2, \ldots, X_n)$ is the same as the joint distribution of $(X_{\pi(1)}, X_{\pi(2)}, \ldots, X_{\pi(n)})$ for any permutation $(\pi(1), \pi(2), \ldots, \pi(n))$ of $(1, 2, \ldots, n)$.

Remark. So, for example, if $n = 2$, and the variables are continuous, then finite exchangeability simply means that the joint density $f(x_1, x_2)$ satisfies $f(x_1, x_2) = f(x_2, x_1) \; \forall \, x_1, x_2$.

An exchangeable multivariate normal sequence has a simple description.

Theorem 5.6. *Let $(X_1, X_2, \ldots, X_n)$ have a multivariate normal distribution. Then, $\{X_1, X_2, \ldots, X_n\}$ is an exchangeable sequence if and only if for some common $\theta, \sigma^2, \rho, E(X_i) = \theta, \text{var}(X_i) = \sigma^2, \forall i, \text{ and } \forall i \neq j, \rho_{X_i, X_j} = \rho$.*

Proof. The theorem follows on observing that if all the variances are equal, and if every pair has the same correlation, then $(X_1, X_2, \ldots, X_n)$ and $(X_{\pi(1)}, X_{\pi(2)}, \ldots, X_{\pi(n)})$ have the same covariance matrix for any permutation $(\pi(1), \pi(2), \ldots, \pi(n))$ of $(1, 2, \ldots, n)$. Because they also have the same mean vector (obviously), and because any multivariate normal distribution is fully determined by only its mean vector and its covariance matrix, it follows that $(X_1, X_2, \ldots, X_n)$ and $(X_{\pi(1)}, X_{\pi(2)}, \ldots, X_{\pi(n)})$ have the same multivariate normal distribution, and hence $\{X_1, X_2, \ldots, X_n\}$ is exchangeable. $\square$

Example 5.6 (Constructing Exchangeable Normal Variables). Let $Z_0, Z_1, \ldots Z_n$ be independent standard normal variables. Now define $X_i = \theta + \sigma(\sqrt{\rho}Z_0 + \sqrt{1-\rho}Z_i), 1 \leq i \leq n$, where $0 \leq \rho \leq 1$. Because **X** admits the representation $X = \mu + B\mathbf{Z}$, where μ is a vector with each coordinate equal to

θ, $\mathbf{X} = (X_1, X_2, \ldots, X_n)$ and $\mathbf{Z} = (Z_1, Z_2, \ldots, Z_n)$, it follows that $\mathbf{X}$ has a multivariate normal distribution. It is clear that $E(X_i) = \theta \; \forall i$. Also, $\text{Var}(X_i) = \sigma^2(\rho + 1 - \rho) = \sigma^2 \; \forall i$. Next, $\text{Cov}(X_i, X_j) = \sigma^2 \rho \, \text{Cov}(Z_0, Z_0) = \rho \sigma^2$, $\forall i \neq j$. So we have proved that $\{X_1, X_2, \ldots, X_n\}$ is exchangeable. We leave it as an easy exercise to explicitly calculate the matrix B cited in the argument.

In general, the CDF of a multivariate normal distribution is an n-dimensional integral, and for $n > 1$, the dimension of the integral can be reduced by one to $n - 1$ by using a conditioning argument. However, in the exchangeable case with the common correlation $\rho \geq 0$, there is a one-dimensional integral representation, given below.

Theorem 5.7 (CDF of Exchangeable Normals). *Let $\mathbf{X} = (X_1, X_2, \ldots, X_n)$ have an exchangeable multivariate normal distribution with common mean θ, common variance σ^2, and common correlation $\rho \geq 0$. Let $(x_1, \ldots, x_n) \in \mathcal{R}^n$, and let $a_i = \frac{x_i - \theta}{\sigma}$. Then,*

$$P(X_1 \leq x_1, \ldots, X_n \leq x_n) = \int_{-\infty}^{\infty} \phi(z) \prod_{i=1}^{n} \Phi\left(\frac{a_i + z\sqrt{\rho}}{\sqrt{1-\rho}}\right) dz.$$

Proof. We use the representation $X_i = \theta + \sigma(\sqrt{\rho} Z_0 + \sqrt{1-\rho} Z_i)$, where $Z_0, \ldots, Z_n$ are all independent standard normals. Now, use the iterated expectation technique as

$$\begin{aligned}
P(X_1 \leq x_1, \ldots, X_n \leq x_n) &= E[I_{X_1 \leq x_1, \ldots, X_n \leq x_n}] \\
&= E_{Z_0} E[I_{X_1 \leq x_1, \ldots, X_n \leq x_n} | Z_0 = z] \\
&= E_{Z_0} E[I_{\sqrt{1-\rho} Z_1 \leq a_1 - \sqrt{\rho} z, \ldots, \sqrt{1-\rho} Z_n \leq a_n - \sqrt{\rho} z} | Z_0 = z] \\
&= E_{Z_0} E[I_{\sqrt{1-\rho} Z_1 \leq a_1 - \sqrt{\rho} z, \ldots, \sqrt{1-\rho} Z_n \leq a_n - \sqrt{\rho} z}]
\end{aligned}$$

(inasmuch as Z_0 is independent of the rest of the Z_i)

$$= \int_{-\infty}^{\infty} \phi(z) \prod_{i=1}^{n} \Phi\left(\frac{a_i - z\sqrt{\rho}}{\sqrt{1-\rho}}\right) dz$$

$$= \int_{-\infty}^{\infty} \phi(z) \prod_{i=1}^{n} \Phi\left(\frac{a_i + z\sqrt{\rho}}{\sqrt{1-\rho}}\right) dz,$$

because $\phi(z) = \phi(-z)$. This proves the theorem. □

Example 5.7. Suppose $X_1, X_2, \ldots, X_n$ are exchangeable normal variables with zero means, unit standard deviations, and correlation ρ. We are interested in the probability $P(\cap_{i=1}^{n} \{X_i \leq 1\})$, as a function of ρ. By the theorem above, this equals the one-dimensional integral $\int_{-\infty}^{\infty} \phi(z) \Phi^n(\frac{1+z\sqrt{\rho}}{\sqrt{1-\rho}}) dz$. Although we cannot simplify this analytically, it is interesting to look at the effect of ρ and n on this probability. Here is a table.

5.4 Sampling Distributions Useful in Statistics

ρ	$n=2$	$n=4$	$n=6$	$n=10$
0	.7079	.5011	.3547	.1777
.25	.7244	.5630	.4569	.3262
.5	.7452	.6267	.5527	.4606
.75	.7731	.6989	.6549	.6004
.95	.8108	.7823	.7665	.7476

The probabilities decrease with n for fixed ρ; this, again, is the *curse of dimensionality*. On the other hand, for fixed n, the probabilities increase with ρ; this is because the event under consideration says that the variables, in some sense, act similarly. The probability of their doing so increases when they have a larger correlation.

5.4 Sampling Distributions Useful in Statistics

Much as in the case of one dimension, structured results are available for functions of a set of n-dimensional independent random vectors, each distributed as a multivariate normal. The applications of most of these results are in statistics. A few major distributional results are collected together in this section.

First, we need some notation. Given N independent random vectors, $\mathbf{X_i}, 1 \leq i \leq N$, each $X_i \sim N_n(\mu, \Sigma)$, we define the *sample mean vector* and the *sample covariance matrix* as

$$\bar{\mathbf{X}} = \frac{1}{N} \sum_{i=1}^{N} \mathbf{X_i},$$

$$S = \frac{1}{N-1} \sum_{i=1}^{N} (\mathbf{X_i} - \bar{\mathbf{X}})(\mathbf{X_i} - \bar{\mathbf{X}})',$$

where $\sum$ in the definitions above is defined as vector addition, and for a vector $\mathbf{u}$, $\mathbf{uu}'$ means a matrix product. Note that in one dimension (i.e., when n = 1) $\bar{X}$ is also distributed as a normal, and $\bar{X}, S$ are independently distributed. Moreover, in one dimension, $(N-1)S$ has a χ^2_{N-1} distribution. Analogues of all of these results exist in this general multivariate case. This part of the multivariate normal theory is very classic.

We need another definition.

Definition 5.4 (Wishart Distribution). Let $W_{p \times p}$ be a symmetric positive definite random matrix with elements $w_{ij}, 1 \leq i, j \leq p$. W is said to have a *Wishart distribution* with k degrees of freedom $(k \geq p)$ and scale matrix Σ, if the joint density of its elements, $w_{ij}, 1 \leq i \leq j \leq p$ is given by

$$f(W) = c|W|^{(k-p-1)/2} e^{-\frac{1}{2} tr(\Sigma^{-1} W)},$$

where the normalizing constant c equals

$$c = \frac{1}{2^{kp/2}|\Sigma|^{k/2}\pi^{p(p-1)/4}\prod_{i=1}^{p}\Gamma\left(\frac{k-i+1}{2}\right)}.$$

We write $W \sim W_p(k, \Sigma)$.

Theorem 5.8. *Let* $\mathbf{X_i}$ *be independent* $N_n(\mu, \Sigma)$ *random vectors,* $1 \leq i \leq N$. *Then,*

(a) $\bar{\mathbf{X}} \sim N_n(\mu, \frac{\Sigma}{N})$.
(b) *For* $N > n$, S *is positive definite with probability one.*
(c) *For* $N > n, (N-1)S \sim W_n(N-1, \Sigma)$.
(d) $\bar{\mathbf{X}}$ *and* S *are independently distributed.*

Part (a) of this theorem follows easily from the representation of a multivariate normal vector in terms of a multivariate standard normal vector. For part (b), see Eaton and Eaton and Perlman (1973) and Dasgupta (1971). Part (c) is classic. Numerous proofs of part (c) are available. Specifically, see Mahalanobis et al. (1937), Olkin and Roy (1954), and Ghosh and Sinha (2002). Part (d) has also been proved by various methods. Classic proofs are available in Tong (1990). The most efficient proof follows from an application of Basu's theorem, *a theorem in statistical inference (Basu (1955); see Chapter 18).*

5.4.1 * Wishart Expectation Identities

A series of elegant and highly useful expectation identities for the Wishart distribution were derived in Haff (1977, 1979a,b, 1981). The identities were derived by clever use of the divergence theorem of multidimensional calculus, and resulted in drastic reduction in the amount of algebraic effort involved in classic derivations of various Wishart expectations and moments. Although they can be viewed as results in multivariate probability, their main concrete applications are in multivariate statistics. A selection of these moment and expectation formulas are collected together in the result below.

Theorem 5.9 (Wishart Identities). *We need some notation to describe the identities. We start with a general Wishart distributed random matrix:*

$S_{p \times p} \sim W_p(k, \Sigma), \Sigma$ *nonsingular.*

$\mathcal{S} =$ *The set of all* $p \times p$ *nnd matrices.*

$f(S), g(S) :$ *Scalar functions on* $\mathcal{S}$.

$\frac{\partial f}{\partial S} =$ *The matrix with elements* $\frac{\partial f}{\partial S_{ij}}$.

5.4 Sampling Distributions Useful in Statistics

$diag(M)$ = Diagonal matrix with the same diagonal elements as those of M.

$M_{(t)} = tM + (1-t)diagM$.

$$\|M\| = \left(\sum_{i,j} m_{ij}^2\right)^{\frac{1}{2}},$$

where m_{ij} are the elements of M. We mention that the identities are also valid for the case $p = 1$, which corresponds to the chi-square case.

Theorem 5.10 (Wishart Identities). Let $S \sim W_p(k, \Sigma)$. Suppose $f(S), g(S)$ are twice differentiable with respect to each s_{ij}. Let Q be a nonrandom real matrix. Then,

(a) If $kp > 4$, $E[f(S)tr\Sigma^{-1}] = (k-p-1)tr(E[f(S)S^{-1}]) + 2tr(E[\frac{\partial f}{\partial S}])$;

(b) If $kp > 4$, $E[g(S)tr(\Sigma^{-1}Q)] = (k-p-1)tr(E[g(S)S^{-1}Q]) + 2tr(E[\frac{\partial g}{\partial S} \cdot Q_{(\frac{1}{2})}])$;

(c) If $kp > 2$, $E[g(S)tr(\Sigma^{-1}S)] = kpE[g(S)] + 2tr(E[\frac{\partial g}{\partial S} \cdot S_{(\frac{1}{2})}])$;

(d) If $kp > 4$, $E[f(S)tr(S^{-1}Q\Sigma^{-1})] = (k-p-2)E[f(S)tr(S^{-2}Q)] - E[f(S)(trS^{-1})(tr(S^{-1}Q))] + 2tr(E[\frac{\partial f}{\partial S} \cdot (S^{-1}Q)_{(\frac{1}{2})}])$.

Example 5.8. In identity (b) above, choose $g(S) = 1$, and $Q = I_p$. Then, the identity gives $tr(\Sigma^{-1}) = (k-p-1)E[trS^{-1}] \Rightarrow E[trS^{-1}] = \frac{1}{k-p-1}tr\Sigma^{-1}$.

Next, in identity (a), choose $f(S) = tr(S^{-1})$. Then, the identity gives

$$E[trS^{-1}]tr\Sigma^{-1} = (k-p-1)E[trS^{-1}]^2 - 2E[trS^{-2}]$$

$$\Rightarrow (k-p-1)E[trS^{-1}]^2 - 2E[trS^{-2}] = \frac{(tr\Sigma^{-1})^2}{k-p-1}$$

$$\Rightarrow \frac{(tr\Sigma^{-1})^2}{k-p-1} \leq (k-p-1)E[trS^{-1}]^2 - \frac{2}{p}E[trS^{-1}]^2$$

$$= (k-p-1-\frac{2}{p})E[trS^{-1}]^2$$

$$\Rightarrow E[trS^{-1}]^2 \geq \frac{(tr\Sigma^{-1})^2}{(k-p-1)(k-p-1-\frac{2}{p})}.$$

Note that we are able to obtain these expectations without doing the hard distributional calculations that accompany classic derivations of such Wishart expectations.

5.4.2 * Hotelling's T^2 and Distribution of Quadratic Forms

The multidimensional analogue of the t statistic is another important statistic in multivariate statistics. Thus, suppose $\mathbf{X}_i, 1 \leq i \leq N$ are independent $N_n(\mu, \Sigma)$

variables, and let $\bar{\mathbf{X}}$ and S be the sample mean vector and covariance matrix. The *Hotelling's T^2 statistic* (proposed in Hotelling (1931)) is the quadratic form

$$T^2 = N(\bar{\mathbf{X}} - \mu)' S^{-1} (\bar{\mathbf{X}} - \mu),$$

assuming that $N > n$, so that S^{-1} exists with probability one. This is an extremely important statistic in multivariate statistical analysis. Its distribution was also worked out in Hotelling (1931).

Theorem 5.11. *Let $\mathbf{X}_i \sim N_n(\mu, \Sigma)$, $1 \leq i \leq N$, and suppose they are independent. Assume that $N > n$. Then*

$$\frac{N-n}{n(N-1)} T^2 \sim F(n, N-n),$$

the F-distribution with n and $N - n$ degrees of freedom.

The T^2 statistic is an example of a *self-normalized statistic*, in that the defining quadratic form has been normalized by S, the sample covariance matrix, which is a random matrix. Also of great importance in statistical problems is the distribution of quadratic forms that are normalized by some nonrandom matrix, that is, distributions of statistics of the form $(\bar{\mathbf{X}} - \mu)' A^{-1} (\bar{\mathbf{X}} - \mu)$, where A is a suitable $n \times n$ nonsingular matrix. It turns out that such quadratic forms are sometimes distributed as chi square, and some classic and very complete results in this direction are available. A particularly important result is the *Fisher Cochran theorem*. The references for the four theorems below are Rao (1973) and Tong (1990).

Theorem 5.12 (Distribution of Quadratic Forms). *Let $\mathbf{X} \sim N_n(0, I_n)$, and $B_{n \times n}$ be a symmetric matrix of rank $r \leq n$. Then $Q = \mathbf{X}' B \mathbf{X}$ has a chi-square distribution if and only if B is idempotent, in which case the degrees of freedom of Q are $r = \text{Rank}(B) = tr(B)$.*

Sketch of Proof: By the spectral decomposition theorem, there is an orthogonal matrix P such that $P' B P = \Lambda$, the diagonal matrix of the eigenvalues of B. Denote the nonzero eigenvalues as $\lambda_1, \ldots, \lambda_r$, where $r = \text{Rank}(B) = tr(B)$. Then, making the orthogonal transformation $\mathbf{Y} = P'\mathbf{X}$, $Q = \mathbf{X}'B\mathbf{X} = \mathbf{Y}'P'BP\mathbf{Y} = \mathbf{Y}'\Lambda\mathbf{Y} = \sum_{i=1}^r \lambda_i Y_i^2$.

$\mathbf{X} \sim N_n(\mu, I_n)$, and $\mathbf{Y} = P'\mathbf{X}$ is an orthogonal transformation, therefore $\mathbf{Y}$ is also $N_n(\mu, I_n)$, and so, the Y_i^2 are independent χ_1^2 variables. This allows one to write the mgf, and hence the cgf (the cumulant generating function) of $Q = \sum_{i=1}^r \lambda_i Y_i^2$. If Q were to be distributed as a chi square, its expectation and hence its degrees of freedom would have to be r. If one compares the cgf of a χ_r^2 distribution with that of $Q = \sum_{i=1}^r \lambda_i Y_i^2$, then by matching the coefficients in the Taylor series expansion of both, one gets that $tr(B^k) = tr(B) = r$ $\forall k \geq 1$, which would cause B to be idempotent.

Conversely, if B is idempotent, then necessarily r eigenvalues of B are one, and the others zero, and so $Q = \sum_{i=1}^r \lambda_i Y_i^2$ is a sum of r independent χ_1^2 variables, and hence, is χ_r^2. □

5.4 Sampling Distributions Useful in Statistics

A generalization to arbitrary covariance matrices is the following.

Theorem 5.13. *Let* $\mathbf{X} \sim N_n(\mu, \Sigma)$, *and* $B_{n \times n}$ *a symmetric matrix of rank r. Then* $Q = (\mathbf{X} - \mu)' B (\mathbf{X} - \mu) \sim \chi_r^2$ *if and only if* $B \Sigma B = B$.

The following theorem is of great use in statistics, and especially so in the area of linear statistical models.

Theorem 5.14 (Fisher Cochran Theorem). *Let* $\mathbf{X} \sim N_n(\mathbf{0}, I_n)$. *Let* B_i, $1 \leq i \leq k$ *be symmetric nonnegative definite matrices of rank* r_i, $1 \leq i \leq k$, *where* $\sum_{i=1}^{k} r_i = n$. *Suppose* $Q_i = \mathbf{X}' B_i \mathbf{X}$, *and suppose* $Q = \mathbf{X}' \mathbf{X}$ *decomposes as* $Q = \sum_{i=1}^{k} Q_i$. *Then, each* $Q_i \sim \chi_{r_i}^2$, *and the* Q_i *are all independent.*

Once again, see Tong (1990) for proofs of the last two theorems. Here is a pretty application of the Fisher Cochran theorem.

Example 5.9 (Independence of Mean and Variance). Suppose $X_1, X_2, \ldots, X_n$ are independent $N(\mu, \sigma^2)$ variables. Consider the algebraic identity

$$\sum_{i=1}^{n}(X_i - \mu)^2 = \sum_{i=1}^{n}(X_i - \bar{X})^2 + n(\bar{X} - \mu)^2$$

$$\Rightarrow \sum_{i=1}^{n} Y_i^2 = \sum_{i=1}^{n}(Y_i - \bar{Y})^2 + n(\bar{Y})^2,$$

where we let $Y_i = \frac{X_i - \mu}{\sigma}$.

Letting $Q = \sum_{i=1}^{n} Y_i^2$, $Q_1 = \sum_{i=1}^{n}(Y_i - \bar{Y})^2$, $Q_2 = n(\bar{Y})^2$, we have the decomposition $Q = Q_1 + Q_2$. Because the Y_i are independent standard normals, $Q \sim \chi_n^2$, and the matrices B_1, B_2 corresponding to the quadratic forms Q_1, Q_2 have ranks $n - 1$ and 1. Thus, all the assumptions of the Fisher Cochran theorem hold, and so, it follows that

$$Q_1 = \sum_{i=1}^{n}(Y_i - \bar{Y})^2 = \frac{\sum_{i=1}^{n}(X_i - \bar{X})^2}{\sigma^2} \sim \chi_{n-1}^2,$$

and that moreover $\sum_{i=1}^{n}(Y_i - \bar{Y})^2$ and $(\bar{Y})^2$ must be independent. On using the symmetry of the Y_i, it follows that $\sum_{i=1}^{n}(Y_i - \bar{Y})^2$ and $\bar{Y}$ also must be independent, which is the same as saying $\sum_{i=1}^{n}(X_i - \bar{X})^2$ and $\bar{X}$ must be independent.

In fact, simple general results on independence of linear forms or of quadratic forms are known in the multivariate standard normal case.

Theorem 5.15 (Independence of Forms). *Let* $\mathbf{X} \sim N_n(\mathbf{0}, I_n)$. *Then,*

(a) *Two linear functions* $\mathbf{c}_1'\mathbf{X}, \mathbf{c}_2'\mathbf{X}$ *are independent if and only if* $\mathbf{c}_1'\mathbf{c}_2 = 0$,
(b) *The linear functions* $\mathbf{c}'\mathbf{X}$ *and the quadratic form* $\mathbf{X}'B\mathbf{X}$ *are independent if and only if* $B\mathbf{c} = \mathbf{0}$.

(c) *The quadratic forms* $\mathbf{X}'A\mathbf{X}$ *and* $\mathbf{X}'B\mathbf{X}$ *are independent if and only if* $AB = \phi$, *the null matrix*.

Part (a) of this theorem is obvious. Standard proofs of parts (b) and (c) use characteristic functions, *which we have not discussed yet*.

5.4.3 * Distribution of Correlation Coefficient

In a bivariate normal distribution, independence of the two coordinate variables is equivalent to their being uncorrelated. As a result, there is an intrinsic interest in knowing about the value of the correlation coefficient in a bivariate normal distribution. In statistical problems, the correlation ρ is taken to be an unknown parameter, and one tries to estimate it from sample data, $(X_i, Y_i), 1 \leq i \leq n$, from the underlying bivariate normal distribution. The usual estimate of ρ is the *sample correlation coefficient* r. It is simply a plug-in estimate; that is, one takes the definition of ρ:

$$\rho = \frac{E(XY) - E(X)E(Y)}{\sigma_X \sigma_Y},$$

and for each quantity in this defining expression, substitutes the corresponding sample statistic, which gives

$$r = \frac{\frac{1}{n}\sum_{i=1}^n X_i Y_i - \bar{X}\bar{Y}}{\sqrt{\frac{1}{n}\sum_{i=1}^n (X_i - \bar{X})^2}\sqrt{\frac{1}{n}\sum_{i=1}^n (Y_i - \bar{Y})^2}} \tag{5.1}$$

$$= \frac{\sum_{i=1}^n X_i Y_i - n\bar{X}\bar{Y}}{\sqrt{\sum_{i=1}^n (X_i - \bar{X})^2 \sum_{i=1}^n (Y_i - \bar{Y})^2}}. \tag{5.2}$$

This is the sample correlation coefficient.

Ronald Fisher worked out the density of r for a general bivariate normal distribution. Obviously, the density depends only on ρ, and not on $\mu_1, \mu_2, \sigma_1, \sigma_2$. Although the density of r is simple if the true $\rho = 0$, Fisher found it to be extremely complex if $\rho \neq 0$ (Fisher (1915)). This is regarded as a classic calculation.

Theorem 5.16. *Let* (X, Y) *have a bivariate normal distribution with general parameters. Then, the sample correlation coefficient* r *has the density*

$$f_R(r) = \frac{n-2}{\sqrt{2}(n-1)B(1/2, n-1/2)} (1-\rho^2)^{(n-1)/2} (1-r^2)^{(n-4)/2} (1-\rho r)^{3/2-n}$$
$$\times {}_2F_1\left(1/2, 1/2; n-1/2; \frac{1+\rho r}{2}\right),$$

where ${}_2F_1$ denotes the ordinary hypergeometric function, ususally denoted by that notation.

5.5 Noncentral Distributions

In particular, if X, Y are independent, then

$$\frac{\sqrt{n-2}\, r}{\sqrt{1-r^2}} \sim t(n-2),$$

a t distribution with $n-2$ degrees of freedom.

5.5 * Noncentral Distributions

The T^2 statistic of Hotelling is centered at the mean μ of the distribution. In statistical problems, it is important also to know what the distribution of the T^2 statistic is when it is instead centered at some other vector, say **a**, rather than the mean vector μ itself. Note that the same question can also be asked of just the one-dimensional t statistic, and of the one-dimensional sample variance. These questions are discussed in this section.

The Noncentral t-Distribution. Suppose $X_1, X_2, \ldots, X_n$ are independent $N(\mu, \sigma^2)$ variables. The *noncentral t statistic* is defined as

$$t(a) = \frac{\sqrt{n}(\bar{X}-a)}{s},$$

where $s^2 = \frac{1}{n-1}\sum_{i=1}^{n}(X_i - \bar{X})^2$ is the sample variance. The distribution of $t(a)$ is given in the following result.

Theorem 5.17. *The statistic $t(a)$ has the* noncentral t-distribution *with $n-1$ degrees of freedom and noncentrality parameter $\delta = \frac{\sqrt{n}(\mu-a)}{\sigma}$, with the density function*

$$f_{t(a)}(x) = c e^{-(n-1)\delta^2/(2(x^2+n-1))} (x^2+n-1)^{-n/2}$$

$$\times \int_0^\infty t^{n-1} e^{-\left(t - \frac{\delta x}{\sqrt{x^2+n-1}}\right)^2 / 2} \, dt, \quad -\infty < x < \infty,$$

where the normalizing constant c equals

$$c = \frac{(n-1)^{(n-1)/2}}{\sqrt{\pi}\, \Gamma(\frac{n-1}{2}) 2^{n/2-1}}.$$

Furthermore, for $n > 2$,

$$E(t(a)) = \delta \sqrt{n-1} \frac{\Gamma(\frac{n}{2}-1)}{\Gamma(\frac{n-1}{2})}.$$

The Noncentral Chi-Square Distribution. *Suppose* $X_1, X_2, \ldots, X_n$ *are independent* $N(\mu, \sigma^2)$ *variables. Then the distribution of* $S_a^2 = \frac{\sum_{i=1}^n (X_i - a)^2}{\sigma^2}$ *is given in the next theorem.*

Theorem 5.18. *The statistic* S_a^2 *has the* noncentral chi-square distribution *with n degrees of freedom and noncentrality parameter* $\lambda = n\frac{(\mu - a)^2}{\sigma^2}$, *with the density function given by a* Poisson mixture of ordinary chi squares

$$f_{S_a^2}(x) = \sum_{k=0}^{\infty} \frac{e^{-\lambda} \lambda^k}{k!} g_{n+2k}(x),$$

where $g_j(x)$ *stands for the density of an ordinary chi-square random variable with j degrees of freedom. Furthermore,*

$$E(S_a^2) = n + \lambda; \quad Var(S_a^2) = 2(n + 2\lambda).$$

5.6 ∗ Some Important Inequalities for Easy Reference

Multivariate normal distributions satisfy a large number of elegant inequalities, many of which are simultaneously intuitive, entirely nontrivial, and also useful. A selection of some of the most prominent such inequalities is presented here for easy reference.

Slepian's Inequality I. Let $\mathbf{X} \sim N_n(\mu, \Sigma)$, and $a_1, \ldots, a_n$ any n real constants. Let $\rho_{r,s}$ be the correlation between X_r, X_s. Let $(i, j), 1 \leq i < j \leq n$ be a fixed pair of indices. Then, each of

$$P(\cap_{i=1}^n X_i \leq a_i), \quad P(\cap_{i=1}^n X_i \geq a_i)$$

is strictly increasing in $\rho_{i,j}$, and $\rho_{r,s}$ is held fixed $\forall (r, s) \neq (i, j)$.

Slepian's Inequality II. Let $\mathbf{X} \sim N_n(\mu, \Sigma)$, and $C \subseteq \mathcal{R}^n$ a convex set, symmetric around μ. Let $\theta \in \mathcal{R}^n$, and $0 \leq s < t < \infty$. then,

$$P(\mathbf{X} + s\theta \in C) \geq P(\mathbf{X} + t\theta \in C).$$

Anderson's Inequality. Let $\mathbf{X} \sim N_n(\mu, \Sigma_1), \mathbf{Y} \sim N_n(\mu, \Sigma_2)$, where $\Sigma_2 - \Sigma_1$ is nnd, and C any convex set symmetric around μ. Then,

$$P(\mathbf{X} \in C) \geq P(\mathbf{Y} \in C).$$

Monotonicity Inequality. Let $\mathbf{X} \sim N_n(\mu, \Sigma)$, and let $\rho_* = \min\{\rho_{i,j}, i \neq j\} \geq 0$. Then,

5.6 Some Important Inequalities for Easy Reference

$$P(\cap_{i=1}^n X_i \leq a_i) \geq \prod_{i=1}^n \Phi\left(\frac{a_i - \mu_i}{\sigma_i}\right);$$

$$P(\cap_{i=1}^n X_i \geq a_i) \geq \prod_{i=1}^n \Phi\left(\frac{\mu_i - a_i}{\sigma_i}\right).$$

Positive Dependence Inequality. Let $\mathbf{X} \sim N_n(\mu, \Sigma)$, and suppose $\sigma^{ij} \leq 0$ for all $i, j, i \neq j$. Then,

$$P(X_i \geq c_i, i = 1, \ldots, n) \geq \prod_{i=1}^n P(X_i \geq c_i).$$

Sidak Inequality. Let $\mathbf{X} \sim N_n(\mu, \Sigma)$. Then,

$$P(|X_i| \leq c_i, i = 1, \ldots, n) \geq \prod_{i=1}^n P(|X_i| \leq c_i),$$

for any constants $c_1, \ldots, c_n$.

Chen Inequality. Let $\mathbf{X} \sim N_n(\mu, \Sigma)$, and let $g(\mathbf{x})$ be a real-valued function having all partial derivatives g_i, such that $E(|g_i(\mathbf{X})|^2) < \infty$. Then,

$$E(\nabla g(X))' \Sigma E(\nabla g(X)) \leq \text{Var}(g(X)) \leq E[(\nabla g(X))' \Sigma (\nabla g(X))].$$

Cirel'son et al. Concentration Inequality. Let $\mathbf{X} \sim N_n(\mathbf{0}, I_n)$, and $f : \mathcal{R}^n \to \mathcal{R}$ a Lipschitz function with Lipschitz norm $= \sup_{x,y} \frac{|f(x) - f(y)|}{||x-y||} \leq 1$. Then, for any $t > 0$,

$$P(f(X) - Ef(X) \geq t) \leq e^{-\frac{t^2}{2}}.$$

Borell Concentration Inequality. Let $\mathbf{X} \sim N_n(\mathbf{0}, I_n)$, and $f : \mathcal{R}^n \to \mathcal{R}$ a Lipschitz function with Lipschitz norm $= \sup_{x,y} \frac{|f(x) - f(y)|}{||x-y||} \leq 1$. Then, for any $t > 0$,

$$P(f(X) - M_f \geq t) \leq e^{-\frac{t^2}{2}},$$

where M_f denotes the median of $f(X)$.

Covariance Inequality. Let $\mathbf{X} \sim N_n(\mu, \Sigma)$, and let $g_1(\mathbf{x}), g_2(\mathbf{x})$ be real-valued functions, monotone nondecreasing in each coordinate x_i. Suppose $\sigma_{ij} \geq 0$ $\forall i \neq j$. Then,

$$\text{Cov}(g_1(\mathbf{X}), g_2(\mathbf{X})) \geq 0.$$

References to each of these inequalities can be seen in DasGupta (2008) and Tong (1990).

Example 5.10. Let $\mathbf{X} \sim N_n(\mu, \Sigma)$, and suppose we want to find a bound for the variance of $\mathbf{X}'\mathbf{X} = \sum_{i=1}^{n} X_i^2$. Using $g(\mathbf{X}) = \mathbf{X}'\mathbf{X}$ in Chen's inequality, $\nabla g(\mathbf{X}) = 2\mathbf{X}$, and so, $(\nabla g(\mathbf{X}))'\Sigma(\nabla g(\mathbf{X})) = 4\mathbf{X}'\Sigma\mathbf{X}$. This gives,

$$\text{Var}(g(\mathbf{X})) \leq 4E[\mathbf{X}'\Sigma\mathbf{X}] = 4E[tr(\Sigma\mathbf{X}\mathbf{X}')]$$
$$= 4tr(\Sigma(\Sigma + \mu\mu'))$$
$$= 4(tr\Sigma^2 + \mu'\Sigma\mu).$$

Note that it would not be very easy to try to find the variance of $\mathbf{X}'\mathbf{X}$ directly.

Exercises

Exercise 5.1. Prove that the density of any multivariate normal distribution is uniformly bounded.

Exercise 5.2. Suppose $X_1, X_1+X_2, X_3-(X_1+X_2)$ are jointly multivariate normal. Prove or disprove that X_1, X_2, X_3 have a multivariate normal distribution.

Exercise 5.3 (Density Contour). Suppose X_1, X_2 have a bivariate normal distribution with means 1, 2, variances 9, 4, and correlation 0.8. Characterize the set of points (x_1, x_2) at which the joint density $f(x_1, x_2) = k$, a given constant, and plot it for $k = .01, .04$.

Exercise 5.4. Suppose X_1, X_2, X_3, X_4 have an exchangeable multivariate normal distribution with general parameters. Find all constants a, b, c, d such that $aX_1 + bX_2$ and $cX_3 + dX_4$ are independent.

Exercise 5.5 (Quadrant Probability). Suppose X_1, X_2, X_3, X_4 have an exchangeable multivariate normal distribution with zero means, and common variance σ^2, and common correlation ρ. Derive a formula for $P(aX_1+bX_2 > 0, cX_3+dX_4 > 0)$.

Exercise 5.6 *(Quadrant Probability). Suppose $\mathbf{X} \sim N_n(\mu, \Sigma)$. If $\rho_{ij} = \frac{1}{2}$ $\forall i, j$, $i \neq j$, show that $P(X_1 > \mu_1, \ldots, X_n > \mu_n) = \frac{1}{n+1}$.

Exercise 5.7. Suppose (X_1, X_2, X_3, X_4) has a multivariate normal distribution with covariance matrix

$$\Sigma = \begin{pmatrix} 2 & 1 & -1 & 0 \\ 1 & 2 & -1 & -1 \\ -1 & -1 & 3 & 0 \\ 0 & -1 & 0 & 2 \end{pmatrix}.$$

(a) Find the two largest possible subsets of X_1, X_2, X_3, X_4 such that one subset is independent of the other subset.
(b) Find the conditional variance of X_4 given X_1.

(c) Find the conditional variance of X_4 given X_1, X_2.
(d) Find the conditional covariance matrix of (X_2, X_4) given (X_1, X_3).
(e) Find the variance of $X_1 + X_2 + X_3 + X_4$.
(f) Find all linear combinations $aX_1 + bX_2 + cX_3 + dX_4$ which have a zero correlation with $X_1 + X_2 + X_3 + X_4$.
(g) Find the value of $\max_{\mathbf{c}: \|\mathbf{c}\|=1} \text{Var}(\mathbf{c}'\mathbf{X})$.

Exercise 5.8 *(A Problem of Statistical Importance).** Suppose given $Y = y$, $X_1, \ldots, X_n$ are independent $N(y, \sigma^2)$ variables, and Y itself is distributed as $N(m, \tau^2)$. Find the conditional distribution of Y given $X_i = x_i, 1 \leq i \leq n$.

Exercise 5.9. Let X, Y be independent standard normal variables. Define $Z=1$ if $Y > 0$; $Z = -1$ if $Y < 0$. Find the marginal distribution of XZ, and show that Z, XZ are independent.

Exercise 5.10 *(A Covariance Matrix Calculation).** Let $X_1, X_2, \ldots, X_n$ be n independent standard normal variables. For each $k, 1 \leq k \leq n$, let $\bar{X}_k = \frac{X_1 + \ldots + X_k}{k}$. Let $Y_k = \bar{X}_k - \bar{X}_n$.

(a) Find the joint density of $(Y_1, Y_2, \ldots, Y_{n-1})$.
(b) Prove that $(Y_1, Y_2, \ldots, Y_{n-1})$ is independent of $\bar{X}_n$.

Exercise 5.11 *(A Joint Distribution Calculation).** Let $X_0, X_1, \ldots, X_n$ be independent standard normal variables. Let $Y_i = \frac{X_i}{X_0}, 1 \leq i \leq n$.

(a) Find the joint density of $(Y_1, Y_2, \ldots, Y_n)$.
(b) Prove that $\sum_{i=1}^{n} c_i Y_i$ has a Cauchy distribution for any n dimensional vector $(c_1, c_2, \ldots, c_n)$.

Exercise 5.12 *(An Interesting Connection).** Let $(X_1, \ldots, X_n)$ be distributed uniformly on the boundary of the n-dimensional unit sphere. Fix $k < n$ and consider $R_k = \sum_{i=1}^{k} X_i^2$. By using a connection between the uniform distribution on the boundary of a sphere and the standard normal distribution, prove that R_k has a Beta distribution; identify these Beta parameters.

Exercise 5.13 *(A Second Interesting Connection).** Let $(X_1, \ldots, X_n)$ be distributed uniformly on the boundary of the n-dimensional unit sphere. Prove that, for any constants $p_1, \ldots, p_n$, such that they are nonnegative, and add to one, the quantity $\sum_{i=1}^{n} \frac{p_i^2}{X_i^2}$ has exactly the same distribution. Hence, what must this distribution be?

Exercise 5.14 (Covariance Between a Linear and a Quadratic Form). Suppose $\mathbf{X} \sim N_2(\mu, \Sigma)$. Find the covariance between $\mathbf{c}'\mathbf{X}$ and $\mathbf{X}'B\mathbf{X}$, for a general vector $\mathbf{c}$ and a general symmetric matrix B. When is it zero?

Exercise 5.15. Let $\mathbf{X} \sim N_n(\mathbf{0}, I_n)$. Find the covariance between $\mathbf{c}'\mathbf{X}$ and $\mathbf{X}'B\mathbf{X}$, for a general vector $\mathbf{c}$ and a general symmetric matrix B. When is it zero?

Exercise 5.16. Let $\mathbf{X} \sim N_2(\mathbf{0}, \Sigma)$. Characterize all symmetric matrices B such that $\mathbf{X}'B\mathbf{X}$ has a chi square distribution.

Exercise 5.17 *(**Noncentral Chi Square MGF**). Show that a noncentral chi-square distribution has the mgf $e^{\frac{t\lambda}{1-2t}}(1-2t)^{-k/2}$, where λ is the noncentrality parameter and k the degrees of freedom.

For what values of t does this formula apply?

Exercise 5.18 *(**Normal Approximation to Noncentral Chi Square**). Show that if the degrees of freedom k is large, then a noncentral chi-square random variable is approximately normal, with mean $k + \lambda$ and variance $2(k + 2\lambda)$, where λ is the noncentrality parameter.

Exercise 5.19 (**Noncentral F Distribution**). Suppose X, Y are independent random variables, and X is noncentral chi square with m degrees of freedom and noncentrality parameter λ, and Y is an ordinary (or central) chi square with n degrees of freedom. Find the density of $\frac{nX}{mY}$.

Remark. The density of $\frac{nX}{mY}$ is called the *noncentral F-distribution* with m and n degrees of freedom, and noncentrality parameter λ.

Exercise 5.20 (**Noncentral F Mean**). Suppose X has a noncentral F-distribution with m and n degrees of freedom, and noncentrality parameter λ. Show that $E(X) = \frac{n(m+\lambda)}{m(n-2)}$.

When is this formula valid?

Exercise 5.21 (**Noncentral F-Distribution**). If X has a noncentral t-distribution, show that X^2 has a noncentral F distribution.

Exercise 5.22 *(**Application of Anderson's Inequality**). Let $\mathbf{X}$ have a bivariate normal distribution with means zero, variances 1, 4, and a correlation of .5. Let $\mathbf{Y}$ have a bivariate normal distribution with means zero, variances 6, 9 and a correlation of $\sqrt{\frac{2}{3}}$. Show that $P(X_1^2 + X_2^2 < c) \geq P(Y_1^2 + Y_2^2 < c)$ for all $c > 0$.

Exercise 5.23. *Let $(X, Y, Z) \sim N_3(\mathbf{0}, \Sigma)$. Show that $P((X-1)^2 + (Y-1)^2 + (Z-1)^2 \leq 1) > P((X-2)^2 + (Y-2)^2 + (Z-2)^2 \leq 1)$.

Hint: Use one of the Slepian inequalities.

Exercise 5.24 (**Normal Marginals with Nonnormal Joint**). Give an example of a random vector (X, Y, Z) such that each of X, Y, Z has a normal distribution, but jointly (X, Y, Z) is not multivariate normal.

References

Basu, D. (1955). On statistics independent of a complete sufficient statistic, *Sankhyá*, 15, 377–380.
DasGupta, A. (2008). *Asymptotic Theory of Statistics and Probability*, Springer-Verlag, New York.
Dasgupta, S. (1971). Nonsingularity of the sample covariance matrix, *Sankhyá, Ser A*, 33, 475–478.
Eaton, M. and Perlman, M. (1973). The nonsingularity of generalized sample covariance matrices, Ann. Stat., 1, 710–717.

References

Fisher, R. A. (1915). Frequency distribution of the values of the correlation coefficient in samples from an indefinitely large population, *Biometrika*, 10, 507–515.

Ghosh, M. and Sinha, B. (2002). A simple derivation of the Wishart distribution, *Amer. Statist.*, 56, 100–101.

Haff, L. (1977). Minimax estimators for a multinormal precision matrix, *J. Mult. Anal.*, 7, 374–385.

Haff, L. (1979a). Estimation of the inverse covariance matrix, *Ann. Stat.*, 6, 1264–1276.

Haff, L. (1979b). An identity for the Wishart distribution with applications, *J. Mult. Anal.*, 9, 531–544.

Haff, L. (1981). Further identities for the Wishart distribution with applications in regression, *Canad. J. Stat.*, 9, 215–224.

Hotelling, H. (1931). The generalization of Student's ratio, *Ann. Math. Statist.*, 2, 360–378.

Mahalanobis, P., Bose, R. and Roy, S. (1937). Normalization of statistical variates and the use of rectangular coordinates in the theory of sampling distributions, *Sankhyā*, 3, 1–40.

Olkin, I. and Roy, S. (1954). On multivariate distribution theory, *Ann. Math. Statist.*, 25, 329–339.

Rao, C. R. (1973). *Linear Statistical Inference and Its Applications*, Wiley, New York.

Tong, Y. (1990). *The Multivariate Normal Distribution,* Springer-Verlag, New York.

Chapter 6
Finite Sample Theory of Order Statistics and Extremes

The ordered values of a sample of observations are called the order statistics of the sample, and the smallest and the largest are called the extremes. Order statistics and extremes are among the most important functions of a set of random variables that we study in probability and statistics. There is natural interest in studying the highs and lows of a sequence, and the other order statistics help in understanding the concentration of probability in a distribution, or equivalently, the diversity in the population represented by the distribution. Order statistics are also useful in statistical inference, where estimates of parameters are often based on some suitable functions of the order statistics. In particular, the median is of very special importance. There is a well-developed theory of the order statistics of a fixed number n of observations from a fixed distribution, as also an asymptotic theory where n goes to infinity. We discuss the case of fixed n in this chapter. A distribution theory for order statistics when the observations are from a discrete distribution is complex, both notationally and algebraically, because of the fact that there could be several observations which are actually equal. These ties among the sample values make the distribution theory cumbersome. We therefore concentrate on the continuous case.

Principal references for this chapter are the books by David (1980), Reiss (1989), Galambos (1987), Resnick (2007), and Leadbetter et al. (1983). Specific other references are given in the sections.

6.1 Basic Distribution Theory

Definition 6.1. Let $X_1, X_2, \ldots, X_n$ be any n real-valued random variables. Let $X_{(1)} \leq X_{(2)} \leq \cdots \leq X_{(n)}$ denote the ordered values of $X_1, X_2, \cdots, X_n$. Then, $X_{(1)}, X_{(2)}, \ldots, X_{(n)}$ are called the *order statistics* of $X_1, X_2, \ldots, X_n$.

Remark. Thus, the minimum among $X_1, X_2, \ldots, X_n$ is the first-order statistic, and the maximum the nth-order statistic. The middle value among $X_1, X_2, \ldots, X_n$ is called the median. But it needs to be defined precisely, because there is really no middle value when n is an even integer. Here is our definition.

Definition 6.2. Let $X_1, X_2, \ldots, X_n$ be any n real-valued random variables. Then, the *median* of $X_1, X_2, \ldots, X_n$ is defined to be $M_n = X_{(m+1)}$ if $n = 2m + 1$ (an odd integer), and $M_n = X_{(m)}$ if $n = 2m$ (an even integer). That is, in either case, the median is the order statistic $X_{(k)}$ where k is the smallest integer $\geq \frac{n}{2}$.

Example 6.1. Suppose .3, .53, .68, .06, .73, .48, .87, .42, .89, .44 are ten independent observations from the $U[0, 1]$ distribution. Then, the order statistics are .06, .3, .42, .44, .48, .53, .68, .73, .87, .89. Thus, $X_{(1)} = .06, X_{(n)} = .89$, and because $\frac{n}{2} = 5, M_n = X_{(5)} = .48$.

An important connection to understand is the connection order statistics have with the *empirical CDF*, a function of immense theoretical and methodological importance in both probability and statistics.

Definition 6.3. Let $X_1, X_2, \ldots, X_n$ be any n real-valued random variables. The *empirical CDF* of $X_1, X_2, \ldots, X_n$, also called the empirical CDF of the sample, is the function
$$F_n(x) = \frac{\#\{X_i : X_i \leq x\}}{n};$$
that is, $F_n(x)$ measures the proportion of sample values that are $\leq x$ for a given x.

Remark. Therefore, by its definition, $F_n(x) = 0$ whenever $x < X_{(1)}$, and $F_n(x) = 1$ whenever $x \geq X_{(n)}$. It is also a constant, namely, $\frac{k}{n}$, for all x-values in the interval $[X_{(k)}, X_{(k+1)})$. So F_n satisfies all the properties of being a valid CDF. Indeed, it is the CDF of a discrete distribution, which puts an equal probability of $\frac{1}{n}$ at the sample values $X_1, X_2, \ldots, X_n$. This calls for another definition.

Definition 6.4. Let P_n denote the discrete distribution that assigns probability $\frac{1}{n}$ to each X_i. Then, P_n is called the *empirical measure of the sample*.

Definition 6.5. Let $Q_n(p) = F_n^{-1}(p)$ be the quantile function corresponding to F_n. Then, $Q_n = F_n^{-1}$ is called the *quantile function* of $X_1, X_2, \ldots, X_n$, or the empirical quantile function.

We can now relate the median and the order statistics to the quantile function F_n^{-1}.

Proposition. Let $X_1, X_2, \ldots, X_n$ be n random variables. Then,

(a) $X_{(i)} = F_n^{-1}(\frac{i}{n})$;
(b) $M_n = F_n^{-1}(\frac{1}{2})$.

We now specialize to the case where $X_1, X_2, \ldots, X_n$ are independent random variables with a common density function $f(x)$ and CDF $F(x)$, and work out the fundamental distribution theory of the order statistics $X_{(1)}, X_{(2)}, \ldots, X_{(n)}$.

Theorem 6.1 (Joint Density of All the Order Statistics). Let $X_1, X_2, \ldots, X_n$ be independent random variables with a common density function $f(x)$. Then, the joint density function of $X_{(1)}, X_{(2)}, \ldots, X_{(n)}$ is given by

$$f_{1,2,\ldots,n}(y_1, y_2, \ldots, y_n) = n! f(y_1) f(y_2) \cdots f(y_n) I_{\{y_1 < y_2 < \cdots < y_n\}}.$$

6.1 Basic Distribution Theory

Proof. A verbal heuristic argument is easy to understand. If $X_{(1)} = y_1, X_{(2)} = y_2, \ldots, X_{(n)} = y_n$, then exactly one of the sample values $X_1, X_2, \ldots, X_n$ is y_1, exactly one is y_2, and so on, but we can put any of the n observations at y_1, any of the other $n-1$ observations at y_2, and so on, and so the density of $X_{(1)}, X_{(2)}, \ldots, X_{(n)}$ is $f(y_1)f(y_2)\cdots f(y_n) \times n(n-1)\cdots 1 = n!f(y_1)f(y_2)\cdots f(y_n)$, and obviously if the inequality $y_1 < y_2 < \cdots < y_n$ is not satisfied, then at such a point the joint density of $X_{(1)}, X_{(2)}, \ldots, X_{(n)}$ must be zero.

Here is a formal proof. The multivariate transformation $(X_1, X_2, \ldots, X_n) \to (X_{(1)}, X_{(2)}, \ldots, X_{(n)})$ is not one-to-one, as any permutation of a fixed $(X_1, X_2, \ldots, X_n)$ vector has exactly the same set of order statistics $X_{(1)}, X_{(2)}, \ldots, X_{(n)}$. However, fix a specific permutation $\{\pi(1), \pi(2), \ldots, \pi(n)\}$ of $\{1, 2, \ldots, n\}$ and consider the subset $A_\pi = \{(x_1, x_2, \ldots, x_n) : x_{\pi(1)} < x_{\pi(2)} < \cdots < x_{\pi(n)}\}$. Then, the transformation $(x_1, x_2, \ldots, x_n) \to (x_{(1)}, x_{(2)}, \ldots, x_{(n)})$ is one-to-one on each such A_π, and indeed, then $x_{(i)} = x_{\pi(i)}, i = 1, 2, \ldots, n$. The Jacobian matrix of the transformation is 1, for each such A_π. A particular vector $(x_1, x_2, \ldots, x_n)$ falls in exactly one A_π, and there are $n!$ such regions A_π, as we exhaust all the $n!$ permutations $\{\pi(1), \pi(2), \ldots, \pi(n)\}$ of $\{1, 2, \ldots, n\}$. By a modification of the Jacobian density theorem, we then get

$$f_{1,2,\ldots,n}(y_1, y_2, \ldots, y_n) = \sum_\pi f(x_1)f(x_2)\cdots f(x_n)$$
$$= \sum_\pi f(x_{\pi(1)})f(x_{\pi(2)})\cdots f(x_{\pi(n)})$$
$$= \sum_\pi f(y_1)f(y_2)\cdots f(y_n)$$
$$= n!f(y_1)f(y_2)\cdots f(y_n). \qquad \square$$

Example 6.2 (Uniform Order Statistics). Let $U_1, U_2, \ldots, U_n$ be independent $U[0, 1]$ variables, and $U_{(i)}, 1 \leq i \leq n$, their order statistics. Then, by our theorem above, the joint density of $U_{(1)}, U_{(2)}, \ldots, U_{(n)}$ is

$$f_{1,2,\ldots,n}(u_1, u_2, \ldots, u_n) = n! I_{0 < u_1 < u_2 < \cdots < u_n < 1}.$$

Once we know the joint density of all the order statistics, we can find the marginal density of any subset by simply integrating out the rest of the coordinates, but being extremely careful in using the correct domain over which to integrate the rest of the coordinates. For example, if we want the marginal density of just $U_{(1)}$, that is, of the sample minimum, then we will want to integrate out $u_2, \ldots, u_n$, and the correct domain of integration would be, for a given u_1, a value of $U_{(1)}$, in $(0,1)$,

$$u_1 < u_2 < u_3 < \cdots < u_n < 1.$$

So, we integrate down in the order $u_n, u_{n-1}, \ldots, u_2$, to obtain

$$f_1(u_1) = n! \int_{u_1}^1 \int_{u_2}^1 \cdots \int_{u_{n-1}}^1 du_n du_{n-1} \cdots du_3 du_2$$
$$= n(1-u_1)^{n-1}, \ 0 < u_1 < 1.$$

Likewise, if we want the marginal density of just $U_{(n)}$, that is, of the sample maximum, then we want to integrate out $u_1, u_2, \ldots, u_{n-1}$, and now the answer is

$$f_n(u_n) = n! \int_0^{u_n} \int_0^{u_{n-1}} \cdots \int_0^{u_2} du_1 du_2 \cdots du_{n-1}$$
$$= n u_n^{n-1}, \quad 0 < u_n < 1.$$

However, it is useful to note that for the special case of the minimum and the maximum, we could have obtained the densities much more easily and directly. Here is why. First take the maximum. Consider its CDF; for $0 < u < 1$:

$$P(U_{(n)} \leq u) = P(\cap_{i=1}^n \{X_i \leq u\}) = \prod_{i=1}^n P(X_i \leq u)$$
$$= u^n,$$

and hence, the density of $U_{(n)}$ is $f_n(u) = \frac{d}{du}[u^n] = n u^{n-1}, 0 < u < 1$.

Likewise, for the minimum, for $0 < u < 1$, the tail CDF is:

$$P(U_{(1)} > u) = P(\cap_{i=1}^n \{X_i > u\}) = (1-u)^n,$$

and so the density of $U_{(1)}$ is

$$f_1(u) = \frac{d}{du}[1 - (1-u)^n] = n(1-u)^{n-1}, \quad 0 < u < 1.$$

For a general r, $1 \leq r \leq n$, the density of $U_{(r)}$ works out to a Beta density:

$$f_r(u) = \frac{n!}{(r-1)!(n-r)!} u^{r-1}(1-u)^{n-r}, \quad 0 < u < 1,$$

which is the Be $(r, n-r+1)$ density.

As a rule, if the underlying CDF F is symmetric about its median, then the sample median will also have a density symmetric about the median of F; see the exercises. When n is even, one has to be careful about this, because there is no universal definition of a sample median when n is even. In addition, the density of the sample maximum will generally be skewed to the right, and that of the sample minimum skewed to the left. For general CDFs, the density of the order statistics usually will not have a simple formula in terms of elementary functions; but approximations for large n are often possible. This is treated in a later chapter. Although such approximations for large n are often available, they may not be very accurate unless n is very large; see Hall (1979).

6.2 More Advanced Distribution Theory

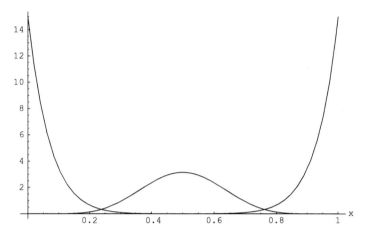

Fig. 6.1 Density of minimum, median, and maximum of U[0, 1] variables; n = 15

Above we have plotted in Fig. 6.1 the density of the minimum, median, and maximum in the $U[0, 1]$ case when $n = 15$. The minimum and the maximum clearly have skewed densities, whereas the density of the median is symmetric about .5.

6.2 More Advanced Distribution Theory

Example 6.3 (Density of One and Two Order Statistics). The joint density of any subset of the order statistics $X_{(1)}, X_{(2)}, \ldots, X_{(n)}$ can be worked out from their joint density, which we derived in the preceding section. The most important case in applications is the joint density of two specific order statistics, say $X_{(r)}$ and $X_{(s)}, 1 \leq r < s \leq n$, or the density of a specific one, say $X_{(r)}$. A verbal heuristic argument helps in understanding the formula for the joint density of $X_{(r)}$ and $X_{(s)}$, and also the density of a specific one $X_{(r)}$.

First consider the density of just $X_{(r)}$. Fix u. To have $X_{(r)} = u$, we must have exactly one observation at u, another $r - 1$ below u, and $n - r$ above u. This will suggest that the density of $X_{(r)}$ is

$$f_r(u) = nf(u) \binom{n-1}{r-1} (F(u))^{r-1} (1 - F(u))^{n-r}$$

$$= \frac{n!}{(r-1)!(n-r)!} (F(u))^{r-1} (1 - F(u))^{n-r} f(u),$$

$-\infty < u < \infty$. This is in fact the correct formula for the density of $X_{(r)}$.

Next, consider the case of the joint density of two order statistics, $X_{(r)}$ and $X_{(s)}$. Fix $0 < u < v < 1$. Then, to have $X_{(r)} = u, X_{(s)} = v$, we must have exactly one observation at u, another $r-1$ below u, one at v, $n-s$ above v, and $s-r-1$ between u and v. This will suggest that the joint density of $X_{(r)}$ and $X_{(s)}$ is

$$f_{r,s}(u,v) = nf(u)\binom{n-1}{r-1}(F(u))^{r-1}(n-r)f(v)\binom{n-r-1}{n-s}(1-F(v))^{n-s}$$
$$(F(v) - F(u))^{s-r-1}$$
$$= \frac{n!}{(r-1)!(n-s)!(s-r-1)!}(F(u))^{r-1}(1-F(v))^{n-s}$$
$$(F(v) - F(u))^{s-r-1}f(u)f(v),$$

$-\infty < u < v < \infty$.

Again, this is indeed the joint density of two specific order statistics $X_{(r)}$ and $X_{(s)}$.

The arguments used in this example lead to the following theorem.

Theorem 6.2 (Density of One and Two Order Statistics and Range). *Let X_1, $X_2, \ldots, X_n$ be independent observations from a continuous CDF $F(x)$ with density function $f(x)$. Then,*

(a) $X_{(n)}$ has the density $f_n(u) = nF^{n-1}(u)f(u), -\infty < u < \infty$.
(b) $X_{(1)}$ has the density $f_1(u) = n(1 - F(u))^{n-1}f(u), -\infty < u < \infty$.
(c) For a general $r, 1 \leq r \leq n$, $X_{(r)}$ has the density

$$f_r(u) = \frac{n!}{(r-1)!(n-r)!}F^{r-1}(u)(1-F(u))^{n-r}f(u), \quad -\infty < u < \infty.$$

(d) For general $1 \leq r < s \leq n$, $(X_{(r)}, X_{(s)})$ have the joint density

$$= \frac{n!}{(r-1)!(n-s)!(s-r-1)!}(F(u))^{r-1}(1-F(v))^{n-s}(F(v)-F(u))^{s-r-1}$$
$$f(u)f(v), \quad -\infty < u < v < \infty.$$

(e) The minimum and the maximum, $X_{(1)}$ and $X_{(n)}$ have the joint density

$$f_{1,n}(u,v) = n(n-1)(F(v) - F(u))^{n-2}f(u)f(v), \quad -\infty < u < v < \infty.$$

(f) **(CDF of Range).** $W = W_n = X_{(n)} - X_{(1)}$ has the CDF

$$F_W(w) = n\int_{-\infty}^{\infty}[F(x+w) - F(x)]^{n-1}f(x)dx.$$

(g) **(Density of Range).** $W = W_n = X_{(n)} - X_{(1)}$ has the density

$$f_W(w) = n(n-1)\int_{-\infty}^{\infty}[F(x+w) - F(x)]^{n-2}f(x)f(x+w)dx.$$

6.2 More Advanced Distribution Theory

Example 6.4 (Moments of Uniform Order Statistics). The general formulas in the above theorem lead to the following moment formulas in the uniform case.

In the $U[0, 1]$ case,

$$E(U_{(1)}) = \frac{1}{n+1}, \quad E(U_{(n)}) = \frac{n}{n+1},$$

$$\text{Var}(U_{(1)}) = \text{Var}(U_{(n)}) = \frac{n}{(n+1)^2(n+2)}; \quad 1 - U_{(n)} \stackrel{\mathcal{L}}{=} U_{(1)};$$

$$\text{Cov}(U_{(1)}, U_{(n)}) = \frac{1}{(n+1)^2(n+2)},$$

$$E(W_n) = \frac{n-1}{n+1}, \quad \text{Var}(W_n) = \frac{2(n-1)}{(n+1)^2(n+2)}.$$

For a general order statistic, it follows from the fact that $U_{(r)} \sim Be(r, n-r+1)$, that

$$E(U_{(r)}) = \frac{r}{n+1}; \quad \text{Var}(U_{(r)}) = \frac{r(n-r+1)}{(n+1)^2(n+2)}.$$

Furthermore, it follows from the formula for their joint density that

$$\text{Cov}(U_{(r)}, U_{(s)}) = \frac{r(n-s+1)}{(n+1)^2(n+2)}.$$

Example 6.5 (Exponential Order Statistics). A second distribution of importance in the theory of order statistics is the exponential distribution. The mean λ essentially arises as just a multiplier in the calculations. So, we treat only the standard exponential case.

Let $X_1, X_2, \ldots, X_n$ be independent standard exponential variables. Then, by the general theorem on the joint density of the order statistics, in this case the joint density of $X_{(1)}, X_{(2)}, \ldots, X_{(n)}$ is

$$f_{1,2,\ldots,n}(u_1, u_2, \ldots, u_n) = n! e^{-\sum_{i=1}^n u_i},$$

$0 < u_1 < u_2 < \cdots < u_n < \infty$. Also, in particular, the minimum $X_{(1)}$ has the density

$$f_1(u) = n(1 - F(u))^{n-1} f(u) = n e^{-(n-1)u} e^{-u} = n e^{-nu},$$

$0 < u < \infty$. In other words, we have the quite remarkable result that the minimum of n independent standard exponentials is itself an exponential with mean $\frac{1}{n}$. Also, from the general formula, the maximum $X_{(n)}$ has the density

$$f_n(u) = n(1 - e^{-u})^{n-1} e^{-u} = n \sum_{i=0}^{n-1} (-1)^i \binom{n-1}{i} e^{-(i+1)u}, \quad 0 < u < \infty.$$

As a result,

$$E(X_{(n)}) = n\sum_{i=0}^{n-1}(-1)^i \binom{n-1}{i}\frac{1}{(i+1)^2} = \sum_{i=1}^{n}(-1)^{i-1}\binom{n}{i}\frac{1}{i},$$

which also equals $1 + \frac{1}{2} + \cdots + \frac{1}{n}$. We show later in the section on spacings that by another argument, it also follows that in the standard exponential case, $E(X_{(n)}) = 1 + \frac{1}{2} + \cdots + \frac{1}{n}$.

Example 6.6 (Normal Order Statistics). Another clearly important case is that of the order statistics of a normal distribution. Because the general $N(\mu, \sigma^2)$ random variable is a location-scale transformation of a standard normal variable, we have the distributional equivalence that $(X_{(1)}, X_{(2)}, \ldots, X_{(n)})$ have the same joint distribution as $(\mu + \sigma Z_{(1)}, \mu + \sigma Z_{(2)}, \ldots, \mu + \sigma Z_{(n)})$. So, we consider just the standard normal case.

Because of the symmetry of the standard normal distribution around zero, for any r, $Z_{(r)}$ has the same distribution as $-Z_{(n-r+1)}$. In particular, $Z_{(1)}$ has the same distribution as $-Z_{(n)}$. From our general formula, the density of $Z_{(n)}$ is

$$f_n(x) = n\Phi^{n-1}(x)\phi(x), \quad -\infty < x < \infty.$$

Again, this is a skewed density. It can be shown, either directly, or by making use of the general theorem on existence of moments of order statistics (see the next section) that every moment, and in particular the mean and the variance of $Z_{(n)}$, exists. Except for very small n, closed-form formulas for the mean or variance are not possible. For small n, integration tricks do produce exact formulas. For example,

$$E(Z_{(n)}) = \frac{1}{\sqrt{\pi}}, \quad \text{if } n = 2; \quad E(Z_{(n)}) = \frac{3}{2\sqrt{\pi}}, \quad \text{if } n = 3.$$

Such formulas are available for $n \leq 5$; see David (1980).

We tabulate the expected value of the maximum for some values of n to illustrate the slow growth.

n	$E(Z_{(n)})$
2	.56
5	1.16
10	1.54
20	1.87
30	2.04
50	2.25
100	2.51
500	3.04
1000	3.24
10000	3.85

More about the expected value of $Z_{(n)}$ is discussed later.

6.3 Quantile Transformation and Existence of Moments

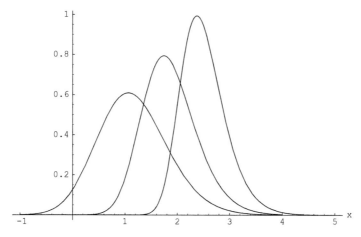

Fig. 6.2 Density of maximum of standard normals; n = 5, 20, 100

The density of $Z_{(n)}$ is plotted in Fig. 6.2 for three values of n. We can see that the density is shifting to the right, and at the same time getting more peaked. Theoretical asymptotic (i.e., as $n \to \infty$) justifications for these visual findings are possible, and we show some of them in a later chapter.

6.3 Quantile Transformation and Existence of Moments

Uniform order statistics play a very special role in the theory of order statistics, because many problems about order statistics of samples from a general density can be reduced, by a simple and common technique, to the case of uniform order statistics. It is thus especially important to understand and study uniform order statistics. The technique that makes helpful reductions of problems in the general continuous case to the case of a uniform distribution on [0,1] is one of making just the quantile transformation. We describe the exact correspondence below.

Theorem 6.3 (Quantile Transformation Theorem). *Let $X_1, X_2, \ldots, X_n$ be independent observations from a continuous CDF $F(x)$ on the real line, and let $X_{(1)}, X_{(2)}, \ldots, X_{(n)}$ denote their order statistics. Let $F^{-1}(p)$ denote the quantile function of F. Let $U_1, U_2, \ldots, U_n$ be independent observations from the $U[0, 1]$ distribution, and let $U_{(1)}, U_{(2)}, \ldots, U_{(n)}$ denote their order statistics. Also let $g(x)$ be any nondecreasing function and let $Y_i = g(X_i), 1 \leq i \leq n$, with $Y_{(1)}, Y_{(2)}, \ldots, Y_{(n)}$ be the order statistics of $Y_1, Y_2, \ldots, Y_n$. Then, the following equalities in distributions hold:*

(a) $F(X_1) \sim U[0, 1]$, that is, $F(X_1)$ and U_1 have the same distribution.
    We write this equality in distribution as $F(X_1) \stackrel{\mathcal{L}}{=} U_1$.
(b) $F^{-1}(U_1) \stackrel{\mathcal{L}}{=} X_1$.

(c) $F(X_{(i)}) \stackrel{\mathcal{L}}{=} U_{(i)}$.

(d) $F^{-1}(U_{(i)}) \stackrel{\mathcal{L}}{=} X_{(i)}$.

(e) $(F(X_{(1)}), F(X_{(2)}), \ldots, F(X_{(n)})) \stackrel{\mathcal{L}}{=} (U_{(1)}, U_{(2)}, \ldots, U_{(n)})$.

(f) $(F^{-1}(U_{(1)}), F^{-1}(U_{(2)}), \ldots, F^{-1}(U_{(n)})) \stackrel{\mathcal{L}}{=} (X_{(1)}, X_{(2)}, \ldots, X_{(n)})$.

(g) $(Y_{(1)}, Y_{(2)}, \ldots, Y_{(n)}) \stackrel{\mathcal{L}}{=} (g(F^{-1}(U_{(1)})), g(F^{-1}(U_{(2)})), \ldots, g(F^{-1}(U_{(n)})))$.

Remark. We are already familiar with parts (a) and (b); they are restated here only to provide the context. The parts that we need to focus on are the last two parts. They say that any question about the set of order statistics $X_{(1)}, X_{(2)}, \ldots, X_{(n)}$ of a sample from a general continuous distribution can be rephrased in terms of the set of order statistics from the $U[0, 1]$ distribution. For this, all we need to do is to substitute $F^{-1}(U_{(i)})$ in place of $X_{(i)}$, where F^{-1} is the quantile function of F.

So, at least in principle, as long as we know how to work skillfully with the joint distribution of the uniform order statistics, we can answer questions about any set of order statistics from a general continuous distribution, because the latter is simply a transformation of the set of order statistics of the uniform. This has proved to be a very useful technique in the theory of order statistics.

As a corollary of part (f) of the above theorem, we have the following connection between order statistics of a general continuous CDF and uniform order statistics.

Corollary. *Let $X_{(1)}, X_{(2)}, \ldots, X_{(n)}$ be the order statistics of a sample from a general continuous CDF F, and $U_{(1)}, U_{(2)}, \ldots, U_{(n)}$ the uniform order statistics. Then, for any $1 \leq r_1 < r_2 < \cdots < r_k \leq n$,*

$$P(X_{(r_1)} \leq u_1, \ldots, X_{(r_k)} \leq u_k) = P(U_{(r_1)} \leq F(u_1), \ldots, U_{(r_k)} \leq F(u_k)),$$

$\forall u_1, \ldots, u_k$.

Several important applications of this quantile transformation method are given below.

Proposition. *Let $X_1, X_2, \ldots, X_n$ be independent observations from a continuous CDF F. Then, for any r, s, $Cov(X_{(r)}, X_{(s)}) \geq 0$.*

Proof. We use the fact that if $g(x_1, x_2, \ldots, x_n), h(x_1, x_2, \ldots, x_n)$ are two functions such that they are coordinatewise nondecreasing in each x_i, then

$Cov(g(U_{(1)}, \ldots, U_{(n)}), h(U_{(1)}, \ldots, U_{(n)})) \geq 0$. By the quantile transformation theorem, $Cov(X_{(r)}, X_{(s)}) = Cov(F^{-1}(U_{(r)}), F^{-1}(U_{(s)})) \geq 0$, as $F^{-1}(U_{(r)})$ is a nondecreasing function of $U_{(r)}$, and $F^{-1}(U_{(s)})$ is a nondecreasing function of $U_{(s)}$, and hence they are also coordinatewise nondecreasing in each order statistic $U_{(1)}, U_{(2)}, \ldots, U_{(n)}$. □

This proposition was first proved in Bickel (1967), but by using a different method. The next application is to existence of moments of order statistics.

Theorem 6.4 (On the Existence of Moments). *Let $X_1, X_2, \ldots, X_n$ be independent observations from a continuous CDF F, and let $X_{(1)}, X_{(2)}, \ldots, X_{(n)}$*

6.3 Quantile Transformation and Existence of Moments

be the order statistics. Let $g(x_1, x_2, \ldots, x_n)$ be a general function. Suppose $E[|g(X_1, X_2, \ldots, X_n)|] < \infty$. Then, $E[|g(X_{(1)}, X_{(2)}, \ldots, X_{(n)})|] < \infty$.

Proof. By the quantile transformation theorem above,

$$E[|g(X_{(1)}, X_{(2)}, \ldots, X_{(n)})|]$$
$$= E[|g(F^{-1}(U_{(1)}), F^{-1}(U_{(2)}), \ldots, F^{-1}(U_{(n)}))|]$$
$$= n! \int_{0<u_1<u_2<\cdots<u_n<1} |g(F^{-1}(u_1), F^{-1}(u_2), \ldots, F^{-1}(u_n))| du_1 du_2 \cdots du_n$$
$$\leq n! \int_{(0,1)^n} |g(F^{-1}(u_1), F^{-1}(u_2), \ldots, F^{-1}(u_n))| du_1 du_2 \cdots du_n$$
$$= n! \int_{(0,1)^n} |g(u_1, u_2, \ldots, u_n)| f(u_1) f(u_2) \cdots f(u_n) du_1 du_2 \cdots du_n$$
$$= n! E[|g(X_1, X_2, \ldots, X_n)|] < \infty. \qquad \square$$

Corollary. *Suppose F is a continuous CDF such that $E_F(|X|^k) < \infty$, for some given k. Then, $E(|X_{(i)}|^k) < \infty \; \forall i \; 1 \leq i \leq n$.* $\qquad \square$

Aside from just the existence of the moment, explicit bounds are always useful. Here is a concrete bound (see Reiss (1989)); approximations for moments of order statistics for certain distributions are derived in Hall (1978).

Proposition. *(a) $\forall r \leq n, E(|X_{(r)}|^k) \leq \frac{n!}{((r-1)!(n-r)!)} E_F(|X|^k)$;*
(b) $E(|X_{(r)}|^k) < \infty \Rightarrow |F^{-1}(p)|^k p^r (1-p)^{n-r+1} \leq C < \infty \; \forall p$;
(c) $|F^{-1}(p)|^k p^r (1-p)^{n-r+1} \leq C < \infty \; \forall p \Rightarrow E(|X_{(s)}|^m) < \infty$, if $1 + \frac{mr}{k} \leq s \leq n - \frac{(n-r+1)m}{k}$.

Example 6.7 (Nonexistence of Every Moment of Every Order Statistic). Consider the continuous CDF $F(x) = 1 - \frac{1}{\log x}, x \geq e$. Setting $1 - \frac{1}{\log x} = p$, we get the quantile function $F^{-1}(p) = e^{\frac{1}{1-p}}$. Fix any n, k, and $r \leq n$. Consider what happens when $p \to 1$.

$$|F^{-1}(p)|^k p^r (1-p)^{n-r+1} = e^{k(\frac{1}{1-p})} p^r (1-p)^{n-r+1}$$
$$\geq C e^{\frac{k}{1-p}} (1-p)^{n-r+1} = C e^{ky} y^{-(n-r+1)},$$

writing y for $\frac{1}{1-p}$. For any $k > 0$, as $y \to \infty (\Leftrightarrow p \to 1), e^{ky} y^{-(n-r+1)} \to \infty$. Thus, the necessary condition of the proposition above is violated, and it follows that for any $r, n, k, E(|X_{(r)}|^k) = \infty$.

Remark. The preceding example and the proposition show that existence of moments of order statistics depends on the tail of the underlying CDF (or, equivalently, the tail of the density). If the tail is so thin that the density has a finite mgf in some neighborhood of zero, then all order statistics will have all moments finite. Evaluating them in closed form is generally impossible, however. If the tail of the underlying density is heavy, then existence of moments of the order statistics, and especially the minimum and the maximum, may be a problem. It is possible for some central order statistics to have a few finite moments, and the minimum or the maximum to have none. In other words, depending on the tail, anything can happen. An especially interesting case is the case of a Cauchy density, notorious for its troublesome tail. The next result describes what happens in that case.

Proposition. Let $X_1, X_2, \ldots, X_n$ be independent $C(\mu, \sigma)$ variables. Then,

(a) $\forall n, E(|X_{(n)}|) = E(|X_{(1)}|) = \infty$;
(b) For $n \geq 3$, $E(|X_{(n-1)}|)$ and $E(|X_{(2)}|)$ are finite;
(c) For $n \geq 5$, $E(|X_{(n-2)}|^2)$ and $E(|X_{(3)}|^2)$ are finite;
(d) In general, $E(|X_{(r)}|^k) < \infty$ if and only if $k < \min\{r, n+1-r\}$.

Example 6.8 (Cauchy Order Statistics). From the above proposition we see that the truly problematic order statistics in the Cauchy case are the two extreme ones, the minimum and the maximum. Every other order statistic has a finite expectation for $n \geq 3$, and all but the two most extremes from each tail even have a finite variance, as long as $n \geq 5$. The table below lists the mean of $X_{(n-1)}$ and $X_{(n-2)}$ for some values of n.

n	$E(X_{(n-1)})$	$E(X_{(n-2)})$
5	1.17	.08
10	2.98	1.28
20	6.26	3.03
30	9.48	4.67
50	15.87	7.90
100	31.81	15.88
250	79.56	39.78
500	159.15	79.57

Example 6.9 (Mode of Cauchy Sample Maximum). Although the sample maximum $X_{(n)}$ never has a finite expectation in the Cauchy case, it always has a unimodal density (see a general result in the exercises). So it is interesting to see what the modal values are for various n. The table below lists the mode of $X_{(n)}$ for some values of n.

n	Mode of $X_{(n)}$
5	.87
10	1.72
20	3.33
30	4.93
50	8.12
100	16.07
250	39.98
500	79.76

By comparing the entries in this table with the previous table, we see that the mode of $X_{(n)}$ is quite close to the mean of $X_{(n-2)}$. It would be interesting to find a theoretical result in this regard.

6.4 Spacings

Another set of statistics helpful in understanding the distribution of probability are the *spacings*, which are the gaps between successive order statistics. They are useful in discerning tail behavior. At the same time, for some particular underlying distributions, their mathematical properties are extraordinarily structured, and in turn lead to results for other distributions. Two instances are the spacings of uniform and exponential order statistics. Some basic facts about spacings are discussed in this section.

Definition 6.6. Let $X_{(1)}, X_{(2)}, \ldots, X_{(n)}$ be the order statistics of a sample of n observations $X_1, X_2, \ldots, X_n$. Then, $W_i = X_{(i+1)} - X_{(i)}, 1 \leq i \leq n-1$ are called the *spacings* of the sample, or the spacings of the order statistics.

6.4.1 Exponential Spacings and Réyni's Representation

The spacings of an exponential sample have the characteristic property that the spacings are all independent exponentials as well. Here is the precise result.

Theorem 6.5. *Let $X_{(1)}, X_{(2)}, \ldots, X_{(n)}$ be the order statistics from an $\mathrm{Exp}(\lambda)$ distribution. Then $W_0 = X_{(1)}, W_1, \ldots, W_{n-1}$ are independent, with $W_i \sim \mathrm{Exp}(\frac{\lambda}{n-i}), i = 0, 1, \ldots, n-1$.*

Proof. The proof follows on transforming to the set of spacings from the set of order statistics, and by applying the Jacobian density theorem. The transformation $(u_1, u_2, \ldots, u_n) \to (w_0, w_1, \ldots, w_{n-1})$, where $w_0 = u_1, w_1 = u_2 - u_1, \ldots, w_{n-1} = u_n - u_{n-1}$ is one to one, with the inverse transformation $u_1 = w_0, u_2 = w_0 + w_1, u_3 = w_0 + w_1 + w_2, \ldots, u_n = w_0 + w_1 + \cdots + w_{n-1}$. The Jacobian matrix is triangular, and has determinant one. From our general theorem, the order statistics $X_{(1)}, X_{(2)}, \ldots, X_{(n)}$ have the joint density

$$f_{1,2,\ldots,n}(u_1, u_2, \ldots, u_n) = n! f(u_1) f(u_2) \cdots f(u_n) I_{\{0 < u_1 < u_2 < \cdots < u_n < \infty\}}.$$

Therefore, the spacings have the joint density

$$g_{0,1,\ldots,n-1}(w_0, w_1, \ldots, w_{n-1})$$
$$= n! f(w_0) f(w_0 + w_1) \cdots f(w_0 + w_1 + \cdots + w_{n-1}) I_{\{w_i > 0 \forall i\}}.$$

This is completely general for any underlying nonnegative variable. Specializing to the standard exponential case, we get

$$g_{0,1,\ldots,n-1}(w_0, w_1, \ldots, w_{n-1}) = n! e^{-\sum_{j=0}^{n-1} \sum_{i=0}^{j} w_i} I_{\{w_i > 0 \forall i\}}$$
$$= n! e^{-nw_0 - (n-1)w_1 - \cdots - w_{n-1}} I_{\{w_i > 0 \forall i\}}.$$

It therefore follows that $W_0, W_1, \ldots, W_{n-1}$ are independent, and also that $W_i \sim$ Exp($\frac{1}{n-i}$). The case of a general λ follows from the standard exponential case. □

Corollary (Réyni). *The joint distribution of the order statistics of an exponential distribution with mean λ have the representation*

$$(X_{(r)})|_{r=1}^n \stackrel{\mathcal{L}}{=} \left(\sum_{i=1}^r \frac{X_i}{n-i+1}\right)\Big|_{r=1}^n,$$

where $X_1, \ldots, X_n$ are themselves independent exponentials with mean λ. □

Remark. Verbally, the order statistics of an exponential distribution are linear combinations of independent exponentials, with a very special sequence of coefficients. In an obvious way, the representation can be extended to the order statistics of a general continuous CDF by simply using the quantile transformation.

Example 6.10 (Moments and Correlations of Exponential Order Statistics). From the representation in the above corollary, we immediately have

$$E(X_{(r)}) = \lambda \sum_{i=1}^r \frac{1}{n-i+1}; \quad \text{Var}(X_{(r)}) = \lambda^2 \sum_{i=1}^r \frac{1}{(n-i+1)^2}.$$

Furthermore, by using the same representation, for $1 \leq r < s \leq n$, $\text{Cov}(X_{(r)}, X_{(s)}) = \lambda^2 \sum_{i=1}^r \frac{1}{n-i+1^2}$, and therefore the correlation

$$\rho_{X_{(r)}, X_{(s)}} = \sqrt{\frac{\sum_{i=1}^r \frac{1}{(n-i+1)^2}}{\sum_{i=1}^s \frac{1}{(n-i+1)^2}}}.$$

In particular,

$$\rho_{X_{(1)}, X_{(n)}} = \frac{\frac{1}{n}}{\sqrt{\sum_{i=1}^n \frac{1}{i^2}}} \approx \frac{\sqrt{6}}{n\pi},$$

for large n. In particular, $\rho_{X_{(1)}, X_{(n)}} \to 0$, as $n \to \infty$. In fact, in large samples the minimum and the maximum are in general approximately independent.

6.4.2 Uniform Spacings

The results on exponential spacings lead to some highly useful and neat representations for the spacings and the order statistics of a uniform distribution. The next result describes the most important properties of uniform spacings and order statistics. Numerous other properties of a more special nature are known. David (1980) and Reiss (1989) are the best references for such additional properties of the uniform order statistics.

Theorem 6.6. Let $U_1, U_2, \ldots, U_n$ be independent $U[0, 1]$ variables, and $U_{(1)}$, $U_{(2)}, \ldots, U_{(n)}$ the order statistics. Let $W_0 = U_{(1)}, W_i = U_{(i+1)} - U_{(i)}, 1 \leq i \leq n-1$, and $V_i = \frac{U_{(i)}}{U_{(i+1)}}, 1 \leq i \leq n-1, V_n = U_{(n)}$. Let also $X_1, X_2, \ldots, X_{n+1}$ be $(n+1)$ independent standard exponentials, independent of the $U_i, i = 1, 2, \ldots, n$. Then,

(a) $V_1, V_2^2, \ldots, V_{n-1}^{n-1}, V_n^n$ are independent $U[0, 1]$ variables, and $(V_1, V_2, \ldots V_{n-1})$ are independent of V_n.

(b) $(W_0, W_1, \ldots, W_{n-1}) \sim \mathcal{D}(\alpha)$, a Dirichlet distribution with parameter vector $\alpha_{n+1 \times 1} = (1, 1, \ldots, 1)$. That is, $(W_0, W_1, \ldots, W_{n-1})$ is uniformly distributed in the n-dimensional simplex.

(c) $(W_0, W_1, \ldots, W_{n-1}) \stackrel{\mathcal{L}}{=} \left(\frac{X_1}{\sum_{j=1}^{n+1} X_j}, \frac{X_2}{\sum_{j=1}^{n+1} X_j}, \ldots, \frac{X_n}{\sum_{j=1}^{n+1} X_j} \right)$.

(d) $(U_{(1)}, U_{(2)}, \ldots, U_{(n)}) \stackrel{\mathcal{L}}{=} \left(\frac{X_1}{\sum_{j=1}^{n+1} X_j}, \frac{X_1+X_2}{\sum_{j=1}^{n+1} X_j}, \ldots, \frac{X_1+X_2+\cdots+X_n}{\sum_{j=1}^{n+1} X_j} \right)$.

Proof. For part (a), use the fact that the negative of the logarithm of a $U[0, 1]$ variable is standard exponential, and then use the result that the exponential spacings are themselves independent exponentials. That $V_1, V_2^2, \ldots, V_{n-1}^{n-1}$ are also uniformly distributed follows from looking at the joint density of $U_{(i)}, U_{(i+1)}$ for any given i. It follows trivially from the density of V_n that $V_n^n \sim U[0, 1]$.

For parts (b) and (c), first consider the joint density of the uniform order statistics, and then transform to the variables $W_i, i = 0, \ldots, n-1$. This is a one-to-one transformation, and so we can apply the Jacobian density theorem. The Jacobian theorem easily gives the joint density of the $W_i, i = 0, \ldots, n-1$, and we simply recognize it to be the density of a Dirichlet with the parameter vector having each coordinate equal to one. Finally, use the representation of a Dirichlet random vector in the form of ratios of Gammas (see Chapter 4). □

Part (d) is just a restatement of part (c).

Remark. Part (d) of this theorem, representing uniform order statistics in terms of independent exponentials is one of the most useful results in the theory of order statistics.

6.5 Conditional Distributions and Markov Property

The conditional distributions of a subset of the order statistics given another subset satisfy some really structured properties. An illustration of such a result is that if we know that the sample maximum $X_{(n)} = x$, then the rest of the order statistics would act like the order statistics of a sample of size $n - 1$ from the original CDF, but truncated on the right at that specific value x. Another prominent property of the conditional distributions is the *Markov property*. Again, a lot is known about the conditional distributions of order statistics, but we present the most significant and easy to state results. The best references for reading more about the conditional distributions are still David (1980) and Reiss (1989). Each of the following theorems follows on straightforward calculations, and we omit the calculations.

Theorem 6.7. Let $X_1, X_2, \ldots, X_n$ be independent observations from a continuous CDF F with density f. Fix $1 \leq i < j \leq n$. Then, the conditional distribution of $X_{(i)}$ given $X_{(j)} = x$ is the same as the unconditional distribution of the ith order statistic in a sample of size $j - 1$ from a new distribution, namely the original F truncated at the right at x. In notation,

$$f_{X_{(j)}|X_{(i)}=x}(u) = \frac{(j-1)!}{(i-1)!(j-1-i)!} \left(\frac{F(u)}{F(x)}\right)^{i-1} \left(1 - \frac{F(u)}{F(x)}\right)^{j-1-i} \frac{f(u)}{F(x)}, \quad u < x.$$

Theorem 6.8. Let $X_1, X_2, \ldots, X_n$ be independent observations from a continuous CDF F with density f. Fix $1 \leq i < j \leq n$. Then, the conditional distribution of $X_{(j)}$ given $X_{(i)} = x$ is the same as the unconditional distribution of the $(j-i)$th order statistic in a sample of size $n-i$ from a new distribution, namely the original F truncated at the left at x. In notation,

$$f_{X_{(j)}|X_{(i)}=x}(u) = \frac{(n-i)!}{(j-i-1)!(n-j)!} \left(\frac{F(u)-F(x)}{1-F(x)}\right)^{j-i-1} \left(\frac{1-F(u)}{1-F(x)}\right)^{n-j}$$
$$\frac{f(u)}{1-F(x)}, \quad u > x.$$

Theorem 6.9 (The Markov Property). Let $X_1, X_2, \ldots, X_n$ be independent observations from a continuous CDF F with density f. Fix $1 \leq i < j \leq n$. Then, the conditional distribution of $X_{(j)}$ given $X_{(1)} = x_1, X_{(2)} = x_2, \ldots, X_{(i)} = x_i$ is the same as the conditional distribution of $X_{(j)}$ given $X_{(i)} = x_i$. That is, given $X_{(i)}$, $X_{(j)}$ is independent of $X_{(1)}, X_{(2)}, \ldots, X_{(i-1)}$.

Theorem 6.10. Let $X_1, X_2, \ldots, X_n$ be independent observations from a continuous CDF F with density f. Then, the conditional distribution of $X_{(1)}, X_{(2)}, \ldots, X_{(n-1)}$ given $X_{(n)} = x$ is the same as the unconditional distribution of the order statistics in a sample of size $n - 1$ from a new distribution, namely the original F truncated at the right at x. In notation,

$$f_{X_{(1)},\ldots,X_{(n-1)}|X_{(n)}=x}(u_1,\ldots,u_{n-1}) = (n-1)! \prod_{i=1}^{n-1} \frac{f(u_i)}{F(x)}, \quad u_1 < \cdots < u_{n-1} < x.$$

Remark. A similar and transparent result holds about the conditional distribution of $X_{(2)}, X_{(3)}, \ldots, X_{(n)}$ given $X_{(1)} = x$.

6.5 Conditional Distributions and Markov Property

Theorem 6.11. *Let $X_1, X_2, \ldots, X_n$ be independent observations from a continuous CDF F with density f. Then, the conditional distribution of $X_{(2)}, \ldots, X_{(n-1)}$ given $X_{(1)} = x, X_{(n)} = y$ is the same as the unconditional distribution of the order statistics in a sample of size $n - 2$ from a new distribution, namely the original F truncated at the left at x, and at the right at y. In notation,*

$$f_{X_{(2)},\ldots,X_{(n-1)} | X_{(1)}=x, X_{(n)}=y}(u_2, \ldots, u_{n-1}) = (n-2)! \prod_{i=2}^{n-1} \frac{f(u_i)}{F(y) - F(x)},$$
$$x < u_2 < \cdots < u_{n-1} < y.$$

Example 6.11 (Mean Given the Maximum). Suppose $X_1, X_2, \ldots, X_n$ are independent $U[0, 1]$ variables. We want to find $E(\bar{X} | X_{(n)} = x)$. We use the theorem above about the conditional distribution of $X_{(1)}, X_{(2)}, \ldots, X_{(n-1)}$ given $X_{(n)} = x$.

$$E(n\bar{X} | X_{(n)} = x) = E\left(\sum_{i=1}^{n} X_i | X_{(n)} = x\right)$$

$$= E\left(\sum_{i=1}^{n} X_{(i)} | X_{(n)} = x\right) = x + E\left(\sum_{i=1}^{n-1} X_{(i)} | X_{(n)} = x\right)$$

$$= x + \sum_{i=1}^{n-1} \frac{ix}{n},$$

because, given $X_{(n)} = x$, $X_{(1)}, X_{(2)}, \ldots, X_{(n-1)}$ act like the order statistics of a sample of size $n - 1$ from the $U[0, x]$ distribution. Now summing the series, we get,

$$E(n\bar{X} | X_{(n)} = x) = x + \frac{(n-1)x}{2} = \frac{n+1}{2} x,$$
$$\Rightarrow E(\bar{X} | X_{(n)} = x) = \frac{n+1}{2n} x.$$

Example 6.12 (Maximum Given the First Half). Suppose $X_1, X_2, \ldots, X_{2n}$ are independent standard exponentials. We want to find $E(X_{(2n)} | X_{(1)} = x_1, \ldots, X_{(n)} = x_n)$. By the theorem on the *Markov property*, this conditional expectation equals $E(X_{(2n)} | X_{(n)} = x_n)$. Now, we further use the representation that

$$(X_{(n)}, X_{(2n)}) \stackrel{\mathcal{L}}{=} \left(\sum_{i=1}^{n} \frac{X_i}{2n - i + 1}, \sum_{i=1}^{2n} \frac{X_i}{2n - i + 1}\right).$$

Therefore,

$$E(X_{(2n)} | X_{(n)} = x_n) = E\left(\sum_{i=1}^{n} \frac{X_i}{2n - i + 1}\right.$$
$$\left. + \sum_{i=n+1}^{2n} \frac{X_i}{2n - i + 1} \middle| \sum_{i=1}^{n} \frac{X_i}{2n - i + 1} = x_n\right)$$

$$= x_n + E\left(\sum_{i=n+1}^{2n} \frac{X_i}{2n-i+1} \Big| \sum_{i=1}^{n} \frac{X_i}{2n-i+1} = x_n\right)$$

$$= x_n + E\left(\sum_{i=n+1}^{2n} \frac{X_i}{2n-i+1}\right)$$

because the X_i are all independent

$$= x_n + \sum_{i=n+1}^{2n} \frac{1}{2n-i+1}.$$

For example, in a sample of size 4, $E(X_{(4)}|X_{(1)}=x, X_{(2)}=y) = E(X_{(4)}|X_{(2)}=y) = y + \sum_{i=3}^{4} \frac{1}{5-i} = y + \frac{3}{2}$.

6.6 Some Applications

Order statistics and the related theory have many interesting and important applications in statistics, in modeling of empirical phenomena, for example, climate characteristics, and in probability theory itself. We touch on a small number of applications in this section for purposes of reference. For further reading on the vast literature on applications of order statistics, we recommend, among numerous possibilities, Lehmann (1975), Shorack and Wellner (1986), David (1980), Reiss (1989), Martynov (1992), Galambos (1987), Falk et al. (1994), Coles (2001), Embrechts et al. (2008), and DasGupta (2008).

6.6.1 * Records

Record values and their timings are used for the purposes of tracking changes in some process, such as temperature, and for preparation for extremal events, such as protection against floods. They are also interesting on their own right.

Let $X_1, X_2, \ldots$, be an infinite sequence of independent observations from a continuous CDF F. We first give some essential definitions.

Definition 6.7. We say that a *record* occurs at time i if $X_i > X_j \ \forall j < i$. By convention, we say that X_1 is a record value, and $i = 1$ is a record time.

Let Z_i be the indicator of the event that a record occurs at time i. The sequence $T_1, T_2, \ldots$ defined as $T_1 = 1; T_j = \min\{i > T_{j-1} : Z_i = 1\}$ is called the sequence of *record times*. The differences $D_{i+1} = T_{i+1} - T_i$ are called the *interarrival times*.

The sequence $X_{T_1}, X_{T_2}, \ldots$, is called the sequence of *record values*.

6.6 Some Applications

Example 6.13. The values 1.46, .28, 2.20, .72, 2.33, .67, .42, .85, .66, .67, 1.54, .76, 1.22, 1.72, .33 are 15 simulated values from a standard exponential distribution. The record values are 1.46, 2.20, 2.33, and the record times are $T_1 = 1, T_2 = 3, T_3 = 5$. Thus, there are three records at time $n = 15$. We notice that no records were observed after the fifth observation in the sequence. In fact, in general, it becomes increasingly more difficult to obtain a record as time passes; justification for this statement is shown in the following theorem.

The following theorem summarizes a number of key results about record values, times, and number of records; this theorem is a superb example of the power of the quantile transformation method, because the results for a general continuous CDF F can be obtained from the $U[0, 1]$ case by making a quantile transformation. The details are worked out in Port (1993, pp. 502–509).

Theorem 6.12. *Let $X_1, X_2, \ldots$ be an infinite sequence of independent observations from a CDF F, and assume that F has the density f. Then,*

(a) *The sequence $Z_1, Z_2, \ldots$ is an infinite sequence of independent Bernoulli variables, with $E(Z_i) = P(Z_i = 1) = \frac{1}{i}, i \geq 1$.*
(b) *Let $N = N_n = \sum_{i=1}^{n} Z_i$ be the number of records at time n. Then,*

$$E(N) = \sum_{i=1}^{n} \frac{1}{i}; \quad Var(N) = \sum_{i=1}^{n} \frac{i-1}{i^2}.$$

(c) *Fix $r \geq 2$. Then D_r has the pmf*

$$P(D_r = k) = \sum_{i=0}^{k-1} (-1)^i \binom{k-1}{i} (i+2)^{-r}, k \geq 1.$$

(d) *The rth record value X_{T_r} has the density*

$$f_r(x) = \frac{[-\log(1 - F(x))]^{r-1}}{(r-1)!} f(x), \quad -\infty < x < \infty.$$

(e) *The first n record values, $X_{T_1}, X_{T_2}, \ldots, X_{T_n}$ have the joint density*

$$f_{12\cdots n}(x_1, x_2, \ldots, x_n) = f(x_n) \prod_{r=1}^{n-1} \frac{f(x_i)}{1 - F(x_i)} I_{\{x_1 < x_2 < \cdots < x_n\}}.$$

(f) *Fix a sequence of reals $t_1 < t_2 < t_3 < \cdots < t_k$, and let for any given real t, $M(t)$ be the total number of record values that are $\leq t$:*

$$M(t) = \#\{i : X_i \leq t \text{ and } X_i \text{ is a record value}\}.$$

Then, $M(t_i) - M(t_{i-1}), 2 \leq i \leq k$ are independently distributed, and

$$M(t_i) - M(t_{i-1}) \sim \text{Poi}\left(\log \frac{1 - F(t_{i-1})}{1 - F(t_i)}\right).$$

Remark. From part (a) of the theorem, we learn that if indeed the sequence of observations keeps coming from the same CDF, then obtaining a record becomes harder as time passes; $P(Z_i = 1) \to 0$. We learn from part (b) that both the mean and the variance of the number of records observed until time n are of the order of $\log n$. The number of records observed until time n is well approximated by a Poisson distribution with mean $\log n$, or a normal distribution with mean and variance equal to $\log n$. We learn from parts (c) and (d) that the interarrival times of the record values do not depend on F, but the *magnitudes* of the record values do. Part (f) is another example of the Poisson distribution providing an approximation in an interesting problem. It is interesting to note the connection between part (b) and part (f). In part (f), if we take $t = F^{-1}(1 - \frac{1}{n})$, then *heuristically*, N_n, the number of records observed up to time n, satisfies $N_n \approx M(X_{(n)}) \approx M(F^{-1}(1 - \frac{1}{n})) \approx \text{Poi}(-\log(1 - F(F^{-1}(1 - \frac{1}{n})))) = \text{Poi}(\log n)$, which is what we mentioned in the paragraph above.

Example 6.14 (Density of Record Values and Times). It is instructive to look at the effect of the tail of the underlying CDF F on the magnitude of the record values. Figure 6.3 gives the density of the third record value for three choices of F, $F = N(0, 1), \text{DoubleExp}(0, 1), C(0, 1)$. Although the modal values are not very different, the effect of the tail of F on the tail of the record density is clear. In particular, for the standard Cauchy case, record values do not have a finite expectation.

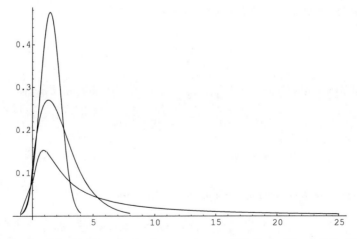

Fig. 6.3 Density of the third record value for, top to bottom, N(0, 1), double exp (0,1), C(0,1) case

6.6 Some Applications

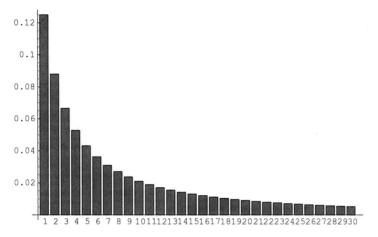

Fig. 6.4 PMF of interarrival time between second and third record

Next, consider the distribution of the gap between the arrival times of the second and the third record. Note the long right tail, akin to an exponential density in Fig. 6.4.

6.6.2 The Empirical CDF

The empirical CDF $F_n(x)$, defined in Section 6.1.2, is a tool of tremendous importance in statistics and probability. The reason for its effectiveness as a tool is that if sample observations arise from some CDF F, then the empirical CDF F_n will be very close to F for large n. So, we can get a very good idea of what the true F is by looking at F_n. Furthermore, because $F_n \approx F$, it can be expected that if $T(F)$ is a nice functional of F, then the empirical version $T(F_n)$ would be close to $T(F)$. Perhaps the simplest example of this is the mean $T(F) = E_F(X)$. The empirical version then is $T(F_n) = E_{F_n}(X) = \frac{\sum_{i=1}^{n} X_i}{n}$, because F_n assigns the equal probability $\frac{1}{n}$ to just the observation values $X_1, \ldots, X_n$. This means that the mean of the sample values should be close to the expected value under the true F. And, this is indeed true under simple conditions, and we have already seen some evidence for it in the form of the central limit theorem. We provide some basic properties and applications of the empirical CDF in this section.

Theorem 6.13. *Let $X_1, X_2, \ldots$ be independent observations from a CDF F. Then,*

(a) *For any fixed x, $nF_n(x) \sim \text{Bin}(n, F(x))$.*
(b) **(DKW Inequality).** *Let $\Delta_n = \sup_{x \in \mathcal{R}} |F_n(x) - F(x)|$. Then, for all $n, \epsilon > 0$, and all F,*

$$P(\Delta_n > \epsilon) \leq 2e^{-2n\epsilon^2}.$$

(c) Assume that F is continuous. For any given n, and $\alpha, 0 < \alpha < 1$, there exist positive constants D_n, independent of F, such that whatever be F,

$$P(\forall x \in \mathcal{R}, F_n(x) - D_n \leq F(x) \leq F_n(x) + D_n) \geq 1 - \alpha.$$

Remark. Part (b), the DKW inequality, was first proved in Dvoretzky et al. (1956), but in a weaker form. The inequality stated here is proved in Massart (1990). Furthermore, the constant 2 in the inequality is the best possible choice of the constant; that is, the inequality is false with any other constant $C < 2$. The inequality says that *uniformly in x*, for large n, the empirical CDF is arbitrarily close to the true CDF with a very high probability, and the probability of the contrary is *sub-Gaussian*. We show more precise consequences of this in a later chapter. Part (c) is important for statisticians, as we show in our next example.

Example 6.15 (Confidence Band for a Continuous CDF). This example is another important application of the quantile transformation method. Imagine a hypothetical sequence of independent $U[0, 1]$ variables, $U_1, U_2, \ldots$, and let G_n denote the empirical CDF of this sequence of uniform random variables; that is,

$$G_n(t) = \frac{\#\{i : U_i \leq t\}}{n}.$$

By the quantile transformation,

$$\Delta_n = \sup_{x \in \mathcal{R}} |F_n(x) - F(x)| \stackrel{\mathcal{L}}{=} \sup_{x \in \mathcal{R}} |G_n(F(x)) - F(x)|$$
$$= \sup_{0 < t < 1} |G_n(t) - t|,$$

which shows that as long as F is a continuous CDF, so that the quantile transformation can be applied, for any n, the distribution of Δ_n is the same for all F. This common distribution is just the distribution of $\sup_{0 < t < 1} |G_n(t) - t|$. Consequently, if D_n is such that $P(\sup_{0 < t < 1} |G_n(t) - t| > D_n) \leq \alpha$, then D_n also satisfies (the apparently stronger statement)

$$P(\forall x \in \mathcal{R}, F_n(x) - D_n \leq F(x) \leq F_n(x) + D_n) \geq 1 - \alpha.$$

The probability statement above provides the assurance that with probability $1 - \alpha$ or more, the true CDF $F(x)$, as a function, is caught between the pair of functions $F_n(x) \pm D_n$. As a consequence, the band $F_n(x) - D_n \leq F(x) \leq F_n(x) + D_n, x \in \mathcal{R}$, is called a $100(1 - \alpha)\%$ *confidence band* for F. This is of great use in statistics, because statisticians often consider the true CDF F to be not known.

The constants D_n have been computed and tabulated for small and moderate n. We tabulate the values of D_n for some selected n for easy reference and use.

n	95th Percentile	99th Percentile
20	.294	.352
21	.287	.344
22	.281	.337
23	.275	.330
24	.269	.323
25	.264	.317
26	.259	.311
27	.254	.305
28	.250	.300
29	.246	.295
30	.242	.290
35	.224	.269
40	.210	.252
>40	$\frac{1.36}{\sqrt{n}}$	$\frac{1.63}{\sqrt{n}}$

6.7 * Distribution of the Multinomial Maximum

The maximum cell frequency in a multinomial distribution is of current interest in several areas of probability and statistics. It is of wide interest in cluster detection, data mining, goodness of fit, and in occupancy problems in probability. It also arises in sequential clinical trials. It turns out that the technique of *Poissonization* can be used to establish, in principle, the exact distribution of the multinomial maximum cell frequency. This can be of substantial practical use. Precisely, if $N \sim \text{Poisson}(\lambda)$, and given $N = n, (f_1, f_2, \ldots, f_k)$ has a multinomial distribution with parameters $(n, p_1, p_2, \ldots, p_k)$, then unconditionally, $f_1, f_2, \ldots, f_k$ are independent and $f_i \sim \text{Poisson}(\lambda p_i)$. It follows that with any given fixed value n, and any given fixed set A in the k-dimensional Euclidean space $\mathcal{R}^k$, the multinomial probability that $(f_1, f_2, \ldots, f_k)$ belongs to A equals $n!c(n)$, with $c(n)$ being the coefficient of λ^n in the power series expansion of $e^\lambda P((X_1, X_2, \ldots, X_k) \in A)$, where now X_i are independent Poisson(λp_i). In the *equiprobable case* (i.e., when the p_i are all equal to $\frac{1}{k}$), this leads to the equality that

$$P(\max\{f_1, f_2, \ldots, f_k\} \geq x) = \frac{n!}{k^n} \times \text{the coefficient of } \lambda^n \text{ in } \left(\sum_{j=0}^{x-1} \frac{\lambda^j}{j!}\right)^k;$$

see Chapter 2.

As a result, we can compute $P(\max\{f_1, f_2, \ldots, f_k\} \geq x)$ exactly whenever we can compute the coefficient of λ^n in the expansion of $(\sum_{j=0}^{x-1} \frac{\lambda^j}{j!})^K$. This is possible to do by using symbolic software; see Ethier (1982) and DasGupta (2009).

Example 6.16 (Maximum Frequency in Die Throws). Suppose a fair six-sided die is rolled 30 times. Should we be surprised if one of the six faces appears 10 times? The usual probability calculation to quantify the surprise is to calculate $P(\max\{f_1, f_2, \ldots, f_6\} \geq 10)$, namely the P-value, where $f_1, f_2, \ldots, f_6$ are the frequencies of the six faces in the 30 rolls. Because of our Poissonization result, we can compute this probability. From the table of exact probabilities below, we can see that it would not be very surprising if some face appeared 10 times in 30 rolls of a fair die; after all $P(\max\{f_1, f_2, \ldots, f_6\} \geq 10) = .1176$, not a very small number, for 30 rolls of a fair die. Similarly, it would not be very surprising if some face appeared 15 times in 50 rolls of a fair die, as can be seen in the table below.

$P(\max\{f_1, f_2, \ldots, f_k\} \geq x)(k = 6)$		
x	$n = 30$	$n = 50$
8	.6014	1
9	.2942	1
10	.1176	.9888
11	.0404	.8663
12	.0122	.6122
13	.0032	.3578
14	.00076	.1816
15	.00016	.0827
16	.00003	.0344

Exercises

Exercise 6.1. Suppose X, Y, Z are three independent $U[0, 1]$ variables. Let U, V, W denote the minimum, median, and the maximum of X, Y, Z.

(a) Find the densities of U, V, W.
(b) Find the densities of $\frac{U}{V}$ and $\frac{V}{W}$, and their joint density.
(c) Find $E(\frac{U}{V})$ and $E(\frac{V}{W})$.

Exercise 6.2. Suppose $X_1, \ldots, X_5$ are independent $U[0, 1]$ variables. Find the joint density of $X_{(2)}, X_{(3)}, X_{(4)}$, and $E(X_{(4)} + X_{(2)} - 2X_{(3)})$.

Exercise 6.3. * Suppose $X_1, \ldots, X_n$ are independent $U[0, 1]$ variables.

(a) Find the probability that all n observations fall within some interval of length at most .9.
(b) Find the smallest n such that $P(X_{(n)} \geq .99, X_{(1)} \leq .01) \geq .99$.

Exercise 6.4 (Correlation Between Order Statistics). Suppose $X_1, \ldots, X_5$ are independent $U[0, 1]$ variables. Find the exact values of $\rho_{X_{(i)}, X_{(j)}}$ for all $1 \leq i < j \leq 5$.

Exercises

Exercise 6.5 * (Correlation Between Order Statistics). Suppose $X_1, \ldots, X_n$ are independent $U[0, 1]$ variables. Find the smallest n such that $\rho_{X_{(1)}, X_{(n)}} < \epsilon, \epsilon = .5, .25, .1$.

Exercise 6.6. Suppose X, Y, Z are three independent standard exponential variables, and let U, V, W be their minimum, median, and maximum. Find the densities of $U, V, W, W - U$.

Exercise 6.7 (Comparison of Mean, Median, and Midrange). Suppose $X_1, X_2, \ldots, X_{2m+1}$ are independent observations from $U[\mu - \sigma, \mu + \sigma]$.

(a) Show that the expectation of each of $\bar{X}, X_{(m+1)}, \frac{X_{(1)}+X_{(n)}}{2}$ is μ.
(b) Find the variance of each $\bar{X}, X_{(m+1)}, \frac{X_{(1)}+X_{(n)}}{2}$. Is there an ordering among their variances?

Exercise 6.8 * (Waiting Time). Peter, Paul, and Mary went to a bank to do some business. Two counters were open, and Peter and Paul went first. Each of Peter, Paul, and Mary will take, independently, an Exp(λ) amount of time to finish their business, from the moment they arrive at the counter.

(a) What is the density of the epoch of the last departure?
(b) What is the probability that Mary will be the last to finish?
(c) What is the density of the total time taken by Mary from arrival to finishing her business?

Exercise 6.9. Let $X_1, \ldots, X_n$ be independent standard exponential variables.

(a) Derive an expression for the CDF of the maximum of the spacings, $W_0 = X_{(1)}, W_i = X_{(i+1)} - X_{(i)}, i = 1, \ldots, n - 1$.
(b) Use it to calculate the probability that among 20 independent standard Exponential observations, no two consecutive observations are more than .25 apart.

Exercise 6.10 * (A Characterization). Let X_1, X_2 be independent observations from a continuous CDF F. Suppose that $X_{(1)}$ and $X_{(2)} - X_{(1)}$ are independent. Show that F must be the CDF of an exponential distribution.

Exercise 6.11 * (Range and Midrange). Let $X_1, \ldots, X_n$ be independent $U[0, 1]$ variables. Let $W_n = (X_{(n)} - X_{(1)}), Y_n = \frac{X_{(n)}+X_{(1)}}{2}$. Find the joint density of W_n, Y_n (be careful about where the joint density is positive). Use it to find the conditional expectation of Y_n given $W_n = w$.

Exercise 6.12 * (Density of Midrange). Let $X_1, \ldots, X_n$ be independent observations from a continuous CDF F with density f. Show that the density of $Y_n = \frac{X_{(n)}+X_{(1)}}{2}$ is given by

$$f_Y(y) = n \int_{-\infty}^{y} [F(2y - x) - F(x)]^{n-1} f(x) dx.$$

Exercise 6.13 * (**Mean Given the Minimum and Maximum**). Let $X_1, \ldots, X_n$ be independent $U[0, 1]$ variables. Derive a formula for $E(\bar{X} \mid X_{(1)}, X_{(n)})$.

Exercise 6.14 * (**Mean Given the Minimum and Maximum**). Let $X_1, \ldots, X_n$ be independent standard exponential variables. Derive a formula for $E(\bar{X} \mid X_{(1)}, X_{(n)})$.

Exercise 6.15 * (**Distance Between Mean and Maximum**). Let $X_1, \ldots, X_n$ be independent $U[0, 1]$ variables. Derive as clean a formula as possible for $E(|\bar{X} - X_{(n)}|)$.

Exercise 6.16 * (**Distance Between Mean and Maximum**). Let $X_1, \ldots, X_n$ be independent standard exponential variables. Derive as clean a formula as possible for $E(|\bar{X} - X_{(n)}|)$.

Exercise 6.17 * (**Distance Between Mean and Maximum**). Let $X_1, \ldots, X_n$ be independent standard normal variables. Derive as clean a formula as possible for $E(|\bar{X} - X_{(n)}|)$.

Exercise 6.18 * (**Relation Between Uniform and Standard Normal**). Let $Z_1, Z_2, \ldots$ be independent standard normal variables. Let $U_1, U_2, \ldots$ be independent $U[0, 1]$ variables. Prove the distributional equivalence:

$$(U_{(r)})|_{r=1}^n \stackrel{\mathcal{L}}{=} \left(\frac{\sum_{i=1}^{2r} Z_i^2}{\sum_{i=1}^{2(n+1)} Z_i^2} \right) \Big|_{r=1}^n.$$

Exercise 6.19 * (**Confidence Interval for a Quantile**). Let $X_1, \ldots, X_n$ be independent observations from a continuous CDF F. Fix $0 < p < 1, 0 < \alpha < 1$, and let $F^{-1}(p)$ be the pth quantile of F. Show that for large enough n, there exist $1 \leq r < s \leq n$ such that $P(X_{(r)} \leq F^{-1}(p) \leq X_{(s)}) \geq 1 - \alpha$.

Do such r, s exist for all n?

Hint: Use the quantile transformation.

Exercise 6.20. Let $X_1, \ldots, X_n$ be independent observations from a continuous CDF F. Find the smallest value of n such that $P(X_{(2)} \leq F^{-1}(\frac{1}{2}) \leq X_{(n-1)}) \geq .95$.

Exercise 6.21. Let $X_1, \ldots, X_n$ be independent observations from a continuous CDF F with a density symmetric about some μ. Show that for all odd sample sizes $n = 2m + 1$, the median $X_{(m+1)}$ has a density symmetric about μ.

Exercise 6.22. Let $X_1, \ldots, X_n$ be independent observations from a continuous CDF F with a density symmetric about some μ. Show that for any r, $X_{(n-r+1)} - \mu \stackrel{\mathcal{L}}{=} \mu - X_{(r)}$.

Exercise 6.23 * (**Unimodality of Order Statistics**). Let $X_1, \ldots, X_n$ be independent observations from a continuous CDF F with a density f. Suppose $\frac{1}{f(x)}$ is convex on the support of f, namely, on $\mathcal{S} = \{x : f(x) > 0\}$. Show that for any r, the density of $X_{(r)}$ is unimodal. You may assume that $\mathcal{S}$ is an interval.

Exercise 6.24. Let $X_1, X_2, \ldots, X_n$ be independent standard normal variables. Prove that the mode of $X_{(n)}$ is the unique root of

$$(n-1)\phi(x) = x\Phi(x).$$

Exercise 6.25 (Conditional Expectation Given the Order Statistics). Let $g(x_1, x_2, \ldots, x_n)$ be a general real-valued function of n variables. Let $X_1, X_2, \ldots, X_n$ be independent observations from a common CDF F. Find as clean an expression as possible for $E(g(X_1, X_2, \ldots, X_n) \mid X_{(1)}, X_{(2)}, \ldots, X_{(n)})$.

Exercise 6.26. Derive a formula for the expected value of the rth record when the sample observations are from an exponential density.

Exercise 6.27 *(Record Values in Normal Case).** Suppose $X_1, X_2, \ldots$ are independent observations from the standard normal distribution. Compute the expected values of the first ten records.

Exercise 6.28. Let $F_n(x)$ be the empirical CDF of n observations from a CDF F. Show that

$$\Delta_n = \sup_{x \in \mathcal{R}} |F_n(x) - F(x)|$$

$$= \max_{1 \leq i \leq n} \max \left\{ \frac{i}{n} - F(X_{(i)}), F(X_{(i)}) - \frac{i-1}{n} \right\}.$$

References

Bickel, P. (1967). Some contributions to order statistics, *Proc. Fifth Berkeley Symp.*, I, 575–591, L. Le Cam and J. Neyman Eds., University of California Press, Berkeley.
Coles, S. (2001). *An Introduction to Statistical Modeling of Extreme Values*, Springer, New York.
DasGupta, A. (2008). *Asymptotic Theory of Statistics and Probability*, Springer, New York.
DasGupta, A. (2009). Exact tail probabilities and percentiles of the multinomial maximum, Technical Report, Purdue University.
David, H. (1980). *Order Statistics*, Wiley, New York.
Dvoretzky, A., Kiefer, J., and Wolfowitz, J. (1956). Asymptotic minimax character of the sample distribution function, *Ann. Math. Statist.*, 27, 3, 642–669.
Embrechts, P., Klüppelberg, C., and Mikosch, T. (2008). *Modelling Extremal Events: For Insurance and Finance*, Springer, New York.
Ethier, S. (1982). Testing for favorable numbers on a roulette wheel, *J. Amer. Statist. Assoc.*, 77, 660–665.
Falk, M., Hüsler, J. and Reiss, R. (1994). *Laws of Small Numbers, Extremes, and Rare Events*, Birkhauser, Basel.
Galambos, J. (1987). *Asymptotic Theory of Extreme Order Statistics*, Academic Press, New York.
Hall, P. (1978). Some asymptotic expansions of moments of order statistics, *Stoch. Proc. Appl.*, 7, 265–275.
Hall, P. (1979). On the rate of convergence of normal extremes, *J. Appl. Prob.*, 16, 2, 433–439.
Leadbetter, M., Lindgren, G., and Rootzén, H. (1983). *Extremes and Related Properties of Random Sequences and Processes*, Springer, New York.

Lehmann, E. (1975). *Nonparametrics: Statistical Methods Based on Ranks*, McGraw-Hill, Columbus, OH.

Martynov, G. (1992). Statistical tests based on empirical processes and related problems, *Soviet J. Math*, 61, 4, 2195–2271.

Massart, P. (1990). The tight constant in the DKW inequality, *Ann. Prob.*, 18, 1269–1283.

Port, S. (1993). *Theoretical Probability for Applications*, Wiley, New York.

Reiss, R. (1989). *Approximation Theorems of Order Statistics*, Springer-Verlag, New York.

Resnick, S. (2007). *Extreme Values, Regular Variation, and Point Processes*, Springer, New York.

Shorack, G. and Wellner, J. (1986). *Empirical Processes with Applications to Statistics*, Wiley, New York.

Chapter 7
Essential Asymptotics and Applications

Asymptotic theory is the study of how distributions of functions of a set of random variables behave, when the number of variables becomes large. One practical context for this is statistical sampling, when the number of observations taken is large. Distributional calculations in probability are typically such that exact calculations are difficult or impossible. For example, one of the simplest possible functions of n variables is their sum, and yet in most cases, we cannot find the distribution of the sum for fixed n in an exact closed form. But the central limit theorem allows us to conclude that in some cases the sum will behave as a normally distributed random variable, when n is large. Similarly, the role of general asymptotic theory is to provide an approximate answer to exact solutions in many types of problems, and often very complicated problems. The nature of the theory is such that the approximations have remarkable unity of character, and indeed nearly unreasonable unity of character. Asymptotic theory is the single most unifying theme of probability and statistics. Particularly, in statistics, nearly every method or rule or tradition has its root in some result in asymptotic theory. No other branch of probability and statistics has such an incredibly rich body of literature, tools, and applications, in amazingly diverse areas and problems. Skills in asymptotics are nearly indispensable for a serious statistician or probabilist.

Numerous excellent references on asymptotic theory are available. A few among them are Bickel and Doksum (2007), van der Vaart (1998), Lehmann (1999), Hall and Heyde (1980), Ferguson (1996), and Serfling (1980). A recent reference is DasGupta (2008). These references have a statistical undertone. Treatments of asymptotic theory with a probabilistic undertone include Breiman (1968), Ash (1973), Chow and Teicher (1988), Petrov (1975), Bhattacharya and Rao (1986), Cramér (1946), and the all-time classic, Feller (1971). Other specific references are given in the sections.

In this introductory chapter, we lay out the basic concepts of asymptotics with concrete applications. More specialized tools are separately treated in subsequent chapters.

7.1 Some Basic Notation and Convergence Concepts

Some basic definitions, notation, and concepts are put together in this section.

Definition 7.1. Let a_n be a sequence of real numbers. We write $a_n = o(1)$ if $\lim_{n \to \infty} a_n = 0$. We write $a_n = O(1)$ if $\exists K < \infty \ni |a_n| \leq K \, \forall n \geq 1$.

More generally, if $a_n, b_n > 0$ are two sequences of real numbers, we write $a_n = o(b_n)$ if $\frac{a_n}{b_n} = o(1)$; we write $a_n = O(b_n)$ if $\frac{a_n}{b_n} = O(1)$.

Remark. Note that the definition allows the possibility that a sequence a_n which is $O(1)$ is also $o(1)$. The converse is always true; that is, $a_n = o(1) \Rightarrow a_n = O(1)$.

Definition 7.2. Let a_n, b_n be two real sequences. We write $a_n \sim b_n$ if $\frac{a_n}{b_n} \to 1$, as $n \to \infty$. We write $a_n \asymp b_n$ if $0 < \liminf \frac{a_n}{b_n} \leq \limsup \frac{a_n}{b_n} < \infty$, as $n \to \infty$.

Example 7.1. Let $a_n = \frac{n}{n+1}$. Then, $|a_n| \leq 1 \, \forall n \geq 1$; so $a_n = O(1)$. But, $a_n \to 1$. as $n \to \infty$; so a_n is not $o(1)$.

However, suppose $a_n = \frac{1}{n}$. Then, again, $|a_n| \leq 1 \, \forall n \geq 1$; so $a_n = O(1)$. But, this time $a_n \to 0$. as $n \to \infty$; so a_n is both $O(1)$ and $o(1)$. But $a_n = O(1)$ is a weaker statement in this case than saying $a_n = o(1)$.

Next, suppose $a_n = -n$. Then $|a_n| = n \to \infty$, as $n \to \infty$; so a_n is not $O(1)$, and therefore also cannot be $o(1)$.

Example 7.2. Let $c_n = \log n$, and $a_n = \frac{c_n}{c_{n+k}}$, where $k \geq 1$ is a fixed positive integer. Thus,

$$a_n = \frac{\log n}{\log(n+k)} = \frac{\log n}{\log n + \log(1 + \frac{k}{n})} = \frac{1}{1 + \frac{1}{\log n}\log(1 + \frac{k}{n})} \to \frac{1}{1+0} = 1.$$

Therefore, $a_n = O(1), a_n \sim 1$, but a_n is not $o(1)$. The statement that $a_n \sim 1$ is stronger than saying $a_n = O(1)$.

Example 7.3. Let $a_n = \frac{1}{\sqrt{n+1}}$. Then,

$$\begin{aligned}
a_n &= \frac{1}{\sqrt{n}} + \frac{1}{\sqrt{n+1}} - \frac{1}{\sqrt{n}} \\
&= \frac{1}{\sqrt{n}} + \frac{\sqrt{n} - \sqrt{n+1}}{\sqrt{n}\sqrt{n+1}} = \frac{1}{\sqrt{n}} - \frac{1}{\sqrt{n}\sqrt{n+1}(\sqrt{n} + \sqrt{n+1})} \\
&= \frac{1}{\sqrt{n}} - \frac{1}{\sqrt{n}\sqrt{n}(1+o(1))(2\sqrt{n}+o(1))} \\
&= \frac{1}{\sqrt{n}} - \frac{1}{n(1+o(1))(2\sqrt{n}+o(1))} \\
&= \frac{1}{\sqrt{n}} - \frac{1}{2n^{3/2}(1+o(1))(1+o(1))}
\end{aligned}$$

7.1 Some Basic Notation and Convergence Concepts

$$= \frac{1}{\sqrt{n}} - \frac{1}{2n^{3/2}(1+o(1))}$$

$$= \frac{1}{\sqrt{n}} - \frac{1}{2n^{3/2}}(1+o(1)) = \frac{1}{\sqrt{n}} - \frac{1}{2n^{3/2}} + o\left(n^{-3/2}\right).$$

In working with asymptotics, it is useful to be skilled in calculations of the type of this example. To motivate the first probabilistic convergence concept, we give an illustrative example.

Example 7.4. For $n \geq 1$, consider the simple discrete random variables X_n with the pmf $P(X_n = \frac{1}{n}) = 1 - \frac{1}{n}, P(X_n = n) = \frac{1}{n}$. Then, for large n, X_n is close to zero with a large probability. Although for any given n, X_n is never equal to zero, the probability of it being far from zero is very small for large n. For example, $P(X_n > .001) \leq .001$ if $n \geq 1000$. More formally, for any given $\epsilon > 0$, $P(X_n > \epsilon) \leq P(X_n > \frac{1}{n}) = \frac{1}{n}$, if we take n to be so large that $\frac{1}{n} < \epsilon$. As a consequence, $P(|X_n| > \epsilon) = P(X_n > \epsilon) \to 0$, as $n \to \infty$. This example motivates the following definition.

Definition 7.3. Let $X_n, n \geq 1$, be an infinite sequence of random variables defined on a common sample space Ω. We say that X_n *converges in probability to* c, a specific real constant, if for any given $\epsilon > 0$, $P(|X_n - c| > \epsilon) \to 0$, as $n \to \infty$. Equivalently, X_n converges in probability to c given any $\epsilon > 0, \delta > 0$, there exists an $n_0 = n_0(\epsilon, \delta)$ such that $P(|X_n - c| > \epsilon) < \delta \, \forall n \geq n_0(\epsilon, \delta)$.

If X_n converges in probability to c, we write $X_n \overset{\mathcal{P}}{\Rightarrow} c$, or sometimes also, $X_n \overset{P}{\Rightarrow} c$.

However, sometimes the sequence X_n may get close to some random variable, rather than a constant. Here is an example of such a situation.

Example 7.5. Let X, Y be two independent standard normal variables. Define a sequence of random variables X_n as $X_n = X + \frac{Y}{n}$. Then, intuitively, we feel that for large n, the $\frac{Y}{n}$ part is small, and X_n is very close to the fixed random variable X. Formally, $P(|X_n - X| > \epsilon) = P(|\frac{Y}{n}| > \epsilon) = P(|Y| > n\epsilon) = 2[1 - \Phi(n\epsilon)] \to 0$, as $n \to \infty$. This motivates a generalization of the previous definition.

Definition 7.4. Let $X_n, X, n \geq 1$ be random variables defined on a common sample space Ω. We say that X_n *converges in probability to X* if given any $\epsilon > 0$, $P(|X_n - X| > \epsilon) \to 0$, as $n \to \infty$.

We denote it as $X_n \overset{\mathcal{P}}{\Rightarrow} X$, or $X_n \overset{P}{\Rightarrow} X$.

Definition 7.5. A sequence of random variables X_n is said to be *bounded in probability* or *tight* if, given $\epsilon > 0$, one can find a constant k such that $P(|X_n| > k) \leq \epsilon$ for all $n \geq 1$.

Notation. If $X_n \overset{P}{\Rightarrow} 0$, then we write $X_n = o_P(1)$. More generally, if $a_n X_n \overset{P}{\Rightarrow} 0$ for some positive sequence a_n, then we write $X_n = o_P(\frac{1}{a_n})$.

If X_n is bounded in probability, then we write $X_n = O_P(1)$. If $a_n X_n = O_P(1)$, we write $X_n = O_P(\frac{1}{a_n})$.

Proposition. *Suppose $X_n = o_P(1)$. Then, $X_n = O_P(1)$. The converse is, in general, not true.*

Proof. If $X_n = o_P(1)$, then by definition of convergence in probability, given $c > 0, P(|X_n| > c) \to 0$, as $n \to \infty$. Thus, given any fixed $\epsilon > 0$, for all large n, say $n \geq n_0(\epsilon)$, $P(|X_n| > 1) < \epsilon$. Next find constants $c_1, c_2, \ldots, c_{n_0}$, such that $P(|X_i| > c_i) < \epsilon, i = 1, 2, \ldots, n_0$. Choose $k = \max\{1, c_1, c_2, c_{n_0}\}$. Then, $P(|X_n| > k) < \epsilon \; \forall n \geq 1$. Therefore, $X_n = O_P(1)$.

To see that the converse is, in general, not true, let $X \sim N(0, 1)$, and define $X_n \equiv X, \forall n \geq 1$. Then, $X_n = O_P(1)$. But $P(|X_n| > 1) \equiv P(|X| > 1)$, which is a fixed positive number, and so, $P(|X_n| > 1)$ does not converge to zero. □

Definition 7.6. *Let $\{X_n, X\}$ be defined on the same probability space. We say that X_n converges almost surely to X (or X_n converges to X with probability 1) if $P(\omega : X_n(\omega) \to X(\omega)) = 1$. We write $X_n \overset{a.s.}{\to} X$ or $X_n \overset{a.s.}{\Rightarrow} X$.*

Remark. If the limit X is a finite constant c with probability one, we write $X_n \overset{a.s.}{\Rightarrow} c$. If $P(\omega : X_n(\omega) \to \infty) = 1$, we write $X_n \overset{a.s.}{\to} \infty$. Almost sure convergence is a stronger mode of convergence than convergence in probability. In fact, a characterization of almost sure convergence is that for any given $\epsilon > 0$,

$$\lim_{m \to \infty} P(|X_n - X| \leq \epsilon \; \forall n \geq m) = 1.$$

It is clear from this characterization that almost sure convergence is stronger than convergence in probability. However, the following relationships hold.

Theorem 7.1. *(a) Let $X_n \overset{P}{\Rightarrow} X$. Then there is a sub-sequence X_{n_k} such that $X_{n_k} \overset{a.s.}{\to} X$.*

(b) Suppose X_n is a monotone nondecreasing sequence and that $X_n \overset{P}{\Rightarrow} X$. Then $X_n \overset{a.s.}{\to} X$.

Example 7.6 *(Pattern in Coin Tossing).* For iid Bernoulli trials with a success probability $p = \frac{1}{2}$, let T_n denote the number of times in the first n trials that a success is followed by a failure. Denoting

$$I_i = I_{i\text{th trial is a success and } (i+1)\text{th trial is a failure}},$$

we have $T_n = \sum_{i=1}^{n-1} I_i$, and therefore $E(T_n) = \frac{n-1}{4}$, and

$$\text{Var}(T_n) = \sum_{i=1}^{n-1} \text{Var}(I_i) + 2\sum_{i=1}^{n-2} \text{Cov}(I_i, I_{i+1}) = \frac{3(n-1)}{16} - \frac{2(n-2)}{16} = \frac{n+1}{16}.$$

It now follows by an application of Chebyshev's inequality that $\frac{T_n}{n} \overset{P}{\Rightarrow} \frac{1}{4}$.

7.1 Some Basic Notation and Convergence Concepts

Example 7.7 (Uniform Maximum). Suppose $X_1, X_2, \ldots$ is an infinite sequence of iid $U[0, 1]$ random variables and let $X_{(n)} = \max\{X_1, \ldots, X_n\}$. Intuitively, $X_{(n)}$ should get closer and closer to 1 as n increases. In fact, $X_{(n)}$ converges almost surely to 1. For $P(|1 - X_{(n)}| \leq \epsilon \ \forall n \geq m) = P(1 - X_{(n)} \leq \epsilon \ \forall n \geq m) = P(X_{(n)} \geq 1 - \epsilon \ \forall n \geq m) = P(X_{(m)} \geq 1 - \epsilon) = 1 - (1 - \epsilon)^m \to 1$ as $m \to \infty$, and hence $X_{(n)} \stackrel{a.s.}{\Rightarrow} 1$.

Example 7.8 (Spiky Normals). Suppose $X_n \sim N(\frac{1}{n}, \frac{1}{n})$ is a sequence of independent variables. The mean and the variance are both converging to zero, therefore one would intuitively expect that the sequence X_n converges to zero in some sense. In fact, it converges almost surely to zero. Indeed, $P(|X_n| \leq \epsilon \ \forall n \geq m) = \prod_{n=m}^{\infty} P(|X_n| \leq \epsilon) = \prod_{n=m}^{\infty} [\Phi(\epsilon\sqrt{n} - \frac{1}{\sqrt{n}}) + \Phi(\epsilon\sqrt{n} + \frac{1}{\sqrt{n}}) - 1] = \prod_{n=m}^{\infty} [1 + O(\frac{\phi(\epsilon\sqrt{n})}{\sqrt{n}})] = 1 + O(\frac{(e^{-\frac{m\epsilon^2}{2}})}{\sqrt{m}}) \to 1$, as $m \to \infty$, implying $X_n \stackrel{a.s.}{\Rightarrow} 0$. In the above, the last but one equality follows on using the tail property of the standard normal CDF that

$$\frac{1 - \Phi(x)}{\phi(x)} = \frac{1}{x} + O\left(\frac{1}{x^3}\right), \quad \text{as } x \to \infty.$$

Next, we introduce the concept of convergence in mean. It often turns out to be a convenient method for establishing convergence in probability.

Definition 7.7. Let $X_n, X, n \geq 1$ be defined on a common sample space Ω. Let $p \geq 1$, and suppose $E(|X_n|^p), E(|X|^p) < \infty$. We say that X_n *converges in pth mean to* X or X_n *converges in* L_p *to* X if $E(|X_n - X|^p) \to 0$, as $n \to \infty$. If $p = 2$, we also say that X_n *converges to* X *in quadratic mean*. If X_n converges in L_p to X, we write $X_n \stackrel{L_p}{\Rightarrow} X$.

Example 7.9 (Some Counterexamples). Let X_n be the sequence of two point random variables with pmf $P(X_n = 0) = 1 - \frac{1}{n}, P(X_n = n) = \frac{1}{n}$. Then X_n converges in probability to zero. But, $E(|X_n|) = 1 \ \forall n$, and hence X_n does not converge in L_1 to zero. In fact, it does not converge to zero in L_p for any $p \geq 1$.

Now take the same sequence X_n as above, and assume moreover that they are independent. Take an $\epsilon > 0$, and positive integers m, k. Then,

$$P(|X_n| < \epsilon \ \forall m \leq n \leq m + k)$$

$$= P(X_n = 0 \ \forall m \leq n \leq m + k) = \prod_{n=m}^{m+k} \left(1 - \frac{1}{n}\right)$$

$$= \frac{m - 1}{m + k}.$$

For any m, this converges to zero as $k \to \infty$. Therefore, $\lim_{m \to \infty} P(|X_n| < \epsilon \ \forall n \geq m)$ cannot be one, and so, X_n does not converge almost surely to zero.

Next, let X_n have the pmf $P(X_n = 0) = 1 - \frac{1}{n}$, $P(X_n = \sqrt{n}) = \frac{1}{n}$. Then, X_n again converges in probability to zero. Furthermore, $E(|X_n|) = \frac{1}{\sqrt{n}} \to 0$, and so X_n converges in L_1 to zero. But, $E(X_n^2) = 1 \;\forall n$, and hence X_n does not converge in L_2 to zero.

The next result says that convergence in L_p is a useful method for establishing convergence in probability.

Proposition. *Let $X_n, X, n \geq 1$ be defined on a common sample space Ω. Suppose X_n converges to X in L_p for some $p \geq 1$. Then $X_n \stackrel{P}{\Rightarrow} X$.*

Proof. Simply observe that, by using Markov's inequality,

$$P(|X_n - X| > \epsilon) = P(|X_n - X|^p > \epsilon^p) \leq \frac{E(|X_n - X|^p)}{\epsilon^p}$$

$\to 0$, by hypothesis. □

Remark. It is easily established that if $p > 1$, then

$$X_n \text{ converges in } L_p \text{ to } X \Rightarrow X_n \text{ converges in } L_1 \text{ to } X.$$

7.2 Laws of Large Numbers

The definitions and the treatment in the previous section are for general sequences of random variables. Averages and sums are sequences of special importance in applications. The classic *laws of large numbers*, which characterize the long run behavior of averages, are given in this section. Truly, the behavior of averages and sums in complete generality is very subtle, and is beyond the scope of this book. Specialized treatments are available in Feller (1971), Révész (1968), and Kesten (1972).

A very useful tool for establishing almost sure convergence is stated first.

Theorem 7.2 (Borel–Cantelli Lemma). *Let $\{A_n\}$ be a sequence of events on a sample space Ω. If*

$$\sum_{n=1}^{\infty} P(A_n) < \infty,$$

then $P(\text{infinitely many } A_n \text{ occur}) = 0$.

If $\{A_n\}$ are pairwise independent, and

$$\sum_{n=1}^{\infty} P(A_n) = \infty,$$

then $P(\text{infinitely many } A_n \text{ occur}) = 1$.

7.2 Laws of Large Numbers

Proof. We prove the first statement. In order that infinitely many among the events $A_n, n \geq 1$, occur, it is necessary and sufficient that given any m, there is at least one event among $A_m, A_{m+1}, \ldots$ that occurs. In other words,

$$\text{Infinitely many } A_n \text{ occur} = \cap_{m=1}^{\infty} \cup_{n=m}^{\infty} A_n.$$

On the other hand, the events $B_m = \cup_{n=m}^{\infty} A_n$ are decreasing in m (i.e., $B_1 \supseteq B_2 \supseteq \ldots$). Therefore,

$$P(\text{infinitely many } A_n \text{ occur}) = P(\cap_{m=1}^{\infty} B_m) = \lim_{m \to \infty} P(B_m)$$

$$= \lim_{m \to \infty} P(\cup_{n=m}^{\infty} A_n) \leq \limsup_{m \to \infty} \sum_{n=m}^{\infty} P(A_n)$$

$$= 0,$$

because, by assumption, $\sum_{n=1}^{\infty} P(A_n) < \infty$. □

Remark. Although pairwise independence suffices for the conclusion of the second part of the Borel–Cantelli lemma, common applications involve cases where the A_n are mutually independent.

The next example gives an application of the Borel–Cantelli lemma.

Example 7.10 (Tail Runs of Arbitrary Length). Most of us do not feel that it is likely that in tossing a coin repeatedly, we are likely to see many tails or many heads in succession. Problems of this kind were discussed in Chapter 1. This example shows that in some sense, that intuition is wrong.

Consider a sequence of independent Bernoulli trials in which success occurs with probability p and failure with probability $q = 1 - p$. Suppose $p > q$, so that successes are more likely than failures. Consider a hypothetical long uninterrupted run of m failures, say $FF \ldots F$, for some fixed m. Break up the Bernoulli trials into nonoverlapping blocks of m trials, and consider A_n to be the event that the nth block consists of only failures. The probability of each A_n is q^m, which is free of n. Therefore, $\sum_{n=1}^{\infty} P(A_n) = \infty$ and it follows from the second part of the Borel–Cantelli lemma that no matter how large p may be, as long as $p < 1$, a string of consecutive failures of any given arbitrary length m reappears infinitely many times in the sequence of Bernoulli trials. In particular, if we keep tossing an ordinary coin, then with certainty, we will see 1000 tails (or, 10,000 tails) in succession, and we will see this occur again and again, infinitely many times, as our coin tosses continue indefinitely.

Here is another important application of the Borel–Cantelli lemma.

Example 7.11 (Almost Sure Convergence of Binomial Proportion). Let $X_1, X_2, \ldots$ be an infinite sequence of independent $Ber(p)$ random variables, where $0 < p < 1$. Let

$$\bar{X}_n = \frac{S_n}{n} = \frac{X_1 + X_2 + \ldots + X_n}{n}.$$

Then, from our previous formula in Chapter 1 for binomial distributions, $E(S_n - np)^4 = np(1-p)[1 + 3(n-2)p(1-p)]$. Thus, by Markov's inequality,

$$\begin{aligned}
P(|\bar{X}_n - p| > \epsilon) &= P(|S_n - np| > n\epsilon) \\
&= P((S_n - np)^4 > (n\epsilon)^4) \leq \frac{E(S_n - np)^4}{(n\epsilon)^4} \\
&= \frac{np(1-p)[1 + 3(n-2)p(1-p)]}{(n\epsilon)^4} \\
&\leq \frac{3n^2(p(1-p))^2 + np(1-p)}{n^4 \epsilon^4} \leq \frac{C}{n^2} + \frac{D}{n^3},
\end{aligned}$$

for finite constants C, D. Therefore,

$$\sum_{n=1}^{\infty} P(|\bar{X}_n - p| > \epsilon) \leq C \sum_{n=1}^{\infty} \frac{1}{n^2} + D \sum_{n=1}^{\infty} \frac{1}{n^3} < \infty.$$

It follows from the Borel–Cantelli lemma that the binomial sample proportion $\bar{X}_n$ converges almost surely to p.

In fact, the convergence of the sample mean $\bar{X}_n$ to $E(X_1)$ (i.e., the common mean of the X_i) holds in general. The general results, due to Khintchine and Kolmogorov, are known as the *laws of large numbers*, stated below.

Theorem 7.3 (Weak Law of Large Numbers). *Suppose $X_1, X_2, \ldots$ are independent and identically distributed (iid) random variables (defined on a common sample space Ω), such that $E(|X_1|) < \infty$, and $E(X_1) = \mu$. Let $\bar{X}_n = \frac{1}{n} \sum_{i=1}^{n} X_i$. Then $\bar{X}_n \stackrel{P}{\Rightarrow} \mu$.*

Theorem 7.4 (Strong Law of Large Numbers). *Suppose $X_1, X_2, \ldots$ are independent and identically distributed random variables (defined on a common sample space Ω). Then, $\bar{X}_n$ has an a.s. (almost sure) limit iff $E(|X_1|) < \infty$, in which case $\bar{X}_n \stackrel{a.s.}{\Rightarrow} \mu = E(X_1)$.*

Remark. It is not very simple to prove either of the two laws of large numbers in the generality stated above. We prove the weak law in Chapter 8, and the strong law in Chapter 14. If the X_i have a finite variance, then Markov's inequality easily leads to the weak law. If X_i have four finite moments, then a careful argument on the lines of our special binomial proportion example above does lead to the strong law. Once again, the Borel–Cantelli lemma does the trick.

It is extremely interesting that existence of an expectation is not necessary for the WLLN (weak law of large numbers) to hold. That is, it is possible that $E(|X|) = \infty$, and yet $\bar{X} \stackrel{P}{\Rightarrow} a$, for some real number a. We describe this more precisely shortly.

The SLLN (strong law of large numbers) already tells us that if $X_1, X_2, \ldots$ are independent with a common CDF F (that is, iid), then the sample mean $\bar{X}$ does not

7.2 Laws of Large Numbers

have any almost sure limit if $E_F(|X|) = \infty$. An obvious question is what happens to $\bar{X}$ in such a case. A great deal of deep work has been done on this question, and there are book-length treatements of this issue. The following theorem gives a few key results only for easy reference.

Definition 7.8. Let x be a real number. The *positive and negative part* of x are defined as

$$x^+ = \max\{x, 0\}; \quad x^- = \max\{-x, 0\}.$$

That is, $x^+ = x$ when $x \geq 0$, and 0 when $x \leq 0$. On the other hand, $x^- = 0$ when $x \geq 0$, and $-x$ when $x \leq 0$. Consequently, for any real number x,

$$x^+, x^- \geq 0; \quad x = x^+ - x^-; \quad |x| = x^+ + x^-.$$

Theorem 7.5 (Failure of the Strong Law). *Let $X_1, X_2, \ldots$ be independent observations from a common CDF F on the real line. Suppose $E_F(|X|) = \infty$.*

(a) *For any sequence of real numbers, a_n,*

$$P(\limsup |\bar{X} - a_n| = \infty) = 1.$$

(b) *If $E(X^+) = \infty$, and $E(X^-) < \infty$, then $\bar{X} \stackrel{a.s.}{\Rightarrow} \infty$.*
(c) *If $E(X^-) = \infty$, and $E(X^+) < \infty$, then $\bar{X} \stackrel{a.s.}{\Rightarrow} -\infty$.*

More refined descriptions of the set of all possible limit points of the sequence of means $\bar{X}$ are worked out in Kesten (1972). See also Chapter 3 in DasGupta (2008).

Example 7.12. Let F be the CDF of the standard Cauchy distribution. Due to the symmetry, we get $E(X^+) = E(X^-) = \frac{1}{\pi} \int_0^\infty \frac{x}{1+x^2} dx = \infty$. Therefore, from part (a) of the above theorem. with probability one, $\limsup |\bar{X}| = \infty$ (i.e., the sequence of sample means cannot remain bounded). Also, from the statement of the strong law itself, the sequence will not settle down near any fixed real number. The four simulated plots in Fig. 7.1 help illustrate these phenomena. In each plot, 1000 standard Cauchy values were simulated, and the sequence of means $\bar{X}_j = \frac{1}{j} \sum_{i=1}^{j} X_i$ were plotted, for $j = 1, 2, \ldots, 1000$.

Now, we consider the issue of a possibility of a WLLN when the expectation does not exist. The answer is that the tail of F should not be too slow. Here is the precise result.

Theorem 7.6 (Weak Law Without an Expectation). *Let $X_1, X_2, \ldots$ be independent observations from a CDF F. Then, there exist constants μ_n such that $\bar{X} - \mu_n \stackrel{P}{\Rightarrow} 0$ if and only if*

$$x(1 - F(x) + F(-x)) \to 0,$$

as $x \to \infty$, in which case the constants μ_n may be chosen as $\mu_n = E_F(XI_{\{|X| \leq n\}})$.

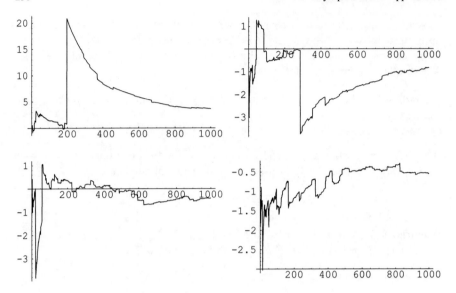

Fig. 7.1 Sequence of sample means for simulated C(0,1) data

In particular, if F is symmetric, $x(1 - F(x)) \to 0$ as $x \to \infty$, and $\int_0^\infty (1 - F(x))dx = \infty$, then $E_F(|X|) = \infty$, where as $\bar{X} \stackrel{P}{\Rightarrow} 0$.

Remark. See Feller (1971, p. 235) for a proof. It should be noted that the two conditions $x(1 - F(x)) \to 0$ and $\int_0^\infty (1 - F(x))dx = \infty$ are not inconsistent. It is easy to find an F that satisfies both conditions.

We close this section with an important result on the uniform closeness of the empirical CDF to the underlying CDF in the iid case.

Theorem 7.7 (Glivenko–Cantelli Theorem). *Let F be any CDF on the real line, and $X_1, X_2, \ldots$ iid with common CDF F. Let $F_n(x) = \frac{\#\{i \leq n : X_i \leq x\}}{n}$ be the sequence of empirical CDFs. Then, $\Delta_n = \sup_{x \in \mathcal{R}} |F_n(x) - F(x)| \stackrel{a.s.}{\Rightarrow} 0$.*

Proof. The main idea of the proof is to discretize the problem, and exploit Kolmogorov's SLLN.

Fix m and define $a_0, a_1, \ldots, a_m, a_{m+1}$ by the relationships $[a_i, a_{i+1}) = \{x : \frac{i}{m} \leq F(x) < \frac{i+1}{m}\}, i = 1, 2, \ldots, m - 1$, and $a_0 = -\infty, a_{m+1} = \infty$. Now fix an i and look at $x \in [a_i, a_{i+1})$. Then

$$F_n(x) - F(x) \leq F_n(a_{i+1}-) - F(a_i) \leq F_n(a_{i+1}-) - F(a_{i-1}-) + \frac{1}{m}.$$

Similarly, for $x \in [a_i, a_{i+1})$,

$$F(x) - F_n(x) \leq F(a_i) - F_n(a_i) + \frac{1}{m}.$$

Therefore, because any x belongs to one of these intervals $[a_i, a_{i+1})$,

$$\sup_{x \in \mathcal{R}} |F_n(x) - F(x)| = \max_i \left\{ |F(a_i) - F_n(a_i)| + |F(a_i-) - F_n(a_i-)| + \frac{1}{m} \right\}.$$

For fixed m, as $n \to \infty$, by the SLLN each of the terms within the absolute value sign above goes almost surely to zero, and so, for any fixed m, almost surely, $\lim_n \sup_{x \in \mathcal{R}} |F_n(x) - F(x)| \leq \frac{1}{m}$. Now let $m \to \infty$, and the result follows. □

7.3 Convergence Preservation

We have already seen the importance of being able to deal with transformations of random variables in Chapters 3 and 4. This section addresses the question of when convergence properties are preserved if we suitably transform a sequence of random variables.

The next important theorem gives some frequently useful results, that are analogous to corresponding results on convergence of sequences in calculus.

Theorem 7.8 (Convergence Preservation).

(a) $X_n \stackrel{P}{\Rightarrow} X, Y_n \stackrel{P}{\Rightarrow} Y \Rightarrow X_n \pm Y_n \stackrel{P}{\Rightarrow} X \pm Y$.

(b) $X_n \stackrel{P}{\Rightarrow} X, Y_n \stackrel{P}{\Rightarrow} Y \Rightarrow X_n Y_n \stackrel{P}{\Rightarrow} XY$, $X_n \stackrel{P}{\Rightarrow} X, Y_n \stackrel{P}{\Rightarrow} Y, P(Y \neq 0) = 1 \Rightarrow \frac{X_n}{Y_n} \stackrel{P}{\Rightarrow} \frac{X}{Y}$.

(c) $X_n \stackrel{a.s.}{\Rightarrow} X, Y_n \stackrel{a.s.}{\Rightarrow} Y \Rightarrow X_n \pm Y_n \stackrel{a.s.}{\Rightarrow} X \pm Y$.

(d) $X_n \stackrel{a.s.}{\Rightarrow} X, Y_n \stackrel{a.s.}{\Rightarrow} Y \Rightarrow X_n Y_n \stackrel{a.s.}{\Rightarrow} XY$, $X_n \stackrel{a.s.}{\Rightarrow} X, Y_n \stackrel{a.s.}{\Rightarrow} Y, P(Y \neq 0) = 1 \Rightarrow \frac{X_n}{Y_n} \stackrel{a.s.}{\Rightarrow} \frac{X}{Y}$.

(e) $X_n \stackrel{L_1}{\Rightarrow} X, Y_n \stackrel{L_1}{\Rightarrow} Y \Rightarrow X_n + Y_n \stackrel{L_1}{\Rightarrow} X + Y$.

(f) $X_n \stackrel{L_2}{\Rightarrow} X, Y_n \stackrel{L_2}{\Rightarrow} Y \Rightarrow X_n + Y_n \stackrel{L_2}{\Rightarrow} X + Y$.

The proofs of each of these parts use relatively simple arguments, such as the triangular, Minkowski, and the Cauchy–Schwarz inequality (see Chapter 1 for their exact statements). We omit the details of these proofs; Chow and Teicher (1988, pp. 254–256) give the details for several parts of this convergence preservation theorem.

Example 7.13. Suppose $X_1, X_2, \ldots$ are independent $N(\mu_1, \sigma_1^2)$ variables, and $Y_1, Y_2, \ldots$ are independent $N(\mu_2, \sigma_2^2)$ variables. For $n, m \geq 1$, let $\bar{X}_n = \frac{1}{n} \sum_{i=1}^n X_i, \bar{Y}_m = \frac{1}{m} \sum_{j=1}^m Y_j$. By the strong law of large numbers, (SLLN), as $n, m \to \infty, \bar{X}_n \stackrel{a.s.}{\Rightarrow} \mu_1$, and $\bar{Y}_m \stackrel{a.s.}{\Rightarrow} \mu_2$. Then, by the theorem above, $\bar{X}_n - \bar{Y}_m \stackrel{a.s.}{\Rightarrow} \mu_1 - \mu_2$.

Also, by the same theorem, $\bar{X}_n \bar{Y}_m \stackrel{a.s.}{\Rightarrow} \mu_1 \mu_2$.

Definition 7.9 (The Multidimensional Case). Let $\mathbf{X_n}, n \geq 1, \mathbf{X}$ be d-dimensional random vectors, for some $1 \leq d < \infty$. We say that $\mathbf{X_n} \stackrel{P}{\Rightarrow} \mathbf{X}$, if $||\mathbf{X_n} - \mathbf{X}|| \stackrel{P}{\Rightarrow} 0$.

We say that $\mathbf{X_n} \stackrel{a.s.}{\Rightarrow} \mathbf{X}$, if $P(\omega : ||\mathbf{X_n}(\omega) - \mathbf{X}(\omega)|| \to 0) = 1$. Here, $||.||$ denotes Euclidean length (norm).

Operationally, the following equivalent conditions are more convenient.

Proposition. (a) $\mathbf{X_n} \stackrel{P}{\Rightarrow} \mathbf{X}$ if and only if $X_{n,i} \stackrel{P}{\Rightarrow} X_i$, for each $i = 1, 2, \ldots, d$. That is, each coordinate of $\mathbf{X_n}$ converges in probability to the corresponding coordinate of $\mathbf{X}$;
(b) $\mathbf{X_n} \stackrel{a.s.}{\Rightarrow} \mathbf{X}$ if and only if $X_{n,i} \stackrel{a.s.}{\Rightarrow} X_i$, for each $i = 1, 2, \ldots, d$.

Theorem 7.9 (Convergence Preservation in Multidimension). Let $\mathbf{X_n}, \mathbf{Y_n}$, $n \geq 1, \mathbf{X}, \mathbf{Y}$ be d-dimensional random vectors. Let A be some $p \times d$ matrix of real elements, where $p \geq 1$. Then,

(a) $\mathbf{X_n} \stackrel{P}{\Rightarrow} \mathbf{X}, \mathbf{Y_n} \stackrel{P}{\Rightarrow} \mathbf{Y} \Rightarrow \mathbf{X_n} \pm \mathbf{Y_n} \stackrel{P}{\Rightarrow} \mathbf{X} \pm \mathbf{Y}$;
$\mathbf{X_n}'\mathbf{Y_n} \stackrel{P}{\Rightarrow} \mathbf{X}'\mathbf{Y}; A\mathbf{X_n} \stackrel{P}{\Rightarrow} A\mathbf{X}$.
(b) *Exactly the same results hold, when convergence in probability is replaced everywhere by almost sure convergence.*

Proof. This theorem follows easily from the convergence preservation theorem in one dimension, and the proposition above, which says that multidimensional convergence is the same as convergence in each coordinate separately. □

The next result is one of the most useful results on almost sure convergence and convergence in probability. It says that convergence properties are preserved if we make smooth transformations. However, the force of the result is partially lost if we insist on the transformations being smooth everywhere. To give the most useful version of the result, we need a technical definition.

Definition 7.10. Let $d, p \geq 1$ be positive integers, and $f : S \subseteq \mathcal{R}^d \to \mathcal{R}^p$ a function. Let $C(f) = \{\mathbf{x} \in S : f \text{ is continuous at } \mathbf{x}\}$. Then $C(f)$ is called the *continuity set of f*.

We can now give the result on preservation of convergence under smooth transformations.

Theorem 7.10 (Continuous Mapping). *Let $\mathbf{X_n}, \mathbf{X}$ be d-dimensional random vectors, and $f : S \subseteq \mathcal{R}^d \to \mathcal{R}^p$ a function. Let $C(f)$ be the continuity set of f. Suppose the random vector $\mathbf{X}$ satisfies the condition*

$$P(\mathbf{X} \in C(f)) = 1.$$

Then,

(a) $\mathbf{X_n} \stackrel{P}{\Rightarrow} \mathbf{X} \Rightarrow f(\mathbf{X_n}) \stackrel{P}{\Rightarrow} f(\mathbf{X})$;
(b) $\mathbf{X_n} \stackrel{a.s.}{\Rightarrow} \mathbf{X} \Rightarrow f(\mathbf{X_n}) \stackrel{a.s.}{\Rightarrow} f(\mathbf{X})$.

Proof. We prove part (b). Let $S_1 = \{\omega \in \Omega : f \text{ is continuous at } X(\omega)\}$. Let $S_2 = \{\omega \in \Omega : \mathbf{X_n}(\omega) \to \mathbf{X}(\omega)\}$. Then, $P(S_1 \cap S_2) = 1$, and for each $\omega \in S_1 \cap S_2$, $f(\mathbf{X_n}(\omega)) \to f(\mathbf{X}(\omega))$. That means that $f(\mathbf{X_n}) \stackrel{a.s.}{\Rightarrow} f(\mathbf{X})$. $\square$

We give two important applications of this theorem next.

Example 7.14 (Convergence of Sample Variance). Let $X_1, X_2, \ldots, X_n$ be independent observations from a common distribution F, and suppose that F has finite mean μ and finite variance σ^2. The sample variance, of immense importance in statistics, is defined as $s^2 = \frac{1}{n-1} \sum_{i=1}^{n} (X_i - \bar{X})^2$. The purpose of this example is to show that $s^2 \stackrel{a.s.}{\Rightarrow} \sigma^2$, as $n \to \infty$.

First note that if we can prove that $\frac{1}{n} \sum_{i=1}^{n} (X_i - \bar{X})^2 \stackrel{a.s.}{\Rightarrow} \sigma^2$, then it follows that s^2 also converges almost surely to σ^2, because $\frac{n}{n-1} \to 1$ as $n \to \infty$. Now,

$$\frac{1}{n} \sum_{i=1}^{n} (X_i - \bar{X})^2 = \frac{1}{n} \sum_{i=1}^{n} X_i^2 - (\bar{X})^2$$

(an algebraic identity). Because F has a finite variance, it also possesses a finite second moment, namely, $E_F(X^2) = \sigma^2 + \mu^2 < \infty$. By applying the strong law of large numbers to the sequence $X_1^2, X_2^2, \ldots$, we get $\frac{1}{n} \sum_{i=1}^{n} X_i^2 \stackrel{a.s.}{\Rightarrow} E_F(X^2) = \sigma^2 + \mu^2$. By applying the SLLN to the sequence $X_1, X_2, \ldots$, we get $\bar{X} \stackrel{a.s.}{\Rightarrow} \mu$, and therefore by the continuous mapping theorem, $(\bar{X})^2 \stackrel{a.s.}{\Rightarrow} \mu^2$. Now, by the theorem on preservation of convergence, we get that $\frac{1}{n} \sum_{i=1}^{n} X_i^2 - (\bar{X})^2 \stackrel{a.s.}{\Rightarrow} \sigma^2 + \mu^2 - \mu^2 = \sigma^2$, which finishes the proof.

Example 7.15 (Convergence of Sample Correlation). Suppose $F(x, y)$ is a joint CDF in $\mathcal{R}^2$, and suppose that $E(X^2), E(Y^2)$ are both finite. Let

$$\rho = \frac{\text{Cov}(X, Y)}{\sqrt{\text{Var}(X) \text{Var}(Y)}} = \frac{E(XY) - E(X)E(Y)}{\sqrt{\text{Var}(X) \text{Var}(Y)}}$$

be the correlation between X and Y. Suppose $(X_i, Y_i), 1 \leq i \leq n$ are n independent observations from the joint CDF F. The sample correlation coefficient is defined as

$$r = \frac{\sum_{i=1}^{n} (X_i - \bar{X})(Y_i - \bar{Y})}{\sqrt{\sum_{i=1}^{n} (X_i - \bar{X})^2 \sum_{i=1}^{n} (Y_i - \bar{Y})^2}}.$$

The purpose of this example is to show that r converges almost surely to ρ.

It is convenient to rewrite r in the equivalent form

$$r = \frac{\frac{1}{n} \sum_{i=1}^{n} X_i Y_i - \bar{X}\bar{Y}}{\sqrt{\frac{1}{n} \sum_{i=1}^{n} (X_i - \bar{X})^2} \sqrt{\frac{1}{n} \sum_{i=1}^{n} (Y_i - \bar{Y})^2}}.$$

By the SLLN, $\frac{1}{n}\sum_{i=1}^{n} X_i Y_i$ converges almost surely to $E(XY)$, and $\bar{X}, \bar{Y}$ converge almost surely to $E(X), E(Y)$. By the previous example, $\frac{1}{n}\sum_{i=1}^{n}(X_i - \bar{X})^2$ converges almost surely to $\text{Var}(X)$, and $\frac{1}{n}\sum_{i=1}^{n}(Y_i - \bar{Y})^2$ converges almost surely to $\text{Var}(Y)$. Now consider the function $f(s, t, u, v, w) = \frac{s - tu}{\sqrt{v}\sqrt{w}}, -\infty < s, t, u < \infty, v, w > 0$. This function is continuous on the set $S = \{(s, t, u, v, w) : -\infty < s, t, u < \infty, v, w > 0, (s - tu)^2 \leq vw\}$. The joint distribution of $\frac{1}{n}\sum_{i=1}^{n} X_i Y_i$, $\bar{X}, \bar{Y}, \frac{1}{n}\sum_{i=1}^{n}(X_i - \bar{X})^2, \frac{1}{n}\sum_{i=1}^{n}(Y_i - \bar{Y})^2$ assigns probability one to the set S. By the continuous mapping theorem, it follows that $r \stackrel{a.s.}{\Rightarrow} \rho$.

7.4 Convergence in Distribution

Studying distributions of random variables is of paramount importance in both probability and statistics. The relevant random variable may be a member of some sequence X_n. Its exact distribution may be cumbersome. But it may be possible to approximate its distribution by a simpler distribution. We can then approximate probabilities for the true distribution of the random variable by probabilities in the simpler distribution. The type of convergence concept that justifies this sort of approximation is called *convergence in distribution* or *convergence in law*. Of all the convergence concepts we are discussing, convergence in distribution is among the most useful in answering practical questions. For example, statisticians are usually much more interested in constructing confidence intervals than just point estimators, and a central limit theorem of some kind is necessary to produce a confidence interval.

We start with an illustrative example.

Example 7.16. Suppose

$$X_n \sim U\left[\frac{1}{2} - \frac{1}{n+1}, \frac{1}{2} + \frac{1}{n+1}\right], n \geq 1.$$

Because the interval $[\frac{1}{2} - \frac{1}{n+1}, \frac{1}{2} + \frac{1}{n+1}]$ is shrinking to the single point $\frac{1}{2}$, intuitively we feel that the distribution of X_n is approaching a distribution concentrated at $\frac{1}{2}$, that is, a *one-point distribution*. The CDF of the distribution concentrated at $\frac{1}{2}$ equals the function $F(x) = 0$ for $x < \frac{1}{2}$, and $F(x) = 1$ for $x \geq \frac{1}{2}$. Consider now the CDF of X_n; call it $F_n(x)$. Fix $x < \frac{1}{2}$. Then, for all large n, $F_n(x) = 0$, and so $\lim_n F_n(x)$ is also zero. Next fix $x > \frac{1}{2}$. Then, for all large n, $F_n(x) = 1$, and so $\lim_n F_n(x)$ is also one. Therefore, if $x < \frac{1}{2}$, or if $x > \frac{1}{2}$, $\lim_n F_n(x) = F(x)$. If x is exactly equal to $\frac{1}{2}$, then $F_n(x) = \frac{1}{2}$. But $F(\frac{1}{2}) = 1$. So $x = \frac{1}{2}$ is a problematic point, and the only problematic point, in that $F_n(\frac{1}{2}) \not\to F(\frac{1}{2})$. Interestingly, $x = \frac{1}{2}$ is also exactly the only point at which F is not continuous. However, we do not want this one problematic point to ruin our intuitive feeling that X_n is approaching the one-point distribution concentrated at $\frac{1}{2}$. That is, we do not take into account any points where the limiting CDF is not continuous.

7.4 Convergence in Distribution

Definition 7.11. Let $X_n, X, n \geq 1$, be real-valued random variables defined on a common sample space Ω. We say that X_n *converges in distribution (in law)* to X if $P(X_n \leq x) \to P(X \leq x)$ as $n \to \infty$, at every point x that is a continuity point of the CDF F of the random variable X.

We denote convergence in distribution by $X_n \stackrel{\mathcal{L}}{\Rightarrow} X$.

If $\mathbf{X_n}, \mathbf{X}$ are d-dimensional random vectors, then the same definition applies by using the joint CDFs of $\mathbf{X_n}, \mathbf{X}$, that is, $\mathbf{X_n}$ converges in distribution to $\mathbf{X}$ if $P(X_{n1} \leq x_1, \ldots, X_{nd} \leq x_d) \to P(X_1 \leq x_1, \ldots, X_d \leq x_d)$ at each point $(x_1, \ldots, x_d)$ that is a continuity point of the joint CDF $F(x_1, \ldots, x_d)$ of the random vector $\mathbf{X}$.

An important point of caution is the following.

Caution. In order to prove that d-dimensional vectors $\mathbf{X_n}$ converge in distribution to $\mathbf{X}$, it is not, in general, enough to prove that each coordinate of $\mathbf{X_n}$ converges in distribution to the corresponding coordinate of $\mathbf{X}$. However, convergence of general linear combinations is enough, which is the content of the following theorem.

Theorem 7.11 (Cramér–Wold Theorem). *Let* $\mathbf{X_n}, \mathbf{X}$ *be* d*-dimensional random vectors. Then* $\mathbf{X_n} \stackrel{\mathcal{L}}{\Rightarrow} \mathbf{X}$ *if and only if* $\mathbf{c}'\mathbf{X_n} \stackrel{\mathcal{L}}{\Rightarrow} \mathbf{c}'\mathbf{X}$ *for all unit* d*-dimensional vectors* $\mathbf{c}$.

The shortest proof of this theorem uses a tool called characteristic functions, which we have not discussed yet. We give a proof in the next chapter by using characteristic functions. Returning to the general concept of convergence in distribution, two basic facts are the following.

Theorem 7.12. *(a) If* $X_n \stackrel{\mathcal{L}}{\Rightarrow} X$, *then* $X_n = O_P(1)$;
(b) If $X_n \stackrel{\mathcal{P}}{\Rightarrow} X$, *then* $X_n \stackrel{\mathcal{L}}{\Rightarrow} X$.

Proof. We sketch a proof of part (b). Take a continuity point x of the CDF F of X, and fix $\epsilon > 0$. Then,

$$F_n(x) = P(X_n \leq x)$$
$$= P(X_n \leq x, |X_n - X| \leq \epsilon) + P(X_n \leq x, |X_n - X| > \epsilon)$$
$$\leq P(X \leq x + \epsilon) + P(|X_n - X| > \epsilon).$$

Now let $n \to \infty$ on both sides of the inequality. Then, we get $\limsup_n F_n(x) \leq F(x + \epsilon)$, because $P(|X_n - X| > \epsilon) \to 0$ by hypothesis. Now, letting $\epsilon \to 0$, we get $\limsup_n F_n(x) \leq F(x)$, because $F(x + \epsilon) \to F(x)$ by right continuity of F.

The proof will be complete if we show that $\liminf_n F_n(x) \geq F(x)$. This is proved similarly, except we now start with $P(X \leq x - \epsilon)$ on the left, and follow the same steps. It should be mentioned that it is in this part that the continuity of F at x is used. □

Remark. The fact that if a sequence X_n of random variables converges in distribution, then the sequence must be $O_P(1)$, tells us that there must be sequences of random variables which do not converge in distribution to anything. For example, take $X_n \sim N(n, 1), n \geq 1$. This sequence X_n is not $O_p(1)$, and therefore cannot converge in distribution. The question arises if the $O_p(1)$ property suffices for convergence. Even that, evidently, is not true; just consider $X_{2n-1} \sim N(0, 1)$, and $X_{2n} \sim N(1, 1)$. However, separately, the odd and the even subsequences do converge. That is, there might be a partial converse to the fact that if a sequence X_n converges in distribution, then it must be $O_p(1)$. This is a famous theorem on convergence in distribution, and is stated below.

Theorem 7.13 (Helly's Theorem). *Let $X_n, n \geq 1$ be random variables defined on a common sample space Ω, and suppose X_n is $O_p(1)$. Then there is a sub-sequence $X_{n_j}, j \geq 1$, and a random variable X (on the same sample space Ω), such that $X_{n_j} \stackrel{\mathcal{L}}{\Rightarrow} X$. Furthermore, $X_n \stackrel{\mathcal{L}}{\Rightarrow} X$ if and only if every convergent sub-sequence X_{n_j} converges in distribution to this same X.*

See Port (1994, p. 625) for a proof. Major generalizations of Helly's theorem to much more general spaces are known. Typically, some sort of a metric structure is assumed in these results; see van der Vaart and Wellner (2000) for such general results on weak compactness.

Example 7.17 (Various Convergence Phenomena Are Possible). This quick example shows that a sequence of discrete distributions can converge in distribution to a discrete distribution, or a continuous distribution, and a sequence of continuous distributions can converge in distribution to a continuous one, or a discrete one.

A good example of discrete random variables converging in distribution to a discrete random variable is the sequence $X_n \sim \text{Bin}(n, \frac{1}{n})$. Although it was not explicitly put in the language of convergence in distribution, we have seen in Chapter 6 that X_n converges to a Poisson random variable with mean one. A familiar example of discrete random variables converging in distribution to a continuous random variable is the de Moivre–Laplace central limit theorem (Chapter 1), which says that if $X_n \sim \text{Bin}(n, p)$, then $\frac{X_n - np}{\sqrt{np(1-p)}}$ converges to a standard normal variable.

Examples of continuous random variables converging to a continuous random variable are immediately available by using the general central limit theorem (also Chapter 1). For example, if X_i are independent $U[-1, 1]$ variables, then $\frac{\sqrt{n}\bar{X}}{\sigma}$, where $\sigma^2 = \frac{1}{3}$, converges to a standard normal variable. Finally, as an example of continuous random variables converging to a discrete random variable, consider $X_n \sim \text{Be}(\frac{1}{n}, \frac{1}{n})$. Visually, the density of X_n for large n is a symmetric U-shaped density, unbounded at both 0 and 1. It is not hard to show that X_n converges in distribution to X, where X is a Bernoulli random variable with parameter $\frac{1}{2}$.

Thus, we see that any types of random variables can indeed converge to the same or the other type.

By definition of convergence in distribution, if $X_n \stackrel{\mathcal{L}}{\Rightarrow} X$, and if X has a continuous CDF F (continuous everywhere), then $F_n(x) \to F(x)$ $\forall x$ where $F_n(x)$ is the CDF of X_n. The following theorem says that much more is true, namely that the convergence is actually uniform; see p. 265 in Chow and Teicher (1988).

7.4 Convergence in Distribution

Theorem 7.14 (Pólya's Theorem). *Let $X_n, n \geq 1$ have CDF F_n, and let X have CDF F. If F is everywhere continuous, and if $X_n \xrightarrow{\mathcal{L}} X$, then*

$$sup_{x \in \mathcal{R}} |F_n(x) - F(x)| \to 0,$$

as $n \to \infty$.

A large number of equivalent characterizations of convergence in distribution are known. Collectively, these conditions are called the **portmanteau theorem**. Note that the parts of the theorem are valid for real valued random variables, or d-dimensional random variables, for any $1 < d < \infty$.

Theorem 7.15 (The Portmanteau Theorem). *Let $\{X_n, X\}$ be random variables taking values in a finite-dimensional Euclidean space. The following are characterizations of $X_n \xrightarrow{\mathcal{L}} X$:*

(a) $E(g(X_n)) \to E(g(X))$ *for all bounded continuous functions g.*
(b) $E(g(X_n)) \to E(g(X))$ *for all bounded uniformly continuous functions g.*
(c) $E(g(X_n)) \to E(g(X))$ *for all bounded Lipschitz functions g. Here a Lipschitz function is such that $|g(x) - g(y)| \leq C ||x - y||$ for some C, and all x, y.*
(d) $E(g(X_n)) \to E(g(X))$ *for all continuous functions g with a compact support.*
(e) $\liminf P(X_n \in G) \geq P(X \in G)$ *for all open sets G.*
(f) $\limsup P(X_n \in S) \leq P(X \in S)$ *for all closed sets S.*
(g) $P(X_n \in B) \to P(X \in B)$ *for all (Borel) sets B such that the probability of X belonging to the boundary of B is zero.*

See Port (1994, p. 614) for proofs of various parts of this theorem.

Example 7.18. Consider $X_n \sim \text{Uniform}\{\frac{1}{n}, \frac{2}{n}, \ldots, \frac{n-1}{n}, 1\}$. Then, it can be shown easily that the sequence X_n converges in law to the $U[0, 1]$ distribution. Consider now the function $g(x) = x^{10}, 0 \leq x \leq 1$. Note that g is continuous and bounded. Therefore, by part (a) of the portmanteau theorem, $E(g(X_n)) = \sum_{k=1}^n (\frac{k^{10}}{n^{11}}) \to E(g(X)) = \int_0^1 x^{10} dx = \frac{1}{11}$.

This can be proved by using convergence of Riemann sums to a Riemann integral. But it is interesting to see the link to convergence in distribution.

Example 7.19 (Weierstrass's Theorem). **Weierstrass's theorem** says that any continuous function on a closed bounded interval can be uniformly approximated by polynomials. In other words, given a continuous function $f(x)$ on a bounded interval, one can find a polynomial $p(x)$ (of a sufficiently large degree) such that $|p(x) - f(x)|$ is uniformly small. Consider the case of the unit interval; the case of a general bounded interval reduces to this case.

Here we show pointwise convergence by using the portmanteau theorem. Laws of large numbers are needed for establishing uniform approximability. We give a constructive proof. Towards this, for $n \geq 1, 0 \leq p \leq 1$, and a given continuous function $g : [0, 1] \to \mathcal{R}$, define the sequence of *Bernstein polynomials*, $B_n(p) = \sum_{k=0}^n g(\frac{k}{n}) \binom{n}{k} p^k (1-p)^{n-k}$. Note that we can think of

$B_n(p)$ as $B_n(p) = E[g(\frac{X}{n}) \mid X \sim \text{Bin}(n, p)]$. As $n \to \infty$, $\frac{X}{n} \xrightarrow{\mathcal{P}} p$, and it follows that $\frac{X}{n} \xRightarrow{\mathcal{L}} \delta_p$, the one-point distribution concentrated at p (we have already seen that convergence in probability implies convergence in distribution). Because g is continuous and hence bounded, it follows from the portmanteau theorem that $B_n(p) \to g(p)$, at any p.

It is not hard to establish that $B_n(p) - g(p)$ converges uniformly to zero, as $n \to \infty$. Here is a sketch. As above, X denotes a binomial random variable with parameters n and p. We need to use the facts that a continuous function on $[0, 1]$ is also uniformly continuous and bounded. Thus, for any given $\epsilon > 0$, we can find $\delta > 0$ such that $|x - y| < \delta \Rightarrow |g(x) - g(y)| \leq \epsilon$, and also we can find a finite C such that $|g(x)| \leq C \ \forall x$. So,

$$
\begin{aligned}
|B_n(p) - g(p)| &= \left| E\left[g\left(\frac{X}{n}\right) \right] - g(p) \right| \\
&\leq E\left[\left| g\left(\frac{X}{n}\right) - g(p) \right| \right] \\
&= E\left[\left| g\left(\frac{X}{n}\right) - g(p) \right| I_{\{|\frac{X}{n} - p| \leq \delta\}} \right] \\
&\quad + E\left[\left| g\left(\frac{X}{n}\right) - g(p) \right| I_{\{|\frac{X}{n} - p| > \delta\}} \right] \\
&\leq \epsilon + 2CP\left(\left|\frac{X}{n} - p\right| > \delta \right).
\end{aligned}
$$

Now, in the last line, just apply Chebyshev's inequality and bound the function $p(1 - p)$ in Chebyshev's inequality by $\frac{1}{4}$. It easily follows then that for all large n, the second term $2CP(|\frac{X}{n} - p| > \delta)$ is also $\leq \epsilon$, which means, that for all large n, uniformly in p, $|B_n(p) - g(p)| \leq 2\epsilon$.

The most important result on convergence in distribution is the *central limit theorem*, which we have already seen in Chapter 1. The proof of the general case is given later in this chapter; it requires some additional development.

Theorem 7.16 (CLT). *Let $X_i, i \geq 1$ be iid with $E(X_i) = \mu$ and $\text{Var}X_i) = \sigma^2 < \infty$. Then*

$$\frac{\sqrt{n}(\bar{X} - \mu)}{\sigma} \xRightarrow{\mathcal{L}} Z \sim N(0, 1).$$

We also write

$$\frac{\sqrt{n}(\bar{X} - \mu)}{\sigma} \xRightarrow{\mathcal{L}} N(0, 1).$$

The multidimensional central limit theorem is stated next. We show that it easily follows from the one-dimensional central limit theorem, by making use of the Cramér–Wold theorem.

Theorem 7.17 (Multivariate CLT). *Let* $X_i, i \geq 1$, *be iid d-dimensional random vectors with* $E(X_1) = \mu$, *and covariance matrix* $Cov(X_1) = \Sigma$. *Then,*

$$\sqrt{n}(\bar{X} - \mu) \stackrel{\mathcal{L}}{\Rightarrow} N_d(0, \Sigma).$$

Remark. If $X_i, i \geq 1$ are iid with mean μ and variance σ^2, then the CLT in one dimension says that $\frac{S_n - n\mu}{\sigma\sqrt{n}} \stackrel{\mathcal{L}}{\Rightarrow} N(0, 1)$, where S_n is the nth partial sum $X_1 + \cdots + X_n$. In particular, therefore, $\frac{S_n - n\mu}{\sigma\sqrt{n}} = O_p(1)$. In other words, in a distributional sense, $\frac{S_n - n\mu}{\sigma\sqrt{n}}$ stabilizes. If we take a large n, then for most sample points ω, $\frac{|S_n(\omega) - n\mu|}{\sigma\sqrt{n}}$ will be, for example, less than 4. But as n changes, this collection of *good sample points* also changes. Indeed, any fixed sample point ω is one of the good sample points for certain values of n, and falls into the category of *bad sample points* for (many) other values of n. The *law of the iterated logarithm* says that if we fix ω and look at $\frac{|S_n(\omega) - n\mu|}{\sigma\sqrt{n}}$ along such *unlucky values of n*, then $\frac{S_n(\omega) - n\mu}{\sigma\sqrt{n}}$ will not appear to be stable. In fact, it will keep growing with n, although at a slow rate. Here is what the law of the iterated logarithm says.

Theorem 7.18 (Law of Iterated Logarithm(LIL)). *Let* $X_i, i \geq 1$ *be iid with mean* μ, *variance* $\sigma^2 < \infty$, *and let* $S_n = \sum_{i=1}^n X_i, n \geq 1$. *Then,*

(a) $\limsup_n \frac{S_n - n\mu}{\sqrt{2n \log \log n}} = \sigma$ a.s.

(b) $\liminf_n \frac{S_n - n\mu}{\sqrt{2n \log \log n}} = -\sigma$ a.s.

(c) *If finite constants* a, τ *satisfy*

$$\limsup_n \frac{S_n - na}{\sqrt{2n \log \log n}} = \tau,$$

then necessarily $Var(X_1) < \infty$, *and* $a = E(X_1), \tau^2 = Var(X_1)$.

See Chow and Teicher (1988, p. 355) for a proof. The main use of the LIL is in proving other strong laws. Because $\sqrt{n \log \log n}$ grows at a very slow rate, the practical use of the LIL is quite limited. We remark that the LIL provides another example of a sequence which converges in probability (to zero), but does not converge almost surely.

7.5 Preservation of Convergence and Statistical Applications

Akin to the results on preservation of convergence in probability and almost sure convergence under various operations, there are similar other extremely useful results on preservation of convergence in distribution. The first theorem is of particular importance in statistics.

7.5.1 Slutsky's Theorem

Theorem 7.19 (Slutsky's Theorem). Let $\mathbf{X_n}, \mathbf{Y_n}$ be d and p-dimensional random vectors for some $d, p \geq 1$. Suppose $\mathbf{X_n} \stackrel{\mathcal{L}}{\Rightarrow} \mathbf{X}$, and $\mathbf{Y_n} \stackrel{\mathcal{P}}{\Rightarrow} \mathbf{c}$. Let $h(x, y)$ be a scalar or a vector-valued jointly continuous function in $(x, y) \in \mathcal{R}^d \times \mathcal{R}^p$. Then $h(\mathbf{X_n}, \mathbf{Y_n}) \stackrel{\mathcal{L}}{\Rightarrow} h(\mathbf{X}, \mathbf{c})$.

Proof. We use the part of the portmanteau theorem which says that a random vector $\mathbf{Z_n} \stackrel{\mathcal{L}}{\Rightarrow} \mathbf{Z}$ if $E[g(\mathbf{Z_n})] \to E[g(\mathbf{Z})]$ for all bounded uniformly continuous functions g. Now, if we simply repeat the proof of the uniform convergence of the Bernstein polynomials in our example on Weierstrass's theorem, the result is obtained. □

The following are some particularly important consequences of Slutsky's theorem.

Corollary. *(a)* Suppose $\mathbf{X_n} \stackrel{\mathcal{L}}{\Rightarrow} \mathbf{X}, \mathbf{Y_n} \stackrel{\mathcal{P}}{\Rightarrow} \mathbf{c}$, where $\mathbf{X_n}, \mathbf{Y_n}$ are of the same order. Then, $\mathbf{X_n} + \mathbf{Y_n} \stackrel{\mathcal{L}}{\Rightarrow} \mathbf{X} + \mathbf{c}$.
(b) Suppose $\mathbf{X_n} \stackrel{\mathcal{L}}{\Rightarrow} \mathbf{X}, Y_n \stackrel{\mathcal{P}}{\Rightarrow} c$, where Y_n are scalar random variables. Then $Y_n \mathbf{X_n} \stackrel{\mathcal{L}}{\Rightarrow} c\mathbf{X}$.
(c) Suppose $\mathbf{X_n} \stackrel{\mathcal{L}}{\Rightarrow} \mathbf{X}, Y_n \stackrel{\mathcal{P}}{\Rightarrow} c \neq 0$, where Y_n are scalar random variables. Then $\frac{\mathbf{X_n}}{Y_n} \stackrel{\mathcal{L}}{\Rightarrow} \frac{\mathbf{X}}{c}$.

Example 7.20 (Convergence of the t to Normal). Let $X_i, i \geq 1$, be iid $N(\mu, \sigma^2)$, $\sigma > 0$, and let $T_n = \frac{\sqrt{n}(\bar{X}-\mu)}{s}$, where $s^2 = \frac{1}{n-1}\sum_{i=1}^{n}(X_i - \bar{X})^2$, namely the sample variance. We saw in Chapter 15 that T_n has the central $t(n-1)$ distribution. Write

$$T_n = \frac{\sqrt{n}(\bar{X}-\mu)/\sigma}{s/\sigma}.$$

We have seen that $s^2 \stackrel{a.s.}{\Rightarrow} \sigma^2$. Therefore, by the *continuous mapping theorem*, $s \stackrel{a.s.}{\Rightarrow} \sigma$, and so $\frac{s}{\sigma} \stackrel{a.s.}{\Rightarrow} 1$. On the other hand, by the central limit theorem, $\frac{\sqrt{n}(\bar{X}-\mu)}{\sigma} \stackrel{\mathcal{L}}{\Rightarrow} N(0, 1)$. Therefore, now by Slutsky's theorem, $T_n \stackrel{\mathcal{L}}{\Rightarrow} N(0, 1)$.

Indeed, this argument shows that whatever the common distribution of the X_i is, if $\sigma^2 = \text{Var}(X_1) < \infty$ and > 0, then $T_n \stackrel{\mathcal{L}}{\Rightarrow} N(0, 1)$, although the exact distribution of T_n is no longer the central t distribution, unless the common distribution of the X_i is normal.

Example 7.21 (A Normal–Cauchy Connection). Consider iid standard normal variables, $X_1, X_2, \ldots, X_{2n}, n \geq 1$. Let

$$R_n = \frac{X_1}{X_2} + \frac{X_3}{X_4} + \cdots + \frac{X_{2n-1}}{X_{2n}}, \quad \text{and} \quad D_n = X_1^2 + X_2^2 + \cdots + X_n^2.$$

Let $T_n = \frac{R_n}{D_n}$.

Recall that the quotient of two independent standard normals is distributed as a standard Cauchy. Thus,
$$\frac{X_1}{X_2}, \frac{X_3}{X_4}, \ldots, \frac{X_{2n-1}}{X_{2n}}$$
are independent standard Cauchy. In the following, we write C_n to denote a random variable with a Cauchy distribution with location parameter zero, and scale parameter n. From our results on convolutions, we know that the sum of n independent standard Cauchy random variables is distributed as C_n; the scale parameter is n. Thus, $R_n \stackrel{\mathcal{L}}{=} C_n \sim C(0, n) \stackrel{\mathcal{L}}{=} nC_1 \sim nC(0, 1)$. Therefore,
$$T_n \stackrel{\mathcal{L}}{=} \frac{nC_1}{\sum_{i=1}^{n} X_i^2} = \frac{C_1}{\frac{1}{n}\sum_{i=1}^{n} X_i^2}.$$

Now, by the WLLN, $\frac{1}{n}\sum_{i=1}^{n} X_i^2 \stackrel{\mathcal{P}}{\Rightarrow} E(X_1^2) = 1$. A sequence of random variables with a distribution identically equal to the fixed $C(0, 1)$ distribution also, tautologically, converges in distribution to C_1, thus by applying Slutsky's theorem, we conclude that $T_n \stackrel{\mathcal{L}}{\Rightarrow} C_1 \sim C(0, 1)$.

7.5.2 Delta Theorem

The next theorem says that convergence in distribution is appropriately preserved by making smooth transformations. In particular, we present a general version of a theorem of fundamental use in statistics, called *the delta theorem*.

Theorem 7.20 (Continuous Mapping Theorem). *(a) Let $\mathbf{X_n}$ be d-dimensional random vectors and let $S \subseteq \mathcal{R}^d$ be such that $P(\mathbf{X_n} \in S) = 1 \; \forall n$. Suppose $\mathbf{X_n} \stackrel{\mathcal{L}}{\Rightarrow} \mathbf{X}$. Let $g : S \to \mathcal{R}^p$ be a continuous function, where p is a positive integer. Then $g(\mathbf{X_n}) \stackrel{\mathcal{L}}{\Rightarrow} g(\mathbf{X})$.*

*(b) (**Delta Theorem of Cramér**). Let $\mathbf{X_n}$ be d-dimensional random vectors and let $S \subseteq \mathcal{R}^d$ be such that $P(\mathbf{X_n} \in S) = 1 \; \forall n$. Suppose for some d-dimensional vector μ, and some sequence of reals $c_n \to \infty$, $c_n(\mathbf{X_n} - \mu) \stackrel{\mathcal{L}}{\Rightarrow} \mathbf{X}$. Let $g : S \to \mathcal{R}^p$ be a function with each coordinate of g once continuously differentiable with respect to every coordinate of $\mathbf{x} \in S$ at $\mathbf{x} = \mu$. Then*
$$c_n(g(\mathbf{X_n}) - g(\mu)) \stackrel{\mathcal{L}}{\Rightarrow} Dg(\mu)\mathbf{X},$$
where $Dg(\mu)$ is the matrix of partial derivatives $((\frac{\partial g_i}{\partial x_j}))|_{\mathbf{x}=\mu}$.

Proof. For part (a), we use the Portmanteau theorem. Denote $g(\mathbf{X_n}) = \mathbf{Y_n}, g(\mathbf{X}) = \mathbf{Y}$, and consider bounded continuous functions $f(\mathbf{Y_n})$. Now, $f(\mathbf{Y_n}) = f(g(\mathbf{X_n})) = h(\mathbf{X_n})$, where $h(.)$ is the composition function $f(g(.))$. Because h is continuous, because f, g are, and h is bounded, because f is, the Portmanteau theorem implies that $E(h(\mathbf{X_n})) \to E(h(\mathbf{X}))$, that is, $E(f(\mathbf{Y_n})) \to E(f(\mathbf{Y}))$. Now the reverse implication in the Portmanteau theorem implies that $\mathbf{Y_n} \stackrel{\mathcal{L}}{\Rightarrow} \mathbf{Y}$.

We prove part (b) for the case $d = p = 1$. First note that it follows from the assumption $c_n \to \infty$ that $X_n - \mu = o_p(1)$. Also, by an application of Taylor's theorem,

$$g(x_0 + h) = g(x_0) + hg'(x_0) + o(h)$$

if g is differentiable at x_0. Therefore,

$$g(X_n) = g(\mu) + (X_n - \mu)g'(\mu) + o_p(X_n - \mu).$$

That the remainder term is $o_p(X_n - \mu)$ follows from our observation that $X_n - \mu = o_p(1)$. Taking $g(\mu)$ to the left and multiplying by c_n, we obtain

$$c_n[g(X_n) - g(\mu)] = c_n(X_n - \mu)g'(\mu) + c_n o_p(X_n - \mu).$$

The term $c_n o_p(X_n - \mu) = o_p(1)$, because $c_n(X_n - \mu) = O_p(1)$. Hence, by an application of Slutsky's theorem, $c_n[g(X_n) - g(\mu)] \stackrel{\mathcal{L}}{\Rightarrow} g'(\mu)X$. □

Example 7.22 (A Quadratic Form). Let $X_i, i \geq 1$ be iid random variables with finite mean μ and finite variance σ^2. By the central limit theorem, $\frac{\sqrt{n}(\bar{X}-\mu)}{\sigma} \stackrel{\mathcal{L}}{\Rightarrow} Z$, where $Z \sim N(0, 1)$. Therefore, by the continuous mapping theorem, if $Q_n = \frac{n}{\sigma^2}(\bar{X}-\mu)^2$, then $Q_n = \left(\frac{\sqrt{n}(\bar{X}-\mu)}{\sigma}\right)^2 \stackrel{\mathcal{L}}{\Rightarrow} Z^2$. But $Z^2 \sim \chi_1^2$. Therefore, $Q_n \stackrel{\mathcal{L}}{\Rightarrow} \chi_1^2$.

Example 7.23 (An Important Statistics Example). Let $X = X_n \sim \text{Bin}(n, p), n \geq 1, 0 < p < 1$. In statistics, p is generally treated as an unknown parameter, and the usual estimate of p is $\hat{p} = \frac{X}{n}$. Define $T_n = |\frac{\sqrt{n}(\hat{p}-p)}{\sqrt{\hat{p}(1-\hat{p})}}|$. The goal of this example is to find the limiting distribution of T_n. First, by the central limit theorem,

$$\frac{\sqrt{n}(\hat{p}-p)}{\sqrt{p(1-p)}} = \frac{X - np}{\sqrt{np(1-p)}} \stackrel{\mathcal{L}}{\Rightarrow} N(0, 1).$$

Next, by the WLLN, $\hat{p} \stackrel{\mathcal{P}}{\Rightarrow} p$, and hence by the continuous mapping theorem for convergence in probability, $\sqrt{\hat{p}(1-\hat{p})} \stackrel{\mathcal{P}}{\Rightarrow} \sqrt{p(1-p)}$. This gives, by Slutsky's theorem, $\frac{\sqrt{n}(\hat{p}-p)}{\sqrt{\hat{p}(1-\hat{p})}} \stackrel{\mathcal{L}}{\Rightarrow} N(0, 1)$. Finally, because the absolute value function is continuous, by the continuous mapping theorem for convergence in distribution,

$$T_n = |\frac{\sqrt{n}(\hat{p}-p)}{\sqrt{\hat{p}(1-\hat{p})}}| \stackrel{\mathcal{L}}{\Rightarrow} |Z|,$$

the absolute value of a standard normal.

7.5 Preservation of Convergence and Statistical Applications

Example 7.24. Suppose $X_i, i \geq 1$, are iid with mean μ and variance $\sigma^2 < \infty$. By the central limit theorem, $\frac{\sqrt{n}(\bar{X}-\mu)}{\sigma} \stackrel{\mathcal{L}}{\Rightarrow} Z$, where $Z \sim N(0,1)$. Consider the function $g(x) = x^2$. This is continuously differentiable, in fact at any x, and $g'(x) = 2x$. If $\mu \neq 0$, then $g'(\mu) = 2\mu \neq 0$. By the delta theorem, we get that $\sqrt{n}(\bar{X}^2 - \mu^2) \stackrel{\mathcal{L}}{\Rightarrow} N(0, 4\mu^2\sigma^2)$. If $\mu = 0$, this last statement is still true, and that means $\sqrt{n}\bar{X}^2 \stackrel{P}{\Rightarrow} 0$, if $\mu = 0$.

Example 7.25 (Sample Variance and Standard Deviation). Suppose again $X_i, i \geq 1$, are iid with mean θ, variance σ^2, and $E(X_1^4) < \infty$. Also let $\mu_j = E(X_1 - \theta)^j$, $1 \leq j \leq 4$. This example has $d = 2, p = 1$. Take

$$\mathbf{X_n} = \begin{pmatrix} \bar{X} \\ \frac{1}{n}\sum_{i=1}^n X_i^2 \end{pmatrix}, \quad \mu = \begin{pmatrix} EX_1 \\ EX_1^2 \end{pmatrix}, \quad \Sigma = \begin{pmatrix} \text{Var}(X_1) & \text{Cov}(X_1, X_1^2) \\ \text{Cov}(X_1, X_1^2) & \text{Var}(X_1^2) \end{pmatrix}$$

By the multivariate central limit theorem, $\sqrt{n}(\mathbf{X_n} - \mu) \stackrel{\mathcal{L}}{\Rightarrow} N(\mathbf{0}, \Sigma)$. Now consider the function $g(u, v) = v - u^2$. This is once continuously differentiable with respect to each of u, v (in fact at any u, v), and the partial derivatives are $g_u = -2u$, $g_v = 1$. Using the delta theorem, with a little bit of matrix algebra calculations, it follows that

$$\sqrt{n}\left(\frac{1}{n}\sum_{i=1}^n (X_i - \bar{X})^2 - \text{Var}(X_1)\right) \stackrel{\mathcal{L}}{\Rightarrow} N(0, \mu_4 - \sigma^4).$$

If we choose $s_n^2 = \sum_{i=1}^n (X_i - \bar{X})^2/(n-1)$ then

$$\sqrt{n}(s_n^2 - \sigma^2) = \frac{\sum_{i=1}^n (X_i - \bar{X})^2}{(n-1)\sqrt{n}} + \sqrt{n}\left(\frac{1}{n}\sum_{i=1}^n (X_i - \bar{X})^2 - \sigma^2\right)$$

$= \sqrt{n}(\frac{1}{n}\sum_{i=1}^n (X_i - \bar{X})^2 - \sigma^2) + o_p(1)$, which also converges in law to $N(0, \mu_4 - \sigma^4)$ by Slutsky's theorem. By another use of the delta theorem, this time with $d = p = 1$, and with the function $g(u) = \sqrt{u}$, one gets

$$\sqrt{n}(s_n - \sigma) \stackrel{\mathcal{L}}{\Rightarrow} N\left(0, \frac{\mu_4 - \sigma^4}{4\sigma^2}\right).$$

Example 7.26 (Sample Correlation). Another use of the delta theorem is the derivation of the limiting distribution of the sample correlation coefficient r for iid bivariate data (X_i, Y_i). We have

$$r_n = \frac{\frac{1}{n}\sum X_i Y_i - \bar{X}\bar{Y}}{\sqrt{\frac{1}{n}\sum(X_i - \bar{X})^2}\sqrt{\frac{1}{n}\sum(Y_i - \bar{Y})^2}}.$$

By taking

$$T_n = \left(\bar{X}, \bar{Y}, \frac{1}{n}\sum X_i^2, \frac{1}{n}\sum Y_i^2, \frac{1}{n}\sum X_i Y_i\right)$$
$$\theta = (EX_1, EY_1, EX_1^2, EY_1^2, EX_1Y_1)$$

and by taking Σ to be the covariance matrix of $(X_1, Y_1, X_1^2, Y_1^2, X_1Y_1)$, and on using the transformation $g(u_1, u_2, u_3, u_4, u_5) = (u_5 - u_1 u_2)/\sqrt{(u_3 - u_1^2)(u_4 - u_2^2)}$ it follows from the delta theorem, with $d = 5, p = 1$, that

$$\sqrt{n}(r_n - \rho) \stackrel{\mathcal{L}}{\Rightarrow} N(0, v^2)$$

for some $v > 0$. It is not possible to write a clean formula for v in general. If (X_i, Y_i) are iid $N_2(\mu_X, \mu_Y, \sigma_X^2, \sigma_Y^2, \rho)$ then the calculation of v^2 can be done in closed-form, and

$$\sqrt{n}(r_n - \rho) \stackrel{\mathcal{L}}{\Rightarrow} N(0, (1-\rho^2)^2).$$

However, convergence to normality is very slow.

7.5.3 Variance Stabilizing Transformations

A major use of the delta theorem is construction of variance stabilizing transformations (VST), a technique that is of fundamental use in statistics. VSTs are useful tools for constructing confidence intervals for unknown parameters. The general idea is the following. Suppose we want to find a confidence interval for some parameter $\theta \in \mathcal{R}$. If $T_n = T_n(X_1, \ldots, X_n)$ is some natural estimate for θ (e.g., sample mean as an estimate of a population mean), then often the CLT, or some generalization of the CLT, will tell us that $T_n - \theta \stackrel{\mathcal{L}}{\Rightarrow} N(0, \sigma^2(\theta))$, for some suitable function $\sigma^2(\theta)$. This implies that in large samples,

$$P_\theta(T_n - z_{\frac{\alpha}{2}} \frac{\sigma(\theta)}{\sqrt{n}} \leq \theta \leq T_n + z_{\frac{\alpha}{2}} \frac{\sigma(\theta)}{\sqrt{n}}) \approx 1 - \alpha,$$

where α is some specified number in $(0,1)$ and $z_{\alpha/2} = \Phi^{-1}(1 - \frac{\alpha}{2})$. Finally, plugging in T_n in place of θ in $\sigma(\theta)$, a confidence interval for θ is $T_n \pm z_{\frac{\alpha}{2}} \frac{\sigma(T_n)}{\sqrt{n}}$. The delta theorem provides an alternative solution that is sometimes preferred. By the delta theorem, if $g(\cdot)$ is once differentiable at θ with $g'(\theta) \neq 0$, then

$$\sqrt{n}(g(T_n) - g(\theta)) \stackrel{\mathcal{L}}{\Rightarrow} N(0, [g'(\theta)]^2 \sigma^2(\theta)).$$

7.5 Preservation of Convergence and Statistical Applications

Therefore, if we set

$$[g'(\theta)]^2 \sigma^2(\theta) = k^2$$

for some constant k, then $\sqrt{n}(g(T_n) - g(\theta)) \stackrel{\mathcal{L}}{\Rightarrow} N(0, k^2)$, and this produces a confidence interval for $g(\theta)$:

$$g(T_n) \pm z_{\frac{\alpha}{2}} \frac{k}{\sqrt{n}}.$$

By *retransforming* back to θ, we get another confidence interval for θ:

$$g^{-1}\left(g(T_n) - z_{\frac{\alpha}{2}} \frac{k}{\sqrt{n}}\right) \leq \theta \leq g^{-1}\left(g(T_n) + z_{\frac{\alpha}{2}} \frac{k}{\sqrt{n}}\right).$$

The reason that this one is sometimes preferred to the first confidence interval, namely, $T_n \pm z_{\frac{\alpha}{2}} \frac{\sigma(T_n)}{\sqrt{n}}$, is that no additional plug-in is necessary to estimate the penultimate variance function in this second confidence interval. The penultimate variance function is already a constant k^2 by choice in this second method. The transformation $g(T_n)$ obtained from its defining property

$$[g'(\theta)]^2 \sigma^2(\theta) = k^2$$

has the expression

$$g(\theta) = k \int \frac{1}{\sigma(\theta)} d\theta,$$

where the integral is to be interpreted as a primitive. The constant k can be chosen as any nonzero real number, and $g(T_n)$ is called a variance stabilizing transformation. Although the delta theorem is certainly available in $\mathcal{R}^d$ even when $d > 1$, unfortunately the concept of VSTs does not generalize to multiparameter cases. It is generally infeasible to find a dispersion-stabilizing transformation when the dimension of θ is more than one. This example is a beautiful illustration of how probability theory leads to useful and novel statistical techniques.

Example 7.27 (VST in Binomial Case). Suppose $X_n \sim \text{Bin}(n, p)$. Then $\sqrt{n}(X_n/n - p) \stackrel{\mathcal{L}}{\Rightarrow} N(0, p(1-p))$. So using the notation used above, $\sigma(p) = \sqrt{p(1-p)}$ and consequently, on taking $k = \frac{1}{2}$,

$$g(p) = \int \frac{1/2}{\sqrt{p(1-p)}} dp = \arcsin(\sqrt{p}).$$

Hence, $g(X_n) = \arcsin(\sqrt{X_n/n})$ is a variance-stabilizing transformation and indeed,

$$\sqrt{n}\left(\arcsin\left(\sqrt{\frac{X_n}{n}}\right) - \arcsin(\sqrt{p})\right) \stackrel{\mathcal{L}}{\Rightarrow} N\left(0, \frac{1}{4}\right).$$

Thus, a confidence interval for p is

$$\sin^2\left(\arcsin\left(\sqrt{\frac{X_n}{n}}\right) \mp \frac{z_{\alpha/2}}{2\sqrt{n}}\right).$$

Example 7.28 (Fisher's z). Suppose (X_i, Y_i), $i = 1, \ldots, n$, are iid bivariate normal with parameters $\mu_X, \mu_Y, \sigma_X^2, \sigma_Y^2, \rho$. Then, as we saw above, $\sqrt{n}(r_n - \rho) \overset{\mathcal{L}}{\Rightarrow} N(0, (1-\rho^2)^2)$, r_n being the sample correlation coefficient. Therefore,

$$g(\rho) = \int \frac{1}{(1-\rho)^2} d\rho = \frac{1}{2}\log\frac{1+\rho}{1-\rho} = \operatorname{arctanh}(\rho)$$

provides a variance-stabilizing transformation for r_n. This is the famous arctanh transformation of Fisher, popularly known as *Fisher's z*. By the delta theorem, $\sqrt{n}(\operatorname{arctanh}(r_n) - \operatorname{arctanh}(\rho))$ converges in distribution to the $N(0, 1)$ distribution. Confidence intervals for ρ are computed from Fisher's z as

$$\tanh\left(\operatorname{arctanh}(r_n) \pm \frac{z_{\alpha/2}}{\sqrt{n}}\right).$$

The arctanh transformation $z = \operatorname{arctanh}(r_n)$ attains approximate normality much more quickly than r_n itself.

Example 7.29 (An Unusual VST). Here is a *nonregular* example on variance stabilization. Suppose we have iid observations $X_1, X_2, \ldots$ from the $U[0, \theta]$ distribution. Then, the usual estimate of θ is the sample maximum $X_{(n)}$, and $n(\theta - X_{(n)}) \overset{\mathcal{L}}{\Rightarrow} \operatorname{Exp}(\theta)$. The asymptotic variance function in the distribution of the sample maximum is therefore simply θ^2, and therefore, a VST is

$$g(\theta) = \int \frac{1}{\theta} d\theta = \log\theta.$$

So, $g(X_{(n)}) = \log X_{(n)}$ is a variance-stabilizing transformation of $X_{(n)}$. In fact, $n(\log\theta - \log X_{(n)}) \overset{\mathcal{L}}{\Rightarrow} \operatorname{Exp}(1)$. However, the interesting fact is that for every n, the distribution of $n(\log\theta - \log X_{(n)})$ is exactly a standard exponential. There is no nontrivial example such as this in the regular cases (although $N(\theta, 1)$ is a trivial example).

7.6 Convergence of Moments

If some sequence of random variables X_n converges in distribution to a random variable X, then sometimes we are interested in knowing whether moments of X_n converge to moments of X. More generally, we may want to find approximations for moments of X_n. Convergence in distribution just by itself cannot ensure convergence of any moment. An extra condition that ensures convergence

7.6 Convergence of Moments

of appropriate moments is *uniform integrability*. There is another side of this story. If we can show that the moments of X_n converge to moments of some recognizable distribution, then we can sometimes show that X_n converges in distribution to that distinguished distribution. Some of these issues are discussed in this section.

7.6.1 Uniform Integrability

Definition 7.12. Let $\{X_n\}$ be a sequence of random variables on some common sample space Ω. The sequence $\{X_n\}$ is called uniformly integrable if $\sup_{n\geq 1} E(|X_n|) < \infty$, and if for any $\epsilon > 0$ there exists a sufficiently small $\delta > 0$, such that whenever $P(A) < \delta$, $\sup_{n\geq 1} \int_A |X_n| dP < \epsilon$.

Remark. We give two results on the link between uniform integrability and convergence of moments.

Theorem 7.21. *Suppose $X, X_n, n \geq 1$ are such that $E(|X|^p) < \infty$, and $E(|X_n|^p) < \infty \; \forall n \geq 1$. Suppose $X_n \stackrel{\mathcal{P}}{\Rightarrow} X$, and $|X_n|^p$ is uniformly integrable. Then, $E(|X_n - X|^p) \to 0$, as $n \to \infty$.*

For proving this theorem, we need two lemmas. The first one is one of the most fundamental results in real analysis. It uses the terminology of Lebesgue integrals and the Lebesgue measure, which we are not treating in this book. Thus, the statement below uses an undefined concept.

Lemma (Dominated Convergence Theorem). *Let f_n, f be functions on $\mathcal{R}^d$, $d < \infty$, and suppose f and each f_n is (Lebesgue) integrable. Suppose $f_n(x) \to f(x)$, except possibly for a set of x values of Lebesgue measure zero. If $|f_n| \leq g$ for some integrable function g, then $\int_{\mathcal{R}^d} f_n(x) dx \to \int_{\mathcal{R}^d} f(x) dx$, as $n \to \infty$.*

Lemma. *Suppose $X, X_n, n \geq 1$ are such that $E(|X|^p) < \infty$, and $E(|X_n|^p) < \infty$ $\forall n \geq 1$. Then $|X_n|^p$ is uniformly integrable if and only if $|X_n - X|^p$ is uniformly integrable.*

Proof of Theorem: Fix $c > 0$, and define $Y_n = |X_n - X|^p$, $Y_{n,c} = Y_n I_{\{|X_n - X| \leq c\}}$. Because, by hypothesis, $X_n \stackrel{\mathcal{P}}{\Rightarrow} X$, by the continuous mapping theorem, $Y_n \stackrel{\mathcal{P}}{\Rightarrow} 0$, and as a consequence, $Y_{n,c} \stackrel{\mathcal{P}}{\Rightarrow} 0$. Furthermore, $|Y_{n,c}| \leq c^p$, and the dominated convergence theorem implies that $E(Y_{n,c}) \to 0$. Now,

$$E(|X_n - X|^p) = E(Y_n) = E(Y_{n,c}) + E(Y_n I_{\{|X_n - X| > c\}})$$

$$\leq E(Y_{n,c}) + \sup_{n \geq 1} E(Y_n I_{\{|X_n - X| > c\}})$$

$$\Rightarrow \limsup_n E(|X_n - X|^p) \leq \sup_{n \geq 1} E(Y_n I_{\{|X_n - X| > c\}})$$

$$\Rightarrow \limsup_n E(|X_n - X|^p) \leq \inf_c \sup_{n \geq 1} E(Y_n I_{\{|X_n - X| > c\}})$$

$$= 0.$$

Therefore, $E(|X_n - X|^p) \to 0$. □

Remark. Sometimes we do not need the full force of the result that $E(|X_n - X|^p) \to 0$, but all we want is that $E(X_n^p)$ converges to $E(X^p)$. In that case, the conditions in the previous theorem can be relaxed, and in fact from a statistical point of view, the relaxed condition is much more natural. The following result gives the relaxed conditions.

Theorem 7.22. *Suppose $X_n, X, n \geq 1$ are defined on a common sample space Ω, that $X_n \stackrel{\mathcal{L}}{\Rightarrow} X$, and that for some given $p \geq 1$, $|X_n|^p$ is uniformly integrable. Then $E(X_n^k) \to E(X^k) \; \forall k \leq p$.*

Remark. To apply these last two theorems, we have to verify that for the appropriate sequence X_n, and for the relevant p, $|X_n|^p$ is uniformly integrable. Direct verification of uniform integrability from definition is often cumbersome. But simple sufficient conditions are available, and these are often satisfied in many applications. The next result lists a few useful sufficient conditions.

Theorem 7.23 (Sufficient Conditions for Uniform Integrability).

(a) *Suppose for some $\delta > 0$, $\sup_n E|X_n|^{1+\delta} < \infty$. Then $\{X_n\}$ is uniformly integrable.*
(b) *If $|X_n| \leq Y, n \geq 1$, and $E(Y) < \infty$, then $\{X_n\}$ is uniformly integrable.*
(c) *If $|X_n| \leq Y_n, n \geq 1$, and Y_n is uniformly integrable, then $\{X_n\}$ is uniformly integrable.*
(d) *If $X_n, n \geq 1$ are identically distributed, and $E(|X_1|) < \infty$, then $\{X_n\}$ is uniformly integrable.*
(e) *If $\{X_n\}$ and $\{Y_n\}$ are uniformly integrable then $\{X_n + Y_n\}$ is uniformly integrable.*
(f) *If $\{X_n\}$ is uniformly integrable and $|Y_n| \leq M < \infty$, then $\{X_n Y_n\}$ is uniformly integrable.*
See Chow and Teicher (1988, p. 94) for further details on the various parts of this theorem.

Example 7.30 (Sample Maximum). We saw in Chapter 6 that if $X_1, X_2, \ldots$ are iid, and if $E(|X_1|^k) < \infty$, then any order statistic $X_{(r)}$ satisfies

$$E(|X_{(r)}|^k) \leq \frac{n!}{(r-1)!(n-r)!} E(|X_1|^k).$$

In particular, for the sample maximum $X_{(n)}$ of n observations,

$$E(|X_{(n)}|) \leq n E(|X_1|) \Rightarrow E\left(\frac{|X_{(n)}|}{n}\right) \leq E(|X_1|).$$

By itself, this does not ensure that $\frac{|X_{(n)}|}{n}$ is uniformly integrable.

However, if we also assume that $E(X_1^2) < \infty$, then the same argument gives $E(|X_{(n)}|^2) \leq n E(X_1)^2$, so that $\sup_n E(\frac{|X_{(n)}|}{n})^2 < \infty$, which is enough to conclude that $\frac{|X_{(n)}|}{n}$ is uniformly integrable.

7.6 Convergence of Moments

However, we do not need the existence of $E(X_1^2)$ for this conclusion. Note that

$$|X_{(n)}| \leq \sum_{i=1}^{n} |X_i| \Rightarrow \frac{|X_{(n)}|}{n} \leq \frac{\sum_{i=1}^{n} |X_i|}{n}.$$

If $E(|X_1|) < \infty$, then in fact $\frac{\sum_{i=1}^{n} |X_i|}{n}$ is uniformly integrable, and as a consequence, $\frac{|X_{(n)}|}{n}$ is also uniformly integrable under just the condition $E(|X_1|) < \infty$.

7.6.2 The Moment Problem and Convergence in Distribution

We remarked earlier that convergence of moments can be useful to establish convergence in distribution. Clearly, however, if we only know that $E(X_n^k)$ converges to $E(X^k)$ for each k, from that alone we cannot conclude that the distributions of X_n converge to the distribution of X. This is because there could, in general, be another random variable Y with a distribution distinct from that of X but with all moments equal to the moments of X. However, if we rule out that possibility, then convergence in distribution follows.

Theorem 7.24. *Suppose for some sequence $\{X_n\}$ and a random variable X, $E(X_n^k) \to E(X^k) \, \forall k \geq 1$. If the distribution of X is determined by its moments, then $X_n \xrightarrow{\mathcal{L}} X$.*

When is a distribution determined by its sequence of moments? This is a hard analytical problem, and is commonly known as the moment problem. There is a huge and sophisticated literature on the moment problem. A few easily understood conditions for determinacy by moments are given in the next result.

Theorem 7.25. *(a) If a random variable X is uniformly bounded, then it is determined by its moments.*
(b) If the mgf of a random variable X exists in a nonempty interval containing zero, then it is determined by its moments.
(c) Let X have a density function $f(x)$. If there exist positive constants c, α, β, k such that $f(x) \leq ce^{-\alpha|x|}|x|^\beta \, \forall x$ such that $|x| > k$, then X is determined by its moments.

Remark. See Feller (1971, pp. 227–228 and p. 251) for the previous two theorems. Basically, part (b) is the primary result here, because if the conditions in (a) or (c) hold, then the mgf exists in an interval containing zero. However, (a) and (c) are useful special sufficient conditions.

Example 7.31 (Discrete Uniforms Converging to Continuous Uniform). Consider random variables X_n with the discrete uniform distribution on $\{\frac{1}{n}, \frac{2}{n}, \ldots, 1\}$. Fix a positive integer k. Then, $E(X_n^k) = \frac{1}{n} \sum_{i=1}^{n} (\frac{i}{n})^k$. This is the upper Riemann sum corresponding to the partition

$$\left(\frac{i-1}{n}, \frac{i}{n}\right], i = 1, 2, \ldots, n$$

for the function $f(x) = x^k$ on $(0, 1]$. Therefore, as $n \to \infty$, $E(X_n^k)$, which is the upper Riemann sum, converges to $\int_0^1 x^k dx$, which is the kth moment of a random variable X having the uniform distribution on the unit interval. Because k is arbitrary, it follows from part (a) of the above theorem that the discrete uniform distribution on $\{\frac{1}{n}, \frac{2}{n}, \ldots, 1\}$ converges to the uniform distribution on the unit interval.

7.6.3 Approximation of Moments

Knowing the limiting value of a moment of some sequence of random variables is only a first-order approximation to the moment. Sometimes we want more refined approximations. Suppose $X_i, 1 \leq i \leq d$, are jointly distributed random variables, and $T_d(X_1, X_2, \ldots, X_d)$ is some function of $X_1, X_2, \ldots, X_d$. To find approximations for a moment of T_d, one commonly used technique is to approximate the function $T_d(x_1, x_2, \ldots, x_d)$ by a simpler function, say $g_d(x_1, x_2, \ldots, x_d)$, and then use the moment of $g_d(X_1, X_2, \ldots, X_d)$ as an approximation to the moment of $T_d(X_1, X_2, \ldots, X_d)$. Note that this is a formal approximation. It does not come with an automatic quantification of the error of the approximation. Such quantification is usually a harder problem, and limited answers are available. We address these two issues in this section. We consider approximation of the mean and variance of a statistic, because of their special importance.

A natural approximation of a smooth function is obtained by expanding the function around some point in a Taylor series. For the formal approximations, we assume that all the moments of X_i that are necessary for the approximation to make sense do exist.

It is natural to expand the statistic around the mean vector μ of $\mathbf{X} = (X_1, \ldots, X_d)$. For notational simplicity, we write t for T_d. Then, the first- and second-order Taylor expansions for $t(X_1, X_2, \ldots, X_d)$ are:

$$t(x_1, x_2, \ldots, x_d) \approx t(\mu_1, \ldots, \mu_d) + \sum_{i=1}^{n}(x_i - \mu_i) t_i(\mu_1, \ldots, \mu_d),$$

and

$$t(x_1, x_2, \ldots, x_d) \approx t(\mu_1, \ldots, \mu_d) + \sum_{i=1}^{d}(x_i - \mu_i) t_i(\mu_1, \ldots, \mu_d)$$
$$+ \frac{1}{2} \sum_{1 \leq i,j \leq d}(x_i - \mu_i)(x_j - \mu_j) t_{ij}(\mu_1, \ldots, \mu_d).$$

If we formally take an expectation on both sides, we get the first- and second-order approximations to $E[T_d(X_1, X_2, \ldots, X_d)]$:

$$E[T_d(X_1, X_2, \ldots, X_d)] \approx T_d(\mu_1, \ldots, \mu_d),$$

7.6 Convergence of Moments

and

$$E[T_d(X_1, X_2, \ldots, X_d)] \approx T_d(\mu_1, \ldots, \mu_d) + \frac{1}{2} \sum_{1 \le i,j \le d} t_{ij}(\mu_1, \ldots, \mu_d)\sigma_{ij},$$

where σ_{ij} is the covariance between X_i and X_j.

Consider now the variance approximation problem. From the first-order Taylor approximation

$$t(x_1, x_2, \ldots, x_d) \approx t(\mu_1, \ldots, \mu_d) + \sum_{i=1}^{d}(x_i - \mu_i)t_i(\mu_1, \ldots, \mu_d),$$

by formally taking the variance of both sides, we get the first-order variance approximation

$$\text{Var}(T_d(X_1, X_2, \ldots, X_d)) \approx \text{Var}\left(\sum_{i=1}^{d}(x_i - \mu_i)t_i(\mu_1, \ldots, \mu_d)\right)$$

$$= \sum_{1 \le i,j \le d} t_i(\mu_1, \ldots, \mu_d)t_j(\mu_1, \ldots, \mu_d)\sigma_{ij}.$$

The second-order variance approximation takes more work. By using the second-order Taylor approximation for $t(x_1, x_2, \ldots, x_d)$, the second-order variance approximation is

$$\text{Var}(T_d(X_1, X_2, \ldots, X_d)) \approx \text{Var}\left(\sum_i (X_i - \mu_i)t_i(\mu_1, \ldots, \mu_d)\right)$$

$$+ \frac{1}{4}\text{Var}\left(\sum_{i,j}(X_i - \mu_i)(X_j - \mu_j)t_{ij}(\mu_1, \ldots, \mu_d)\right)$$

$$+ \text{Cov}\left(\sum_i (X_i - \mu_i)t_i(\mu_1, \ldots, \mu_d),\right.$$

$$\left.\sum_{j,k}(X_j - \mu_j)(X_k - \mu_k)t_{jk}(\mu_1, \ldots, \mu_d)\right).$$

If we denote $E(X_i - \mu_i)(X_j - \mu_j)(X_k - \mu_k) = m_{3,ijk}$, and $E(X_i - \mu_i)(X_j - \mu_j)(X_k - \mu_k)(X_l - \mu_l) = m_{4,ijkl}$, then the second-order variance approximation becomes

$$\text{Var}(T_d(X_1, X_2, \ldots, X_d)) \approx \sum_{i,j} t_i(\mu_1, \ldots, \mu_d)t_j(\mu_1, \ldots, \mu_d)\sigma_{ij}$$

$$+ \sum_{i,j,k} t_i(\mu_1, \ldots, \mu_d)t_{jk}(\mu_1, \ldots, \mu_d)m_{3,ijk}$$

$$+ \frac{1}{4} \sum_{i,j,k,l} t_{ij}(\mu_1, \ldots, \mu_d) t_{kl}(\mu_1, \ldots, \mu_d)$$

$$[m_{4,ijkl} - \sigma_{ij}\sigma_{kl}].$$

For general d, this is a complicated expression. For $d = 1$, it reduces to the reasonably simple approximation

$$\text{Var}(T(X)) \approx [t'(\mu)]^2 \sigma^2 + t'(\mu) t''(\mu) E(X - \mu)^3 + \frac{1}{4}[t''(\mu)]^2 [E(X - \mu)^4 - \sigma^4].$$

Example 7.32. Let X, Y be two jointly distributed random variables, with means μ_1, μ_2, variances σ_1^2, σ_2^2, and covariance σ_{12}. We work out the second-order approximation to the expectation of $T(X, Y) = XY$. Writing t for T as above, the various relevant partial derivatives are $t_x = y, t_y = x, t_{xx} = t_{yy} = 0, t_{xy} = 1$. Plugging into the general formula for the second-order approximation to the mean, we get $E(XY) \approx \mu_1 \mu_2 + \frac{1}{2}[\sigma_{12} + \sigma_{21}] = \mu_1 \mu_2 + \sigma_{12}$. Thus, in this case, the second-order approximation reproduces the exact mean of XY.

Example 7.33 (A Multidimensional Example). Let $\mathbf{X} = (X_1, X_2, \ldots, X_d)$ have mean vector μ and covariance matrix Σ. Assume that μ is not the null vector. We find a second-order approximation to $E(||\mathbf{X}||)$. Denoting $T(x_1, \ldots, x_d) = ||\mathbf{x}||$, the successive partial derivatives are

$$t_i(\mu) = \frac{\mu_i}{||\mu||}, \quad t_{ii}(\mu) = \frac{1}{||\mu||} - \frac{\mu_i^2}{||\mu||^3}, \quad t_{ij}(\mu) = -\frac{\mu_i \mu_j}{||\mu||^3} (i \neq j).$$

Plugging these into the general formula for the second-order approximation of the expectation, on some algebra, we get the approximation

$$E(||\mathbf{x}||) \approx ||\mu|| + \frac{tr\Sigma}{2||\mu||} - \frac{\sum_i \mu_i^2 \sigma_{ii}}{2||\mu||^3} - \frac{\sum_{i \neq j} \mu_i \mu_j \sigma_{ij}}{2||\mu||^3}$$

$$= ||\mu|| + \frac{1}{2||\mu||}\left[tr\Sigma - \frac{\mu'\Sigma\mu}{\mu'\mu}\right].$$

The ratio $\frac{\mu'\Sigma\mu}{\mu'\mu}$ varies between the minimum and the maximum eigenvalue of Σ, where as $tr\Sigma$ equals the sum of all the eigenvalues. Thus, $tr\Sigma - \frac{\mu'\Sigma\mu}{\mu'\mu} \geq 0$, Σ being a nnd matrix, which implies that the approximation $||\mu|| + \frac{1}{2||\mu||}[tr\Sigma - \frac{\mu'\Sigma\mu}{\mu'\mu}]$ is $\geq ||\mu||$. This is consistent with the bound $E(||\mathbf{X}||) \geq ||\mu||$, as is implied by Jensen's inequality.

The second-order variance approximation is difficult to work out in this example. However, the first-order approximation is easily worked out, and gives

$$\text{Var}(||\mathbf{X}||) \approx \frac{\mu'\Sigma\mu}{\mu'\mu}.$$

7.6 Convergence of Moments

Note that no distributional assumption about **X** was made in deriving the approximations.

Example 7.34 (Variance of the Sample Variance). It is sometimes necessary to calculate the variance of a centered sample moment $m_k = \frac{1}{n}\sum_{i=1}^{n}(X_i - \bar{X})^k$, for iid observations $X_1, \ldots, X_n$ from some one-dimensional distribution. Particularly, the case $k = 2$ is of broad interest in statistics. Because we are considering centered moments, we may assume that $E(X_i) = 0$, so that $E(X_i^k)$ for any k will equal the population centered moment $\mu_k = E(X_i - \mu)^k$. We also recall that $E(m_2) = \frac{n-1}{n}\sigma^2$.

Using the algebraic identity $\sum_{i=1}^{n}(X_i - \bar{X})^2 = \sum_{i=1}^{n} X_i^2 - (\sum_{i=1}^{n} X_i)^2/n$, one can make substantial algebraic simplification towards calculating the variance of m_2. Indeed,

$$\text{Var}(m_2) = E[m_2^2] - [E(m_2)]^2$$

$$= \frac{1}{n^2} E\left[\sum_{i=1}^{n} X_i^4 + \sum_{i \neq j} X_i^2 X_j^2 + \frac{8}{n^2} \sum_{i \neq j} X_i^2 X_j^2 + \frac{4}{n^2} \sum_{i \neq j \neq k} X_i^2 X_j X_k\right.$$

$$\left. - \frac{4}{n} \sum_{i \neq j} X_i^3 X_j\right] - [E(m_2)]^2.$$

The expectation of each term above can be found by using the independence of the X_i and the zero mean assumption, and interestingly, in fact the variance of m_2 can be thus found exactly for any n, namely,

$$\text{Var}(m_2) = \frac{1}{n^3}[(n-1)^2(\mu_4 - \sigma^4) + 2(n-1)\sigma^4].$$

The approximate methods would have produced the answer

$$\text{Var}(m_2) \approx \frac{\mu_4 - \sigma^4}{n}.$$

It is useful to know that the approximate methods would likewise produce the general first-order variance approximation

$$\text{Var}(m_r) \approx \frac{\mu_{2r} + r^2\sigma^2\mu_{r-1}^2 - 2r\mu_{r-1}\mu_{r+1} - \mu_r^2}{n}.$$

The formal approximations described above may work well in some cases, however, it is useful to have some theoretical quantification of the accuracy of the approximations. This is difficult in general, and we give one result in a special case, with $d=1$.

Theorem 7.26. *Suppose $X_1, X_2, \ldots$ are iid observations with a finite fourth moment. Let $E(X_1) = \mu$, and $\text{Var}(X_1) = \sigma^2$. Let g be a scalar function with four uniformly bounded derivatives. Then*

(a) $E[g(\overline{X})] = g(\mu) + \frac{g^{(2)}(\mu)\sigma^2}{2n} + O(n^{-2})$;
(b) $\text{Var}[g(\overline{X})] = \frac{(g'(\mu))^2 \sigma^2}{n} + O(n^{-2})$.

See Bickel and Doksum (2007) for a proof of this theorem.

7.7 Convergence of Densities and Scheffé's Theorem

Suppose $X_n, X, n \geq 1$ are continuous random variables with densities f_n, f, and that X_n converges in distribution to X. It is natural to ask whether that implies that f_n converges to f pointwise. Simple counterexamples show that this need not be true. We show an example below. However, convergence of densities, when true, is very useful. It ensures a mode of convergence much stronger than convergence in distribution. We discuss convergence of densities in general, and for sample means of iid random variables in particular, in this section.

First, we give an example to show that convergence in distribution does not imply convergence of densities.

Example 7.35 (Convergence in Distribution Is Weaker Than Convergence of Density). Suppose X_n is a sequence of random variables on $[0, 1]$ with density $f_n(x) = 1 + \cos(2\pi n x)$. Then, $X_n \stackrel{\mathcal{L}}{\Rightarrow} U[0, 1]$ by a direct verification of the definition using CDFs. Indeed, $F_n(x) = x + \frac{\sin(2n\pi x)}{2n\pi} \to x\; \forall x \in (0, 1)$. However, note that the densitities f_n do not converge to the uniform density 1 as $n \to \infty$.

Convergence of densities is useful to have when true, because it ensures a much stronger form of convergence than convergence in distribution. Suppose X_n have CDF F_n and density f_n, and X has CDF F and density f. If $X_n \stackrel{\mathcal{L}}{\Rightarrow} X$, then we can only assert that $F_n(x) = P(X_n \leq x) \to F(x) = P(X \leq x)\; \forall x$. However, if we have convergence of the densities, then we can make the much stronger assertion that for any event A, $P(X_n \in A) \to P(X \in A)$, not just for events A of the form $A = (-\infty, x]$. This is explained below.

Definition 7.13. *Let X, Y be two random variables defined on a common sample space Ω. The total variation distance between the distributions of X and Y is defined as $d_{TV}(X, Y) = \sup_A |P(X \in A) - P(Y \in A)|$.*

Remark. Again, actually the set A is not completely arbitrary. We do need the restriction that A be a Borel set, a concept in measure theory. However, we make no further mention of this qualification.

The relation between total variation distance and densities when the random variables X, Y are continuous is described by the following result.

7.7 Convergence of Densities and Scheffé's Theorem

Lemma. *Let X, Y be continuous random variables with densities f, g. Then $d_{TV}(X, Y) = \frac{1}{2} \int_{-\infty}^{\infty} |f(x) - g(x)| dx$.*

Proof. The proof is based on two facts:

$$\int_{-\infty}^{\infty} |f(x) - g(x)| dx = 2 \int_{-\infty}^{\infty} (f - g)^+ dx,$$

and, for any set A,

$$|P(X \in A) - P(Y \in A)| \leq \int_{x: f(x) > g(x)} (f(x) - g(x)) dx.$$

Putting these two together,

$$d_{TV}(X, Y) = \sup_A |P(X \in A) - P(Y \in A)| \leq \frac{1}{2} \int_{-\infty}^{\infty} |f(x) - g(x)| dx.$$

However, for the particular set $A_0 = \{x : f(x) > g(x)\}$, $|P(X \in A_0) - P(Y \in A_0)| = \frac{1}{2} \int_{-\infty}^{\infty} |f(x) - g(x)| dx$, and so, that proves that $\sup_A |P(X \in A) - P(Y \in A)|$ exactly equals $\frac{1}{2} \int_{-\infty}^{\infty} |f(x) - g(x)| dx$. □

Example 7.36 (Total Variation Distance Between Two Normals). Total variation distance is usually hard to find in closed analytical form. The absolute value sign makes closed-form calculations difficult. It is, however, possible to write a closed-form formula for the total variation distance between two arbitrary normal distributions in one dimension. No such formula would be possible in higher dimensions.

Let $X \sim N(\mu_1, \sigma_1^2), Y \sim N(\mu_2, \sigma_2^2)$. We use the result that $d_{TV}(X, Y) = \frac{1}{2} \int_{-\infty}^{\infty} |f(x) - g(x)| dx$, where f, g are the densities of X, Y. To evaluate the integral of $|f(x) - g(x)|$, we need to find the set of all values of x for which $f(x) \geq g(x)$. We assume that $\sigma_1 > \sigma_2$, and use the notation

$$c = \frac{\sigma_1}{\sigma_2}, \qquad \Delta = \frac{\mu_1 - \mu_2}{\sigma_2},$$

$$A = \frac{\sqrt{(c^2 - 1) 2 \log c + \Delta^2} - c\Delta}{c^2 - 1}, \qquad B = -\frac{\sqrt{(c^2 - 1) 2 \log c + \Delta^2} + c\Delta}{c^2 - 1}.$$

The case $\sigma_1 = \sigma_2$ is commented on below.

With this notation, by making a change of variable,

$$\int_{-\infty}^{\infty} |f(x) - g(x)| dx = \int_{-\infty}^{\infty} |\phi(z) - c\phi(\Delta + cz)| dz,$$

and $\phi(z) \leq c\phi(\Delta + cz)$ if and only if $A \leq z \leq B$. Therefore,

$$\int_{-\infty}^{\infty} |f(x) - g(x)|dx = \int_{A}^{B} [c\phi(\Delta + cz) - \phi(z)]dz$$
$$+ \int_{-\infty}^{A} [\phi(z) - c\phi(\Delta + cz)]dz$$
$$+ \int_{B}^{\infty} [\phi(z) - c\phi(\Delta + cz)]dz$$
$$= 2[(\Phi(\Delta + cB) - \Phi(B)) - (\Phi(\Delta + cA) - \Phi(A))],$$

where the quantities $\Delta + cB, \Delta + cA$ work out to

$$\Delta + cB = \frac{c\sqrt{(c^2-1)2\log c + \Delta^2} - \Delta}{c^2 - 1},$$

$$\Delta + cA = -\frac{c\sqrt{(c^2-1)2\log c + \Delta^2} + \Delta}{c^2 - 1}.$$

When $\sigma_1 = \sigma_2$, the expression reduces to $\Phi(\frac{|\Delta|}{2}) - \Phi(-\frac{|\Delta|}{2})$. In applying the formula we have derived, it is important to remember that the larger of the two variances has been called σ_1. Finally, now,

$$d_{TV}(X, Y) = (\Phi(\Delta + cB) - \Phi(B)) - (\Phi(\Delta + cA) - \Phi(A)),$$

with $\Delta + cA, \Delta + cB, A, B$ as given above explicitly.

We see from the formula that $d_{TV}(X, Y)$ depends on both individual variances, and on the difference between the means. When the means are equal, the total variation distance reduces to the simpler expression

$$2\left[\Phi\left(\frac{c\sqrt{2\log c}}{\sqrt{c^2-1}}\right) - \Phi\left(\frac{\sqrt{2\log c}}{\sqrt{c^2-1}}\right)\right]$$
$$\approx \frac{1}{2\sqrt{2\pi e}}(3-c)(c-1),$$

for $c \approx 1$.

The next fundamental result asserts that if X_n, X are continuous random variables with densities f_n, f, and if $f_n(x) \to f(x) \,\forall x$, then $d_{TV}(X_n, X) \to 0$. This means that pointwise convergence of densities, when true, ensures an extremely strong mode of convergence, namely convergence in total variation.

Theorem 7.27 (Scheffé's Theorem). *Let f_n, f be nonnegative integrable functions. Suppose:*

$$f_n(x) \to f(x) \,\forall x; \quad \int_{-\infty}^{\infty} f_n(x)dx \to \int_{-\infty}^{\infty} f(x)dx.$$

Then $\int_{-\infty}^{\infty} |f_n(x) - f(x)|dx \to 0$.

7.7 Convergence of Densities and Scheffé's Theorem

In particular, if f_n, f are all density functions, and if $f_n(x) \to f(x) \; \forall x$, then $\int_{-\infty}^{\infty} |f_n(x) - f(x)| dx \to 0$.

Proof. The proof is based on the pointwise algebraic identity

$$|f_n(x) - f(x)| = f_n(x) + f(x) - 2\min\{f_n(x), f(x)\}.$$

Now note that $\min\{f_n(x), f(x)\} \to f(x) \; \forall x$, as $f_n(x) \to f(x) \; \forall x$, and $\min\{f_n(x), f(x)\} \leq f(x)$. Therefore, by the dominated convergence theorem (see the previous section), $\int_{-\infty}^{\infty} \min\{f_n(x), f(x)\} dx \to \int_{-\infty}^{\infty} f(x) dx$. The pointwise algebraic identity now gives that

$$\int_{-\infty}^{\infty} |f_n(x) - f(x)| dx \to \int_{-\infty}^{\infty} f(x) dx + \int_{-\infty}^{\infty} f(x) dx - 2\int_{-\infty}^{\infty} f(x) dx = 0,$$

which completes the proof. □

Remark. As we remarked before, convergence in total variation is very strong, and should not be expected, without some additional structure. The following theorems exemplify the kind of structure that may be necessary. The first theorem below is a general theorem: no assumptions are made on the structural form of the statistic. In the second theorem below, convergence in total variation is considered for sample means of iid random variables: there is a restriction on the structural form of the underlying statistic.

Theorem 7.28 (Ibragimov). *Suppose X_n, X are continuous random variables with densities f_n, f that are unimodal. Then $X_n \stackrel{\mathcal{L}}{\Rightarrow} X$ if and only if $\int_{-\infty}^{\infty} |f_n(x) - f(x)| dx \to 0$.*

See Reiss Aui (1989) for this theorem. The next result for sample means of iid random variables was already given in Chapter 1; we restate it for completeness.

Theorem 7.29 (Gnedenko). *Let $X_i, i \geq 1$ be iid continuous random variables with density $f(x)$, mean μ, and finite variance σ^2. Let $Z_n = \frac{\sqrt{n}(\bar{X}-\mu)}{\sigma}$, and let f_n denote the density of Z_n. If f is uniformly bounded, then f_n converges uniformly to the standard normal density $\phi(x)$ on $(-\infty, \infty)$, and $\int_{-\infty}^{\infty} |f_n(x) - \phi(x)| dx \to 0$.*

Remark. This is an easily stated result covering many examples. But better results are available. Feller (1971) is an excellent reference for some of the better results, which, however, involve more complex concepts.

Example 7.37. Let $X_n \sim N(\mu_n, \sigma_n^2), n \geq 1$. For X_n to converge in distribution, each of μ_n, σ_n^2 must converge. This is because, if X_n does converge in distribution, then there is a CDF $F(x)$ such that $\Phi(\frac{x-\mu_n}{\sigma_n}) \to F(x)$ at all continuity points of F. This implies, by selecting two suitable continuity points of F, that each of μ_n, σ_n must converge. If σ_n converges to zero, then X_n will converge to a one-point distribution. Otherwise, $\mu_n \to \mu, \sigma_n \to \sigma$, for some $\mu, \sigma, -\infty < \mu < \infty, 0 < \sigma < \infty$.

It follows that $P(X_n \leq x) \to \Phi(\frac{x-\mu}{\sigma})$ for any fixed x, and so, X_n converges in distribution to another normal, namely to $X \sim N(\mu, \sigma^2)$. Now, either by direct verification, or from Ibragimov's theorem, we have that X_n also converges to X in total variation. The converse is also true. That is, if $X_n \sim N(\mu_n, \sigma_n^2), n \geq 1$, then X_n can either converge to a one-point distribution, or to another normal distribution, say $N(\mu, \sigma^2)$, in which case $\mu_n \to \mu, \sigma_n \to \sigma$, and convergence in total variation also holds. Conversely, if $\mu_n \to \mu, \sigma_n \to \sigma > 0$, then X_n converges in total variation to $X \sim N(\mu, \sigma^2)$.

Exercises

Exercise 7.1. (a) Show that $X_n \xrightarrow{2} c$ (i.e., X_n converges in quadratic mean to c) if and only if $E(X_n - c)$ and $\text{Var}(X_n)$ both converge to zero.
(b) Show by an example (different from text) that convergence in probability does not necessarily imply almost sure convergence.

Exercise 7.2. (a) Suppose $E|X_n - c|^\alpha \to 0$, where $0 < \alpha < 1$. Does X_n necessarily converge in probability to c?
(b) Suppose $a_n(X_n - \theta) \xrightarrow{\mathcal{L}} N(0, 1)$. Under what condition on a_n can we conclude that $X_n \xrightarrow{P} \theta$?
(c) $o_p(1) + O_p(1) = ?$
(d) $o_p(1) O_p(1) = ?$
(e) $O_p(1) + o_p(1) O_p(1) = ?$
(f) Suppose $X_n \xrightarrow{\mathcal{L}} X$. Then, $o_p(1) X_n = ?$

Exercise 7.3 (Monte Carlo). Consider the purely mathematical problem of finding a definite integral $f(x) dx$ for some (possibly complicated) function $f(x)$. Show that the SLLN provides a method for approximately finding the value of the integral by using appropriate averages $\frac{1}{n} \sum_{i=1}^{n} f(X_i)$.
Numerical analysts call this Monte Carlo integration.

Exercise 7.4. Suppose $X_1, X_2, \ldots$ are iid and that $E(X_1) = \mu \neq 0, \text{Var}(X_1) = \sigma^2 < \infty$. Let $S_{m,p} = \sum_{i=1}^{m} X_i^p, m \geq 1, p = 1, 2$.

(a) Identify with proof the almost sure limit of $\frac{S_{m,1}}{S_{n,1}}$ for fixed m, and $n \to \infty$.
(b) Identify with proof the almost sure limit of $\frac{S_{n-m,1}}{S_{n,1}}$ for fixed m, and $n \to \infty$.
(c) Identify with proof the almost sure limit of $\frac{S_{n,1}}{S_{n,2}}$ as $n \to \infty$.
(d) Identify with proof the almost sure limit of $\frac{S_{n,1}}{S_{n^2,2}}$ as $n \to \infty$.

Exercise 7.5. Let $A_n, n \geq 1, A$ be events with respect to a common sample space Ω.

(a) Prove that $I_{A_n} \overset{\mathcal{L}}{\Rightarrow} I_A$ if and only if $P(A_n) \to P(A)$.
(b) Prove that $I_{A_n} \overset{2}{\Rightarrow} I_A$ if and only if $P(A \Delta A_n) \to 0$.

Exercise 7.6. Suppose $g : \mathcal{R}_+ \to \mathcal{R}$ is continuous and bounded. Show that

$$e^{-n\lambda} \sum_{k=0}^{\infty} g(\frac{k}{n}) \frac{(n\lambda)^k}{k!} \to g(\lambda)$$

as $n \to \infty$.

Exercise 7.7 *(Convergence of Medians).** Suppose X_n is a sequence of random variables converging in probability to a random variable X; X is absolutely continuous with a strictly positive density. Show that the medians of X_n converge to the median of X.

Exercise 7.8. Suppose $\{A_n\}$ is an infinite sequence of independent events. Show that $P(\text{infinitely many } A_n \text{ occur}) = 1 \Leftrightarrow P(\bigcup A_n) = 1$.

Exercise 7.9 *(Almost Sure Limit of Mean Absolute Deviation).** Suppose $X_i, i \geq 1$ are iid random variables from a distribution F with $E_F(|X|) < \infty$.

(a) Prove that the mean absolute deviation $\frac{1}{n} \sum_{i=1}^n |X_i - \bar{X}|$ has a finite almost sure limit.
(b) Evaluate this limit explicitly when F is standard normal.

Exercise 7.10. *Let X_n be any sequence of random variables. Prove that one can always find a sequence of numbers c_n such that $\frac{X_n}{c_n} \overset{a.s.}{\Rightarrow} 0$.

Exercise 7.11 (Sample Maximum). Let $X_i, i \geq 1$ be an iid sequence, and $X_{(n)}$ the maximum of $X_1, \ldots, X_n$. Let $\xi(F) = \sup\{x : F(x) < 1\}$, where F is the common CDF of the X_i. Prove that $X_{(n)} \overset{a.s.}{\Rightarrow} \xi(F)$.

Exercise 7.12. Suppose $\{A_n\}$ is an infinite sequence of events. Suppose that $P(A_n) \geq \delta \; \forall n$. Show that $P(\text{infinitely many } A_n \text{ occur}) \geq \delta$.

Exercise 7.13. Let X_i be independent $N(\mu, \sigma_i^2)$ variables.

(a) Find the BLUE (best linear unbiased estimate) of μ.
(b) Suppose $\sum_{i=1}^{\infty} \sigma_i^{-2} = \infty$. Prove that the BLUE converges almost surely to μ.

Exercise 7.14. Suppose X_i are iid standard Cauchy. Show that

(a) $P(|X_n| > n \text{ infinitely often}) = 1$,
(b) * $P(|S_n| > n \text{ infinitely often}) = 1$.

Exercise 7.15. Suppose X_i are iid standard exponential. Show that $\limsup_n \frac{X_n}{\log n} = 1$ with probability 1.

Exercise 7.16 *(Coupon Collection). Cereal boxes contain independently and with equal probability exactly one of n different celebrity pictures. Someone having the entire set of n pictures can cash them in for money. Let W_n be the minimum number of cereal boxes one would need to purchase to own a complete set of the pictures. Find a sequence a_n such that $\frac{W_n}{a_n} \stackrel{P}{\Rightarrow} 1$.
Hint : Approximate the mean of W_n.

Exercise 7.17. Let $X_n \sim \text{Bin}(n, p)$. Show that $(X_n/n)^2$ and $X_n(X_n - 1)/n(n-1)$ both converge in probability to p^2. Do they also converge almost surely?

Exercise 7.18. Suppose $X_1, \ldots, X_n$ are iid standard exponential variables, and let $S_n = X_1 + \ldots + X_n$. Apply the Chernoff–Bernstein inequality (see Chapter 1) to show that for $c > 1$,
$$P(S_n > cn) \leq e^{-n(c-1-\ln c)}$$
and hence that $P(S_n > cn) \to 0$ exponentially fast.

Exercise 7.19. Let $X_1, X_2, \ldots$ be iid nonnegative random variables. Show that $\frac{X_{(n)}}{n} \stackrel{P}{\Rightarrow} 0$ if and only if $nP(X_1 > n) \to 0$.
Is this true in the normal case?

Exercise 7.20 (Failure of Weak Law). Let $X_1, X_2, \ldots$ be a sequence of independent variables, with $P(X_i = i) = P(X_i = -i) = 1/2$. Show that $\bar{X}$ does not converge in probability to the common mean $\mu = 0$.

Exercise 7.21. Let $X_1, X_2, X_3, \ldots$ be iid $U[0, 1]$. Let
$$G_n = (X_1 X_2 \ldots X_n)^{1/n}.$$
Find c such that $G_n \stackrel{P}{\Rightarrow} c$.

Exercise 7.22 *(Uniform Integrability of Sample Mean). Suppose $X_i, i \geq 1$ are iid from some CDF F with mean zero and variance one. Find a sufficient condition on F for $E(\sqrt{n}\bar{X})^k$ to exist and converge to $E(Z^k)$, where k is fixed, and $Z \sim N(0, 1)$.

Exercise 7.23 *(Sufficient Condition for Uniform Integrability). Let $\{X_n\}, n \geq 1$ be a sequence of random variables, and suppose for some function $f : \mathcal{R}_+ \to \mathcal{R}_+$ such that f is nondecreasing and $\frac{f(x)}{x} \to \infty$ as $x \to \infty$, we know that $\sup_n E[f(|X_n|)] < \infty$. Show that $\{X_n\}$ is uniformly integrable.

Exercise 7.24 (Uniform Integrability of IID Sequence). Suppose $\{X_n\}$ is an iid sequence with $E(|X_1|) < \infty$. Show that $\{X_n\}$ is uniformly integrable.

Exercise 7.25. Give an example of a sequence $\{X_n\}$, and an X such that $X_n \overset{\mathcal{L}}{\Rightarrow} X$, $E(X_n) \to E(X)$, but $E(|X_n|)$ does not converge to $E(|X|)$.

Exercise 7.26. Suppose X_n has a normal distribution with mean μ_n and variance σ_n^2. Let $\mu_n \to \mu$ and $\sigma_n \to \sigma$ as $n \to \infty$. What is the limiting distribution of X_n?

Exercise 7.27 (Delta Theorem). Suppose $X_1, X_2, \ldots$ are iid with mean μ and variance σ^2, a finite fourth moment, and let $Z \sim N(0, 1)$.

(a) Show that $\sqrt{n}(\overline{X}^2 - \mu^2) \overset{\mathcal{L}}{\Rightarrow} 2\mu\sigma Z$.

(b) Show that $\sqrt{n}(e^{\overline{X}} - e^\mu) \overset{\mathcal{L}}{\Rightarrow} e^\mu Z$.

(c) Show that $\sqrt{n}(\log(1/n \sum_{i=1}^n (X_i - \overline{X})^2) - \log \sigma^2) \overset{\mathcal{L}}{\Rightarrow} (1/\sigma^2)(EX_1^4)^{1/2} Z$.

Exercise 7.28 (Asymptotic Variance and True Variance). Let $X_1, X_2, \ldots$ be iid observations from a CDF F with four finite moments. For each of the following cases, find the exact variance of $m_2 = \frac{1}{n}\sum_{i=1}^n (X_i - \bar{X})^2$ by using the formula in the text, and also find the asymptotic variance by using the formula in the text. Check when the true variance is larger than the asymptotic variance.

(a) $F = N(\mu, \sigma^2)$.
(b) $F = \text{Exp}(\lambda)$.
(c) $F = \text{Poi}(\lambda)$.

Exercise 7.29 (All Distributions as Limits of Discrete). Show that any distribution on R^d is the limit in distribution of distributions on R^d that are purely discrete with finitely many values.

Exercise 7.30 (Conceptual). Suppose $X_n \overset{\mathcal{L}}{\Rightarrow} X$, and also $Y_n \overset{\mathcal{L}}{\Rightarrow} X$. Does this mean that $X_n - Y_n$ converge in distribution to (the point mass at) zero?

Exercise 7.31. (a) Suppose $a_n(X_n - \theta) \to N(0, \tau^2)$; what can be said about the limiting distribution of $|X_n|$, when $\theta \neq 0, \theta = 0$?

(b) *Suppose X_i are iid Bernoulli(p); what can be said about the limiting distribution of the sample variance s^2 when $p = \frac{1}{2}; p \neq \frac{1}{2}$?

Exercise 7.32 (Delta Theorem). Suppose $X_1, X_2, \ldots$ are iid Poi(λ). Find the limiting distribution of $e^{-\bar{X}}$.

Remark. It is meant that on suitable centering and norming, you will get a nondegenerate limiting distribution.

Exercise 7.33 (Delta Theorem). Suppose $X_1, X_2, \ldots$ are iid $N(\mu, 1)$. Find the limiting distribution of $\Phi(\bar{X})$, where Φ, as usual, is the standard normal CDF.

Exercise 7.34 *(Delta Theorem with Lack of Smoothness). Suppose $X_1, X_2, \ldots$ are iid $N(\mu, 1)$. Find the limiting distribution of $|\bar{X}|$ when

(a) $\mu \neq 0$.
(b) $\mu = 0$.

Exercise 7.35 (**Delta Theorem**). For each F below, find the limiting distributions of $\frac{\bar{X}}{s}$ and $\frac{s}{\bar{X}}$:

(i) $F = U[0,1]$, (ii) $F = \text{Exp}(\lambda)$, (iii) $F = \chi^2(p)$.

Exercise 7.36 *(**Delta Theorem**). Suppose $X_1, X_2, \ldots$ are iid $N(\mu, \sigma^2)$. Let

$$b_1 = \frac{\frac{1}{n}\sum (X_i - \bar{X})^3}{\left[\frac{1}{n}\sum (X_i - \bar{X})^2\right]^{3/2}} \quad \text{and} \quad b_2 = \frac{\frac{1}{n}\sum (X_i - \bar{X})^4}{\left[\frac{1}{n}\sum (X_i - \bar{X})^2\right]^2} - 3$$

be the sample skewness and kurtosis coefficients. Find the joint limiting distribution of (b_1, b_2).

Exercise 7.37 *(**Slutsky**). Let X_n, Y_m be independent Poisson with means $n, m, m, n \geq 1$. Find the limiting distribution of $\frac{X_n - Y_m - (n-m)}{\sqrt{X_n + Y_m}}$ as $n, m \to \infty$.

Exercise 7.38 (**Approximation of Mean and Variance**). Let $X \sim \text{Bin}(n, p)$. Find the first- and the second-order approximation to the mean and variance of $\frac{X}{n-X}$.

Exercise 7.39 (**Approximation of Mean and Variance**). Let $X \sim \text{Poi}(\lambda)$. Find the first- and the second-order approximation to the mean and variance of e^{-X}. Compare to the exact mean and variance by consideration of the mgf of X.

Exercise 7.40 (**Approximation of Mean and Variance**). Let $X_1, \ldots, X_n$ be iid $N(\mu, \sigma^2)$. Find the first- and the second-order approximation to the mean and variance of $\Phi(\bar{X})$.

Exercise 7.41 *(**Approximation of Mean and Variance**). Let $X \sim \text{Bin}(n, p)$. Find the first- and the second-order approximation to the mean and variance of $\arcsin(\sqrt{\frac{X}{n}})$.

Exercise 7.42 *(**Approximation of Mean and Variance**). Let $X \sim \text{Poi}(\lambda)$. Find the first- and the second-order approximation to the mean and variance of $\sqrt{X}$.

Exercise 7.43 *(**Multidimensional Approximation of Mean**). Let $\mathbf{X}$ be a d-dimensional random vector. Find a first- and second-order approximation to the mean of $\sqrt{\sum_{i=1}^d X_i^4}$.

Exercise 7.44 *(**Expected Length of Poisson Confidence Interval**). In Chapter 1, the approximate 95% confidence interval $X + 1.92 \pm \sqrt{3.69 + 3.84X}$ for a Poisson mean λ was derived. Find a first- and second-order approximation to the expected length of this confidence interval.

Exercise 7.45 *(**Expected Length of the t Confidence Interval**). The modified t confidence interval for a population mean μ has the limits $\bar{X} \pm z_{\alpha/2}\frac{s}{\sqrt{n}}$, where $\bar{X}$ and s are the mean and the standard deviation of an iid sample of size n, and $z_{\alpha/2} = \Phi^{-1}(\frac{\alpha}{2})$. Find a first- and a second-order approximation to the expected length of the modified t confidence interval when the population distribution is

(a) $N(\mu, \sigma^2)$.
(b) $\text{Exp}(\mu)$.
(c) $U[\mu - 1, \mu + 1]$.

Exercise 7.46 * **(Coefficient of Variation).** Given a set of positive iid random variables $X_1, X_2, \ldots, X_n$, the coefficient of variation (CV) is defined as $CV = \frac{s}{\bar{X}}$. Find a second-order approximation to its mean, and a first-order approximation to its variance, in terms of suitable moments of the distribution of the X_i. Make a note of how many finite moments you need for each approximation to make sense.

Exercise 7.47 * **(Variance-Stabilizing Transformation).** Let $X_i, i \geq 1$ be iid $\text{Poi}(\lambda)$.

(a) Show that for each a, b, $\sqrt{\frac{\sum_{i=1}^n X_i + a}{n+b}}$ is a variance stabilizing transformation.
(b) Find the first- and the second-order approximation to the mean of $\sqrt{\frac{\sum_{i=1}^n X_i + a}{n+b}}$.
(c) Are there some particular choices of a, b that make the approximation

$$E\left(\sqrt{\frac{\sum_{i=1}^n X_i + a}{n+b}}\right) \approx \sqrt{\lambda}$$

more accurate? Justify your answer.

Exercise 7.48 * **(Variance-Stabilizing Transformation).** Let $X_i, i \geq 1$ be iid $\text{Ber}(p)$.

(a) Show that for each a, b, $\arcsin\left(\sqrt{\frac{\sum_{i=1}^n X_i + a}{n+b}}\right)$ is a variance stabilizing transformation.
(b) Find the first- and the second-order approximation to the mean of $\arcsin\left(\sqrt{\frac{\sum_{i=1}^n X_i + a}{n+b}}\right)$.
(c) Are there some particular choices of a, b that make the approximation

$$E\left[\arcsin\left(\sqrt{\frac{\sum_{i=1}^n X_i + a}{n+b}}\right)\right] \approx \arcsin\left(\sqrt{p}\right)$$

more accurate? Justify your answer.

Exercise 7.49. For each of the following cases, evaluate the total variation distance between the indicated distributions:

(a) $N(0, 1)$ and $C(0, 1)$.
(b) $N(0, 1)$ and $N(0, 10^4)$.
(c) $C(0, 1)$ and $C(0, 10^4)$.

Exercise 7.50 (**Plotting the Variation Distance**)**.** Calculate and plot (as a function of μ) $d_{TV}(X,Y)$ if $X \sim N(0,1), Y \sim N(\mu,1)$.

Exercise 7.51 (**Convergence of Densities**)**.** Let $Z \sim N(0,1)$ and Y independent of Z. Let $X_n = Z + \frac{Y}{n}, n \geq 1$.

(a) Prove by direct calculation that the density of X_n converges pointwise to the standard normal density in each of the following cases.

  (i) $Y \sim N(0,1)$.
  (ii) $Y \sim U[0,1]$.
  (iii) $Y \sim \text{Exp}(1)$.

(b) Hence, or by using Ibragimov's theorem prove that $X_n \to Z$ in total variation.

Exercise 7.52. Show that $d_{TV}(X,Y) \leq P(X \neq Y)$.

Exercise 7.53. Suppose $X_1, X_2, \ldots$ are iid Exp(1). Does $\sqrt{n}(\bar{X} - 1)$ converge to standard normal in total variation? Prove or disprove.

Exercise 7.54 *(**Minimization of Variation Distance**)**.** Let $X \sim U[-a,a]$ and $Y \sim N(0,1)$. Find a that minimizes $d_{TV}(X,Y)$.

Exercise 7.55. Let X,Y be integer-valued random variables. Show that $d_{TV}(X,Y) = \frac{1}{2} \sum_k |P(X=k) - P(Y=k)|$.

References

Ash, R. (1973). *Real Analysis and Probability*, Academic Press, New York.
Bhattacharya, R. and Rao, R. (1986). *Normal Approximation and Asymptotic Expansions*, Wiley, New York.
Bickel, P. and Doksum (2007). *Mathematical Statistics: Basic Ideas and Selected Topics*, Prentice Hall, Upper Saddle River, NJ.
Breiman, L. (1968). *Probability*, Addison-Wesley, Reading, MA.
Chow, Y. and Teicher, H. (1988). *Probability Theory*, 3rd ed., Springer, New York.
Cramér, H. (1946). *Mathematical Methods of Statistics*, Princeton University Press, Princeton, NJ.
DasGupta, A. (2008). *Asymptotic Theory of Statistics and Probability*, Springer, New York.
Feller, W. (1971). *Introduction to Probability Theory with Applications*, Wiley, New York.
Ferguson, T. (1996). *A Course in Large Sample Theory*, Chapman and Hall, New York.
Hall, P. (1997). *Bootstrap and Edgeworth Expansions*, Springer, New York.
Hall, P. and Heyde, C. (1980). *Martingale Limit Theory and Its Applications*, Academic Press, New York.
Kesten, H. (1972). Sums of independent random variables, *Ann. Math. Statist.*, 43, 701–732.
Lehmann, E. (1999). *Elements of Large Sample Theory*, Springer, New York.
Petrov, V. (1975). *Limit Theorems of Probability Theory*, Oxford University Press, London.
Port, S. (1994). *Theoretical Probability for Applications*, Wiley, New York.
Reiss, R. (1989). *Approximate Distribution of Order Statistics*, Springer-Verlag, New York.
Révész, P. (1968). *The Laws of Large Numbers*, Academic Press, New York.
Sen, P. K. and Singer, J. (1993). *Large Sample Methods in Statistics*, Chapman and Hall, New York.
Serfling, R. (1980). *Approximation Theorems of Mathematical Statistics*, Wiley, New York.
van der Vaart, Aad (1998). *Asymptotic Statistics*, Cambridge University Press, Cambridge, UK.

Chapter 8
Characteristic Functions and Applications

Characteristic functions were first systematically studied by Paul Lévy, although they were used by others before him. It provides an extremely powerful tool in probability in general, and in asymptotic theory in particular. The power of the characteristic function derives from a set of highly convenient properties. Like the mgf, it determines a distribution. But unlike mgfs, existence is not an issue, and it is a bounded function. It is easily transportable for common functions of random variables, such as convolutions. And it can be used to prove convergence of distributions, as well as to recognize the name of a limiting distribution. It is also an extremely handy tool in proving characterizing properties of distributions. For instance, the Cramér–Levy theorem (see Chapter 1), which characterizes a normal distribution, has so far been proved by only using characteristic function methods. There are two disadvantages in working with characteristic functions. First, it is a complex-valued function, in general, and so, familiarity with basic complex analysis is required. Second, characteristic function proofs usually do not lead to any intuition as to why a particular result should be true. All things considered, knowledge of basic characteristic function theory is essential for statisticians, and certainly for students of probability.

We introduce some notation.

The set of complex numbers is denoted by $\mathbb{C}$. Given a complex number $z = x + iy$, where $i = \sqrt{-1}$, x is referred to as the real part of z, and y the complex part of z, and written as $\Re z, \Im z$. The complex conjugate of z is $x - iy$ and denoted as $\bar{z}$. The absolute value is denoted as $|z|$ and equals $\sqrt{x^2 + y^2}$. We have $z\bar{z} = |z|^2$. We recall *Euler's formula* $e^{it} = \cos t + i \sin t \ \forall t \in \mathcal{R}$, and $|e^{it}| = 1 \ \forall t \in \mathcal{R}$. Any $z \in \mathbb{C}$ may be represented as $z = r(\cos\theta + i \sin\theta)$, where $r = |z|$, and $\theta \in (-\pi, \pi]$; θ is called the argument of z. Note the formal equivalence of $\mathbb{C}$ to $\mathcal{R}^2$ by identifying the real and the imaginary part of a complex number to $r \cos\theta$ and $r \sin\theta$, where r, θ are the polar coordinates of the point $(x, y) \in \mathcal{R}^2$. We also recall *de Moivre's formula* $(\cos\theta + i \sin\theta)^n = \cos n\theta + i \sin n\theta \ \forall n \geq 1$.

Although we mostly limit ourselves to the case of real-valued random variables, we define a characteristic function (cf) for general d-vectors.

Definition 8.1. Let $\mathbf{X}$ be a d-dimensional random vector. Then the cf of $\mathbf{X}, \psi = \psi_\mathbf{X} : \mathcal{R}^d \to \mathbb{C}$ is defined as $\psi(\mathbf{t}) = E(e^{i\mathbf{t}'\mathbf{X}}) = E(\cos \mathbf{t}'\mathbf{X}) + iE(\sin \mathbf{t}'\mathbf{X})$.

The simplest properties of a cf in the case $d = 1$ are given first.

Proposition.

(a) For any real-valued random variable X, $\psi(t)$ exists for any t, $|\psi(t)| \leq 1$, and $\psi(0) = 1$.
(b) $\psi_{-X}(t) = \psi_X(-t) = \overline{\psi_X(t)}$.
(c) For any real-valued random variables X, Y such that X, Y are independent, $\psi_{X+Y}(t) = \psi_X(t)\psi_Y(t)$.
(d) If $Y = a + bX$, then $\psi_Y(t) = e^{ita}\psi_X(bt)$.
(e) If $\psi_X(t) = \psi_Y(t)$ $\forall t$, then X, Y have the same distribution; that is, a cf determines the distribution.
(f) A random variable X has a distribution symmetric about zero if and only if its characteristic function is real and even.
(g) The cf of any real-valued random variable X is continuous, and even uniformly continuous on the whole real line.

Proof. $|\psi(t)| = |E(e^{itX})| \leq E(|e^{itX}|) = E(1) = 1$; it is obvious that $\psi(0) = 1$. Next, $\psi_{-X}(t) = E(e^{it(-X)}) = E(e^{-itX}) = E(\cos tX) - iE(\sin tx) = \overline{\psi_X(t)} = \psi_X(-t)$. If X, Y are independent, $\psi_{X+Y}(t) = E(e^{it(X+Y)}) = E(e^{itX}e^{itY}) = E(e^{itX})E(e^{itY}) = \psi_X(t)\psi_Y(t)$. Part (d) is obvious. Part (e) is proved later. For part (f), X has a distribution symmetric about zero if and only if X and $-X$ have the same distribution if and only if $\psi_X(t) = \overline{\psi_X(t)} = \psi_X(-t)$. Part (g) can be proved by using simple inequalities on the exponential function and at the same time, the dominated convergence theorem. We leave this as a short exercise for the reader. □

8.1 Characteristic Functions of Standard Distributions

Characteristic functions of many standard distributions are collected together in the following table for convenient reference. Some special cases are worked out as examples.

Distribution	Density/pmf	cf
Point mass at a		e^{ita}
Bernoulli(p)	$p^x(1-p)^{1-x}$	$1 - p + pe^{it}$
Binomial	$\binom{n}{x}p^x(1-p)^{n-x}$	$(1 - p + pe^{it})^n$
Poisson	$e^{-\lambda}\frac{\lambda^x}{x!}$	$e^{-\lambda(e^{it}-1)}$
Geometric	$p(1-p)^{x-1}$	$\frac{pe^{it}}{1-(1-p)e^{it}}$
Negative binomial	$\binom{x-1}{n-1}p^n(1-p)^{x-n}$	$[\frac{pe^{it}}{1-(1-p)e^{it}}]^n$
Uniform$[-a, a]$	$\frac{1}{2a}$	$\frac{\sin at}{at}$
Triangular$[-2a, 2a]$	$\frac{1}{2a}(1 - \frac{|x|}{2a})$	$(\frac{\sin at}{at})^2$

(continued)

8.1 Characteristic Functions of Standard Distributions

(continued)

Distribution	Density/pmf	cf
Beta(α, α)	$\frac{\Gamma(2\alpha)}{[\Gamma(\alpha)]^2} x^{\alpha-1}(1-x)^{\alpha-1}$	$\frac{\sqrt{\pi}\Gamma(2\alpha)}{\Gamma(\alpha)} \frac{J_{\alpha-\frac{1}{2}}(\frac{t}{2})}{t^{\alpha-\frac{1}{2}}}$ $\times[\cos(\frac{t}{2}) + i\sin(\frac{t}{2})], t > 0$
Exponential	$\frac{1}{\lambda} e^{-x/\lambda}$	$\frac{1}{1-i\lambda t}$
Gamma	$\frac{e^{-x/\lambda} x^{\alpha-1}}{\lambda^\alpha \Gamma(\alpha)}$	$[\frac{1}{1-i\lambda t}]^\alpha$
Double exponential	$\frac{1}{2\sigma} e^{-|x|/\sigma}$	$\frac{1}{1+\sigma^2 t^2}$
Normal	$\frac{1}{\sigma\sqrt{2\pi}} e^{-(x-\mu)^2/(2\sigma^2)}$	$e^{it\mu - t^2\sigma^2/2}$
Cauchy	$\frac{1}{\sigma\pi\left(1 + \frac{(x-\mu)^2}{\sigma^2}\right)}$	$e^{it\mu - \sigma|t|}$
t	$\frac{\Gamma(\frac{a+1}{2})}{\sqrt{a\pi}\Gamma(\frac{a}{2})(1+\frac{x^2}{a})^{(a+1)/2}}$	$\frac{a^{a/4}}{2^{a/2-1}\Gamma(\frac{a}{2})} |t|^{a/2} K_{\frac{a}{2}}(\sqrt{a}\,|t|)$
Multivariate Normal	$\frac{1}{(2\pi)^{d/2}|\Sigma|^{\frac{1}{2}}} e^{-\frac{1}{2}(x-\mu)'\Sigma^{-1}(x-\mu)}$	$e^{it'\mu - \frac{t'\Sigma t}{2}}$
Uniform in n-dimensional unit ball	$\frac{\Gamma(\frac{n}{2}+1)}{\pi^{\frac{n}{2}}}$	$2^{\frac{n}{2}}\Gamma(\frac{n}{2}+1)\frac{J_{\frac{n}{2}}(\|t\|)}{\|t\|^{\frac{n}{2}}}$
Multinomial	$\frac{n!}{\prod_{j=1}^k x_j!}\prod_{j=1}^k p_j^{x_j}$	$\left(\sum_{j=1}^k p_j e^{it_j}\right)^n$

Example 8.1 (Binomial, Normal, Poisson). In this example we work out the cf of general binomial, Normal, and Poisson distributions.

If $X \sim \text{Ber}(p)$, then immediately from definition its cf is $\psi_X(t) = (1-p+pe^{it})$. because the general binomial random variable can be written as the sum of n iid bernoulli variables, the cf of the $\text{Bin}(n, p)$ distribution is $(1 - p + pe^{it})^n$.

For the Normal case, first consider the standard Normal case. By virtue of symmetry, from part (f) of the above theorem (or, directly from definition), if $Z \sim N(0, 1)$, then its cf is $\psi_Z(t) = \int_{-\infty}^{\infty} (\cos tz)\phi(z)dz$. Therefore,

$$\frac{d}{dt}\psi_Z(t) = \frac{d}{dt}\int_{-\infty}^{\infty} (\cos tz)\phi(z)dz$$
$$= \int_{-\infty}^{\infty} \left(\frac{d}{dt}\cos tz\right)\phi(z)dz = \int_{-\infty}^{\infty}(\sin tz)(-z\phi(z))dz$$
$$= \int_{-\infty}^{\infty} (\sin tz)\left(\frac{d}{dz}\phi(z)\right)dz = -t\int_{-\infty}^{\infty}(\cos tz)\phi(z)dz$$
$$= -t\psi_Z(t),$$

where the interchange of the derivative and the integral is permitted by the dominated convergence theorem, and integration by parts has been used in the final step of the calculation. Because $\psi_Z(0)$ must be one, $\frac{d}{dt}\psi_Z(t) = -t\psi_Z(t) \Rightarrow \psi_Z(t) = e^{-\frac{t^2}{2}}$.

If $X \sim N(\mu, \sigma^2)$, by part (d) of the above theorem, its cf equals $e^{it\mu - \frac{t^2\sigma^2}{2}}$. If $X \sim \text{Poi}(\lambda)$, then

$$\psi_X(t) = e^{-\lambda} \sum_{x=0}^{\infty} e^{itx} \frac{\lambda^x}{x!}$$

$$= e^{-\lambda} \sum_{x=0}^{\infty} \frac{(\lambda e^{it})^x}{x!}$$

$$= e^{-\lambda} e^{\lambda e^{it}} = e^{\lambda(e^{it}-1)}.$$

The power series representation of the exponential function on $\mathbb{C}$ has been used in summing the series $\sum_{x=0}^{\infty} \frac{(\lambda e^{it})^x}{x!}$.

Example 8.2 (Exponential, Double Exponential, and Cauchy). Let X have the standard exponential density. Then its cf is $\psi_X(t) = \int_0^{\infty} e^{itx} e^{-x} dx$. One can use methods of integration of complex-valued functions, and get the answer. Alternatively, one can separately find the real integrals $\int_0^{\infty} (\cos tx) e^{-x} dx$, and $\int_0^{\infty} (\sin tx) e^{-x} dx$, and thus obtain the cf as $\psi_X(t) = \int_0^{\infty} (\cos tx) e^{-x} dx + i \int_0^{\infty} (\sin tx) e^{-x} dx$. Each integral can be evaluated by various methods. For example, one can take the power series expansion $\sum_{k=0}^{\infty} (-1)^k \frac{(tx)^{2k}}{(2k)!}$ for $\cos(tx)$ and integrate $\int_0^{\infty} (\cos tx) e^{-x} dx$ term by term, and then sum the series, and likewise for the integral $\int_0^{\infty} (\sin tx) e^{-x} dx$. Or, one can evaluate the integrals by repeated integration by parts. Any of these gives the formula $\psi_X(t) = \frac{1}{1-it}$.

If X has the standard double exponential density, then simply use the fact that $X \stackrel{\mathcal{L}}{=} X_1 - X_2$, where X_1, X_2 are iid standard exponential. Then, by parts (b) and (c) of the above theorem,

$$\psi_X(t) = \frac{1}{1-it} \frac{1}{1+it} = \frac{1}{1+t^2}.$$

Note the very interesting outcome that the cf of the standard double exponential density is the renormalized standard Cauchy density. By a theorem (the inversion theorem in the next section) that we have not discussed yet, this implies that the cf of the standard Cauchy density is $e^{-|t|}$, the renormalized standard double exponential density.

Example 8.3 (Uniform Distribution in n-Dimensional Unit Ball). This example well illustrates the value of a skilled calculation. We work out the cf of the uniform distribution in B_n, the n-dimensional unit ball, with the constant density

$$f(x) = \frac{1}{\text{Vol}(B_n)} I_{\{x \in B_n\}} = \frac{\Gamma(\frac{n}{2}+1)}{\pi^{\frac{n}{2}}} I_{\{x \in B_n\}}.$$

We write $f(x) = c_n I_{\{x \in B_n\}}$. Before we start the derivation, it is important to note that the constants in the calculation do not have to be explicitly carried along the steps, and a final constant can be found at the end by simply forcing the cf to equal one at $\mathbf{t} = \mathbf{0}$.

First note that by virtue of symmetry (around the origin) of the uniform density in the ball, the cf equals $\psi(\mathbf{t}) = c_n \int_{B_n} \cos(\mathbf{t}'\mathbf{x}) d\mathbf{x}$. Let P be an orthogonal matrix such that $P\mathbf{t} = ||\mathbf{t}|| \mathbf{e}_1$, where $\mathbf{e}_1$ is the first n-dimensional basis unit vector

8.1 Characteristic Functions of Standard Distributions

$(1, 0, \ldots, 0)'$. Because $||P\mathbf{x}|| = ||\mathbf{x}||$ (P being an orthogonal matrix), and because $|P| = 1$, we get

$$\psi(\mathbf{t}) = c_n \int_{B_n} \cos(||\mathbf{t}|| x_1) d\mathbf{x}.$$

Now make the n-dimensional polar transformation $(x_1, x_2, \ldots, x_n) \to (\rho, \theta_1, \ldots, \theta_{n-1})$ (consult Chapter 4), which we recall has the Jacobian $\rho^{n-1} \sin^{n-2} \theta_1 \sin^{n-3} \theta_2 \cdots \sin \theta_{n-2}$. Thus,

$$\psi(\mathbf{t}) = c_n \int_0^1 \int_0^\pi \int_0^\pi \cdots \int_0^{2\pi} \rho^{n-1} \cos(\rho ||\mathbf{t}|| \cos \theta_1)(\sin \theta_1)^{n-2} \sin^{n-3} \theta_2 \cdots$$
$$\times \sin \theta_{n-2}$$

$d\theta_{n-1} \cdots d\theta_2 d\theta_1 d\rho$

$$= k_{1n} \int_0^1 \int_0^\pi \rho^{n-1} \cos(\rho ||\mathbf{t}|| \cos \theta_1)(\sin \theta_1)^{n-2} d\theta_1 d\rho$$

(it is not important to know right now what the constant k_{1n} is)

$$= k_{2n} \int_0^1 \int_{-1}^1 \rho^{n-1} \cos(\rho ||\mathbf{t}|| u)(1 - u^2)^{\frac{n-3}{2}} du d\rho$$

(on making the change of variable $\cos \theta_1 = u$, a monotone transformation on $(0, \pi)$)

$$= k_{3n} \int_0^1 \rho^{n-1} (\rho ||\mathbf{t}||)^{1-\frac{n}{2}} J_{\frac{n}{2}-1}(\rho ||\mathbf{t}||) d\rho$$

(on using the known integral representation of $J_\alpha(x)$)

$$= k_{3n} ||\mathbf{t}||^{1-\frac{n}{2}} \int_0^1 \rho^{\frac{n}{2}} J_{\frac{n}{2}-1}(\rho ||\mathbf{t}||) d\rho$$

$$= k_{4n} ||\mathbf{t}||^{1-\frac{n}{2}} \frac{J_{\frac{n}{2}}(||\mathbf{t}||)}{||\mathbf{t}||}$$

(on using the known formula for the integral in the above line)

$$= k_{4n} \frac{J_{\frac{n}{2}}(||\mathbf{t}||)}{||\mathbf{t}||^{\frac{n}{2}}}.$$

Now, at this final step use the fact that $\lim_{u \to 0} \frac{J_{\frac{n}{2}}(u)}{u^{\frac{n}{2}}} = \frac{1}{2^{\frac{n}{2}}(\Gamma \frac{n}{2} + 1)}$, and this gives the formula that

$$\psi(t) = 2^{\frac{n}{2}} \Gamma\left(\frac{n}{2} + 1\right) \frac{J_{\frac{n}{2}}(||\mathbf{t}||)}{||\mathbf{t}||^{\frac{n}{2}}}.$$

8.2 Inversion and Uniqueness

Given the cf of a random variable, it is possible to recover the CDF at its points of continuity; additionally, if the cf is absolutely integrable, then the random variable must have a density, and the density too can be recovered from the cf. The first result on the recovery of the CDF leads to the distribution determining property of a cf.

Theorem 8.1 (Inversion Theorems). *Let X have the CDF F and cf $\psi(t)$.*

(a) *For any $-\infty < a < b < \infty$, such that a, b are both continuity points of F,*

$$F(b) - F(a) = \lim_{T \to \infty} \frac{1}{2\pi} \int_{-T}^{T} \frac{e^{-iat} - e^{-ibt}}{it} \psi(t) dt;$$

(b) *If $\int_{-\infty}^{\infty} |\psi(t)| dt < \infty$, then X has a density, say $f(x)$, and*

$$f(x) = \frac{1}{2\pi} \int_{-\infty}^{\infty} e^{-itx} \psi(t) dt.$$

Furthermore, in this case, f is bounded and continuous.

(c) **(Plancherel's Identity)** *If a random variable X has a density f and the cf ψ, then*

$$\int_{-\infty}^{\infty} |\psi(t)|^2 dt < \infty \Leftrightarrow \int_{-\infty}^{\infty} |f(x)|^2 dx < \infty,$$

and $\int_{-\infty}^{\infty} |\psi(t)|^2 dt = \int_{-\infty}^{\infty} |f(x)|^2 dx$.

Proof. We prove part (b), by making use of part (a).

From part (a), if a, b are continuity points of F, then because $\int_{-\infty}^{\infty} |\psi(t)| dt < \infty$, the dominated convergence theorem gives the simpler expression

$$\begin{aligned}
F(b) - F(a) &= \frac{1}{2\pi} \int_{-\infty}^{\infty} \frac{e^{-iat} - e^{0-ibt}}{it} \psi(t) dt \\
&= \frac{1}{2\pi} \int_{-\infty}^{\infty} \left[\int_{a}^{b} e^{-itx} dx \right] \psi(t) dt \\
&= \int_{a}^{b} \left[\frac{1}{2\pi} \int_{-\infty}^{\infty} e^{-itx} \psi(t) dt \right] dx
\end{aligned}$$

(the interchange of the order of integration justified by Fubini's theorem). Let $a \to -\infty$; noting that we can indeed approach $-\infty$ through continuity points, we obtain

$$F(b) = \int_{-\infty}^{b} \left[\frac{1}{2\pi} \int_{-\infty}^{\infty} e^{-itx} \psi(t) dt \right] dx,$$

at continuity points b. To make the proof completely rigorous from here, some measure theory is needed; but we have basically shown that the CDF can be written as

8.2 Inversion and Uniqueness

the integral of the function $\frac{1}{2\pi}\int_{-\infty}^{\infty} e^{-itx}\psi(t)dt$, and so X must have this function as its density. It is bounded and continuous for the same reason that a cf is bounded and continuous.

For part (c), notice that $|\psi(t)|^2$ is the cf of $X - Y$, where X, Y are iid with cf $\psi(t)$. But the density of $X - Y$, by the density convolution formula, is $g(x) = \int_{-\infty}^{\infty} f(x)f(x+y)dy$. If $|\psi(t)|^2$ is integrable, the inversion formula of part (b) applies to this cf, namely, $|\psi(t)|^2$. If we do apply the inversion formula, we get that $\int_{-\infty}^{\infty}|\psi(t)|^2 dt = g(0)$. But $g(0) = \int_{-\infty}^{\infty}|f(x)|^2 dx$. Therefore, we must have $\int_{-\infty}^{\infty}|\psi(t)|^2 dt = \int_{-\infty}^{\infty}|f(x)|^2 dx$. We leave the converse part as an easy exercise. □

The inversion theorem leads to the distribution determining property.

Theorem 8.2 (Distribution Determining Property). *Let X, Y have characteristic functions $\psi_X(t), \psi_Y(t)$. If $\psi_X(t) = \psi_Y(t)$ $\forall t$, then X, Y have the same distribution.*

The proof uses part (a) of the inversion theorem, which applies to any type of random variables. Applying part (a) separately to X and Y, one gets that the CDFs F_X, F_Y must be the same at all points b that are continuity points of both F_X, F_Y. This forces F_X and F_Y to be equal everywhere, because one can approach an arbitrary point from above through common continuity points of F_X, F_Y. □

Remark. It is important to note that the assumption that $\psi_X(t) = \psi_Y(t)$ $\forall t$ cannot really be relaxed in this theorem, because it is actually possible for random variables X, Y with different distributions to have identical characteristic functions at a lot of points, for example, over arbitrarily long intervals. In fact, we show such an example later.

We had stated in Chapter 1 that the sum of any number of independent Cauchy random variables also has a Cauchy distribution. Aided by cfs, we can now prove it.

Example 8.4 (Sum of Cauchys). Let $X_i, 1 \leq i \leq n$, be independent, and suppose $X_i \sim C(\mu_i, \sigma_i^2)$. Let $S_n = X_1 + \cdots + X_n$. Then, the cf of S_n is

$$\psi_{S_n}(t) = \prod_{i=1}^{n} \psi_{X_i}(t) = \prod_{i=1}^{n}[e^{it\mu_i - \sigma_i|t|}]$$
$$= e^{it(\sum_{i=1}^{n}\mu_i) - |t|(\sum_{i=1}^{n}\sigma_i)}.$$

This coincides with the cf of a $C\left(\sum_{i=1}^{n}\mu_i, \left(\sum_{i=1}^{n}\sigma_i\right)^2\right)$ distribution, therefore by the distribution determining property of cfs, one has

$$S_n \sim C\left(\sum_{i=1}^{n}\mu_i, \left(\sum_{i=1}^{n}\sigma_i\right)^2\right)$$

$$\Leftrightarrow \bar{X}_n = \frac{S_n}{n} \sim C\left(\frac{\sum_{i=1}^{n}\mu_i}{n}, \left(\frac{\sum_{i=1}^{n}\sigma_i}{n}\right)^2\right).$$

In particular, if $X_1, \ldots, X_n$ are iid $C(\mu, \sigma^2)$, then $\bar{X}_n$ is also distributed as $C(\mu, \sigma^2)$ for all n. No matter how large n is, the distribution of the sample mean is exactly the same as the distribution of one observation. This is a remarkable stability property, and note the contrast to situations where the laws of large numbers and the central limit theorem hold.

Analogous to the inversion formula for the density case, there is a corresponding inversion formula for integer-valued random variables, and actually more generally, for lattice-valued random variables. We need two definitions for presenting those results.

Definition 8.2. A real number x is called an *atom* of the distribution of a real-valued random variable X if $P(\{x\}) = P(X = x) > 0$.

Definition 8.3. A real-valued random variable X is called a *lattice-valued random variable* and its distribution a *lattice distribution* if there exist numbers $a, h > 0$ such that $P(X = a + nh$ for some $n \in \mathbb{Z}) = 1$, where $\mathbb{Z}$ denotes the set of integers.

In other words, X can only take values $a, a \pm h, a \pm 2h, \ldots$, for some $a, h > 0$. The largest number h satisfying this property is called the *span of the lattice*.

Example 8.5. This example helps illustrate the concepts of an atom and a lattice.

If X has a density, then it cannot have any atoms. It also cannot be a lattice random variable, because any random variable with a density assigns zero probability to any countable set. Now consider a mixture distribution such as $p\delta_0 + (1-p)N(0, 1), 0 < p < 1$. This distribution has one atom, namely the value $x = 0$. Consider a Poisson distribution. This has all nonnegative integers as its atoms, and it is also a lattice distribution, with $a = 0, h = 1$. Moreover, this distribution is purely atomic. Now take $Y \sim \text{Bin}(n, p)$ and let $X = \frac{Y}{n}$. Then X has the atoms $0, \frac{1}{n}, \frac{2}{n}, \ldots, 1$. This is also a lattice distribution, with $a = 0, h = \frac{1}{n}$. This distribution is also purely atomic. Lastly, let $Z \sim N(0, 1)$ and let $X = 1 + 2\lfloor Z \rfloor$. Then, the atoms of X are the integers $\pm 1, \pm 3, \ldots$ and again, X has a lattice distribution, with $a = 1$ and $h = 2$.

We now give the inversion formulas for lattice-valued random variables.

Theorem 8.3 (Inversion for Lattice Random Variables).

(a) Let X be an integer-valued random variable with cf $\psi(t)$. Then for any integer n,
$$P(X = n) = \frac{1}{2\pi} \int_{-\pi}^{\pi} e^{-int} \psi(t) dt.$$

(b) More generally, if X is a lattice random variable, then for any integer n,
$$P(X = a + nh) = \frac{h}{2\pi} \int_{-\frac{\pi}{h}}^{\frac{\pi}{h}} e^{-i(a+nh)t} \psi(t) dt.$$

8.2 Inversion and Uniqueness

(c) Given a random variable X with cf $\psi(t)$, let $\mathcal{A}$ be the (countable) set of all the atoms of X. Then

$$\sum_{x \in \mathcal{A}} [P(X = x)]^2 = \lim_{T \to \infty} \frac{1}{2T} \int_{-T}^{T} |\psi(t)|^2 dt.$$

An interesting immediate corollary of part (c) is the following.

Corollary. *Suppose X has a cf $\psi(t)$ that is square integrable. Then X cannot have any atoms. In particular, if $\psi(t)$ is absolutely integrable, then X cannot have any atoms, and in fact must have a density.*

Because there is a one-to-one correspondence between characteristic functions and CDFs, it seems natural to expect that closeness of characteristic functions should imply closeness of the CDFs, and distant characteristic functions should imply some distance between the CDFs. In other words, there should be some relationship of implication between nearby laws and nearby characteristic functions. Indeed, there are many such results available. Perhaps the most well known among them is *Esseen's lemma*, which is given below, together with a reverse inequality, and a third result for a pair of integer-valued random variables.

Theorem 8.4 (Esseen's Lemma). *(a) Let F, G be two CDFs, of which G has a uniformly bounded density g, $g \leq K < \infty$. Let F, G have cfs ψ, ξ. Then for any $T > 0$, and $b > \frac{1}{2\pi}$, there exists a finite positive constant $C = C(b)$ such that*

$$\sup_x |F(x) - G(x)| \leq b \int_{-T}^{T} \frac{|\psi(t) - \xi(t)|}{|t|} dt + \frac{CK}{T}.$$

*(b) **(Reversal)** Let F, G be any two CDFs, with cfs ψ, ξ. Then*

$$\sup_x |F(x) - G(x)| \geq \frac{1}{2} \left| \int_{-\infty}^{\infty} [\psi(t) - \xi(t)] \phi(t) dt \right|,$$

where ϕ is the standard normal density.

*(c) **(Integer-Valued Case)** Let F, G be CDFs of two integer-valued random variables, with cfs ψ, ξ. Then*

$$\sup_x |F(x) - G(x)| \leq \frac{1}{4} \int_{-\pi}^{\pi} \frac{|\psi(t) - \xi(t)|}{|t|} dt.$$

See Chapter 5 in Petrov (1975, pp. 142–147 and pp. 186–187) for each part of this theorem.

8.3 Taylor Expansions, Differentiability, and Moments

Unlike mgfs, characteristic functions need not be differentiable, even once. We have already seen such an example, namely the cf of the standard Cauchy distribution, which is $e^{-|t|}$, and therefore continuous but not differentiable. The tail of the Cauchy distribution is causing the lack of differentiability of its cf. If the tail tapers off rapidly, then the cf will be differentiable, and could even be infinitely differentiable. Thus, thin tails of the distribution go hand in hand with smoothness of the cf. Conversely, erratic tails of the cf go hand in hand with a CDF that is not sufficiently smooth. Inasmuch as a thin tail of the CDF is helpful for existence of moments, these three attributes are linked together, namely,

- Does the CDF F have a thin tail?
  - Is the cf differentiable enough number of times?
  - Does F have enough number of finite moments, and if so, how does one recover them directly from the cf?

  Conversely, these two attributes are linked together:
- Does the cf taper off rapidly?
  - Is the CDF F very smooth, for example, differentiable a (large) number of times?

These issues, together with practical applications in the form of Taylor expansions for the cf are discussed next. It should be remarked that Taylor expansions for the cf and its logarithm form hugely useful tools in various problems in asymptotics. For example, the entire area of *Edgeworth expansions* in statistics uses these Taylor expansions as the most fundamental tool.

Theorem 8.5. *Let X have cf $\psi(t)$.*

(a) *If $E(|X|^k) < \infty$, then ψ is k times continuously differentiable at any point, and moreover,*
$$E(X^m) = (-i)^m \psi^{(m)}(0) \; \forall m \leq k.$$

(b) *If for some even integer $k = 2m$, ψ is k times differentiable at zero, then $E(X^k) < \infty$.*

(c) *If for some odd integer $k = 2m + 1$, ψ is k times differentiable at zero, then $E(X^{k-1}) < \infty$.*

(d) **(Riemann–Lebesgue Lemma)** *If X has a density, then $\psi(t) \to 0$ as $t \to \pm\infty$.*

(e) *If X has a density and the density itself has n derivatives, each of which is absolutely integrable, then $\psi(t) = o(|t|^{-n})$ as $t \to \pm\infty$.*

Proof. We outline the proof of parts (a) and (e) of this theorem. See Port (1994, pp. 658–663 and p. 670) for the remaining parts. For part (a), for notational simplicity, assume that X has a density f, and formally differentiate $\psi(t)$ m times inside the integral sign. We get the expression $i^m \int_{-\infty}^{\infty} x^m e^{itx} f(x) dx$. The absolute value of the integrand $|x^m e^{itx} f(x)|$ is bounded by $|x^m f(x)|$, which is integrable, because

8.4 Continuity Theorems

$m \leq k$ and the kth moment exists by hypothesis. This means that the dominated convergence theorem is applicable, and that at any t the cf is m times differentiable, with $\psi^{(m)}(t) = i^m \int_{-\infty}^{\infty} x^m e^{itx} f(x) dx$. Putting $t = 0$, $\psi^{(m)}(0) = i^m E(X^m)$.

For part (e), suppose X has a density f and that f' exists and is absolutely integrable. Then, by integration by parts, $\psi(t) = -\frac{i}{t} \int_{-\infty}^{\infty} e^{itx} f'(x) dx$. Now apply the Riemann–Lebesgue lemma, namely part (d), to conclude that $|\psi(t)| = o\left(\frac{1}{|t|}\right)$ as $t \to \pm\infty$. For general n, use this same argument by doing repeated integration by parts. □

The practically useful Taylor expansions for the cf and its logarithm are given next. For completeness, we recall the definition of cumulants of a random variable.

Definition 8.4. Suppose $E(|X|^n) < \infty$, and let $c_j = E(X^j)$, $1 \leq j \leq n$. The first cumulant is defined to be $\kappa_1 = c_1$, and for $2 \leq j \leq n$, κ_j defined by the recursion relation $\kappa_j = c_j - \sum_{k=1}^{j-1} \binom{j-1}{k-1} c_{j-k} \kappa_k$ is the jth cumulant of X. In particular, $\kappa_2 = \mathrm{Var}(X)$, $\kappa_3 = E(X - \mu)^3$, $\kappa_4 = E(X - \mu)^4 - 3[\mathrm{Var}(X)]^2$, where $\mu = c_1 = E(X)$.

Equivalently, if $E(|X|^n) < \infty$, then for $1 \leq j \leq n$, $\kappa_j = \frac{1}{i^j} \left[\frac{d^j}{dt^j} \log \psi(t) \right]\big|_{t=0}$.

Theorem 8.6 (Taylor Expansion of Characteristic Functions). *Let X have the cf $\psi(t)$.*

(a) If $E(|X|^n) < \infty$, then $\psi(t)$ admits the Taylor expansion

$$\psi(t) = 1 + \sum_{j=1}^{n} \frac{(it)^j}{j!} E(X^j) + o(|t|^n),$$

as $t \to 0$.

(b) If $E(|X|^n) < \infty$, then $\log \psi(t)$ admits the Taylor expansion

$$\log \psi(t) = \sum_{j=1}^{n} \frac{(it)^j}{j!} \kappa_j + o(|t|^n),$$

as $t \to 0$. See Port 1994 Port (1994, p. 660) for a derivation of the expansion.

8.4 Continuity Theorems

One of the principal uses of characteristic functions is in establishing convergence in distribution. Suppose Z_n is some sequence of random variables, and we suspect that Z_n converges in distribution to some specific random variable Z (e.g., standard normal). Then, a standard method of attack is to calculate the cf of Z_n and show that it converges (pointwise) to the cf of the specific Z we have in mind. Another case is where we are not quite sure what the limiting distribution of Z_n is. But, still, if we can calculate the cf of Z_n and obtain a pointwise limit for it, say $\psi(t)$, and if we can also establish that this function $\psi(t)$ is indeed a valid cf (it is not automatically guaranteed), then the limiting distribution will be whatever distribution has $\psi(t)$ as its cf. Characteristic functions thus make a particularly effective tool in asymptotic theory. In fact, we later give a proof of the CLT in the iid case, without making restrictive assumptions such as the existence of the mgf, by using characteristic functions.

Theorem 8.7. *(a) Let $X_n, n \geq 1$, X have cfs $\psi_n(t), \psi(t)$. Then $X_n \stackrel{\mathcal{L}}{\Rightarrow} X$ if and only if $\psi_n(t) \to \psi(t)$ $\forall t$.*
(b) (Lévy) Let $X_n, n \geq 1$ have cfs $\psi_n(t)$, and suppose $\psi_n(t)$ converges pointwise to (some function) $\psi(t)$. If ψ is continuous at zero, then $\psi(t)$ must be a cf, in which case $X_n \stackrel{\mathcal{L}}{\Rightarrow} X$, where X has $\psi(t)$ as its cf.

Proof. We prove only part (a). Let F_n, F denote the CDFs of X_n, X. Let

$$\mathcal{B} = \{b : b \text{ is a continuity point of each } F_n \text{ and } F\}.$$

Note that the complement of $\mathcal{B}$ is at most countable. Suppose $\psi_n(t) \to \psi(t)$ $\forall t$. By the inversion theorem, for any $a, b (a < b) \in \mathcal{B}$,

$$\begin{aligned}
F(b) - F(a) &= \lim_{T \to \infty} \frac{1}{2\pi} \int_{-T}^{T} \psi(t) \frac{e^{-iat} - e^{-ibt}}{it} dt \\
&= \lim_{T \to \infty} \frac{1}{2\pi} \int_{-T}^{T} \left[\lim_{n} \psi_n(t) \frac{e^{-iat} - e^{-ibt}}{it} \right] dt \\
&= \lim_{T \to \infty} \lim_{n} \frac{1}{2\pi} \int_{-T}^{T} \psi_n(t) \frac{e^{-iat} - e^{-ibt}}{it} dt
\end{aligned}$$

(by the dominated convergence theorem)

$$\begin{aligned}
&= \lim_{n} \lim_{T \to \infty} \frac{1}{2\pi} \int_{-T}^{T} \psi_n(t) \frac{e^{-iat} - e^{-ibt}}{it} dt \\
&= \lim_{n} [F_n(b) - F_n(a)].
\end{aligned}$$

This implies that $X_n \stackrel{\mathcal{L}}{\Rightarrow} X$.

Conversely, suppose $X_n \stackrel{\mathcal{L}}{\Rightarrow} X$. Then, by the portmanteau theorem, for any t, each of $E[\cos(tX_n)]$, $E[\sin(tX_n)]$ converges to $E[\cos(tX)]$, $E[\sin(tX)]$, and so, $\psi_n(t) \to \psi(t)$ $\forall t$. □

8.5 Proof of the CLT and the WLLN

Perhaps the most major application of part (a) of the continuity theorem is in supplying a proof of the CLT in the iid case, without making any assumptions other than what the CLT says. Characteristic functions also provide a very efficient proof of the weak law of large numbers and the Cramér–Wold theorem. These three proofs are provided below.

Theorem 8.8 (CLT in the IID Case). *Let $X_i, i \geq 1$ be iid variables with mean μ, and variance $\sigma^2(< \infty)$. Let $Z_n = \frac{\sqrt{n}(\bar{X}-\mu)}{\sigma}$. Let $Z \sim N(0, 1)$. Then $Z_n \stackrel{\mathcal{L}}{\Rightarrow} Z$.*

Proof. Without any loss of generality, we may assume that $\mu = 0, \sigma = 1$. Let $\psi(t)$ denote the cf of the X_i. Then, the cf of Z_n is

$$\psi_n(t) = \left[\psi\left(\frac{t}{\sqrt{n}}\right)\right]^n$$
$$= \left(1 - \frac{t^2}{2n} + o\left(\frac{1}{n}\right)\right)^n$$

(by the Taylor expansion for cfs)

$$\to e^{-\frac{t^2}{2}},$$

and hence, by part (a) of the continuity theorem, $Z_n \stackrel{\mathcal{L}}{\Rightarrow} Z$. □

Characteristic functions also provide a quick proof of Khintchine's weak law of large numbers.

Theorem 8.9 (WLLN). *Suppose $X_i, i \geq 1$ are iid, with $E(|X_i|) < \infty$. Let $E(X_i) = \mu$. Let $S_n = X_1 + \cdots + X_n, n \geq 1$. Then $\bar{X} = \frac{S_n}{n} \stackrel{P}{\Rightarrow} \mu$.*

Proof. We may assume without loss of generality that $\mu = 0$. Because $E(|X_i|) < \infty$, the cf of X_i admits the Taylor expansion $\psi(t) = 1 + o(|t|), t \to 0$. Now, the cf of $\bar{X}$ is

$$\psi_n(t) = \left(\psi\left(\frac{t}{n}\right)\right)^n = \left(1 + o\left(\frac{1}{n}\right)\right)^n$$
$$\to 1 \;\forall t.$$

Therefore, $\bar{X}$ converges in distribution to (the point mass at) zero, and so, also converges in probability to zero. □

A third good application of characteristic functions is writing a proof of the Cramér–Wold theorem.

Theorem 8.10 (Cramér–Wold Theorem). *Let* $\mathbf{X}_n, \mathbf{X}$ *be d-dimensional vectors. Then* $\mathbf{X}_n \stackrel{\mathcal{L}}{\Rightarrow} \mathbf{X}$ *if and only if* $\mathbf{c}'\mathbf{X}_n \stackrel{\mathcal{L}}{\Rightarrow} \mathbf{c}'\mathbf{X}$ *for all unit vectors* $\mathbf{c}$.

Proof. The *only if* part is an immediate consequence of the continuous mapping theorem for convergence in distribution. For the *if* part, if $\mathbf{c}'\mathbf{X}_n \stackrel{\mathcal{L}}{\Rightarrow} \mathbf{c}'\mathbf{X}$ for all unit vectors $\mathbf{c}$, then by the continuity theorem for characteristic functions, for all real λ, and for all unit vectors $\mathbf{c}$, $E(e^{i\lambda \mathbf{c}'\mathbf{X}_n}) \to E(e^{i\lambda \mathbf{c}'\mathbf{X}})$. But that is the same as saying $E(e^{i\mathbf{t}'\mathbf{X}_n}) \to E(e^{i\mathbf{t}'\mathbf{X}})$ for all d-dimensional vectors $\mathbf{t}$, and so, once again by the continuity theorem for characteristic functions, $\mathbf{X}_n \stackrel{\mathcal{L}}{\Rightarrow} \mathbf{X}$. □

8.6 * Producing Characteristic Functions

An interesting question is which functions can be characteristic functions of some probability distribution. We only address the case of one-dimensional random variables. It turns out that there is a classic necessary and sufficient condition. However, the condition is not practically very useful, because of the difficulty verifying it in specific examples. Simple sufficient conditions are thus also very useful. In this section, we first describe the necessary and sufficient condition, and then give some examples and sufficient conditions.

Theorem 8.11 (Bochner's Theorem). *Let ψ be a complex-valued function defined on the real line. Then ψ is the characteristic function of some probability distribution on $\mathcal{R}$ if and only if*

(a) ψ *is continuous.*
(b) $\psi(0) = 1$.
(c) $|\psi(t)| \leq 1 \; \forall t$.
(d) *For all $n \geq 1$, for all reals $t_1, \ldots, t_n$, and for all $\omega_1, \ldots, \omega_n \in \mathbb{C}$,*

$$\sum_{i=1}^{n} \sum_{j=1}^{n} \omega_i \bar{\omega}_j \psi(t_i - t_j) \geq 0.$$

See Port (1994, p. 663) for a proof of Bochner's theorem.

Example 8.6. We give some simple consequences of Bochner's theorem. Suppose $\psi_i, i \geq 1$ are all characteristic functions, and let $p_i \geq 0, \sum_{i=1}^{\infty} p_i = 1$. Then, because each ψ_i satisfies the four conditions of Bochner's theorem, so does $\sum_{i=1}^{\infty} p_i \psi_i$, and therefore $\sum_{i=1}^{\infty} p_i \psi_i$ is also a characteristic function. However, we really do not need Bochner's theorem for this, because we can easily recognize $\sum_{i=1}^{\infty} p_i \psi_i$ to be the characteristic function of $\sum_{i=1}^{\infty} p_i F_i$ where ψ_i is the characteristic function of F_i. Similarly, if ψ satisfies the four conditions of Bochner's theorem, so does $\bar{\psi}$, by an easy verification of the last condition, and the first three are obvious. But again, that $\bar{\psi}$ is a characteristic function is directly clear on observing that it is the cf of $-X$ if ψ is the cf of X.

8.6 Producing Characteristic Functions

We can now conclude more. If ψ_1, ψ_2 are both cfs, then so must be their product $\psi_1 \psi_2$, by the convolution theorem for cfs. Applied to the present situation, this means that if $\psi(t)$ is a cf, so must be $\psi(t)\overline{\psi(t)} = |\psi(t)|^2$. In fact, this is the cf of $X_1 - X_2$, with X_1, X_2 being iid with the cf ψ. Recall that this is just the *symmetrization* of X_1.

To conclude this example, because $\overline{\psi}$ is a cf whenever ψ is, the special convex combination $\frac{1}{2}\psi + \frac{1}{2}\overline{\psi}$ is also a cf. But, $\frac{1}{2}\psi + \frac{1}{2}\overline{\psi}$ is simply the real part of ψ. Therefore, if ψ is any cf, then so is its real part $\Re(\psi)$. There is a simple interpretation for it. For example, if X has a density $f(x)$ and the cf ψ, then $\Re(\psi)$ is just the cf of the mixture density $\frac{1}{2}f(x) + \frac{1}{2}f(-x)$.

The following sufficient condition is among the most practically useful methods for constructing valid characteristic functions. We also show some examples of its applications.

Theorem 8.12 (Polýa's Criterion). *Let ψ be a real and even function. Suppose $\psi(0) = 1$, and suppose that $\psi(t)$ is nonincreasing and convex for $t > 0$, and converges to zero as $t \to \infty$. Then ψ is the characteristic function of some distribution having a density.*

See Feller (1971, p. 509) for a proof.

Example 8.7 (Stable Distributions). Let $\psi(t) = e^{-|t|^\alpha}, 0 < \alpha \leq 1$. Then, by simple calculus, $\psi(t)$ satisfies the convexity condition in Polýa's criterion, and the other conditions are obviously satisfied. Therefore, for $0 < \alpha \leq 1, e^{-|t|^\alpha}$ is a valid characteristic function. In fact, these are the characteristic functions of some very special distributions. Distributions F with the property that for any n, if $X_1, \ldots X_n$ are iid with CDF F, then their sum $S_n = X_1 + \cdots + X_n$ is distributed as $a_n X_1 + b_n$ for some sequences a_n, b_n called *stable distributions*. It turns out that the sequence a_n must be of the form $n^{\frac{1}{\alpha}}$ for some $\alpha \in (0, 2]$; α is called the index of the stable distribution. A symmetric stable law has cfs of the form $e^{-c|t|^\alpha}, c > 0$. Therefore, we have arrived here at the cfs of symmetric stable laws of index ≤ 1. Polýa's criterion breaks down if $1 < \alpha \leq 2$. So, although $e^{-c|t|^\alpha}, c > 0$ are also valid cfs for $1 < \alpha \leq 2$, it cannot be proved by using Polýa's criterion. Note that the case $\alpha = 2$ corresponds to mean zero normal distributions, and $\alpha = 1$ corresponds to centered Cauchy distributions. The case $\alpha = \frac{1}{2}$ arises as the limiting distribution of $\frac{n\nu_r}{r^2}$, where ν_r is the time of the rth return to the starting point zero of a *simple symmetric random walk*; see Chapter 11. It can be proved that stable distributions must be continuous, and have a density. However, except for $\alpha = \frac{1}{2}, 1$, and 2, a stable density cannot be written in a simple form using elementary functions. Nevertheless, stable distributions are widely used in modeling variables in economics, finance, extreme event modeling, and generally, whenever densities with heavy tails are needed.

Example 8.8 (An Example Where the Inversion Theorem Fails). The inversion theorem says that if the cf is absolutely integrable, then the density can be found by using the inversion formula. This example shows that the cf need not be absolutely integrable for the distribution to possess a density. For this, consider the function $\psi(t) = \frac{1}{1+|t|}$. Then ψ clearly satisfies every condition in Polýa's criterion, and therefore is a cf corresponding to a density. However, at the same time,

$\int_{-\infty}^{\infty} \frac{1}{1+|t|} dt = \infty$, and therefore the density cannot be found by using the inversion theorem.

Example 8.9 (Two Characteristic Functions That Coincide in an Interval). We remarked earlier that unlike mgfs, cfs of different distributions can coincide over nonempty intervals. We give such an example now. Towards this, define

$$\psi_1(t) = \left(1 - \frac{|t|}{T}\right) I_{\{|t| \leq T\}};$$

$$\psi_2(t) = \left(1 - \frac{|t|}{T}\right) I_{\{|t| \leq \frac{T}{2}\}} + \frac{T}{4|t|} I_{\{|t| > \frac{T}{2}\}}.$$

Each of ψ_1, ψ_2 is a cf by Polýa's criterion. Note that, however, $\psi_1(t) = \psi_2(t) \; \forall t \in [-\frac{T}{2}, \frac{T}{2}]$. Because T is arbitrary, this shows that two different cfs can coincide on arbitrarily large intervals. Also note that at the same time this example provides a cf with a bounded support.

Perhaps a little explanation is useful: mgfs can be extended into analytic functions defined on $\mathbb{C}$, and there is a theorem in complex analysis that two analytic functions cannot coincide on any subset of $\mathbb{C}$ that has a limit point. Thus, two different mgfs cannot coincide on a nonempty real interval. However, unlike mgfs, characteristic functions are not necessarily analytic. That leaves an opening for finding nonanalytic cfs which do coincide over nonempty intervals.

8.7 Error of the Central Limit Theorem

Suppose a sequence of CDFs F_n converges in distribution to F, for some F. Such a weak convergence result is usually used to approximate the true value of $F_n(x)$ at some fixed n and x by $F(x)$. However, the weak convergence result, by itself, says absolutely nothing about the accuracy of approximating $F_n(x)$ by $F(x)$ for that particular value of n. To approximate $F_n(x)$ by $F(x)$ for a given finite n is a leap of faith, unless we have some idea of the error committed (i.e., of the magnitude of $|F_n(x) - F(x)|$). More specifically, for a sequence of iid random variables $X_i, i \geq 1$, with mean μ and variance σ^2, we need some quantification of the error of the normal approximation $|P\left(\frac{\sqrt{n}(\bar{X}-\mu)}{\sigma}\right) \leq x) - \Phi(x)|$, both for a fixed x and uniformly over all x.

The first result for the iid case in this direction is the classic *Berry–Esseen theorem* (Berry (1941), Esseen (1945)). Typically, these accuracy measures give bounds on the error in the central limit theorem approximation for all fixed n, in terms of moments of the X_i. Good general references on this topic are Petrov (1975), Feller (1971), Hall (1992), and Bhattacharya et al. (1986). Other specific references are given later.

In the canonical iid case with a finite variance, the CLT says that $\frac{\sqrt{n}(\bar{X}-\mu)}{\sigma}$ converges in law to the $N(0, 1)$ distribution. By Polya's theorem, the uniform error

8.7 Error of the Central Limit Theorem

$\Delta_n = \sup_{-\infty < x < \infty} |P\left(\frac{\sqrt{n}(\bar{X}-\mu)}{\sigma} \leq x\right) - \Phi(x)| \to 0$, as $n \to \infty$. Bounds on Δ_n for any given n are called *uniform bounds*. Here is the classic Berry–Esseen uniform bound.

Theorem 8.13 (Berry–Esseen Theorem). *Let $X_i, i \geq 1$, be iid with $E(X_1) = \mu$, $Var(X_1) = \sigma^2$, and $\beta_3 = E|X_1 - \mu|^3 < \infty$. Then there exists a universal constant C, not depending on n or the distribution F of the X_i, such that*

$$\Delta_n \leq \frac{C\beta_3}{\sigma^3 \sqrt{n}}.$$

The proof of the Berry–Esseen theorem uses two technical inequalities on characteristic functions. One is Esseen's lemma, and the other a lemma that further bounds a term in Esseen's lemma itself. It is the second lemma that needs elaborate arguments involving Taylor expansions for the characteristic function of $\frac{\sqrt{n}(\bar{X}-\mu)}{\sigma}$, and its logarithm. A detailed proof of this second lemma can be seen in Petrov (1975, pp. 142–147).

Lemma. *Let $X_1, X_2, \ldots, X_n$ be iid random variables with mean μ, variance σ^2, and $\beta_3 = E(|X_1 - \mu|^3) < \infty$. Let $L_n = \frac{\beta_3}{\sigma^3 \sqrt{n}}$, and let $\phi_n(t)$ denote the characteristic function of $Z_n = \frac{\sqrt{n}(\bar{X}-\mu)}{\sigma}$. Then,*

$$|\phi_n(t) - e^{-\frac{t^2}{2}}| \leq 16 L_n |t|^3 e^{-\frac{t^2}{3}} \; \forall t \text{ such that } |t| \leq \frac{1}{4L_n}.$$

Sketch of Proof of Berry–Esseen Theorem. Denote the CDF of Z_n by F_n. In Esseen's lemma, use $F = F_n, G = \Phi, b = \frac{1}{4\pi}$, and $T = \frac{1}{4L_n}$. Then, Esseen's lemma gives

$$\sup_{-\infty < x < \infty} |F_n(x) - \Phi(x)| \leq \frac{1}{\pi} \int_{t:|t| \leq \frac{1}{4L_n}} \left|\frac{\phi_n(t) - e^{-\frac{t^2}{2}}}{t}\right| dt + C_1 L_n,$$

on using the simple inequality that the standard normal density $\phi(x) \leq \frac{1}{\sqrt{2\pi}} \; \forall x$.

Now, in the first term on the right, use the pointwise bound for $|\phi_n(t) - e^{-\frac{t^2}{2}}|$ from the above lemma, and integrate. Some algebra then gives the Berry–Esseen theorem with a new universal constant C. □

Remark. The universal constant C may be taken as $C = 0.8$. Fourier analytic proofs give the best available constant; direct proofs have so far not succeeded in producing good constants. The constant C cannot be taken to be smaller than $\frac{3+\sqrt{10}}{6\sqrt{2\pi}} \doteq 0.41$. Better values of the constant C can be found for specific types of the underlying CDF, for example, if it is known that the samples are iid from an exponential distribution. However, no systematic studies in this direction seem to have been done. Also for some specific underlying CDFs F, better rates of convergence in the CLT may be possible. For example, under suitable additional conditions on F, one can have $\Delta_n = \frac{C(F)}{n} + o(n^{-1})$.

The main use of the Berry–Esseen theorem is that it establishes the facts that, in general, the rate of convergence in the central limit theorem for sums is $O\left(\frac{1}{\sqrt{n}}\right)$, and that the accuracy of the CLT approximation is linked to the third moment. It need not, necessarily, give accurate practical bounds on the error of the CLT in specific applications. We take an example below.

Example 8.10. Suppose $X_i, 1 \leq i \leq n$ are iid Bernoulli variables with parameter p. Take n to be 100. We want to investigate if the Berry–Esseen theorem can let us conclude that the error of the CLT approximation is at most $\epsilon = .005$. In the Bernoulli case, $\beta_3 = E|X_i - \mu|^3 = pq(1 - 2pq)$, where $q = 1 - p$. Using $C = 0.8$, the uniform Berry–Esseen bound is

$$\Delta_n \leq \frac{0.8pq(1-2pq)}{(pq)^{3/2}\sqrt{n}}.$$

This is less than or equal to the prescribed $\epsilon = 0.005$ iff $pq > 0.4784$, which does not hold for any $p, 0 < p < 1$. Even for $p = .5$, the Berry–Esseen bound is less than or equal to $\epsilon = 0.005$ only when $n > 25{,}000$, which is a very large sample size.

Various refinements of the Berry–Esseen theorem are available. They replace the third absolute central moment β_3 by an expectation of some more general function. Petrov (1975) and van der Vaart et al. (1996) are good references for more general Berry–Esseen type bounds on the error of the CLT for sums. Here is an important refinement that does not assume the existence of the third moment, or that the variables are iid. The zero mean assumption in the next theorem is not a restriction, because we can make the means zero by writing $X_i - \mu_i$ in place of X_i.

Theorem 8.14. Let $X_1, \ldots, X_n$ be independent with $E(X_i) = 0$, $\text{Var}(X_i) = \sigma_i^2$ and let $B_n = \sum_{i=1}^n \sigma_i^2$. Let $g : \mathcal{R} \to \mathcal{R}_+$ be such that

(a) g is even.
(b) $g(x)$ and $\frac{x}{g(x)}$ are nondecreasing for $x > 0$.
(c) $E(X_i^2 g(X_i)) < \infty$ for each $i = 1, 2, \ldots, n$.

Then,

$$\sup_x \left| P\left(\frac{\bar{X} - E(\bar{X})}{\sqrt{\text{Var}(\bar{X})}} \leq x\right) - \Phi(x) \right| \leq C \frac{\sum_{i=1}^n E(X_i^2 g(X_i))}{g(\sqrt{B_n}) B_n},$$

for some universal constant $C, 0 < C < \infty$.

See Petrov (1975, p. 151) for a proof. An important corollary, obtained by using the function $g(x) = |x|^\delta (0 < \delta \leq 1)$, is the following uniform bound in the iid case without assuming the existence of the third moment.

Corollary. Let $X_1, \ldots, X_n$ be iid with mean μ, variance σ^2, and suppose for some $\delta, 0 < \delta \leq 1$, $E(|X_1|^{2+\delta}) < \infty$. Then,

$$\Delta_n = \sup_x |P\left(\frac{\sqrt{n}(\bar{X} - \mu)}{\sigma} \leq x\right) - \Phi(x)| \leq C \frac{E\left(|X_1 - \mu|^{2+\delta}\right)}{\sigma^{2+\delta} n^{\frac{\delta}{2}}},$$

for some universal constant C.

Bounds on the error of the central limit theorem at a fixed x are called *local bounds*; Petrov (1975), Serfling (1980), and DasGupta (2008) discuss local bounds. The problem of deriving bounds on the error of the central limit theorem in the multidimensional case is much harder. No uniform bounds can be obtained with the class of all possible (Borel) sets. Bounds have been derived for some special classes of sets, such as the class of all closed balls, and the class of closed convex sets. DasGupta (2008) gives a detailed presentation of the results.

8.8 Lindeberg–Feller Theorem for General Independent Case

The central limit theorem generalizes far beyond the iid situation. The general rule is that if we add a large number of independent summands, none of which dominates the rest, then the sum should still behave as does a normally distributed random variable. There are several theorems in this regard, of which the *Lindeberg–Feller theorem* is usually considered to be the last word. A weaker result, which is easier to apply in many problems, is *Lyapounov's theorem*, which we present first.

Theorem 8.15 (Lyapounov's Theorem). *Suppose $\{X_n\}$ is a sequence of independent variables, with $E(X_i) = \mu_i$, $Var(X_i) = \sigma_i^2$. Let $s_n^2 = \sum_{i=1}^n \sigma_i^2$.*
If for some $\delta > 0$,

$$\frac{1}{s_n^{2+\delta}} \sum_{j=1}^n E|X_j - \mu_j|^{2+\delta} \to 0 \text{ as } n \to \infty,$$

then

$$\frac{\sum_{i=1}^n (X_i - \mu_i)}{s_n} \stackrel{\mathcal{L}}{\Rightarrow} N(0, 1).$$

Corollary. *If $s_n \to \infty$ and $|X_j - \mu_j| \leq C < \infty \; \forall j$, then*

$$\frac{\sum_{i=1}^n (X_i - \mu_i)}{s_n} \stackrel{\mathcal{L}}{\Rightarrow} N(0, 1).$$

Sketch of Proof of Lyapounov's Theorem. We explain the idea of the proof when the condition of the theorem holds with $\delta = 1$. Here is the idea, first in nontechnical terms.

Assume without any loss of generality that $\mu_i = 0, \sigma_i^2 = 1 \; \forall i \geq 1$. Denote $\frac{\sum_{i=1}^n X_i}{s_n}$ by Z_n and its characteristic function by $\psi_n(t)$. The idea is to approximate the logarithm of $\psi_n(t)$ and to show that it is approximately equal to $-\frac{t^2}{2}$ for each fixed t. This means that $\psi_n(t)$ itself is approximately equal to $e^{-\frac{t^2}{2}}$, and hence by the continuity theorem for characteristic functions, $Z_n \stackrel{\mathcal{L}}{\Rightarrow} N(0, 1)$.

Let the cumulants of Z_n be denoted by $\kappa_{j,n}$. Note that $\kappa_{1,n} = 0, \kappa_{2,n} = 1$. We approximate the logarithm o f $\psi_n(t)$ by using our previously given Taylor expansion:

$$\log \psi_n(t) \approx \sum_{j=1}^{3} \kappa_{j,n} \frac{(it)^j}{j!}$$

$$= -\frac{t^2}{2} - \frac{i}{6}\kappa_{3,n} t^3.$$

Now, $\kappa_{3,n} = E(Z_n^3) = \frac{1}{s_n^3} E[(\sum_{i=1}^{n} X_i)^3]$. By an expansion of $(\sum_{i=1}^{n} X_i)^3$, and on using the independence and zero mean property of the X_i, and on using the triangular inequality, one gets $|\kappa_{3,n}| \leq \frac{\sum_{i=1}^{n} E(|X_i|^3)}{s_n^3}$, which by the Lyapounov condition goes to zero, when $\delta = 1$. Thus, $\log \psi_n(t) \approx -\frac{t^2}{2}$, which is what we need. □

As remarked earlier, a central limit theorem for independent but not iid summands holds under conditions weaker than the Lyapounov condition. The Lindeberg–Feller theorem is usually considered the best possible result on this, in the sense that the conditions of the theorem are not only sufficient, but also necessary under some natural additional restrictions. However, the conditions of the Lindeberg–Feller theorem involve calculations with the moments of more complicated functions of the summands. Thus, in applications, using the Lindeberg–Feller theorem gives a CLT under weaker conditions than Lyapounov's theorem, but typically at the expense of more cumbersome calculations to verify the Lindeberg–Feller conditions.

Theorem 8.16 (Lindeberg–Feller Theorem). *With the same notation as in Lyapounov's theorem, assume that*

$$\forall \delta > 0, \quad \frac{1}{s_n^2} \sum_{j=1}^{n} E\left(X_j^2 I_{\{|X_j| > \delta s_n\}}\right) \to 0.$$

Then,

$$\frac{\sum_{i=1}^{n}(X_i - \mu_i)}{s_n} \stackrel{\mathcal{L}}{\Rightarrow} N(0, 1).$$

Proof. We give a characteristic function proof. Merely for notational simplicity, we assume that each X_i has a density f_i. This assumption is not necessary, and the proof goes through verbatim in general, with integrals replaced by the expectation notation.

Denote the cf of X_i by ψ_i, and without any loss of generality, we assume that $\mu_i = 0 \; \forall i$, so that we have to show that

8.8 Lindeberg–Feller Theorem for General Independent Case

$$\frac{\sum_{i=1}^{n} X_i}{s_n} \stackrel{\mathcal{L}}{\Rightarrow} N(0,1)$$

$$\Leftrightarrow \prod_{k=1}^{n} \psi_k\left(\frac{t}{s_n}\right) \to e^{-t^2/2}.$$

The last statement is equivalent to

$$\sum_{k=1}^{n}\left[\psi_k\left(\frac{t}{s_n}\right) - 1\right] \to -\frac{t^2}{2}$$

$$\Leftrightarrow \sum_{k=1}^{n}\left[\psi_k\left(\frac{t}{s_n}\right) - 1\right] + \frac{t^2}{2} \to 0,$$

by using a two-term Taylor expansion for the logarithm of each ψ_k (see the section on characteristic functions).

Now, because each X_i has mean zero, the quantity on the left in this latest expression can be rewritten as

$$\sum_{k=1}^{n}\int_{-\infty}^{\infty}\left[e^{i\frac{t}{s_n}x} - 1 - i\frac{t}{s_n}x + \frac{t^2}{2s_n^2}x^2\right]f_k(x)dx$$

$$= \sum_{k=1}^{n}\int_{|x|\leq \delta s_n}\left[e^{i\frac{t}{s_n}x} - 1 - i\frac{t}{s_n}x + \frac{t^2}{2s_n^2}x^2\right]f_k(x)dx$$

$$+ \sum_{k=1}^{n}\int_{|x|>\delta s_n}\left[e^{i\frac{t}{s_n}x} - 1 - i\frac{t}{s_n}x + \frac{t^2}{2s_n^2}x^2\right]f_k(x)dx$$

We bound each term in this expression. First, we bound the integrand. In the first term, that is when $|x| \leq \delta s_n$, the integrand is bounded above by $|\frac{t}{s_n}x|^3 < \delta|t|^3 x^2/s_n^2$. In the second term, that is when $|x| > \delta s_n$, the integrand is bounded above by $t^2\frac{x^2}{s_n^2}$. Therefore,

$$\sum_{k=1}^{n}\int_{|x|\leq \delta s_n}\left[e^{i\frac{t}{s_n}x} - 1 - i\frac{t}{s_n}x + \frac{t^2}{2s_n^2}x^2\right]f_k(x)dx$$

$$+ \sum_{k=1}^{n}\int_{|x|>\delta s_n}\left[e^{i\frac{t}{s_n}x} - 1 - i\frac{t}{s_n}x + \frac{t^2}{2s_n^2}x^2\right]f_k(x)dx$$

$$\leq \delta|t|^3\frac{\sum_{k=1}^{n}\int_{|x|\leq \delta s_n}x^2 f_k(x)dx}{s_n^2}$$

$$+ \frac{t^2}{s_n^2}\sum_{k=1}^{n}\int_{|x|>\delta s_n}x^2 f_k(x)dx.$$

Hold δ fixed and let $n \to \infty$. Then, the second term $\to 0$ by hypothesis of the Lindeberg–Feller theorem. Now, notice that the expression we started with, namely,

$$\sum_{k=1}^{n} \int_{-\infty}^{\infty} \left[e^{i \frac{t}{s_n} x} - 1 - i \frac{t}{s_n} x + \frac{t^2}{2s_n^2} x^2 \right] f_k(x) dx$$

has no δ in it, and so, now letting $\delta \to 0$, even the first term is handled, and we conclude that

$$\sum_{k=1}^{n} \int_{-\infty}^{\infty} \left[e^{i \frac{t}{s_n} x} - 1 - i \frac{t}{s_n} x + \frac{t^2}{2s_n^2} x^2 \right] f_k(x) dx \to 0,$$

which is what we needed. □

Example 8.11 (Linear Combination of IID Variables). Let $U_i, i \geq 1$ be iid variables, and assume that U_i has mean μ and variance σ^2. Without loss of generality, we may assume that $\mu = 0, \sigma = 1$. Let also $c_n, n \geq 1$ be a sequence of constants, such that it satisfies the *uniform asymptotic negligibility condition*

$$r_n = \frac{\max\{|c_1|, \ldots, |c_n|\}}{\sqrt{\sum_{i=1}^{n} c_i^2}} \to 0, \quad \text{as } n \to \infty.$$

We want to show that $\frac{\sum_{i=1}^{n} c_i U_i}{\sqrt{\sum_{i=1}^{n} c_i^2}} \stackrel{\mathcal{L}}{\Rightarrow} N(0, 1)$. We do this by verifying the Lindeberg–Feller condition.

Denote $X_i = c_i U_i, s_n = \sqrt{\sum_{i=1}^{n} c_i^2}$. We need to show that for any $\delta > 0$,

$$\frac{1}{s_n^2} \sum_{j=1}^{n} E\left(X_j^2 I_{\{|X_j| > \delta s_n\}}\right) \to 0.$$

Fix j, then,

$$E\left(X_j^2 I_{\{|X_j| > \delta s_n\}}\right) = c_j^2 E\left(U_j^2 I_{\{|U_j| > \delta s_n/|c_j|\}}\right)$$
$$\leq c_j^2 E\left(U_j^2 I_{\{|U_j| > \delta s_n/\max\{|c_1|,\ldots,|c_n|\}\}}\right).$$

By the uniform asymptotic negligibility condition, $s_n/\max\{|c_1|, \ldots, |c_n|\} \to \infty$, and therefore $I_{\{|U_1| > \delta s_n/\max\{|c_1|,\ldots,|c_n|\}\}}$ goes almost surely to zero. Furthermore, $U_1^2 I_{\{|U_1| > \delta s_n/\max\{|c_1|,\ldots,|c_n|\}\}}$ is bounded above by U_1^2, and we have assumed that $E U_1^2 < \infty$. This implies by the dominated convergence theorem that

$$E\left(U_1^2 I_{\{|U_1| > \delta s_n/\max\{|c_1|,\ldots,|c_n|\}\}}\right) \to 0.$$

But this implies $\frac{1}{s_n^2}\sum_{j=1}^n E(X_j^2 I_{\{|X_j|>\delta s_n\}})$ goes to zero too, because $\frac{c_j^2}{s_n^2}\leq 1$. This completes the verification of the Lindeberg–Feller condition.

Example 8.12 (Failure of Lindeberg–Feller Condition). It is possible for standardized sums of independent variables to converge in distribution to $N(0, 1)$ without the Lindeberg–Feller condition being satisfied. Basically, one of the variables has to have an undue influence on the sum. To keep the distributional calculation simple, take the X_i to be independent normal with mean 0 and variance rapidly increasing so that the nth variable dominates the sum of the first n, for example, with $\text{Var}(X_i) = 2^i$. Then, for each n, $\frac{\sum_{i=1}^n X_i}{s_n} \sim N(0, 1)$, and so it is $N(0, 1)$ in the limit. However, the Lindeberg–Feller condition fails, which can be shown by an analysis of the standard normal tail CDF, or by invoking the first part of the following theorem.

Theorem 8.17 (Necessity of Lindeberg–Feller Condition). *(a) In order that the Lindeberg–Feller condition holds, it must be the case that $\frac{\max\{\sigma_1^2,\ldots,\sigma_n^2\}}{s_n^2} \to 0$.*
(b) Conversely, if $\frac{\max\{\sigma_1^2,\ldots,\sigma_n^2\}}{s_n^2} \to 0$, then for $\frac{\sum_{i=1}^n (X_i-\mu_i)}{s_n}$ to converge in distribution to $N(0, 1)$, the Lindeberg–Feller condition must hold.

Part (b) is attributed to Feller; see Port (1994, p. 704). It shows that if variances exist, then one cannot do away with the Lindeberg–Feller condition, except in unusual cases where the summands are not uniformly negligible relative to the sum.

8.9 * Infinite Divisibility and Stable Laws

Infinitely divisible distributions were introduced by de Finetti in 1929 and the most fundamental results were developed by Kolmogorov, Lévy, and Khintchine in the thirties. The area has undergone tremendous growth and a massive literature now exists. Stable distributions form a subclass of infinitely divisible distributions and are quite extensively used in modeling heavy tail data in various applied fields.

The origin of infinite divisibility and stable laws was in connection with characterizing possible limit distributions of centered and normed partial sums of independent random variables. If $X_1, .X_2, \ldots$ is an iid sequence with mean μ and a finite variance σ^2, then we know that with $a_n = n\mu$ and $b_n = \sigma\sqrt{n}$, $\frac{S_n-a_n}{b_n} \stackrel{\mathcal{L}}{\Rightarrow} N(0, 1)$. But, if $X_1, .X_2, \ldots$ are iid standard Cauchy, then with $a_n = 0$ and $b_n = n$, $\frac{S_n-a_n}{b_n} \stackrel{\mathcal{L}}{\Rightarrow} C(0, 1)$. It is natural to ask what are all the possible limit laws for suitably centered and normed partial sums of iid random variables. One can remove the iid assumption and keep just independence, and ask the same question. It turns out that stable and infinitely divisible distributions arise as the class of all possible limits in these cases. Precise statements are given in the theorems below. But, first we need to define infinite divisibility and stable laws.

Definition 8.5. A real-valued random variable X with cumulative distribution function F and characteristic function ϕ is said to be infinitely divisible if for each $n > 1$, there exist iid random variables $X_{1n}, \ldots, X_{nn}$ with cdf, say F_n, and characteristic function ϕ_n, such that X has the same distribution as $X_{1n} + \cdots + X_{nn}$, or, equivalently, $\phi = \phi_n^n$.

Example 8.13. Let X be $N(\mu, \sigma^2)$. For any $n > 1$, let $X_{1n}, \ldots, X_{nn}$ be iid $N(\mu/n, \sigma^2/n)$. Then X has the same distribution as $X_{1n} + \cdots + X_{nn}$, and so X is infinitely divisible.

Example 8.14. Let X have a Poisson distribution with mean λ. For a given n, let $X_{1n}, \ldots, X_{nn}$ be iid Poisson variables with mean λ/n. Then X has the same distribution as $X_{1n} + \cdots + X_{nn}$, and so X is infinitely divisible.

Example 8.15. Let X have the continuous $U[0, 1]$ distribution. Then X is not infinitely divisible. For if it is, then for any n, there exist iid random variables $X_{1n}, \ldots, X_{nn}$ with some distribution F_n such that X has the same distribution as $X_{1n} + \cdots + X_{nn}$. Evidently, the supremum of the support of F_n is at most $1/n$. This implies $\mathrm{Var}(X_{1n}) \leq 1/n^2$ and hence $\mathrm{Var}(X) \leq 1/n$, a contradiction.

In fact, a random variable X with a bounded support cannot be infinitely divisible, and the uniform case proof applies in general.

The following important property of the class of infinitely divisible distributions describes the connection of infinite divisibility to possible weak limits of partial sums of *triangular arrays* of independent random variables.

Theorem 8.18. *A random variable X is infinitely divisible if and only if for each n, there is an iid sequence $Z_{1n}, \ldots, Z_{nn}$, such that $\sum_{i=1}^{n} Z_{in} \stackrel{\mathcal{L}}{\Rightarrow} X$.*

See Feller (1971, p. 303) for its proof. The result above allows triangular arrays of independent random variables, with possibly different common distributions H_n for the different rows. An important special case is that of just one iid sequence $X_1, X_2, \ldots$ with a common cdf H. If the partial sums $S_n = \sum_{i=1}^{n} X_i$, possibly after suitable centering and norming, converge in distribution to some random variable Z, then Z belongs to a subclass of the class of infinitely divisible distributions. This class is the so-called stable family. We first give a more direct definition of a stable distribution that better explains the reason for the name stable.

Definition 8.6. A random variable X, or its CDF F, is said to be *stable* if for every $n \geq 1$, there exist constants b_n and a_n such that $S_n = X_1 + X_2 + \cdots + X_n$ and $b_n X_1 + a_n$ have the same distribution, where $X_1, X_2, \ldots, X_n$ are iid with the common distribution F.

It turns out that b_n has to be $n^{1/\alpha}$ for some $0 < \alpha \leq 2$. The constant α is said to be the *index* of the stable distribution F. The case $\alpha = 2$ corresponds to the normal case, and $\alpha = 1$ corresponds to the Cauchy case. Generally, stable distributions are heavy tailed. For instance, the only stable laws with a finite variance are the normal distributions. The following theorem is often useful.

8.10 Some Useful Inequalities

Theorem 8.19. *If X is stable with an index $0<\alpha<2$, then for any $p>0$, $E|X|^p<\infty$ if and only if $0<p<\alpha$.*

Thus, stable laws with an index $\alpha \leq 1$ cannot even have a finite mean; see Feller *(1971, pp. 578–579) for a proof of the above theorem.*

Stable distributions are necessarily absolutely continuous, and therefore have densities. However, except for $\alpha = \frac{1}{2}, 1, \frac{3}{2}$, and 2, it is not possible to write a stable density analytically using elementary functions. See Chapter 11 and Chapter 18 for examples of situations where the stable law with $\alpha = \frac{1}{2}$ naturally arises. There are various infinite series and integral representations for them. The general stable distribution is parametrized by four parameters; a location parameter μ, a scale parameter σ, a skewness parameter β, and the index parameter, which we have called α. For instance, for the standard normal distribution, which is stable, $\mu = 0$, $\sigma = 1, \beta = 0$, and $\alpha = 2$. By varying μ, σ, β, and α, one can fit a lot of heavy-tailed data by using a stable law. However, fitting the four parameters from observed data is a very hard statistical problem. Feller (1971) is an excellent reference for more details on stable and infinitely divisible distributions. Infinite divisibility and stable laws are also treated in more detail in DasGupta (2008).

Here is the result connecting stable laws to limits of partial sums of iid random variables; see Feller (1971, p. 172).

Theorem 8.20. *Let $X_1, X_2, \ldots$ be iid with the common cdf H. Suppose for constant sequences $\{a_n\}, \{b_n\}, \frac{S_n - a_n}{b_n} \stackrel{\mathcal{L}}{\Rightarrow} Z \sim F$. Then F is stable.*

8.10 ∗ Some Useful Inequalities

References to the inequalities below are given in DasGupta (2008), Chapter 34.

Bikelis Nonuniform Inequality. Given independent random variables $X_1, \ldots, X_n$ with mean zero, and finite third absolute moments, and $F_n(x) = P(\sqrt{n}\bar{X} \leq x)$,

$$|F_n(x) - \Phi(x)| \leq A \frac{\sum_{i=1}^{n} E|X_i|^3}{B_n^3} \frac{1}{(1+|x|)^3},$$

for all real x, where $B_n^2 = \sum_{i=1}^{n} \text{Var}(X_i)$.

Reverse Berry–Esseen Inequality of Hall and Barbour. Given independent random variables $X_1, \ldots, X_n$ with mean zero, variances σ_i^2 scaled so that $\sum_{i=1}^{n} \sigma_i^2 = 1, \delta_n = \sum_{i=1}^{n} E[(X_i^3 + X_i^4)I_{|X_i|\leq 1}] + \sum_{i=1}^{n} E[X_i^2 I_{|X_i|>1}]$,

$$\sup_{-\infty<x<\infty}|F_n(x) - \Phi(x)| \geq \frac{1}{392}\left(\delta_n - 121\sum_{i=1}^{n}\sigma_i^4\right).$$

Kolmogorov's Maximal Inequality. Given independent random variables $X_1, \ldots, X_n$ with mean zero, variances σ_i^2, $S_k = \sum_{i=1}^{k} X_i$, and a positive number λ,

$$P\left(\max_{1 \leq k \leq n} |S_k| \geq \lambda\right) \leq \frac{\sum_{i=1}^{n} \sigma_i^2}{\lambda^2}.$$

Hoeffding's Inequality. Given independent random variables $X_1, \ldots, X_n$ with mean zero such that $a_i \leq X_i \leq b_i, 1 = 1, 2, \ldots, n$, $S_n = \sum_{i=1}^{n} X_i$, and a positive number t,

$$P(S_n \geq nt) \leq e^{-\frac{2n^2 t^2}{\sum_{i=1}^{n}(b_i - a_i)^2}}.$$

$$P(S_n \leq -nt) \leq e^{-\frac{2n^2 t^2}{\sum_{i=1}^{n}(b_i - a_i)^2}}.$$

$$P(|S_n| \geq nt) \leq 2e^{-\frac{2n^2 t^2}{\sum_{i=1}^{n}(b_i - a_i)^2}}.$$

Partial Sums Moment Inequality. Given n random variables $X_1, \ldots, X_n$, $p > 1$,

$$E\left|\sum_{i=1}^{n} X_i\right|^p \leq n^{p-1} \sum_{k=1}^{n} E|X_k|^p.$$

Given n independent random variables $X_1, \ldots, X_n$, $p \geq 2$,

$$E\left|\sum_{i=1}^{n} X_i\right|^p \leq c(p) n^{p/2 - 1} \sum_{k=1}^{n} E|X_k|^p,$$

for some finite universal constant $c(p)$.

von Bahr–Esseen Inequality. Given independent random variables $X_1, \ldots, X_n$ with mean zero, $1 \leq p \leq 2$,

$$E\left|\sum_{i=1}^{n} X_i\right|^p \leq \left(2 - \frac{1}{n}\right) \sum_{k=1}^{n} E|X_k|^p.$$

Rosenthal Inequality. Given independent random variables $X_1, \ldots, X_n$, $p \geq 1$, $E(X_i) = 0$, $E(|X_i|^p) < \infty \; \forall i$, there exists a finite constant $C(p)$ independent of n such that

$$E\left[\left|\sum_{i=1}^{n} X_i\right|^p\right] \leq C(p) E\left[\left(\sum_{i=1}^{n} X_i^2\right)^{\frac{p}{2}}\right].$$

Hence, if $p \geq 2$, and X_i are iid, $E(X_i) = 0, E(|X_i|^p) < \infty$, then

$$E\left[\left|\sum_{i=1}^{n} X_i\right|^p\right] \leq C(p) n^{\frac{p}{2}} E\left(|X_1|^p\right).$$

Rosenthal Inequality II. Given independent random variables $X_1, \ldots, X_n$, $p > 1$,

$$E\left(\left|\sum_{i=1}^n X_i\right|^p\right) \leq 2^{p^2} \max\left\{\sum_{k=1}^n E|X_k|^p, \left(\sum_{k=1}^n E|X_k|\right)^p\right\}.$$

Given independent random variables $X_1, \ldots, X_n$, symmetric about zero, $2 < p < 4$, $E(|X_i|^p) < \infty \, \forall i$,

$$\left(E\left[\left|\sum_{i=1}^n X_i\right|^p\right]\right)^{1/p} \leq \left(1 + \frac{2^{p/2}}{\sqrt{\pi}} \Gamma\left(\frac{p+1}{2}\right)\right)^{1/p}$$

$$\cdot \max\left\{\left(E\left[\left|\sum_{i=1}^n X_i\right|^2\right]\right)^{1/2}, \left(\sum_{i=1}^n E|X_i|^p\right)^{1/p}\right\}.$$

Exercises

Exercise 8.1 (Symmetrization of $U[0,1]$). Let X, Y be iid $U[0,1]$. Calculate the characteristic function of $X - Y$. Is $X - Y$ uniformly distributed on $[-1, 1]$?

Exercise 8.2 (Limit of Characteristic Functions). Let $X_n \sim U[-n, n], n \geq 1$, and let ψ_n be the cf of X_n. Show that there is a function ψ such that $\psi_n(t) \to \psi(t)$, but that ψ is not a cf.

Exercise 8.3 (Limit of Characteristic Functions). Let $X_n \sim N(n, n^2), n \geq 1$, and let ψ_n be the cf of X_n.

(a) Does ψ_n have a pointwise limit?
(b) Does X_n converge in distribution?
(c) Does Helly's theorem apply to the sequence $\{X_n\}$?
(d) Is there any sequence c_n such that $c_n X_n$ converges in distribution? If so, identify it and also identify the limit in distribution of $c_n X_n$.

Exercise 8.4. Calculate the cf of the *standard logistic distribution* with the CDF $F(x) = \frac{1}{1+e^{-x}}, -\infty < x < \infty$.

Exercise 8.5 * (Normal–Cauchy Convolution). Show that for appropriate constants, $ae^{-b(|t|+c)^2}$ is a characteristic function. Explicitly describe what a, b, c can be.

Exercise 8.6 * (Decimal Expansion). Let $X_k, k \geq 1$ be iid discrete uniform on $\{0, 1, \ldots, 9\}$.

(a) Calculate the cf of $\sum_{k=1}^n \frac{X_k}{10^k}$.
(b) Find its limit as $n \to \infty$.
(c) Recognize the limit and make an interesting conclusion.

Exercise 8.7. For each of the $Bin(n, p)$, $Geo(p)$, and $Poi(\lambda)$, use the cf formula given in text and show that there exists $t \neq 0$ at which $|\psi(t)| = 1$.

Remark. This is true of any lattice distribution.

Exercise 8.8 * **(Random Sums).** Suppose $X_i, i \geq 1$ are iid, and $N \sim Poi(\lambda)$ is independent of $\{X_1, X_2, \cdots\}$. Let $S_n = X_1 + \ldots + X_n, n \geq 1, S_0 = 0$. Derive a formula for the cf of S_N.

Exercise 8.9 * **(Characteristic Function of Products).**

(a) Write a general expression for the cf of XY if X, Y are independent.
(b) Find the cf of the product of two independent standard normal variable s.

Exercise 8.10 * **(Sum of Dice Rolls).** Let $X_i, 1 \leq i \leq n$ be iid integer-valued random variables, with cf $\psi(t)$. Let $S_n = X_1 + \cdots + X_n, n \geq 1$. By using the inversion formula for integer-valued random variables, derive an expression for $P(S_n = k)$.

Hence, derive the pmf of the sum of n independent rolls of a fair die.

Exercise 8.11 * **(A Characterization of Normal Distribution).** Let X, Y be iid random variables. Give a characteristic function proof that if $X + Y, X - Y$ are independent, then X, Y must be normal.

Exercise 8.12 (Characteristic Function of Compact Support). Consider the example $\psi(t) = (1 - |t|)I_{\{|t| \leq 1\}}$ given in the text. Find the density function corresponding to this cf.

Exercise 8.13 (Practice with the Inversion Formula). For each of the following cfs, find the corresponding density function:

$$(\cosh t)^{-1}; (\cosh t)^{-2}; t(\sinh t)^{-1}; (-1)^n H_{2n}(t)e^{-\frac{t^2}{2}},$$

where $H_r(x)$ is the rth Hermite polynomial.

Exercise 8.14. Suppose X, Y are independent, and that $X + Y \stackrel{\mathcal{L}}{=} X$. Show that $Y = 0$ with probability one.

Exercise 8.15 ($e^{-|t|^4}$ **Is Not a Characteristic Function).**

(a) Suppose a cf ψ is twice differentiable at zero, and $\psi''(0) = 0$. Show that the corresponding random variable equals zero with probability one.
(b) Prove that for any $\alpha > 2$, $e^{-|t|^\alpha}$ cannot be a characteristic function.

Exercise 8.16 * **(A Remarkable Fact).** Let $X_i, 1 \leq i \leq 4$ be iid standard normal. Prove that $X_1 X_2 - X_3 X_4$ has a standard double exponential distribution.

Exercise 8.17 * **(Convergence in Total Variation and Characteristic Functions).** Let $X_n, n \geq 1, X$ have densities f_n, f, and cfs ψ_n, ψ. Suppose $|\psi_n|$ and $|\psi|$ are all integrable. Show that if $\int_{-\infty}^{\infty} |\psi_n(t) - \psi(t)| dt \to 0$, then X_n converges to X in total variation.

Exercise 8.18. Suppose $X_i \stackrel{\text{indep.}}{\sim} U[-a_i, a_i], i \geq 1, a_i \leq a < \infty \; \forall i$.

(a) Give a condition on $\{a_i\}$ such that the Lindeberg–Feller condition holds.
(b) Give $\{a_i\}$ such that the Lindeberg–Feller condition does not hold.
(c) * Prove that the Lindeberg–Feller condition holds if and only if $\sum_{i=1}^{\infty} a_i^2 = \infty$.

Exercise 8.19. Let $X_i \stackrel{\text{indep}}{\sim} \text{Poi}(\lambda_i), i \geq 1$. Find a sufficient condition on $\{\lambda_i\}$ so that the Lindeberg–Feller condition holds. Next, find a sufficient condition for Lyapounov's condition to hold with $\delta = 1$.

Exercise 8.20. (Lindeberg–Feller for Independent Bernoullis). Let $X_{ni} \stackrel{\text{indep.}}{\sim}$ $\text{Bin}(1, t_{ni}), 1 \leq i \leq n$, and suppose $\sum_{i=1}^{n} t_{ni}(1 - t_{ni}) \to \infty$. Show that

$$\frac{\sum_{i=1}^{n} X_{ni} - \sum_{i=1}^{n} t_{ni}}{\sqrt{\sum_{i=1}^{n} t_{ni}(1 - t_{ni})}} \stackrel{\mathcal{L}}{\Rightarrow} N(0, 1).$$

Exercise 8.21 (Lindeberg–Feller for Independent Exponentials). Let $X_i \stackrel{\text{indep}}{\sim}$ $\text{Exp}(\lambda_i), i \geq 1$, and suppose $\frac{\max_{1 \leq i \leq n} \lambda_i}{\sum_{i=1}^{n} \lambda_i} \to 0$. Show that, on standardization to zero mean and unit variance, $\bar{X}$ converges in distribution to $N(0, 1)$.

Exercise 8.22 (Poisson Limit in Bernoulli Case). Let $X_{ni} \stackrel{\text{indep.}}{\sim} \text{Bin}(1, t_{ni}), 1 \leq i \leq n$, and suppose $\sum_{i=1}^{n} t_{ni} \to 0 < \lambda < \infty$. Where does $\sum_{i=1}^{n} X_{ni}$ converge in law?
Hint: Look at characteristic functions.

Exercise 8.23. Verify, for which of the following cases, the Lindeberg–Feller condition holds:

(a) $P(X_n = \pm \frac{1}{n}) = \frac{1}{2}$.
(b) $P(X_n = \pm n) = \frac{1}{2}$.
(c) $P(X_n = \pm 2^{-n}) = \frac{1}{2}$.
(d) $P(X_n = \pm 2^{-n}) = 2^{-n-1} \cdot P(X_n = \pm 1) = \frac{1}{2} - 2^{-n-1}$.

Exercise 8.24. Let $X \sim \text{Bin}(n, p)$. Use the Berry–Esseen theorem to bound $P(X \leq k)$ in terms of the standard normal CDF Φ uniformly in k. Is it possible to give any nontrivial bounds that are uniform in both k, p?

Exercise 8.25. Suppose $X_1, \ldots, X_{20}$ are the scores in 20 independent rolls of a fair die. Obtain an upper and a lower bound on $P(X_1 + X_2 + \cdots + X_{20} \leq 75)$ by using the Berry–Esseen inequality.

Exercise 8.26 * (**Hall–Barbour Reverse Bound in Binomial Case**). By using the Hall–Barbour inequality given in the text, find a lower bound on the maximum error of the CLT in the iid Bernoulli case. Plot this bound as a function of p for some selected values of n.

Exercise 8.27 * (**Bikelis Local Bound**). For which values of p, does the Bikelis local bound imply that $\int_{-\infty}^{\infty} |F_n(x) - \Phi(x)|^p dx \to 0$ as $n \to \infty$?

Exercise 8.28 (**Using the Generalized Berry–Esseen**). Suppose $X_i \stackrel{indep}{\sim}$ Poisson $(i), i \geq 1$.

(a) Does $\bar{X}$, on standardization, converge to $N(0, 1)$?
(b) Obtain an explicit upper bound on the error of the CLT by a suitable choice of the function g in the generalized Berry–Esseen inequality given in text.

Exercise 8.29. Point out a defect of the Bikelis local bound.

Exercise 8.30 (**Applying the Rosenthal Inequalities**). Suppose $X_1, X_2, \ldots, X_n$ are iid $U[-1, 1]$. Derive bounds on the expected value of $|\sum_{i=1}^{n} X_i|^3$ by using both Rosenthal inequalities given in text.

References

Berry, A. (1941). The accuracy of the Gaussian approximation to the sum of independent variates, *Trans. Amer. Math. Soc.*, 49, 122–136.
Bhattacharya, R.N., and Rao, R.R. (1986). *Normal Approximation and Asymptotic Expansions*, John Wiley, New York.
DasGupta, A. (2008). *Asymptotic Theory of Statistics and Probability*, Springer, New York.
Esseen, C. (1945). Fourier analysis of distribution functions: A mathematical study, *Acta Math.*, 77, 1–125.
Feller, W. (1971). *An Introduction to Probability Theory and Its Applications*, Vol II, Wiley, New York.
Hall, P. (1992). *The Bootstrap and the Edgeworth Expansion*, Springer, New York.
Petrov, V. (1975). *Limit Theorems of Probability Theory*, Oxford University Press, London.
Port, S. (1994). *Theoretical Probability for Applications*, Wiley, New York.
Serfling, R. (1980). *Approximation Theorems of Mathematical Statistics*, Wiley, New York.
van der Vaart, Aad and Wellner, J. (1996). *Weak Convergence and Empirical Processes: With Applications to Statistics*, Springer-Verlag, New York.

Chapter 9
Asymptotics of Extremes and Order Statistics

We discussed the importance of order statistics and sample percentiles in detail in Chapter 6. The exact distribution theory of one or several order statistics was presented there. Although closed-form in principle, the expressions are complicated, except in some special cases, such as the uniform and the exponential. However, once again it turns out that just like sample means, order statistics and sample percentiles also have a very structured asymptotic distribution theory. We present a selection of the fundamental results on the asymptotic theory for order statistics and sample percentiles, including the sample extremes. Principal references for this chapter are David (1980), Galambos (1987), Serfling (1980), Reiss (1989), de Haan (2006), and DasGupta (2008); other references are given in the sections. First, we recall some notation for convenience.

Suppose $X_i, i \geq 1$, are iid random variables with CDF F. We denote the order statistics of $X_1, X_2, \ldots, X_n$ by $X_{1:n}, X_{2:n}, \ldots, X_{n:n}$, or sometimes, simply as $X_{(1)}, X_{(2)}, \ldots, X_{(n)}$. The empirical CDF is denoted as $F_n(x) = \frac{\#\{i : X_i \leq x\}}{n}$ and $F_n^{-1}(p) = \inf\{x : F_n(x) \geq p\}$ denotes the empirical quantile function. The population quantile function is $F^{-1}(p) = \inf\{x : F(x) \geq p\}$. $F^{-1}(p)$ is also denoted as ξ_p.

Consider the kth-order statistic $X_{k:n}$ where $k = k_n$. Three distinguishing cases are:

(a) $\sqrt{n}(\frac{k_n}{n} - p) \to 0$, for some $0 < p < 1$. This is called *the central case*.
(b) $n - k_n \to \infty$ and $\frac{k_n}{n} \to 1$. This is called *the intermediate case*.
(c) $n - k_n = O(1)$. This is called *the extreme case*.

Different asymptotics apply to the three cases; the case of central order statistics is considered first.

9.1 Central-Order Statistics

9.1.1 Single-Order Statistic

Example 9.1 (Uniform Case). As a simple motivational example, consider the case of iid $U[0, 1]$ observations $U_1, U_2, \ldots$. Fix n, and $p, 0 < p < 1$, and assume for convenience that $k = np$ is an integer. The goal of this example is to show why the kth-order statistic of $U_1, U_2, \ldots, U_n$ is asymptotically normal. More precisely, if $q = 1 - p$, and $\sigma^2 = pq$, then we want to show that $\frac{\sqrt{n}(U_{(k)} - p)}{\sigma}$ converges in distribution to the standard normal.

This can be established in a number of ways. For example, this can be proved by using the representation of $U_{(k)}$ as $U_{(k)} = \frac{\sum_{i=1}^{k} X_i}{\sum_{i=1}^{n+1} X_i}$, where the X_i are iid standard exponential (see Chapter 6). However, we find the following heuristic argument more illuminating than using the above representation. We let $F_n(u)$ denote the empirical CDF of $U_1, \ldots, U_n$, and recall that $nF_n(u) \sim \text{Bin}(n, F(u))$, which is $\text{Bin}(n, u)$ in the uniform case.

Denote $Z_k = \frac{\sqrt{n}(U_{(k)} - p)}{\sigma}$. Then,

$$P(Z_k \leq z) = P\left(U_{(k)} \leq p + \frac{\sigma z}{\sqrt{n}}\right)$$

$$= P\left(nF_n\left(p + \frac{\sigma z}{\sqrt{n}}\right) \geq k\right)$$

$$\approx 1 - \Phi\left(\frac{k - np - \sigma z\sqrt{n}}{\sqrt{n\left(p + \frac{\sigma z}{\sqrt{n}}\right)\left(q - \frac{\sigma z}{\sqrt{n}}\right)}}\right)$$

$$\approx 1 - \Phi\left(\frac{-\sigma z\sqrt{n}}{\sqrt{npq}}\right)$$

$$= 1 - \Phi(-z) = \Phi(z).$$

The steps of this argument can be made rigorous and this shows that in the $U[0,1]$ case, $\frac{\sqrt{n}(U_{(k)} - p)}{\sqrt{pq}}$ converges in distribution to $N(0, 1)$. This is an important example, because the case of a general continuous CDF will follow from the uniform case by making the very convenient quantile transformation (see Chapter 1).

If the observations are iid according to some general continuous CDF F, with a density f, then, by the quantile transformation, we can write $X_{(k)} \stackrel{\mathcal{L}}{=} F^{-1}(U_{(k)})$. The transformation $u \to F^{-1}(u)$ is a continuously differentiable function at $u = p$ with the derivative $\frac{1}{f(F^{-1}(p))}$, provided $0 < f(F^{-1}(p)) < \infty$. By the delta theorem, it then follows that $\sqrt{n}(X_{(k)} - F^{-1}(p)) \stackrel{\mathcal{L}}{\Rightarrow} N(0, \frac{pq}{[f(F^{-1}(p))]^2}$. This is an important result. However, the best result is quite a bit stronger. We do not have to assume that $k = k_n$ is exactly equal to np for some fixed p, and we do not require the CDF F to have a density everywhere. In other words, F need not be differentiable at all x; all we need is enough local smoothness at the limiting value of $\frac{k}{n}$. Here is the exact theorem. Note that this theorem applies if we take $k = \lfloor np \rfloor$, the integer part of np for some fixed p.

Theorem 9.1 (Single-Order Statistic). *Let $X_i, i \geq 1$, be iid with CDF F. Let $\sqrt{n}(\frac{k_n}{n} - p) \to 0$ for some $0 < p < 1$, as $n \to \infty$. Denote $\xi_p = F^{-1}(p)$. Then*

(a)
$$\lim_{n\to\infty} P_F(\sqrt{n}(X_{k_n:n} - \xi_p) \leq t) = P\left(N\left(0, \frac{p(1-p)}{(F'(\xi_p-))^2}\right) \leq t\right)$$

for $t < 0$, provided the left derivative $F'(\xi_p-)$ exists and is > 0.

(b)
$$\lim_{n\to\infty} P_F(\sqrt{n}(X_{k_n:n} - \xi_p) \leq t) = P\left(N\left(0, \frac{p(1-p)}{(F'(\xi_p+))^2}\right) \leq t\right)$$

for $t > 0$, provided the right derivative $F'(\xi_p+)$ exists and is > 0.

(c) If $F'(\xi_p-) = F'(\xi_p+) \stackrel{say}{=} f(\xi_p) > 0$, then

$$\sqrt{n}(X_{k_n:n} - \xi_p) \stackrel{\mathcal{L}}{\Rightarrow} N\left(0, \frac{p(1-p)}{f^2(\xi_p)}\right).$$

Remark. Part (c) of the theorem is the most ubiquitous version and most used in applications. The same results hold with $X_{k_n:n}$ replaced by $F_n^{-1}(p)$.

9.1.2 Two Statistical Applications

Asymptotic distributions of central order statistics are useful in statistics in various ways. One common use is to consider estimators of parameters based on central order statistics and compare their performance with the performance of a more traditional estimator. Usually, such comparisons would have to be done asymptotically, because we cannot do the fixed sample size calculations in closed-form. Another use is to write confidence intervals for a percentile of a distribution (i.e., a population), when only minimal assumptions have been made about the nature of the distribution. We give an illustrative example of each kind below.

Example 9.2 (Asymptotic Relative Efficiency). Suppose $X_1, X_2, \ldots$ are iid $N(\mu, 1)$. Let $M_n = X_{\lfloor \frac{n}{2} \rfloor:n}$ denote the sample median. Because the standard normal density $f(0)$ at zero equals $\frac{1}{\sqrt{2\pi}}$, it follows from the above theorem that $\sqrt{n}(M_n - \mu) \stackrel{\mathcal{L}}{\Rightarrow} N(0, \frac{\pi}{2})$. On the other hand, $\sqrt{n}(\bar{X} - \mu) \stackrel{\mathcal{L}}{\Rightarrow} N(0, 1)$. The ratio of the variances in the two asymptotic distributions, namely, $1/\frac{\pi}{2} = \frac{2}{\pi}$ is called the *ARE* (asymptotic relative efficiency) of M_n relative to $\bar{X}$. The idea is that asymptotically the sample median has a larger variance than the sample mean by a factor of $\frac{\pi}{2}$, and therefore is only $\frac{2}{\pi}$ as efficient as the sample mean in the normal case. For other distributions symmetric about some μ (e.g., $C(\mu, 1)$), the preference between the mean and the median as an estimate of the point of symmetry μ can reverse.

Example 9.3. This is an example where the CDF F possesses left and right derivatives at the median, but they are unequal. Suppose

$$F(x) = \begin{cases} x, & \text{for } 0 \leq x \leq \frac{1}{2}, \\ 2x - \frac{1}{2}, & \text{for } \frac{1}{2} \leq x \leq \frac{3}{4}. \end{cases}$$

By the previous theorem, $P_F(\sqrt{n}(X_{\lfloor \frac{n}{2} \rfloor:n} - \frac{1}{2}) \leq t)$ can be approximated by $P(N(0, \frac{1}{4}) \leq t)$ when $t < 0$ and by $P(N(0, \frac{1}{16}) \leq t)$ when $t > 0$. Separate approximations are necessary because F changes slope at $x = \frac{1}{2}$.

Here is an important statistical application.

Example 9.4 (Confidence Interval for a Quantile). Suppose $X_1, X_2, \ldots, X_n$ are iid observations from some CDF F, and to keep things simple, assume that F has a density f. Suppose we wish to estimate $\xi_p = F^{-1}(p)$ for some fixed $p, 0 < p < 1$. Let $k = k_n = \lfloor np \rfloor$, and let $\hat{\xi}_p = X_{k:n}$. Then, from the above theorem,

$$\sqrt{n}(\hat{\xi}_p - \xi_p) \stackrel{\mathcal{L}}{\Rightarrow} N\left(0, \frac{p(1-p)}{(f(\xi_p))^2}\right).$$

Thus, a confidence interval for ξ_p can be constructed as

$$\hat{\xi}_p \pm \frac{z_{\alpha/2}}{\sqrt{n}} \frac{\sqrt{p(1-p)}}{f(\xi_p)}.$$

The interval has a simplistic appeal and is computed much more easily than an *exact interval* based on order statistics.

9.1.3 Several Order Statistics

Just as a single central order statistic is asymptotically normal under very mild conditions, any fixed number of central order statistics are jointly asymptotically normal, under similar conditions. Furthermore, any two of them are positively correlated, and there is a simple explicit description of the asymptotic covariance matrix. We present that result next; a detailed proof can be found in Serfling (1980) and Reiss (1989).

Theorem 9.2 (Several Order Statistics). *Let $X_{k_i:n}, 1 \leq i \leq r$, be r specified order statistics, and suppose for some $0 < q_1 < q_2 < \cdots < q_r < 1$, $\sqrt{n}(\frac{k_i}{n} - q_i) \to 0$ as $n \to \infty$. Then*

$$\sqrt{n}[(X_{k_1:n}, X_{k_2:n}, \ldots, X_{k_r:n}) - (\xi_{q_1}, \xi_{q_2}, \ldots, \xi_{q_r})] \stackrel{\mathcal{L}}{\Rightarrow} N_r(0, \Sigma),$$

where for $i \leq j$,

$$\sigma_{ij} = \frac{q_i(1-q_j)}{F'(\xi_{q_i})F'(\xi_{q_j})},$$

provided $F'(\xi_{q_i})$ exists and is > 0 for each $i = 1, 2, \ldots, r$.

9.1 Central-Order Statistics

Remark. Note that $\sigma_{ij} > 0$ for $i \neq j$ in the above theorem. However, as we proved in Chapter 6, what is true is that if $X_i, 1 \leq i \leq n$, are iid from any CDF F, then provided the covariance exists, $\text{Cov}(X_{i:n}, X_{j:n}) \geq 0$ for any i, j and any $n \geq 2$.

An important consequence of the joint asymptotic normality of a finite number of order statistics is that linear combinations of a finite number of order statistics will be asymptotically univariate normal. A precise statement is as follows.

Corollary. *Let $c_1, c_2, \ldots, c_r$ be r fixed real numbers, and let $\mathbf{c}' = (c_1, \ldots, c_r)$. Under the assumptions of the above theorem,*

$$\sqrt{n}\left(\sum_{i=1}^{r} c_i X_{k_i:n} - \sum_{i=1}^{r} c_i \xi_{q_i}\right) \stackrel{\mathcal{L}}{\Rightarrow} N(0, \mathbf{c}'\Sigma\mathbf{c}).$$

The corollary is a simple consequence of the continuous mapping theorem. Here is an important statistical application.

Example 9.5 (The Interquartile Range). The 25th and the 75th percentiles of a set of sample observations are called the first and the third quartile of the sample. The difference between them gives information about the spread in the distribution from which the sample values are coming. The difference is called the *interquartile range*, and we denote it as IQR. In statistics, suitable multiples of the IQR are often used as measures of spread, and are then compared to traditional measures of spread, such as the sample standard deviation s.

Let $k_1 = \lfloor \frac{n}{4} \rfloor, k_2 = \lfloor \frac{3n}{4} \rfloor$. Then $IQR = X_{k_2:n} - X_{k_1:n}$. It follows on some calculation from the above corollary that

$$\sqrt{n}(IQR - (\xi_{\frac{3}{4}} - \xi_{\frac{1}{4}})) \stackrel{\mathcal{L}}{\Rightarrow} N\left(0, \frac{1}{16}\left[\frac{3}{f^2(\xi_{\frac{3}{4}})} + \frac{3}{f^2(\xi_{\frac{1}{4}})} - \frac{2}{f(\xi_{\frac{1}{4}})f(\xi_{\frac{3}{4}})}\right]\right).$$

Here, the notation f means the derivative of F at the particular point. In most cases, f is simply the density of F.

Specializing to the case when F is the CDF of $N(\mu, \sigma^2)$, on some algebra and computation, for normally distributed iid observations,

$$\sqrt{n}(IQR - 1.35\sigma) \stackrel{\mathcal{L}}{\Rightarrow} N(0, 2.48\sigma^2)$$

$$\Rightarrow \sqrt{n}\left(\frac{IQR}{1.35} - \sigma\right) \stackrel{\mathcal{L}}{\Rightarrow} N\left(0, \frac{2.48}{1.35^2}\sigma^2\right) = N(0, 1.36\sigma^2).$$

On the other hand, $\sqrt{n}(s - \sigma) \stackrel{\mathcal{L}}{\Rightarrow} N(0, .5\sigma^2)$ (we have previously solved this problem in general for any distribution with four finite moments). The ratio of the asymptotic variances, namely, $\frac{.5}{1.36} = .37$ is the *ARE* of the IQR-based estimate relative to s. Thus, for normal data, one is better off using s. For populations with thicker tails, IQR-based estimates can be more efficient. DasGupta and Haff (2006) work out the general asymptotic theory for comparison between estimates based on IQR and s.

9.2 Extremes

The asymptotic theory of sample extremes for iid observations is completely different from that of central order statistics. For one thing, the limiting distributions of extremes are never normal. Sample extremes are becoming increasingly useful in various statistical problems, such as financial modeling, climate studies, multiple testing, and disaster planning. As such, the asymptotic theory of extremes is gaining in importance. General references for this section are Galambos (1987), Reiss (1989), Sen and Singer (1993), and DasGupta (2008).

We start with a familiar easy example that illustrates the different kind of asymptotics that extremes have, compared to central order statistics.

Example 9.6. Let $U_1, \ldots, U_n \stackrel{iid}{\sim} U[0,1]$. Then

$$P(n(1 - U_{n:n}) > t) = P\left(1 - U_{n:n} > \frac{t}{n}\right)$$
$$= P\left(U_{n:n} < 1 - \frac{t}{n}\right) = \left(1 - \frac{t}{n}\right)^n, \quad \text{if } 0 < t < n.$$

So $P(n(1 - U_{n:n}) > t) \to e^{-t}$ for all real t, which implies that $n(1 - U_{n:n}) \stackrel{\mathcal{L}}{\Rightarrow}$ Exp(1). Notice two key things: the limit is nonnormal and the norming constant is n, not $\sqrt{n}$. The norming constant in general depends on the tail of the underlying CDF F.

It turns out that if $X_1, X_2, \ldots$ are iid from some F, then the limit distributions of $X_{n:n}$, if it at all exists, can be only one of three types. Characterizations are available and were obtained rigorously by Gnedenko (1943), although some of his results were previously known to Frechet, Fisher, Tippett, and von Mises.

The usual characterization result, called *the convergence of types theorem*, is somewhat awkward to state and can be difficult to verify. Therefore, we first present more easily verifiable sufficient conditions. We make the assumption until further notified that F is continuous.

9.2.1 Easily Applicable Limit Theorems

We start with a few definitions.

Definition 9.1. A CDF F on an interval with $\xi(F) = \sup\{x : F(x) < 1\} < \infty$ is said to have *terminal contact of order m* if $F^{(j)}(\xi(F)-) = 0$ for $j = 1, \ldots, m$ and $F^{(m+1)}(\xi(F)-) \neq 0$.

Example 9.7. Consider the Beta density $f(x) = (m+1)(1-x)^m, 0 < x < 1$. Then the CDF $F(x) = 1 - (1-x)^{m+1}$. For this distribution, $\xi(F) = 1$, and $F^{(j)}(1) = 0$ for $j = 1, \ldots, m$, whereas $F^{(m+1)}(1) = (m+1)!$. Thus, F has terminal contact of order m.

9.2 Extremes

Definition 9.2. A CDF F with $\xi(F) = \infty$ is said to be of an exponential type if F is absolutely continuous and infinitely differentiable, and if for each fixed

$$j \geq 2, \frac{F^{(j)}(x)}{F^{(j-1)}(x)} \sim (-1)^{j-1}\frac{f(x)}{1-F(x)}, \quad \text{as } x \to \infty,$$

where $\sim$ means that the ratio converges to 1.

Example 9.8. Suppose $F(x) = \Phi(x)$, the $N(0,1)$ CDF. Then

$$F^{(j)}(x) = (-1)^{j-1} H_{j-1}(x)\phi(x),$$

where $H_j(x)$ is the j-th Hermite polynomial and is of degree j (see Chapter 12). Therefore, for every j, $\frac{F^{(j)}(x)}{F^{(j-1)}(x)} \sim (-1)^{j-1}x$. Thus, $F = \Phi$ is a CDF of the exponential type.

Definition 9.3. A CDF F with $\xi(F) = \infty$ is said to be of a polynomial type of order k if $x^k(1-F(x)) \to c$ for some $0 < c < \infty$, as $x \to \infty$.

Example 9.9. All t distributions, including therefore the Cauchy, are of polynomial type. Consider the t-distribution with α degrees of freedom and with median zero. Then, it is easily seen that $x^\alpha(1-F(x))$ has a finite nonzero limit. Hence, a t_α-distribution is of the polynomial type of order α.

We now present our sufficient conditions for weak convergence of the maximum to three different types of limit distributions. The first three theorems below are proved in de Haan (2006, pp. 15–19); also see Sen and Singer (1993). The first result handles cases such as the uniform on a bounded interval.

Theorem 9.3. *Suppose $X_1, X_2, \ldots$ are iid from a CDF with m^{th} order terminal contact at $\xi(F) < \infty$. Then for suitable a_n, b_n,*

$$\frac{X_{n:n} - a_n}{b_n} \stackrel{\mathcal{L}}{\Rightarrow} G,$$

where

$$G(t) = \begin{cases} e^{-(-t)^{m+1}} & t \leq 0 \\ 1 & t > 0. \end{cases}$$

Moreover, a_n can be chosen to be $\xi(F)$ and one can choose

$$b_n = \left\{\frac{(-1)^m(m+1)!}{nF^{(m+1)}(\xi(F)-)}\right\}^{\frac{1}{m+1}}.$$

The second result handles cases such as the t-distribution.

Theorem 9.4. *Suppose $X_1, X_2, \ldots$ are iid from a CDF F of a polynomial type of order k. Then for suitable a_n, b_n,*

$$\frac{X_{n:n} - a_n}{b_n} \stackrel{\mathcal{L}}{\Rightarrow} G,$$

where

$$G(t) = \begin{cases} e^{-t^{-k}} & t \geq 0 \\ 0 & t < 0. \end{cases}$$

Moreover, a_n can be chosen to be 0 and one can choose $b_n = F^{-1}(1 - \frac{1}{n})$. The last result handles in particular the important Normal case.

Theorem 9.5. *Suppose $X_1, X_2, \ldots$ are iid from a CDF F of an exponential type. Then for suitable a_n, b_n,*

$$\frac{X_{n:n} - a_n}{b_n} \stackrel{\mathcal{L}}{\Rightarrow} G,$$

where $G(t) = e^{-e^{-t}}$, $-\infty < t < \infty$.

Remark. The choice of a_n, b_n is discussed later.

Example 9.10. Suppose $X_1, X_2, \ldots$ are iid from a triangular density on $[0, \theta]$. So,

$$F(x) = \begin{cases} \frac{2x^2}{\theta^2} & \text{if } 0 \leq x \leq \frac{\theta}{2} \\ \frac{4x}{\theta} - \frac{2x^2}{\theta^2} - 1 & \text{if } \frac{\theta}{2} \leq x \leq \theta. \end{cases}$$

It follows that $1 - F(\theta) = 0$ and $F^{(1)}(\theta-) = 0$, $F^{(2)}(\theta-) = -\frac{4}{\theta^2} \neq 0$. Thus F has terminal contact of order $(m+1)$ at θ with $m = 1$. It follows that $\frac{\sqrt{2n}(X_{n:n} - \theta)}{\theta} \stackrel{\mathcal{L}}{\Rightarrow} G$, where $G(t) = e^{-t^2}$, $t \leq 0$.

Example 9.11. Suppose $X_1, X_2, \ldots$ are iid standard Cauchy. Then

$$1 - F(x) = \frac{1}{\pi} \int_x^\infty \frac{1}{1+t^2} dt = \frac{1}{2} - \frac{\arctan(x)}{\pi} \sim \frac{1}{\pi x},$$

as $x \to \infty$; that is, $F(x) \sim 1 - \frac{1}{\pi x}$. Therefore, $F^{-1}(1 - \frac{1}{n}) \sim \frac{n}{\pi}$. Hence, it follows that $\frac{\pi X_{n:n}}{n} \stackrel{\mathcal{L}}{\Rightarrow} G$, where $G(t)$ is the CDF of the reciprocal of an *exponential*(1) random variable.

The derivation of the sequences a_n, b_n for the asymptotic distribution of the sample maximum in the all-important normal case is surprisingly involved. The normal case is so special that we present it as a theorem.

Theorem 9.6. *Suppose $X_1, X_2, \ldots$ are iid $N(0, 1)$. Then*

$$\sqrt{2 \log n} \left(X_{n:n} - \sqrt{2 \log n} + \frac{\log \log n + \log 4\pi}{2\sqrt{2 \log n}} \right) \stackrel{\mathcal{L}}{\Rightarrow} G,$$

where $G(t) = e^{-e^{-t}}$, $-\infty < t < \infty$.

9.2 Extremes

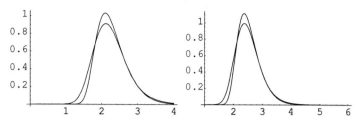

Fig. 9.1 True and asymptotic density of maximum in N(0, 1) case; n = 50, 100

See de Haan (2006, pp. 11–12) or Galambos (1987) for a proof. The distribution with the CDF $e^{-e^{-t}}$ is generally known as the **Gumbel distribution** or the extreme value distribution.

Example 9.12 (Sample Maximum in Normal Case). The density of the Gumbel distribution is $g(t) = e^{-t}e^{-e^{-t}}, -\infty < t < \infty$. This distribution has mean $m = C$ (the Euler constant), and variance $v^2 = \frac{\pi^2}{6}$. The asymptotic distribution gives us a formal approximation for the density of $X_{n:n}$:

$$\hat{f}_n(x) = \sqrt{2\log n}\, g\left(\sqrt{2\log n}\left(x - \sqrt{2\log n} + \frac{\log\log n + \log 4\pi}{2\sqrt{2\log n}}\right)\right).$$

Of course, the true density of $X_{n:n}$ is $n\phi(x)\Phi^{n-1}(x)$. The asymptotic and the true density are plotted in Fig. 9.1 for $n = 50, 100$. The asymptotic density is more peaked at the center, and although it is fairly accurate at the upper tail, it is badly inaccurate at the center and the lower tail. Its lower tail dies too quickly. Hall (1979) shows that the rate of convergence of the asymptotic distribution is extremely slow, in a uniform sense.

Formal approximations to the mean and variance of $X_{n:n}$ are also obtained from the asymptotic distribution. Uniform integrability arguments are required to make these formal approximations rigorous.

$$E(X_{n:n}) \approx \sqrt{2\log n} - \frac{\log\log n}{2\sqrt{2\log n}} + \frac{C - \frac{1}{2}\log 4\pi}{\sqrt{2\log n}},$$

$$\mathrm{Var}(X_{n:n}) \approx \frac{\pi^2}{12\log n}.$$

The moment approximations are not as inaccurate as the density approximation. We evaluated the exact means of $X_{n:n}$ in Chapter 6 for selected values of n. We reproduce those values with the approximate mean given by the above formula for comparison.

n	Exact $E(X_{n:n})$	Approximated Value
10	1.54	1.63
30	2.04	2.11
50	2.25	2.31
100	2.51	2.56
500	3.04	3.07

9.2.2 The Convergence of Types Theorem

We now present the famous *trichotomy result* of asymptotics for the sample maximum in the iid case. Either the sample maximum $X_{n:n}$ cannot be centered and normalized in any way at all to have a nondegenerate limit distribution, or, it can be, in which case the limit distribution must be one of exactly three types. Which type applies to a specific example depends on the support and the upper tail behavior of the underlying CDF. The three possible types of limit distributions are the following.

$$\left. \begin{aligned} G_{1,r}(x) &= e^{-x^{-r}}, & x &> 0 \\ &= 0, & x &\leq 0 \\ G_{2,r}(x) &= e^{-(-x)^r}, & x &\leq 0 \\ &= 1, & x &> 0 \\ G_3(x) &= e^{-e^{-x}}, & -\infty &< x < \infty \end{aligned} \right\},$$

where $r > 0$.

To identify which type applies in a specific case, we need a few definitions that are related to the upper tail behavior of the CDF F.

Definition 9.4. A function g is called slowly varying at ∞ if for each $t > 0$, $\lim_{x \to \infty} \frac{g(tx)}{g(x)} = 1$ and is said to be of regular variation of index r if for each $t > 0$, $\lim_{x \to \infty} \frac{g(tx)}{g(x)} = t^r$.

Definition 9.5. Suppose $X_1, X_2, \ldots$ are iid with CDF F. We say that F is in the *domain of maximal attraction* of a CDF G and write $F \in \mathcal{D}(G)$, if for some a_n, b_n,

$$\frac{X_{n:n} - a_n}{b_n} \stackrel{\mathcal{L}}{\Rightarrow} G.$$

We now state the three main theorems that make up the trichotomy result. See Chapter 1 in de Haan (2006) for a proof of these three theorems.

Theorem 9.7. $F \in \mathcal{D}(G_{1,r})$ iff $\xi(F) = \infty$ and F is of regular variation at $\xi(F) = \infty$. In this case, a_n can be chosen to be zero, and b_n such that $1 - F(b_n) \sim \frac{1}{n}$.

Theorem 9.8. $F \in \mathcal{D}(G_{2,r})$ iff $\xi(F) < \infty$ and $\widetilde{F} \in \mathcal{D}(G_{1,r})$, where $\widetilde{F}(x)$ is the CDF $\widetilde{F}(x) = F(\xi(F) - \frac{1}{x})$, $x > 0$. In this case, a_n may be chosen to be $\xi(F)$ and b_n such that $1 - F(a - b_n) \sim \frac{1}{n}$.

Theorem 9.9. $F \in \mathcal{D}(G_3)$ iff there is a function $u(t) > 0$, such that

$$\lim_{t \to \xi(F)} \frac{1 - F(t + xu(t))}{1 - F(t)} = e^{-x}, \quad \text{for all } x.$$

Remark. In this last case, a_n, b_n can be chosen as $a_n = F^{-1}(1 - \frac{1}{n})$ and $b_n = u(a_n)$. However, the question of choosing such a function $u(t)$ is the most complicated part of the trichotomy phenomenon. We have a limited discussion on it below. A detailed discussion is available in DasGupta (2008). We recall a definition.

Definition 9.6. Given a CDF F, the *mean residual life* is the function $L(t) = E_F(X - t | X > t)$.

One specific important result covering some important special cases with unbounded upper end points for the support of F is the following.

Theorem 9.10. *Suppose $\xi(F) = \infty$. If $L(t)$ is of regular variation at $t = \infty$, and if $\lim_{t \to \infty} \frac{L(t)}{t} = 0$, then $F \in \mathcal{D}(G_3)$.*

Here are two important examples.

Example 9.13. Let $X \sim \text{Exp}(\lambda)$. Then trivially, $L(t) = $ constant. Any constant function is obviously slowly varying and so is of regular variation. Also, obviously, $\lim_{t \to \infty} \frac{L(t)}{t} = 0$. Therefore, for any exponential distribution, the CDF $F \in \mathcal{D}(G_3)$.

Example 9.14. Let $X \sim N(0, 1)$. Then, from Chapter 14,

$$L(t) = \frac{1}{R(t)} - t \sim \frac{1}{t},$$

where $R(t)$ is the Mills ratio. Therefore, $L(t)$ is of regular variation at $t = \infty$, and also, obviously, $\lim_{t \to \infty} \frac{L(t)}{t} = 0$. It follows that $\Phi \in \mathcal{D}(G_3)$, and as a consequence the CDF of any normal distribution also belongs to $\mathcal{D}(G_3)$.

9.3 * Fisher–Tippett Family and Putting it Together

The three different types of distributions that can at all arise as limit distributions of centered and normalized sample extremes can be usefully unified in a single one-parameter family of distributions, called the *Fisher–Tippett distributions*. We do so in this section, together with some additional simplifications.

Suppose for some real sequences a_n, b_n, we have the convergence result that $\frac{X_{(n)}-a_n}{b_n} \stackrel{\mathcal{L}}{\Rightarrow} G$. Then, from the definition of convergence in distribution, we have that at each continuity point x of G,

$$P\left(\frac{X_{(n)}-a_n}{b_n} \leq x\right) \to G(x) \Leftrightarrow P(X_{(n)} \leq a_n + b_n x) \to G(x)$$
$$\Leftrightarrow F^n(a_n + b_n x) \to G(x)$$
$$\Leftrightarrow n \log F(a_n + b_n x) \to \log G(x)$$
$$\Leftrightarrow n \log[1 - (1 - F(a_n + b_n x))] \to \log G(x)$$
$$\Leftrightarrow n[1 - F(a_n + b_n x)] \to -\log G(x)$$
$$\Leftrightarrow \frac{1}{1 - F(a_n + b_n x)} \to \frac{1}{-\log G(x)}.$$

Now, if we consider the case where F is strictly monotone, and let $U(t)$ denote the inverse function of $\frac{1}{1-F(x)}$, then the last convergence assertion above is equivalent to

$$\frac{U(nx) - a_n}{b_n} \to G^{-1}\left(e^{-\frac{1}{x}}\right),$$

$x > 0$.

We can appreciate the role of this function $U(t)$ in determining the limit distribution of the maximum in a given case. Not only that, when this is combined with the Fisher–Tippett result that the possible choices of G are a very precisely defined one-parameter family, the statement takes an even more aesthetically pleasing form. The Fisher–Tippett result and a set of equivalent characterizations for a given member of the Fisher–Tippett family to be the correct limiting distribution in a specific problem are given below. See pp. 6–8 in de Haan (2006) for a proof of the Fisher–Tippett theorem.

Theorem 9.11 (Fisher–Tippett Theorem). *Let $X_1, X_2, \ldots$ be iid with the common CDF F. Suppose for some sequences a_n, b_n, and some CDF G, $\frac{X_{(n)}-a_n}{b_n} \stackrel{\mathcal{L}}{\Rightarrow} G$. Then G must be a location-scale shift of some member of the one parameter family*

$$\{G_\gamma : G_\gamma(x) = e^{-(1+\gamma x)^{-1/\gamma}}, 1 + \gamma x > 0, \gamma \in \mathcal{R}\};$$

in the above, G_0 is defined to be $G_0(x) = e^{-e^{-x}}, -\infty < x < \infty$. See de Haan (2006) for a proof of this theorem.

We can reconcile the Fisher–Tippett theorem with the trichotomy result we have previously described. Indeed,

(a) For $\gamma > 0$, using the particular location-scale shift $G_\gamma(\frac{x-1}{\gamma})$, and denoting $\frac{1}{\gamma}$ as α, we end up with $G_{1,\alpha}$.
(b) For $\gamma = 0$, we directly arrive at the Gumbel law G_3.
(c) For $\gamma < 0$, using the particular location-scale shift $G_\gamma\left(\frac{-(x+1)}{\gamma}\right)$, and denoting $-\frac{1}{\gamma}$ as α, we end up with $G_{2,\alpha}$.

If we combine the Fisher–Tippett theorem with the convergence condition $\frac{U(nx)-a_n}{b_n} \to G^{-1}(e^{-\frac{1}{x}})$, we get the following neat and useful theorem.

Theorem 9.12. *Let $X_1, X_2, \ldots$ be iid with the common CDF F. Then the following are all equivalent.*

(a) *For some real sequences a_n, b_n, and some real γ, $\frac{X_{(n)}-a_n}{b_n} \stackrel{\mathcal{L}}{\Rightarrow} G_\gamma$.*
(b) *$F^n(a_n + b_n x) \to G_\gamma(x)$ for all real x.*
(c) *$\frac{1}{n[1-F(a_n+b_n x)]} \to (1+\gamma x)^{1/\gamma}$ for all real x.*
(d) *For some positive function $a(t)$, and any $x > 0$, $\frac{U(tx)-U(t)}{a(t)} \to \frac{x^\gamma - 1}{\gamma}$ as $t \to \infty$.*

Here, for $\gamma = 0$, $\frac{x^\gamma - 1}{\gamma}$ is defined to be $\log x$. Furthermore, if the condition in this part holds with a specific function $a(t)$, then in part (a), one may choose $a_n = a(n)$, and $b_n = U(n)$.

Exercises

Exercise 9.1 (Sample Maximum). Let $X_i, i \geq 1$ be an iid sequence, and $X_{(n)}$ the maximum of $X_1, \ldots, X_n$. Let $\xi(F) = \sup\{x : F(x) < 1\}$, where F is the common CDF of the X_i. Prove that $X_{(n)} \stackrel{a.s.}{\Rightarrow} \xi(F)$.

Exercise 9.2 *(Records). Let $X_i, i \geq 1$ be an iid sequence with a common density $f(x)$, and let N_n denote the number of records observed up to time n. By using the results in Chapter 6, and the Borel–Cantelli lemma, prove that $N_n \stackrel{a.s.}{\Rightarrow} \infty$.

Exercise 9.3 (Asymptotic Relative Efficiency of Median). For each of the following cases, derive the asymptotic relative efficiency of the sample median with respect to the sample mean:

(a) Double exponential$(\mu, 1)$.
(b) $U[\mu - \sigma, \mu + \sigma]$.
(c) Beta(α, α).

Exercise 9.4 *(Percentiles with Large and Small Variance). Consider the standard normal, standard double exponential, and the standard Cauchy densities. For $0 < p < 1$, find expressions for the variance in the limiting normal distribution of $F_n^{-1}(p)$, and plot them as functions of p. Which percentiles are the most variable, and which the least?

Exercise 9.5 (Interquartile Range). Find the limiting distribution of the interquartile range for sampling from the normal, double exponential, and Cauchy distributions.

Exercise 9.6 (Quartile Ratio). Find the limiting distribution of the quartile ratio defined as $X_{\lfloor \frac{3n}{4} \rfloor:n} / X_{\lfloor \frac{n}{4} \rfloor:n}$ for the exponential, Pareto, and uniform distributions.

Exercise 9.7 * **(Best Linear Combination).** Suppose $X_1, X_2, \ldots$ are iid with density $f(x - \mu)$. For each of the following cases, find the estimate of the form

$$pX_{\lfloor n\alpha_1 \rfloor:n} + pX_{\lfloor n-n\alpha_1 \rfloor:n} + (1 - 2p)X_{\lfloor \frac{n}{2} \rfloor:n}$$

which has the smallest asymptotic variance:

(a) $f(x) = \frac{1}{\sqrt{2\pi}} e^{-\frac{x^2}{2}}$;

(b) $f(x) = \frac{1}{2} e^{-|x|}$;

(c) $f(x) = \frac{1}{\pi(1+x^2)}$.

Exercise 9.8 * **(Poisson Median Oscillates).** Suppose X_i are iid from a Poisson distribution with mean 1. How would the sample median behave asymptotically? Specifically, does it converge in probability to some number? Does it converge in distribution on any centering and norming?

Exercise 9.9 * **(Position of the Mean Among Order Statistics).** Let X_i be iid standard Cauchy. Let N_n be defined as the number of observations among $X_1, \ldots, X_n$ that are less than or equal to $\bar{X}$. Let $p_n = \frac{N_n}{n}$. Show that p_n converges to the $U[0, 1]$ distribution.

Hint: Use the Glivenko–Cantelli theorem.

Exercise 9.10 * **(Position of the Mean Among Order Statistics).** Let X_i be iid standard normal. Let N_n be defined as the number of observations among $X_1, \ldots, X_n$ that are less than or equal to $\bar{X}$. Let $p_n = \frac{N_n}{n}$. Show that on suitable centering and norming, p_n converges to a normal distribution, and find the variance of this limiting normal distribution.

Exercise 9.11 (Domain of Attraction for Sample Maximum). In what domain of (maximal) attraction are the following distributions. (a) $F = t_\alpha, \alpha > 0$; (b) $F = \chi_k^2$; (c) $F = (1 - \epsilon)N(0, 1) + \epsilon C(0, 1)$.

Exercise 9.12 (Maximum of the Absolute Values). Let X_i be iid standard normal, and let $\xi_n = \max_{1 \leq i \leq n} |X_i|, n \geq 1$. Find constants a_n, b_n and a CDF G such that $\frac{\xi_n - a_n}{b_n} \stackrel{\mathcal{L}}{\Rightarrow} G$.

Exercise 9.13 (Limiting Distribution of the Second Largest).

(a) Let X_i be iid $U[0, 1]$ random variables. Find constants a_n, b_n and a CDF G such that $\frac{X_{n-1:n} - a_n}{b_n} \stackrel{\mathcal{L}}{\Rightarrow} G$;

(b) * Let X_i be iid standard Cauchy random variables. Find constants a_n, b_n, and a CDF G such that $\frac{X_{n-1:n} - a_n}{b_n} \stackrel{\mathcal{L}}{\Rightarrow} G$.

Exercise 9.14. Let $X_1, X_2, \ldots$ be iid Exp(1) samples.

(a) Find a sequence a_n such that $X_{1:n} - a_n \stackrel{a.s.}{\Rightarrow} 0$.

(b) * Find a sequence c_n such that $\frac{X_{1:n}}{c_n} \stackrel{a.s.}{\Rightarrow} 1$.

Exercise 9.15 (**Unusual Behavior of Sample Maximum**). Suppose $X_1, X_2, \ldots, X_n$ are iid Bin(N, p), where N, p are considered fixed. Show, by using the Borel–Cantelli lemma that $X_{n:n} = N$ with probability one for all large n.

Remark. However, typically, $E(X_{n:n}) \ll N$, and estimating N is a very difficult statistical problem. The problem is treated in detail in DasGupta and Rubin (2005).

Exercise 9.16 * (**Limiting Distribution of Sample Range**). Let $X_i, i \geq 1$ be iid $N(0, 1)$ and $W_n = X_{n:n} - X_{1:n}$. Identify constants α_n, β_n, and a CDF G such that $\beta_n(W_n - \alpha_n) \stackrel{\mathcal{L}}{\Rightarrow} G$.

Hint: Use without proof that $X_{1:n}, X_{n:n}$ are asymptotically independent and the result in the text on the limiting distribution of $X_{n:n}$ in the $N(0, 1)$ case. Finally, calculate in closed-form the appropriate convolution density.

Exercise 9.17 * (**Limiting Distribution of Sample Range**). Let $X_i, i \geq 1$ be iid $C(0, 1)$ and $W_n = X_{n:n} - X_{1:n}$. Identify constants α_n, β_n, and a CDF G such that $\beta_n(W_n - \alpha_n) \stackrel{\mathcal{L}}{\Rightarrow} G$.

Hint: Use the same hint as in the previous exercise.

Exercise 9.18 * (**Mean and Variance of Maximum**). Suppose $X_1, X_2, \ldots$ are iid Exp(1). Do the mean and the variance of $X_{n:n} - \log n$ converge to those of the Gumbel law G_3?

Hint: They do. Either use direct calculations or use uniform integrability arguments.

Exercise 9.19 (**Uniform Integrability of Order Statistics**). Suppose $X_i, i \geq 1$ are iid from some CDF F and suppose F has an mgf in an interval around zero. Given $k \geq 1$, find a sequence c_n such that $\frac{X_{n-k:n}}{c_n}$ is uniformly integrable.

Hint: Use the moment inequalities in Chapter 6.

Exercise 9.20 * (**Asymptotic Independence of Minimum and Maximum**). Let $X_i, i \geq 1$ be iid random variables with a common CDF F. Find the weakest condition on F such that

$$P(X_{(1)} > x, X_{(n)} \leq y) - P(X_{(1)} > x)P(X_{(n)} \leq y) \to 0$$
$$\text{as } n \to \infty \; \forall x, y, \; x < y.$$

Exercise 9.21. Suppose $X_i, i \geq 1$ are iid $U[0, 1]$. Does $E[n(1 - X_{n:n})]^k$ have a limit for all k?

References

DasGupta, A. (2008). *Asymptotic Theory of Statistics and Probability*, Springer, New York.

DasGupta, A. and Haff, L. (2006). Asymptotic expansions for correlations between different measures of spread, *JSPI*, 136, 2197–2212.

DasGupta, A. and Rubin, H. (2005). Estimation of binomial parameters when N, p are both unknown, *JSPI, Felicitation Volume for Herman Chernoff*, 130, 391–404.

David, H.A. (1980). *Order Statistics*, Wiley, New York.

de Haan, L. (2006). *Extreme Value Theory: An Introduction*, Springer, New York.

Galambos, J. (1987). *Asymptotic Theory of Extreme Order Statistics*, Wiley, New York.

Gnedenko, B.V. (1943). Sur la distribution limité du terme maximum d'une serie aleatoire, *Annals od Math.*, 44, 423–453.

Hall, P. (1979). On the rate of convergence of normal extremes, *J. Appl. Prob.*, 16(2), 433–439.

Reiss, R. (1989). *Approximate Distributions of Order Statistics*, Springer, New York.

Sen, P.K. and Singer, J. (1993). *Large Sample Methods in Statistics*, Chapman and Hall, New York.

Serfling, R. (1980). *Approximation Theorems of Mathematical Statistics*, Wiley, New York.

Chapter 10
Markov Chains and Applications

In many applications, successive observations of a process, say, $X_1, X_2, \ldots$, have an inherent time component associated with them. For example, the X_i could be the state of the weather at a particular location on the ith day, counting from some fixed day. In a simplistic model, the state of the weather could be "dry" or "wet", quantified as, say, 0 and 1. It is hard to believe that in such an example, the sequence $X_1, X_2, \ldots$ could be mutually independent. The question then arises how to model the dependence among the X_i. Probabilists have numerous dependency models. A particular model that has earned a very special status is called a *Markov chain*. In a Markov chain model, we assume that the future, given the entire past and the present state of a process, depends only on the present state. In the weather example, suppose we want to assign a probability that tomorrow, say March 10, will be dry, and suppose that we have available to us the precipitation history for each of January 1 to March 9. The Markov chain model would entail that our probability that March 10 will be dry will depend only on the state of the weather on March 9, even though the entire past precipitation history was available to us. As simple as it sounds, Markov chains are enormously useful in applications, perhaps more than any other specific dependency model. They also are independently relevant to statistical computing in very important ways. The topic has an incredibly rich and well-developed theory, with links to many other topics in probability theory. Familiarity with basic Markov chain terminology and theory is often considered essential for anyone interested in studying statistics and probability. We present an introduction to basic Markov chain theory in this chapter.

Feller (1968), Freedman (1975), and Isaacson and Madsen (1976) are classic references for Markov chains. Modern treatments are available in Bhattacharya and Waymire (2009), Brémaud (1999), Meyn and Tweedie (1993), Norris (1997), Seneta (1981), and Stirzaker (1994). Classic treatment of the problem of gambler's ruin is available in Feller (1968) and Kemperman (1950). Numerous interesting examples at more advanced levels are available in Diaconis (1988); sophisticated applications at an advanced level are also available in Bhattacharya and Waymire (2009).

10.1 Notation and Basic Definitions

Definition 10.1. A sequence of random variables $\{X_n\}, n \geq 0$, is said to be a *Markov chain* if for some countable set $S \subseteq \mathcal{R}$, and any $n \geq 1, x_{n+1}, x_n, \ldots, x_0 \in S$,

$$P(X_{n+1} = x_{n+1} | X_0 = x_0, \ldots, X_n = x_n) = P(X_{n+1} = x_{n+1} | X_n = x_n).$$

The set S is called the *state space of the chain*. If S is a finite set, the chain is called a *finite state Markov chain*. X_0 is called the *initial state*.

Without loss of generality, we can denote the elements of S as $1, 2, \ldots$, although in some examples we may use the original labeling of the states to avoid confusion.

Definition 10.2. The distribution of the initial state X_0 is called the *initial distribution* of the chain. We denote the pmf of the initial distribution as $\lambda_i = P(X_0 = i)$.

Definition 10.3. A Markov chain $\{X_n\}$ is called *homogeneous* or *stationary* if $P(X_{n+1} = y | X_n = x)$ is independent of n for any x, y.

Definition 10.4. Let $\{X_n\}$ be a stationary Markov chain. Then the probabilities $p_{ij} = P(X_{n+1} = j | X_n = i)$ are called the *one-step transition probabilities*, or simply transition probabilities. The matrix $P = ((p_{ij}))$ is called the *transition probability matrix*.

Definition 10.5. Let $\{X_n\}$ be a stationary Markov chain. Then the probabilities $p_{ij}(n) = P(X_{n+m} = j | X_m = i) = P(X_n = j | X_0 = i)$ are called the *n-step transition probabilities*, and the matrix $P^{(n)} = ((p_{ij}(n)))$ is called the *n-step transition probability matrix*.

Remark. If the state space of the chain is finite and has, say t elements, then the transition probability matrix P is a $t \times t$ matrix. Note that $\sum_{j \in S} p_{ij}$ is always one. A matrix with this property is called a *stochastic matrix*.

Definition 10.6. A $t \times t$ square matrix P is called a *stochastic matrix* if for each $i, \sum_{j=1}^{t} p_{ij} = 1$. It is called *doubly stochastic* or *bistochastic* if, in addition, for every $j, \sum_{i=1}^{t} p_{ij} = 1$. Thus, a transition probability matrix is always a stochastic matrix.

10.2 Examples and Various Applications as a Model

Markov chains are widely used as models for discrete time sequences that exhibit local dependence. Part of the reason for this popularity of Markov chains as a model is that a coherent, complete, and elegant theory exists for how a chain evolves. We describe below examples from numerous fields where a Markov chain is either the correct model, or is chosen as a model.

10.2 Examples and Various Applications as a Model

Example 10.1 (Weather Pattern). Suppose that in some particular city, any day is either dry or wet. If it is dry on some day, it remains dry the next day with probability α, and will be wet with the residual probability $1 - \alpha$. On the other hand, if it is wet on some day, it remains wet the next day with probability β, and becomes dry with probability $1 - \beta$. Let $X_0, X_1, \ldots$ be the sequence of states of the weather, with X_0 being the state on the initial day (on which observation starts). Then $\{X_n\}$ is a two-state stationary Markov chain with the transition probability matrix

$$P = \begin{pmatrix} \alpha & 1-\alpha \\ 1-\beta & \beta \end{pmatrix}.$$

Example 10.2 (Voting Preference). Suppose that in a presidential election, voters can vote for either the labor party, or the conservative party, or the Independent party. Someone who has voted for the Labor candidate in this election will vote Labor again with 80% probability, will switch to Conservative with 5% probability, and vote Independent with 15% probability. Someone who has voted for the Conservative candidate in this election will vote Conservative again with 90% probability, switch to Labor with 3% probability, and vote Independent with 7% probability. Someone who has voted for the Independent candidate in this election will vote Independent again with 80% probability, or switch to one of the other parties with 10% probability each. This is a three-state stationary Markov chain with state space $S = \{1, 2, 3\} \equiv \{\text{Labor, Conservative, Independent}\}$ and the transition matrix

$$P = \begin{pmatrix} .8 & .05 & .15 \\ .03 & .9 & .07 \\ .1 & .1 & .8 \end{pmatrix}.$$

Example 10.3 (An Urn Model Example). Two balls, say A, B are initially in urn 1, and two others, say C, D are in urn 2. In each successive trial, one ball is chosen at random from urn 1, and one independently and also at random from urn 2, and these balls switch urns. We let X_n denote the vector of locations of the four balls A, B, C, D, in that order of the balls, after the nth trial. Thus, $X_{10} = 1122$ means that after the tenth trial, A, B are located in urn 1, and C, D in urn 2, and so on. Note that $X_0 = 1122$. Two of the four balls are always in urn 1, and two in urn 2. Thus, the possible number of states is $\binom{4}{2} = 6$. They are $1122, 1212, 1221, 2112, 2121, 2211$. Then $\{X_n\}$ is a six-state stationary Markov chain. What are the transition probabilities?

For notational convenience, denote the above six states as $1, 2, \ldots, 6$ respectively. For the state of the chain to move from state 1 to state 2 in one trial, B, C have to switch their urns. This will happen with probability $.5 \times .5 = .25$. As another example, for the state of the chain to move from state 1 to state 6, all of the four balls must switch their urns. This is not possible. Therefore, this transition probability is zero. Also, note that if the chain is in some state now, it cannot remain in that same state after the next trial. Thus, all diagonal elements in the transition probability matrix must be zero. Indeed, the transition probability matrix is

$$P = \begin{pmatrix} 0 & .25 & .25 & .25 & .25 & 0 \\ .25 & 0 & .25 & .25 & 0 & .25 \\ .25 & .25 & 0 & 0 & .25 & .25 \\ .25 & .25 & 0 & 0 & .25 & .25 \\ .25 & 0 & .25 & .25 & 0 & .25 \\ 0 & .25 & .25 & .25 & .25 & 0 \end{pmatrix}.$$

Notice that there are really three distinct rows in P, each occuring twice. It is easy to argue that that is how it should be in this particular urn experiment. Also note the very interesting fact that in each row, as well as in each column, there are two zeroes, and the nonzero entries obviously add to one. This is an example of a transition probability matrix that is *doubly stochastic*. Markov chains with a doubly stochastic transition probability matrix show a unified long run behavior. By definition, initially the chain is in state 1, and so $P(X_0 = 1) = 1, P(X_0 = i) = 0 \ \forall i \neq 0$. However, after many trials, the state of the chain would be any of the six possible states with essentially an equal probability; that is, $P(X_n = i) \approx \frac{1}{6}$ for each possible state i for large n. This unifying long run behavior of Markov chains with a doubly stochastic transition probability matrix is a significant result of wide applications in Markov chain theory.

Example 10.4 (Urn Model II: Ehrenfest Model). This example has wide applications in the theory of heat transfers. The mathematical model is that we initially have m balls, some in one urn, say urn I, and the rest in another urn, say urn II. At each subsequent time $n = 1, 2, \ldots$, one ball among the m balls is selected at random. If the ball is in urn I, with probability α it is transferred to urn II, and with probability $1 - \alpha$ it continues to stay in urn I. If the ball is in urn II, with probability β it is transferred to urn I, and with probability $1 - \beta$ it continues to stay in urn II.

Let X_0 be the number of balls initially in urn I, and X_n the number of balls in urn I after time n. Then $\{X_n\}$ is a stationary Markov chain with state space $S = \{0, 1, \ldots, m\}$. If there are, say i balls in urn I at a particular time, then at the next instant urn I could lose one ball, gain one ball, or neither lose nor gain any ball. It loses a ball if one of its i balls gets selected for possible transfer, and then the transfer actually happens. So $p_{i,i-1} = \frac{i}{m}\alpha$. Using this simple argument, we get the one-step transition probabilities as

$$p_{i,i-1} = \frac{i}{m}\alpha; \quad p_{i,i+1} = cm - im\beta; \quad p_{ii} = 1 - \frac{i}{m}\alpha - \frac{m-i}{m}\beta,$$

and $p_{ij} = 0$ if $j \neq i-1, i, i+1$.

As a specific example, suppose $m = 7$, and $\alpha = \beta = \frac{1}{2}$. Then the transition matrix on the state space $S = \{0, 1, \ldots, 7\}$ can be worked out by using the formulas given just above, and it is

10.2 Examples and Various Applications as a Model

$$P = \begin{pmatrix} \frac{1}{2} & \frac{1}{2} & 0 & 0 & 0 & 0 & 0 & 0 \\ \frac{1}{14} & \frac{1}{2} & \frac{3}{7} & 0 & 0 & 0 & 0 & 0 \\ 0 & \frac{1}{7} & \frac{1}{2} & \frac{5}{14} & 0 & 0 & 0 & 0 \\ 0 & 0 & \frac{3}{14} & \frac{1}{2} & \frac{2}{7} & 0 & 0 & 0 \\ 0 & 0 & 0 & \frac{2}{7} & \frac{1}{2} & \frac{3}{14} & 0 & 0 \\ 0 & 0 & 0 & 0 & \frac{5}{14} & \frac{1}{2} & \frac{1}{7} & 0 \\ 0 & 0 & 0 & 0 & 0 & \frac{3}{7} & \frac{1}{2} & \frac{1}{14} \\ 0 & 0 & 0 & 0 & 0 & 0 & \frac{1}{2} & \frac{1}{2} \end{pmatrix}.$$

Example 10.5 (Machine Maintenance). Of the machines in a factory, a certain number break down or are identified to be in need of maintenance on a given day. They are sent to a maintenance shop the next morning. The maintenance shop is capable of finishing its maintenance work on some k machines on any given day. We are interested in the sequence $\{X_n\}$, where X_n denotes the number of machines in the maintenance shop on the nth day, $n \geq 0$. We may take $X_0 = 0$.

Let Z_0 machines break down on day zero. Then, $X_1 = Z_0$. Of these, up to k machines can be fixed by the shop on that day, and these are returned. But now, on day 1, some Z_1 machines break down at the factory, so that $X_2 = \max\{X_1 - k, 0\} + Z_1 = \max\{Z_0 - k, 0\} + Z_1$, of which up to k machines can be fixed by the shop on the second day itself and those are returned to the factory. We then have $X_3 = \max\{X_2 - k, 0\} + Z_2$

$$\begin{aligned} &= Z_0 + Z_1 + Z_2 - 2k, && \text{if } Z_0 \geq k, Z_0 + Z_1 \geq 2k; \\ &= Z_2, && \text{if } Z_0 \geq k, Z_0 + Z_1 < 2k; \\ &= Z_1 + Z_2 - k, && \text{if } Z_0 < k, Z_1 \geq k; \\ &= Z_2, && \text{if } Z_0 < k, Z_1 < k, \end{aligned}$$

and so on.

If $Z_i, i \geq 0$ are iid, then $\{X_n\}$ forms a stationary Markov chain. The state space of this chain is $\{0, 1, 2, \ldots\}$. What is the transition probability matrix? For simplicity, take $k = 1$. For example, $P(X_2 = 1 \mid X_1 = 0) = P(Z_1 = 1 \mid Z_0 = 0) = P(Z_1 = 1) = p_1$ (say). On the other hand, as another example, $P(X_2 = 2 \mid X_1 = 4) = 0$. If we denote the common mass function of the Z_i by $P(Z_i = j) = p_j, j \geq 0$, then the transition probability matrix is

$$P = \begin{pmatrix} p_0 & p_1 & p_2 & p_3 & \cdots \\ p_0 & p_1 & p_2 & p_3 & \cdots \\ 0 & p_0 & p_1 & p_2 & \cdots \\ 0 & 0 & p_0 & p_1 & \cdots \\ 0 & 0 & 0 & p_0 & \cdots \\ & & \vdots & & \end{pmatrix}.$$

Example 10.6 (Hopping Mosquito). Suppose a mosquito makes movements between the forehead, the left cheek, and the right cheek of an individual, which we designate as states 1, 2, 3, according to the following rules. If at some time n, the mosquito is sitting on the forehead, then it will definitely move to the left cheek at the next time $n + 1$; if it is sitting on the left cheek, it will stay there, or move to the right cheek with probability .5 each, and if it is on the right cheek, it will stay there, or move to the forehead with probability .5 each.

Then the sequence of locations of the mosquito form a three-state Markov chain with the one-step transition probability matrix

$$P = \begin{pmatrix} 0 & 1 & 0 \\ 0 & .5 & .5 \\ .5 & 0 & .5 \end{pmatrix}.$$

Example 10.7 (An Example from Genetics). Many traits in organisms, for example, humans, are determined by genes. For example, eye color in humans is determined by a pair of genes. Genes can come in various forms or versions, which are called *alleles*. An offspring receives one allele from each parent. A parent contributes one of his or her alleles with equal probability to an offspring, and the parents make their contributions independently. Certain alleles dominate over others. For example, the allele for blue eye color is dominated by the allele for brown eye color. The allele for blue color would be called *recessive*, and the allele for brown eye color would be called *dominant*. If we denote these as b, B respectively, then a person may have the pair of alleles BB, Bb, or bb. They are called the dominant, hybrid, and recessive *genotypes*. We denote them as d, h, r, respectively. Consider now the sequence of genotypes of descendants of an initial individual, and denote the sequence as $\{X_n\}$; for any n, X_n must be one of d, h, r (we may call them 1, 2, 3).

Consider now a person with an unknown genotype (X_0) mating with a known hybrid. Suppose he has genotype d. He will necessarily contribute B to the offspring. Therefore, the offspring can only have genotype d or h, and not r. It will be d if the offspring also gets the B allele from the mother, and it will be h if the offspring gets b from the mother. The chance of each is $\frac{1}{2}$. Therefore, the transition probability $P(X_1 = d \mid X_0 = d) = P(X_1 = h \mid X_0 = d) = \frac{1}{2}$, and $P(X_1 = r \mid X_0 = d) = 0$.

Suppose $X_0 = h$. Then the father contributes B or b with probability $\frac{1}{2}$ each, and so does the mother, who was assumed to be a hybrid. So the probabilities that $X_1 = d, h, r$ are, respectively, $\frac{1}{4}, \frac{1}{2}, \frac{1}{4}$.

If $X_0 = r$, then X_1 can only be h or r, with probability $\frac{1}{2}$ each. So, if we assume this same mating scheme over generations, then $\{X_n\}$ forms a three-state stationary Markov chain with the transition probability matrix

$$P = \begin{pmatrix} .5 & .5 & 0 \\ .25 & .5 & .25 \\ 0 & .5 & .5 \end{pmatrix}.$$

Example 10.8 (Simple Random Walk). Consider a particle starting at the origin at time zero, and making independent movements of one step to the right or one step to the left at each successive time instant $1, 2, \ldots$. Assume that the particle moves to the right at any particular time with probability p, and to the left with probability $q = 1 - p$. The mathematical formulation is that the successive movements are iid random variables $X_1, X_2, \ldots$, with common pmf $P(X_i = 1) = p$, $P(X_i = -1) = q$. The particle's location after the nth step has been taken is denoted as $S_n = X_0 + X_1 + \cdots + X_n = X_1 + \ldots + X_n$, assuming that $X_0 = 0$ with probability one. At each time the particle can move by just one unit, thus $\{S_n\}$ is a stationary Markov chain with state space $S = \mathbb{Z} = \{\ldots, -2, -1, 0, 1, 2, \ldots\}$, and with the transition probabilities

$$\begin{aligned} p_{ij} &= P(S_{n+1} = j \mid S_n = i) \\ &= p \text{ if } j = i + 1; \\ &= q \text{ if } j = i - 1; \\ &= 0 \text{ if } |j - i| > 1, \quad i, j \in \mathbb{Z}. \end{aligned}$$

By virtue of the importance of random walks in theory and applications of probability, this is an important example of a stationary Markov chain. Note that the chain is stationary because the individual steps X_i are iid. This is also an example of a Markov chain with an infinite state space.

10.3 Chapman–Kolmogorov Equation

The Chapman–Kolmogorov equation provides a simple method for obtaining the higher-order transition probabilities of a Markov chain in terms of lower-order transition probabilities. Carried to its most convenient form, the equation describes how to calculate by a simple and explicit method all higher-order transition probabilities in terms of the one-step transition probabilities. Because we always start analyzing a chain with the one-step probabilities, it is evidently very useful to know how to calculate all higher-order transition probabilities using just the knowledge of the one-step transition probabilities.

Theorem 10.1 (Chapman–Kolmogorov Equation). *Let $\{X_n\}$ be a stationary Markov chain with the state space S. Let $n, m \geq 1$. Then,*

$$p_{ij}(m+n) = P(X_{m+n} = j \mid X_0 = i) = \sum_{k \in S} p_{ik}(m) p_{kj}(n).$$

Proof. A verbal proof is actually the most easily understood. In order to get to state j from state i in $m + n$ steps, the chain must go to some state $k \in S$ in m steps, and then travel from that k to the state j in the next n steps. By adding over all possible $k \in S$, we get the Chapman–Kolmogorov equation.

An extremely important corollary is the following result.

Corollary. Let $P^{(n)}$ denote the n-step transition probability matrix. Then, for all $n \geq 2$, $P^{(n)} = P^n$, where P^n denotes the usual nth power of P.

Proof. From the Chapman–Kolmogorov equation, by using the definition of matrix product, for all $m, n \geq 1$, $P^{(m+n)} = P^{(m)} P^{(n)} \Rightarrow P^{(2)} = PP = P^2$. We now finish the proof by induction. Suppose $P^{(n)} = P^n \; \forall n \leq k$. Then, $P^{(k+1)} = P^{(k)} P = P^k P = P^{k+1}$, which finishes the proof. □

A further important consequence is that we can now write an explicit formula for the pmf of the state of the chain at a given time n.

Proposition. Let $\{X_n\}$ be a stationary Markov chain with the state space S, and one-step transition probability matrix P. Fix $n \geq 1$. Then, $\lambda_n(i) = P(X_n = i) = \sum_{k \in S} p_{ki}(n) P(X_0 = k)$. In matrix notation, if $\lambda = (\lambda_1, \lambda_2, \ldots)'$ denotes the vector of the initial probabilities $P(X_0 = k), k = 1, 2, \cdots$, and if λ_n denotes the row vector of probabilities $P(X_n = i), i = 1, 2, \ldots$, then $\lambda_n = \lambda P^n$.

This is an important formula, because it lays out how to explicitly find the distribution of X_n from the initial distribution λ and the one-step transition matrix P.

Example 10.9 (Weather Pattern). Consider once again the weather pattern example with the one-step transition probability matrix

$$P = \begin{pmatrix} \alpha & 1-\alpha \\ 1-\beta & \beta \end{pmatrix}.$$

We let the states be 1, 2 (1 = dry; 2 = wet). We use the Chapman–Kolmogorov equation to answer two questions. First, suppose it is Wednesday today, and it is dry. We want to know what the probability is that Saturday will be dry. In notation, we want to find $p_{11}^{(3)} = P(X_3 = 1 | X_0 = 1)$. In order to get a concrete numerical answer at the end, let us take $\alpha = \beta = .8$ Now, by direct matrix multiplication,

$$P^3 = \begin{pmatrix} .608 & .392 \\ .392 & .608 \end{pmatrix}.$$

Therefore, the probability that Saturday will be dry if Wednesday is dry is $p_{11}^{(3)} = .608$.

Next suppose that we want to know what the probability is that Saturday and Sunday will both be dry if Wednesday is dry. In notation, we now want to find

$$P(X_3 = 1, X_4 = 1 | X_0 = 1)$$
$$= P(X_3 = 1 | X_0 = 1) P(X_4 = 1 | X_3 = 1, X_0 = 1)$$
$$= P(X_3 = 1 | X_0 = 1) P(X_4 = 1 | X_3 = 1) = .608 \times .8$$
$$= .4864.$$

10.3 Chapman–Kolmogorov Equation

Coming now to evaluating the pmf of X_n itself, we know how to calculate it, namely, $P(X_n = i) = \sum_{k \in S} p_{ki}(n) P(X_0 = k)$. Denote $P(\text{The initial day was dry}) = \lambda_1$, $P(\text{The initial day was wet}) = \lambda_2$, $\lambda_1 + \lambda_2 = 1$. Let us evaluate the probabilities that it will be dry one week, two weeks, three weeks, four weeks from the initial day. This requires calculation of, respectively, $P^7, P^{14}, P^{21}, P^{28}$. For example,

$$P^7 = \begin{pmatrix} .513997 & .486003 \\ .486003 & .513997 \end{pmatrix}.$$

Therefore,

$$P(\text{It will be dry one week from the initial day}) = .513997\lambda_1 + .486003\lambda_2$$
$$= .5 + .013997(\lambda_1 - \lambda_2).$$

Similarly, we can compute P^{14} and show that

$$P(\text{It will be dry two weeks from the initial day}) = .500392\lambda_1 + .499608\lambda_2$$
$$= .5 + .000392(\lambda_1 - \lambda_2).$$
$$P(\text{It will be dry three weeks from the initial day}) = .5 + .000011(\lambda_1 - \lambda_2).$$
$$P(\text{It will be dry four weeks from the initial day}) = .5.$$

We see that convergence to .5 has occurred, regardless of λ_1, λ_2. That is, regardless of the initial distribution, eventually you will put 50–50 probability that a day far into the future will be dry or wet. Is this always the case? The answer is no. In this case, convergence to .5 occurred because the one-step transition matrix P has the doubly stochastic characteristic: each row as well as each column of P adds to one. We discuss more about this later.

Example 10.10 (Voting Preferences). Consider the example previously given on voting preferences. Suppose we want to know what the probabilities are that a Labor voter in this election will vote, respectively, Labor, Conservative, or Independent two elections from now. Denoting the states as 1, 2, 3, in notation, we want to find $P(X_2 = i | X_0 = 1), i = 1, 2, 3$. We can answer this by simply computing P^2. Because

$$P = \begin{pmatrix} .80 & .05 & .15 \\ .03 & .90 & .07 \\ .1 & .1 & .8 \end{pmatrix},$$

by direct computation,

$$P^2 = \begin{pmatrix} .66 & .1 & .24 \\ .06 & .82 & .12 \\ .16 & .18 & .66 \end{pmatrix}.$$

Hence, the probabilities that a Labor voter in this election will vote Labor, Conservative, or Independent two elections from now are 66%, 10%, and 24%. We also see from P^2 that a Conservative voter will vote Conservative two elections from now with 82% probability and has a chance of just 6% to switch to Labor, and so on.

Example 10.11 (Hopping Mosquito). Consider again the hopping mosquito example previously introduced. The goal of this example is to find the n-step transition probability matrix P^n for a general n. We describe a general method for finding P^n using a linear algebra technique, known as *diagonalization of a matrix*. If a square matrix P (not necessarily symmetric) of order $t \times t$ has t distinct eigenvalues, say $\delta_1, \ldots, \delta_t$, which are complex numbers in general, and if $\mathbf{u}_1, \ldots, \mathbf{u}_t$ are a set of t eigenvectors of P corresponding to the eigenvalues $\delta_1, \ldots, \delta_t$, then define a matrix U as $U = (\mathbf{u}_1, \ldots, \mathbf{u}_t)$; that is, U has $\mathbf{u}_1, \ldots, \mathbf{u}_t$ as its t columns. Then, U has the property that $U^{-1}PU = L$, where L is the diagonal matrix with the diagonal elements $\delta_1, \ldots, \delta_t$. Now, just note that

$$U^{-1}PU = L \Rightarrow P = ULU^{-1} \Rightarrow P^n = UL^nU^{-1},$$

$\forall n \geq 2$.

Therefore, we only need to compute the eigenvalues of P, and the matrix U of a set of eigenvectors. As long as the eigenvalues are distinct, the n-step transition matrix will be provided by the unified formula $P^n = UL^nU^{-1}$.

The eigenvalues of our P are

$$\delta_1 = -\frac{i}{2}, \quad \delta_2 = \frac{i}{2}, \quad \delta_3 = 1;$$

note that they are distinct. The eigenvectors (one set) turn out to be:

$$\mathbf{u}_1 = \left(-1-i, \frac{i-1}{2}, 1\right)', \quad \mathbf{u}_2 = \left(i-1, -\frac{i+1}{2}, 1\right)', \quad \mathbf{u}_3 = (1,1,1)'.$$

Therefore,

$$U = \begin{pmatrix} -i-1 & i-1 & 1 \\ \frac{i-1}{2} & -\frac{i+1}{2} & 1 \\ 1 & 1 & 1 \end{pmatrix},$$

$$U^{-1} = \begin{pmatrix} \frac{3i-1}{10} & -\frac{2i+1}{5} & \frac{i+3}{10} \\ -\frac{3i+1}{10} & \frac{2i-1}{5} & \frac{3-i}{10} \\ \frac{1}{5} & \frac{2}{5} & \frac{2}{5} \end{pmatrix}.$$

This leads to

$$P^n = U \begin{pmatrix} (-\frac{i}{2})^n & 0 & 0 \\ 0 & (\frac{i}{2})^n & 0 \\ 0 & 0 & 1 \end{pmatrix} U^{-1},$$

with U, U^{-1} as above.

10.4 Communicating Classes

For example,

$$p_{11}(n) = (-i-1)\frac{3i-1}{10}\left(-\frac{i}{2}\right)^n + (i-1)\left(-\frac{3i+1}{10}\right)\left(\frac{i}{2}\right)^n + 1\left(\frac{1}{5}\right)1$$

$$= \frac{1}{5} + \frac{2-i}{5}\left(-\frac{i}{2}\right)^n + \frac{2+i}{5}\left(\frac{i}{2}\right)^n;$$

this is the probability that the mosquito will be back on the forehead after n moves if it started at the forehead. If we take $n = 2$, we get, on doing the complex multiplication, $p_{11}(2) = 0$. We can logically verify that $p_{11}(2)$ must be zero by just looking at the one-step transition matrix P. However, if we take $n = 3$, then the formula will give $p_{11}(3) = \frac{1}{4} > 0$. Indeed, if we take $n = 3$, we get

$$P^3 = \begin{pmatrix} \frac{1}{4} & \frac{1}{4} & \frac{1}{2} \\ \frac{1}{4} & \frac{3}{8} & \frac{3}{8} \\ \frac{1}{8} & \frac{1}{2} & \frac{3}{8} \end{pmatrix}.$$

We notice that every element in P^3 is strictly positive. That is, no matter where the mosquito was initially seated, by the time it has made three moves, we cannot rule out any location for where it will be: it can now be anywhere. In fact, this property of a transition probability matrix is so important in Markov chain theory, that it has a name. It is the first definition in our next section.

10.4 Communicating Classes

Definition 10.7. Let $\{X_n\}$ be a stationary Markov chain with transition probability matrix P. It is called a *regular chain* if there exists a universal $n_0 > 0$ such that $p_{ij}(n_0) > 0 \; \forall i, j \in S$.

So, what we just saw in the last example is that the mosquito is engaged in movements according to a regular Markov chain.

A weaker property is that of *irreducibility*.

Definition 10.8. Let $\{X_n\}$ be a stationary Markov chain with transition probability matrix P. It is called an *irreducible chain* if for any $i, j \in S, i \neq j$, there exists $n_0 > 0$, possibly depending on i, j such that $p_{ij}(n_0) > 0$.

Irreducibility means that it is possible to travel from any state to any other state, however many steps it might take, depending on which two states are involved.

Another terminology also commonly used is that of *communicating*.

Definition 10.9. Let $\{X_n\}$ be a stationary Markov chain with transition probability matrix P. Let i, j be two specific states. We say that i *communicates with* $j (i \leftrightarrow j)$

if there exists $n_0 > 0$, possibly depending on i, j such that $p_{ij}(n_0) > 0$, and also, there exists $n_1 > 0$, possibly depending on i, j such that $p_{ji}(n_1) > 0$.

In words, a pair of specific states i, j are *communicating states* if it is possible to travel back and forth between i, j, however many steps it might take, depending on i, j, and possibly even depending on the direction of the journey, that is, whether the direction is from i to j, or from j to i.

By convention, we say that $i \leftrightarrow i$. Thus, $\leftrightarrow$ defines an *equivalence relation* on the state space S:

$$i \leftrightarrow i; i \leftrightarrow j \Rightarrow j \leftrightarrow i; i \leftrightarrow j, j \leftrightarrow k \Rightarrow i \leftrightarrow k.$$

Therefore, like all equivalence relations, $\leftrightarrow$ partitions the state space S into mutually exclusive subsets of S, say $C_1, C_2, \ldots$. These partitioning subsets $C_1, C_2, \ldots$ are called the *communicating classes* of the chain.

Here is an example to help illustrate the notion.

Example 10.12 (Identifying Communicating Classes). Consider the one-step transition matrix

$$P = \begin{pmatrix} .75 & .25 & 0 & 0 & 0 & 0 \\ 0 & 0 & 1 & 0 & 0 & 0 \\ .25 & 0 & 0 & .25 & .5 & 0 \\ 0 & 0 & 0 & .75 & .25 & 0 \\ 0 & 0 & 0 & 0 & 0 & 1 \\ 0 & 0 & 0 & 0 & 1 & 0 \end{pmatrix}.$$

Inspecting the transition matrix, we see that $1 \leftrightarrow 2$, because it is possible to go from 1 to 2 in just one step, and conversely, starting at 2, one will always go to 3, and it is possible to then go from 3 to 1. Likewise, $2 \leftrightarrow 3$, because if we are at 2, we will always go to 3, and conversely, if we are at 3, then we can first go to 1, and then from 1 to 2. Next, state 5 and state 6 are communicative, but they are clearly not communicative with any other state, because once at 5, we can only go to 6, and once at 6, we can only go to 5. Finally, if we are at 4, then we can go to 5, and from 5 to 6, but 6 does not communicate with 4. So, the communicating classes in this example are

$$C_1 = \{1, 2, 3\}, \quad C_2 = \{4\}, \quad C_3 = \{5, 6\}.$$

Note that they are disjoint, and that $C_1 \cup C_2 \cup C_3 = \{1, 2, 3, 4, 5, 6\} = S$. As a further interesting observation, if we are in C_3, then we cannot make even one-way trips to any state outside of C_3. Such a communicating class is called *closed*. In this example, C_3 is the only closed communicating class. For example, $C_1 = \{1, 2, 3\}$ is not a closed class because it is possible to make one-way trips from 1 to 4, 5. The reader can verify trivially that $C_2 = \{4\}$ is also not a closed class.

10.4 Communicating Classes

We can observe more interesting things about the chain from the transition matrix. Consider, for example, state 5. If you are in state 5, then your transitions would have to be 565656⋯. So, starting at 5, you can return to 5 only at times $n = 2k, k \geq 1$. In such a case, we call the state *periodic* with period equal to two. Likewise, state 6 is also periodic with period equal to two. An exercise asks to show that all states within the same communicating class always have the same period. It is useful to have a formal definition, because there is an element of subtlety about the exact meaning of the period of a state.

Definition 10.10. A state $i \in S$ is said to have the period $d(> 1)$ if the greatest common divisor of all positive integers n for which $p_{ii}(n) > 0$ is the given number d. If a state i has no period $d > 1$, it is called an *aperiodic state*. If every state of a chain is aperiodic, the chain itself is called an *aperiodic chain*.

Example 10.13 (Computing the Period). Consider the hopping mosquito example again. First, let us look at state 1. Evidently, we can go to 1 from 1 in any number of steps: $p_{11}(n) > 0 \ \forall n \geq 1$. So the set of integers n for which $p_{11}(n) > 0$ is $\{1, 2, 3, 4, \ldots\}$, and the gcd (greatest common divisor) of these integers is one. So 1 is an aperiodic state. Because $\{1, 2, 3\}$ is a communicating class, we then must have that 3 is also an aperiodic state. Let us see it. Note that in fact we cannot go to 3 from 3 in one step. But we can go from 3 to 1, then from 1 to 2, and then from 2 to 3. That takes three steps. But we can also go from 3 to 3 in n steps for any $n > 3$, because once we go from 3 to 1, we can stay there with a positive probability for any number of times, and then go from 1 to 2, and from 2 to 3. So the set of integers n for which $p_{33}(n) > 0$ is $\{3, 4, 5, 6, \ldots\}$, and we now see that 3 is an aperiodic state. Similarly, one can verify that 2 is also an aperiodic state.

Remark. It is important to note the subtle point that just because a state i has period d, it does not mean that $p_{ii}(d) > 0$. Suppose, for example, that we can travel from i back to i in steps $6, 9, 12, 15, 18, \ldots$, which have gcd equal to 3, and yet $p_{ii}(3)$ is not greater than zero.

A final definition for now is that of an *absorbing state*. Absorption means that once you have gotten there, you will remain there forever. The formal definition is as follows.

Definition 10.11. A state $i \in S$ is called an *absorbing state* if $p_{ij}(n) = 0$ for all n and for all $j \neq i$. Equivalently, $i \in S$ is an absorbing state if $p_{ii} = 1$; that is, the singleton set $\{i\}$ is a closed class.

Remark. Plainly, if a chain has an absorbing state, then it cannot be regular, and cannot even be irreducible. Absorption is fundamentally interesting in gambling scenarios. A gambler may decide to quit the game as soon as his net fortune becomes zero. If we let X_n denote the gambler's net fortune after the nth play, then zero will be an absorbing state for the chain $\{X_n\}$. For chains that have absorbing states, time taken to get absorbed is considered to be of basic interest.

10.5 Gambler's Ruin

The problem of the *gambler's ruin* is a classic and entertaining example in the theory of probability. It is an example of a Markov chain with absorbing states. Answers to numerous interesting questions about the problem of the gambler's ruin have been worked out; this is all very classic. We provide an introductory exposition to this interesting problem.

Imagine a gambler who goes to a casino with $\$a$ in his pocket. He will play a game that pays him one dollar if he wins the game, or has him pay one dollar to the house if he loses the game. He will play repeatedly until he either goes broke, or his total fortune increases from the initial amount a he entered with to a prespecified larger amount $b (b > a)$. The idea is that he is forced to quit if he goes broke, and he leaves of his own choice if he wins handsomely and is happy to quit. We can ask numerous interesting questions. But let us just ask what is the probability that he will leave because he goes broke.

This is really a simple random walk problem again. Let the gambler's initial fortune be $S_0 = a$. Then, the gambler's net fortune after n plays is $S_n = S_0 + X_1 + X_2 + \cdots + X_n$, where the X_i are iid with the distribution $P(X_i = 1) = p, P(X_i = -1) = q = 1 - p$. We make the realistic assumption that $p < q \Leftrightarrow p < \frac{1}{2}$; that is, the game is favorable to the house and unfavorable to the player. Let p_a denote the probability that the player will leave broke if he started with $\$a$ as his initial fortune. In the following argument, we hold b fixed, and consider p_a as a function of a, with a varying between 0 and the fixed $b; 0 \leq a \leq b$. Note that $p_0 = 1$ and $p_b = 0$. Then, p_a satisfies the recurrence relation

$$p_a = p p_{a+1} + (1-p) p_{a-1}, \quad 1 \leq a < b.$$

The argument is that if the player wins the very first time, which would happen with probability p, then he can eventually go broke with probability p_{a+1}, because the first win increases his fortune by one dollar from a to $a + 1$; but, if the player loses the very first time, which would happen with probability $1 - p$, then he can eventually go broke with probability p_{a-1}, because the first loss will decrease his fortune by one dollar from a to $a - 1$.

Rewrite the above equation in the form

$$p_{a+1} - p_a = \frac{1-p}{p}(p_a - p_{a-1}).$$

If we iterate this identity, we get

$$p_{a+1} - p_a = \left(\frac{1-p}{p}\right)^a (p_1 - 1);$$

here, we have used the fact that $p_0 = 1$.

10.5 Gambler's Ruin

Now use this to find an expression for p_{a+1} as follows:

$$p_{a+1} - 1 = [p_{a+1} - p_a] + [p_a - p_{a-1}] + \cdots + [p_1 - p_0]$$

$$= (p_1 - 1)\left[\left(\frac{1-p}{p}\right)^a + \left(\frac{1-p}{p}\right)^{a-1} + \cdots + 1\right]$$

$$= (p_1 - 1)\frac{\left(\frac{1-p}{p}\right)^a - 1}{\frac{1-p}{p} - 1}$$

$$\Rightarrow p_{a+1} = 1 + (p_1 - 1)\frac{\left(\frac{1-p}{p}\right)^a - 1}{\frac{1-p}{p} - 1}.$$

However, we can find p_1 explicitly by using the last equation with the choice $a = b - 1$, which gives

$$0 = p_b = 1 + (p_1 - 1)\frac{\left(\frac{1-p}{p}\right)^{b-1} - 1}{\frac{1-p}{p} - 1}.$$

Substituting the expression we get for p_1 from here into the formula for p_{a+1}, we have

$$p_{a+1} = \frac{(q/p)^b - (q/p)^{a+1}}{(q/p)^b - 1}.$$

This last formula actually gives an expression for p_x for a general $x \le b$; we can use it with $x = a$ in order to write the final formula

$$p_a = \frac{(q/p)^b - (q/p)^a}{(q/p)^b - 1}.$$

Note that this formula does give $p_0 = 1, p_b = 0$, and that $\lim_{b \to \infty} p_a = 1$, on using the important fact that $\frac{q}{p} > 1$. The practical meaning of $\lim_{b \to \infty} p_a = 1$ is that if the gambler is targeting too high, then actually he will certainly go broke before he reaches that high target.

To summarize, this is an example of a stationary Markov chain with two distinct absorbing states, and we have worked out here the probability that the chain reaches one absorbing state (the gambler going broke) before it reaches the other absorbing state (the gambler leaving as a winner on his terms).

10.6 First Passage, Recurrence, and Transience

Recurrence, transience, and first passage times are fundamental to understanding the long run behavior of a Markov chain. Recurrence is also linked to the stationary distribution of a chain, one of the most important things to study in analyzing and using a Markov chain.

Definition 10.12. Let $\{X_n\}, n \geq 0$ be a stationary Markov chain. Let D be a given subset of the state space S. Suppose the initial state of the chain is state i. The *first passage time* to the set D, denoted as T_{iD}, is defined to be the first time that the chain enters the set D; formally,

$$T_{iD} = \inf\{n > 0 : X_n \in D\},$$

with $T_{iD} = \infty$ if $X_n \in D^c$, the complement of D, for every $n > 0$. If D is a singleton set $\{j\}$, then we denote the first passage time to j as just T_{ij}. If $j = i$, then the first passage time T_{ii} is just the first time the chain returns to its initial state i. We use the simpler notation T_i to denote T_{ii}.

Example 10.14 (Simple Random Walk). Let $X_i, i \geq 1$ be iid random variables, with $P(X_i = \pm 1) = \frac{1}{2}$, and let $S_n = X_0 + \sum_{i=1}^n X_i, n \geq 0$, with the understanding that $X_0 = 0$. Then $\{S_n\}, n \geq 0$ is a stationary Markov chain with initial state as zero, and state space $S = \{\ldots, -2, -1, 0, 1, 2, \ldots\}$.

A graph of the first 50 steps of a simulated random walk is given in Fig. 10.1. By carefully reading the plot, we see that the first passage to zero, the initial state, occurs at $T_0 = 4$. We can also see from the graph that the walk returns to zero a total of nine times within these first 50 steps. The first passage to $j = 5$ occurs at $T_{05} = 9$. The first passage to the set $D = \{\ldots, -9, -6, -3, 3, 6, 9, \ldots\}$ occurs at

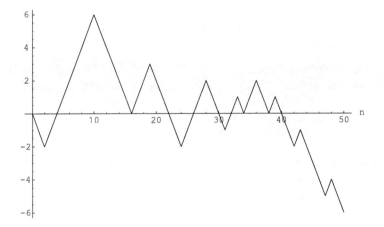

Fig. 10.1 First 50 steps of a simple symmetric random walk

10.6 First Passage, Recurrence, and Transience

$T_{0D} = 7$. The walk goes up to a maximum of 6 at the tenth step. So, we can say that $T_{07} > 50$; in fact, we can make a stronger statement about T_{07} by looking at where the walk is at time $n = 50$. The reader is asked to find the best statement we can make about T_{07} based on the graph.

Example 10.15 (*Infinite Expected First Passage Times*). Consider the three-state Markov chain with state space $S = \{1, 2, 3\}$ and transition probability matrix

$$P = \begin{pmatrix} x & y & z \\ p & q & 0 \\ 0 & 0 & 1 \end{pmatrix},$$

where $x + y + z = p + q = 1$.

First consider the recurrence time T_1. Note that for the chain to return at all to state 1, having started at 1, it cannot ever land in state 3, because 3 is an absorbing state. So, if $T_1 = t$, then the chain spends $t - 1$ time instants in state 2, and then returns to 1. In other words, $P(T_1 = 1) = x$, and for $t > 1$, $P(T_1 = t) = yq^{t-2}p$. From here, we can compute $P(T_1 < \infty)$. Indeed,

$$P(T_1 < \infty) = x + \frac{py}{q^2} \sum_{t=2}^{\infty} q^t$$

$$= x + \frac{py}{q^2} \frac{q^2}{p} = x + y = 1 - z.$$

Therefore, $P(T_1 = \infty) = z$, and if $z > 0$, then obviously $E(T_1) = \infty$, because T_1 itself can be ∞ with a positive probability. If $z = 0$, then

$$E(T_1) = x + \frac{py}{q^2} \sum_{t=2}^{\infty} tq^t$$

$$= x + \frac{py}{q^2} \frac{2q^2 - q^3}{p^2 q} = \frac{1 + p - x(1 + p^2)}{p(1 - p)}.$$

We now define the properties of recurrence and transience of a state. At first glance, it would appear that there could be something in between recurrence and transience; but, in fact, a state is either recurrent or transient. The mathematical meanings of recurrence and transience would really correspond to their dictionary meanings. A recurrent state is one that you keep coming back to over and over again with certainty; a transient state is one that you will ultimately leave behind forever with certainty. Below, we are going to use the simpler notation $P_i(A)$ to denote the conditional probability $P(A|X_0 = i)$, where A is a generic event. Here are the formal definitions of recurrence and transience.

Definition 10.13. A state $i \in S$ is called *recurrent* if $P_i(X_n = i$ for infinitely many $n \geq 1) = 1$. The state $i \in S$ is called *transient* if $P_i(X_n = i$ for infinitely many $n \geq 1) = 0$.

Remark. Note that if a stationary chain returns to its original state i (at least) once with probability one, it will then also return infinitely often with probability one. So, we could also think of recurrence and transience of a state in terms of the following questions.

(a) Is $P_i(X_n = i$ for some $n \geq 1) = 1$?
(b) Is $P_i(X_n = i$ for some $n \geq 1) < 1$?

Here is another way to think about it. Consider our previously defined recurrence time T_i (still with the understanding that the initial state is i). We can think of recurrence in terms of whether $P_i(T_i < \infty) = 1$.

Needless to say that just because $P_i(T_i < \infty) = 1$, it does not follow that its expectation $E_i(T_i) < \infty$. It is a key question in Markov chain theory whether $E_i(T_i) < \infty$ for every state i. Not only is it of practical value to compute $E_i(T_i)$, the finiteness of $E_i(T_i)$ for every state i crucially affects the long run behavior of the chain. If we want to predict where the chain will be after it has run for a long time, our answers will depend on these expected values $E_i(T_i)$, provided they are all finite. The relationship of $E_i(T_i)$ to the limiting value of $P(X_n = i)$ is made clear in the next section. Because of the importance of the issue of finiteness of $E_i(T_i)$, the following are important definitions.

Definition 10.14. A state i is called *null recurrent* if $P_i(T_i < \infty) = 1$, but $E_i(T_i) = \infty$. The state i is called *positive recurrent* if $E_i(T_i) < \infty$. The Markov chain $\{X_n\}$ is called positive recurrent if every state i is positive recurrent.

Recurrence and transience can be discussed at various levels of sophistication, and the treatment and ramifications can be confusing. So a preview is going to be useful.

Preview.

(a) You can verify recurrence or transience of a given state i by verifying whether $\sum_{i=0}^{\infty} p_{ii}(n) = \infty$ or $< \infty$.
(b) You can also try to directly verify whether $P_i(T_i < \infty) = 1$ or < 1.
(c) Chains with a finite state space are more easily handled as regards settling recurrence or transience issues. For finite chains, there must be at least one recurrent state; that is, not all states can be transient, if the chain has a finite state space.
(d) Recurrence is a *class property*; that is, states within the same communicating class have the same recurrence status. If one of them is recurrent, so are all the others.
(e) In identifying exactly which communicating classes have the recurrence property, you can identify which of the communicating classes are *closed*.
(f) Even if a state i is recurrent, $E_i(T_i)$ can be infinite: the state i can be *null recurrent*. However, if the state space is finite, and if the chain is regular, then you do not have to worry about it. As a matter of fact, for any set D, T_{iD} will

10.6 First Passage, Recurrence, and Transience

be finite with probability one, and even $E_i(T_{iD})$ will be finite. So, for a finite regular chain, you have a very simple recurrence story; every state is not just recurrent, but even *positive recurrent*.

(g) For chains with an infinite state space, it is possible that every state is transient, and it is also possible that every state is recurrent, or it could also be something in between. Whether the chain is *irreducible* is going to be a key factor in sorting out the exact recurrence structure.

Some of the major results on recurrence and transience are now given.

Theorem 10.2. *Let $\{X_n\}$ be a stationary Markov chain. If $\sum_{n=0}^{\infty} p_{ii}(n) = \infty$, then i is a recurrent state, and if $\sum_{n=0}^{\infty} p_{ii}(n) < \infty$, then i is a transient state.*

Proof. Introduce the variable $V_i = \sum_{n=0}^{\infty} I_{\{X_n=i\}}$; thus, V_i is the total number of visits of the chain to state i. Let also $p_i = P_i(T_i < \infty)$. By using the Markov property of $\{X_n\}$, it follows that $P_i(V_i > m) = p_i^m$ for any $m \geq 0$. Suppose now $p_i < 1$. Then, by the tailsum formula for expectations,

$$E_i(V_i) = \sum_{m=0}^{\infty} P_i(V_i > m)$$

$$= \sum_{m=0}^{\infty} p_i^m = \frac{1}{1-p_i} < \infty.$$

But also,

$$E_i(V_i) = E_i\left[\sum_{n=0}^{\infty} I_{\{X_n=i\}}\right]$$

$$= \sum_{n=0}^{\infty} E[I_{\{X_n=i\}}] = \sum_{n=0}^{\infty} P_i(X_n = i)$$

$$= \sum_{n=0}^{\infty} p_{ii}(n).$$

So, if $p_i < 1$, then we must have $\sum_{n=0}^{\infty} p_{ii}(n) < \infty$, which is the same as saying if $\sum_{n=0}^{\infty} p_{ii}(n) = \infty$, then p_i must be equal to 1, and so i must be a recurrent state.

Suppose on the other hand that $p_i = 1$. Then, for any m, $P_i(V_i > m) = 1$, and so, with probability one, $V_i = \infty$. So, $E_i(V_i) = \infty$, which implies that $\sum_{n=0}^{\infty} p_{ii}(n) = E_i(V_i) = \infty$. So, if $p_i = 1$, then $\sum_{n=0}^{\infty} p_{ii}(n)$ must be ∞, which is the same as saying that if $\sum_{n=0}^{\infty} p_{ii}(n) < \infty$, then $p_i < 1$, which would mean that i is a transient state. □

The next theorem formalizes the intuition that if you keep coming back to some state over and over again, and that state communicates with some other state, then you will be visiting that state over and over again as well. That is, recurrence is a class property, and that implies that transience is also a class property.

Theorem 10.3. *Let C be any communicating class of states of a stationary Markov chain $\{X_n\}$. Then, either all states in C are recurrent, or all states in C are transient.*

Proof. The theorem is proved if we can show that if i, j both belong to a common communicating class, and i is transient, then j must also be transient. If we can prove this, it follows that if j is recurrent, then i must also be recurrent, for if it were not, it would be transient, and so that would make j transient, which would be a contradiction.

So, suppose $i \in C$, and assume that i is transient. By virtue of the transience of i, we know that $\sum_{r=0}^{\infty} p_{ii}(r) < \infty$, and so, $\sum_{r=R}^{\infty} p_{ii}(r) < \infty$ for any fixed R. This is useful to us in the proof.

Now consider another state $j \in C$. Because C is a communicating class, there exist k, n such that $p_{ij}(k) > 0, p_{ji}(n) > 0$. Take such k, n and hold them fixed.

Now observe that for any m, we have the inequality

$$p_{ii}(k+m+n) \geq p_{ij}(k) p_{jj}(m) p_{ji}(n)$$

$$\Rightarrow p_{jj}(m) \leq \frac{1}{p_{ij}(k) p_{ji}(n)} p_{ii}(k+m+n)$$

$$\Rightarrow \sum_{m=0}^{\infty} p_{jj}(m) \leq \frac{1}{p_{ij}(k) p_{ji}(n)} \sum_{m=0}^{\infty} p_{ii}(k+m+n) < \infty,$$

because $p_{ij}(k), p_{ji}(n)$ are two fixed positive numbers, and $\sum_{m=0}^{\infty} p_{ii}(k+m+n) = \sum_{r=k+n}^{\infty} p_{ii}(r) < \infty$. But, if $\sum_{m=0}^{\infty} p_{jj}(m) < \infty$, then we already know that j must be transient, which is what we want to prove. □

If a particular communicating class C consists of (only) recurrent states, we call C a *recurrent class*. The following are two important consequences of the above theorem.

Theorem 10.4. *(a) Let $\{X_n\}$ be a stationary irreducible Markov chain with a finite state space. Then every state of $\{X_n\}$ must be recurrent.*
(b) For any stationary Markov chain with a finite state space, a communicating class is recurrent if and only if it is closed.

Example 10.16 (Various Illustrations.). We revisit some of the chains in our previous examples and examine their recurrence structure.

In the weather pattern example,

$$P = \begin{pmatrix} \alpha & 1-\alpha \\ 1-\beta & \beta \end{pmatrix}.$$

If $0 < \alpha < 1$ and also $0 < \beta < 1$, then clearly the chain is irreducible, and it obviously has a finite state space. And so, each of the two states is recurrent. If $\alpha = \beta = 1$, then each state is an absorbing state, and clearly, $\sum_{n=0}^{\infty} p_{ii}(n) = \infty$

10.7 Long Run Evolution and Stationary Distributions

for both $i = 1, 2$. So, each state is recurrent. If $\alpha = \beta = 0$, then the chain evolves either as $121212\ldots$, or $212121\cdots$. Each state is periodic and recurrent.

In the hopping mosquito example,

$$P = \begin{pmatrix} 0 & 1 & 0 \\ 0 & .5 & .5 \\ .5 & 0 & .5 \end{pmatrix}.$$

In this case, some elements of P are zero. However, we have previously seen that every element in P^3 is strictly positive. Hence, the chain is again irreducible. Once again, each of the three states is recurrent.

Next consider the chain with the transition matrix

$$P = \begin{pmatrix} .75 & .25 & 0 & 0 & 0 & 0 \\ 0 & 0 & 1 & 0 & 0 & 0 \\ .25 & 0 & 0 & .25 & .5 & 0 \\ 0 & 0 & 0 & .75 & .25 & 0 \\ 0 & 0 & 0 & 0 & 0 & 1 \\ 0 & 0 & 0 & 0 & 1 & 0 \end{pmatrix}.$$

We have previously proved that the communicating classes of this chain are $\{1, 2, 3\}, \{4\}, \{5, 6\}$, of which $\{5, 6\}$ is the only closed class. Therefore, $5, 6$ are the only recurrent states of this chain.

10.7 Long Run Evolution and Stationary Distributions

A natural human instinct is to want to predict the future. It is not surprising that we often want to know exactly where a Markov chain will be after it has evolved for a fairly long time. Of course, we cannot say with certainty where it will be. But perhaps we can make probabilistic statements. In notation, suppose a stationary Markov chain $\{X_n\}$ started at some initial state $i \in S$. A natural question is what can we say about $P(X_n = j | X_0 = i)$ for arbitrary $j \in S$, if n is large. Again, a short preview might be useful.

Preview. For chains with a finite state space, the answers are concrete, extremely structured, and furthermore, convergence occurs rapidly. That is, under some reasonable conditions on the chain, regardless of what the initial state i is, $P(X_n = j | X_0 = i)$ has a limit π_j, and $P(X_n = j | X_0 = i) \approx \pi_j$ for quite moderate values of n. In addition, the marginal probabilities $P(X_n = j)$ are also well approximated by the same π_j, and there is an explicit formula for determining the limiting probability π_j for each $j \in S$. Somewhat different versions of these results are often presented in different texts, under different sets of conditions on the chain.

Our version balances the ease of understanding the results with the applicability of the conditions assumed. But, first let us see two illustrative examples.

Example 10.17. Consider first the weather pattern example, and for concreteness, take the one-step transition probability matrix to be

$$P = \begin{pmatrix} .8 & .2 \\ .2 & .8 \end{pmatrix}.$$

Then, by direct computation,

$$P^{10} = \begin{pmatrix} .50302 & .49698 \\ .49698 & .50302 \end{pmatrix}; \quad P^{15} = \begin{pmatrix} .50024 & .49976 \\ .49976 & .50024 \end{pmatrix};$$

$$P^{20} = \begin{pmatrix} .50018 & .49982 \\ .49982 & .50018 \end{pmatrix}; \quad P^{25} = \begin{pmatrix} .50000 & .50000 \\ .50000 & .50000 \end{pmatrix}.$$

We notice that P^n appears to converge to a limiting matrix with each row of the limiting matrix being the same, namely, $(.5, .5)$. That is, regardless of the initial state i, $P(X_n = j \,|X_0 = i)$ appears to converge to $\pi_j = .5$. Thus, if indeed $\alpha = \beta = .8$ in the weather pattern example, then in the long run the chances of a dry or wet day would be just 50–50, and the effect of the weather on the initial day is going to wash out.

On the other hand, consider a chain with the one-step transition matrix

$$P = \begin{pmatrix} x & y & z \\ p & q & 0 \\ 0 & 0 & 1 \end{pmatrix}.$$

Notice that this chain has an absorbing state; once you are in state 3, you can never leave from there. To be concrete, take $x = .25, y = .75, p = q = .5$. Then, by direct computing,

$$P^{10} = \begin{pmatrix} .400001 & .599999 & 0 \\ .4 & .6 & 0 \\ 0 & 0 & 1 \end{pmatrix}; \quad P^{20} = \begin{pmatrix} .4 & .6 & 0 \\ .4 & .6 & 0 \\ 0 & 0 & 1 \end{pmatrix}.$$

This time it appears that P^n converges to a limiting matrix whose first two rows are the same, but the third row is different. Specifically, the first two rows of P^n seem to be converging to $(.4, .6, 0)$ and the third row is $(0, 0, 1)$, the same as the third row in P itself. Thus, the limiting behavior of $P(X_n = j \,|X_0 = i)$ seems to depend on the initial state i.

The difference between the two chains in this example is that the first chain is regular, whereas the second chain has an absorbing state and cannot be regular. Indeed, regularity of the chain is going to have a decisive effect on the limiting behavior of $P(X_n = j \,|X_0 = i)$. An important theorem is the following.

10.7 Long Run Evolution and Stationary Distributions

Theorem 10.5 (Fundamental Theorem for Finite Markov Chains). Let $\{X_n\}$ be a stationary Markov chain with a finite state space S, consisting of t elements. Assume furthermore that $\{X_n\}$ is regular. Then, there exist $\pi_j, j = 1, 2, \ldots, t$ such that

(a) For any initial state i, $P(X_n = j \mid X_0 = i) \to \pi_j, j = 1, 2, \ldots, t$.

(b) $\pi_1, \pi_2, \ldots, \pi_t$ are the unique solutions of the system of equations $\pi_j = \sum_{i=1}^{t} \pi_i p_{ij}, j = 1, 2, \ldots, t$, $\sum_{j=1}^{t} \pi_j = 1$, where p_{ij} denotes the (i, j)th element in the one-step transition matrix P.

Equivalently, the row vector $\pi = (\pi_1, \pi_2, \ldots, \pi_t)$ is the unique solution of the equations $\pi P = \pi$, $\pi \mathbf{1}' = 1$, where $\mathbf{1}$ is a row vector with each coordinate equal to one.

(c) The chain $\{X_n\}$ is positive recurrent; that is, for any state i, the mean recurrence time $\mu_i = E_i(T_i) < \infty$, and furthermore $\mu_i = \frac{1}{\pi_i}$.

The vector $\pi = (\pi_1, \pi_2, \ldots, \pi_t)$ is called the stationary distribution of the regular finite chain $\{X_n\}$. It is also sometimes called the equilibrium distribution or the invariant distribution of the chain. The difference in terminology can be confusing.

Suppose now that a stationary chain has a stationary distribution π. If we use this π as the initial distribution of the chain, then we observe that

$$P(X_1 = j) = \sum_{k \in S} P(X_1 = j \mid X_0 = k)\pi_k = \pi_j,$$

by the fact that π is a stationary distribution of the chain. Indeed, it now follows easily by induction that for any n, $P(X_n = j) = \pi_j, j \in S$. Thus, if a chain has a stationary distribution, and the chain starts out with that distribution, then at all subsequent times, the distribution of the state of the chain remains exactly the same, namely the stationary distribution. This is why a chain that starts out with its stationary distribution is sometimes described to be in steady-state.

We now give a proof of parts (a) and (b) of the *fundamental theorem of Markov chains*. For this, we use a famous result in linear algebra, which we state as a lemma; see Seneta (1981) for a proof.

Lemma (Perron–Frobenius Theorem). Let P be a real $t \times t$ square matrix with all elements p_{ij} strictly positive. Then,

(a) P has a positive real eigenvalue λ_1 such that for any other eigenvalue λ_j of P, $|\lambda_j| < \lambda_1, j = 2, \ldots, t$.

(b) λ_1 satisfies
$$\min_i \sum_j p_{ij} \leq \lambda_1 \leq \max_i \sum_j p_{ij}.$$

(c) There exist left and right eigenvectors of P, each having only strictly positive elements, corresponding to the eigenvalue λ_1; that is, there exist vectors π, ω, with both π, ω having only strictly positive elements, such that $\pi P = \lambda_1 \pi$; $P\omega = \lambda_1 \omega$.

(d) The algebraic multiplicity of λ_1 is one and the dimension of the set of left as well as right eigenvectors corresponding to λ_1 equals one.

(e) For any i, and any vector $(c_1, c_2, \ldots, c_t)$ with each $c_j > 0$,

$$\lim_n \frac{1}{n} \log \left[\sum_j p_{ij}(n) c_j \right] = \lim_n \frac{1}{n} \log \left[\sum_j p_{ji}(n) c_j \right] = \log \lambda_1.$$

Proof of Fundamental Theorem. Because for a transition probability matrix of a Markov chain, the row sums are all equal to one, it follows immediately from the Perron–Frobenius theorem that if every element of P is strictly positive, then $\lambda_1 = 1$ is an eigenvalue of P and that there is a left eigenvector π with only strictly positive elements such that $\pi P = \pi$. We can always normalize π so that its elements add to exactly one, and so the renormalized π is a stationary distribution for the chain, by the definition of a stationary distribution. If the chain is regular, then in general we can only assert that every element of P^n is strictly positive for some n. Then the Perron–Frobenius theorem applies to P^n and we have a left eigenvector π satisfying $\pi P^n = \pi$. It can be proved from here that the same vector π satisfies $\pi P = \pi$, and so the chain has a stationary distribution. The uniqueness of the stationary distribution is a consequence of part (d) of the Perron–Frobenius theorem.

Coming to part (a), note that part (a) asserts that every row of P^n converges to the vector π; that is,

$$P^n \to \begin{pmatrix} \pi \\ \pi \\ \vdots \\ \pi \end{pmatrix}.$$

We prove this by the diagonalization argument we previously used in working out a closed-form formula for P^n in the hopping mosquito example. Thus, consider the case where the eigenvalues of P are distinct, remembering that one eigenvalue is one, and the rest less than one in absolute value. Let $U^{-1} P U = L = \text{diag}\{1, \lambda_2, \cdots, \lambda_t\}$, where

$$U = \begin{pmatrix} 1 & u_{12} & u_{13} & \cdots \\ 1 & u_{22} & u_{23} & \cdots \\ & \vdots & \vdots & \\ 1 & u_{t2} & u_{t3} & \cdots \end{pmatrix}; \quad U^{-1} = \begin{pmatrix} \pi_1 & \pi_2 & \cdots & \pi_t \\ u^{21} & u^{22} & \cdots & u^{2t} \\ \vdots & \vdots & & \\ u^{t1} & u^{t2} & \cdots & u^{tt} \end{pmatrix}.$$

This implies $P = ULU^{-1} \Rightarrow P^n = UL^n U^{-1}$. Because each λ_j for $j > 1$ satisfies $|\lambda_j| < 1$, we have $|\lambda_j|^n \to 0$, as $n \to \infty$. This fact, together with the explicit forms of U, U^{-1} given immediately above leads to the result that each row of $UL^n U^{-1}$ converges to the fixed row vector π, which is the statement in part (a). □

10.7 Long Run Evolution and Stationary Distributions

We assumed that our chain is regular for the fundamental theorem. An exercise asks us to show that regularity is not necessary for the existence of a stationary distribution. Regular chains are of course irreducible. But irreducibility alone is not enough for the existence of a stationary distribution. More is said of the issue of existence of a stationary distribution a bit later. For finite chains, irreducibility plus aperiodicity is enough for the validity of the fundamental theorem because of the simple reason that such chains are regular in the finite case. It is worth mentioning this as a formal result.

Theorem 10.6. *Let $\{X_n\}$ be a stationary Markov chain with a finite-state space S. If $\{X_n\}$ is irreducible and aperiodic, then the fundamental theorem holds.*

Example 10.18 (Weather Pattern). Consider the two-state Markov chain with the transition probability matrix

$$P = \begin{pmatrix} \alpha & 1-\alpha \\ 1-\beta & \beta \end{pmatrix}.$$

Assume $0 < \alpha, \beta < 1$, so that the chain is regular. The stationary probabilities π_1, π_2 are to be found from the equation

$$(\pi_1, \pi_2)P = (\pi_1, \pi_2)$$
$$\Rightarrow \alpha\pi_1 + (1-\beta)\pi_2 = \pi_1;$$
$$\Rightarrow (1-\alpha)\pi_1 = (1-\beta)\pi_2 \Rightarrow \pi_2 = \frac{1-\alpha}{1-\beta}\pi_1.$$

Substituting this into $\pi_1 + \pi_2 = 1$ gives $\pi_1 + \frac{1-\alpha}{1-\beta}\pi_1 = 1$, and so $\pi_1 = \frac{1-\beta}{2-\alpha-\beta}$, which then gives $\pi_2 = 1 - \pi_1 = \frac{1-\alpha}{2-\alpha-\beta}$. For example, if $\alpha = \beta = .8$, then we get $\pi_1 = \pi_2 = \frac{1-.8}{2-.8-.8} = .5$, which is the numerical limit we saw in our example by computing P^n explicitly for large n. For general $0 < \alpha, \beta < 1$, each of the states is positive recurrent. For instance, if $\alpha = \beta = .8$, then $E_i(T_i) = \frac{1}{.5} = 2$ for each of $i = 1, 2$.

Example 10.19. With the row vector $\pi = (\pi_1, \pi_2, \ldots, \pi_t)$ denoting the vector of stationary probabilities of a chain, π satisfies the vector equation $\pi P = \pi$, and taking a transpose on both sides, $P'\pi' = \pi'$. That is, the column vector π' is a right eigenvector of P', the transpose of the transition matrix.

For example, consider the voting preferences example with

$$P = \begin{pmatrix} .8 & .05 & .15 \\ .03 & .9 & .07 \\ .1 & .1 & .8 \end{pmatrix}.$$

The transpose of P is

$$P' = \begin{pmatrix} .8 & .03 & .1 \\ .05 & .9 & .1 \\ .15 & .07 & .8 \end{pmatrix}.$$

A set of its three eigenvectors is

$$\begin{pmatrix} .38566 \\ .74166 \\ .54883 \end{pmatrix}, \quad \begin{pmatrix} .44769 \\ -.81518 \\ .36749 \end{pmatrix}, \quad \begin{pmatrix} -.56867 \\ -.22308 \\ .79174 \end{pmatrix}.$$

Of these, the last two cannot be the eigenvector we are looking for, because they contain negative elements. The first eigenvector contains only nonnegative (actually, strictly positive) elements, and when normalized to give elements that add to one, results in the stationary probability vector $\pi = (.2301, .4425, .3274)$. We could have also obtained it by the method of elimination as in our preceding example, but the eigenvector method is a general clean method, and is particularly convenient when the number of states t is not small.

Example 10.20 (Ehrenfest Urn). Consider the symmetric version of the Ehrenfest urn model in which a certain number among m balls are initially in urn I, the rest in urn II, and at each successive time one of the m balls is selected completely at random and transferred to the other urn with probability $\frac{1}{2}$ (and left in the same urn with probability $\frac{1}{2}$). The one-step transition probabilities are

$$p_{i,i-1} = \frac{i}{2m}, \quad p_{i,i+1} = \frac{m-i}{2m}, \quad p_{ii} = \frac{1}{2}.$$

A stationary distribution π would satisfy the equations

$$\pi_j = \frac{m-j+1}{2m}\pi_{j-1} + \frac{j+1}{2m}\pi_{j+1} + \frac{\pi_j}{2}, \quad 1 \leq j \leq m-1;$$

$$\pi_0 = \frac{\pi_0}{2} + \frac{\pi_1}{2m}; \quad \pi_m = \frac{\pi_m}{2} + \frac{\pi_{m-1}}{2m}.$$

These are equivalent to the equations

$$\pi_0 = \frac{\pi_1}{m}; \quad \pi_m = \frac{\pi_{m-1}}{m}; \quad , \pi_j = \frac{m-j+1}{m}\pi_{j-1}$$

$$+ \frac{j+1}{m}\pi_{j+1}, \quad 1 \leq j \leq m-1.$$

Starting with π_1, one can solve these equations by just successive substitution, leaving π_0 as an undetermined constant to get $\pi_j = \binom{m}{j}\pi_0$. Now use the fact that

$\sum_{j=0}^{m} \pi_j$ must equal one. This forces $\pi_0 = \frac{1}{2^m}$, and hence, $\pi_j = \frac{\binom{m}{j}}{2^m}$. We now realize that these are exactly the probabilities in a binomial distribution with parameters m and $\frac{1}{2}$. That is, in the symmetric Ehrenfest urn problem, there is a stationary distribution and it is the $\text{Bin}(m, \frac{1}{2})$ distribution. In particular, after the process has evolved for a long time, we would expect close to half the balls to be in each urn. Each state is positive recurrent, that is, the chain is sure to return to its original configuration with a finite expected value for the time it takes to return to that configuration. As a specific example, suppose $m = 10$ and that initially, there were five balls in each urn. Then, the stationary probability $\pi_5 = \frac{\binom{10}{5}}{2^{10}} = \frac{63}{256} = .246$, so that we can expect that after about four transfers, the urns will once again have five balls each.

Example 10.21 (Asymmetric Random Walk). Consider a random walk $\{S_n\}, n \geq 0$ starting at zero, and taking independent steps of length one at each time, either to the left or to the right, with the respective probabilities depending on the current position of the walk. Formally, S_n is a Markov chain with initial state zero, and with the one-step transition probabilities $p_{i,i+1} = \alpha_i$, $p_{i,i-1} = \beta_i$, $\alpha_i + \beta_i = 1$ for any $i \geq 0$. In order to restrict the state space of the chain to just the nonnegative integers $S = \{0, 1, 2, \ldots\}$, we assume that $\alpha_0 = 1$. Thus, if you ever reach zero, then you start over.

If a stationary distribution π exists, by virtue of the matrix equation $\pi = \pi P$, it satisfies the recursion

$$\pi_j = \pi_{j-1} \alpha_{j-1} + \pi_{j+1} \beta_{j+1},$$

with the initial equation

$$\pi_0 = \pi_1 \beta_1.$$

This implies, by successive substitution,

$$\pi_1 = \frac{1}{\beta_1} \pi_0 = \frac{\alpha_0}{\beta_1} \pi_0; \quad \pi_2 = \frac{\alpha_0 \alpha_1}{\beta_1 \beta_2} \pi_0; \ldots,$$

and for a general $j > 1$,

$$\pi_j = \frac{\alpha_0 \alpha_1 \cdots \alpha_{j-1}}{\beta_1 \beta_2 \cdots \beta_j} \pi_0.$$

Because each $\pi_j, j \geq 0$ is clearly nonnegative, the only issue is whether they constitute a probability distribution, that is, whether $\pi_0 + \sum_{j=1}^{\infty} \pi_j = 1$. This is equivalent to asking whether $(1 + \sum_{j=1}^{\infty} c_j) \pi_0 = 1$, where $c_j = \frac{\alpha_0 \alpha_1 \cdots \alpha_{j-1}}{\beta_1 \beta_2 \cdots \beta_j}$. In other words, the chain has a stationary distribution if and only if the infinite series $\sum_{j=1}^{\infty} c_j$ converges to some positive finite number δ, in which case $\pi_0 = \frac{1}{1+\delta}$ and for $j \geq 1$, $\pi_j = \frac{c_j}{1+\delta}$.

Consider now the special case when $\alpha_i = \beta_i = \frac{1}{2}$ for all $i \geq 1$. Then, for any $j \geq 1, c_j = \frac{1}{2}$, and hence $\sum_{j=1}^{\infty} c_j$ diverges. Therefore, the case of the symmetric random walk does not possess a stationary distribution, in the sense that no stationary distribution exists that is a valid probability distribution.

The stationary distribution of a Markov chain is not just the limit of the n-step transition probabilities; it also has important interpretations in terms of the marginal distribution of the state of the chain. Suppose the chain has run for a long time, and we want to know what the chances are that the chain is now in some state j. It turns out that the stationary probability π_j approximates that probability too. The approximations are valid in a fairly strong sense, made precise below. Even more, π_j is approximately equal to the fraction of the time so far that the chain has spent visiting state j. To describe these results precisely, we need a little notation.

Given a stationary chain $\{X_n\}$, we denote $f_n(j) = P(X_n = j)$. Also let $I_k(j) = I_{\{X_k = j\}}$, and $V_n(j) = \sum_{k=1}^{n} I_k(j)$. Thus, $V_n(j)$ counts the number of times up to time n that the chain has been in state j, and $\delta_n(j) = \frac{V_n(j)}{n}$ measures the fraction of times up to time n that the chain has been in state j. Then, the following results hold.

Theorem 10.7 (Weak Ergodic Theorem). *Let $\{X_n\}$ be a regular Markov chain with a finite state space and the stationary distribution $\pi = (\pi_1, \pi_2, \ldots, \pi_t)$. Then,*

(a) Whatever the initial distribution of the chain is, for any $j \in S$, $P(X_n = j) \to \pi_j$ as $n \to \infty$.
(b) For any $\epsilon > 0$, and for any $j \in S$, $P(|\delta_n(j) - \pi_j| > \epsilon) \to 0$ as $n \to \infty$.
(c) More generally, given any bounded function g, and any $\epsilon > 0$, $P(|\frac{1}{n}\sum_{k=1}^{n} g(X_k) - \sum_{j=1}^{t} g(j)\pi_j| > \epsilon) \to 0$ as $n \to \infty$.

Remark. See Norris (1997, p. 53) for a proof of this theorem. Also see Section 19.3.1 in this text, where an even stronger version is proved. The theorem provides a basis for estimating the stationary probabilities of a chain by following its trajectory for a long time. Part (c) of the theorem says that time averages of a general bounded function will ultimately converge to the state space average of the function with respect to the stationary distribution. In fact, a stronger convergence result than the one we state here holds and is commonly called the *ergodic theorem for stationary Markov chains*; see Brémaud 1999 or Norris (1997).

Exercises

Exercise 10.1. A particular machine is either in working order or broken on any particular day. If it is in working order on some day, it remains so the next day with probability .7, whereas if it is broken on some day, it stays broken the next day with probability .2.

(a) If it is in working order on Monday, what is the probability that it is in working order on Saturday?
(b) If it is in working order on Monday, what is the probability that it remains in working order all the way through Saturday?

Exercise 10.2. Consider the voting preferences example in text with the transition probability matrix

$$P = \begin{pmatrix} .8 & .05 & .15 \\ .03 & .9 & .07 \\ .1 & .1 & .8 \end{pmatrix}.$$

Suppose a family consists of the two parents and a son. The three follow the same Markov chain described above in deciding their votes. Assume that the family members act independently, and that in this election the father voted Conservative, the mother voted Labor, and the son voted Independent.

(a) Find the probability that they will all vote the same parties in the next election as they did in this election.
(b) * Find the probability that as a whole, the family will split their votes among the three parties, one member for each party, in the next election.

Exercise 10.3. Suppose $\{X_n\}$ is a stationary Markov chain. Prove that for all n, and all $x_i, i = 0, 1, \ldots, n + 2, P(X_{n+2} = x_{n+2}, X_{n+1} = x_{n+1} | X_n = x_n, X_{n-1} = x_{n-1}, \ldots, X_0 = x_0) = P(X_{n+2} = x_{n+2}, X_{n+1} = x_{n+1} | X_n = x_n)$.

Exercise 10.4 *(What the Markov Property Does Not Mean). Give an example of a stationary Markov chain with a small number of states such that $P(X_{n+1} = x_{n+1} | X_n \leq x_n, X_{n-1} \leq x_{n-1}, \ldots, X_0 \leq x_0) = P(X_{n+1} = x_{n+1} | X_n \leq x_n)$ is not true for arbitrary $x_0, x_1, \ldots, x_{n+1}$.

Exercise 10.5 (Ehrenfest Urn). Consider the Ehrenfest urn model when there are only two balls to distribute.

(a) Write the transition probability matrix P.
(b) Calculate P^2, P^3.
(c) Find general formulas for P^{2k}, P^{2k+1}.

Exercise 10.6 (The Cat and Mouse Chain). In one of two adjacent rooms, say room 1, there is a cat, and in the other one, room 2, there is a mouse. There is a small hole in the wall through which the mouse can travel between the rooms, and there is a larger hole through which the cat can travel between the rooms. Each minute, the cat and the mouse decide the room they want to be in by following a stationary Markov chain with the transition probability matrices

$$P_1 = \begin{pmatrix} .5 & .5 \\ .5 & .5 \end{pmatrix}; \quad P_2 = \begin{pmatrix} .1 & .9 \\ .6 & .4 \end{pmatrix}.$$

Let X_n be the room in which the cat is at time n, and Y_n the room in which the mouse is at time n. Assume that the chains $\{X_n\}$ and $\{Y_n\}$ are independent.

(a) Write the transition matrix for the chain $Z_n = (X_n, Y_n)$.
(b) Let $p_n = P(X_n = Y_n)$. Compute p_n for $n = 1, 2, 3, 4, 5$, taking the initial time to be $n = 0$.
(c) The very first time that they end up in the same room, the cat will eat the mouse. Let q_n be the probability that the cat eats the mouse at time n. Compute q_n for $n = 1, 2, 3$.

Exercise 10.7 (Diagonalization in the Two-State Case). Consider a two-state stationary chain with the transition probability matrix

$$P = \begin{pmatrix} \alpha & 1-\alpha \\ 1-\beta & \beta \end{pmatrix}.$$

(a) Find the eigenvalues of P. When are they distinct?
(b) Diagonalize P when the eigenvalues are distinct.
(c) Hence find a general formula for $p_{11}(n)$.

Exercise 10.8. A flea is initially located on the top face of a cube, which has six faces, top and bottom, left and right, and front and back. Every minute it moves from its current location to one of the other five faces chosen at random.

(a) Find the probability that after four moves it is back to the top face.
(b) Find the probability that after n moves it is on the top face; on the bottom face.
(c) * Find the probability that the next five moves are distinct. This is the same as the probability that the first six locations of the flea are the six faces of the cube, each location chosen exactly once.

Exercise 10.9 (Subsequences of Markov Chains). Suppose $\{X_n\}$ is a stationary Markov chain. Let $Y_n = X_{2n}$. Prove or disprove that $\{Y_n\}$ is a stationary Markov chain. How about $\{X_{3n}\}$? $\{X_{kn}\}$ for a general k?

Exercise 10.10. Let $\{X_n\}$ be a three-state stationary Markov chain with the transition probability matrix

$$P = \begin{pmatrix} 0 & x & 1-x \\ y & 1-y & 0 \\ 1 & 0 & 0 \end{pmatrix}.$$

Define a function g as $g(1) = 1, g(2) = g(3) = 2$ and let $Y_n = g(X_n)$. Is $\{Y_n\}$ a stationary Markov chain?

Give an example of a function g such that $g(X_n)$ is not a Markov chain.

Exercise 10.11 (An IID Sequence). Let $X_i, i \geq 1$ be iid Poisson random variables with some common mean λ. Prove or disprove that $\{X_n\}$ is a staionary Markov chain. If it is, describe the transition probability matrix.

How important is the Poisson assumption? What happens if $X_i, i \geq 1$ are independent, but not iid?

Exercises

Exercise 10.12. Let $\{X_n\}$ be a stationary Markov chain with transition matrix P, and g a one-to-one function. Define $Y_n = g(X_n)$. Prove that $\{Y_n\}$ is a Markov chain, and characterize as well as you can the transition probability matrix of $\{Y_n\}$.

Exercise 10.13 * **(Loop Chains).** Suppose $\{X_n\}$ is a stationary Markov chain with state space S and transition probability matrix P.

(a) Let $Y_n = (X_n, X_{n+1})$. Show that Y_n is also a stationary Markov chain.
(b) Find the transition probability matrix of Y_n.
(c) How about $Y_n = (X_n, X_{n+1}, X_{n+2})$? Is this also a stationary Markov chain?
(d) How about $Y_n = (X_n, X_{n+1}, \cdots, X_{n+d})$ for a general $d \geq 1$?

Exercise 10.14 (Dice Experiments). Consider the experiment of rolling a fair die repeatedly. Define

(a) $X_n = $ # sixes obtained up to the nth roll.
(b) $X_n = $ number of rolls, at time n, that a six has not been obtained since the last six.

Prove or disprove that each $\{X_n\}$ is a Markov chain, and if they are, obtain the transition probability matrices.

Exercise 10.15. Suppose $\{X_n\}$ is a regular stationary Markov chain with transition probability matrix P. Prove that there exists $m \geq 1$ such that every element in P^n is strictly positive for all $n \geq m$.

Exercise 10.16 (Communicating Classes). Consider a finite-state stationary Markov chain with the transition matrix

$$P = \begin{pmatrix} 0 & .5 & 0 & .5 & 0 \\ 0 & 0 & 1 & 0 & 0 \\ .5 & 0 & 0 & 0 & .5 \\ 0 & .25 & .25 & .25 & .25 \\ .5 & 0 & 0 & 0 & .5 \end{pmatrix}.$$

(a) Identify the communicating classes of this chain.
(b) Identify those classes that are closed.

Exercise 10.17 * **(Periodicity and Simple Random Walk).** Consider the Markov chain corresponding to the simple random walk with general step probabilities $p, q, p + q = 1$.

(a) Identify the periodic states of the chain and the periods.
(b) Find the communicating classes.
(c) Are there any communicating classes that are not closed? If there are, identify them. If not, prove that there are no communicating classes that are not closed.

Exercise 10.18 *(Gambler's Ruin).* Consider the Markov chain corresponding to the problem of the gambler's ruin, with initial fortune a, and absorbing states at 0 and b.

(a) Identify the periodic states of the chain and the periods.
(b) Find the communicating classes.
(c) Are there any communicating classes that are not closed? If there are, identify them.

Exercise 10.19. Prove that a stationary Markov chain with a finite-state space has at least one closed communicating class.

Exercise 10.20 *(Chain with No Closed Classes).* Give an explicit example of a stationary Markov chain with no closed communicating classes.

Exercise 10.21 (Skills Exercise). Consider the stationary Markov chains corresponding to the following transition probability matrices:

$$P = \begin{pmatrix} \frac{1}{3} & \frac{2}{3} & 0 \\ 0 & \frac{1}{3} & \frac{2}{3} \\ \frac{2}{3} & 0 & \frac{1}{3} \end{pmatrix}; \quad P = \begin{pmatrix} \frac{1}{2} & 0 & 0 & 0 & \frac{1}{2} \\ 0 & \frac{1}{2} & 0 & \frac{1}{2} & 0 \\ 0 & \frac{3}{4} & \frac{1}{8} & \frac{1}{8} & 0 \\ \frac{1}{2} & 0 & 0 & 0 & \frac{1}{2} \end{pmatrix}.$$

(a) Are the chains irreducible?
(b) Are the chains regular?
(c) For each chain, find the communicating classes.
(d) Are there any periodic states? If there are, identify them.
(e) Do both chains have a stationary distribution? Is there anything special about the stationary distribution of either chain? If so, what is special?

Exercise 10.22 *(Recurrent States).* Let $Z_i, i \geq 1$ be iid Poisson random variables with mean one. For each of the sequences

$$X_n = \sum_{i=1}^{n} Z_i, \ X_n = \max\{Z_1, \ldots, Z_n\}, \ X_n = \min\{Z_1, \ldots, Z_n\},$$

(a) Prove or disprove that $\{X_n\}$ is a stationary Markov chain.
(b) For those that are, write the transition probability matrix.
(c) Find the recurrent and the transient states of the chain.

Exercise 10.23 (Irreducibility and Aperiodicity). For stationary Markov chains with the following transition probability matrices, decide whether the chains are irreducible and aperiodic.

$$P = \begin{pmatrix} 0 & 1 \\ p & 1-p \end{pmatrix}; \quad P = \begin{pmatrix} \frac{1}{4} & \frac{1}{2} & \frac{1}{4} \\ 0 & \frac{1}{2} & \frac{1}{2} \\ 1 & 0 & 0 \end{pmatrix}; \quad P = \begin{pmatrix} 0 & 1 & 0 \\ 0 & 0 & 1 \\ p & 1-p & 0 \end{pmatrix}.$$

Exercise 10.24 (Irreducibility of the Machine Maintenance Chain). Consider the machine maintenance example given in the text. Prove that the chain is irreducible if and only if $p_0 > 0$ and $p_0 + p_1 < 1$.

Do some numerical computing that reinforces this theoretical result.

Exercise 10.25 * (Irreducibility of Loop Chains). Let $\{X_n\}$ be a stationary Markov chain and consider the *loop chain* defined by $Y_n = (X_n, X_{n+1})$. Prove that if $\{X_n\}$ is irreducible, then so is $\{Y_n\}$.

Do you think this generalizes to $Y_n = (X_n, X_{n+1}, \ldots, X_{n+d})$ for general $d \geq 1$?

Exercise 10.26 * (Functions of a Markov Chain). Consider the Markov chain $\{X_n\}$ corresponding to the simple random walk with general step probabilities $p, q, p + q = 1$.

(a) If $f(.)$ is any strictly monotone function defined on the set of integers, show that $\{f(X_n)\}$ is a stationary Markov chain.
(b) Is this true for a general chain $\{Y_n\}$? Prove it or give a counterexample.
(c) Show that $\{|X_n|\}$ is a stationary Markov chain, although $x \to |x|$ is not a strictly monotone function.
(d) Give an example of a function f such that $\{f(X_n)\}$ is not a Markov chain.

Exercise 10.27 (A Nonregular Chain with a Stationary Distribution). Consider a two-state stationary Markov chain with the transition probability matrix

$$P = \begin{pmatrix} 0 & 1 \\ 1 & 0 \end{pmatrix}.$$

(a) Show that the chain is not regular.
(b) Prove that nevertheless, the chain has a unique stationary distribution and identify it.

Exercise 10.28 * (Immigration–Death Model). At time $n, n \geq 1, U_n$ particles enter into a box. $U_1, U_2, \ldots$ are assumed to be iid with some common distribution F. The lifetimes of all the particles are assumed to be iid with common distribution G. Initially, there are no particles in the box. Let X_n be the number of particles in the box just after time n.

(a) Take F to be a Poisson distribution with mean two, and G to be geometric with parameter $\frac{1}{2}$. That is, G has the mass function $\frac{1}{2^x}, x = 1, 2, \ldots$. Write the transition probability matrix for $\{X_n\}$.
(b) Does $\{X_n\}$ have a stationary distribution? If it does, find it.

Exercise 10.29 * (Betting on the Basis of a Stationary Distribution). A particular stock either retains the value that it had at the close of the previous day, or gains a point, or loses a point, the respective states denoted as $1, 2, 3$. Suppose X_n is the state of the stock on the nth day; thus, X_n takes the values 1, 2, or 3. Assume that $\{X_n\}$ forms a stationary Markov chain with the transition probability matrix

$$P = \begin{pmatrix} 0 & \frac{1}{2} & \frac{1}{2} \\ \frac{1}{3} & \frac{1}{3} & \frac{1}{3} \\ \frac{1}{2} & \frac{3}{8} & \frac{1}{8} \end{pmatrix}.$$

A friend offers you the following bet: if the stock goes up tomorrow, he pays you 15 dollars, and if it goes down you pay him 10 dollars. If it remains the same as where it closes today, a fair coin will be tossed and he will pay you 10 dollars if a head shows up, and you will pay him 15 dollars if a tail shows up. Will you accept this bet? Justify with appropriate calculations.

Exercise 10.30 * **(Absent-Minded Professor).** A mathematics professor has two umbrellas, both of which were originally at home. The professor walks back and forth between home and office, and if it is raining when he starts a journey, he carries an umbrella with him unless both his umbrellas are at the other location. If it is clear when he starts a journey, he does not take an umbrella with him. We assume that at the time of starting a journey, it rains with probability p, and that the states of weather are mutually independent.

(a) Find the limiting proportion of journeys in which the professor gets wet.
(b) What if the professor had three umbrellas to begin with, all of which were originally at home?
(c) Is the limiting proportion affected by how many were originally at home?

Exercise 10.31 * **(Wheel of Fortune).** A pointed arrow is set on a circular wheel marked with m positions labeled as $0, 1, \ldots, m-1$. The hostess turns the wheel at each game, so that the arrow either remains where it was before the wheel was turned, or it moves to a different position. Let X_n denote the position of the arrow after n turns.

(a) Suppose at any turn, the arrow has an equal probability $\frac{1}{m}$ of ending up at any of the m positions. Does $\{X_n\}$ have a stationary distribution? If it does, identify it.
(b) Suppose at each turn, the hostess keeps the arrow where it was, or moves it one position clockwise or one position counterclockwise, each with an equal probability $\frac{1}{3}$. Does $\{X_n\}$ have a stationary distribution? If it does, identify it.
(c) Suppose again that each turn, the hostess keeps the arrow where it was, or moves it one position clockwise or one position counterclockwise, but now with respective probabilities $\alpha, \beta, \gamma, \alpha + \beta + \gamma = 1$. Does $\{X_n\}$ have a stationary distribution? If it does, identify it.

Exercise 10.32 (Wheel of Fortune Continued). Consider again the Markov chains corresponding to the wheel of fortune. Prove or disprove that they are irreducible and aperiodic.

Exercises

Exercise 10.33 * **(Stationary Distribution in Ehrenfest Model).** Consider the general Ehrenfest chain defined in the text, with m balls, and transfer probabilities $\alpha, \beta, 0 < \alpha, \beta < 1$. Identify a stationary distribution, if it exists.

Exercise 10.34 * **(Time Till Breaking Away).** Consider a general stationary Markov chain $\{X_n\}$ and let $T = \min\{n \geq 1 : X_n \neq X_0\}$.

(a) Can T be equal to ∞ with a positive probability?
(b) Give a simple necessary and sufficient condition for $P(T < \infty) = 1$.
(c) For each of the weather pattern, Ehrenfest urn, and the cat and mouse chains, compute $E(T \mid X_0 = i)$ for a general i in the corresponding state space S.

Exercise 10.35 ** **(Constructing Examples).** Construct an example of each of the following phenomena.

(a) A Markov chain with only absorbing states.
(b) A Markov chain that is irreducible but not regular.
(c) A Markov chain that is irreducible but not aperiodic.
(d) A Markov chain on an infinite state space that is irreducible and aperiodic, but not regular.
(e) A Markov chain in which there is at least one null recurrent state.
(f) A Markov chain on an infinite state space such that every state is transient.
(g) A Markov chain such that each first passage time T_{ij} has all moments finite.
(h) A Markov chain without a proper stationary distribution.
(i) Independent irreducible chains $\{X_n\}, \{Y_n\}$, such that $Z_n = (X_n, Y_n)$ is not irreducible.
(j) Markov chains $\{X_n\}, \{Y_n\}$ such that $Z_n = (X_n, Y_n)$ is not a Markov chain.

Exercise 10.36 * **(Reversibility of a Chain).** A stationary chain $\{X_n\}$ with transition probabilities p_{ij} is called reversible if there is a function $m(x)$ such that $p_{ij}m(i) = p_{ji}m(j)$ for all $i, j \in S$. Give a simple sufficient condition in terms of the function m which ensures that a reversible chain has a proper stationary distribution. Then, identify the stationary distribution.

Exercise 10.37. Give a physical interpretation for the property of reversibility of a Markov chain.

Exercise 10.38 (Reversibility). Give an example of a Markov chain that is reversible, and of one that is not.

Exercise 10.39 (Use Your Computer: Cat and Mouse). Take the cat and mouse chain and simulate it to find how long it takes for the cat and the mouse to end up in the same room. Repeat the simulation and estimate the expected time until the cat and the mouse end up in the same room. Vary the transition matrix and examine how the expected value changes.

Exercise 10.40 (Use Your Computer: Ehrenfest Urn). Take the symmetric Ehrenfest urn; that is, take $\alpha = \beta = .5$. Put all the m balls in the second urn. Simulate the chain and find how long it takes for the urns to have an equal number of balls for the first time. Repeat the simulation and estimate the expected time until both urns have an equal number of balls. Take $m = 10, 20$.

Exercise 10.41 (Use Your Computer: Gambler's Ruin). Take the gambler's ruin problem with $p = .4, .49$. Simulate the chain using $a = 10, b = 25$, and find the proportion of times that the gambler goes broke by repeating the simulation. Compare your empirical proportion with the exact theoretical value of the probability that the gambler will go broke.

References

Bhattacharya, R.N. and Waymire, E. (2009). *Stochastic Processes with Applications*, Siam, Philadelphia.
Brémaud, P. (1999). *Markov Chains, Gibbs Fields, Monte Carlo, and Queues*, Springer, New York.
Diaconis, P. (1988). *Group Representations in Probability and Statistics*, IMS, Lecture Notes and Monographs Series, Hayward, CA.
Feller, W. (1968). *An Introduction to Probability Theory, With Applications*, Wiley, New York.
Freedman, D. (1975). *Markov Chains*, Holden Day, San Francisco.
Isaacson, D. and Madsen, R. (1976). *Markov Chains, Theory and Applications*, Wiley, New York.
Kemperman, J. (1950). *The General One-Dimensional Random Walk with Absorbing Barriers*, Geboren Te, Amsterdam.
Meyn, S. and Tweedie, R. (1993). *Markov Chains and Stochastic Stability*, Springer, New York.
Norris, J. (1997). *Markov Chains*, Cambridge University Press, Cambridge, UK.
Seneta, E. (1981). *Nonnegative Matrices and Markov Chains*, Springer-Verlag, New York.
Stirzaker, D. (1994). *Elementary Probability*, Cambridge University Press, Cambridge, UK.

Chapter 11
Random Walks

We have already encountered the simple random walk a number of times in the previous chapters. Random walks occupy an extremely important place in probability because of their numerous applications, and because of their theoretical connections in suitable limiting paradigms to other important random processes in time. Random walks are used to model the value of stocks in economics, the movement of the molecules of a particle in a liquid medium, animal movements in ecology, diffusion of bacteria, movement of ions across cells, and numerous other processes that manifest random movement in time in response to some external stimuli. Random walks are indirectly of interest in various areas of statistics, such as sequential statistical analysis and testing of hypotheses. They also help a student of probability simply to understand randomness itself better.

We present a treatment of the theory of random walks in one or more dimensions in this section, focusing on the asymptotic aspects that relate to the long run probabilistic behavior of a particle performing a random walk. We recommend Feller (1971), Rényi (1970), and Spitzer (2008) for classic treatment of random walks; Spitzer (2008), in particular, provides a comprehensive coverage of the theory of random walks with numerous examples in setups far more general than we consider in this chapter.

11.1 Random Walk on the Cubic Lattice

Definition 11.1. A particle is said to perform a d-dimensional *cubic lattice random walk*, starting at the origin, if at each time instant $n, n \geq 1$, one of the d locational coordinates of the particle is changed by either $+1$ or -1, the direction of movement being equally likely to be either of these $2d$ directions.

As an example, in three dimensions, the particle's initial position is $\mathbf{S}_0 = (0, 0, 0)$ and the position at time $n = 1$ is $\mathbf{S}_1 = \mathbf{S}_0 + \mathbf{X}_1$, where $\mathbf{X}_1$ takes one of the values $(1, 0, 0), (-1, 0, 0), (0, 1, 0), (0, -1, 0), (0, 0, 1), (0, 0, -1)$ with an equal probability, and so on for the successive steps of the walk. In the cubic lattice random walk, at any particular time, the particle can take a unit step to the right, or left, or front, or back, or up, or down, choosing one of these six options with an equal probability. In d dimensions, it chooses one of the $2d$ options with an equal probability.

We show two simulations of 100 steps of a simple symmetric random walk on the line; this is just the cubic lattice random walk when $d = 1$. We use these two simulated plots to illustrate a number of important and interesting variables connected to a random walk. One example of such an interesting variable is the number of times the walk returns to its starting position in the first n steps (in the simulated plots, $n = 100$).

We now give some notation:

$$\mathbf{S}_n = d - \text{dimensional cubic lattice random walk}, \quad n \geq 0, \mathbf{S}_0 = \mathbf{0};$$
$$\mathcal{S} = \text{State space of } \{\mathbf{S}_n\} = \{(i_1, i_2, \ldots, i_d) : i_1, i_2, \ldots, i_d \in \mathbb{Z}\};$$
$$P_{d,n} = P(\mathbf{S}_n = \mathbf{0}); \quad P_n \equiv P_{1,n};$$
$$Q_{d,n} = P(\mathbf{S}_k \neq \mathbf{0} \,\forall 1 \leq k < n, \mathbf{S}_n = \mathbf{0});$$
$$= P(\text{The random walk returns to its starting point } \mathbf{0} \text{ for the first time at time n});$$
$$Q_d = \sum_{n=1}^{\infty} Q_{d,n} = \sum_{k=1}^{\infty} Q_{d,2k} = P(\text{The random walk ever returns to its starting point } \mathbf{0}).$$

And, for $d = 1$,

$v_i = $ The time of the ith return of the walk to 0;
$\theta_n = $ Number of times the walk returns to 0 in the first n steps;
$$\pi_n = \sum_{k=1}^{n} I_{\{S_k > 0\}}$$
$=$ Number of times in the first n steps that the walk takes a positive value.

Example 11.1 (Two Simulated Walks). We refer to the plots in Fig. 11.1 and Fig. 11.2 of the first 100 steps of the two simulated simple random walks. First, note that in both plots, the walk does not at all stay above and below the axis about 50% of the time. In the first simulation, the walk spends most of its time on the negative side, and in the second simulation, it does the reverse. This is borne out by theory, although at first glance it seems counter to intuition of most people. We give a table providing the values of the various quantities defined above corresponding to the two simulations.

Walk	θ_n	v_i	π_n
1	3	2, 14, 16	11
2	8	2, 4, 6, 8, 10, 22, 24, 34	69

11.1 Random Walk on the Cubic Lattice

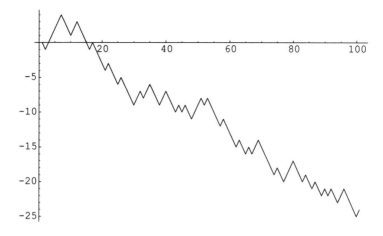

Fig. 11.1 A simulated random walk

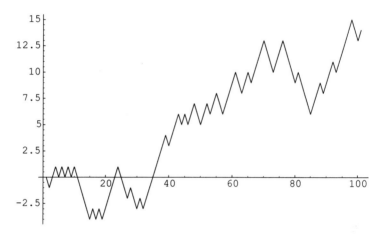

Fig. 11.2 A simulated random walk

A matter of key interest is whether the random walk returns to its origin, and if so how many times. More generally, we may ask how many times the random walk visits a given state $x \in \mathcal{S}$. These considerations lead to the following fundamental definition.

Definition 11.2. A state $x \in \mathcal{S}$ is said to be a *recurrent state* if

$$P(S_n = x \text{ infinitely often}) = 1.$$

The random walk $\{S_n\}$ is said to be *recurrent* if every $x \in \mathcal{S}$ is a recurrent state. $\{S_n\}$ is said to be *transient* if it is not recurrent.

11.1.1 Some Distribution Theory

We now show some exact and asymptotic distribution theory. Part of the asymptotic distribution theory has elements of surprise.

It is possible to write combinatorial formulas for $P_{d,n}$. These simplify to simple expressions for $d = 1, 2$. Note first that the walk cannot return to the origin at odd times $2n + 1$. For the walk to return to the origin at an even time $2n$, in each of the d coordinates, the walk has to take an equal number of positive and negative steps. We can think of this as a multinomial trial, where the $2n$ times $1, 2, \ldots, 2n$ are thought of as $2n$ balls, and they are dropped into $2d$ cells, with the restriction that pairs of cells receive equal number of balls $n_1, n_1, n_2, n_2, \ldots, n_d, n_d$. Thus,

$$P_{d,2n} = \frac{1}{(2d)^{2n}} \sum_{n_1,\ldots,n_d \geq 0, n_1+\cdots+n_d=n} \frac{(2n)!}{(n_1!)^2 \cdots (n_d!)^2}$$

$$= \frac{\binom{2n}{n}}{(2d)^{2n}} \sum_{n_1,\ldots,n_d \geq 0, n_1+\cdots+n_d=n} \left(\frac{n!}{n_1! \cdots n_d!}\right)^2.$$

In particular,

$$P_{1,2n} = \frac{\binom{2n}{n}}{(4)^n};$$

$$P_{2,2n} = \frac{[\binom{2n}{n}]^2}{(16)^n};$$

$$P_{3,2n} = \frac{\binom{2n}{n}}{(36)^n} \sum_{k,l \geq 0, k+l \leq n} \left[\frac{n!}{k!l!(n-k-l)!}\right]^2.$$

Apart from the fact that simply computable exact formulas are always attractive, these formulas have other very important implications, as we demonstrate shortly.

Example 11.2. We give a plot of $P_{d,2n}$ in Fig. 11.3 for $d = 1, 2, 3$ for $n \leq 25$. There are two points worth mentioning. The return probabilities for $d = 2$ and $d = 3$ cross each other. As $n \to \infty$, $P_{3,n} \to 0$ at a faster rate. This is shown later. The second point is that the return probabilities decrease the most when the dimension jumps from one to two.

In addition, here are some actual values of the return probabilities.

11.1 Random Walk on the Cubic Lattice

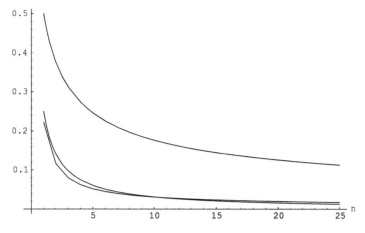

Fig. 11.3 Probability that d-dimensional lattice walk $= 0$ at time 2n, d $= 1, 2, 3$

n	$P_{1,n}$	$P_{2,n}$	$P_{3,n}$
2	.5	.25	.222
4	.375	.145	.116
6	.3125	.098	.080
10	.246	.061	.052
20	.176	.031	.031
50	.112	.013	.017

11.1.2 Recurrence and Transience

It is possible to make deeper conclusions about the lattice random walk by using the above exact formulas for $P_{d,n}$. Plainly, by using Stirling's approximation,

$$P_{1,2n} \sim \frac{1}{\sqrt{n\pi}}; \quad P_{2,n} \sim \frac{1}{n\pi},$$

where recall that the $\sim$ notation means that the ratio converges to one, as $n \to \infty$. Establishing the asymptotic order of $P_{d,n}$ for $d > 2$ takes a little more effort. This can be done by other more powerful means, but we take a direct approach. We use two facts:

(a) $\frac{1}{d^n} \sum_{n_1,\ldots,n_d \geq 0, n_1 + \cdots + n_d = n} \frac{n!}{n_1! \cdots n_d!} = 1$.
(b) The multinomial coefficient $\frac{n!}{n_1! \cdots n_d!}$ is maximized essentially when $n_1, \ldots, n_d$ are all equal.

Now using the exact expression from above,

$$\begin{aligned}
P_{d,2n} &= \frac{\binom{2n}{n}}{(2d)^{2n}} \sum_{n_1,\ldots,n_d \geq 0, n_1+\cdots+n_d=n} \left(\frac{n!}{n_1!\cdots n_d!}\right)^2 \\
&= \frac{\binom{2n}{n}}{(4d)^n} \sum_{n_1,\ldots,n_d \geq 0, n_1+\cdots+n_d=n} \frac{1}{d^n} \frac{n!}{n_1!\cdots n_d!} \frac{n!}{n_1!\cdots n_d!} \\
&\leq \frac{\binom{2n}{n}}{(4d)^n} \max_{n_1,\ldots,n_d \geq 0, n_1+\cdots+n_d=n} \frac{n!}{n_1!\cdots n_d!} \\
&\approx \frac{\binom{2n}{n}}{(4d)^n} \frac{n!}{[\Gamma(\frac{n}{d}+1)]^d} \\
&= O\left(\frac{1}{n^{d/2}}\right),
\end{aligned}$$

by applying Stirling's approximation to the factorials and also the Gamma function.

This proves that $P_{d,2n} = O(\frac{1}{n^{d/2}})$, but it does not prove that $P_{d,2n}$ is of the exact order of $\frac{1}{n^{d/2}}$. In fact, it is. We state the following exact asymptotic result.

Theorem 11.1. *For the cubic lattice random walk in d dimensions, $1 \leq d < \infty$,*

$$P_{d,2n} = \frac{1}{2^{d-1}} \left(\frac{d}{n\pi}\right)^{d/2} + o(n^{-d/2}).$$

See Rényi (1970, p. 503). An important consequence of this theorem is the following result.

Corollary. *The d-dimensional cubic lattice random walk is transient for $d \geq 3$.*

Proof. From the above theorem, if $d \geq 3$, $\sum_{n=1}^{\infty} P_{d,n} = \sum_{n=1}^{\infty} P_{d,2n} < \infty$, because $\sum_{n=1}^{\infty} n^{-d/2} < \infty \; \forall d \geq 3$, and if $\epsilon_n = o(n^{-d/2})$ for any $d \geq 3$, then $\sum_{n=1}^{\infty} \epsilon_n < \infty$. Therefore, by the Borel–Cantelli lemma, the probability that the random walk takes the zero value infinitely often is zero. Hence, for $d \geq 3$, the walk must be transient. □

How about the case of one and two dimensions? In those two cases, the cubic lattice random walk is in fact recurrent. To show this, note that it is enough to show that the walk returns to zero at least once with probability one (if it returns to zero at least once with certainty, it will return infinitely often with certainty). But, using our previous notation,

$$P(\text{The random walk returns to zero at least once}) = \sum_{n=1}^{\infty} Q_{d,2n}.$$

This can be proved to be one for $d = 1, 2$ in several ways. One possibility is to show, by using generating functions, that, always,

$$\sum_{n=1}^{\infty} Q_{d,2n} = \frac{\sum_{n=1}^{\infty} P_{d,2n}}{1 + \sum_{n=1}^{\infty} P_{d,2n}},$$

11.1 Random Walk on the Cubic Lattice

and therefore, $\sum_{n=1}^{\infty} Q_{d,2n} = 1$ if $d = 1, 2$. Other more sophisticated methods using more analytical means are certainly available; one of these is to use the Chung–Fuchs (1951) theorem, which is described in a subsequent section. There is also an elegant Markov chain argument for proving the needed recurrence result, and we outline that argument below. In fact, the cubic lattice random walk visits every state infinitely often with probability one when $d = 1, 2$; we record this formally.

Theorem 11.2 (Recurrence and Transience). *The d-dimensional cubic lattice random walk is recurrent for $d = 1, 2$ and transient for $d \geq 3$.*

We now outline a Markov chain argument for establishing the recurrence properties of a general Markov chain. This then automatically applies to the lattice random walk, because it is a Markov chain. For notational brevity, we treat only the case $d = 1$. But the Markov chain argument will in fact work in any number of dimensions. Here is the Markov chain theorem that we use (this is the same as Theorem 10.2).

Theorem 11.3. *Let $\{X_n\}, n \geq 0$ be a Markov chain with the state space S equal to the set of all integers. Assume that $X_0 = 0$. Let*

$$p_{00}^{(k)} = P(X_k = 0); \quad \rho = P(\text{The chain always returns to zero});$$
$$N = \text{Total number of visits of the chain to zero}.$$

Then,

$$\sum_{k=1}^{\infty} p_{00}^{(k)} = \sum_{n=1}^{\infty} \rho^n = E(N).$$

The consequence we are interested in is the following.

Corollary 11.1. *The Markov chain $\{X_n\}$ returns to zero infinitely often with probability one if and only if $\sum_{k=1}^{\infty} p_{00}^{(k)} = \infty$.*

It is now clear why the lattice random walk is recurrent in one dimension. Indeed, we have already established that for the one-dimensional lattice random walk, $p_{00}^{(k)} = 0$ if k is odd, and $p_{00}^{(2n)} = P_{1,2n} \sim \frac{1}{\sqrt{n\pi}}$. Consequently, $\sum_{k=1}^{\infty} p_{00}^{(k)} = \sum_{n=1}^{\infty} P_{1,2n} = \infty$, and therefore, by the general Markov chain theorem we stated above, the chain returns to zero infinitely often with probability one in one dimension.

Exactly the same Markov chain argument also establishes the lattice random walk's recurrence in two dimensions, and transience in dimensions higher than two. This is an elegant and shorter derivation of the recurrence structure of the lattice random walk than the alternative direct method we provided first in this section. The disadvantage of the shorter proof is that we must appeal to Markov chain theory to give the shorter proof.

11.1.3 * Pólya's Formula for the Return Probability

We just saw that in dimensions three or higher, the probability that the cubic lattice random walk returns to zero infinitely often is zero. In notation, if $d \geq 3$, $P(S_n = 0$ for infinitely many $n) = 0$. But this does not mean that $P(S_n = 0$ for some $n) = 0$ also. This latter probability is the probability that the walk returns to the origin at least once. We know that for $d \geq 3$, it is not 1; but it need not be zero. Indeed, it is something in between. In 1921, Pólya gave a pretty formula for the probability that the cubic lattice random walk returns to the origin at least once.

Theorem 11.4 (Pólya's Formula). *Let S_n be the cubic lattice random walk starting at 0. Then,*

$$Q_{d,n} = P(\text{The walk returns to } 0 \text{ at least once})$$
$$= \frac{\sum_{n=1}^{\infty} P_{d,2n}}{1 + \sum_{n=1}^{\infty} P_{d,2n}},$$

where

$$\sum_{n=1}^{\infty} P_{d,2n} = \text{The expected number of returns to the origin}$$

$$= \frac{d}{(2\pi)^d} \int_{-\pi}^{\pi} \cdots \int_{-\pi}^{\pi} \left(d - \sum_{k=1}^{d} \cos \theta_k \right)^{-1} d\theta_1 \cdots d\theta_d - 1$$

$$= \int_0^{\infty} e^{-t} \left[I_0 \left(\frac{t}{d} \right) \right]^d dt - 1.$$

In particular, for $d = 3$,

$$\sum_{n=1}^{\infty} P_{3,2n} = \frac{3}{(2\pi)^3} \int_{-\pi}^{\pi} \int_{-\pi}^{\pi} \int_{-\pi}^{\pi} \frac{1}{3 - \cos x - \cos y - \cos z} dx dy dz - 1$$

$$= \frac{\sqrt{6}}{32\pi^3} \Gamma\left(\frac{1}{24}\right) \Gamma\left(\frac{5}{24}\right) \Gamma\left(\frac{7}{24}\right) \Gamma\left(\frac{11}{24}\right) - 1$$

$$= .5164$$

⇒ P(The three-dimensional cubic lattice random walk returns at least once to the origin) $= .3405$.

In addition to Pólya (1921), see Finch (2003, p. 322) for these formulas.

Example 11.3 (Expected Number and Probability of Return). For computational purposes, the integral representation using the Bessel I_0 function is the most convenient, although careful numerical integration is necessary to get accurate values. Some values are given in the following table.

d	Expected Number of Returns	Probability of Return
3	.5164	.3405
4	.2394	.1931
5	.1562	.1351
8	.0785	.0728
10	.0594	.0561
15	.0371	.0358
50	.0102	.0101
51	.0100	.0099

We note that it takes 50 dimensions for the probability of a return to become 1%.

11.2 First Passage Time and Arc Sine Law

For the simple symmetric random walk on the real line starting at zero, consider the number of steps necessary for the walk to reach a given integer j for the first time:

$$T_j = \min\{n > 0 : S_n = j\}.$$

Also let $T_{j,r}, r \geq 1$ denote the successive visit times of the random walk of the integer j; note that the one-dimensional symmetric walk we are considering visits every integer j infinitely often with probability one, and so it is sensible to talk of $T_{j,r}$.

Definition 11.3. $T_j = T_{j,1}$ is called the *first passage time* to j, and the sequence $\{T_{j,r}\}$ the *recurrence times* of j.

For the special case $j = 0$, we denote the recurrence times as $v_1, v_2, v_3, \ldots$, instead of using the more complicated notation $T_{0,1}, T_{0,2}, \ldots$. Our goal is to write the distribution of v_1, and from there conclude the asymptotic distribution of v_r as $r \to \infty$. It turns out that although the random walk returns to zero infinitely often with probability one, the expected value of v_1 is infinite! This precludes Gaussian asymptotics for v_r as $r \to \infty$. Recall also that θ_n denotes the number of times in the first n steps that the random walk returns to zero.

Theorem 11.5 (Distribution of Return Times).

(a) $v_1, v_2 - v_1, v_3 - v_2, \ldots$ are independent and identically distributed.
(b) The generating function of v_1 is $E(t^{v_1}) = 1 - \sqrt{1 - t^2}, |t| \leq 1$.
(c) $P(v_1 = 2k + 1) = 0 \; \forall k; \; P(v_1 = 2k) = \frac{\binom{2k-2}{k-1}}{k \, 2^{2k-1}}$.
(d) $E(v_1) = \infty$.
(e) The characteristic function of v_1 is $\psi_1(t) = 1 - \sqrt{1 - e^{2it}}$.

(f) $P(\frac{v_r}{r^2} \leq x) \to 2[1 - \Phi(\frac{1}{\sqrt{x}})]\ \forall x > 0,\ as\ r \to \infty.$

(g) $P(\frac{\theta_n}{\sqrt{n}} \leq x) \to 2\Phi(x) - 1\ \forall x > 0,\ as\ n \to \infty.$

Detailed Sketch of Proof. Part (a) is a consequence of the observation that each time the random walk returns to zero, it simply starts over again. A formal proof establishes that the Markov property is preserved at first passage times.

For part (b), first obtain the generating function

$$G(t) = \sum_{n=1}^{\infty} P_{1,n} t^n$$

$$= \sum_{n=1}^{\infty} P_{1,2n} t^{2n} = \sum_{n=1}^{\infty} \frac{\binom{2n}{n}}{4^n} t^{2n}$$

$$= \frac{1}{\sqrt{1-t^2}} - 1,\quad |t| < 1.$$

Now, notice that the sequences $P_{1,n}, Q_{1,n}$ satisfy the recursion relation

$$P_{1,2n} = Q_{1,2n} + \sum_{k=1}^{n-1} P_{1,2k} Q_{1,2n-2k}.$$

This results in the functional identity

$$E(t^{\nu_1}) = \sum_{n=1}^{\infty} Q_{1,n} t^n$$

$$= \frac{G(t)}{1 + G(t)},\ |t| < 1,$$

which produces the expression of part (b). Part (c) now comes out of part (b) on differentiation of the generating function $E(t^{\nu_1})$. Part (d) follows on directly showing that $\sum_{k=1}^{\infty} k \frac{\binom{2k-2}{k-1}}{k \, 2^{2k-1}} = \infty$.

Parts (e) and (f) are connected. The characteristic function formula follows by a direct evaluation of the complex power series $\psi_1(t) = \sum_{n=1}^{\infty} Q_{1,n} e^{itn} = \sum_{n=1}^{\infty} Q_{1,2n} e^{2itn} = \sum_{n=1}^{\infty} e^{2itn} \frac{\binom{2n-2}{n-1}}{n 2^{2n-1}}$. Alternatively, it can also be deduced by the argument that led to the generating function formula of part (b).

For part (f), by the iid property of $\nu_1, \nu_2 - \nu_1, \ldots$, we can represent ν_r as a sum of iid variables

$$\nu_r = \nu_1 + (\nu_2 - \nu_1) + \cdots + (\nu_r - \nu_{r-1}).$$

Therefore, by virtue of part (e), we can write the characteristic function $\psi_r(t)$ of ν_r, and we get

$$\lim_{r \to \infty} \psi_r\left(\frac{t}{r^2}\right) = \lim_{r \to \infty} \left[\psi\left(\frac{t}{r^2}\right)\right]^r = e^{-\sqrt{-2it}}.$$

11.2 First Passage Time and Arc Sine Law

We now use the inversion of technique to produce a density, which, by the continuity theorem for characteristic functions, must be the density of the limiting distribution of $\frac{v_r}{r^2}$. The inversion formula gives

$$f(x) = \frac{1}{2\pi} \int_{-\infty}^{\infty} e^{-itx} e^{-\sqrt{-2it}} dt$$

$$= \frac{e^{-\frac{1}{2x}} x^{-\frac{3}{2}}}{\sqrt{2\pi}}, x > 0.$$

The CDF corresponding to this density f is $2[1 - \Phi(\frac{1}{\sqrt{x}})], x > 0$, which is what part (f) says.

Finally, part (g) is actually a restatement of part (f) because of the identity $P(v_r \leq n) = P(\theta_n \geq r)$. □

Example 11.4 (Returns to Origin and Stable Law of Index $\frac{1}{2}$). There are several interesting things about the density $f(x)$ of the limiting distribution of $\frac{v_r}{r^2}$. First, by inspection, we see that it is the density of $\frac{1}{Z^2}$ where $Z \sim N(0, 1)$. It clearly does not have a finite expectation (and neither does v_r for any r). Moreover, the characteristic function of f matches the form of the characteristic function of a one-sided (positive) stable law of index $\alpha = \frac{1}{2}$. This is an example of a density with tails even heavier than that of the absolute value of a Cauchy random variable. Although the recurrence times do not have a finite expectation, either for finite r or asymptotically, it is interesting to know that the median of the asymptotic distribution of $\frac{v_r}{r^2}$ is $[\frac{1}{\Phi^{-1}(\frac{3}{4})}]^2 = 2.195$. So, for large r, with probability about 50%, the rth return to zero will come within about $2.2r^2$ steps of the random walk.

A plot of the asymptotic density in Fig. 11.4 shows the extremely sharp peak near zero, accompanied by a very flat tail. The lack of a finite expectation is due to this tail.

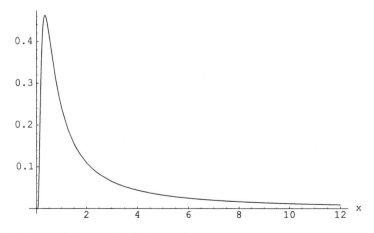

Fig. 11.4 Asymptotic density of scaled return times

We next turn to the question of studying the proportion of time that a random walk spends above the horizontal axis. In notation, we are interested in the distribution of $\frac{\pi_n}{n}$, where recall that $\pi_n = \sum_{k=1}^n I_{\{S_k > 0\}}$. The steps each have a symmetric distribution, and they are independent, thus each S_k has a symmetric distribution. Intuitively, one might expect that a trajectory of the random walk, meaning the graph of the points (n, S_n), is above and below the axis close to 50% of the time. The arc sine law, which we now present, says that this is not true. It is more likely that the proportion $\frac{\pi_n}{n}$ will either be near zero or near one, than that it will be near the naively expected value $\frac{1}{2}$. We provide formulas for the exact and asymptotic distribution of π_n.

Theorem 11.6 (Arc Sine Law).

(a) For each $n \geq 1$, $P(\pi_n = k) = \frac{\binom{2k}{k}\binom{2n-2k}{n-k}}{4^n}$, $k = 0, 1, \ldots, n$, with $\binom{0}{0} = 1$.

(b) For each $x \in [0, 1]$, $P(\frac{\pi_n}{n} \leq x) \to \frac{2}{\pi} \arcsin(\sqrt{x})$ as $n \to \infty$.

The proof of part (a) is a careful combinatorial exercise, and we omit it. Feller (1971) gives a very careful presentation of the argument. However, granted part (a), part (b) then follows by a straightforward application of Stirling's approximation, using the exact formula of part (a) with $k = k_n = \lfloor nx \rfloor$.

Example 11.5. The density of the CDF $\frac{2}{\pi} \arcsin(\sqrt{x})$ is the Beta density $f(x) = \frac{1}{\pi\sqrt{x(1-x)}}, 0 < x < 1$. The density is unbounded as $x \to 0, 1$, and has its minimum (rather than the maximum) at $x = \frac{1}{2}$. Consider the probability that the random walk takes a positive value between 45% and 55% of the times in the first n steps. The arc sine law implies, by just integrating the Beta density, that this probability is .0637 for large n. Now, consider the probability that the random walk takes a positive value either more than 95% or less than 5% of the times in the first n steps. By integrating the Beta density, we find this probability to be .2871, more than four times the .0637 value. We can see the tendency of a trajectory to spend most of the time either above or below the axis, rather than dividing its time on a close to 50–50 basis above and below the axis. This seems counter to intuition largely because we automatically expect a random variable to concentrate near its expectation, which in this case is $\frac{1}{2}$. The arc sine law of random walks says that this sort of intuitive expectation can land us in trouble.

We provide some values on the exact distribution of $\frac{\pi_n}{n}$ for certain selected n.

n	$P(.45 \leq \frac{\pi_n}{n} \leq .55)$	$P(\frac{\pi_n}{n} \leq .05$ or $\geq .95)$
10	.0606	.5379
25	.0500	.4269
50	.0631	.3518
100	.0698	.3077
250	.0636	.3011

11.3 ∗ The Local Time

Consider the simple symmetric random walk starting at zero, with iid steps having the common distribution $P(X_i = 1) = P(X_i = -1) = \frac{1}{2}$, and fix any integer x. The family of random variables

$$\xi(x,n) = \#\{k : 0 \leq k \leq n, S_k = x\}$$

is called the *local time* of the random walk. Note, therefore, that $\xi(0,n)$ is the same as θ_n. The local time of the random walk answers the following interesting question of how much time does the random walk spend at x up to the time n? It turns out that the distribution of $\xi(x,n)$ can be written in a simple closed-form, and therefore we can compute it.

Note that $\xi(0,2n) = \xi(0,2n+1)$, because the simple symmetric walk starting at zero can visit the origin only at even numbered times. Therefore, in the case of $\xi(0,n)$, it is enough to know the distribution of $\xi(0,2n)$ for a general n.

Theorem 11.7. *For any $n \geq 1, 0 \leq k \leq n$,*

$$P(\xi(0,2n) = k) = P(\xi(0,2n+1) = k) = \frac{\binom{2n-k}{n}}{2^{2n-k}}.$$

Proof. For the proof of this formula, we require a few auxiliary combinatorial facts, which we state together as a lemma.

Lemma. *For the simple symmetric random walk, with the same notation as in the previous section,*

(a) *For any $k \geq 1$, $P(v_1 = 2k \mid X_1 = 1) = P(v_1 = 2k \mid X_1 = -1) = P(T_1 = 2k-1)$.*

(b) *For any $k \geq 1$, T_k and $v_k - k$ have the same distribution.*

(c) *For any n, and $0 \leq k \leq n$, $P(M_n = k) = \frac{\binom{n}{\lfloor \frac{n-k}{2} \rfloor}}{2^n}$.*

Part (c) of the lemma is not very easy to prove, and we do not present its proof here. Part (a) is immediate from the symmetry of the distribution of X_1 and from the total probability formula

$$P(v_1 = 2k) = \frac{1}{2}P(v_1 = 2k \mid X_1 = 1) + \frac{1}{2}P(v_1 = 2k \mid X_1 = -1)$$
$$= \frac{1}{2}P(T_1 = 2k - 1) + \frac{1}{2}P(T_1 = 2k - 1) = P(T_1 = 2k - 1).$$

Therefore, $v_1 - 1$ and T_1 have the same distribution. It therefore follows that for any $k \geq 1$, $v_k - k$ and T_k have the same distribution. As a consequence, we have the important identity

$$P(v_k > n + k) = P(v_k - k > n) = P(T_k > n) = P(M_n < k).$$

Now, observe that

$$P(\xi(0,2n) = k) = P(v_k \leq 2n, v_{k+1} > 2n)$$

$$= P(M_{2n-k} \geq k, M_{2n-k-1} \leq k) = P(M_{2n-k} = k) = \frac{\binom{2n-k}{n}}{2^{2n-k}},$$

by part (c) of the lemma. This proves the formula given in the theorem.

It turns out that once we have the distribution of $\xi(0, 2n)$, that of $\xi(x, n)$ can be found by cleverly conditioning on the value of the first time instant that the random walk hits the number x. Precisely, for $x > 0$, write

$$P(\xi(x,n) = k) = \sum_{j=x}^{n-2k} P(\xi(x,n) = k \mid T_x = j) P(T_x = j)$$

$$= \sum_{j=x}^{n-2k} P(\xi(0, n-j) = k-1) P(T_x = j),$$

and from here, on some algebra, the distribution of $\xi(x, n)$ works out to the following formula. □

Theorem 11.8. *For $x, k > 0$,*

$$P(\xi(x,n) = k) = \frac{\binom{n-k+1}{\frac{n+x}{2}}}{2^{n-k+1}},$$

if $(n + x)$ is even;

$$P(\xi(x,n) = k) = \frac{\binom{n-k}{\frac{n+x-1}{2}}}{2^{n-k}},$$

if $(n + x)$ is odd.

Remark. In the above, combinatorial coefficients $\binom{r}{s}$ are to be taken as zero if $r < s$. It is a truly rewarding consequence of clever combinatorial arguments that in the symmetric case, the distribution of the local time can in fact be fully written down.

Example 11.6 (The Local Time at Zero). Consider the simple symmetric random walk starting at zero. Because zero is the starting point of the random walk, the local time at zero is of special interest. In this example, we want to investigate the local time at zero in some additional detail.

First, what exactly does the distribution look like as a function of n? By using our analytical formula,

$$P(\xi(0, 2n) = k) = \frac{\binom{2n-k}{n}}{2^{2n-k}},$$

we can easily compute the pmf of the local time for (small) given n. An inspection of the pmf will help us in understanding the distribution. Here is a small table.

	$P(\xi(0, 2n) = k)$		
k	$n = 3$	$n = 6$	$n = 10$
0	.3125	.2256	.1762
1	.3125	.2256	.1762
2	.25	.2051	.1669
3	.125	.1641	.1484
4		.1094	.1222
5		.0547	.0916
6		.0156	.0611
7			.0349
8			.0161
9			.0054
10			.0010

We see from this table that the distribution of the local time at zero has a few interesting properties: it is monotone nondecreasing, and it has its maximum value at $k = 0$ and $k = 1$. Thus the random walk spends more time near its original home than at another location. Both of these properties can be proved analytically. Also, from the table of the pmf, we can easily compute the mean of the local time at zero. For $n = 3, 6, 10$, $E[\xi(0, 2n)]$ equals 1.1875, 1.93, and 2.70. For somewhat larger n, say $n = 100$, $E[\xi(0, 2n)]$ equals 10.33. The mean local time grows at the rate of $\sqrt{n}$. In fact, when normalized by $\sqrt{n}$, the local time has a limiting distribution, and the limiting distribution is a very interesting one, namely the absolute value of a standard normal. This can be proved by using Stirling's approximation in the exact formula for the pmf of the local time. Here is the formal result.

Theorem 11.9. *Let* $Z \sim N(0, 1)$. *Then,*

$$\frac{\xi(0, n)}{\sqrt{n}} \stackrel{\mathcal{L}}{\Rightarrow} |Z|.$$

11.4 Practically Useful Generalizations

Difficult and deep work by a number of researchers, including in particular, Sparre-Andersen, Erdös, Kac, Spitzer, and Kesten, give fairly sweeping generalizations of many of the results and phenomena of the previous section to random walks more general than the cubic lattice random walk. Feller (1971) and Rényi (1970) are classic and readable accounts of the theory of random walks, as it stood at that time. These generalizations are of particular interest to statisticians, because of their methodological potential in fields such as testing of hypotheses. For the sake of reference, we collect two key generalizations in this section. Another major general result is postponed till the next section.

Definition 11.4. Let $X_i, i \geq 1$ be iid random variables with common CDF F, and let $S_n = \sum_{i=1}^{n} X_i, n \geq 1$. Let x be a real number and $S_{n,x} = x + S_n$. Then $\{S_{n,x}\}, n \geq 1$ is called a random walk with step distribution F starting at x.

Theorem 11.10. *(a) (Sparre-Andersen). If $d = 1$, and the common distribution F of X_i has a density symmetric about zero, then the distribution of π_n is as in the case of the simple symmetric random walk; that is,*

$$P(\pi_n = k) = \frac{\binom{2k}{k}\binom{2n-2k}{n-k}}{4^n}, \quad k = 0, 1, \ldots, n.$$

(b) (Erdös–Kac). If X_i are independent, have zero mean, and satisfy the conditions of the Lindeberg–Feller theorem, but are not necessarily iid, then the arc sine law holds:

$$P\left(\frac{\pi_n}{n} \leq x\right) \to \frac{2}{\pi} \arcsin(\sqrt{x}), \quad \forall x \in [0, 1].$$

Part (a) is also popularly known as Spitzer's identity. If, in part (a), we choose $k = n$, we get

$$\begin{aligned} P(\pi_n = n) &= P(S_1 > 0, S_2 > 0, \ldots, S_n > 0) \\ &= P(S_1 \geq 0, S_2 \geq 0, \ldots, S_n \geq 0) \\ &= \frac{\binom{2n}{n}}{4^n} = P(S_{2n} = 0), \end{aligned}$$

for every CDF F with a symmetric density function. To put it another way, if T marks the first time that the random walk enters negative territory, then $P(T > n)$ is completely independent of the underlying CDF, as long as it has a symmetric density. For the simple symmetric random walk, the same formula also holds, and is commonly known as the ballot theorem. A Fourier analytic proof of Spitzer's identity is given in Dym and McKean (1972, pp. 184–187).

11.5 Wald's Identity

Returning to the random variable T, namely the first time the random walk enters negative territory, an interesting general formula for its expectation can be given, for random walks more general than the simple symmetric random walk. We have to be careful about talking about $E(T)$, because it need not be finite. But if it is, then it is an interesting number to know. Consider, for example, a person gambling in a casino, and repeatedly playing a specific game. We may assume that the game is (at least slightly) favorable to the house, and unfavorable to the player. So, intuitively, the player already knows that eventually she will be sunk. But how long can she continue without being sunk? If the player expects that she can hang around for a long time, it may well add to the excitement of the game.

11.5 Wald's Identity

We do in fact deal with a more general result, known as *Wald's identity*. It was proved in the context of sequential testing of hypotheses in statistics. To describe the identity, we need a definition. We caution the reader that although the meaning of the definition is clear, it is not truly rigorous because of our not using measure-theoretic terminology.

Definition 11.5. Let $X_1, X_2, \ldots$ be an infinite sequence of random variables defined on a common sample space Ω. A nonnegative integer-valued random variable N, also defined on Ω, is called a *stopping time* if whether or not $N \leq n$ for a given n can be determined by only knowing $X_1, X_2, \ldots, X_n$, and if, moreover, $P(N < \infty) = 1$.

Example 11.7. Suppose a fair die is rolled repeatedly (and independently), and let the sequence of rolls be $X_1, X_2, \ldots$. Let N be the first throw at which the sum of the rolls exceeds 10. In notation, $N = \min\{n : S_n = X_1 + \cdots + X_n > 10\}$. Clearly, N cannot be more than 11 and so $P(N < \infty) = 1$, and also, whether the sum has exceeded 10 within the first n rolls can be decided by knowing the values of only the first n rolls. So N is a valid stopping time.

Example 11.8. Suppose $X_1, X_2, \ldots$ are iid $N(0, 1)$, and let $W_n = X_{n:n} - X_{1:n}$ be the range of the $X_1, X_2, \ldots, X_n$. Suppose N is the first time the range W_n exceeds five. Because $X_{n:n} \stackrel{a.s.}{\Rightarrow} \infty$, and $X_{1:n} \stackrel{a.s.}{\Rightarrow} -\infty$, we have that $W_n \stackrel{a.s.}{\Rightarrow} \infty$. Therefore, $P(N < \infty) = 1$. Also, evidently, whether $W_n > 5$ can be decided by knowing the values of only $X_1, X_2, \ldots, X_n$. So N is a valid stopping time.

Theorem 11.11 (Wald's Identity). *Let $X_i, i \geq 1$ be iid random variables, with $E(|X_1|) < \infty$. Let $S_n = X_1 + \cdots + X_n, n \geq 1, S_0 = 0$ and let N be a stopping time with a finite expectation. Then $E(S_N) = E(N)E(X_1)$.*

Proof. The proof is not completely rigorous, and we should treat it as a sketch of the proof. First, note that

$$S_N = \sum_{n=1}^{\infty} S_N I_{\{N=n\}} = \sum_{n=1}^{\infty} S_n I_{\{N=n\}}$$

$$= \sum_{n=1}^{\infty} S_n [I_{\{N>n-1\}} - I_{\{N>n\}}]$$

$$= \sum_{n=1}^{\infty} X_n I_{\{N>n-1\}}$$

$$\Rightarrow E(S_N) = E\left[\sum_{n=1}^{\infty} X_n I_{\{N>n-1\}}\right]$$

$$= \sum_{n=1}^{\infty} E[X_n I_{\{N>n-1\}}]$$

$$= \sum_{n=1}^{\infty} E[X_n] E[I_{\{N>n-1\}}]$$

(because N is assumed to be a stopping time, and so whether $N > n - 1$ depends only on $X_1, X_2, \ldots, X_{n-1}$, and so $I_{\{N>n-1\}}$ is independent of X_n)

$$= \sum_{n=1}^{\infty} E[X_1]E[I_{\{N>n-1\}}]$$

(because the X_i all have the same expectation)

$$= E[X_1] \sum_{n=1}^{\infty} P(N > n - 1)$$

$$= E(X_1)E(N). \qquad \square$$

Example 11.9 (Time to Enter Negative Territory). Consider a one-dimensional random walk on $\mathbb{Z}$, but not the symmetric one. The random walk is defined as follows. Suppose $X_i, i \geq 1$ are iid with the common distribution $P(X_i = 1) = p, P(X_i = -1) = 1 - p, p < \frac{1}{2}$. Let $S_0 = 0, S_n = X_1 + \cdots + X_n, n \geq 1$. This corresponds to a gambler betting repeatedly on something, where he wins one dollar with probability p and loses one dollar to the house with probability $1 - p$. Then, S_n denotes the player's net fortune after the nth play. Now consider the first time his net fortune becomes negative. This is of interest to him if he has decided to pack it in as soon as his net fortune becomes negative. In notation, we are looking at the stopping time $T = \min\{n \geq 1 : S_n < 0\}$. The random walk moves by just one step at a time, therefore note that $S_T = -1$. Provided that $E(T) < \infty$ (which we have not proved), by Wald's identity,

$$E(S_T) = E(-1) = -1 = E(X_1)E(T) = (2p - 1)E(T)$$
$$\Rightarrow E(T) = \frac{1}{1 - 2p}.$$

This is in fact correct, as it can be shown that $E(T) < \infty$.

Suppose now the game is just slightly favorable to the house, say, $p = .49$. Then we get, $E(T) = 50$. If each play takes just one minute, the gambler can expect to hang in for about an hour, with a minimal loss.

11.6 * Fate of a Random Walk

Suppose $X_1, X_2, \ldots$ is a sequence of iid real random variables, and for each $n \geq 1$, let $S_n = \sum_{i=1}^{n} X_i$; set also $S_0 = 0$. This is a *general one-dimensional random walk* with iid steps $X_1, X_2, \ldots$. A very natural question is what does such a general random walk do in the long run? Excluding the trivial case that the X_i are degenerate at zero, the random walk $S_n, n \geq 1$ can do one of three things. It can drift off to $+\infty$,

11.6 Fate of a Random Walk

or drift off to $-\infty$, or it can oscillate. If it oscillates, what can we say about the nature of its oscillation? For example, under certain conditions on the common distribution of the X_i, does it come arbitrarily close to any real number over and over again? Clearly, the answer depends on the common distribution of the X_i. For example, if the X_i can only take the values ± 1, then obviously the random walk cannot come arbitrarily close to any real number over and over again. The answer in general is that either the random walk does not visit neighborhoods of any number over and over again, or that it visits neighborhoods of every number over and over again, or perhaps that it visits neighborhoods of certain distinguished numbers over and over again. We give two formal definitions and a theorem for the one-dimensional case first. The case of random walks in general dimensions is treated following the one-dimensional case.

Definition 11.6. Let $S_n, n \geq 1$ be a general random walk on $\mathcal{R}$. A specific real number x is called a *possible value* for the random walk S_n if for any given $\epsilon > 0$, $P(|S_n - x| \leq \epsilon) > 0$ for some $n \geq 1$.

Definition 11.7. Let $S_n, n \geq 1$ be a general random walk on $\mathcal{R}$. A specific real number x is called a *recurrent value* for the random walk S_n if

for any given $\epsilon > 0$, $P(|S_n - x| \leq \epsilon$ infinitely often$) = 1$.

Definition 11.8. Let $\mathcal{X} = \{x \in \mathcal{R} : x \text{ is a recurrent value of } S_n\}$. Then $\mathcal{X}$ is called the *recurrent set* of S_n.

Then, we have the following neat dichotomy result on the recurrence structure of a general nontrivial random walk; see Chung (1974) or Chung (2001, p. 279) for a proof.

Theorem 11.12. Let $S_n, n \geq 1$ be a general random walk on $\mathcal{R}$. Assume that $P(X_i = 0) \neq 1$. Then the recurrent set $\mathcal{X}$ of S_n is either empty, or the entire real line, or a countable lattice set of the form $\{\pm n x_0 : n = 0, 1, 2, \ldots\}$ for some specific real number x_0.

Remark. The recurrent set will be empty when the random walk drifts off to one of $+\infty$ or $-\infty$. For example, an asymmetric simple random walk will do so. The simple symmetric random walk will have a countable lattice set as its recurrent set. On the other hand, as we later show, a random walk with iid standard normal steps will have the entire real line as its recurrent set.

Although we shortly present a handy and effective all at one time theorem for verifying whether a particular random walk in the general dimension has a specific point, say the origin, in its recurrent set, the following intuitively plausible result in the one-dimensional case is worth knowing; see Chung (2001, pp. 279–283) for a proof.

Theorem 11.13. (a) If for some $\epsilon > 0$, $\sum_{n=1}^{\infty} P(|S_n| \leq \epsilon) = \infty$, then $0 \in \mathcal{X}$.
(b) If for some $\epsilon > 0$, $\sum_{n=1}^{\infty} P(|S_n| \leq \epsilon) < \infty$, then $0 \notin \mathcal{X}$.

Although part (b) of this theorem easily follows by an application of the Borel–Cantelli lemma, the proof of part (a) is nontrivial; see Chung (1974).

A generalization of this is assigned as a chapter exercise. Let us see a quick example of application of this theorem.

Example 11.10 (Standard Normal Random Walk). Suppose $X_1, X_2, \ldots$ are iid $N(0, 1)$ and consider the random walk $S_n = \sum_{i=1}^{n} X_i, n \geq 1$. Then, $S_n \sim N(0, n)$ and therefore $\frac{S_n}{\sqrt{n}} \sim N(0, 1)$. Fix any $\epsilon > 0$. Then,

$$P(|S_n| \leq \epsilon) = P\left(\left|\frac{S_n}{\sqrt{n}}\right| \leq \frac{\epsilon}{\sqrt{n}}\right) = 2\Phi\left(\frac{\epsilon}{\sqrt{n}}\right) - 1,$$

where Φ denotes the standard normal CDF. Now,

$$\Phi\left(\frac{\epsilon}{\sqrt{n}}\right) \approx \Phi(0) + \frac{\epsilon}{\sqrt{n}}\phi(0),$$

and therefore, $2\Phi(\frac{\epsilon}{\sqrt{n}}) - 1 \approx \frac{2\epsilon\phi(0)}{\sqrt{n}}$. Because $\phi(0) > 0$, and $\sum_{n=1}^{\infty} \frac{1}{\sqrt{n}}$ diverges, by the theorem above, 0 is seen to be a recurrent state for S_n. The rough approximation argument we gave can be made rigorous easily, by using lower bounds on the standard normal CDF. In fact, any real number x is a recurrent state for S_n in this case, and it is shown shortly.

11.7 Chung–Fuchs Theorem

We show a landmark application of characteristic functions to the problem of recurrence or transience of general d-dimensional random walks.

Consider $\mathbf{X}_i, i \geq 1$, iid with the common CDF F, and assume that (each) $\mathbf{X}_i$ has a distribution symmetric around zero. Still defining $\mathbf{S}_0 = 0, \mathbf{S}_n = \mathbf{X}_1 + \cdots + \mathbf{X}_n$, $\mathbf{S}_n$ is called a *random walk with steps driven by* F. A question of very great interest is whether $\mathbf{S}_n$ will revisit neighborhoods of the origin at infinitely many future time instants n. In a landmark article, Chung and Fuchs (1951) completely settled the problem for random walks in a general dimension $d < \infty$. We first define the necessary terminology.

Definition 11.9. Let $\mathbf{X}_i, i \geq 1$ be iid d-dimensional vectors with common CDF F. Assume that F is symmetric, in the sense that $\mathbf{X}_i \stackrel{\mathcal{L}}{=} -\mathbf{X}_i$. Let $\mathbf{S}_0 = 0, \mathbf{S}_n = \mathbf{X}_1 + \cdots + \mathbf{X}_n, n \geq 1$. Then $\mathbf{S}_n, n \geq 0$ is called the d-dimensional random walk driven by F.

Definition 11.10. The d-dimensional random walk driven by F is called *recurrent* if for every open set $C \subseteq \mathcal{R}^d$ containing the origin, $P(\mathbf{S}_n \in C \text{ infinitely often}) = 1$. If $\mathbf{S}_n$ is not recurrent, then it is called *transient*.

Some authors use the terminology *neighborhood recurrent* for the above definition. Here is the Chung–Fuchs result.

11.7 Chung–Fuchs Theorem

Theorem 11.14 (Chung–Fuchs). *Let S_n be the d-dimensional random walk driven by F. Then S_n is recurrent if and only if*

$$\int_{(-1,1)^d} \frac{1}{1-\psi(\mathbf{t})} d\mathbf{t} = \infty,$$

where $\psi(\mathbf{t})$ is the characteristic function of F.

The proof is somewhat technical to repeat here and we omit it. However, let us see some interesting examples.

Example 11.11 (One-Dimensional Simple, Gaussian, and Cauchy Random Walks). Let F be, respectively, the CDF of the symmetric two-point distribution on ± 1, the standard normal distribution, and the standard Cauchy distribution. We show that the random walk driven by each is recurrent.

If $P(X_i = \pm 1) = \frac{1}{2}$, then the cf equals $\psi(t) = \cos t \geq 1 - \frac{t^2}{2} \Rightarrow \frac{1}{1-\psi(t)} \geq \frac{2}{t^2}$, and therefore $\int_{-1}^{1} \frac{1}{1-\psi(t)} dt = \infty$. Therefore, by the Chung–Fuchs theorem, the one-dimensional simple random walk is recurrent.

If $X_i \sim N(0,1)$, then the cf equals $\psi(t) = e^{-t^2/2} \geq 1 - \frac{t^2}{2}$, and so again, the random walk driven by F is recurrent.

If $X_i \sim C(0,1)$, then the cf equals $\psi(t) = e^{-|t|} \geq 1 - |t| \Rightarrow \frac{1}{1-\psi(t)} \geq \frac{1}{|t|}$, and hence, $\int_{-1}^{1} \frac{1}{1-\psi(t)} dt \geq 2\int_0^1 \frac{1}{t} dt = \infty$, and so the random walk driven by F is recurrent.

Example 11.12 (One-Dimensional Stable Random Walk). Let F be the CDF of a symmetric stable distribution with index $\alpha < 1$. We have previously seen that the cf of F equals $\psi(t) = e^{-c|t|^\alpha}, c > 0$. Take $c = 1$, for simplicity. For

$$|t| < 1, \quad e^{-|t|^\alpha} < 1 - \frac{|t|^\alpha}{2} \Rightarrow \frac{1}{1-\psi(t)} < \frac{2}{|t|^\alpha}.$$

But for $0 < \alpha < 1$, $\int_0^1 \frac{2}{t^\alpha} dt < \infty$, and therefore, by the reverse part of the Chung–Fuchs theorem, one-dimensional stable random walks with index smaller than one are transient.

Example 11.13 (d-Dimensional Gaussian Random Walk). Consider the d-dimensional random walk driven by the CDF F of the $N_d(\mathbf{0}, I_d)$ distribution. The cf of F equals $\psi(t) = e^{-t't/2}$. Therefore, locally near $\mathbf{t} = \mathbf{0}$, $1 - \psi(t) \sim t't/2$. Now, with S_d denoting the surface area of the unit d-dimensional ball, $\int_{t:t't \leq 1} \frac{1}{t't} dt = S_d \int_0^1 r^{d-3} dr$ (by transforming to the d-dimensional polar coordinates; see Chapter 4) $< \infty$ if and only if $d \geq 3$. Now note that $\int_{(-1,1)^d} \frac{1}{1-\psi(t)} dt < \infty$ if and only if $\int_{t:t't \leq 1} \frac{1}{1-\psi(t)} dt < \infty$. Therefore, the d-dimensional Gaussian random walk is recurrent in one and two dimensions, and transient in all dimensions higher than two.

A number of extremely important general results follow from the Chung–Fuchs theorem. Among them, the following three stand out.

Theorem 11.15 (Three General Results). *(a) Suppose $d = 1$, and that $E(|X_i|) < \infty$, with $E(X_i) = 0$. Then the random walk S_n is recurrent.*
(b) More generally, suppose $\frac{S_n}{n} \overset{P}{\Rightarrow} 0$. Then S_n is recurrent.
(c) Consider a general $d \geq 3$, and suppose F is not supported on any two-dimensional subset of $\mathcal{R}^d$. Then $\mathbf{S}_n$ must be transient.

Part (c) is the famous result that there are no recurrent random walks beyond two dimensions. Beyond two dimensions, a random walker has too many directions in which to wander off, and does not return to its original position recurrently, that is, infinitely often.

Proof. Parts (a) and (c) are in fact not very hard to derive from the Chung–Fuchs theorem. For example, for part (a), finiteness of $E(|X_i|)$ allows us to write a Taylor expansion for $\psi(t)$, the characteristic function of F (as we have already discussed). Furthermore, that $E(X_i) = 0$ allows us to conclude that locally, near the origin, $1 - \psi(t)$ is $o(|t|)$. This leads to divergence of the integral $\int_{-1}^{1} \frac{1}{1-\psi(t)} dt$, and hence recurrence of the random walk by the Chung–Fuchs theorem. Part (c) is also proved by a Taylor expansion, but we do not show the argument, because we have not discussed Taylor expansions of characteristic functions for $d > 1$. Part (b) is harder to prove, but it too follows from the Chung–Fuchs theorem. □

11.8 ∗ Six Important Inequalities

Inequalities are of tremendous value for proving convergence of suitable things, for obtaining rates of convergences, and for finding concrete bounds on useful functions and sequences. We collect a number of classic inequalities on the moments and distributions of partial sums in this section. References to each inequality are given in DasGupta (2008).

Kolmogorov's Maximal Inequality. For independent random variables $X_1, \ldots, X_n$, $E(X_i) = 0$, $\mathrm{Var}(X_i) < \infty$, for any $t > 0$,

$$P\left(\max_{1 \leq k \leq n} |X_1 + \cdots + X_k| \geq t\right) \leq \frac{\sum_{k=1}^{n} \mathrm{Var}(X_k)}{t^2}.$$

Hájek–Rényi Inequality. For independent random variables $X_1, \ldots, X_n$, $E(X_i) = 0$, $\mathrm{Var}(X_i) = \sigma_i^2$, and a nonincreasing positive sequence c_k,

$$P\left(\max_{m \leq k \leq n} c_k |(X_1 + \cdots + X_k)| \geq \epsilon\right) \leq \frac{1}{\epsilon^2}\left(c_m^2 \sum_{k=1}^{m} \sigma_k^2 + \sum_{k=m+1}^{n} c_k^2 \sigma_k^2\right).$$

Lévy Inequality. Given independent random variables $X_1, \ldots, X_n$, each symmetric about zero,

$$P\left(\max_{1 \leq k \leq n} |X_k| \geq x\right) \leq 2P(|X_1 + \cdots + X_n| \geq x);$$

$$P\left(\max_{1 \leq k \leq n} |X_1 + \cdots + X_k| \geq x\right) \leq 2P(|X_1 + \cdots + X_n| \geq x).$$

Doob–Klass Prophet Inequality. Given iid mean zero random variables $X_1, \ldots, X_n$,

$$E\left[\max_{1 \leq k \leq n} (X_1 + \cdots + X_k)\right] \leq 3E|X_1 + \cdots + X_n|.$$

Bickel Inequality. Given independent random variables $X_1, \ldots, X_n$, each symmetric about $E(X_k) = 0$, and a nonnegative convex function g,

$$E\left[\max_{1 \leq k \leq n} g(X_1 + \cdots + X_k)\right] \leq 2E[g(X_1 + \cdots + X_n)].$$

General Prophet Inequality of Bickel (Private Communication). Given independent random variables $X_1, \ldots, X_n$, each with mean zero,

$$E\left[\max_{1 \leq k \leq n} (X_1 + \cdots + X_k)\right] \leq 4E[|X_1 + \cdots + X_n|].$$

Exercises

Exercise 11.1 (Simple Random Walk). By evaluating its generating function, calculate the probabilities that the second return to zero of the simple symmetric random walk occurs at the fourth step; at the sixth step.

Exercise 11.2 (Simple Random Walk). For the simple symmetric random walk, find $\lim_{n \to \infty} \frac{E(\theta_n)}{\sqrt{n}}$, where θ_n is the number of returns to zero by step n.

Exercise 11.3. Let $S_n, T_n, n \geq 1$ be two independent simple symmetric random walks. Show that $S_n - T_n$ is also a random walk, and find its step distribution.

Exercise 11.4. Consider two particles starting out at specified integer values x, y. At each subsequent time, one of the two particles is selected at random, and moved one unit to the right or one unit to the left, with probabilities $p, q, p + q = 1$. Calculate the probability that the two particles will eventually meet.

Exercise 11.5 (On the Local Time). For $n = 10, 20$, and 30, calculate the probability that the simple symmetric random walk spends zero time at the state x for $x = -2, -1, 1, 2, 5, 10$.

Exercise 11.6 (On the Local Time). Find and plot the mass function of the local time $\xi(x, n)$ of the simple symmetric random walk for $x = 5, 10, 15$ for $n = 25$.

Exercise 11.7 (On the Local Time). Calculate the expected value of the local time $\xi(x, n)$ of the simple symmetric random walk for $x = 2, 5, 10, 15$ and $n = 25, 50, 100$. Comment on the patterns that emerge.

Exercise 11.8 (Quartiles of Number of Positive Terms). For the simple symmetric random walk, compute the quartiles of π_n, the number of positive values among $S_1, \ldots, S_n$, for $n = 5, 8, 10, 15$.

Exercise 11.9 * (Range of a Random Walk). Consider the simple symmetric random walk and let R_n be the number of distinct states visited by the walk up to time n (i.e., R_n is the number of distinct elements in the set of numbers $\{S_0, S_1, \ldots, S_n\}$.

(a) First derive a formula for $P(S_k \neq S_j \ \forall j, 0 \leq j \leq k-1)$ for any given $k \geq 1$.
(b) Hence derive a formula for $E(R_n)$ for any n.
(c) Compute this formula for $n = 2, 5, 10, 20$.

Exercise 11.10 * (Range of a Random Walk). In the notation of the previous exercise, show that $\frac{R_n}{n} \stackrel{P}{\Rightarrow} 0$.

Exercise 11.11 * (A Different Random Walk on $\mathbb{Z}^d$). For $d > 1$, let $\mathbf{X}_i = (X_{i1}, \ldots, X_{id})$, where $X_{i1}, \ldots, X_{id}$ are iid and take the values ± 1 with probability $\frac{1}{2}$ each. The $\mathbf{X}_i$ are themselves independent. Let $\mathbf{S}_0 = \mathbf{0}, \mathbf{S}_n = \mathbf{X}_1 + \cdots + \mathbf{X}_n, n \geq 1$. Show that $P(\mathbf{S}_n = \mathbf{0} \text{ infinitely often}) = 0$ if $d \geq 3$.

Exercise 11.12. Show that for a general one-dimensional random walk, a recurrent value must be a possible value.

Exercise 11.13. Consider the one-dimensional random walk with iid steps having the common distribution as the standard exponential. Characterize its recurrent class $\mathcal{X}$.

Exercise 11.14. Consider the one-dimensional random walk with iid steps having the $U[-1, 1]$ distribution as the common distribution of the steps. Characterize its recurrent class $\mathcal{X}$.

Exercise 11.15. Consider the one-dimensional random walk with iid steps having the common distribution given by $P(X_i = \pm 2) = .2, P(X_i = \pm 1) = .3$. Characterize its recurrent class $\mathcal{X}$.

Exercise 11.16. Consider the one-dimensional random walk with iid steps having the common distribution given by $P(X_i = \pm 1) = .4995, P(X_i = .001) = 001$. Characterize its recurrent class $\mathcal{X}$.

Exercise 11.17 * (Two-Dimensional Cauchy Random Walk). Consider the two-dimensional random walk driven by the CDF F of the two-dimensional Cauchy density $\frac{c}{(1+\|x\|^2)^{3/2}}$, where c is the normalizing constant. Verify whether the random walk driven by F is recurrent.

Exercise 11.18 (The Asymmetric Cubic Lattice Random Walk). Consider the cubic lattice random walk in d dimensions, but with the change that a coordinate changes by ± 1 with respective probabilities $p, q, p + q = 1$.

Derive a formula for $P_{d,2n}$ in this general situation, and then verify whether $\sum_{n=1}^{\infty} P_{d,2n}$ converges or diverges for given $d = 1, 2, 3, \ldots$. Make a conclusion about the recurrence or transience of such an asymmetric random walk for each $d \geq 1$.

Exercise 11.19 * **(A Random Walk with Poisson Steps).** Let $X_i, i \geq 1$ be iid $Poi(\lambda)$ and let $S_n = \sum_{i=1}^{n} X_i, n \geq 1, S_0 = 0$. Show that the random walk $\{S_n\}, n \geq 1$ is transient.

Exercise 11.20 * **(Recurrence Time in an Asymmetric Random Walk).** Consider again the simple random walk on $\mathbb{Z}$, but with the change that the steps take the values ± 1 with respective probabilities $p, q, p + q = 1$. Let ν_1 denote the first instant at which the walk returns to zero; that is, $\nu_1 = \min\{n \geq 1 : S_n = 0\}$.

Find a formula for $P(\nu_1 < \infty)$.

Exercise 11.21 * **(Conditional Distribution of Recurrence Times).** For the simple symmetric random walk on $\mathbb{Z}$,

(a) Find $P(\nu_2 \leq k \mid \nu_1 = m)$, where ν_i denotes the time of the ith return of the random walk to zero. Compute your formula for $k = 4, 6, 8, 10$ when $m = 2$.
(b) Prove or disprove: $E(\nu_2 \mid \nu_1 = m) = \infty \; \forall m$.

Exercise 11.22 (Ratio of Recurrence Times). For the simple symmetric random walk on $\mathbb{Z}$,

(a) Find an explicit formula for $E(\frac{\nu_r}{\nu_s}), 1 \leq r < s < \infty$, where ν_i denotes the time of the ith return of the random walk to zero.
(b) Find an explicit answer for $E(\frac{\nu_s}{\nu_r}), 1 \leq r < s < \infty$.

Exercise 11.23 * **(Number of Positive Terms Given the Last Term).** Consider a one-dimensional random walk driven by $F = \Phi$, the standard normal CDF. Derive a formula for $E(\pi_n \mid S_n = c)$, that is, the conditional expectation of the number of positive terms given the value of the last term.

Exercise 11.24 (Expected Number of Returns). Consider the simple symmetric random walk and consider the quantities $\nu_{m,n} = E[\#\{j : m \leq j \leq m+n, S_j = 0\}]$.

Prove that $\nu_{m,n}$ is always maximized at $m = 0$, whatever n is.

Exercise 11.25 (Application of Wald's Identity). In repeated independent rolls of a fair die, show that the expected number of throws necessary for the total to become 10 or more is less than 4.3.

Exercise 11.26 (Application of Wald's Identity). Let X have the negative binomial distribution with parameters r, p. Prove by using only Wald's identity that $E(X) = \frac{r}{p}$.

Exercise 11.27 (Application of Lévy and Kolmogorov's Inequalities). Consider a random walk with step distribution F starting at zero. Let m be a fixed integer. For each of the following cases, find a lower bound on the probability that the random walk does not cross either of the boundaries $y = -m$, $y = +m$ within the first n steps.

(a) $F = Ber(.5)$.
(b) $F = N(0, 1)$.
(c) $F = C(0, 1)$.

Exercise 11.28 (Applying Prophet Inequalities). Suppose $X_1, X_2, \ldots, X_n$ are iid $N(0, 1)$ and let $S_n = X_1 + \cdots + X_n$.

(a) Calculate $E(S_n^+)$ exactly.
(b) Find an upper bound on $E(\max_{1 \leq k \leq n} S_k^+)$.
(c) Repeat part (b) with $\max_{1 \leq k \leq n} |S_k|$.

References

Chung, K.L. (1974). *A Course in Probability*, Academic Press, New York.
Chung, K.L. (2001). *A Course in Probability*, 3rd Edition, Academic Press, New York.
Chung, K.L. and Fuchs, W. (1951). On the distribution of values of sums of random variables, *Mem. Amer. Math. Soc.*, 6, 12.
Dym, H. and McKean, H. (1972). *Fourier Series and Integrals*, Academic Press, New York.
Erdös, P. and Kac, M. (1947). On the number of positive sums of independent random variables, *Bull. Amer. Math. Soc.*, 53, 1011–1020.
Feller, W. (1971). *Introduction to Probability Theory with Applications*, Wiley, New York.
Finch, S. (2003). *Mathematical Constants*, Cambridge University Press, Cambridge, UK.
Pólya, G. (1921). Uber eine Aufgabe der Wahrsch einlichkeitstheorie betreffend die Irrfahrt im Strasseenetz, *Math. Annalen*, 84, 149–160.
Rényi, A. (1970). *Probability Theory*, Nauka, Moscow.
Sparre-Andersen, E. (1949). On the number of positive sums of random variables, *Aktuarietikskr*, 32, 27–36.
Spitzer, F. (2008). *Principles of Random Walk*, Springer, New York.

Chapter 12
Brownian Motion and Gaussian Processes

We started this text with discussions of a single random variable. We then proceeded to two and more generally, a finite number of random variables. In the last chapter, we treated the random walk, which involved a countably infinite number of random variables, namely the positions of the random walk S_n at times $n = 0, 1, 2, 3, \ldots$. The time parameter n for the random walks we discussed in the last chapter belongs to the set of nonnegative integers, which is a countable set. We now look at a special *continuous time stochastic process*, which corresponds to an uncountable family of random variables, indexed by a *time parameter t* belonging to a suitable uncountable time set T. The process we mainly treat in this chapter is *Brownian motion*, although some other *Gaussian processes* are also treated briefly.

Brownian motion is one of the most important continuous-time stochastic processes and has earned its special status because of its elegant theoretical properties, its numerous important connections to other continuous-time stochastic processes, and due to its real applications and its physical origin. If we look at the path of a random walk when we run the clock much faster, and the steps of the walk are also suitably smaller, then the random walk converges to Brownian motion. This is an extremely important connection, and it is made precise later in this chapter. Brownian motion arises naturally in some form or other in numerous statistical inference problems. It is also used as a real model for modeling stock market behavior.

The process owes its name to the Scottish botanist Robert Brown, who noticed under a microscope that pollen particles suspended in fluid engaged in a zigzag and eccentric motion. It was, however, Albert Einstein who in 1905 gave Brownian motion a formal physical formulation. Einstein showed that Brownian motion of a large particle visible under a microscope could be explained by assuming that the particle gets ceaselessly bombarded by invisible molecules of its surrounding medium. The theoretical predictions made by Einstein were later experimentally verified by various physicists, including Jean Baptiste Perrin who was awarded the Nobel prize in physics for this work. In particular, Einstein's work led to the determination of *Avogadro's constant*, perhaps the first major use of what statisticians call a *moment estimate*. The existence and construction of Brownian motion was first explicitly established by Norbert Wiener in 1923, which accounts for the other name *Wiener process* for a Brownian motion.

There are numerous excellent references at various technical levels on the topics of this chapter. Comprehensive and lucid mathematical treatments are available in Freedman (1983), Karlin and Taylor (1975), Breiman (1992), Resnick (1992), Revuz and Yor (1994), Durrett (2001), Lawler (2006), and Bhattacharya and Waymire (2009). Elegant and unorthodox treatment of Brownian motion is given in Mörters and Peres (2010). Additional specific references are given in the sections.

12.1 Preview of Connections to the Random Walk

We remarked in the introduction that random walks and Brownian motion are interconnected in a suitable asymptotic paradigm. It would be helpful to understand this connection in a conceptual manner before going into technical treatments of Brownian motion.

Consider then the usual simple symmetric random walk defined by $S_0 = 0, S_n = X_1 + X_2 + \cdots + X_n, n \geq 1$, where the X_i are iid with common distribution $P(X_i = \pm 1) = \frac{1}{2}$. Consider now a random walk that makes its steps at much smaller time intervals, but the jump sizes are also smaller. Precisely, with the $X_i, i \geq 1$ still as above, define

$$S_n(t) = \frac{S_{\lfloor nt \rfloor}}{\sqrt{n}}, \quad 0 \leq t \leq 1,$$

where $\lfloor x \rfloor$ denotes the integer part of a nonnegative real number x. This amounts to joining the points

$$(0,0), \quad \left(\frac{1}{n}, \frac{X_1}{\sqrt{n}}\right), \quad \left(\frac{2}{n}, \frac{X_1 + X_2}{\sqrt{n}}\right), \ldots$$

by linear interpolation, thereby obtaining a curve. The simulated plot of $S_n(t)$ for $n = 1000$ in Fig. 12.1 shows the zigzag path of the scaled random walk. We can see that the plot is rather rough, and the function takes the value zero at $t = 0$; that is, $S_n(0) = 0$, and $S_n(1) \neq 0$.

It turns out that in a suitable precise sense, the graph of $S_n(t)$ on [0, 1] for large n should mimic the graph of a *random function* called *Brownian motion* on [0, 1]. Brownian motion is a special *stochastic process*, which is a collection of infinitely many random variables, say $W(t), 0 \leq t \leq 1$, each $W(t)$ for a fixed t being a normally distributed random variable, with other additional properties for their joint distributions. They are introduced formally and analyzed in greater detail in the next sections.

The question arises of why is the connection between a random walk and the Brownian motion of any use or interest to us. A short nontechnical answer to that question is that because $S_n(t)$ acts like a realization of a Brownian motion, by using known properties of Brownian motion, we can approximately describe properties of $S_n(t)$ for large n. This is useful, because the stochastic process $S_n(t)$ arises in

12.2 Basic Definitions

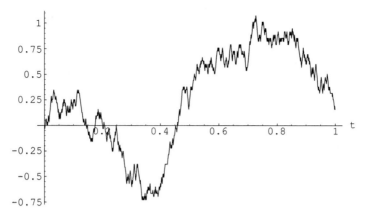

Fig. 12.1 Simulated plot of a scaled random walk

numerous problems of interest to statisticians and probabilists. By simultaneously using the connection between $S_n(t)$ and Brownian motion, and known properties of Brownian motion, we can assert useful things concerning many problems in statistics and probability that would be nearly impossible to assert in a simple direct manner. That is why the connections are not just mathematically interesting, but also tremendously useful.

12.2 Basic Definitions

Our principal goal in the subsequent sections is to study Brwonian motion and Brownian bridge due to their special importance among Gaussian processes.

The Brownian bridge is closely related to Brownian motion, and shares many of the same properties as Brownian motion. They both arise in many statistical applications. It should also be understood that the Brownian motion and bridge are of enormous independent interest in the study of probability theory, regardless of their connections to problems in statistics.

We caution the reader that it is not possible to make all the statements in this chapter mathematically rigorous without using measure theory. This is because we are now dealing with uncountable collections of random variables, and problems of measure zero sets can easily arise. However, the results are accurate and they can be practically used without knowing exactly how to fix the measure theory issues.

We first give some general definitions for future use.

Definition 12.1. A *stochastic process* is a collection of random variables $\{X(t), t \in T\}$ taking values in some finite-dimensional Euclidean space $\mathcal{R}^d$, $1 \leq d < \infty$, where the indexing set T is a general set.

Definition 12.2. A real-valued stochastic process $\{X(t), -\infty < t < \infty\}$ is called *weakly stationary* if

(a) $E(X(t)) = \mu$ is independent of t.
(b) $E[X(t)]^2 < \infty$ for all t, and $\text{Cov}(X(t), X(s)) = \text{Cov}(X(t+h), X(s+h))$ for all s, t, h.

Definition 12.3. A real-valued stochastic process $\{X(t), -\infty < t < \infty\}$ is called *strictly stationary* if for every $n \geq 1, t_1, t_2, \ldots, t_n$, and every h, the joint distribution of $(X_{t_1}, X_{t_2}, \ldots, X_{t_n})$ is the same as the joint distribution of $(X_{t_1+h}, X_{t_2+h}, \ldots, X_{t_n+h})$.

Definition 12.4. A real-valued stochastic process $\{X(t), -\infty < t < \infty\}$ is called *a Markov process* if for every $n \geq 1, t_1 < t_2 < \cdots < t_n$,

$$P(X_{t_n} \leq x_{t_n} | X_{t_1} = x_{t_1}, \ldots, X_{t_{n-1}} = x_{t_{n-1}}) = P(X_{t_n} \leq x_{t_n} | X_{t_{n-1}} = x_{t_{n-1}});$$

that is, the distribution of the future values of the process given the entire past depends only on the most recent past.

Definition 12.5. A stochastic process $\{X(t), t \in T\}$ is called a *Gaussian process* if for every $n \geq 1, t_1, t_2, \ldots, t_n$, the joint distribution of $(X_{t_1}, X_{t_2}, \ldots, X_{t_n})$ is a multivariate normal.

This is often stated as a process is a Gaussian process if all its *finite-dimensional distributions* are Gaussian.

With these general definitions at hand, we now define Brownian motion and the Brownian bridge. Brownian motion is intimately linked to the simple symmetric random walk, and partial sums of iid random variables. The Brownian bridge is intimately connected to the *empirical process* of iid random variables. We focus on the properties of Brownian motion in this chapter, and postpone discussion of the empirical process and the Brownian bridge to a later chapter. However, we define both Brownian processes right now.

Definition 12.6. A stochastic process $W(t)$ defined on a probability space $(\Omega, \mathcal{A}, P), t \in [0, \infty)$ is called a standard Wiener process or the standard Brownian motion starting at zero if:

(i) $W(0) = 0$ with probability one.
(ii) For $0 \leq s < t < \infty, W(t) - W(s) \sim N(0, t - s)$.
(iii) Given $0 \leq t_0 < t_1 < \ldots < t_k < \infty$, the random variables $W(t_{j+1}) - W(t_j), 0 \leq j \leq k - 1$ are mutually independent.
(iv) The sample paths of $W(.)$ are almost all continuous; that is except for a set of sample points of probability zero, as a function of $t, W(t, \omega)$ is a continuous function.

Remark. Property (iv) actually can be proved to follow from the other three properties. But it is helpful to include it in the definition to emphasize the importance of the continuity of Brownian paths. Property (iii) is the celebrated *independent increments*

12.2 Basic Definitions

property and lies at the heart of numerous further properties of Brownian motion. We often just omit the word *standard* when referring to standard Brownian motion.

Definition 12.7. If $W(t)$ is a standard Brownian motion, then $X(t) = x + W(t)$, $x \in \mathcal{R}$, is called Brownian motion starting at x, and $Y(t) = \sigma W(t), \sigma > 0$ is called Brownian motion with scale coefficient or diffusion coefficient σ.

Definition 12.8. Let $W(t)$ be a standard Wiener process on $[0, 1]$. The process $B(t) = W(t) - tW(1)$ is called a standard Brownian bridge on $[0, 1]$.

Remark. Note that the definition implies that $B(0) = B(1) = 0$ with probability one. Thus, the Brownian bridge on $[0, 1]$ starts and ends at zero. Hence the name *tied down Wiener process*. The Brownian bridge on $[0, 1]$ can be defined in various other equivalent ways. The definition we adopt here is convenient for many calculations.

Definition 12.9. Let $1 < d < \infty$, and let $W_i(t), 1 \leq i \leq d$, be independent Brownian motions on $[0, \infty)$. Then $\mathbf{W}_d(t) = (W_1(t), \ldots, W_d(t))$ is called d-*dimensional Brownian motion*.

Remark. In other words, if a particle performs independent Brownian movements along d different coordinates, then we say that the particle is engaged in d-dimensional Brownian motion. Figure 12.2 demonstrates the case $d = 2$. When the dimension is not explicitly mentioned, it is understood that $d = 1$.

Example 12.1 (Some Illustrative Processes). We take a few stochastic processes, and try to understand some of their basic properties. The processes we consider are the following.

(a) $X_1(t) \equiv X$, where $X \sim N(0, 1)$.
(b) $X_2(t) = tX$, where $X \sim N(0, 1)$.

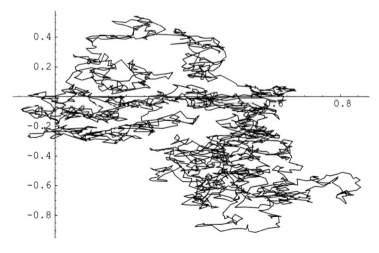

Fig. 12.2 State visited by a planar Brownian motion

(c) $X_3(t) = A\cos\theta t + B\sin\theta t$, where θ is a fixed positive number, $t \geq 0$, and A, B are iid $N(0, 1)$.
(d) $X_4(t) = \int_0^t W(u)du, t \geq 0$, where $W(u)$ is standard Brownian motion on $[0, \infty)$, starting at zero.
(e) $X_5(t) = W(t+1) - W(t), t \geq 0$, where $W(t)$ is standard Brownian motion on $[0, \infty)$, starting at zero.

Each of these processes is a Gaussian process on the time domain on which it is defined. The mean function of each process is the zero function.

Coming to the covariance functions, for $s \leq t$,

$\text{Cov}(X_1(s), X_1(t)) \equiv 1$.
$\text{Cov}(X_2(s), X_2(t)) = st$.
$\text{Cov}(X_3(s), X_3(t)) = \cos\theta s \cos\theta t + \sin\theta s \sin\theta t = \cos\theta(s-t)$.
$\text{Cov}(X_4(s), X_4(t)) = E\left[\int_0^s W(u)du \int_0^t W(v)dv\right] = E\left[\int_0^t \int_0^s W(u)W(v)du\,dv\right]$
$= \int_0^t \int_0^s E[W(u)W(v)]du\,dv = \int_0^t \int_0^s \min(u,v)du\,dv$
$= \int_0^s \int_0^s \min(u,v)du\,dv + \int_s^t \int_0^s \min(u,v)du\,dv$
$= \int_0^s \int_0^v \min(u,v)du\,dv + \int_0^s \int_v^s \min(u,v)du\,dv$
$\quad + \int_s^t \int_0^s \min(u,v)du\,dv$
$= \frac{s^3}{6} + \left[\frac{s^3}{2} - \frac{s^3}{3}\right] + \frac{s^2}{2}(t-s)$
$= \frac{s^2 t}{2} - \frac{s^3}{6} = \frac{s^2}{6}(3t-s)$.

$\text{Cov}(X_5(s), X_5(t)) = \text{Cov}(W(s+1) - W(s), W(t+1) - W(t))$
$= s + 1 - \min(s+1, t) - s + s$
$= 0$, if $s - t \leq -1$, and $= s - t + 1$ if $s - t > -1$.

The two cases are combined into the single formula $\text{Cov}(W(s+1) - W(s), W(t+1) - W(t)) = (s - t + 1)_+$. The covariance functions of $X_1(t), X_3(t)$, and $X_5(t)$ depend only on $s - t$, and these are stationary.

12.2.1 Condition for a Gaussian Process to be Markov

We show a simple and useful result on characterizing Gaussian processes that are Markov. It turns out that there is a simple way to tell if a given Gaussian process is Markov by simply looking at its correlation function. Because we only need to consider finite-dimensional distributions to decide if a stochastic process is Markov, it is only necessary for us to determine when a finite sequence of jointly normal variables has the Markov property. We start with that case.

Definition 12.10. Let $X_1, X_2, \ldots, X_n$ be n jointly distributed continuous random variables. The n-dimensional random vector $(X_1, \ldots, X_n)$ is said to have the Markov property if for every $k \leq n$, the conditional distribution of X_k given $X_1, \ldots, X_{k-1}$ is the same as the conditional distribution of X_k given X_{k-1} alone.

Theorem 12.1. *Let $(X_1, \ldots, X_n)$ have a multivariate normal distribution with means zero and the correlations $\rho_{X_j, X_k} = \rho_{jk}$. Then $(X_1, \ldots, X_n)$ has the Markov property if and only if for $1 \leq i \leq j \leq k \leq n$, $\rho_{ik} = \rho_{ij}\rho_{jk}$.*

Proof. We may assume that each X_i has variance one. If $(X_1, \ldots, X_n)$ has the Markov property, then for any k, $E(X_k|X_1, \ldots, X_{k-1}) = E(X_k|X_{k-1}) = \rho_{k-1,k} X_{k-1}$ (see Chapter 5). Therefore, $X_k - \rho_{k-1,k} X_{k-1} = X_k - E(X_k|X_1, \ldots, X_{k-1})$ is independent of the vector $(X_1, \ldots, X_{k-1})$. In particular, each covariance $\text{Cov}(X_k - \rho_{k-1,k} X_{k-1}, X_i)$ must be zero for all $i \leq k - 1$. This leads to $\rho_{ik} = \rho_{i,k-1}\rho_{k-1,k}$, and to the claimed identity $\rho_{ik} = \rho_{ij}\rho_{jk}$ by simply applying $\rho_{ik} = \rho_{i,k-1}\rho_{k-1,k}$ repeatedly.

Conversely, suppose the identity $\rho_{ik} = \rho_{ij}\rho_{jk}$ holds for all $1 \leq i \leq j \leq k \leq n$. Then, it follows from the respective formulas for $E(X_k|X_1, \ldots, X_{k-1})$ and $\text{Var}(X_k|X_1, \ldots, X_{k-1})$ (see Chapter 5) that $E(X_k|X_1, \ldots, X_{k-1}) = \rho_{k-1,k} X_{k-1} = E(X_k|X_{k-1})$, and $\text{Var}(X_k|X_1, \ldots, X_{k-1}) = \text{Var}(X_k|X_{k-1})$. All conditional distributions for a multivariate normal distribution are also normal, therefore it must be the case that the distribution of X_k given $X_1, \ldots, X_{k-1}$ and distribution of X_k given just X_{k-1} are the same. This being true for all k, the full vector $(X_1, \ldots, X_n)$ has the Markov property. □

Because the Markov property for a continuous-time stochastic process is defined in terms of finite-dimensional distributions, the above result gives us the following simple and useful result as a corollary.

Corollary. *A Gaussian process $X(t), t \in \mathcal{R}$ is Markov if and only if $\rho_{X(s), X(u)} = \rho_{X(s), X(t)} \times \rho_{X(t), X(u)}$ for all $s, t, u, s \leq t \leq u$.*

12.2.2 * Explicit Construction of Brownian Motion

It is not a priori obvious that an uncountable collection of random variables with the defining properties of Brownian motion can be constructed on a common probability space (a measure theory terminology). In other words, that Brownian motion exists requires a proof. Various proofs of the existence of Brownian motion can be given. We provide two explicit constructions, of which one is more classic in nature. But the second construction is also useful.

Theorem 12.2 (Karhunen–Loéve Expansion). *(a) Let $Z_1, Z_2, \ldots$ be an infinite sequence of iid standard normal variables. Then, with probability one, the infinite series*

$$W(t) = \sqrt{2} \sum_{m=1}^{\infty} \frac{\sin\left(\left[m - \frac{1}{2}\right]\pi t\right)}{\left[m - \frac{1}{2}\right]\pi} Z_m$$

converges uniformly in t on $[0, 1]$ and the process $W(t)$ is a Brownian motion on $[0, 1]$.

The infinite series $B(t) = \sqrt{2} \sum_{m=1}^{\infty} \frac{\sin(m\pi t)}{m\pi} Z_m$ converges uniformly in t on $[0, 1]$ and the process $B(t)$ is a Brownian Bridge on $[0, 1]$.

(b) For $n \geq 0$, let I_n denote the set of odd integers in $[0, 2^n]$. Let $Z_{n,k}, n \geq 0, k \in I_n$ be a double array of iid standard normal variables. Let $H_{n,k}(t), n \geq 0, k \in I_n$ be the sequence of Haar wavelets *defined as*

$$H_{n,k}(t) = 0 \text{ if } t \notin \left[\frac{k-1}{2^n}, \frac{k+1}{2^n}\right], \quad \text{and } H_{n,k}(t) = 2^{(n-1)/2}$$

$$\text{if } t \in \left[\frac{k-1}{2^n}, \frac{k}{2^n}\right), \quad \text{and} \quad -2^{(n-1)/2} \text{ if } t \in \left[\frac{k}{2^n}, \frac{k+1}{2^n}\right].$$

Let $S_{n,k}(t)$ be the sequence of Schauder functions *defined as $S_{n,k}(t) = \int_0^t H_{n,k}(s)ds, 0 \leq t \leq 1, n \geq 0, k \in I_n$.*

Then the infinite series $W(t) = \sum_{n=0}^{\infty} \sum_{k \in I_n} Z_{n,k} S_{n,k}(t)$ converges uniformly in t on $[0, 1]$ and the process $W(t)$ is a Brownian motion on $[0, 1]$.

Remark. See Bhattacharya and Waymire (2007, p. 135) for a proof. Both constructions of Brownian motion given above can be heuristically understood by using ideas of Fourier theory. If the sequence $f_0(t) = 1, f_1(t), f_2(t), \ldots$ forms an orthonormal basis of $L_2[0, 1]$, then we can expand a square integrable function, say $w(t)$, as an infinite series $\sum_i c_i f_i(t)$, where c_i equals the inner product $\int_0^1 w(t) f_i(t) dt$. Thus, $c_0 = 0$ if the integral of $w(t)$ is zero. The Karhunen–Loéve expansion can be heuristically explained as a random orthonormal expansion of $W(t)$. The basis functions $f_i(t)$ chosen do depend on the process $W(t)$, specifically the covariance function. The inner products $\int_0^1 W(t) f_i(t) dt, i \geq 1$ form a sequence of iid standard normals. This is very far from a proof, but provides a heuristic context for the expansion. The second construction is based similarly on expansions using a wavelet basis instead of a Fourier basis.

12.3 Basic Distributional Properties

Distributional properties and formulas are always useful in doing further calculations and for obtaining concrete answers to questions. The most basic distributional properties of the Brownian motion and bridge are given first.

12.3 Basic Distributional Properties

Throughout this chapter, the notation $W(t)$ and $B(t)$ mean a (standard) Brownian motion and Brownian bridge. The phrase *standard* is often deleted for brevity.

Proposition. *(a)* $Cov(W(s), W(t)) = \min(s, t); Cov(B(s), B(t)) = \min(s, t) - st$.
(b) **(The Markov Property).** *For any given n and $t_0 < t_1 < \cdots < t_n$, the conditional distribution of $W(t_n)$ given that $W(t_0) = x_0, W(t_1) = x_1, \ldots, W(t_{n-1}) = x_{n-1}$ is the same as the conditional distribution of $W(t_n)$ given $W(t_{n-1}) = x_{n-1}$.*
(c) *Given $s < t$, the conditional distribution of $W(t)$ given $W(s) = w$ is $N(w, t-s)$.*
(d) *Given $t_1 < t_2 < \cdots < t_n$, the joint density of $W(t_1), W(t_2), \ldots, W(t_n)$ is given by the function*

$$f(x_1, x_2, \ldots, x_n) = p(x_1, t_1) p(x_2 - x_1, t_2 - t_1) \cdots p(x_n - x_{n-1}, t_n - t_{n-1}),$$

where $p(x, t)$ is the density of $N(0, t)$; that is, $p(x, t) = \frac{1}{\sqrt{2\pi t}} e^{-\frac{x^2}{2t}}$.

Each part of this proposition follows on simple and direct calculation by using the definition of a Brownian motion and Brownian bridge. It is worth mentioning that the Markov property is extremely important and is a consequence of the independent increments property. Alternatively, one can simply use our previous characterization that a Gaussian process is Markov if and only if its correlation function satisfies $\rho_{X(s), X(u)} = \rho_{X(s), X(t)} \times \rho_{X(t), X(u)}$ for all $s \leq t \leq u$.

The Markov property can be strengthened to a very useful property, known as the strong Markov property. For instance, suppose you are waiting for the process to reach a level b for the first time. The process will reach that level at some random time, say τ. At this point, the process will simply start over, and $W(t) - b$ will act like a path of a standard Brownian motion from that point onwards. For the general description of this property, we need a definition.

Definition 12.11. A nonnegative random variable τ is called a *stopping time* for the process $W(t)$ if for any $s > 0$, whether $\tau \leq s$ depends only on the values of $W(t)$ for $t \leq s$.

Example 12.2. For $b > 0$, consider the first passage time $T_b = \inf\{t > 0 : W(t) = b\}$. Then, $T_b > s$ if and only if $W(t) < b$ for all $t \leq s$. Therefore, T_b is a stopping time for the process $W(t)$.

Example 12.3. Let X be a $U[0, 1]$ random variable independent of the process $W(t)$. Then the nonnegative random variable $\tau = X$ is not a stopping time for the process $W(t)$.

Theorem 12.3 (Strong Markov Property). *If τ is a stopping time for the process $W(t)$, then $W(\tau + t) - W(\tau)$ is also a Brownian motion on $[0, \infty)$ and is independent of $\{W(s), s \leq \tau\}$.*

See Bhattacharya and Waymire (2007, p. 153) for its proof.

12.3.1 Reflection Principle and Extremes

It is important in applications to be able to derive the distribution of special functionals of Brownian processes. They can be important because a Brownian process is used directly as a statistical model in some problem, or they can be important because the functional arises as the limit of some suitable sequence of statistics in a seemingly unrelated problem. Examples of the latter kind are seen in applications of the so-called *invariance principle*. For now, we provide formulas for the distribution of certain extremes and first passage times of a Brownian motion. The following notation is used:

$$M(t) = \sup_{0 < s \leq t} W(s); \quad T_b = \inf\{t > 0 : W(t) = b\}.$$

Theorem 12.4 (Reflection Principle). *(a) For $b > 0$, $P(M(t) > b) = 2P(W(t) > b)$.*

(b) For $t > 0$, $M(t) = \sup_{0 < s \leq t} W(s)$ has the density

$$\sqrt{\frac{2}{t\pi}} e^{-x^2/(2t)}, \quad x > 0.$$

(c) For $b > 0$, the first passage time T_b has the density

$$\frac{\phi\left(\frac{b}{\sqrt{t}}\right)}{t^{3/2}},$$

where ϕ denotes the standard normal density function.

(d) (First Arcsine Law). Let τ be the point of maxima of $W(t)$ on $[0, 1]$. Then τ is almost surely unique, and $P(\tau \leq t) = \frac{2}{\pi}\arcsin(\sqrt{t})$.

(e) (Reflected Brownian Motion). Let $X(t) = \sup_{0 \leq s \leq t}|W(s)|$. Then $X(1) = \sup_{0 \leq s \leq 1}|W(s)|$ has the CDF

$$G(x) = \frac{4}{\pi} \sum_{m=0}^{\infty} \frac{(-1)^m}{2m+1} e^{-(2m+1)^2 \pi^2/(8x^2)}, \quad x \geq 0.$$

(f) (Maximum of a Brownian Bridge). Let $B(t)$ be a Brownian bridge on $[0, 1]$. Then, $\sup_{0 \leq t \leq 1} |B(t)|$ has the CDF

$$H(x) = 1 - \sum_{k=-\infty}^{\infty} (-1)^{k-1} e^{-2k^2 x^2}, \quad x \geq 0.$$

(g) (Second Arcsine Law). Let $L = \sup\{t \in [0, 1] : W(t) = 0\}$. Then $P(L \leq t) = \frac{2}{\pi}\arcsin(\sqrt{t})$.

(h) Given $0 < s < t$, $P(W(t)$ has at least one zero in the time interval $(s, t)) = \frac{2}{\pi}\arccos(\sqrt{\frac{s}{t}})$.

12.3 Basic Distributional Properties

Proof of the Reflection Principle: The reflection principle is of paramount importance and we provide a proof of it. The reflection principle follows from two observations, the first of which is obvious, and the second needs a clever argument. The observations are:

$$P(T_b < t) = P(T_b < t, W(t) > b) + P(T_b < t, W(t) < b),$$

and,

$$P(T_b < t, W(t) > b) = P(T_b < t, W(t) < b).$$

Because $P(T_b < t, W(t) > b) = P(W(t) > b)$ (because $W(t) > b$ implies that $T_b < t$), if we accept the second identity above, then we immediately have the desired result $P(M(t) > b) = P(T_b < t) = 2P(W(t) > b)$. Thus, only the second identity needs a proof. This is done by a clever argument.

The event $\{T_b < t, W(t) < b\}$ happens if and only if at some point $\tau < t$, the process reaches the level b, and then at time t drops to a lower level $l, l < b$. However, once at level b, the process could as well have taken the path reflected along the level b, which would have caused the process to end up at level $b + (b - l) = 2b - l$ at time t. We now observe that $2b - l > b$, meaning that corresponding to every path in the event $\{T_b < t, W(t) < b\}$, there is a path in the event $\{T_b < t, W(t) > b\}$, and so $P(T_b < t, W(t) < b)$ must be equal to $P(T_b < t, W(t) > b)$.

This is the famous *reflection principle* for Brownian motion. An analytic proof of the identity $P(T_b < t, W(t) < b) = P(T_b < t, W(t) > b)$ can be given by using the strong Markov property of Brownian motion.

Note that both parts (b) and (c) of the theorem are simply restatements of part (a). Many of the remaining parts follow on calculations that also use the reflection principle. Detailed proofs can be seen, for example, in Karlin and Taylor (1975, pp. 345–354). □

Example 12.4 (Density of Last Zero Before T). Consider standard Brownian motion $W(t)$ on $[0, \infty)$ starting at zero and fix a time $T > 0$. We want to find the density of the last zero of $W(t)$ before the time T. Formally, let $\tau = \tau_T = \sup\{t < T : W(t) = 0\}$. Then, we want to find the density of τ.

By using part (g) of the previous theorem,

$$P(\tau > s) = P(\text{There is at least one zero of W in } (s, T)) = \frac{2}{\pi} \arccos\left(\sqrt{\frac{s}{T}}\right).$$

Therefore, the density of τ is

$$f_\tau(s) = -\frac{d}{ds}\left[\frac{2}{\pi} \arccos\left(\sqrt{\frac{s}{T}}\right)\right] = \frac{1}{\pi\sqrt{s(T-s)}}, \quad 0 < s < T.$$

Notice that the density is symmetric about $\frac{T}{2}$, and therefore $E(\tau) = \frac{T}{2}$. A calculation shows that $E(\tau^2) = \frac{3}{8}T^2$, and therefore $\text{Var}(\tau) = \frac{3}{8}T^2 - \frac{T^2}{4} = \frac{1}{8}T^2$.

12.3.2 Path Properties and Behavior Near Zero and Infinity

A textbook example of a nowhere differentiable and yet everywhere continuous function is Weierstrass's function $f(t) = \sum_{n=0}^{\infty} 2^{-n}\cos(b^n \pi t), -\infty < t < \infty$, for $b > 2 + 3\pi$. Constructing another example of such a function is not trivial. A result of notoriety is that almost all sample paths of Brownian motion are functions of this kind; that is, as a function of t, $W(t)$ is continuous at every t, and differentiable at no t! The paths are extremely crooked. The Brownian bridge shares the same property. The sample paths show other evidence of extreme oscillation; for example, in any arbitrarily small interval containing the starting time $t = 0$, $W(t)$ changes its sign infinitely often. The various important path properties of Brownian motion are described and discussed below.

Theorem 12.5. *Let $W(t), t > 0$ be a Brownian motion on $[0, \infty)$. Then,*

(a) **(Scaling).** *For $c > 0$, $X(t) = c^{-\frac{1}{2}} W(ct)$ is a Brownian motion on $[0, \infty)$.*
(b) **(Time reciprocity).** *$X(t) = tW(\frac{1}{t})$, with the value being defined as zero at $t = 0$ is a Brownian motion on $[0, \infty)$.*
(c) **(Time Reversal).** *Given $0 < T < \infty$, $X_T(t) = W(T) - W(T - t)$ is a Brownian motion on $[0, T]$.*

Proof. Only part (b) requires a proof, the others being obvious. First note that for $s \leq t$, the covariance function is

$$\text{Cov}\left(sW\left(\frac{1}{s}\right), tW\left(\frac{1}{t}\right)\right) = st\left[\min\left\{\frac{1}{s}, \frac{1}{t}\right\}\right] = st\frac{1}{t} = s = \min\{s, t\}.$$

It is obvious that $X(t) - X(s) \sim N(0, t - s)$. Next, for $s < t < u$, $\text{Cov}(X(t) - X(s), X(u) - X(t)) = t - s - t + s = 0$, and the independent increments property holds. The sample paths are continuous (including at $t = 0$) because $W(t)$ has continuous sample paths, and $X(0) = 0$. Thus, all the defining properties of a Brownian motion are satisfied, and hence $X(t)$ must be a Brownian motion. □

Part (b) leads to the following useful property.

Proposition. *With probability one, $\frac{W(t)}{t} \to 0$ as $t \to \infty$.*

The behavior of Brownian motion near $t = 0$ is quite a bit more subtle, and we postpone its discussion till later. We next describe a series of classic results that illustrate the extremely rough nature of the paths of a Brownian motion. The results essentially tell us that at any instant, it is nearly impossible to predict what a particle performing a Brownian motion will do next. Here is a simple intuitive explanation for why the paths of a Brownian motion are so rough.

12.3 Basic Distributional Properties

Take two time instants $s, t, s < t$. We then have the simple moment formula $E[(W(t) - W(s)]^2 = (t - s)$. Writing $t = s + h$, we get

$$E[W(s+h) - W(s)]^2 = h \Leftrightarrow E\left[\frac{W(s+h) - W(s)}{h}\right]^2 = \frac{1}{h}.$$

If the time instants s, t are close together, then $h \approx 0$, and so $\frac{1}{h}$ is large. We can see that the increment $\frac{W(s+h) - W(s)}{h}$ is blowing up in magnitude. Thus, differentiability is going to be a problem. In fact, not only is the path of a Brownian motion guaranteed to be nondifferentiable at any prespecified t, it is guaranteed to be non-differentiable *simultaneously* at all values of t. This is a much stronger roughness property than lack of differentiability at a fixed t.

The next theorem is regarded as one of the most classic ones in probability theory. We first need a few definitions.

Definition 12.12. Let f be a real-valued continuous function defined on some open subset $\mathcal{T}$ of $\mathcal{R}$. The *upper and the lower Dini right derivatives* of f at $t \in \mathcal{T}$ are defined as

$$D^+ f(t) = \limsup_{h \downarrow 0} \frac{f(t+h) - f(t)}{h}, \quad D_+ f(t) = \liminf_{h \downarrow 0} \frac{f(t+h) - f(t)}{h}.$$

Definition 12.13. Let f be a real-valued function defined on some open subset $\mathcal{T}$ of $\mathcal{R}$. The function f is said to be *Holder continuous* of order $\gamma > 0$ at t if for some finite constant C (possibly depending on t), $|f(t+h) - f(t)| \leq C|h|^\gamma$ for all sufficiently small h. If f is Holder continuous of order γ at every $t \in \mathcal{T}$ with a universal constant C, it is called Holder continuous of order γ in $\mathcal{T}$.

Theorem 12.6 (Crooked Paths and Unbounded Variation). *(a) Given any $T > 0$, $P(\sup_{t \in [0,T]} W(t) > 0, \inf_{t \in [0,T]} W(t) < 0) = 1$. Hence, with probability one, in any nonempty interval containing zero, $W(t)$ changes sign at least once, and therefore infinitely often.*

(b) **(Nondifferentiability Everywhere).** *With probability one, $W(t)$ is (simultaneously) nondifferentiable at all $t > 0$; that is,*

$$P(\text{For each } t > 0, W(t) \text{ is not differentiable at } t) = 1.$$

(c) **(Unbounded Variation).** *For every $T > 0$, with probability one, $W(t)$ has an unbounded total variation as a function of t on $[0, T]$.*

(d) With probability one, no nonempty time interval $W(t)$ can be monotone increasing or monotone decreasing.

(e) $P(\text{For all } t > 0, D^+ W(t) = \infty \text{ or } D_+ W(t) = -\infty \text{ or both}) = 1$.

(f) **(Holder Continuity).** *Given any finite $T > 0$ and $0 < \gamma < \frac{1}{2}$, with probability one, $W(t)$ is Holder continuous on $[0, T]$ of order γ.*

(g) For any $\gamma > \frac{1}{2}$, with probability one, $W(t)$ is nowhere Holder continuous of order γ.

(h) **(Uniform Continuity in Probability).** *Given any* $\epsilon > 0$, *and* $0 < T < \infty$, $P(\sup_{t,s, 0 \leq t,s \leq T, |t-s| < h} |W(t) - W(s)| > \epsilon) \to 0$ *as* $h \to 0$.

Proof. Each of parts (c) and (d) would follow from part (b), because of results in real analysis that monotone functions or functions of bounded variation must be differentiable almost everywhere. Part (e) is a stronger version of the nondifferentiability result in part (b); see Karatzas and Shreve (1991, pp. 106–111) for parts (e)–(h). Part (b) itself is proved in many standard texts on stochastic processes; the proof involves quite a bit of calculation. We show here that part (a) is a consequence of the reflection principle.

Clearly, it is enough just to show that for any $T > 0$, $P(\sup_{t \in [0,T]} W(t) > 0) = 1$. This will imply that $P(\inf_{t \in [0,T]} W(t) < 0) = 1$, because $-W(t)$ is a Brownian motion if $W(t)$ is, and hence it will imply all the other statements in part (a). Fix $c > 0$. Then,

$$P(\sup_{t \in [0,T]} W(t) > 0) \geq P(\sup_{t \in [0,T]} W(t) > c) = 2P(W(T) > c) \text{ (reflection principle)}$$

$\to 1$ as $c \to 0$, and therefore $P(\sup_{t \in [0,T]} W(t) > 0) = 1$. □

Remark. It should be noted that the set of points at which the path of a Brownian motion is Holder continuous of order $\frac{1}{2}$ is not empty, although in some sense such points are rare.

The oscillation properties of the paths of a Brownian motion are further illustrated by the laws of the iterated logarithm for Brownian motion. The path of a Brownian motion is a random function. Can we construct suitable deterministic functions, say $u(t)$ and $l(t)$, such that for large t the Brownian path $W(t)$ will be bounded by the envelopes $l(t), u(t)$? What are the tightest such envelope functions? Similar questions can be asked about small t. The law of the iterated logarithm answers these questions precisely. However, it is important to note that in addition to the intellectual aspect of just identifying the tightest envelopes, the iterated logarithm laws have other applications.

Theorem 12.7 (LIL). *Let* $f(t) = \sqrt{2t \log |\log t|}, t > 0$. *With probability one,*

(a) $\limsup_{t \to \infty} \frac{W(t)}{f(t)} = 1$; $\liminf_{t \to \infty} \frac{W(t)}{f(t)} = -1$.
(b) $\limsup_{t \to 0} \frac{W(t)}{f(t)} = 1$; $\liminf_{t \to 0} \frac{W(t)}{f(t)} = -1$.

Remark on Proof: Note that the lim inf statement in part (a) follows from the lim sup statement because $-W(t)$ is also a Brownian motion if $W(t)$ is. On the other hand, the two statements in part (b) follow from the corresponding statements in part (a) by the time reciprocity property that $tW(\frac{1}{t})$ is also a Brownian motion if $W(t)$ is. For a proof of part (a), see Karatzas and Shreve (1991), or Bhattacharya and Waymire (2007, p. 143). □

12.3.3 * Fractal Nature of Level Sets

For a moment, let us consider a general question. Suppose $\mathcal{T}$ is a subset of the real line, and $X(t), t \in \mathcal{T}$ a real-valued stochastic process. Fix a number u, and ask how many times does the path of $X(t)$ cut the line drawn at level u; that is, consider $N_{\mathcal{T}}(u) = \#\{t \in \mathcal{T} : X(t) = u\}$. It is not a priori obvious that $N_{\mathcal{T}}(u)$ is finite. Indeed, for Brownian motion, we already know that in any nonempty interval containing zero, the path hits zero infinitely often with probability one. One might guess that this lack of finiteness is related to the extreme oscillatory nature of the paths of a Brownian motion. Indeed, that is true. If the process $X(t)$ is a bit more smooth, then the number of level crossings will be finite. However, investigations into the distribution of $N_{\mathcal{T}}(u)$ will still be a formidable problem. For the Brownian motion, it is not the number of level crossings, but the geometry of the set of times at which it crosses a given level u that is of interest. In this section, we describe the fascinating properties of these *level sets* of the path of a Brownian motion. We also give a very brief glimpse into what we can expect for processes whose paths are more smooth, to draw the distinction from the case of Brownian motion.

Given $b \in \mathcal{R}$, let
$$\mathcal{C}_b = \{t \geq 0 : W(t) = b\}.$$

Note that $\mathcal{C}_b$ is a *random set*, in the sense that different sample paths will hit the level b at different sets of times. We only consider the case $b = 0$ here, although most of the properties of $\mathcal{C}_0$ extend in a completely evident way to the case of a general b.

Theorem 12.8. *With probability one, $\mathcal{C}_0$ is an uncountable, unbounded, closed set of Lebesgue measure zero, and has no isolated points; that is, in any neighborhood of an element of $\mathcal{C}_0$, there is at least one other element of $\mathcal{C}_0$.*

Proof. It follows from an application of the reflection principle that $P(\sup_{t \geq 0} W(t) = \infty, \inf_{t \geq 0} W(t) = -\infty) = 1$ (check it!). Therefore, given any $T > 0$, there must be a time instant $t > T$ such that $W(t) = 0$. For if there were a finite last time that $W(t) = 0$, then for such a sample path, the supremum and the infimum cannot simultaneously be infinite. This means that the zero set $\mathcal{C}_0$ is unbounded. It is closed because the paths of Brownian motion are continuous. We have not defined what Lebesgue measure means, therefore we cannot give a rigorous proof that $\mathcal{C}_0$ has zero Lebesgue measure. Think of Lebesgue measure of a set C as its total length $\lambda(C)$. Then, by Fubini's theorem,

$$E[\lambda(\mathcal{C}_0)] = E\left[\int_{\mathcal{C}_0} dt\right] = E\left[\int_{[0,\infty)} I_{\{W(t)=0\}} dt\right]$$
$$= \int_{[0,\infty)} P(W(t) = 0) dt = 0.$$

If the expected length is zero, then the length itself must be zero with probability one. That $\mathcal{C}_0$ has no isolated points is entirely nontrivial to prove and we omit the proof. Finally, by a result in real analysis that any closed set with no isolated points must be uncountable unless it is empty, we have that C_0 is an uncountable set. □

Remark. The implication is that the set of times at which Brownian motion returns to zero is a topologically large set marked by holes, and collectively the holes are big enough that the zero set, although uncountable, has length zero. Such sets in one dimension are commonly called *Cantor sets*. Corresponding sets in higher dimensions often go by the name *fractals*.

12.4 The Dirichlet Problem and Boundary Crossing Probabilities

The *Dirichlet problem* on a *domain* in $\mathcal{R}^d$, $1 \leq d < \infty$ was formulated by Gauss in the mid-nineteenth century. It is a problem of special importance in the area of partial differential equations with boundary value constraints. The Dirichlet problem can also be interpreted as a problem in the physical theory of diffusion of heat in a d-dimensional domain with controlled temperature at the boundary points of the domain. According to the laws of physics, the temperature as a function of the location in the domain would have to be a *harmonic function*. The Dirichlet problem thus asks for finding a function $u(x)$ such that

$$u(x) = g(x) \text{ (specified)}, \quad x \in \partial U; u(.) \text{ harmonic in } U,$$

where U is a specified domain in $\mathcal{R}^d$. In this generality, solutions to the Dirichlet problem need not exist. We need the boundary value function g as well as the domain U to be sufficiently nice. The interesting and surprising thing is that solutions to the Dirichlet problem have connections to the d-dimensional Brownian motion. Solutions to the Dirichlet problem can be constructed by solving suitable problems (which we describe below) about d-dimensional Brownian motion. Conversely, these problems on the Brownian motion can be solved if we can directly find solutions to a corresponding Dirichlet problem, perhaps by inspection, or by using standard techniques in the area of partial differential equations. Thus, we have an altruistic connection between a special problem on partial differential equations and a problem on Brownian motion. It turns out that these connections are more than intellectual curiosities. For example, these connections were elegantly exploited in Brown (1971) to solve certain otherwise very difficult problems in the area of statistical decision theory.

We first provide the necessary definitions. We remarked before that the Dirichlet problem is not solvable on arbitrary domains. The domain must be such that it does not contain any *irregular* boundary points. These are points $x \in \partial U$ such that a Brownian motion starting at x immediately falls back into U. A classic example is that of a disc, from which the center has been removed. Then, the center is an irregular boundary point of the domain. We refer the reader to Karatzas and Shreve (1991, pp. 247–250) for the exact regularity conditions on the domain.

12.4 The Dirichlet Problem and Boundary Crossing Probabilities

Definition 12.14. A set $U \subseteq \mathcal{R}^d$, $1 \leq d < \infty$ is called a *domain* if U is connected and open.

Definition 12.15. A twice continuously differentiable real-valued function $u(x)$ defined on a domain $U \subseteq \mathcal{R}^d$ is called *harmonic* if its Laplacian $\Delta u(x) = \sum_{i=1}^d \frac{\partial^2}{\partial x_i^2} u(x) \equiv 0$ for all $x = (x_1, x_2, \ldots, x_d) \in U$.

Definition 12.16. Let U be a bounded regular domain in $\mathcal{R}^d$, and g a real-valued continuous function on ∂U. The *Dirichlet problem* on the domain $U \subseteq \mathcal{R}^d$ with boundary value constraint g is to find a function $u : \overline{U} \to \mathcal{R}$ such that u is harmonic in U, and $u(x) = g(x)$ for all $x \in \partial U$, where ∂U denotes the boundary of U and $\overline{U}$ the closure of U.

Theorem 12.9. *Let $U \subseteq \mathcal{R}^d$ be a bounded regular domain. Fix $x \in U$. Consider $\mathbf{X}_d(t), t \geq 0$, d-dimensional Brownian motion starting at x, and let $\tau = \tau_U = \inf\{t > 0 : \mathbf{X}_d(t) \notin U\} = \inf\{t > 0 : \mathbf{X}_d(t) \in \partial U\}$. Define the function u pointwise on $\overline{U}$ by*

$$u(x) = E_x[g(\mathbf{X}_d(\tau))], \quad x \in U; \quad u = g \text{ on } \partial U.$$

Then u is continuous on $\overline{U}$ and is the unique solution, continuous on $\overline{U}$, to the Dirichlet problem on U with boundary value constraint g.

Remark. When $\mathbf{X}_d(t)$ exits from U having started at a point inside U, it can exit through different points on the boundary ∂U. If it exits at the point $y \in \partial U$, then $g(\mathbf{X}_d(\tau))$ will equal $g(y)$. The exit point y is determined by chance. If we average over y, then we get a function that is harmonic inside U and equals g on ∂U. We omit the proof of this theorem, and refer the reader to Karatzas and Shreve (1991, p. 244), and Körner (1986, p. 55).

Example 12.5 (Dirichlet Problem on an Annulus). Consider the Dirichlet problem on the d-dimensional annulus $U = \{z : r < ||z|| < R\}$, where $0 < r < R < \infty$. Specifically, suppose we want a function u such that

$$u \text{ harmonic on } \{z : r < ||z|| < R\}, \quad u = 1 \text{ on } \{z : ||z|| = R\},$$
$$u = 0 \text{ on } \{z : ||z|| = r\}.$$

A continuous solution to this can be found directly. The solution is

$$u(z) = \frac{|z| - r}{R - r} \quad \text{for } d = 1;$$

$$u(z) = \frac{\log ||z|| - \log r}{\log R - \log r} \quad \text{for } d = 2;$$

$$u(z) = \frac{r^{2-d} - ||z||^{2-d}}{r^{2-d} - R^{2-d}} \quad \text{for } d > 2.$$

We now relate this solution to the Dirichlet problem on U with d-dimensional Brownian motion. Consider then $\mathbf{X}_d(t)$, d-dimensional Brownian motion that started at a given point x inside U; $r < ||x|| < R$. Because the function g corresponding to the boundary value constraint in this example is $g(z) = I_{\{z:||z||=R\}}$, by the above theorem, $u(x)$ equals

$$u(x) = E_x[g(\mathbf{X}_d(\tau))]$$
$$= P_x(\mathbf{X}_d(t) \text{ reaches the sphere} ||z|| = R \text{ before it reaches the sphere} ||z|| = r).$$

For now, let us consider the case $d = 1$. Fix positive numbers r, R and suppose a one-dimensional Brownian motion starts at a number x between r and R, $0 < r < x < R < \infty$. Then the probability that it will hit the line $z = R$ before hitting the line $z = r$ is $\frac{x-r}{R-r}$. The closer the starting point x is to R, the larger is the probability that it will first hit the line $z = R$. Furthermore, the probability is a very simple linear function. We revisit the case $d > 1$ when we discuss recurrence and transience of d-dimensional Brownian motion in the next section.

12.4.1 Recurrence and Transience

We observed during our discussion of the lattice random walk (Chapter 11) that it is recurrent in dimensions $d = 1, 2$ and transient for $d > 2$. That is, in one and two dimensions the lattice random walk returns to any integer value x at least once (and hence infinitely often) with probability one, but for $d > 2$, the probability that the random walk returns at all to its starting point is less than one. For the Brownian motion, when the dimension is more than one, the correct question is not to ask if it returns to particular points x. The correct question to ask is if it returns to any fixed neighborhood of a particular point, however small. The answers are similar to the case of the lattice random walk; that is, in one dimension, Brownian motion returns to any point x infinitely often with probability one, and in two dimensions, Brownian motion returns to any given neighborhood of a point x infinitely often with probability one. But when $d > 2$, it diverges off to infinity. We can see this by using the connection of Brownian motion to the Dirichlet problem on discs. We first need two definitions.

Definition 12.17. For $d > 1$, a d-dimensional stochastic process $\mathbf{X}_d(t), t \geq 0$ is called *neighborhood recurrent* if with probability one, it returns to any given ball $B(x, \epsilon)$ infinitely often.

Definition 12.18. For any d, a d-dimensional stochastic process $\mathbf{X}_d(t), t \geq 0$ is called *transient* if with probability one, $\mathbf{X}_d(t)$ diverges to infinity.

We now show how the connection of the Brownian motion to the solution of the Dirichlet problem will help us establish that Brownian motion is transient for $d > 2$. That is, if we let B be the event that $\lim_{t \to \infty} ||\mathbf{W}_d(t)|| \neq \infty$, then we show that $P(B)$ must be zero for $d > 2$. Indeed, to be specific, take $d = 3$, pick a point

$x \in \mathcal{R}^3$ with $||x|| > 1$, suppose that our Brownian motion is now sitting at the point x, and ask what the probability is that it will reach the unit ball B_1 before it reaches the disk $||z|| = R$. Here $R > ||x||$. We have derived this probability. The Markov property of Brownian motion gives this probability to be exactly equal to $1 - \frac{1 - \frac{1}{||x||}}{1 - \frac{1}{R}}$. This clearly converges to $\frac{1}{||x||}$ as $R \to \infty$. Imagine now that the process has evolved for a long time, say T, and that it is now sitting at a very distant x (i.e., $||x||$ is large). The LIL for Brownian motion guarantees that we can pick such a large T and such a distant x. Then, the probability of ever returning from x to the unit ball would be the small number $\epsilon = \frac{1}{||x||}$. We can make ϵ arbitrarily small by choosing $||x||$ sufficiently large, and what that means is that the probability of the process returning infinitely often to the unit ball B_1 is zero. The same argument works for B_k, the ball of radius k for any $k \geq 1$, and therefore, $P(B) = P(\cup_{k=1}^\infty B_k) = 0$; that is, the process drifts off to infinity with probability one. The same argument works for any $d > 2$, not just $d = 3$. The case of $d = 1, 2$ is left as a chapter exercise. We then have the following theorem. □

Theorem 12.10. *Brownian motion visits every real x infinitely often with probability one if $d = 1$, is neighborhood recurrent if $d = 2$, and transient if $d > 2$. Moreover, by its neighborhood recurrence for $d = 2$, the graph of a two-dimensional Brownian path on $[0, \infty)$ is dense in the two-dimensional plane.*

12.5 * The Local Time of Brownian Motion

For the simple symmetric random walk in one dimension, we derived the distribution of the local time $\xi(x, n)$, which is the total time the random walk spends at the integer x up to the time instant n. It would not be interesting to ask exactly the same question about Brownian motion, because the number of time points t up to some time T at which the Brownian motion $W(t)$ equals a given x is zero or infinity. Paul Lévy gave the following definition for the local time of a Brownian motion. Fix a set A in the real line and a general time instant $T, T > 0$. Now ask what is the total size of the times t up to T at which the Brownian motion has resided in the given set A. That is, denoting *Lebesgue measure* on $\mathcal{R}$ by λ, look at the following *kernel*

$$H(A, T) = \lambda\{t \leq T : W(t) \in A\}.$$

Using this, Lévy formulated the local time of the Brownian motion at a given x as

$$\eta(x, T) = \lim_{\epsilon \downarrow 0} \frac{H([x - \epsilon, x + \epsilon], T)}{2\epsilon},$$

where the limit is supposed to mean a pointwise almost sure limit. It is important to note that the existence of the almost sure limit is nontrivial.

Instead of the clumsy notation T, we eventually simply use the notation t, and thereby obtain a new stochastic process $\eta(x,t)$, indexed simultaneously by two parameters x and t. We can regard (x,t) together as a vector-valued time parameter, and call $\eta(x,t)$ a *random field*. This is called the *local time of one-dimensional Brownian motion*. The local time of Brownian motion is generally regarded to be an analytically difficult process to study. We give a relatively elementary exposition to the local time of Brownian motion in this section.

Recall now the previously introduced maximum process of standard Brownian motion, namely $M(t) = \sup_{0 \leq s \leq t} W(s)$. The following major theorem on the distribution of the local time of Brownian motion at zero was proved by Paul Lévy.

Theorem 12.11. *Let $W(s), s \geq 0$ be standard Brownian motion starting at zero. Consider the two stochastic processes, $\{\eta(0,t), t \geq 0\}$, and $\{M(t), t \geq 0\}$. These two processes have the same distribution.*

In particular, for any given fixed t, and $y > 0$,

$$P\left(\frac{\eta(0,t)}{\sqrt{t}} \leq y\right) = \sqrt{\frac{2}{\pi}} \int_0^y e^{-z^2/2} dz = 2\Phi(y) - 1$$

$$\Leftrightarrow P(\eta(0,t) \leq y) = \sqrt{\frac{2}{t\pi}} \int_0^y e^{-z^2/(2t)} dz.$$

For a detailed proof of this theorem, we refer to Mörters and Peres (2010, p. 160). A sketch of the proof can be seen in Révész (2005).

For a general level x, the corresponding result is as follows, and it follows from the case $x = 0$ treated above.

Theorem 12.12.

$$P(\eta(x,t) \leq y) = 2\Phi\left(\frac{y+|x|}{\sqrt{t}}\right) - 1, \quad -\infty < x < \infty, t, y > 0.$$

It is important to note that if the level $x \neq 0$, then the local time $\eta(x,t)$ can actually be exactly equal to zero with a positive probability, and this probability is simply the probability that Brownian motion does not reach x within time t, and equals $2\Phi(\frac{|x|}{\sqrt{t}}) - 1$. This is not the case if the level is zero, in which case the local time $\eta(0,t)$ possesses a density function.

The theorem above also says that the local time of Brownian motion grows at the rate of $\sqrt{t}$ for any level x. The expected value follows easily by evaluating the integral $\int_0^\infty [1 - P(\eta(x,t) \leq y)] dy$, and one gets

$$E[\eta(x,t)] = 4\sqrt{t}\phi\left(\frac{x}{\sqrt{t}}\right)\left[1 - \Phi\left(\frac{|x|}{\sqrt{t}}\right)\right] - 4|x|\left[1 - \Phi\left(\frac{|x|}{\sqrt{t}}\right)\right]^2.$$

The limit of this as $x \to 0$ equals $\sqrt{\frac{2}{\pi}}\sqrt{t}$, which agrees with $E[\eta(0,t)]$. The expected local time is plotted in Fig. 12.3.

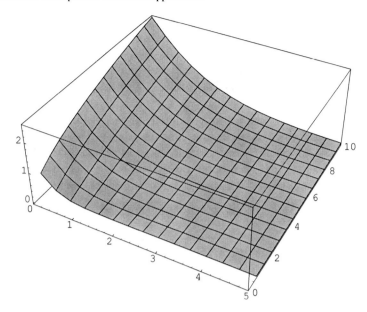

Fig. 12.3 Plot of the expected local time as function of (x,t)

12.6 Invariance Principle and Statistical Applications

We remarked in the first section of this chapter that scaled random walks mimic the Brownian motion in a suitable asymptotic sense. As a matter of fact, if $X_1, X_2, \ldots$ is any iid sequence of one-dimensional random variables satisfying some relatively simple conditions, then the sequence of partial sums $S_n = \sum_{i=1}^{n} X_i, n \geq 1$, when appropriately scaled, mimics Brownian motion in a suitable asymptotic sense. Why is this useful? This is useful because in many concrete problems of probability and statistics, suitable functionals of the sequence of partial sums arise as the objects of direct importance. The *invariance principle* allows us to conclude that if the sequence of partial sums S_n mimics $W(t)$, then any nice functional of the sequence of partial sums will also mimic the same functional of $W(t)$. So, if we can figure out how to deal with the distribution of the needed functional of the $W(t)$ process, then we can use it in practice to approximate the much more complicated distribution of the original functional of the sequence of partial sums. It is a profoundly useful fact in the asymptotic theory of probability that all of this is indeed a reality. This section treats the invariance principle for the partial sum process of one-dimensional iid random variables. We recommend Billingsley (1968), Hall and Heyde (1980), and Csörgo and Révész (1981) for detailed and technical treatments; Erdös and Kac (1946), Donsker (1951), Komlós et al. (1975, 1976, Major (1978), Whitt (1980), and Csörgo and Hall (1984) for invariance principles for the partial sum process; and Pyke (1984) and Csörgo (2002)) for lucid reviews. Also, see Dasgupta (2008) for references to various significant extensions, such as the multidimensional and dependent cases.

Although the invariance principle for partial sums of iid random variables is usually credited to Donsker (1951), Erdös and Kac (1946) contained the basic idea behind the invariance principle and also worked out the asymptotic distribution of a number of key and interesting functionals of the partial sum process. Donsker (1951) provided the full generalization of the Erdös–Kac technique by providing explicit embeddings of the discrete sequence $\frac{S_k}{\sqrt{n}}, k = 1, 2, \ldots, n$ into a continuous-time stochastic process $S_n(t)$ and by establishing the limiting distribution of a general continuous functional $h(S_n(t))$. In order to achieve this, it is necessary to use a continuous mapping theorem for metric spaces, as consideration of Euclidean spaces is no longer enough. It is also useful to exploit a property of the Brownian motion known as the *Skorohod embedding theorem*. We first describe this necessary background material.

Define

$C[0, 1]$ = *Class of all continuous real valued functions on* $[0, 1]$, *and*

$D[0, 1]$ = *Class of all real-valued functions on* $[0, 1]$ *that are right continuous and have a left limit at every point in* $[0, 1]$.

Given two functions f, g in either $C[0, 1]$ or $D[0, 1]$, let $\rho(f, g) = \sup_{0 \leq t \leq 1} |f(t) - g(t)|$ denote the supremum distance between f and g. We refer to ρ as the *uniform metric*. Both $C[0, 1]$ and $D[0, 1]$ are (complete) metric spaces with respect to the uniform metric ρ.

Suppose $X_1, X_2, \ldots$ is an iid sequence of real valued random variables with mean zero and variance one. Two common embeddings of the discrete sequence $\frac{S_k}{\sqrt{n}}, k = 1, 2, \ldots, n$ into a continuous time process are the following.

$$S_{n,1}(t) = \frac{1}{\sqrt{n}}[S_{\lfloor nt \rfloor} + \{nt\}X_{\lfloor nt \rfloor+1}],$$

and

$$S_{n,2}(t) = \frac{1}{\sqrt{n}}S_{\lfloor nt \rfloor},$$

$0 \leq t \leq 1$. Here, $\lfloor . \rfloor$ denotes the integer part and $\{.\}$ the fractional part of a positive real.

The first one simply continuously interpolates between the values $\frac{S_k}{\sqrt{n}}$ by drawing straight lines, but the second one is only right continuous, with jumps at the points $t = \frac{k}{n}, k = 1, 2, \ldots, n$. For certain specific applications, the second embedding is more useful. It is because of these jump discontinuities that Donsker needed to consider weak convergence in $D[0, 1]$. It led to some additional technical complexities.

The main idea from this point on is not difficult. One can produce a version of $S_n(t)$, say $\tilde{S}_n(t)$, such that $\tilde{S}_n(t)$ is close to a sequence of Wiener processes $W_n(t)$. Because $\tilde{S}_n(t) \approx W_n(t)$, if $h(.)$ is a continuous functional with respect to the uniform metric, then one can expect that $h(\tilde{S}_n(t)) \approx h(W_n(t)) = h(W(t))$ in distribution. $\tilde{S}_n(t)$ being a version of $S_n(t), h(S_n(t)) = h(\tilde{S}_n(t))$ in distribution, and so, $h(S_n(t))$ should be close to the fixed Brownian functional $h(W(t))$ in distribution, which is the question we wanted to answer.

The results leading to Donsker's theorem are presented below.

Theorem 12.13 (Skorohod Embedding). *Given a random variable X with mean zero and a finite variance σ^2, we can construct (on the same probability space) a standard Brownian motion $W(t)$ starting at zero, and a stopping time τ with respect to $W(t)$ such that $E(\tau) = \sigma^2$ and X and the stopped Brownian motion $W(\tau)$ have the same distribution.*

Theorem 12.14 (Convergence of the Partial Sum Process to Brownian Motion). *Let $S_n(t) = S_{n,1}(t)$ or $S_{n,2}(t)$ as defined above. Then there exists a common probability space on which one can define Wiener processes $W_n(t)$ starting at zero, and a sequence of processes $\{\tilde{S}_n(t)\}, n \geq 1$, such that*

(a) For each n, $S_n(t)$ and $\tilde{S}_n(t)$ are identically distributed as processes.

(b) $\sup_{0 \leq t \leq 1} |\tilde{S}_n(t) - W_n(t)| \overset{P}{\Rightarrow} 0$.

We prove the last theorem, assuming the Skorohod embedding theorem. A proof of the Skorohod embedding theorem may be seen in Csörgo and Révész (1981), or in Bhattacharya and Waymire (2007, p. 160).

Proof. We treat only the linearly interpolated process $S_{n,1}(t)$, and simply call it $S_n(t)$. To reduce notational clutter, we write as if the version $\tilde{S}_n$ of S_n is S_n itself. Thus, the $\tilde{S}_n$ notation is dropped in the proof of the theorem. Without loss of generality, we take $E(X_1) = 0$ and $\text{Var}(X_1) = 1$. First, by using the Skorohod embedding theorem, construct a stopping time τ_1 with respect to the process $W(t), t \geq 0$ such that $E(\tau_1) = 1$ and such that $W(\tau_1) \overset{\mathcal{L}}{=} X_1$. Using the strong Markov property of Brownian motion, $W(t + \tau_1) - W(\tau_1)$ is also a Brownian motion on $[0, \infty)$, independent of $(\tau_1, W(\tau_1))$, and we can now pick a stopping time, say τ_2' with respect to this process, with the two properties $E(\tau_2') = 1$ and $W(\tau_2') \overset{\mathcal{L}}{=} X_2$. Therefore, if we define $\tau_2 = \tau_1 + \tau_2'$, then we have obtained a stopping time with respect to the original Brownian motion, with the properties that its expectation is 2, and $\tau_2 - \tau_1$ and τ_1 are independent. Proceeding in this way, we can construct an infinite sequence of stopping times $0 = \tau_0 \leq \tau_1 \leq \tau_2 \leq \tau_3 \leq \cdots$, such that $\tau_k - \tau_{k-1}$ are iid with mean one, and the two discrete time processes S_k and $W(\tau_k)$ have the same distribution. Moreover, by the usual SLLN,

$$\frac{\tau_n}{n} = \frac{1}{n}\sum_{k=1}^{n}[\tau_k - \tau_{k-1}] \overset{\text{a.s.}}{\to} 1,$$

from which it follows that

$$\frac{\max_{0 \leq k \leq n} |\tau_k - k|}{n} \overset{P}{\to} 0.$$

Set $W_n(t) = \frac{W(nt)}{\sqrt{n}}, n \geq 1$. Therefore, in this notation, $W(\tau_k) = \sqrt{n} W_n(\frac{\tau_k}{n})$. Now fix $\epsilon > 0$ and consider the event $B_n = \{\sup_{0 \leq t \leq 1} |S_n(t) - W_n(t)| > \epsilon\}$. We need to show that $P(B_n) \to 0$.

Now, because $S_n(t)$ is defined by linear interpolation, in order that B_n happens, at some t in $[0, 1]$ we must have one of

$$|S_k/\sqrt{n} - W_n(t)| \text{ and } |S_{k-1}/\sqrt{n} - W_n(t)|$$

larger than ϵ, where k is the unique k such that $\frac{k-1}{n} \leq t < \frac{k}{n}$. Our goal is to show that the probability of the union of these two events is small. Now use the fact that in distribution, $S_k = W(\tau_k) = \sqrt{n} W_n(\frac{\tau_k}{n})$, and so it will suffice to show that the probability of the union of the two events $\{|W_n(\frac{\tau_k}{n}) - W_n(t)| > \epsilon\}$ and $\{|W_n(\frac{\tau_{k-1}}{n}) - W_n(t)| > \epsilon\}$ is small. However, the union of these two events can only happen if either W_n differs by a large amount in a small interval, or one of the two time instants $\frac{\tau_k}{n}$ and $\frac{\tau_{k-1}}{n}$ are far from t. The first possibility has a small probability by the uniform continuity of paths of a Brownian motion (on any compact interval), and the second possibility has a small probability by our earlier observation that $\frac{\max_{0 \leq k \leq n} |\tau_k - k|}{n} \xrightarrow{P} 0$. This implies that $P(B_n)$ is small for all large n, as we wanted to show.

This theorem implies the following important result by an application of the continuous mapping theorem, continuity being defined through the uniform metric on the space $C[0, 1]$.

Theorem 12.15 (Donsker's Invariance Principle). *Let h be a continuous functional with respect to the uniform metric on $C[0, 1]$ and let $S_n(t)$ be defined as either $S_{n,1}(t)$ or $S_{n,2}(t)$. Then $h(S_n(t)) \xrightarrow{\mathcal{L}} h(W(t))$, as $n \to \infty$.*

Example 12.6 (CLT Follows from Invariance Principle). The central limit theorem for iid random variables having a finite variance follows as a simple consequence of Donsker's invariance principle. Suppose $X_1, X_2, \ldots$ are iid random variables with mean zero and variance 1. Let $S_k = \sum_{i=1}^k X_i, k \geq 1$. Define the functional $h(f) = f(1)$ on $C[0, 1]$. This is obviously a continuous functional on $C[0, 1]$ with respect to the uniform metric $\rho(f, g) = \sup_{0 \leq t \leq 1} |f(t) - g(t)|$. Therefore, with $S_n(t)$ as the linearly interpolated partial sum process, it follows from the invariance principle that

$$h(S_n) = S_n(1) = \frac{\sum_{i=1}^n X_i}{\sqrt{n}} \xrightarrow{\mathcal{L}} h(W) = W(1) \sim N(0, 1),$$

which is the central limit theorem.

Example 12.7 (Maximum of a Random Walk). We apply the Donsker invariance principle to the problem of determination of the limiting distribution of a functional of a random walk. Suppose $X_1, X_2, \ldots$ are iid random variables with mean zero and variance 1. Let $S_k = \sum_{i=1}^k X_i, k \geq 1$. We want to derive the limiting distribution of

$G_n = \frac{\max_{1 \le k \le n} S_k}{\sqrt{n}}$. To derive its limiting distribution, define the functional $h(f) = \sup_{0 \le t \le 1} f(t)$ on $C[0,1]$. This is a continuous functional on $C[0,1]$ with respect to the uniform metric $\rho(f,g) = \sup_{0 \le t \le 1} |f(t) - g(t)|$. Further notice that our statistic G_n can be represented as $G_n = h(S_n)$, where S_n is the linearly interpolated partial sum process. Therefore, by Donsker's invariance principle, $G_n = h(S_n) \overset{\mathcal{L}}{\Rightarrow} h(W) = \sup_{0 \le t \le 1} W(t)$, where $W(t)$ is standard Brownian motion on $[0,1]$. We know its CDF explicitly, namely, for $x > 0$, $P(\sup_{0 \le t \le 1} W(t) \le x) = 2\Phi(x) - 1$. Thus, $P(G_n \le x) \to 2\Phi(x) - 1$ for all $x > 0$.

Example 12.8 (Sums of Powers of Partial Sums). Consider once again iid random variables $X_1, X_2, \ldots$ with zero mean and a unit variance. Fix a positive integer m and consider the statistic $T_{m,n} = n^{-1-m/2} \sum_{k=1}^{n} S_k^m$. By direct integration of the polygonal curve $[S_n(t)]^m$, we find that $T_{m,n} = \int_0^1 [S_n(t)]^m dt$. This guides us to the functional $h(f) = \int_0^1 f^m(t) dt$. Because $[0,1]$ is a compact interval, it is easy to verify that h is a continuous functional on $C[0,1]$ with respect to the uniform metric. Indeed, the continuity of $h(f)$ follows from simply the algebraic identity $|x^m - y^m| = |x - y||x^{m-1} + x^{m-2}y + \cdots + y^{m-1}|$. It therefore follows from Donsker's invariance principle that $T_{m,n} \overset{\mathcal{L}}{\Rightarrow} \int_0^1 W^m(t) dt$. At first glance it seems surprising that a nondegenerate limit distribution for partial sums of S_k^m can exist with only two moments.

12.7 Strong Invariance Principle and the KMT Theorem

In addition to the weak invariance principle described above, there are also *strong invariance principles*. The first strong invariance principle for partial sums was obtained in Strassen (1964). Since then, a lot of literature has developed, including for the multidimensional case. Good sources for information are Strassen (1967), Komlós et al. (1976), Major (1978), Csörgo and Révész (1981), and Einmahl (1987).

It would be helpful to first understand exactly what a strong invariance principle is meant to achieve. Suppose $X_1, X_2, \ldots$ is a zero mean unit variance iid sequence of random variables. For $n \ge 1$, let S_n denote the partial sum $\sum_{i=1}^n X_i$, and $S_n(t)$ the interpolated partial sum process with the special values $S_n(\frac{k}{n}) = \frac{S_k}{\sqrt{n}}$ for each n and $1 \le k \le n$. In the process of proving Donsker's invariance principle, we have shown that we can construct (on a common probability space) a process $\tilde{S}_n(t)$ (which is equivalent to the original process $S_n(t)$ in distribution) and a single Wiener process $W(t)$ such that $\sup_{0 \le t \le 1} |\tilde{S}_n(t) - \frac{1}{\sqrt{n}} W(nt)| \overset{P}{\to} 0$. Therefore,

$$|\tilde{S}_n(1) - \frac{1}{\sqrt{n}} W(n)| \overset{P}{\to} 0$$

$$\Rightarrow \frac{|\tilde{S}_n - W(n)|}{\sqrt{n}} \overset{P}{\to} 0.$$

The strong invariance principle asks if we can find suitable functions $g(n)$ such that we can make the stronger statement $\frac{|\tilde{S}_n - W(n)|}{g(n)} \stackrel{a.s.}{\to} 0$, and as a next step, what is the best possible choice for such a function g.

The exact statements of the strong invariance principle results require us to say that we can construct an *equivalent process* $\tilde{S}_n(t)$ and a Wiener process $W(t)$ on some probability space such that $\frac{|\tilde{S}_n - W(n)|}{g(n)} \stackrel{a.s.}{\to} 0$ for some suitable function g. Due to the clumsiness in repeatedly having to mention these qualifications, we drop the $\tilde{S}_n$ notation and simply say $S_n(t)$, and we also do not mention that the processes have all been constructed on some new probability space. The important thing for applications is that we can use the approximations on the original process itself, by simply adopting the equivalent process on the new probability space.

Paradoxically, the strong invariance principle does not imply the weak invariance principle (i.e., Donsker's invariance principle) in general. This is because under the assumption of just the finiteness of the variance of the X_i, the best possible $g(n)$ increases faster than $\sqrt{n}$. On the other hand, if the common distribution of the X_i satisfies more stringent moment conditions, then we can make $g(n)$ a lot slower, and even slower than $\sqrt{n}$. The array of results that is available is bewildering and they are all difficult to prove. We prefer to report a few results of great importance, including in particular the *KMT theorem*, due to Komlós et al. (1976).

Theorem 12.16. *Let $X_1, X_2, \ldots$ be an iid sequence with $E(X_i) = 0$, $Var(X_i) = 1$.*

(a) *There exists a Wiener process $W(t), t \geq 0$, starting at zero such that $\frac{S_n - W(n)}{\sqrt{n \log \log n}} \stackrel{a.s.}{\to} 0$.*

(b) *The $\sqrt{n \log \log n}$ rate of part (a) cannot be improved in the sense that given any nondecreasing sequence $a_n \to \infty$ (however slowly), there exists a CDF F with zero mean and unit variance, such that with probability one, $\limsup_n a_n \frac{S_n - W(n)}{\sqrt{n \log \log n}} = \infty$, for any iid sequence X_i following the CDF F, and any Wiener process $W(t)$.*

(c) *If we make the stronger assumption that X_i has a finite mgf in some open neighborhood of zero, then the statement of part (a) can be improved to $|S_n - W(n)| = O(\log n)$ with probability one.*

(d) **(KMT Theorem)** *Specifically, if we make the stronger assumption that X_i has a finite mgf in some open neighborhood of zero, then we can find suitable positive constants C, K, λ such that for any real number x and any given n,*

$$P(\max_{1 \leq k \leq n} |S_k - W(k)| \geq C \log n + x) \leq K e^{-\lambda x},$$

where the constants C, K, λ depend only on the common CDF of the X_i.

Remark. The KMT theorem is widely regarded as one of the most major advances in the area of invariance principles and central limit problems. One should note that the inequality given in the above theorem has a qualitative nature attached to it, as we can only use the inequality with constants C, K, λ that are known to exist, depending on the underlying F. Refinements of the version of the inequality given above are available. We refer to Csörgo and Révész (1981) for such refinements and general detailed treatment of the strong invariance principle.

12.8 Brownian Motion with Drift and Ornstein–Uhlenbeck Process

We finish with two special processes derived from standard Brownian motion. Both are important in applications.

Definition 12.19. Let $W(t), t \geq 0$ be standard Brownian motion starting at zero. Fix $\mu \in \mathcal{R}$ and $\sigma > 0$. Then the process $X(t) = t\mu + \sigma W(t), t \geq 0$ is called *Brownian motion with drift μ and diffusion coefficient σ*. It is clear that it inherits the major path properties of standard Brownian motion, such as nondifferentiablity at all t with probability one, the independent increments property, and the Markov property. Also, clearly, for fixed t, $X(t) \sim N(t\mu, \sigma^2 t)$.

12.8.1 Negative Drift and Density of Maximum

There are, however, also some important differences when a drift is introduced. For example, unless $\mu = 0$, the reflection principle no longer holds, and consequently one cannot derive the distribution of the running maximum $M_\mu(t) = \sup_{0 \leq s \leq t} X(s)$ by using symmetry arguments. If $\mu \geq 0$, then it is not meaningful to ask for the distribution of the maximum over all $t > 0$. However, if $\mu < 0$, then the process has a tendency to drift off towards negative values, and in that case the maximum in fact does have a nontrivial distribution. We derive the distribution of the maximum when $\mu < 0$ by using a result on a particular first passage time of the process.

Theorem 12.17. *Let $X(t), t \geq 0$ be Brownian motion starting at zero, and with drift $\mu < 0$ and diffusion coefficient σ. Fix $a < 0 < b$, and let*

$$T_{a,b} = \min[\inf\{t > 0 : X(t) = a\}, \inf\{t > 0 : X(t) = b\}],$$

the first time $X(t)$ reaches either a or b. Then,

$$P(X_{T_{a,b}} = b) = \frac{e^{-2a\mu/\sigma^2} - 1}{e^{-2a\mu/\sigma^2} - e^{-2b\mu/\sigma^2}}.$$

A proof of this theorem can be seen in Karlin and Taylor (1975, p 361). By using this result, we can derive the distribution of $\sup_{t>0} X(t)$ in the case $\mu < 0$.

Theorem 12.18 (The Maximum of Brownian Motion with a Negative Drift). *If $X(t), t \geq 0$ is Brownian motion starting at zero, and with drift $\mu < 0$ and diffusion coefficient σ, then, $\sup_{t>0} X(t)$ is distributed as an exponential with mean $-\frac{\sigma^2}{2\mu}$.*

Proof. In the theorem stated above, by letting $a \to -\infty$, we get

$$P(X(t) \text{ ever reaches the level } b > 0) = e^{2b\mu/\sigma^2}.$$

But this is the probability that an exponential variable with mean $-\frac{\sigma^2}{2\mu}$ is larger than b. On the other hand, $P(X(t)$ ever reaches the level $b > 0)$ is the same as $P(\sup_{t>0} X(t) \geq b)$. Therefore, $\sup_{t>0} X(t)$ must have an exponential distribution with mean $-\frac{\sigma^2}{2\mu}$. □

Example 12.9 (Probability That Brownian Motion Does Not Hit a Line). Consider standard Brownian motion $W(t)$ starting at zero on $[0, \infty)$, and consider a straight line L with the equation $y = a + bt, a, b > 0$. Because $W(0) = 0, a > 0$, and paths of $W(t)$ are continuous, the probability that $W(t)$ does not hit the line L is the same as $P(W(t) < a + bt \, \forall t > 0)$. However, if we define a new Brownian motion (with drift) $X(t)$ as $X(t) = W(t) - bt$, then

$$P(W(t) < a + bt \, \forall t > 0) = P\left(\sup_{t>0} X(t) \leq a\right) = 1 - e^{-2ab},$$

by our theorem above on the maximum of a Brownian motion with a negative drift. We notice that the probability that $W(t)$ does not hit L is monotone increasing in each of a, b, as it should be.

12.8.2 * Transition Density and the Heat Equation

If we consider Brownian motion starting at some number x, and with drift $\mu < 0$ and diffusion coefficient σ, then by simple calculations, the conditional distribution of $X(t)$ given that $X(0) = x$ is $N(x + t\mu, \sigma^2 t)$, which has the density

$$p_t(x, y) = \frac{1}{\sqrt{2\pi}\sigma\sqrt{t}} e^{-\frac{(y-x-t\mu)^2}{2\sigma^2 t}}.$$

This is called the *transition density* of the process. The transition density satisfies a very special partial differential equation, which we now prove.

By direct differentiation,

$$\frac{\partial}{\partial t} p_t(x, y) = \frac{(x-y)^2 - \mu^2 t^2 - \sigma^2 t}{2\sqrt{2\pi}\sigma^3 t^{5/2}} e^{-\frac{(y-x-t\mu)^2}{2\sigma^2 t}};$$

$$\frac{\partial}{\partial y} p_t(x, y) = \frac{x - y + t\mu}{\sqrt{2\pi}\sigma^3 t^{3/2}} e^{-\frac{(y-x-t\mu)^2}{2\sigma^2 t}};$$

$$\frac{\partial^2}{\partial y^2} p_t(x, y) = \frac{(x-y+t\mu)^2 - \sigma^2 t}{\sqrt{2\pi}\sigma^5 t^{5/2}} e^{-\frac{(y-x-t\mu)^2}{2\sigma^2 t}}.$$

12.8 Brownian Motion with Drift and Ornstein–Uhlenbeck Process

On using these three expressions, it follows that the transition density $p_t(x, y)$ satisfies the partial differential equation

$$\frac{\partial}{\partial t} p_t(x, y) = -\mu \frac{\partial}{\partial y} p_t(x, y) + \frac{\sigma^2}{2} \frac{\partial^2}{\partial y^2} p_t(x, y).$$

This is the *drift-diffusion equation* in one dimension. In the particular case that $\mu = 0$ (no drift), and $\sigma = 1$, the equation reduces to the celebrated *heat equation*

$$\frac{\partial}{\partial t} p_t(x, y) = \frac{1}{2} \frac{\partial^2}{\partial y^2} p_t(x, y).$$

Returning to the drift-diffusion equation for the transition density in general, if we now take a general function $f(x, y)$ that is twice continuously differentiable in y and is bounded above by $Ke^{c|y|}$ for some finite $K, c > 0$, then integration by parts in the drift-diffusion equation produces the following expectation identity, which we state as a theorem.

Theorem 12.19. *Let x, μ be any real numbers, and $\sigma > 0$. Suppose $Y \sim N(x + t\mu, \sigma^2 t)$, and $f(x, y)$ twice continuously differentiable in y such that for some $0 < K, c < \infty, |f(x, y)| \leq Ke^{c|y|}$ for all y. Then,*

$$\frac{\partial}{\partial t} E_x[f(x, Y)] = \mu E_x\left[\frac{\partial}{\partial y} f(x, Y)\right] + \frac{\sigma^2}{2} E_x\left[\frac{\partial^2}{\partial y^2} f(x, Y)\right].$$

This identity and a multidimensional version of it has been used in Brown et al. (2006) to derive various results in statistical decision theory.

12.8.3 * The Ornstein–Uhlenbeck Process

The covariance function of standard Brownian motion $W(t)$ is $\text{Cov}(W(s), W(t)) = \min(s, t)$. Therefore, if we scale by $\sqrt{t}$, and let $X(t) = \frac{W(t)}{\sqrt{t}}, t > 0$, we get that $\text{Cov}(X(s), X(t)) = \sqrt{\frac{\min(s,t)}{\max(s,t)}} = \sqrt{\frac{s}{t}}$, if $s \leq t$. Therefore, the covariance is a function of only the time lag on a logarithmic time scale. This motivates the definition of the *Ornstein–Uhlenbeck process* as follows.

Definition 12.20. Let $W(t)$ be standard Brownian motion starting at zero, and let $\alpha > 0$ be a fixed constant. Then $X(t) = e^{-\frac{\alpha t}{2}} W(e^{\alpha t}), -\infty < t < \infty$ is called the *Ornstein–Uhlenbeck process*. The most general Ornstein–Uhlenbeck process is defined as

$$X(t) = \mu + \frac{\sigma}{\sqrt{\alpha}} e^{-\frac{\alpha t}{2}} W(e^{\alpha t}), \quad -\infty < \mu < \infty, \quad \alpha, \sigma > 0.$$

In contrast to the Wiener process, the Ornstein–Uhlenbeck process has a locally time-dependent drift. If the present state of the process is larger than μ, the global mean, then the drift drags the process back towards μ, and if the present state of the process is smaller than μ, then it does the reverse. The α parameter controls this tendency to return to the grand mean. The third parameter σ controls the variability.

Theorem 12.20. *Let $X(t)$ be a general Ornstein–Uhlenbeck process. Then, $X(t)$ is a stationary Gaussian process with $E[X(t)] = \mu$, and $Cov(X(s), X(t)) = \frac{\sigma^2}{\alpha} e^{-\frac{\alpha}{2}|s-t|}$.*

Proof. It is obvious that $E[X(t)] = \mu$. By definition of $X(t)$,

$$Cov(X(s), X(t)) = \frac{\sigma^2}{\alpha} e^{-\frac{\alpha}{2}(s+t)} \min(e^{\alpha s}, e^{\alpha t}) = \frac{\sigma^2}{\alpha} e^{-\frac{\alpha}{2}(s+t)} e^{\alpha \min(s,t)}$$

$$= \frac{\sigma^2}{\alpha} e^{\frac{\alpha}{2} \min(s,t)} e^{-\frac{\alpha}{2} \max(s,t)} = \frac{\sigma^2}{\alpha} e^{-(\alpha/2)|s-t|},$$

and inasmuch as $Cov(X(s), X(t))$ is a function of only $|s - t|$, it follows that it is stationary. □

Example 12.10 (Convergence of Integrated Ornstein–Uhlenbeck to Brownian Motion). Consider an Ornstein–Uhlenbeck process $X(t)$ with parameters μ, α, and σ^2. In a suitable asymptotic sense, the integrated Ornstein–Uhlenbeck process converges to a Brownian motion with drift μ and an appropriate diffusion coefficient; the diffusion coefficient can be adjusted to be one. Towards this, define $Y(t) = \int_0^t X(u) du$. This is clearly also a Gaussian process. We show in this example that if $\sigma^2, \alpha \to \infty$ in such a way that $\frac{4\sigma^2}{\alpha^2} \to c^2, 0 < c < \infty$, then $Cov(Y(s), Y(t)) \to c^2 \min(s, t)$. In other words, in the asymptotic paradigm where $\sigma, \alpha \to \infty$, but are of comparable order, the integrated Ornstein–Uhlenbeck process $Y(t)$ is approximately the same as a Brownian motion with some drift and some diffusion coefficient, in the sense of distribution.

We directly calculate $Cov(Y(s), Y(t))$. There is no loss of generality in taking μ to be zero. Take $0 < s \leq t < \infty$. Then

$$Cov(Y(s), Y(t)) = \int_0^t \int_0^s E[X(u)X(v)] du dv = \frac{\sigma^2}{\alpha} \int_0^t \int_0^s e^{-\frac{\alpha}{2}|u-v|} du dv$$

$$= \frac{\sigma^2}{\alpha} \left[\int_0^s \int_0^v e^{-\frac{\alpha}{2}|u-v|} du dv + \int_0^s \int_v^s e^{-\frac{\alpha}{2}|u-v|} du dv \right.$$

$$\left. + \int_s^t \int_0^s e^{-\frac{\alpha}{2}|u-v|} du dv \right]$$

$$= \frac{\sigma^2}{\alpha} \left[\int_0^s \int_0^v e^{-\frac{\alpha}{2}(v-u)} du dv + \int_0^s \int_v^s e^{-\frac{\alpha}{2}(u-v)} du dv \right.$$

$$\left. + \int_s^t \int_0^s e^{-\frac{\alpha}{2}(v-u)} du dv \right]$$

$$= \frac{\sigma^2}{\alpha} \frac{4}{\alpha^2} \left[\alpha s + e^{-\alpha s/2} + e^{-\alpha t/2} - e^{-\alpha(t-s)/2} \right],$$

on doing the three integrals in the line before, and on adding them.

If $\alpha, \sigma^2 \to \infty$, and $\frac{4\sigma^2}{\alpha^2}$ converges to some finite and nonzero constant c^2, then for any $s, t, 0 < s < t$, the derived expression for $\text{Cov}(Y(s), Y(t)) \to c^2 s = c^2 \min(s, t)$, which is the covariance function of Brownian motion with diffusion coefficient c.

The Ornstein–Uhlenbeck process enjoys another important property besides stationarity. It is also a Markov process. It is the only stationary and Markov Gaussian process with paths that are smooth. This property explains some of the popularity of the Ornstein-Uhlenbeck process in fitting models to real data.

Exercises

Exercise 12.1 (Simple Processes). Let $X_0, X_1, X_2, \ldots$ be a sequence of iid standard normal variables, and $W(t), t \geq 0$ a standard Brownian motion independent of the X_i sequence, starting at zero. Determine which of the following processes are Gaussian, and which are stationary.

(a) $X(t) \equiv \frac{X_1 + X_2}{\sqrt{2}}$.

(b) $X(t) \equiv |\frac{X_1 + X_2}{\sqrt{2}}|$.

(c) $X(t) = \frac{tX_1 X_2}{\sqrt{X_1^2 + X_2^2}}$.

(d) $X(t) = \sum_{j=0}^{k}[X_{2j} \cos \theta_j t + X_{2j+1} \sin \theta_j t]$.

(e) $X(t) = t^2 W(\frac{1}{t^2}), t > 0$, and $X(0) = 0$.

(f) $X(t) = W(t|X_0|)$.

Exercise 12.2. Let $X(t) = \sin \theta t$, where $\theta \sim U[0, 2\pi]$.

(a) Suppose the time parameter t belongs to $T = \{1, 2, \ldots, \}$. Is $X(t)$ stationary?
(b) Suppose the time parameter t belongs to $T = [0, \infty)$. Is $X(t)$ stationary?

Exercise 12.3. Suppose $W(t), t \geq 0$ is a standard Brownian motion starting at zero, and $Y \sim N(0, 1)$, independent of the $W(t)$ process. Let $X(t) = Yf(t) + W(t)$, where f is a deterministic function. Is $X(t)$ stationary?

Exercise 12.4 (Increments of Brownian Motion). Suppose $W(t), t \geq 0$ is a standard Brownian motion starting at zero, and Y is a positive random variable independent of the $W(t)$ process. Let $X(t) = W(t + Y) - W(t)$. Is $X(t)$ stationary?

Exercise 12.5. Suppose $W(t), t \geq 0$ is a standard Brownian motion starting at zero. Let $X(n) = W(1) + W(2) + \ldots + W(n), n \geq 1$. Find the covariance function of the process $X(n), n \geq 1$.

Exercise 12.6 (Moments of the Hitting Time). Suppose $W(t), t \geq 0$ is a standard Brownian motion starting at zero. Fix $a > 0$ and let T_a be the first time $W(t)$ hits a. Characterize all α such that $E[T_a^\alpha] < \infty$.

Exercise 12.7 (Hitting Time of the Positive Quadrant). Suppose $W(t), t \geq 0$ is a standard Brownian motion starting at zero. Let $T = \inf\{t > 0 : W(t) > 0\}$. Show that with probability one, $T = 0$.

Exercise 12.8. Suppose $W(t), t \geq 0$ is standard Brownian motion starting at zero. Fix $z > 0$ and let T_z be the first time $W(t)$ hits z. Let $0 < a < b < \infty$. Find $E(T_b | T_a = t)$.

Exercise 12.9 (Running Maximum of Brownian Motion). Let $W(t), t \geq 0$ be standard Brownian motion on $[0, \infty)$ and $M(t) = \sup_{0 \leq s \leq t} W(s)$. Evaluate $P(M(1) = M(2))$.

Exercise 12.10. Let $W(t), t \geq 0$ be standard Brownian motion on $[0, \infty)$. Let $T > 0$ be a fixed finite time instant. Find the density of the first zero of $W(t)$ after the time $t = T$. Does it have a finite mean?

Exercise 12.11 (Integrated Brownian Motion). Let $W(t), t \geq 0$ be standard Brownian motion on $[0, \infty)$. Let $X(t) = \int_0^t W(s)ds$. Identify explicit positive constants K, α such that for any $t, c > 0$, $P(|X(t)| \geq c) \leq \frac{Kt^\alpha}{c}$.

Exercise 12.12 (Integrated Brownian Motion). Let $W(t), t \geq 0$ be standard Brownian motion on $[0, \infty)$. Let $X(t) = \int_0^t W(s)ds$. Prove that for any fixed t, $X(t)$ has a finite mgf everywhere, and use it to derive the fourth moment of $X(t)$.

Exercise 12.13 (Integrated Brownian Motion). Let $W(t), t \geq 0$ be standard Brownian motion on $[0, \infty)$. Let $X(t) = \int_0^t W(s)ds$. Find

(a) $E(X(t)|W(t) = w)$.
(b) $E(W(t)|X(t) = x)$.
(c) The correlation between $X(t)$ and $W(t)$.
(d) $P(X(t) > 0, W(t) > 0)$ for a given t.

Exercise 12.14 (Application of Reflection Principle). Let $W(t), t \geq 0$ be standard Brownian motion on $[0, \infty)$ and $M(t) = \sup_{0 \leq s \leq t} W(s)$. Prove that $P(W(t) \leq w, M(t) \geq x) = 1 - \Phi(\frac{2x-w}{\sqrt{t}}), x \geq w, x \geq 0$. Hence, derive the joint density of $W(t)$ and $M(t)$.

Exercise 12.15 (Current Value and Current Maximum). Let $W(t), t \geq 0$ be standard Brownian motion on $[0, \infty)$ and $M(t) = \sup_{0 \leq s \leq t} W(s)$. Find $P(W(t) = M(t))$ and find its limit as $t \to \infty$.

Exercise 12.16 (Current Value and Current Maximum). Let $W(t), t \geq 0$ be standard Brownian motion on $[0, \infty)$ and $M(t) = \sup_{0 \leq s \leq t} W(s)$. Find $E(M(t)|W(t) = w)$.

Exercises

Exercise 12.17 (Predicting the Next Value). Let $W(t), t \geq 0$ be standard Brownian motion on $[0, \infty)$ and let $\bar{W}(t) = \frac{1}{t}\int_0^t W(s)ds$ the current running average.

(a) Find $\hat{W}(t) = E(W(t)|\bar{W}(t) = w)$.
(b) Find the prediction error $E[|W(t) - \hat{W}(t)|]$.

Exercise 12.18 (Zero-Free Intervals). Let $W(t), t \geq 0$ be standard Brownian motion, and $0 < s < t < u < \infty$. Find the conditional probability that $W(t)$ has no zeroes in $[s, u]$ given that it has no zeroes in $[s, t]$.

Exercise 12.19 (Application of the LIL). Let $W(t), t \geq 0$ be standard Brownian motion, and $0 < s < t < u < \infty$. Let $X(t) = \frac{W(t)}{\sqrt{t}}, t > 0$. Let K, M be any two positive numbers. Show that infinitely often, with probability one, $X(t) > K$ and $< -M$.

Exercise 12.20. Let $W(t), t \geq 0$ be standard Brownian motion, and $0 < s < t < u < \infty$. Find the conditional expectation of $X(t)$ given $X(s) = x, X(u) = y$.

Hint: Consider first the conditional expectation of $X(t)$ given $X(0) = X(1) = 0$.

Exercise 12.21 (Reflected Brownian Motion Is Markov). Let $W(t), t \geq 0$ be standard Brownian motion starting at zero. Show that $|W(t)|$ is a Markov process.

Exercise 12.22 (Adding a Function to Brownian Motion). Let $W(t)$ be standard Brownian motion on $[0, 1]$ and f a general continuous function on $[0, 1]$. Show that with probability one, $X(t) = W(t) + f(t)$ is everywhere nondifferentiable.

Exercise 12.23 (No Intervals of Monotonicity). Let $W(t), t \geq 0$ be standard Brownian motion, and $0 < a < b < \infty$ two fixed positive numbers. Show, by using the independent increments property, that with probability one, $W(t)$ is nonmonotone on $[a, b]$.

Exercise 12.24 (Two-Dimensional Brownian Motion). Show that two-dimensional standard Brownian motion is a Markov process.

Exercise 12.25 (An Interesting Connection to Cauchy Distribution). Let $W_1(t), W_2(t)$ be two independent standard Brownian motions on $[0, \infty)$ starting at zero. Fix a number $a > 0$ and let T_a be the first time $W_1(t)$ hits a. Find the distribution of $W_2(T_a)$.

Exercise 12.26 (Time Spent in a Nonempty Set). Let $\mathbf{W}_2(t), t \geq 0$ be a two-dimensional standard Brownian motion starting at zero, and let C be a nonempty open set of $\mathcal{R}^2$. Show that with probability one, the *Lebesgue measure* of the set of times t at which $W(t)$ belongs to C is infinite.

Exercise 12.27 (Difference of Two Brownian Motions). Let $W_1(t), W_2(t), t \geq 0$ be two independent Brownian motions, and let c_1, c_2 be two constants. Show that $X(t) = c_1 W_1(t) + c_2 W_2(t)$ is another Brownian motion. Identify any drift and diffusion parameters.

Exercise 12.28 (Intersection of Brownian Motions). Let $W_1(t), W_2(t), t \geq 0$ be two independent standard Brownian motions starting at zero. Let $C = \{t > 0 : W_1(t) = W_2(t)\}$.

(a) Is C nonempty with probability one?
(b) If C is nonempty, is it a finite set, or is it an infinite set with probability one?
(c) If C is an infinite set with probability one, is its *Lebesgue measure* zero or greater than zero?
(d) Does C have accumulation points? Does it have accumulation points with probability one?

Exercise 12.29 (The L_1 Norm of Brownian Motion). Let $W(t), t \geq 0$ be standard Brownian motion starting at zero. Show that with probability one, $I = \int_0^\infty |W(t)| dt = \infty$.

Exercise 12.30 (Median Local Time). Find the median of the local time $\eta(x, t)$ of a standard Brownian motion on $[0, \infty)$ starting at zero.
Caution: For $x \neq 0$, the local time has a mixed distribution.

Exercise 12.31 (Monotonicity of the Mean Local Time). Give an analytical proof that the expected value of the local time $\eta(x, t)$ of a standard Brownian motion starting at zero is strictly decreasing in the spatial coordinate x.

Exercise 12.32 (Application of Invariance Principle). Let $X_1, X_2, \ldots$ be iid variables with the common distribution $P(X_i = \pm 1) = \frac{1}{2}$. Let $S_k = \sum_{i=1}^{k} X_i, k \geq 1$, and $\Pi_n = \frac{1}{n}\#\{k : S_k > 0\}$. Find the limiting distribution of Π_n by applying Donsker's invariance principle.

Exercise 12.33 (Application of Invariance Principle). Let $X_1, X_2, \ldots$ be iid variables with zero mean and a finite variance σ^2. Let $S_k = \sum_{i=1}^{k} X_i, k \geq 1$, and $M_n = n^{-1/2} \max_{1 \leq k \leq n} S_k$. Find the limiting distribution of M_n by applying Donsker's invariance principle.

Exercise 12.34 (Application of Invariance Principle). Let $X_1, X_2, \ldots$ be iid variables with zero mean and a finite variance σ^2. Let $S_k = \sum_{i=1}^{k} X_i, k \geq 1$, and $A_n = n^{-1/2} \max_{1 \leq k \leq n} |S_k|$. Find the limiting distribution of A_n by applying Donsker's invariance principle.

Exercise 12.35 (Application of Invariance Principle). Let $X_1, X_2, \ldots$ be iid variables with zero mean and a finite variance σ^2. Let $S_k = \sum_{i=1}^{k} X_i, k \geq 1$, and $T_n = n^{-3/2} \sum_{k=1}^{n} |S_k|$. Find the limiting distribution of T_n by applying Donsker's invariance principle.

Exercise 12.36 (Distributions of Some Functionals). Let $W(t), t \geq 0$ be standard Brownian motion starting at zero. Find the density of each of the following functionals of the $W(t)$ process:

(a) $\sup_{t>0} W^2(t)$;

(b) $\dfrac{\int_0^1 W(t)dt}{W(\frac{1}{2})}$;

Hint: The terms in the quotient are jointly normal with zero means.

(c) $\sup_{t>0} \dfrac{W(t)}{a+bt}, a, b > 0$.

Exercise 12.37 (Ornstein–Uhlenbeck Process). Let $X(t)$ be a general Ornstein–Uhlenbeck process and $s < t$ two general times. Find the expected value of $|X(t) - X(s)|$.

Exercise 12.38. Let $X(t)$ be a general Ornstein–Uhlenbeck process and $Y(t) = \int_0^t X(u)du$. Find the correlation between $Y(s)$ and $Y(t)$ for $0 < s < t < \infty$, and find its limit when $\sigma, \alpha \to \infty$ and $\frac{\sigma}{\alpha} \to 1$.

Exercise 12.39. Let $W(t), t \geq 0$ be standard Brownian motion starting at zero, and $0 < s < t < \infty$ two general times. Find an expression for $P(W(t) > 0 \,|\, W(s) > 0)$, and its limit when s is held fixed and $t \to \infty$.

Exercise 12.40 (Application of the Heat Equation). Let $Y \sim N(0, \sigma^2)$ and $f(Y)$ a twice continuously differentiable convex function of Y. Show that $E[f(Y)]$ is increasing in σ, assuming that the expectation exists.

References

Bhattacharya, R.N. and Waymire, E. (2007). *A Basic Course in Probability Theory*, Springer, New York.
Bhattacharya, R.N. and Waymire, E. (2009). *Stochastic Processes with Applications*, SIAM, Philadelphia.
Billingsley, P. (1968). *Convergence of Probability Measures*, John Wiley, New York.
Breiman, L. (1992). *Probability*, Addison-Wesley, New York.
Brown, L. (1971). Admissible estimators, recurrent diffusions, and insoluble boundary value problems, *Ann. Math. Statist.*, 42, 855–903.
Brown, L., DasGupta, A., Haff, L.R., and Strawderman, W.E. (2006). The heat equation and Stein's identity: Connections, *Applications*, 136, 2254–2278.
Csörgo, M. (2002). A glimpse of the impact of Pal Erdös on probability and statistics, *Canad. J. Statist.*, 30, 4, 493–556.
Csörgo, M. and Révész, P. (1981). *Strong Approximations in Probability and Statistics*, Academic Press, New York.
Csörgo, S. and Hall, P. (1984). The KMT approximations and their applications, *Austr. J. Statist.*, 26, 2, 189–218.
Dasgupta, A. (2008), Asymptotic Theory of Statistics and Probability Springer, New York.
Donsker, M. (1951). An invariance principle for certain probability limit theorems, *Mem. Amer. Math. Soc.*, 6.
Durrett, R. (2001). *Essentials of Stochastic Processes*, Springer, New York.
Einmahl, U. (1987). Strong invariance principles for partial sums of independent random vectors, *Ann. Prob.*, 15, 4, 1419-1440.
Erdös, P. and Kac, M. (1946). On certain limit theorems of the theory of probability, *Bull. Amer. Math. Soc.*, 52, 292–302.

Freedman, D. (1983). *Brownian Motion and Diffusion,* Springer, New York.
Hall, P. and Heyde, C. (1980). *Martingale Limit Theory and Its Applications,* Academic Press, New York.
Karatzas, I. and Shreve, S. (1991). *Brownian Motion and Stochastic Calculus,* Springer, New York.
Karlin, S. and Taylor, H. (1975). *A First Course in Stochastic Processes,* Academic Press, New York.
Komlós, J., Major, P., and Tusnady, G. (1975). An approximation of partial sums of independent rvs and the sample df :I, *Zeit für Wahr. Verw. Geb.,* 32, 111–131.
Komlós, J., Major, P. and Tusnady, G. (1976). An approximation of partial sums of independent rvs and the sample df :II, *Zeit für Wahr. Verw. Geb.,* 34, 33–58.
Körner, T. (1986). *Fourier Analysis,* Cambridge University Press, Cambridge, UK.
Lawler, G. (2006). *Introduction to Stochastic Processes,* Chapman and Hall, New York.
Major, P. (1978). On the invariance principle for sums of iid random variables, *Mult. Anal.,* 8, 487-517.
Mörters, P. and Peres, Y. (2010). *Brownian Motion,* Cambridge University Press, Cambridge, UK.
Pyke, R. (1984). Asymptotic results for empirical and partial sum processes: A review, *Canad. J. Statist.,* 12, 241–264.
Resnick, S. (1992). *Adventures in Stochastic Processes,* Birkhäuser, Boston.
Révész, P. (2005). *Random Walk in Random and Nonrandom Environments,* World Scientific Press, Singapore.
Revuz, D. and Yor, M. (1994). *Continuous Martingales and Brownian Motion,* Springer, Berlin.
Strassen, V. (1964). An invariance principle for the law of the iterated logarithm, *Zeit. Wahr. verw. Geb.,* 3, 211–226.
Strassen, V. (1967). Almost sure behavior of sums of independent random variables and martingales, *Proc. Fifth Berkeley Symp.,* 1, 315–343, University of California Press, Berkeley.
Whitt, W. (1980). Some useful functions for functional limit theorems, *Math. Opem. Res.,* 5, 67–85.

Chapter 13
Poisson Processes and Applications

A single theme that binds together a number of important probabilistic concepts and distributions, and is at the same time a major tool to the applied probabilist and the applied statistician is the *Poisson process*. The Poisson process is a probabilistic model of situations where events occur completely at random at intermittent times, and we wish to study the number of times the particular event has occurred up to a specific time instant, or perhaps the waiting time till the next event, and so on. Some simple examples are receiving phone calls at a telephone call center, receiving an e-mail from someone, arrival of a customer at a pharmacy or some other store, catching a cold, occurrence of earthquakes, mechanical breakdown in a computer or some other machine, and so on. There is no end to how many examples we can think of, where an event happens, then nothing happens for a while, and then it happens again, and it keeps going like this, apparently at random. It is therefore not surprising that the Poisson process is such a valuable tool in the probabilist's toolbox. It is also a fascinating feature of Poisson processes that it is connected in various interesting ways to a number of special distributions, including the Poisson, exponential, Gamma, Beta, uniform, binomial, and the multinomial. These embracing connections and wide applications make the Poisson process a very special topic in probability.

The examples we mentioned above all correspond to events occurring at time instants taking values on the real line. In all of these examples, we can think of the time parameter as real physical time. However, Poisson processes can also be discussed in two, three, or indeed any number of dimensions. For example, Poisson processes are often used to model spatial distribution of trees in a forest. This would be an example of a Poisson process in two dimensions. Poisson processes are used to model galaxy distributions in space. This would be an example of a Poisson process in three dimensions. The fact is, Poisson processes can be defined and treated in considerably more generality than just the linear Poisson process. We start with the case of the linear Poisson process, and then consider the higher-dimensional cases. Kingman (1993) is a classic reference on Poisson processes. Some other recommended references are Parzen (1962), Karlin and Taylor (1975), and Port (1994). A concise well-written treatment is given in Lawler (2006).

13.1 Notation

The Poisson process is an example of a stochastic process with pure jumps indexed by a running label t. We call t the time parameter, and for the purpose of our discussion here, t belongs to the infinite interval $[0, \infty)$. For each $t \geq 0$, there is a nonnegative random variable $X(t)$, which counts how many events have occurred up to and including time t. As we vary t, we can think of $X(t)$ as a function. It is a *random function*, because each $X(t)$ is a random variable. Like all functions, $X(t)$ has a graph. The graph of $X(t)$ is called a *path* of $X(t)$. It is helpful to look at a typical path of a Poisson process; Fig. 13.1 gives an example.

We notice that the path is a nondecreasing function of the time parameter t, and that it increases by jumps of size one. The time instants at which these jumps occur are called the *renewal or arrival times of the process*. Thus, we have an infinite sequence of arrival times, say $Y_1, Y_2, Y_3, \ldots$; the first arrival occurs exactly at time Y_1, the second arrival occurs at time Y_2, and so on. We define Y_0 to be zero. The gaps between the arrival times, namely, $Y_1 - Y_0, Y_2 - Y_1, Y_3 - Y_2, \ldots$ are called the *interarrival times*. Writing $Y_n - Y_{n-1} = T_n$, we see that the interarrival times and the arrival times are related by the simple identity

$$Y_n = (Y_n - Y_{n-1}) + (Y_{n-1} - Y_{n-2}) + \cdots + (Y_2 - Y_1) + (Y_1 - Y_0) = T_1 + T_2 + \cdots + T_n.$$

A special property of a Poisson process is that these interarrival times are independent. So, for instance, if T_3, the time you had to wait between the second and the third event, were large, then you would have no right to believe that T_4 should be small, because T_3 and T_4 are actually independent for a Poisson process.

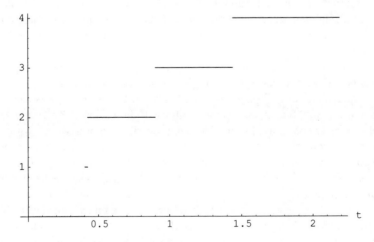

Fig. 13.1 Path of a Poisson process

13.2 Defining a Homogeneous Poisson Process

The Poisson process can be arrived at in a number of ways. All of these apparently different definitions are actually equivalent. Here are some equivalent definitions of a Poisson process.

Definition # 1. One possibility is to start with the interarrival times, and assume that they are iid exponential with some mean μ. Then the number of arrivals up to and including time t is a Poisson process with a constant arrival rate $\lambda = \frac{1}{\mu}$.

Definition # 2. Or, we may assume that the number of arrivals in a general time interval $[t_1, t_2]$ is a Poisson variable with mean $\lambda(t_2 - t_1)$ for some fixed positive number λ, and that the number of arrivals over any finite number of disjoint intervals is mutually independent Poisson variables. This is equivalent to the first definition given in the paragraph above.

Definition # 3. A third possibility is to use a neat result due to Alfred Rényi. Rényi proved that if $X(t)$ satisfies the Poisson property that the number of arrivals $X(B)$ within any set of times B, not necessarily a set of the form of an interval, is a Poisson variable with mean $\lambda |B|$, where $|B|$ denotes the *Lebesgue measure* of B, then $X(t)$ must be a Poisson process. Note that there is no mention of independence over disjoint intervals in this definition. Independence falls out as a consequence of Rényi's condition, and Rényi's result is also a perfectly correct definition of a Poisson process; see Kingman (1993, p. 33).

Definition # 4. From the point of view of physical motivation and its original history, it is perhaps best to define a Poisson process in terms of some characteristic physical properties of the process. In other words, if we make a certain number of specific assumptions about how the process $X(t)$ behaves, then those assumptions serve as a definition of a Poisson process. If you do not believe in one or more of these assumptions in a particular problem, then the Poisson process is not the right model for that problem. Here are the assumptions.

(a) $X(0) = 0$.
(b) The rate of arrival of the events remains constant over time, in the sense there is a finite positive number λ such that for any $t \geq 0$,

$$(i) \text{ For } h \to 0, P(X(t+h) = X(t) + 1) = \lambda h + o(h).$$

(c) The number of events over nonoverlapping time intervals are independent; that is,
   given disjoint time intervals $[a_i, b_i]$, $i = 1, 2, \ldots, n$,
the random variables $X(b_i) - X(a_i), i = 1, 2, \ldots, n$ are mutually independent.
(d) More than one event cannot occur at exactly the same time instant.

Precisely,

(i) For $h \to 0$, $P(X(t+h) = X(t)) = 1 - \lambda h + o(h)$.
(ii) For $h \to 0$, $P(X(t+h) = X(t) + 1) = \lambda h + o(h)$.
(iii) For $h \to 0$, $P(X(t+h) > X(t) + 1) = o(h)$.

The important point is that all of these definitions are equivalent. Depending on taste, one may choose any of these as the definition of a homogeneous or stationary Poisson process on the real line.

13.3 Important Properties and Uses as a Statistical Model

Starting with these assumptions in our Definition #4 about the physical behavior of the process, one can use some simple differential equation methods and probability calculations to establish various important properties of a Poisson process. We go over some of the most important properties next.

Given $k \geq 0$, let $p_k(t) = P(X(t) = k)$, and let $f_k(t) = e^{\lambda t} p_k(t)$. By the total probability formula, as $h \to 0$,

$$p_k(t+h) = P(X(t+h) = k \,|\, X(t) = k) p_k(t)$$
$$+ P(X(t+h) = k \,|\, X(t) = k-1) p_{k-1}(t) + o(h)$$
$$\Rightarrow p_k(t+h) = (1 - \lambda h) p_k(t) + \lambda h p_{k-1}(t) + o(h)$$
$$\Rightarrow \frac{p_k(t+h) - p_k(t)}{h} = -\lambda [p_k(t) - p_{k-1}(t)] + o(1)$$
$$\Rightarrow p_k'(t) = -\lambda [p_k(t) - p_{k-1}(t)],$$

for $k \geq 1, t > 0$, and when $k = 0$, $p_0'(t) = -\lambda p_0(t), t > 0$. Because $X(0) = 0$, the last equation immediately gives $p_0(t) = e^{-\lambda t}$. For $k \geq 1$, the system of differential equations $p_k'(t) = -\lambda [p_k(t) - p_{k-1}(t)]$ is equivalent to

$$f_k'(t) = \lambda f_{k-1}(t), \quad t > 0, \, f_k(0) = 0;$$

note that $f_k(0) = 0$ because $P(X(0) = k) = 0$ if $k \geq 1$.

This last system of differential equations $f_k'(t) = \lambda f_{k-1}(t), t > 0, f_k(0) = 0$ is easy to solve recursively and the solutions are

$$f_k(t) = \frac{(\lambda t)^k}{k!}, \quad k \geq 1, t > 0$$
$$\Rightarrow p_k(t) = P(X(t) = k) = e^{-\lambda t} \frac{(\lambda t)^k}{k!}.$$

This is true also for $k = 0$. So, we have proved the following theorem, which accounts for the name *Poisson process*.

13.3 Important Properties and Uses as a Statistical Model

Theorem 13.1. *If $X(t), t \geq 0$ is a Poisson process with the constant arrival rate λ, then for any $t > 0$, $X(t)$ is distributed as a Poisson with mean λt. More generally, the number of arrivals in an interval $(s, t]$ is distributed as a Poisson with mean $\lambda(t - s)$.*

Example 13.1 (A Medical Example). Suppose between the months of May and October, you catch allergic rhinitis at the constant average rate of once in six weeks. Assuming that the incidences follow a Poisson process, let us answer some simple questions.

First, what is the expected total number of times that you will catch allergic rhinitis between May and October in one year? Take the start of May 1 as $t = 0$, and $X(t)$ as the number of fresh incidences up to(and including) time t. Note that time is being measured in some implicit unit, say weeks. Then, the arrival rate of the Poisson process for $X(t)$ is $\lambda = \frac{1}{6}$. There are 24 weeks between May and October, and $X(24)$ is distributed as a Poisson with mean $24\lambda = 4$, which is the expected total number of times that you will catch allergic rhinitis between May and October.

Next, what is the probability that you will catch allergic rhinitis at least once before the start of August and at least once after the start of August? This is the same as asking what is $P(X(12) \geq 1, X(24) - X(12) \geq 1)$. By the property of independence of $X(12)$ and $X(24) - X(12)$, this probability equals

$$P(X(12) \geq 1)P(X(24) - X(12) \geq 1) = [P(X(12) \geq 1)]^2$$
$$= [1 - P(X(12) = 0)]^2 = [1 - e^{-\frac{12}{6}}]^2 = .7476.$$

A key property of the Poisson process is that the sequence of interarrival times $T_1, T_2, \ldots$ is iid exponential with mean $\frac{1}{\lambda}$. We do not rigorously prove this here, but as the simplest illustration of how the exponential density enters into the picture, we show that T_1 has the exponential density. Indeed, $P(T_1 > h) = P(X(h) = 0) = e^{-\lambda h}$, because $X(h)$ has a Poisson distribution with mean λh. It follows that T_1 has the density $f_{T_1}(h) = \lambda e^{-\lambda h}, h > 0$.

As a further illustration, consider the joint distribution of T_1 and T_2. Here is a very heuristic explanation for why T_1, T_2 are iid exponentials. Fix two positive numbers t, u. The event $\{T_1 > t, T_2 > u\}$ is just the event that the first arrival time Y_1 is at some time later than t, and counting from Y_1, no new further events occur for another time interval of length u. But the intervals $[0, t]$ and $(Y_1, Y_1 + u)$ are nonoverlapping if the first arrival occurs after t, and the probability of zero events in both of these intervals would then factor as $e^{-\lambda t} e^{-\lambda u}$. This means $P(T_1 > t) = e^{-\lambda t}$ and $P(T_2 > u) = e^{-\lambda u}$, and that T_1, T_2 are iid exponentials with the same density function.

Because the sum of iid exponentials has a Gamma distribution (see Chapter 4), it also follows that for a Poisson process, the nth arrival time Y_n has a Gamma distribution. All of these are recorded in the following theorem.

Theorem 13.2. *Let $X(t), t \geq 0$ be a Poisson process with constant arrival rate λ. Then,*

(a) $T_1, T_2, \ldots$ are iid exponential with the density function $f_{T_i}(t) = \lambda e^{-\lambda t}, t > 0$.
(b) Let $n \geq 1$. Then Y_n has the Gamma density $f_{Y_n}(y) = \frac{\lambda^n e^{-\lambda y} y^{n-1}}{(n-1)!}, y > 0$.

See Kingman (1993, p. 39) for a rigorous proof of this key theorem.

Example 13.2 (Geiger Counter). Geiger counters, named after Hans Geiger, are used to detect radiation-emitting particles, such as *beta* and *alpha* particles, or low-energy gamma rays. The counter does so by recording a current pulse when a radioactive particle or ray hits the counter. Poisson processes are standard models for counting particle hits on the counter.

Suppose radioactive particles hit such a Geiger counter at the constant average rate of one hit per 30 seconds. Therefore, the arrival rate of our Poisson process is $\lambda = \frac{1}{30}$. Let $Y_1, Y_2, \ldots, Y_n$ be the times of the first n hits on the counter, and $T_1, T_2, \ldots, T_n$ the time gaps between the successive hits. We ask a number of questions about these arrival and interarrival times in this example.

First, by our previous theorem, Y_n has the Gamma distribution with density

$$f_{Y_n}(y) = \frac{e^{-\frac{y}{30}} y^{n-1}}{(30)^n (n-1)!}, y > 0.$$

Suppose we want to calculate the probability that the hundredth hit on the Geiger counter occurs within the first hour. This is equal to $P(Y_{100} \leq 3600)$, there being $60 \times 60 = 3600$ seconds in an hour. We can try to integrate the Gamma density for Y_n and evaluate $P(Y_{100} \leq 3600)$. This will require a computer. On the other hand, the needed probability is also equal to $P(X(3600) \geq 100)$, where $X(3600) \sim$ Poi(120), because with $t = 3600, \lambda t = \frac{3600}{30} = 120$. However, this calculation is also clumsy, because the Poisson mean is such a large number. We can calculate the probability approximately by using the central limit theorem. Indeed,

$$P(Y_{100} \leq 3600) \approx P\left(Z \leq \frac{3600 - 100 \times 30}{\sqrt{100 \times 900}}\right) = P(Z \leq 2) = .9772,$$

where Z denotes a standard normal variable.

Next, suppose we wish to find the probability that at least once, within the first hundred hits, two hits come within a second of each other. At first glance, it might seem that this is unlikely, because the average gap between two successive hits is as large as 30 seconds. But, actually the probability is very high.

If we denote the order statistics of the first hundred interarrival times by $T_{(1)} < T_{(2)} < \ldots < T_{(100)}$, then we want to find $P(T_{(1)} \leq 1)$. Recall that the interarrival times $T_1, T_2, \ldots$ themselves are iid exponential with mean 30. Therefore, the minimum $T_{(1)}$ also has an exponential distribution, but the mean is $\frac{30}{100} = .3$. Therefore, our needed probability is $\int_0^1 \frac{1}{.3} e^{-\frac{x}{.3}} dx = .9643$.

13.3 Important Properties and Uses as a Statistical Model

Example 13.3 (Poisson Process and the Beta Distribution). It was mentioned in the chapter introduction that the Poisson process has a connection to the Beta distribution. We show this connection in this example.

Suppose customers stream in to a drugstore at the constant average rate of 15 per hour. The pharmacy opens its doors at 8:00 AM and closes at 8:00 PM. Given that the hundredth customer on a particular day walked in at 2:00 PM, we want to know what is the probability that the fiftieth customer came before noon. Write n for 100 and m for 50, and let Y_j be the arrival time of the jth customer on that day. Then, we are told that $Y_n = 6$, and we want to calculate $P(Y_m < 4 | Y_n = 6)$.

We can attack this problem more generally by drawing a connection of a Poisson process to the Beta distribution. For this, we recall the result from Chapter 4 that if X, Y are independent positive random variables, with densities

$$f_X(x) = \frac{\lambda^\alpha e^{-\lambda x} x^{\alpha-1}}{\Gamma(\alpha)}, \quad f_Y(y) = \frac{\lambda^\beta e^{-\lambda x} x^{\beta-1}}{\Gamma(\beta)}, \quad \alpha, \beta, \lambda > 0,$$

then $U = \frac{X}{X+Y}$ and $V = X + Y$ are independent, and U has the Beta density

$$f_U(u) = \frac{\Gamma(\alpha+\beta)}{\Gamma(\alpha)\Gamma(\beta)} u^{\alpha-1}(1-u)^{\beta-1}, \quad 0 < u < 1.$$

Returning now to our problem, $Y_m = T_1 + T_2 + \ldots + T_m$, and $Y_n = Y_m + T_{m+1} + \ldots + T_n$, where $T_1, T_2, \ldots, T_n$ are iid exponentials. We can write the ratio $\frac{Y_m}{Y_n}$ as $\frac{Y_m}{Y_n} = \frac{Y_m}{Y_m + (Y_n - Y_m)}$. Because Y_m and $Y_n - Y_m$ are independent Gamma variables, it follows from the above paragraph that $U = \frac{Y_m}{Y_n}$ has the Beta density

$$f_U(u) = \frac{\Gamma(n)}{\Gamma(m)\Gamma(n-m)} u^{m-1}(1-u)^{n-m-1}, \quad 0 < u < 1,$$

and furthermore, U and Y_n are independent. This is useful to us. Indeed,

$$P(Y_m < 4 | Y_n = 6) = P\left(\frac{Y_m}{Y_n} < \frac{4}{6} \Big| Y_n = 6\right) = P\left(\frac{Y_m}{Y_n} < \frac{4}{6}\right)$$

(inasmuch as $U = \frac{Y_m}{Y_n}$ and Y_n are independent)

$$= \int_0^{\frac{4}{6}} f_U(u) du = \frac{\Gamma(100)}{[\Gamma(50)]^2} \int_0^{\frac{4}{6}} u^{49}(1-u)^{49} du = .9997.$$

Example 13.4 (Filtered Poisson Process). Suppose the webpage of a certain text is hit by a visitor according to a Poisson process $X(t)$ with an average rate of 1.5 per day. However, 30% of the time, the person visiting the page does not purchase the book. We assume that customers make their purchase decisions independently, and independently of the $X(t)$ process. Let $Y(t)$ denote the number of copies of the book sold up to and including the tth day. We assume, by virtue of the Poisson process assumption, that more than one hit is not made at exactly the same time, and that a visitor does not purchase more than one book. What kind of a process is the process $Y(t), t \geq 0$?

We can imagine a sequence of iid Bernoulli variables $U_1, U_2, \ldots$, where $U_i = 1$ if the ith visitor to the webpage actually purchases the book. Each U_i is a Bernoulli with parameter $p = .7$. Also let $X(t)$ denote the number of hits made on the page up to time t. Then, $Y(t) = \sum_{i=1}^{X(t)} U_i$, where the sum over an empty set is defined, as usual, to be zero. Then,

$$P(Y(t) = k) = \sum_{x=k}^{\infty} P(Y(t) = k \mid X(t) = x) P(X(t) = x)$$

$$= \sum_{x=k}^{\infty} \binom{x}{k} p^k (1-p)^{x-k} \frac{e^{-\lambda t} (\lambda t)^x}{x!} = e^{-\lambda t} \frac{\left(\frac{p}{1-p}\right)^k}{k!} \sum_{x=k}^{\infty} \frac{(\lambda t (1-p))^x}{(x-k)!}$$

$$= e^{-\lambda t} (\lambda t (1-p))^k \frac{\left(\frac{p}{1-p}\right)^k}{k!} e^{\lambda t (1-p)}$$

$$= e^{-p\lambda t} \frac{(p\lambda t)^k}{k!}.$$

Therefore, for each t, $Y(t)$ is also a Poisson random variable, but the mean has changed to $p\lambda t$.

This is not enough to prove that $Y(t)$ is also a Poisson process. One needs to show, in addition, the *time homogeneity* property, namely, regardless of s, $Y(t+s) - Y(s)$ is a Poisson variable with mean $p\lambda t$, and the *independent increments* property; that is, over disjoint intervals $(a_i, b_i]$, $Y(b_i) - Y(a_i)$ are independent Poisson variables. Verification of these is just a calculation, and is left as an exercise.

To summarize, a *filtered Poisson process* is also a *Poisson process*.

Example 13.5 (Inspection Paradox). Suppose that buses arrive at a certain bus stop according to a Poisson process with some constant average rate λ, say once in 30 minutes. Thus, the average gap between any two arrivals of the bus is 30 minutes. Suppose now that out of habit, you always arrive at the bus stop at some fixed time t, say 5:00 PM. The term *inspection paradox* refers to the mathematical fact that the average time gap between the last arrival of a bus before 5:00 PM and the first arrival of a bus after 5:00 PM is larger than 30 minutes. It is as if by simply showing up at your bus stop at a fixed time, you can make the buses tardier than they are on an average! We derive this peculiar mathematical fact in this example.

We need some notation. Given a fixed time t, let δ_t denote the time that has elapsed since the last event; in symbols, $\delta_t = t - Y_{X(t)}$. Also let γ_t denote the time until the next event; in symbols, $\gamma_t = Y_{X(t)+1} - t$. The functions δ_t and γ_t are commonly known as *current life* and *residual life* in applications. We then have

$$P(\delta_t > h) = P(\text{No events between } t - h \text{ and } t) = e^{-\lambda h}, \quad 0 \leq h \leq t.$$

13.3 Important Properties and Uses as a Statistical Model

It is important that we note that $P(\delta_t > t) = e^{-\lambda t}$, and that for $h > t$, $P(\delta_t > h) = 0$. Thus, the function $P(\delta_t > h)$ has a discontinuity at $h = t$. Likewise,

$$P(\gamma_t > h) = P(\text{No events between } t \text{ and } t + h) = e^{-\lambda h}, \quad h \geq 0.$$

Therefore,

$$E[\delta_t + \gamma_t] = \left[\int_0^t h(\lambda e^{-\lambda h})dh + te^{-\lambda t}\right] + \frac{1}{\lambda}$$
$$= \frac{2}{\lambda} - \frac{1}{\lambda}e^{-\lambda t},$$

on actually doing the integration.

Now, clearly, $\frac{2}{\lambda} - \frac{1}{\lambda}e^{-\lambda t} > \frac{1}{\lambda}$, because $t > 0$, and so, we have the apparent conundrum that the average gap between the two events just prior to and just succeeding any fixed time t is larger than the average gap between all events in the process.

Example 13.6 (Compound Poisson Process). In some applications, a cluster of individuals of a random size arrive at the arrival times of an independent Poisson process. Here is a simple example. Tourist buses arrive at a particular location according to a Poisson process, and each arriving bus brings with it a random number of tourists. We want to study the total number of arriving tourists in some given time interval, for example, from 0 to t.

Then let $X(t)$ be the underlying Poisson process with average constant rate λ for the arrival of the buses, and assume that the number of tourists arriving in the buses forms an iid sequence $W_1, W_2, \ldots$ with the mass function $P(W_i = k) = p_k, k \geq 0$. The sequence $W_1, W_2, \ldots$ is assumed to be independent of the entire $X(t)$ process. Let $Y(t)$ denote the total arriving number of tourists up to time t. Then, we have $Y(t) = \sum_{i=1}^{X(t)} W_i$. Such a process $Y(t)$ is called a *compound Poisson process*.

We work out the generating function of $Y(t)$ for a general t; see Chapter 1 for the basic facts about the generating function of a nonnegative integer-valued random variable. We recall that for any nonnegative integer-valued random variable Z, the generating function is defined as $G_Z(s) = E(s^Z)$, and for any $k \geq 0$, $P(Z = k) = \frac{G_Z^{(k)}(0)}{k!}$, and also, provided that the kth moment of Z exists, $E[Z(Z-1)\cdots(Z-k+1)] = G_Z^{(k)}(1)$.

The generating function of $Y(t)$ then equals

$$G_Y(s) = E\left[s^{Y(t)}\right] = E\left[s^{\sum_{i=1}^{X(t)} W_i}\right]$$
$$= \sum_{x=0}^{\infty} E\left[s^{\sum_{i=1}^{X(t)} W_i} \mid X(t) = x\right] e^{-\lambda t}\frac{(\lambda t)^x}{x!} = e^{-\lambda t}\sum_{x=0}^{\infty} E\left[s^{\sum_{i=1}^{x} W_i}\right]\frac{(\lambda t)^x}{x!}$$

(because it has been assumed that $W_1, W_2, \ldots$ are independent of the $X(t)$ process)

$$= e^{-\lambda t} \sum_{x=0}^{\infty} \left(\prod_{i=1}^{x} E\left[s^{W_i}\right] \right) \frac{(\lambda t)^x}{x!}$$

$$= e^{-\lambda t} \sum_{x=0}^{\infty} [G_W(s)]^x \frac{(\lambda t)^x}{x!} = e^{-\lambda t} e^{\lambda t G_W(s)},$$

where, in the last two lines we have used our assumption that $W_1, W_2, \ldots$ are iid. We have thus derived a fully closed-form formula for the generating function of $Y(t)$. We can, in principle, derive $P(Y(t) = k)$ for any k from this formula. We can also derive the mean and the variance, for which we need the first two derivatives of this generating function. The first two derivatives are

$$G'_Y(s) = \lambda t e^{-\lambda t} e^{\lambda t G_W(s)} G'_W(s),$$

and

$$G''_Y(s) = e^{-\lambda t} e^{\lambda t G_W(s)} ([\lambda t G'_W(s)]^2 + \lambda t G''_W(s)).$$

If the W_i have a finite mean and variance, then we know from Chapter 1 that $E(W) = G'_W(1)$ and $E[W(W-1)] = G''_W(1)$. Plugging these into our latest expressions for $G'_Y(s)$ and $G''_Y(s)$, and some algebra, we find

$$E[Y(t)] = \lambda t E(W), \text{ and } E[Y(t)]^2 = [\lambda t E(W)]^2 + \lambda t E(W^2).$$

Combining, we get $\text{Var}(Y(t)) = \lambda t E(W^2)$.

A final interesting property of a Poisson process that we present is a beautiful connection to the uniform distribution. It is because of this property of the Poisson process that the Poisson process is equated in probabilistic folklore to *complete randomness of pattern*. The result says that if we are told that n events in a Poisson process have occurred up to the time instant t for some specific t, then the actual arrival times of those n events are just uniformly scattered in the time interval $(0, t]$. Here is a precise statement of the result and its proof.

Theorem 13.3 (Complete Randomness Property). *Let $X(t), t \geq 0$ be a Poisson process with a constant arrival rate, say λ. Let $t > 0$ be fixed. Then the joint conditional density of $Y_1, Y_2, \ldots, Y_n$ given that $X(t) = n$ is the same as the joint density of the n-order statistics of an iid sample from the $U[0, t]$ distribution.*

Proof. Recall the notation that the arrival times are denoted as $Y_1, Y_2, \ldots$, and the interarrival times by $T_1, T_2, \ldots$, and so on, so that $Y_1 = T_1, Y_2 = T_1 + T_2, Y_3 = T_1 + T_2 + T_3, \ldots$, Fix $0 < u_1 < u_2 < \cdots < u_n < t$. We show that

$$\frac{\partial^n}{\partial u_1 \partial u_2 \cdots \partial u_n} P(Y_1 \leq u_1, Y_2 \leq u_2, \ldots, Y_n \leq u_n | X(t) = n) = \frac{n!}{t^n},$$

$0 < u_1 < u_2 < \cdots < u_n < t$. This completes our proof because the mixed partial derivative of the joint CDF gives the joint density in general(see Chapter 3), and the function of n variables $u_1, u_2, \ldots, u_n$ given by $\frac{n!}{t^n}$ on $0 < u_1 < u_2 < \cdots < u_n < t$ indeed is the joint density of the order statistics of n iid $U[0, t]$ variables.

For ease of presentation, we show the proof for $n = 2$; the proof for a general n is exactly the same. Assume then that $n = 2$. The key fact to use is that the interarrival times T_1, T_2, T_3 are iid exponential. Therefore, the joint density of T_1, T_2, T_3 is $\lambda^3 e^{-\lambda(t_1+t_2+t_3)}, t_1, t_2, t_3 > 0$. Now make the linear transformation $Y_1 = T_1, Y_2 = T_1 + T_2, Y_3 = T_1 + T_2 + T_3$. This is a one-to-one transformation with a Jacobian equal to one. Therefore, by the Jacobian method (see Chapter 4), the joint density of Y_1, Y_2, Y_3 is

$$f_{Y_1,Y_2,Y_3}(y_1, y_2, y_3) = \lambda^3 e^{-\lambda y_3},$$

$0 < y_1 < y_2 < y_3$. Consequently,

$$P(Y_1 \leq u_1, Y_2 \leq u_2 \,|\, X(t) = 2)$$
$$= \frac{P(Y_1 \leq u_1, Y_2 \leq u_2, Y_3 > t)}{P(X(t) = 2)}$$
$$= \frac{\int_0^{u_1} \int_{y_1}^{u_2} \int_t^{\infty} \lambda^3 e^{-\lambda y_3} dy_3 dy_2 dy_1}{e^{-\lambda t} \frac{(\lambda t)^2}{2!}}$$
$$= \frac{\lambda^2 e^{-\lambda t} \left(u_1 u_2 - \frac{u_1^2}{2}\right)}{e^{-\lambda t} \frac{(\lambda t)^2}{2}}$$

(by just doing the required integration in the numerator)

$$= \frac{2}{t^2}\left(u_1 u_2 - \frac{u_1^2}{2}\right).$$

Therefore,

$$\frac{\partial^2}{\partial u_1 \partial u_2} P(Y_1 \leq u_1, Y_2 \leq u_2 \,|\, X(t) = 2) = \frac{2}{t^2},$$

$0 < u_1 < u_2 < t$, which is what we wanted to prove. □

Example 13.7 (E-Mails). Suppose that between 9:00 AM and 5:00 PM, you receive e-mails at the average constant rate of one per 10 minutes. You left for lunch at 12 noon, and when you returned at 1:00 PM, you found nine new e-mails waiting for you. We answer a few questions based on the complete randomness property, assuming that your e-mails arrive according to a Poisson process with rate $\lambda = 6$, using an hour as the unit of time.

First, what is the probability that the fifth e-mail arrived before 12:30? From the complete randomness property, given that $X(t) = 9$, with $t = 1$, the arrival times of the nine e-mails are jointly distributed as the order statistics of $n = 9$ iid

$U[0, 1]$ variables. Hence, given that $X(t) = 9, Y_5$, the time of the fifth arrival has the Beta(5, 5) density. From the symmetry of the Beta(5, 5) density, it immediately follows that $P(Y_5 \leq .5 | X(t) = 9) = .5$. So, there is a 50% probability that the fifth e-mail arrived before 12:30.

Next, what is the probability that at least three e-mails arrived after 12:45? This probability is the same as $P(Y_7 > .75 | X(t) = 9)$. Once again, this follows from a Beta distribution calculation, because given that $X(t) = 9, Y_7$ has the Beta(7, 3) density. Hence, the required probability is

$$\frac{\Gamma(10)}{\Gamma(7)\Gamma(3)} \int_{.75}^{1} x^6(1-x)^2 dx = .3993.$$

Finally, what is the expected gap between the times that the third and the seventh e-mail arrived? That is, what is $E(Y_7 - Y_3 | X(t) = 9)$? because $E(Y_7 | X(t) = 9) = \frac{7}{10} = .7$, and $E(Y_3 | X(t) = 9) = \frac{3}{10} = .3$, we have that $E(Y_7 - Y_3 | X(t) = 9) = .7 - .3 = .4$. Hence the expected gap between the arrival of the third and the seventh e-mail is 24 minutes.

13.4 * Linear Poisson Process and Brownian Motion: A Connection

The model in the compound Poisson process example in the previous section can be regarded as a model for displacement of a particle in a medium subject to collisions with other particles or molecules of the surrounding medium. Recall that this was essentially Einstein's derivation of Brownian motion for the movement of a physical particle in a fluid or gaseous medium.

How does the model of the compound Poisson process example become useful in such a context, and what exactly is the connection to Brownian motion? Suppose that a particle immersed in a medium experiences random collisions with other particles or the medium's molecules at random times which are the event times of a homogeneous Poisson process $X(t)$ with constant rate λ. Assume moreover that at each collision, our particle moves a distance of a units linearly to the right, or a units linearly to the left, with an equal probability. In other words, the sequence of displacements caused by the successive collisions form an iid sequence of random variables $W_1, W_2, \ldots$, with $P(W_i = \pm a) = \frac{1}{2}$. Then, the total displacement of the particle up to a given time t equals $W(t) = \sum_{i=1}^{X(t)} W_i$, with the empty sum $W(0)$ being zero. If we increase the rate of collisions λ and simultaneously decrease the displacement a caused by a single collision in just the right way, then the $W(t)$ process is approximately a one-dimensional Brownian motion. That is the connection to Brownian motion in this physical model for how a particle moves when immersed in a medium.

We provide an explanation for the emergence of Brownian motion in such a model. First, recall from our calculations for the compound Poisson example that

13.4 Linear Poisson Process and Brownian Motion: A Connection

the mean of $W(t)$ as defined above is zero for all t, and $\text{Var}(W(t)) = E[W(t)^2] = a^2\lambda t$. Therefore, to have any chance of approximating the $W(t)$ process by a Brownian motion, we should let $a^2\lambda$ converge to some finite constant σ^2. We now look at the characteristic function of $W(t)$ for any fixed t, and show that if $a \to 0, \lambda \to \infty$ in such a way that $a^2\lambda \to \sigma^2$, then the characteristic function of $W(t)$ itself converges to the characteristic function of the $N(0, \sigma^2 t)$ distribution. This is be one step towards showing that the $W(t)$ process mimics a Brownian motion with zero drift and diffusion coefficient σ if $a^2\lambda \to \sigma^2$.

The characteristic function calculation is exactly similar to the generating function calculation that we did in our compound Poisson process example. Indeed, the characteristic function of $W(t)$ equals $\psi_{W(t)}(s) = e^{\lambda t [\psi_W(s) - 1]}$, where $\psi_W(s)$ denotes the common characteristic function of $W_1, W_2, \ldots$, the sequence of displacements of the particle. Therefore, by a Taylor expansion of $\psi_W(s)$ (see Chapter 8),

$$\psi_{W(t)}(s) = e^{\lambda t \left[1 + is E(W_1) - \frac{s^2}{2} E(W_1^2) - 1 + \theta |s|^3 E(|W_1|^3)\right]}$$

(for some θ with $|\theta| \leq 1$)

$$= e^{\lambda t \left[-\frac{s^2}{2} a^2 + \theta |s|^3 a^3\right]} = e^{-\frac{s^2}{2}\sigma^2 t}(1 + o(1)),$$

if $a \to 0, \lambda \to \infty$, and $a^2\lambda \to \sigma^2$. Therefore, for any s, the characteristic function of $W(t)$ converges to that of the $N(0, \sigma^2 t)$ distribution in this specific asymptotic paradigm, namely when $a \to 0, \lambda \to \infty$, and $a^2\lambda \to \sigma^2$. This is clearly not enough to assert that the process $W(t)$ itself is approximately a Brownian motion. We need to establish that the $W(t)$ process has independent increments, that the increments $W(t) - W(s)$ have distribution depending only on the difference $t - s$, that for each $t, E(W(t)) = 0$, and $W(0) = 0$. Of these, the last two do not need a proof, as they are obvious. The independent increments property follows from the independent increments property of the Poisson process $X(t)$ and the independence of the W_i sequence. The stationarity of the increments follows by a straight calculation of the characteristic function of $W(t) - W(s)$, and by explicitly exhibiting that it is a function of $t - s$. Indeed, writing $D = X(t) - X(s)$,

$$E\left[e^{iu[W(t) - W(s)]}\right] = \sum_{d=0}^{\infty} \left(E\left[e^{iu[W(t) - W(s)]}\right] | D = d\right) \frac{e^{-\lambda(t-s)}(\lambda(t-s))^d}{d!}$$

$$= e^{\lambda(t-s)[\psi_W(u) - 1]},$$

on using the facts that conditional on $D = d$, $W(t) - W(s)$ is the sum of d iid variables, which are jointly independent of the $X(t)$ process, and then on some algebra. This shows that the increments $W(t) - W(s)$ are stationary.

13.5 Higher-Dimensional Poisson Point Processes

We remarked in the introduction that Poisson processes are also quite commonly used to model a random scatter of points in a planar area or in space, such as trees in a forest, or galaxies in the universe. In the case of the one-dimensional Poisson process, there was a random sequence of points too, say Π, and these were just the arrival times of the events. We then considered how many elements of Π belonged to a *test set*, such as an interval $[s, t]$. The number of elements in any particular interval was Poisson with mean depending on the length of the interval, and the number of elements in disjoint intervals were independent. In higher dimensions, say $\mathcal{R}^d$, we similarly have a random countable set of points Π of $\mathcal{R}^d$, and we consider how many elements of Π belong to a suitable test set, for example, a d-dimensional rectangle $A = \prod_{i=1}^{d} [a_i, b_i]$. Poisson processes would now have the properties that the counts over disjoint test sets would be independent, and the count of any particular test set A, namely $N(A) = \#\{x \in \Pi : x \in A\}$, will have a Poisson distribution with some appropriate mean depending on the *size* of A.

Definition 13.1. A random countable set $\Pi \subseteq \mathcal{R}^d$ is called a d-dimensional Poisson point process with *intensity or mean measure* μ if

(a) For a general set $A \subseteq \mathcal{R}^d$, $N(A) \sim \text{Poi}(\mu(A))$, where the set function $\mu(A) = \int_A \lambda(x) dx$ for some fixed nonnegative function $\lambda(x)$ on $\mathcal{R}^d$, with the property that $\mu(A) < \infty$ for any bounded set A.
(b) For any $n \geq 2$, and any n disjoint rectangles $A_j = \prod_{i=1}^{d} [a_{i,j}, b_{i,j}]$, the random variables $N(A_1), N(A_2), \ldots, N(A_n)$ are mutually independent.

Remark. If $\lambda(x) \equiv \lambda > 0$, then $\mu(A) = \lambda[\text{vol}(A)]$, and in that case the Poisson process is called *stationary or homogeneous* with the constant intensity λ. Notice that this coincides with the definition of a homogeneous Poisson process in one dimension that was given in the previous sections, because in one dimension volume would simply correspond to length.

According to the definition, given a set $A \subseteq \mathcal{R}^d$, $N(A) \sim \text{Poi}(\mu(A))$, and therefore, the probability that a given set A contains no points of the random countable set Π is $P(N(A) = 0) = e^{-\mu(A)}$. It is clear, therefore, that if we know just these *void probabilities* $P(N(A) = 0)$, then the intensity measure and the distribution of the full process is completely determined. Indeed, one could use this as a definition of a d-dimensional Poisson process.

Definition 13.2. A random countable set Π in $\mathcal{R}^d$ is a Poisson process with intensity measure μ if for each bounded $A \subseteq \mathcal{R}^d$, $P(\Pi \cap A = \phi) = e^{-\mu(A)}$.

Remark. This definition is saying that if the scientist understands the void probabilities, then she understands the entire Poisson process.

It is not obvious that a Poisson process with a given intensity measure exists. This requires a careful proof, within the rigorous measure-theoretic paradigm of probability. We only state the existence theorem here. A proof may be seen in many

13.5 Higher-Dimensional Poisson Point Processes

places, for instance, Kingman (1993). We do need some restrictions on the intensity measure for an existence result. The restrictions we impose are not the weakest possible, but we choose simplicity of the existence theorem over the greatest generality.

Definition 13.3. Suppose $\lambda(x)$ is a nonnegative function on $\mathcal{R}^d$ with the property that for any bounded set A in $\mathcal{R}^d$, $\int_A \lambda(x)dx < \infty$. Then a Poisson process with the intensity measure μ defined by $\mu(A) = \int_A \lambda(x)dx$ exists.

Example 13.8 (Distance to the Nearest Event Site). Consider a Poisson process Π in $\mathcal{R}^d$ with intensity measure μ, and fix a point $\tilde{x} \in \mathcal{R}^d$. We are interested in $D = D(\tilde{x}, \Pi)$, the Euclidean distance from $\tilde{x}$ to the nearest element of the Poisson process Π. The survival function of D is easily calculated. Indeed, if $B(\tilde{x}, r)$ is the ball of radius r centered at $\tilde{x}$, then

$$P(D > r) = P(N(B(\tilde{x}, r)) = 0) = e^{-\mu(B(\tilde{x}, r))}.$$

In particular, if Π is a homogeneous Poisson process with constant intensity $\lambda > 0$, then

$$P(D > r) = e^{-\frac{\lambda \pi^{d/2} r^d}{\Gamma(\frac{d}{2}+1)}}.$$

Differentiating, the density of D is

$$f_D(r) = \frac{\lambda d \pi^{d/2} r^{d-1}}{\Gamma(\frac{d}{2}+1)} e^{-\frac{\lambda \pi^{d/2} r^d}{\Gamma(\frac{d}{2}+1)}}.$$

Or, equivalently, $\frac{\lambda \pi^{d/2}}{\Gamma(\frac{d}{2}+1)} D^d$ has a standard exponential density. In the special case of two dimensions, this means that $\lambda \pi D^2$ has a standard exponential distribution. This result is sometimes used to statistically test whether a random countable set in some specific application is a homogeneous Poisson process.

Example 13.9 (Multinomial Distribution and Poisson Process). Suppose the intensity measure μ of a Poisson process is finite; that is, $\mu(\mathcal{R}^d) < \infty$. In that case, the cardinality of Π itself, that is, $N(\mathcal{R}^d) < \infty$ with probability one, because $N(\mathcal{R}^d)$ has a Poisson distribution with the finite mean $M = \mu(\mathcal{R}^d)$. Suppose in a particular realization, the total number of events $N(\mathcal{R}^d) = n$ for some finite n. We want to know how these n events are distributed among the members of a partition $A_1, A_2, \ldots, A_k$ of $\mathcal{R}^d$. It turns out that the joint distribution of $N(A_1), N(A_2), \ldots, N(A_k)$ is a multinomial distribution.

Indeed, given $n_1, n_2, \ldots, n_k \geq 0$ such that $\sum_{i=1}^{k} n_i = n$,

$$P(N(A_1) = n_1, N(A_2) = n_2, \cdots, N(A_k) = n_k \mid N(\mathcal{R}^d) = n)$$
$$= \frac{P(N(A_1) = n_1, N(A_2) = n_2, \cdots, N(A_k) = n_k)}{P(N(\mathcal{R}^d) = n)}$$

$$= \frac{\prod_{i=1}^{k} \frac{e^{-\mu(A_i)}(\mu(A_i))^{n_i}}{n_i!}}{\frac{e^{-M} M^n}{n!}}$$

$$= \frac{n!}{\prod_{i=1}^{k} n_i!} \prod_{i=1}^{k} \left(\frac{\mu(A_i)}{M}\right)^{n_i},$$

which shows that conditional on $N(\mathcal{R}^d) = n, (N(A_1), N(A_2), \ldots, N(A_k))$ is jointly distributed as a multinomial with parameters $(n, p_1, p_2, \ldots, p_k)$, where $p_i = \frac{\mu(A_i)}{M}$.

13.5.1 The Mapping Theorem

If a random set Π is a Poisson process in $\mathcal{R}^d$ with some intensity measure, and if the points are mapped into some other space by a transformation f, then the mapped points will often form another Poisson process in the new space with a suitable new intensity measure. This is useful, because we are often interested in a scatter induced by an original scatter, and we can view the induced scatter as a mapping of a Poisson process.

If Π is a Poisson process with intensity measure μ, and $f:\mathcal{R}^d \to \mathcal{R}^k$ is a mapping, denote the image of Π under f by Π^*; that is, $\Pi^* = f(\Pi)$. Let $f^{-1}(A) = \{x \in \mathcal{R}^d : f(x) \in A\}$. Then,

$$N^*(A) = \#\{y \in \Pi^* : y \in A\} = \#\{x \in \Pi : x \in f^{-1}(A)\} = N(f^{-1}(A)).$$

Therefore, $N^*(A) \sim \text{Poi}(\mu(f^{-1}(A)))$. If we denote $\mu^*(A) = \mu(f^{-1}(A))$, then $\mu^*(A)$ acts as the intensity measure of the Π^* process. However, in order for Π^* to be a Poisson process, we also need to show the independence property for disjoint sets $A_1, A_2, \ldots, A_n$, for any $n \geq 2$, and we need to ensure that singleton sets do not carry positive weight under the new intensity measure μ^*. The independence over disjoint sets is inherited from the independence over disjoint sets for the original Poisson process Π; the requirement that $\mu^*(\{y\})$ should be zero for any singleton set $\{y\}$ is usually verifiable on a case-by-case basis in applications. The exact statement of the mapping theorem then says the following.

Theorem 13.4. *Let $d, k \geq 1$, and let $f : \mathcal{R}^d \to \mathcal{R}^k$ be a transformation. Suppose $\Pi \subseteq \mathcal{R}^d$ is a Poisson process with the intensity measure μ, and let μ^* be defined by the relation $\mu^*(A) = \mu(f^{-1}(A))$, $A \subseteq \mathcal{R}^k$. Assume that $\mu^*(\{y\}) = 0$ for all $y \in \mathcal{R}^k$. Then the transformed set $\Pi^* \subseteq \mathcal{R}^k$ is a Poisson process with the intensity measure μ^*.*

An important immediate consequence of the mapping theorem is that lower-dimensional projections of a Poisson process are Poisson processes in the corresponding lower dimensions.

Corollary. *Let Π be a Poisson process in $\mathcal{R}^d$ with the intensity measure μ given by $\mu(A) = \int_A \lambda(x_1, x_2, \ldots, x_d) dx_1 dx_2 \ldots dx_d$. Then, for any $p < d$, the projection of Π to $\mathcal{R}^p$ is a Poisson process with the intensity measure $\mu_p(B) = \int_B \lambda_p(x_1, \ldots, x_p) dx_1 \ldots dx_p$, where the intensity function $\lambda_p(x_1, \ldots, x_p)$ is defined by $\lambda_p(x_1, \ldots, x_p) = \int_{\mathcal{R}^{d-p}} \lambda(x_1, x_2, \ldots, x_d) dx_{p+1} \ldots dx_d$.*

In particular, we have the useful result that all projections of a homogeneous Poisson process are homogeneous Poisson processes in the corresponding lower dimensions.

13.6 One-Dimensional Nonhomogeneous Processes

Nonhomogeneous Poisson processes are clearly important from the viewpoint of applications. But an additional fact of mathematical convenience is that the mapping theorem implies that a nonhomogeneous Poisson process in one dimension can be transformed into a homogeneous Poisson process. As a consequence, various formulas and results for the nonhomogeneous case can be derived rather painlessly, by simply applying the mapping technique suitably.

Suppose $X(t), t \geq 0$ is a Poisson process on the real line with intensity function $\lambda(x)$, and intensity measure μ, so that for any given $t, \mu([0,t]) = \int_0^t \lambda(x)dx$. We denote $\mu([0,t])$ as $\Lambda(t)$. Thus, $\Lambda(t)$ is a nonnegative, continuous, and nondecreasing function for $t \geq 0$, with $\Lambda(0) = 0$.

We show how this general Poisson process can be transformed to a homogeneous one by simply changing the clock. For this, think of the Poisson process as a random countable set and simply apply the mapping theorem. Using the same notation as in the mapping theorem, consider the particular mapping $f(x) = \Lambda(x)$. Then, by the mapping theorem, $\Pi^* = \Lambda(\Pi)$ is another Poisson process with the intensity measure $\mu^*([0,t]) = \mu(f^{-1}([0,t])) = \Lambda(\Lambda^{-1}(t)) = t$. In other words, mapping the original Poisson process by using the transformation $f(x) = \Lambda(x)$ converts the general Poisson process with the intensity measure μ into a homogeneous Poisson process with constant intensity equal to one. An equivalent way to state this is to say that if we define a new process $Z(t)$ as $Z(t) = X(\Lambda^{-1}(t)), t \geq 0$, then $Z(t)$ is a homogeneous Poisson process with constant rate (intensity) equal to one. This amounts to simply measuring time according to a new clock.

This transformation result is useful in deriving important distributional properties for a general nonhomogeneous Poisson process on the real line. We first need some notation. Let $T_{i,X}, i = 1, 2, \ldots$ denote the sequence of interarrival times for our Poisson process $X(t)$ with intensity function $\lambda(x)$, and let $T_{i,Z}$ denote the sequence of interarrival times for the transformed process $Z(t) = X(\Lambda^{-1}(t))$. Let also, $Y_{n,X}$ be the nth arrival time in the $X(t)$ process, and $Y_{n,Z}$ the nth arrival time in the $Z(t)$ process.

Consider the simplest distributional question, namely what is the distribution of $T_{1,X}$. By applying the mapping theorem,

$$P(T_{1,X} > t) = P(T_{1,Z} > \Lambda(t)) = e^{-\Lambda(t)}.$$

Therefore, by differentiation, $T_{1,X}$ has the density

$$f_{T_{1,X}}(t) = \lambda(t)e^{-\Lambda(t)}, \quad t > 0.$$

Similarly, once again, by applying the mapping theorem,

$$\begin{aligned}P(T_{2,X} > t \,|\, T_{1,X} = s) &= P(\Lambda^{-1}(Y_{2,Z}) > s+t \,|\, \Lambda^{-1}(Y_{1,Z}) = s)\\ &= P(Y_{2,Z} > \Lambda(t+s) \,|\, T_{1,Z} = \Lambda(s))\\ &= P(T_{2,Z} > \Lambda(t+s) - \Lambda(s) \,|\, T_{1,Z} = \Lambda(s))\\ &= P(T_{2,Z} > \Lambda(t+s) - \Lambda(s)) = e^{-[\Lambda(s+t) - \Lambda(s)]}.\end{aligned}$$

By differentiating, the conditional density of $T_{2,X}$ given $T_{1,X} = s$ is

$$f_{T_{2,X} \,|\, T_{1,X}}(t\,|\,s) = \lambda(s+t)e^{-[\Lambda(s+t) - \Lambda(s)]}.$$

The mapping theorem similarly leads to the main distributional results for the general nonhomogeneous Poisson process on the real line, which are collected in the theorem below.

Theorem 13.5. *Let $X(t), t \geq 0$ be a Poisson process with the intensity function $\lambda(x)$. Let $\Lambda(t) = \int_0^t \lambda(x) dx$. Let the interarrival times be $T_i, i \geq 1$, and the arrival times $Y_i, i \geq 1$. Then,*

(a) *For $n \geq 1$, Y_n has the density*

$$f_{Y_n}(t) = \frac{\lambda(t)e^{-\Lambda(t)}(\Lambda(t))^{n-1}}{(n-1)!}, \quad t > 0.$$

(b) *For $n \geq 2$ the conditional density of T_n given $Y_{n-1} = w$ is*

$$f_{T_n \,|\, Y_{n-1}}(t\,|\,w) = \lambda(t+w)e^{-[\Lambda(t+w) - \Lambda(w)]}, t, w > 0.$$

(c) *T_1 has the density*

$$f_{T_1}(t) = \lambda(t)e^{-\Lambda(t)}, \quad t > 0.$$

(d) *For $n \geq 2$, T_n has the marginal density*

$$f_{T_n}(t) = \frac{1}{(n-2)!} \int_0^\infty \lambda(w)\lambda(t+w)e^{-\Lambda(t+w)}(\Lambda(w))^{n-2} dw.$$

Example 13.10 (Piecewise Linear Intensity). Suppose printing jobs arrive at a computer printer according to the piecewise linear periodic intensity function

$$\lambda(x) = x, \text{ if } 0 \leq x \leq .1; \quad = .1, \text{ if } .1 \leq x \leq .5;$$
$$= .2(1-x), \text{ if } .5 \leq x \leq 1.$$

We assume that the unit of time is a day, so that for the first few hours in the morning the arrival rate increases steadily, and then for a few hours it remains constant. As the

13.6 One-Dimensional Nonhomogeneous Processes

day winds up, the arrival rate decreases steadily. The intensity function is periodic with a period equal to one.

By direct integration, in the interval [0, 1],

$$\Lambda(t) = \frac{t^2}{2}, \text{ if } t \leq .1; \quad = .005 + .1(t - .1), \text{ if } .1 \leq t \leq .5;$$

$$= .045 + .2(t - \frac{t^2}{2} - .375) = .2\left(-\frac{t^2}{2}\right) - .03, \text{ if } .5 \leq t \leq 1.$$

The intensity function $\lambda(x)$ and the mean function $\Lambda(t)$ are plotted in Figs. 13.2 and 13.3. Note that $\Lambda(t)$ would have been a linear function if the process were homogeneous.

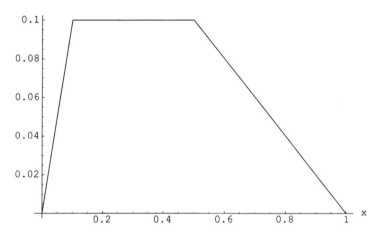

Fig. 13.2 Plot of the intensity function lambda

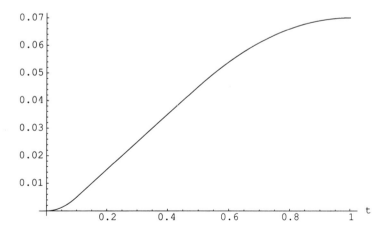

Fig. 13.3 Plot of the mean function lambda

Suppose as an example that we want to know what the probability is that at least one printing job arrives at the printer in the first ten hours of a particular day. Because $\Lambda(\frac{10}{24}) = \Lambda(.4167) = .0367$, the number of jobs to arrive at the printer in the first ten hours has a Poisson distribution with mean .0367, and so the probability we seek equals $1 - e^{-.0367} = .0360$.

13.7 * Campbell's Theorem and Shot Noise

Suppose $f(y_1, y_2, \ldots, y_d)$ is a real-valued function of d variables. Campbell's theorem tells us how to find the distribution and summary measures, such as the mean and the variance, of sums of the form $\sum_{Y \in \Pi} f(Y)$, where the countable set Π is a Poisson process in $\mathcal{R}^d$. A more general form is a sum $\sum_{n=1}^{\infty} X_n f(Y_n)$, where $Y_1, Y_2, \ldots$ are the points of the Poisson process Π, and $X_1, X_2, \ldots$ is some suitable auxiliary sequence of real-valued random variables. Sums of these types occur in important problems in various applied sciences. An important example of this type is Nobel Laureate S. Chandrasekhar's description of the gravitational field in a Poisson stellar ensemble.

Other applications include the so-called *shot effects* in signal detection problems. Imagine radioactive rays hitting a counter at the event times of a Poisson process, $Y_1, Y_2, \ldots$. A ray hitting the counter at time Y_i produces a lingering shot effect in the form of an electrical impulse, say $\omega(t - Y_i)$, the effect being a function of the time lag between the fixed instant t and the event time Y_i. Then the total impulse produced is the sum $\sum_{i=1}^{\infty} \omega(t - Y_i)$, which is a function of the type described above.

We first need some notation and definitions. We follow Kingman (1993) closely. Varadhan (2007) also has an excellent treatment of the topic in this section. Given a function $f : \mathcal{R}^d \to \mathcal{R}$, and a Poisson process Π in $\mathcal{R}^d$, let

$$\Sigma_f = \sum_{Y \in \Pi} f(Y); \quad F(y) = e^{-f(y)}.$$

For $f \geq 0$, let

$$\psi(f) = E(e^{-\Sigma_f}).$$

Definition 13.4. Let Π be a Poisson process in $\mathcal{R}^d$, and f a nonnegative function on $\mathcal{R}^d$. Then $\psi(f)$ is called the *characteristic functional* of Π.

The convergence of the sum Σ_f is clearly an issue of importance. A necessary and sufficient condition for the convergence of Σ_f is given in the theorem below. The trick is first to prove it for functions that take only a finite number of values and are of a compact support (i.e., they are zero outside of a compact set), and then to approximate a general nonnegative function by functions of this type. See Kingman (1993) for the details; these proofs and the outline of our arguments below involve notions of integrals with respect to general measures. This is unavoidable, and some readers may want to skip directly to the statement of the theorem below.

13.7 Campbell's Theorem and Shot Noise

The mean of Σ_f can be calculated by essentially applying *Fubini's theorem for measures*. More specifically, if we define a counting function (measure) $N(A)$ as the number of points of Π in (a general) set A, then the sum Σ_f can be viewed as the mean of the function f with respect to this random measure N. On the other hand, the mean of the random measure N is the intensity measure μ of the Poisson process Π, in the sense $E[N(A)] = \mu(A)$ for any A. Then, heuristically, by interchanging the order of integration, $E(\Sigma_f) = E(\int f dN) = \int f d\mu$. This is a correct formula.

In fact, a similar approximation argument leads to an expression for the characteristic function of Σ_f, or more generally, for $E(e^{\theta \Sigma_f})$ for a general complex number θ, when this expectation exists. Suppose our function f takes only k distinct values $f_1, f_2, \ldots, f_k$ on the sets $A_1, A_2, \ldots, A_k$. Then,

$$E\left(e^{\theta \Sigma_f}\right) = E\left(e^{\theta \sum_{j=1}^{k} f_j N(A_j)}\right)$$

$$= \prod_{j=1}^{k} E\left[e^{\theta f_j N(A_j)}\right] = \prod_{j=1}^{k} e^{(e^{\theta f_j}-1)\mu(A_j)}$$

(because $N(A_j)$ are independent Poisson random variables with means $\mu(A_j)$),

$$= e^{\sum_{j=1}^{k} (e^{\theta f_j}-1)\mu(A_j)} = e^{\sum_{j=1}^{k} \int_{A_j} (e^{\theta f(x)}-1) d\mu(x)}$$

$$= e^{\int_{\mathcal{R}^d} (e^{\theta f(x)}-1) d\mu(x)}.$$

One now approximates a general nonnegative function f by a sequence of such functions taking just a finite number of distinct values, and it turns out that the above formula for $E(e^{\theta \Sigma_f})$ holds generally, when this expectation is finite. This is the cream of what is known as *Campbell's theorem*.

If we take this formula, namely, $E(e^{\theta \Sigma_f}) = e^{\int_{\mathcal{R}^d} (e^{\theta f(x)}-1) d\mu(x)}$, and differentiate it twice with respect to θ (or simply expand the expression in powers of θ and look at the coefficient of θ^2), then we get the second moment, and hence the variance of Σ_f, and it is seen to be $\text{Var}(\Sigma_f) = \int_{\mathcal{R}^d} f^2(x) d\mu(x)$.

Theorem 13.6 (Campbell's Theorem). *Let Π be a Poisson process in $\mathcal{R}^d$ with intensity function $\lambda(x)$, and $f : \mathcal{R}^d \to \mathcal{R}$ a nonnegative function. Then,*

(a) $\Sigma_f < \infty$ *if and only if* $\int_{\mathcal{R}^d} \min(1, f(x))\lambda(x) dx < \infty$.
(b) $E(\Sigma_f) = \int_{\mathcal{R}^d} f(x)\lambda(x) dx$, *provided the integral exists.*
(c) *If* $\int_{\mathcal{R}^d} f(x)\lambda(x) dx$ *and* $\int_{\mathcal{R}^d} f^2(x)\lambda(x) dx$ *are both finite, then* $\text{Var}(\Sigma_f) = \int_{\mathcal{R}^d} f^2(x)\lambda(x) dx$.
(d) *Let θ be any complex number for which* $I(\theta) = \int_{\mathcal{R}^d} (1 - e^{\theta f(x)})\lambda(x) dx < \infty$. *Then,* $E(e^{\theta \Sigma_f}) = e^{-I(\theta)}$.
(e) *The intensity function of Π is completely determined by the characteristic functional $\psi(f)$ on the class of functions f that take a finite number of distinct values on $\mathcal{R}^d$.*
(f) $E[\prod_{Y \in \Pi} F(Y)] = e^{-\int_{\mathcal{R}^d} [1 - F(x)]\lambda(x) dx}$.

For a formal proof of this theorem, see Kingman (1993, p. 28–31).

13.7.1 Poisson Process and Stable Laws

Stable distributions, which we have previously encountered in Chapters 8 and 11 arise as the distributions of $\sum_{Y \in \Pi} f(Y)$ for suitable functions $f(y)$ when Π is a one-dimensional homogeneous Poisson process with some constant intensity λ. We demonstrate this interesting connection below.

Example 13.11 (Shot Noise and Stable Laws). Consider the particular function $f : \mathcal{R} \to \mathcal{R}$ defined by $f(x) = c|x|^{-\beta} \text{sgn}(x), 0 < \beta < \infty$. We eventually take $\beta > \frac{1}{2}$. The integral $\int_{\mathcal{R}} \min(1, f(x)) dx$ exists as a principal value, by taking limit through $\int_{-M}^{M} \min(1, f(x)) dx$. We evaluate the characteristic function of Σ_f, that is, we evaluate $E(e^{\theta \Sigma_f})$ for $\theta = iu$, where u is a real. We evaluate the characteristic function by using Campbell's theorem.

For this, we first evaluate $I(\theta)$ with $\theta = iu$. Because of the skew-symmetry of the function $f(x)$, only the cosine part in the characteristic function remains. In other words, if $\theta = iu, u > 0$, then

$$I(\theta) = \lambda \int_{-\infty}^{\infty} [1 - \cos(uf(x))] dx = 2\lambda \int_{0}^{\infty} [1 - \cos(uf(x))] dx$$

$$= 2\lambda \int_{0}^{\infty} [1 - \cos(ucx^{-\beta})] dx = \frac{2\lambda(uc)^{\frac{1}{\beta}}}{\beta} \int_{0}^{\infty} [1 - \cos(y)] y^{-\frac{1}{\beta} - 1} dy.$$

The integral $\int_{0}^{\infty} [1 - \cos(y)] y^{-\frac{1}{\beta}-1} dy$ converges if and only if $\beta > \frac{1}{2}$, in which case denote $\int_{0}^{\infty} [1 - \cos(y)] y^{-\frac{1}{\beta}-1} dy = k(\beta)$. Thus, for $\beta > \frac{1}{2}, u > 0, I(\theta) = \sigma(\beta) u^{\frac{1}{\beta}}$ for some constant $\sigma(\beta)$. As a function of u, $I(\theta)$ is symmetric in u, and so ultimately we get that if $\beta > \frac{1}{2}$, then the characteristic function of Σ_f is

$$E\left(e^{iu \Sigma_f}\right) = e^{-\sigma |u|^{\frac{1}{\beta}}},$$

for some positive constant σ. This is the *characteristic function of a stable law with exponent* $\alpha = \frac{1}{\beta} < 2$ (and, of course, $\alpha > 0$). We thus find here an interesting manner in which stable distributions arise from consideration of shot noises and Campbell's theorem.

Exercises

Exercise 13.1 (Poisson Process for Catching a Cold). Suppose that you catch a cold according to a Poisson process of once every three months.

(a) Find the probability that between the months of July and October, you will catch at least four colds.

Exercises

(b) Find the probability that between the months of May and July, and also between the months of July and October, you will catch at least four colds.

(c) * Find the probability that you will catch more colds between the months of July and October than between the months of May and July.

Exercise 13.2 (Events up to a Random Time). Jen has two phones on her desk. On one number, she receives internal calls according to a Poisson process at the rate of one in two hours. On the other number, she receives external calls according to a Poisson process at the rate of one per hour. Assume that the two processes run independently.

(a) What is the expected number of external calls by the time the second internal call arrives?

(b) * What is the distribution of the number of external calls by the time the second internal call arrives?

Exercise 13.3 (An Example Due to Emanuel Parzen). Certain machine parts in a factory fail according to a Poisson process at the rate of one in every six weeks. Two such parts are in the factory's inventory. The next supply will come in two months (eight weeks). What is the probability that production will be stopped for a week or more due to the lack of this particular machine part?

Exercise 13.4 (An Example Due to Emanuel Parzen). Customers arrive at a newspaper stall according to a Poisson process at the rate of one customer per two minutes.

(a) Find the probability that five minutes have passed since the last customer arrived.

(b) Find the probability that five minutes have passed since the next to the last customer arrived.

Exercise 13.5 (Spatial Poisson Process). Stars are distributed in a certain part of the sky according to a three-dimensional Poisson process with constant intensity λ. Find the mean and the variance of the separation between a particular star and the star nearest to it.

Exercise 13.6 (Compound Poisson Process). E-mails arrive from a particular friend according to a Poisson process at the constant rate of twice per week. When she writes, the number of lines in her e-mail is exponentially distributed with a mean of 10 lines. Find the mean, variance, and the generating function of $X(t)$, the total number of lines of correspondence received in t weeks.

Exercise 13.7 (An Example Due to Emanuel Parzen: Cry Baby). A baby cries according to a Poisson process at the constant rate of three times per 30 minutes. The parents respond only to every third cry. Find the probability that 20 minutes will elapse between two successive responses of the parents; that 60 minutes will elapse between two successive responses of the parents.

Exercise 13.8 (Nonhomogeneous Poisson Process). Let $X(t)$ be a one-dimensional Poisson process with an intensity function $\lambda(x)$. Give a formula for $P(X(t) = k)$ for a general $k \geq 0$. Here $X(t)$ counts the number of arrivals in the time interval $[0, t]$.

Exercise 13.9 (Nonhomogeneous Poisson Process). A one-dimensional nonhomogeneous Poisson process has an intensity function $\lambda(x) = cx$ for some $c > 0$. Find the density of the nth interarrival time, and the nth arrival time.

Exercise 13.10 (Nonhomogeneous Poisson Process). A one-dimensional Poisson process has an intensity function $\lambda(x)$. Give a necessary and sufficient condition that the interarrival times are mutually independent.

Exercise 13.11 (Nonhomogeneous Poisson Process). A one-dimensional Poisson process has an intensity function $\lambda(x)$. Give a necessary and sufficient condition that $E(T_n)$, the mean of the nth interarrival time, remains bounded as $n \to \infty$.

Exercise 13.12 (Conditional Distribution of Interarrival Times). A one-dimensional Poisson process $X(t)$ has an intensity function $\lambda(x)$. Show that the conditional distribution of the n arrival times given that $X(t) = n$ is still the same as the joint distribution of n-order statistics from a suitable density, say $g(x)$. Identify this g.

Exercise 13.13 (Projection of a Poisson Process). Suppose Π is a Poisson point process in $\mathcal{R}^d$, with the intensity function $\lambda(x_1, x_2, \ldots, x_d) = e^{-\sum_{i=1}^{d} |x_i|}$.

(a) Are the projections of Π to the individual dimensions also Poisson processes?
(b) If they are, find the intensity functions.

Exercise 13.14 (Polar Coordinates of a Poisson Point Process). Let Π be a two-dimensional Poisson point process with a constant intensity function $\lambda(x_1, x_2) = \lambda$. Prove that the set of polar coordinates of the points of Π form another Poisson point process, and identify its intensity function.

Exercise 13.15 (Distances from Origin of a Poisson Point Process). Let Π be a two-dimensional Poisson point process with a constant intensity function $\lambda(x_1, x_2) = \lambda$. Consider the set of distances of the points of Π from the origin.
Is this another Poisson process? If it is, identify its intensity function.

Exercise 13.16 (Deletion of SPAM). Suppose that you receive e-mails according to a Poisson process with a constant arrival rate of one per 10 minutes. Mutually independently, each mail has a 20% probability of being deleted by your spam filter. Find the probability that the tenth mail after you first log in shows up in your mailbox within two hours.

Exercise 13.17 (Production on Demand). Orders for a book arrive at a publisher's office according to a Poisson process with a constant rate of 4 per week. Production for a received order is assumed to start immediately, and the actual time to produce a copy is uniformly distributed between 5 and 10 days. Let $X(t)$ be the number of orders in actual production at a given time t. Find the mean and the variance of $X(t)$.

Exercises

Exercise 13.18 (Compound Poisson Process). An auto insurance agent receives claims at the average constant rate of one per day. The amount of claims are iid, with an equal probability of being 250, 500, 1000, or 1500 dollars.

(a) Find the generating function of the total claim made in a 30-day period.
(b) Hence find the mean and the variance of the total claim made in a 30-day period.

Exercise 13.19 (Connection of a Poisson Process to Binomial Distribution). Suppose $X(t), t \geq 0$ is a Poisson process with constant average rate λ. Given that $X(t) = n$, show that the number of events up to the time u, where $u < t$, has a Binomial distribution. Identify the parameters of this binomial distribution.

Exercise 13.20 * (Inspection Paradox). The city bus arrives at a certain bus stop at the constant average rate of once per 30 minutes. Suppose that you arrive at the bus stop at a fixed time t. Find the probability that the time till the next bus is larger than the time since the last bus.

Exercise 13.21 (Correlation in a Poisson Process). Suppose $X(t)$ is a Poisson process with average constant arrival rate λ. Let $0 < s < t < \infty$. Find the correlation between $X(s)$ and $X(t)$.

Exercise 13.22 (A Strong Law). Suppose $X(t)$ is a Poisson process with constant arrival rate λ. Show that $\frac{X(t)}{t}$ converges almost surely to λ as $t \to \infty$.

Exercise 13.23 (Two Poisson Processes). Suppose $X(t), Y(t), t \geq 0$ are two Poisson processes with rates λ_1, λ_2. Assume that the processes run independently.

(a) Prove or disprove: $X(t) + Y(t)$ is also a Poisson process.
(b) * Prove or disprove: $|X(t) - Y(t)|$ is also a Poisson process.

Exercise 13.24 (Two Poisson Processes; Continued). Suppose $X(t), Y(t), t \geq 0$ are two Poisson processes with rates λ_1, λ_2. Assume that the processes run independently.

(a) Find the probability that the first event in the X process occurs before the first event in the Y process.
(b) * Find the density function, mean, and variance of the minimum of the first arrival time in the X process and the first arrival time in the Y process.
(c) Find the distribution of the number of events in the Y process in the time interval between the first and the second event in the X process.

Exercise 13.25 (Displaced Poisson Process). Let Π be a Poisson point process on $\mathcal{R}^d$ with intensity function $\lambda(x)$ constantly equal to λ. Suppose the points Y of Π are randomly shifted to new points $Y + W$, where the shifts W are iid, and are independent of Π. Let Π^* denote the set of points $Y + W$.

(a) Is Π^* also a Poisson point process?
(b) If it is, what is its intensity function?

Exercise 13.26 (The Midpoint Process). Let Π be a one-dimensional Poisson point process. Consider the point process Π^* of the midpoints of the points in Π.

(a) Is Π^* also a Poisson point process?
(b) If it is, what is its intensity function?

Exercise 13.27 (kth Nearest Neighbor). Let Π be a Poisson point process on $\mathcal{R}^d$ with intensity function $\lambda(x)$ constantly equal to λ. For $k \geq 1$, let D_k be the distance from the origin of the kth nearest point of Π to the origin.
  Find the density, mean, and variance of D_k.

Exercise 13.28 (Generalized Campbell's Theorem). Let $X(t)$ be a homogeneous Poisson process on the real line with constant arrival rate λ, and let $Y_i, i \geq 1$ be the arrival times in the $X(t)$ process. Let $S(t) = \sum_{i=1}^{\infty} Z_i \omega(t - Y_i), t \geq 0$, where $w(s) = e^{-cs} I_{s \geq 0}$ for some positive constant c, and $Z_i, i \geq 1$ is a positive iid sequence, which you may assume to be independent of the $X(t)$ process.

(a) Find the characteristic function of $S(t)$ for a fixed t.
(b) Hence, find the mean and variance of $S(t)$, when they exist.

Exercise 13.29 (Poisson Processes and Geometric Distribution). Suppose $X_i(t)$, $i = 1, 2$ are two homogeneous Poisson processes, independent of each other, and with respective arrival rates λ, μ. Show that the distribution of the number of arrivals N in the $X_2(t)$ process between two successive arrivals in the $X_1(t)$ process is given by $P(N = k) = p(1-p)^k, k = 0, 1, 2, \cdots$ for some $0 < p < 1$. Identify p.

Exercise 13.30 (Nonhomogeneous Shot Noise). Let $X(t)$ be a one dimensional Poisson process with the intensity function $\lambda(x)$. Let $S(t) = \sum_{i=1}^{\infty} \omega(t - Y_i)$, where $Y_i, i \geq 1$ are the arrival times of the $X(t)$ process, and $\omega(s) = 0$ for $s < 0$. Derive formulas for the characteristic function, mean, and variance of $S(t)$ for a given t.

References

Karlin, S. and Taylor, H.M. (1975). *A First Course in Stochastic Processes*, Academic Press, New York.
Kingman, J.F.C. (1993). *Poisson Processes*, Oxford University Press, Oxford, UK.
Lawler, G. (2006). *Introduction to Stochastic Processes*, Chapman and Hall, New York.
Parzen, E. (1962). *Stochastic Processes*, Holden-Day, San Francisco.
Port, S. (1994). *Theoretical Probability for Applications*, Wiley, New York.

Chapter 14
Discrete Time Martingales and Concentration Inequalities

For an independent sequence of random variables $X_1, X_2, \ldots$, the conditional expectation of the present term of the sequence given the past terms is the same as its unconditional expectation. Martingales let the conditional expectation depend on the past terms, but in a special way. Thus, similar to Markov chains, martingales act as natural models for incorporating dependence into a sequence of observed data. But the value of the theory of martingales is much more than simply its modeling value. Martingales arise, as natural byproducts of the mathematical analysis in an amazing variety of problems in probability and statistics. Therefore, results from martingale theory can be immediately applied to all these situations in order to make deep and useful conclusions about numerous problems in probability and statistics. A particular modern set of applications of martingale methods is in the area of concentration inequalities, which place explicit bounds on probabilities of large deviations of functions of a set of variables from their mean values. This chapter gives a glimpse into some important concentration inequalities, and explains how martingale theory enters there. Martingales form a nearly indispensable tool for probabilists and statisticians alike.

Martingales were introduced into the probability literature by Paul Lévy, who was interested in finding situations beyond the iid case where the strong law of large numbers holds. But its principal theoretical studies were done by Joseph Doob. Two extremely lucid expositions on martingales are Doob (1971) and Heyde (1972). Some other excellent references for this chapter are Karlin and Taylor (1975), Chung (1974), Hall and Heyde (1980), Williams (1991), Karatzas and Shreve (1991), Fristedt and Gray (1997), and Chow and Teicher (2003). Other references are provided in the sections.

14.1 Illustrative Examples and Applications in Statistics

We start with a simple example, which nevertheless captures the spirit of the idea of a martingale sequence of random variables.

Example 14.1 (Gambler's Fortune). Consider a gambler repeatedly playing a fair game in a casino. Thus, a fair coin is tossed. If heads show, the player wins $1; if it is tails, the house wins $1. He plays repeatedly. Let $X_1, X_2, \ldots$ be the players's sequence of wins. Thus, the X_i are iid with the common distribution $P(X_i = \pm 1) = \frac{1}{2}$. The player's fortune after n plays is $S_n = S_0 + \sum_{i=1}^n X_i, n \geq 1$. If we take the player's initial fortune S_0 to be just zero, then $S_n = \sum_{i=1}^n X_i$. Suppose now the player has finished playing for n times, and he is looking ahead to what his fortunes will be after he plays the next time. In other words, he wants to find $E(S_{n+1} | S_1, \ldots, S_n)$. But,

$$E(S_{n+1} | S_1, \ldots, S_n)$$
$$= E(S_n + X_{n+1} | S_1, \ldots, S_n) = S_n + E(X_{n+1} | S_1, \ldots, S_n)$$
$$= S_n + E(X_{n+1}) = S_n + 0 = S_n.$$

In the above, $E(X_{n+1} | S_1, \ldots, S_n)$ equals the unconditional expectation of X_{n+1} because X_{n+1} is independent of $(X_1, X_2, \ldots, X_n)$, and hence, independent of $(S_1, \ldots, S_n)$.

Notice that the sequence of fortunes $S_1, S_2, \ldots$ is not an independent sequence. There is information in the past sequence of fortunes for predicting the current fortune. But the players's forecast for what his fortune will be after the next round of play is simply what his fortunes are right now, no more and no less. This is basically what the martingale property means, and is the reason for equating martingales with *fair games*.

Here is the definition. Rigorous treatment of martingales requires use of measure theory. For the most part, our treatment avoids measure-theory terminology.

Definition 14.1. Let $X_n, n \geq 1$ be a sequence of random variables defined on a common sample space Ω such that $E(|X_n|) < \infty$ for all $n \geq 1$. The sequence $\{X_n\}$ is called a *martingale adapted to itself* if for each $n \geq 1$, $E(X_{n+1} | X_1, X_2, \ldots, X_n) = X_n$ with probability one.

The sequence $\{X_n\}$ is called a *supermartingale* if for each $n \geq 1$, $E(X_{n+1} | X_1, X_2, \ldots, X_n) \leq X_n$ with probability one. The sequence $\{X_n\}$ is called a *submartingale* if for each $n \geq 1$, $E(X_{n+1} | X_1, X_2, \ldots, X_n) \geq X_n$ with probability one.

Remark. We generally do not mention the *adapted to itself* qualification when that is indeed the case. It is sometimes useful to talk about the martingale property with respect to a different sequence of random variables. This concept is defined below and Example 14.8 is an example of such a martingale sequence.

Note that X_n is a submartingale if and only if $-X_n$ is a supermartingale, and that it is a martingale if and only if it is both a supermartingale and a submartingale. Also notice that for a martingale sequence X_n, $E(X_{n+m}) = E(X_n)$ for all $n, m \geq 1$; in other words, $E(X_n) = E(X_1)$ for all n.

Definition 14.2. Let $X_n, n \geq 1$ and $Y_n, n \geq 1$ be sequences of random variables defined on a common sample space Ω such that $E(|X_n|) < \infty$ for all $n \geq 1$.

The sequence $\{X_n\}$ is called a *martingale adapted to the sequence* $\{Y_n\}$ if for each $n \geq 1$, X_n is a function of $Y_1, \ldots, Y_n$, and $E(X_{n+1} | Y_1, Y_2, \ldots, Y_n) = X_n$ with probability one.

Some elementary examples are given first.

Example 14.2 (Partial Sums). Let $Z_1, Z_2, \ldots$ be independent zero mean random variables, and let S_n denote the partial sum $\sum_{i=1}^{n} Z_i$. Then, clearly, $E(S_{n+1} | S_1, \ldots, S_n) = S_n + E(Z_{n+1} | S_1, \ldots, S_n) = S_n + E(Z_{n+1}) = S_n$, and so $\{S_n\}$ forms a martingale. More generally, if the common mean of the Z_i is some number μ, then $S_n - n\mu$ is a martingale.

Example 14.3 (Sums of Squares). Let $Z_1, Z_2, \ldots$ be iid $N(0, 1)$ random variables, and let $X_n = (Z_1 + \cdots + Z_n)^2 - n = S_n^2 - n$, where $S_n = Z_1 + \cdots + Z_n$. Then,

$$E(X_{n+1} | X_1, X_2, \ldots, X_n)$$
$$= E[(Z_1 + \cdots + Z_n)^2 + 2Z_{n+1}(Z_1 + \cdots + Z_n)$$
$$+ Z_{n+1}^2 | X_1, X_2, \ldots, X_n] - (n+1)$$
$$= X_n + n + 2(Z_1 + \cdots + Z_n) E(Z_{n+1} | X_1, X_2, \ldots, X_n)$$
$$+ E(Z_{n+1}^2 | X_1, X_2, \ldots, X_n) - (n+1)$$
$$= X_n + n + 0 + 1 - (n+1) = X_n,$$

and so $\{X_n\}$ forms a martingale sequence.

Actually, we did not use the normality of the Z_i at all, and the martingale property holds without the normality assumption. That is, if $Z_1, Z_2, \ldots$ are iid with mean zero and variance σ^2, then $S_n^2 - n\sigma^2$ is a martingale.

Example 14.4. Suppose $X_1, X_2, \ldots$ are iid $N(0, 1)$ variables and $S_n = \sum_{i=1}^{n} X_i$. Because $S_n \sim N(0, n)$, its mgf $E(e^{tS_n}) = e^{nt^2/2}$. Now let $Z_n = e^{tS_n - nt^2/2}$, where t is a fixed real number. Then, $E(Z_{n+1} | Z_1, \ldots, Z_n) = e^{-(n+1)t^2/2} E(e^{tS_n} e^{tX_{n+1}} | S_n) = e^{-(n+1)t^2/2} e^{tS_n} e^{t^2/2} = Z_n$. Therefore, for any real t, the sequence $e^{tS_n - nt^2/2}$ forms a martingale.

Once again, a generalization beyond the normal case is possible; see the chapter exercises for a general result.

Example 14.5 (Matching Problem). Consider the matching problem. For example, suppose N people, each wearing a hat, have gathered in a party and at the end of the party, the N hats are returned to them at random. Those that get their own hats back then leave the room. The remaining hats are distributed among the remaining guests at random, and so on. The process continues until all the hats have been given away. Let X_n denote the number of guests still present after the nth round of this hat returning process.

At each round, we expect one person to get his own hat back and leave the room. In other words, $E(X_n - X_{n+1}) = 1 \quad \forall n$. In fact, with a little calculation, we even have

$$E(X_{n+1}|X_1,\ldots,X_n) = E(X_{n+1} - X_n + X_n|X_1,\ldots,X_n)$$
$$= E(X_{n+1} - X_n|X_1,\ldots,X_n) + X_n = -1 + X_n.$$

This immediately implies that $E(X_{n+1}+n+1|X_1,\ldots,X_n) = -1+(n+1)+X_n = X_n + n$. Hence the sequence $\{X_n + n\}$ is a martingale.

Example 14.6 (Pólya's Urn). The Pólya urn scheme is defined as follows. Initially, an urn contains a white and b black balls, a total of $a + b$ balls. One ball is drawn at random from among all the balls in the urn. It, together with c more balls of its color is returned to the urn, so that after the first draw, the urn has $a + b + c$ balls. This process is repeated.

Suppose X_i is the indicator of the event A_i that a white ball is drawn at the ith trial, and for given $n \geq 1, S_n = X_1 + \cdots + X_n$, which is the total number of times that a white ball has been drawn in the first n trials. For the sake of notational simplicity, we take $c = 1$. Then, the proportion of white balls in the urn just after the nth trial has been completed is $R_n = \frac{a+S_n}{a+b+n}$.

Elementary arguments show that

$$P(X_{n+1} = 1|X_1 = x_1,\ldots, X_n = x_n) = \frac{a + x_1 + \cdots + x_n}{a + b + n}.$$

Thus,

$$E(S_{n+1}|S_1,\ldots,S_n) = E(S_{n+1}|S_n) = S_n + \frac{a + S_n}{a + b + n}$$

$$\Rightarrow E(R_{n+1}|R_1,\ldots,R_n) = \frac{a}{a+b+n+1} + \frac{1}{a+b+n+1}$$
$$[(a+b+n)R_n - a + R_n] = R_n.$$

We therefore have the interesting result that in the Pólya urn scheme, the sequence of proportions of white balls in the urn forms a martingale.

Example 14.7 (The Wright–Fisher Markov Chain). Consider the stationary Markov chain $\{X_n\}$ on the state space $\{0, 1, 2, \ldots, N\}$ with the one-step transition probabilities

$$p_{ij} = \binom{N}{j}\left(\frac{i}{N}\right)^j\left(1 - \frac{i}{N}\right)^{N-j}.$$

14.1 Illustrative Examples and Applications in Statistics

This is the Wright–Fisher chain in population genetics (see Chapter 10). We show that X_n is a martingale adapted to itself. Indeed, by direct calculation,

$$E(X_{n+1}|X_1,\ldots,X_n) = E(X_{n+1}|X_n)$$

$$= \sum_{j=0}^{N} j \binom{N}{j} \left(\frac{X_n}{N}\right)^j \left(1 - \frac{X_n}{N}\right)^{N-j} = N\frac{X_n}{N} = X_n.$$

Example 14.8 (Likelihood Ratios). Suppose $X_1, X_2, \ldots, X_n$ are iid with a common density function f, which is one of f_0, and f_1, two different density functions. The statistician is supposed to choose from the two densities f_0, f_1, the one that is truly generating the observed data $x_1, x_2, \ldots, x_n$. One therefore has the *null hypothesis* H_0 that $f = f_0$, and the *alternate hypothesis* that $f = f_1$. The statistician's decision is commonly based on the *likelihood ratio*

$$\Lambda_n = \prod_{i=1}^{n} \frac{f_1(X_i)}{f_0(X_i)}.$$

If Λ_n is large for the observed data, then one concludes that the data values come from a high-density region of f_1 and a low-density region of f_0, and therefore concludes that the true f generating the observed data is f_1.

Suppose now the null hypothesis is actually true; that is, truly, $X_1, X_2, \ldots$ are iid with the common density f_0. Now,

$$E_{f_0}[\Lambda_{n+1}|\Lambda_1,\ldots,\Lambda_n] = E_{f_0}\left[\frac{f_1(X_{n+1})}{f_0(X_{n+1})}\Lambda_n \Big| \Lambda_1,\ldots,\Lambda_n\right]$$

$$= \Lambda_n E_{f_0}\left[\frac{f_1(X_{n+1})}{f_0(X_{n+1})} \Big| \Lambda_1,\ldots,\Lambda_n\right]$$

$$= \Lambda_n E_{f_0}\left[\frac{f_1(X_{n+1})}{f_0(X_{n+1})}\right]$$

(because the sequence $X_1, X_2, \ldots$ are independent)

$$= \Lambda_n \int_{\mathcal{R}} \frac{f_1(x)}{f_0(x)} f_0(x)dx = \Lambda_n \int_{\mathcal{R}} f_1(x)dx$$

$$= \Lambda_n \times 1 = \Lambda_n.$$

Therefore, the sequence of likelihood ratios forms a martingale under the null hypothesis (i.e., if the true f is f_0).

Example 14.9 (Bayes Estimates). Suppose random variables $Y, X_1, X_2, \ldots$ are defined on a common sample space Ω. For given $n \geq 1, (X_1, X_2, \ldots, X_n)$ has the joint conditional distribution $P_{\theta,n}$ given that $Y = \theta$. From a statistical point of view, Y is supposed to stand for an unknown parameter, which is formally treated as a random variable, and $X^{(n)} = (X_1, X_2, \ldots, X_n)$ for some specific n, namely the actual sample size, is the data that the statistician has available to estimate the unknown parameter. The *Bayes estimate* of the unknown paramter is the posterior mean $E(Y \mid X^{(n)})$ (see Chapter 3).

Denote for each $n \geq 1$, $E(Y \mid X^{(n)}) = Z_n$. We show that Z_n forms a martingale sequence with respect to the sequence $X^{(n)}$; that is, $E(Z_{n+1} \mid X^{(n)}) = Z_n$. However, this follows on simply observing that by the iterated expectation formula,

$$Z_n = E\left(Y \mid X^{(n)}\right) = E_{X_{n+1} \mid X^{(n)}} \left[E\left(Y \mid X^{(n)}, X_{n+1}\right) \right] = E\left(Z_{n+1} \mid X^{(n)}\right).$$

Example 14.10 (Square of a Martingale). Suppose X_n, defined on some sample space Ω is a positive submartingale sequence. For simplicity, let us consider the case when it is adapted to itself. Thus, for any $n \geq 1$, $E(X_{n+1} \mid X_1, \ldots, X_n) \geq X_n$ (with probability one). Therefore, for any $n \geq 1$,

$$E(X_{n+1}^2 \mid X_1, \ldots, X_n) \geq [E(X_{n+1} \mid X_1, \ldots, X_n)]^2$$
$$\geq X_n^2.$$

Therefore, if we let $Z_n = X_n^2$, then Z_n is a submartingale sequence.

If we inspect this example carefully, then we realize that we have only used a very special case of Jensen's inequality to establish the needed submartingale property for the Z_n sequence. Furthermore, if the original $\{X_n\}$ sequence is a martingale, rather than a submartingale, then the positivity restriction on the X_n is no longer necessary. Thus, by simply following the steps of this example, we in fact have the following simple but widely useful general result.

Theorem 14.1 (Convex Function Theorem). *Let $X_n, n \geq 1$ be defined on a common sample space Ω. Let f be a convex function on $\mathcal{R}$, and let $Z_n = f(X_n)$.*

(a) *Suppose $\{X_n\}$ is a martingale adapted to some sequence $\{Y_n\}$. Then $\{Z_n\}$ is a submartingale adapted to $\{Y_n\}$.*
(b) *Suppose $\{X_n\}$ is a submartingale adapted to some sequence $\{Y_n\}$. Assume that f is in addition nondecreasing. Then $\{Z_n\}$ is a submartingale adapted to $\{Y_n\}$.*

14.2 Stopping Times and Optional Stopping

The *optional stopping theorem* is one of the most useful results in martingale theory. It can be explained in gambling terms. Consider a gambler playing a fair game repeatedly, so that her sequence of fortunes forms a martingale. One might think that by gaining experience as the game proceeds, and by quitting at a cleverly chosen opportune time based on the gambler's experience, a fair game could be turned into a favorable game. The optional stopping theorem says that this is in fact not possible, if the gambler does not have unlimited time on her hands and the house

14.2 Stopping Times and Optional Stopping

has limits on how much she can put up on the table. Mathematical formulation of the optional stopping theorem requires use of *stopping times*, which were introduced in Chapter 11 in the context of random walks. We redefine stopping times and give additional examples below before introducing optional stopping.

14.2.1 Stopping Times

Definition 14.3. Let $X_1, X_2, \ldots$ be a sequence of random variables, all defined on a common sample space Ω. Let τ be a nonnegative integer-valued random variable, also defined on Ω. We call τ a *stopping time adapted to the sequence* $\{X_n\}$ if $P(\tau < \infty) = 1$, and if for each $n \geq 1$, $I_{\{\tau \leq n\}}$ is a function of only $X_1, X_2, \ldots, X_n$.

In other words, τ is a stopping time adapted to $\{X_n\}$ if for any $n \geq 1$, whether or not $\tau \leq n$ can be determined by only knowing $X_1, X_2, \ldots, X_n$, and provided that τ cannot be infinite with a positive probability.

We have seen some examples of stopping times in Chapter 11. We start with a few more illustrative examples.

Example 14.11 (Sequential Tests in Statistics). Suppose to start with we have an infinite sequence of random variables $X_1, X_2, \ldots$ on a common sample space Ω, and let S_n denote the nth partial sum, $S_n = \sum_{i=1}^{n} X_i, n \geq 1$. The X_n need not be independent. Fix numbers $-\infty < l < u < \infty$. Then τ defined as

$$\tau = \inf\{n : S_n < l \text{ or } S_n > u\},$$

and $\tau = \infty$ if $l \leq S_n \leq u \, \forall n \geq 1$, is a stopping time adapted to the sequence $\{S_n\}$.

A particular case of this arises in sequential testing of hypotheses in statistics. Suppose an original sequence $Z_1, Z_2, \ldots$ is iid from some density f, which equals either f_0 or f_1. Then, as we have seen above, the *likelihood ratio* is

$$\Lambda_n = \frac{\prod_{i=1}^{n} f_1(Z_i)}{\prod_{i=1}^{n} f_0(Z_i)}.$$

The *Wald sequential probability ratio test* (SPRT) continues sampling as long as Λ_n remains between two specified numbers a and b, $a < b$, and stops and decides in favor of f_1 or f_0 the first time $\Lambda_n > b$ or $< a$. If we denote $l = \log a, u = \log b$, then Wald's test waits till the first time $\log \Lambda_n = \sum_{i=1}^{n} \log \frac{f_1(Z_i)}{f_0(Z_i)} = \sum_{i=1}^{n} X_i$ (say) goes above u or below l, and thus the *sampling number* of Wald's SPRT is a stopping time.

Example 14.12 (Combining Stopping Times). This example shows a few ways that we can make new stopping times out of given ones. Suppose τ is a stopping time (adapted to some sequence $\{X_n\}$) and n is a prespecified positive integer. Then $\tau_n = \min(\tau, n)$ is a stopping time (adapted to the same sequence). This is because

$$\{\tau_n \leq k\} = \{\tau \leq k\} \cup \{n \leq k\},$$

and therefore, τ being a stopping time adapted to $\{X_n\}$, for any given k, deciding whether $\tau_n \leq k$ requires the knowledge of only $X_1, \ldots, X_k$.

Suppose τ_1, τ_2 are both stopping times, adapted to some sequence $\{X_n\}$. Then $\tau_1 + \tau_2$ is also a stopping time adapted to the same sequence. To prove this, note that

$$\{\tau_1 + \tau_2 \leq k\} = \cup_{i=0}^{k} \cup_{j=0}^{i} \{\tau_1 = j, \tau_2 = i - j\} = \cup_{i=0}^{k} \cup_{j=0}^{i} A_{ij}.$$

and whether any A_{ij} occurs depends only on $X_1, \ldots, X_k$.

For the sake of reference, we collect a set of such facts about stopping times in the next result. They are all easy to prove.

Theorem 14.2. *(a) Let τ be a stopping time adapted to some sequence $\{X_n\}$. Then, for any given $n \geq 1$, $\min(\tau, n)$ is also a stopping time adapted to $\{X_n\}$.*
(b) Let τ_1, τ_2 be stopping times adapted to $\{X_n\}$. Then each of $\tau_1 + \tau_2, \min(\tau_1, \tau_2), \max(\tau_1, \tau_2)$ is a stopping time adapted to $\{X_n\}$.
(c) Let $\{\tau_k, k \geq 1\}$ be a countable family of stopping times, each adapted to $\{X_n\}$. Let

$$\underline{\tau} = \inf_{k} \tau_k; \overline{\tau} = \sup_{k} \tau_k; \tau = \lim_{k \to \infty} \tau_k,$$

where $\underline{\tau}, \overline{\tau}$, and τ are defined pointwise, and it is assumed that the limit τ exists almost surely. Then each of $\underline{\tau}, \overline{\tau}$ and τ is a stopping time adapted to $\{X_n\}$.

14.2.2 Optional Stopping

The most significant derivative of introducing the concept of stopping times is the *optional stopping theorem*. At the expense of using some potentially hard to verify conditions, stronger versions of our statement of the optional stopping theorem can be stated. We choose to opt for simplicity of the statement over greater generality, and refer to more general versions (which are useful!). The main message of the optional stopping theorem is that a gambler cannot convert a fair game into a favorable one by using clever quitting strategies.

Theorem 14.3 (Optional Stopping Theorem). *Let $\{X_n, n \geq 0\}$ be a submartingale adapted to some sequence $\{Y_n\}$, and τ a stopping time adapted to the same sequence. For $n \geq 0$, let $\tau_n = \min(\tau, n)$. Then $\{X_{\tau_n}\}$ is also a submartingale adapted to $\{Y_n\}$, and for each $n \geq 0$,*

$$E(X_0) \leq E(X_{\tau_n}) \leq E(X_n).$$

In particular, if

$\{X_n\}$ *is a martingale,* $E(|X_\tau|) < \infty$, *and* $\lim_{n \to \infty} E(X_{\tau_n}) = E(X_\tau)$,

14.2 Stopping Times and Optional Stopping

then
$$E(X_\tau) = E(X_0).$$

Remark. It is of course unsatisfactory to simply demand that $E(|X_\tau|) < \infty$ and $\lim_{n \to \infty} E(X_{\tau_n}) = E(X_\tau)$. What we need are simple sufficient conditions that a user can verify relatively easily. This is addressed following the proof of the above theorem.

Proof of Theorem. For simplicity, we give the proof only for the case when $\{X_n\}$ is adapted to itself. The main step involved is to notice the identity

$$W_n = X_{\tau_n} = \sum_{i=0}^{n-1} X_i I_{\{\tau=i\}} + X_n I_{\{\tau \geq n\}}, \quad (*)$$

for all $n \geq 0$. It follows from this identity and the submartingale property of the $\{X_n\}$ sequence that

$$E(W_{n+1} | X_0, \ldots, X_n)$$
$$= \sum_{i=0}^{n} E(X_i I_{\{\tau=i\}} | X_0, \ldots, X_n) + E(X_{n+1} I_{\{\tau>n\}} | X_0, \ldots, X_n)$$
$$= \sum_{i=0}^{n} X_i I_{\{\tau=i\}} + I_{\{\tau>n\}} E(X_{n+1} | X_0, \ldots, X_n)$$
$$\geq \sum_{i=0}^{n} X_i I_{\{\tau=i\}} + X_n I_{\{\tau>n\}} = X_{\tau_n} = W_n.$$

Thus, as claimed, $W_n = \{X_{\tau_n}\}$ is a submartinagle adapted to the original $\{X_n\}$ sequence. It follows that

$$E(X_{\tau_n}) = E(W_n) \geq E(W_0) = E(X_0).$$

To complete the proof of the theorem, we need the reverse inequality $E(W_n) \leq E(X_n)$. This too follows from the same identity $(*)$ given at the beginning of the proof of this theorem, and on using the additional inequality

$$E(X_n I_{\{\tau=i\}} | X_0, \ldots, X_i)$$
$$= I_{\{\tau=i\}} E(X_n | X_0, \ldots, X_i) \geq I_{\{\tau=i\}} X_i,$$

because $\{X_n\}$ is a submartingale. If this bound on $X_i I_{\{\tau=i\}}$ is plugged into our basic identity $(*)$ above, the reverse inequality follows.

The remaining claim, when $\{X_n\}$ is in fact a martingale, follows immediately from the two inequalities $E(X_0) \leq E(W_n) \leq E(X_n)$. □

14.2.3 Sufficient Conditions for Optional Stopping Theorem

Easy examples show that the assertion $E(X_\tau) = E(X_0)$ for a martingale sequence $\{X_n\}$ cannot hold without some control on the stopping time τ. We first provide a simple example where the assertion of the optional stopping theorem fails. In looking for such counterexamples, it is useful to construct the stopping time in a way that when we stop, the value of the stopped martingale is a constant; that is, X_τ is a constant.

Example 14.13 (An Example Where the Optional Stopping Theorem Fails). Consider again the gambling example, or what really is the simple symmetric random walk, with X_i iid having the common distribution $P(X_i = \pm 1) = \frac{1}{2}$, and $S_n = \sum_{i=1}^n X_i, n \geq 1$. We define $S_0 = 0$. We know S_n to be a martingale. Consider now the stopping time

$$\tau = \inf\{n > 0 : S_n = 1\}.$$

We know from Chapter 11 that the one-dimensional simple symmetric random walk is recurrent; thus, $P(\tau < \infty) = 1$. Note that $S_\tau = 1$, and so, $E(S_\tau) = 1$. However, $E(S_0) = E(S_n) = 0$. So, the assertion of the optional stopping theorem does not hold.

What is going on in this example is that we do not have enough control on the stopping time τ. Although the random walk visits all its states (infinitely often) with probability one, the recurrence times are infinite on the average. Thus, τ can be uncontrollably large. Indeed, the assumption

$$\lim_{n \to \infty} E(S_{\min(\tau,n)}) = E(S_\tau)(= 1)$$

does not hold. Roughly speaking, $P(\tau > n)$ goes to zero at the rate $\frac{1}{\sqrt{n}}$ and if the random walk still has not reached positive territory by time n, then it has traveled to some distance roughly of the order of $-\sqrt{n}$. These two now exactly balance out, so that $E(S_{\min(\tau,n)})I_{\{\tau>n\}}$ does not go to zero. This causes the assumption $\lim_{n \to \infty} E(S_{\min(\tau,n)}) = E(S_\tau) = 1$ to fail.

Thus, our search for sufficient conditions in the optional stopping theorem should be directed at finding nice enough conditions that ensure that the stopping time τ cannot get too large with a high probability. The next two theorems provide such a set of aesthetically attractive sufficient conditions. It is not hard to prove these two theorems. We refer the reader to Fristedt and Gray (1997, Chapter 24) for proofs of these two theorems.

Theorem 14.4. *Suppose $\{X_n, n \geq 0\}$ is a martingale, adapted to itself, and τ a stopping time adapted to the same sequence. Suppose any one of the following conditions holds.*

(a) *For some $n < \infty$, $P(\tau \leq n) = 1$.*
(b) *For some nonnegative random variable Z with $E(Z) < \infty$, the martingale sequence $\{X_n\}$ satisfies $|X_n| \leq Z$ for all $n \geq 0$.*

14.2 Stopping Times and Optional Stopping

(c) *For some positive and finite* c, $|X_{n+1} - X_n| \leq c$ *for all* $n \geq 0$, $E(|X_0|) < \infty$, *and* $E(\tau) < \infty$.

(d) *For some finite constant* c, $E(X_n^2) \leq c$ *for all* $n \geq 0$.

Then $E(X_\tau) = E(X_0)$.

Remark. It is important to keep in mind that none of these four conditions is necessary for the equality $E(X_\tau) = E(X_0)$ to hold. We recall from our discussion of *uniform integrability* in Chapter 7 that conditions (b) and (d) in Theorem 14.4 each imply that the sequence $\{X_n\}$ is uniformly integrable. In fact, it may be shown that under the *weaker condition* that our martingale sequence $\{X_n\}$ is uniformly integrable, the equality $E(X_\tau) = E(X_0)$ holds. The important role played by uniform integrability in martingale theory reappears when we discuss convergence of martingales.

An important case where the equality holds with essentially the minimum requirements is the special case of a random walk. We precisely describe this immediately below. The point is that the four sufficient conditions are all-purpose conditions. But if the martingale has a special structure, then the conditions can sometimes be weakened. Here is such a result for a special martingale, namely the random walk.

Theorem 14.5. Let $Z_1, Z_2, \ldots$ be an iid sequence such that $E(|Z_1|) < \infty$. Let $S_n = \sum_{i=1}^n Z_i$, $n \geq 1$. Let τ be any stopping time adapted to $\{S_n\}$ such that $E(\tau) < \infty$. Consider the martingale sequence $X_n = S_n - n\mu$, $n \geq 1$, where $\mu = E(Z_1)$. Then the equality $E(X_\tau) = E(X_1) = 0$ holds.

Remark. The special structure of the random walk martingale allows us to conclude the assertion of the optional stopping theorem, without requiring the bounded increments condition $|X_{n+1} - X_n| \leq c$, which was included in the all-purpose sufficient condition in Theorem 14.4.

Example 14.14 (Weighted Rademacher Series). Let $X_1, X_2, \ldots$ be a sequence of iid *Rademacher variables* with common distribution $P(X_i = \pm 1) = \frac{1}{2}$. For $n \geq 1$, let $S_n = \sum_{i=1}^n \frac{X_i}{i^\alpha}$, where $\alpha > \frac{1}{2}$. Because X_i are independent and $E(\frac{X_i}{i^\alpha}) = 0$ for all i, S_n forms a martingale sequence (see Example 14.2). On the other hand,

$$E(S_n^2) = \text{Var}(S_n) = \sum_{i=1}^n \frac{\text{Var}(X_i)}{i^{2\alpha}}$$

$$= \sum_{i=1}^n \frac{1}{i^{2\alpha}} \leq \sum_{i=1}^\infty \frac{1}{i^{2\alpha}} = \zeta(2\alpha) < \infty,$$

where $\zeta(z)$ is the Riemann zeta function $\zeta(z) = \sum_{n=1}^\infty \frac{1}{n^z}$, $z > 1$. Therefore, if $\alpha > \frac{1}{2}$, $E(S_n^2) \leq c = \zeta(2\alpha)$ for all n, and hence by our theorem above, $E(S_\tau) = 0$ holds for any stopping time τ adapted to $\{S_n\}$.

Example 14.15 (The Simple Random Walk). Consider the one-dimensional random walk with iid steps X_i, having the common distribution $P(X_i = 1) = p, P(X_i = -1) = q, 0 < p < 1, p + q = 1$. Then, $E(X_i) = p - q = \mu$ (say), and $S_n - n\mu$, where $S_n = \sum_{i=1}^{n} X_i$, is a martingale. We also have, for any n,

$$|S_{n+1} - (n+1)\mu - (S_n - n\mu)| = |X_{n+1} - \mu| \leq 2.$$

Furthermore, $E(|S_1 - \mu|)$ is clearly finite. Therefore, for any stopping time τ with a finite expectation, by using our theorem above, the equality $E(S_\tau - \mu\tau) = 0$, or equivalently, $E(S_\tau) = \mu E(\tau)$ holds. Recall from Chapter 11 that this is a special case of *Wald's identity*. Wald's identity is revisited in the next section.

14.2.4 Applications of Optional Stopping

We provide a few applications of the optional stopping theorem. The optional stopping theorem also has important applications to martingale inequalities, which is our topic in the next section.

Perhaps the two best general applications of the optional stopping theorem are two identities, known as *Wald identities*. Of these, the first Wald identity is already known to us; see Chapter 11. We connect that identity to martingale theory and present a second identity, which was not presented in Chapter 11.

Theorem 14.6 (Wald's First and Second Identity). *Let $X_1, X_2, \ldots$ be a sequence of iid random variables, defined on a common sample space Ω. Let $S_n = \sum_{i=1}^{n} X_i, n \geq 1$. Let τ be a stopping time adapted to the sequence $\{S_n\}$ and suppose that $E(\tau) < \infty$.*

(a) *Suppose $E(|X_1|) < \infty$ and $E(X_1) = \mu$ (which need not be zero). Then $E(S_\tau) = \mu(E\tau)$.*
(b) *Suppose $E(X_1) = 0, E(X_1^2) = \sigma^2 < \infty$. Then $\text{Var}(S_\tau) = \sigma^2(E\tau)$.*

Proof. Both parts of this theorem follow from Theorem 14.5. For part (a), apply Theorem 14.5 to the martingale sequence $S_n - n\mu$ to conclude that $E(S_\tau - \tau\mu) = 0 \Rightarrow E(S_\tau) = \mu(E\tau)$. For part (b), because $\mu = E(X_1)$ has now been assumed to be zero, by applying part (a) of this theorem,

$$\text{Var}(S_\tau) = E(S_\tau - E[S_\tau])^2 = E(S_\tau - 0)^2 = E(S_\tau^2).$$

Next note that because the X_i are independent,

$$\text{Var}(S_{n+1}|S_1, \ldots, S_n) = \text{Var}(X_{n+1}) = \sigma^2$$
$$\Rightarrow E(S_{n+1}^2|S_1, \ldots, S_n) = S_n^2 + \sigma^2$$
$$\Rightarrow E(S_{n+1}^2 - (n+1)\sigma^2|S_1, \ldots, S_n) = S_n^2 - n\sigma^2;$$

14.2 Stopping Times and Optional Stopping

that is, $S_n^2 - n\sigma^2$ is a martingale sequence adapted to the S_n sequence. From here, it follows that $E(S_\tau^2 - \tau\sigma^2) = E(S_1^2 - \sigma^2) = 0$, which means

$$\mathrm{Var}(S_\tau) = E(S_\tau^2) = \sigma^2(E\tau),$$

which is what part (b) says. □

Example 14.16 (Expected Hitting Times for a Random Walk). The Wald identity may be used to evaluate the expected hitting time of a given level by a random walk. Specifically, let S_n be the one-dimensional simple symmetric random walk with the iid steps having the common distribution $P(X_i = \pm 1) = \frac{1}{2}$. Let x be any given positive integer and consider the first passage time

$$\tau_x = \inf\{n > 0 : S_n = x\}.$$

We know from general random walk theory (Chapter 11) that $P(\tau_x < \infty) = 1$. Also, obviously $E(|X_1|) = 1 < \infty$, and $\mu = E(X_1) = 0$. Therefore, if $E(\tau_x)$ is finite, Wald's identity $E(S_{\tau_x}) = \mu E(\tau_x)$ will hold. However, $S_{\tau_x} = x$ with probability one, and hence, $E(S_{\tau_x}) = x$. It follows that the equality $x = 0 \times E(\tau_x)$ cannot hold for any finite value of $E(\tau_x)$. In other words, for any positive x, the expected hitting time of x must be infinite for the simple symmetric random walk. The same argument also works for negative x.

Example 14.17 (Gambler's Ruin). Now let us revisit the so-called *gambler's ruin problem*, wherein the gambler quits when he either goes broke, or attains a prespecified amount of fortune (see Chapter 10). In other words, the gambler waits for the random walk S_n to hit one of two integers $0, b, b > 0$. Suppose $a < b$ is the amount of money with which the gambler walked in, so that the gambler's sequence of fortunes is the random walk $S_n = \sum_{i=1}^n X_i + S_0$, where $S_0 = a$, and the steps are still iid with $P(X_i = \pm 1) = \frac{1}{2}$. Formally, let

$$\tau = \tau_{\{a,b\}} = \inf\{n > 0 : S_n \in \{0, b\}\}.$$

By applying the optional stopping theorem,

$$E(S_\tau) = 0 \times P(S_\tau = 0) + b[1 - P(S_\tau = 0)] = E(S_0) = a;$$

note that we have implicitly assumed the validity of the optional stopping theorem in the last step (which is true in this example; why?). Rearranging terms, we deduce that $P(S_\tau = 0) = \frac{b-a}{b}$, or equivalently, $P(S_\tau = b) = \frac{a}{b}$.

Example 14.18 (Generalized Wald Identity). The two identities of Wald given above assume only the existence of the first and the second moment of X_i, respectively. If we make the stronger assumption that the X_i have a finite mgf, then a more embracing martingale identity can be proved, from which the two Wald identities given above fall out as special cases. This generalized Wald identity is presented in this example.

The basic idea is the same as before, which is to think of a suitable martingale, and apply the optional stopping theorem to it. Suppose then that $X_1, X_2, \ldots$ is an iid sequence, with the mgf $\psi(t) = E(e^{tX_i})$, which we assume to be finite in some nonempty interval containing zero. The martingale that works for our purpose in this example is

$$Z_n = [\psi(t)]^{-n} e^{tS_n}, \quad n \geq 0,$$

where, as usual, $S_n = \sum_{i=1}^{n} X_i$, and we take $S_0 = 0$. The number t is fixed, and is often cleverly chosen in specific applications.

The special normal case of this martingale was seen in Example 14.4. Exactly the same proof works in order to show that Z_n as defined above is a martingale in general, not just the normal case. Formally, therefore, whenever we have a stopping time τ such that the optional stopping theorem is valid for this martingale sequence Z_n, we have the identity

$$E(Z_\tau) = E[(\psi(t))^{-\tau} e^{tS_\tau}] = E(Z_0) = 1.$$

Once we have this general identity, we can manipulate it for special stopping times τ to make useful conclusions in specific applications.

Example 14.19 (Error Probabilities of Wald's SPRT). As a specific application of historical importance in statistics, consider again the example of Wald's SPRT (Example 14.11). The setup is that we are acquiring iid observations $Z_1, Z_2, \ldots$ from a parametric family of densities $f(x|\theta)$, and we have to decide between the two hypotheses $H_0 : \theta = \theta_0$ (the null hypothesis), and $H_1 : \theta = \theta_1$ (the alternative hypothesis). As was explained in Example 14.11, we continue sampling as long as $l < S_n < u$ for some $l, u, l < u$, and stop and decide in favor of H_1 or H_0 when for the first time $S_n \geq u$ or $S_n \leq l$; here, S_n is the log likelihood ratio

$$S_n = \log \Lambda_n = \log \frac{\prod_{i=1}^{n} f(Z_i|\theta_1)}{\prod_{i=1}^{n} f(Z_i|\theta_0)}$$
$$= \sum_{i=1}^{n} \log \frac{f(Z_i|\theta_1)}{f(Z_i|\theta_0)} = \sum_{i=1}^{n} X_i \text{ say.}$$

Therefore, in this particular case, the relevant stopping time is

$$\tau = \inf\{n > 0 : S_n \notin (l, u)\}.$$

The *type I error probability* of our test is the probability that the test would reject H_0 if H_0 happened to be true. Denoting the type I error probability as α, we have $\alpha = P_{\theta=\theta_0}(S_\tau \geq u)$. We use Wald's generalized identity to approximate α. Exact calculation of α is practically impossible except in stray cases.

To proceed with this approximation, suppose there is a number $t \neq 0$ such that $E_{\theta=\theta_0}(e^{tX_i}) = 1$. In our notation for the generalized Wald identity, this makes $\psi(t) = 1$ for this judiciously chosen t. If we now make the assumption (of some faith) that when S_n leaves the interval (l, u), it does not overshoot the limits l, u by too much, we should have

$$S_\tau \approx u I_{\{S_\tau \geq u\}} + l I_{\{S_\tau \leq l\}}.$$

Therefore, by applying Wald's generalized identity,

$$1 = E_{\theta=\theta_0}(e^{tS_\tau}) \approx e^{tu}\alpha + e^{tl}(1 - \alpha)$$
$$\Rightarrow \alpha \approx \frac{1 - e^{tl}}{e^{tu} - e^{tl}}.$$

This is the classic *Wald approximation to the type I error probability of the SPRT (sequential probability ratio test)*. A similar approximation exists for the type II error probability of the SPRT, which is the probability that the test will accept H_0 if H_0 happens to be false.

14.3 Martingale and Concentration Inequalities

The optional stopping theorem is also the main tool in proving a collection of important inequalities involving martingales. To provide a little context for such inequalities, consider the special martingale of a random walk, namely $S_n = \sum_{i=1}^n X_i, n \geq 1$, where we assume the X_i to be iid mean zero random variables with a finite variance σ^2. If we take any fixed n, and any fixed $\lambda > 0$, then simply by Chebyshev's inequality, $P(|S_n| \geq \lambda) \leq \frac{E(S_n^2)}{\lambda^2}$. Kolmogorov's inequality (see Chapter 8) makes the stronger assertion $P(\max_{1 \leq k \leq n} |S_k| \geq \lambda) \leq \frac{E(S_n^2)}{\lambda^2}$. A fundamental inequality in martingale theory says that such an inequality holds for more general martingales, and not just the special martingale of a random walk.

14.3.1 Maximal Inequality

Theorem 14.7 (Martingale Maximal Inequality).

(a) Let $\{X_n, n \geq 0\}$ be a nonnegative submartingale adapted to some sequence $\{Y_n\}$, and λ any fixed positive number. Then, for any $n \geq 0$,

$$P\left(\max_{0 \leq k \leq n} X_k \geq \lambda\right) \leq \frac{E(X_n)}{\lambda}.$$

(b) Let $\{X_n, n \geq 0\}$ be a martingale adapted to some sequence $\{Y_n\}$, and λ any fixed positive number. Suppose $p \geq 1$ is such that $E(|X_k|^p) < \infty$ for any $k \geq 0$. Then, for any $n \geq 0$,

$$P\left(\max_{0 \leq k \leq n} |X_k| \geq \lambda\right) \leq \frac{E\left(|X_n|^p I_{\{\max_{0 \leq k \leq n} |X_k| \geq \lambda\}}\right)}{\lambda^p} \leq \frac{E(|X_n|^p)}{\lambda^p}.$$

Proof. Note that the final inequality in part (b) follows from part (a) by use of Theorem 14.1 because $f(z) = |z|^p$ is a nonnegative convex function, and therefore if $\{X_n\}$ is a martingale adapted to some sequence $\{Y_n\}$, then for $p \geq 1$, $\{|X_n|^p\}$ is a nonnegative submartingale (adapted to that same sequence $\{Y_n\}$). The first inequality in part (b) is proved by partitioning the event $\{\max_{0 \leq k \leq n} |X_k| \geq \lambda\}$ into disjoint events of the form $\{|X_0| < \lambda, \ldots |X_i| < \lambda, |X_{i+1}| \geq \lambda\}$, and then using simple bounds on each of these partitioning sets. This is left as an exercise.

For proving part (a) of this theorem, define the stopping time

$$\tau = \inf\{k \geq 0 : X_k > \lambda\},$$

and $\tau_n = \min(\tau, n)$.

Then, by the optional stopping theorem,

$$E(X_n) \geq E(X_{\tau_n}) = E\left(X_{\tau_n} I_{\{\max_{0 \leq k \leq n} X_k \geq \lambda\}}\right) + E\left(X_{\tau_n} I_{\{\max_{0 \leq k \leq n} X_k < \lambda\}}\right)$$

$$\geq E\left(X_{\tau_n} I_{\{\max_{0 \leq k \leq n} X_k \geq \lambda\}}\right)$$

(since the $\{X_n\}$ sequence has been assumed to be nonnegative)

$$\geq \lambda E\left[I_{\{\max_{0 \leq k \leq n} X_k \geq \lambda\}}\right] = \lambda P\left(\max_{0 \leq k \leq n} X_k \geq \lambda\right),$$

which is what part (a) of this theorem says. □

Part (a) of the theorem above assumes the submartingale $\{X_n\}$ to be nonnegative. This assumption is in fact not needed. In addition, the inequality itself can be somewhat strengthened. The following improved version of the maximal inequality can be proved by minor modifications of the argument given above; we record the stronger version, which is important for applications.

Theorem 14.8 (A Better Maximal Inequality). *Let $\{X_n, n \geq 0\}$ be a submartingale adapted to some sequence $\{Y_n\}$, and λ any fixed positive number. Then, for any $n \geq 0$,*

$$P\left(\max_{0 \leq k \leq n} X_k \geq \lambda\right) \leq \frac{E(X_n^+)}{\lambda} \leq \frac{E(|X_n|)}{\lambda},$$

where for any real number x, $x^+ = \max(x, 0) \leq |x|$.

14.3 Martingale and Concentration Inequalities

Example 14.20 (Sharper Bounds Near Zero). The bounds in Theorem 14.7 and Theorem 14.8 are not useful unless λ is large, because the upper bounds blow up as $\lambda \to 0$. However, if we work a little harder, then useful bounds can be derived at least in some cases even when λ is near zero. This example illustrates such a calculation.

Let $\{X_n\}$ be a zero mean martingale, and suppose $\sigma_k^2 = \text{Var}(X_k) < \infty$ for all k. For $n \geq 0$, denote $M_n = \max_{0 \leq k \leq n} X_k$. Fix a constant $c > 0$; the constant c is chosen later suitably. By Theorem 14.1, $\{(X_k + c)^2\}$ is a submartingale, and therefore, by Theorem 14.8,

$$P(M_n \geq \lambda) = P(M_n + c \geq \lambda + c) = P\left(\max_{0 \leq k \leq n} (X_k + c) \geq \lambda + c\right)$$

$$\leq \frac{E(X_n + c)^2}{(\lambda + c)^2} = \frac{c^2 + \sigma_n^2}{c^2 + 2c\lambda + \lambda^2}.$$

Therefore,

$$P(M_n \geq \lambda) \leq \inf_{c > 0} \frac{c^2 + \sigma_n^2}{c^2 + 2c\lambda + \lambda^2}.$$

The function $\frac{c^2 + \sigma_n^2}{c^2 + 2c\lambda + \lambda^2}$ is uniquely minimized at the root of the derivative equation

$$\frac{c}{c^2 + \sigma_n^2} - \frac{c + \lambda}{c^2 + 2c\lambda + \lambda^2} = 0$$

$$\Leftrightarrow c^2\lambda + c(\lambda^2 - \sigma_n^2) - \lambda\sigma_n^2 = 0 \Leftrightarrow c = \frac{\sigma_n^2}{\lambda}.$$

Plugging this value of c, we get

$$P(M_n \geq \lambda) \leq \inf_{c > 0} \frac{c^2 + \sigma_n^2}{c^2 + 2c\lambda + \lambda^2}$$

$$= \frac{\sigma_n^2}{\lambda^2 + \sigma_n^2},$$

for any $\lambda > 0$. Clearly, this bound is strictly smaller than one for any $\lambda > 0$.

Example 14.21 (Bounds on the Moments of the Maximum). Here is a clever application of Theorem 14.7 to bounding the moments of $M_n = \max_{0 \leq k \leq n} |X_k|$ in terms of the same moment of $|X_n|$ for a martingale sequence $\{X_n\}$. The example is a very nice illustration of the art of putting simple things together to get a pretty end result.

Suppose that $\{X_n, n \geq 0\}$ is a martingale sequence, and $p > 1$ is such that $E(|X_k|^p) < \infty$ for every k. The proof of the result in this example makes use of Holder's inequality $E(|XY|) \leq (E|X|^\alpha)^{1/\alpha}(E|Y|^\beta)^{1/\beta}$, where $\alpha, \beta > 1$, and $\beta = \frac{\alpha}{\alpha - 1}$ (see Chapter 1).

Proceeding,

$$E(M_n^p) = \int_0^\infty p\lambda^{p-1} P(M_n > \lambda) d\lambda$$
$$\leq \int_0^\infty p\lambda^{p-1} \frac{E\left(|X_n|I_{\{M_n \geq \lambda\}}\right)}{\lambda} d\lambda$$

(by using part (b) of Theorem 14.7)

$$= \int_0^\infty p\lambda^{p-2} E\left(|X_n|I_{\{M_n \geq \lambda\}}\right) d\lambda = E\left[p|X_n|\left(\int_0^{M_n} \lambda^{p-2} d\lambda\right)\right]$$

(by Fubini's theorem)

$$= E\left[p|X_n|\frac{M_n^{p-1}}{p-1}\right] = \frac{p}{p-1} E\left(|X_n|M_n^{p-1}\right)$$
$$\leq \frac{p}{p-1}[E|X_n|^p]^{1/p}[E(M_n^p)]^{(p-1)/p}$$

(by using Holder's inequality with $\alpha = p, \beta = \frac{p}{p-1}$).

Transferring $[E(M_n^p)]^{(p-1)/p}$ to the left side,

$$[E(M_n^p)]^{1/p} \leq \frac{p}{p-1}[E|X_n|^p]^{1/p}.$$

In particular, for a square integrable martingale, by using $p = 2$ in the inequality we just derived,

$$[E(M_n^2)]^{1/2} \leq 2[E(X_n^2)]^{1/2} \Rightarrow E(M_n^2) \leq 4E(X_n^2),$$

a very pretty and useful inequality.

14.3.2 * Inequalities of Burkholder, Davis, and Gundy

The previous two examples indicated applications of various versions of the maximal inequality to obtaining bounds on the moments of the maximum $M_n = \max_{0 \leq k \leq n} |X_k|$ for a martingale sequence $\{X_n\}$. The maximal inequality tells us how to obtain bounds on the moments from bounds on the tail probability. In particular, if the martingale is square integrable, that is, if $E(X_k^2) < \infty$ for any k, then the maximal inequality leads to a bound on the second moment of M_n in terms of the second moment of the last term, namely $E(X_n^2)$.

14.3 Martingale and Concentration Inequalities

There is a useful connection between $E(X_n^2)$ and $E(D_n^2)$ for a general square integrable martingale $\{X_n\}$, where $D_n^2 = \sum_{i=1}^n (X_i - X_{i-1})^2$. The connection, which we prove below, is the neat identity $E(X_n^2) - E(X_0^2) = E(D_n^2)$, so that if $X_0 = 0$, then $E(X_n^2)$ and $E(D_n^2)$ are equal. Therefore, we can think of the maximal inequality and the implied moment bounds in terms of $E(D_n^2)$, because $E(D_n^2)$ and $E(X_n^2)$ are, after all, equal. It was shown in Burkholder (1973), Davis (1970), and Burkholder, Davis, and Gundy (1972) that one can bound expectations of far more general functions of M_n in terms of expectations of the same functions of D_n; in particular, one can bound the pth moment of M_n from both directions by multiples of the pth moment of D_n for general $p \geq 1$. In some sense, the moments of M_n and the moments of D_n grow in the same order; if one can control the increments of the martingale sequence, then one can control the maximum. Three such important bounds are presented in this section for reference and completeness. But first, we demonstrate the promised connection between $E(X_n^2)$ and $E(D_n^2)$, an interesting result in its own right.

Proposition. *Suppose $\{X_n, n \geq 0\}$ is a martingale. Let $V_i = X_i - X_{i-1}, i \geq 1$, and $D_n^2 = \sum_{i=1}^n V_i^2$. Suppose $E(X_k^2) < \infty$ for each $k \geq 0$. Then, for any $n \geq 1$,*

$$E(D_n^2) = E(X_n^2) - E(X_0^2).$$

Proof.

$$E(D_n^2) = \sum_{i=1}^n E[(X_i - X_{i-1})^2] = \sum_{i=1}^n E[X_i(X_i - X_{i-1}) - X_{i-1}(X_i - X_{i-1})]$$

$$= \sum_{i=1}^n E(E[X_i(X_i - X_{i-1}) | X_0, \ldots, X_{i-1}])$$

$$- \sum_{i=1}^n E(E[X_{i-1}(X_i - X_{i-1}) | X_0, \ldots, X_{i-1}])$$

$$= \sum_{i=1}^n \{E(E[X_i^2 | X_0, \ldots, X_{i-1}]) - E(X_{i-1}E[X_i | X_0, \ldots, X_{i-1}])\}$$

$$- \sum_{i=1}^n E(X_{i-1}E[X_i | X_0, \ldots, X_{i-1}] - X_{i-1}^2)$$

$$= \sum_{i=1}^n \{E(X_i^2) - E(X_{i-1}^2)\} - \sum_{i=1}^n E(X_{i-1}^2 - X_{i-1}^2)$$

$$= E(X_n^2) - E(X_0^2). \qquad \square$$

Remark. In view of this result, we can restate part (b) of Theorem 14.7 for the case $p = 2$ in the following manner.

Theorem 14.9. *Let $\{X_n, n \geq 0\}$ be a martingale such that $X_0 = 0$ and $E(X_k^2) < \infty$ for all $k \geq 1$. Let λ be any fixed positive number, and for any $n \geq 1$, $M_n = \max_{0 \leq k \leq n} |X_k|$. Then,*

$$P(M_n \geq \lambda) \leq \frac{E(D_n^2)}{\lambda^2}.$$

The inequalities of Burkholder, Davis, and Gundy show how to establish bounds on moments of M_n in terms of the same moments of D_n. To describe some of these bounds, we first need a little notation.

Given a real-valued random variable X, and a positive number p, the L_p norm of X is defined as $\|X\|_p = [E(|X|^p)]^{\frac{1}{p}}$, assuming that $E(|X|^p) < \infty$. Obviously, if X is already a nonnegative random variable, then $\|X\|_p = [E(X^p)]^{\frac{1}{p}}$. Here are two specific bounds on the L_p norms of M_n in terms of the L_p norms of D_n. Of these, the case $p > 1$ was considered in works of Donald Burkholder (e.g., Burkholder (1973)); the case $p = 1$ needed a separate treatment, and was dealt with in Davis (1970).

Theorem 14.10. *(a) Suppose $\{X_n, n \geq 0\}$ is a martingale, with $X_0 = 0$. Suppose for some given $p > 1$, $\|X_k\|_p < \infty$ for all $k \geq 1$. Then, for any $n \geq 1$,*

$$\frac{p-1}{18p^{3/2}}\|D_n\|_p \leq \|M_n\|_p \leq \frac{18p^{3/2}}{(p-1)^{3/2}}\|D_n\|_p.$$

(b) There exist universal positive constants c_1, c_2 such that

$$c_1\|D_n\|_1 \leq \|M_n\|_1 \leq c_2\|D_n\|_1.$$

Moreover, the constant c_2 may be taken to be $\sqrt{3}$.

For $p \geq 1$, the functions $x \to |x|^p$ are convex. It was shown in Burkholder, Davis, and Gundy (1972) that bounds of the same nature as in the theorem above hold for general convex functions. The exact result says the following.

Theorem 14.11. *Suppose $\{X_n, n \geq 0\}$ is a martingale with $X_0 = 0$ and $\phi : \mathcal{R} \to \mathcal{R}$ a convex function. Then there exist universal positive constants $c_\phi, C_\phi, c_\phi \leq C_\phi$, depending only on the function ϕ, such that for any $n \geq 1$,*

$$c_\phi E(\phi(D_n)) \leq E(\phi(M_n)) \leq C_\phi E(\phi(D_n)).$$

Remark. Note that apart from the explicit constants, both parts of Theorem 14.10 follow as special cases of this theorem. To our knowledge, no explicit choices of c_ϕ, C_ϕ are known.

14.3.3 Inequalities of Hoeffding and Azuma

The classical inequality of Hoeffding (Hoeffding (1963); see Chapter 8) gives bounds on the probability of a large deviation of a partial sum of bounded iid random variables from its mean value. The message of that inequality is that if the iid summands can be controlled, then the deviations of the sum from its mean can be controlled. Inequalities on probabilities of the form $P(|f(X_1, X_2, \ldots, X_n) - E(f(X_1, X_2, \ldots, X_n))| > t)$ are called *concentration inequalities*. An equally classic concentration inequality of K. Azuma (Azuma (1967)) shows that a Hoeffding type inequality holds for a martingale sequence, provided that the increments $X_k - X_{k-1}$ vary in bounded intervals. The analogy between the iid case and the martingale case is then clear. In the iid case, we can control $S_n = \sum_{i=1}^{n} X_i$ if we can control the summands X_i; in the martingale case, we can control $X_n - X_0 = \sum_{i=1}^{n}(X_i - X_{i-1})$ if we can control the summands $X_i - X_{i-1}$. Here is Azuma's inequality in its classic form; a more general form is given afterwards.

Theorem 14.12 (Azuma's Inequality). *Suppose $\{X_n, n \geq 0\}$ is a martingale such that $V_i = |X_i - X_{i-1}| \leq c_i$, where c_i are positive constants. Then, for any positive number t and any $n \geq 1$,*

(a) $P(X_n - X_0 \geq t) \leq e^{-\frac{t^2}{2\sum_{i=1}^{n} c_i^2}}$.

(b) $P(X_n - X_0 \leq -t) \leq e^{-\frac{t^2}{2\sum_{i=1}^{n} c_i^2}}$.

(c) $P(|X_n - X_0| \geq t) \leq 2e^{-\frac{t^2}{2\sum_{i=1}^{n} c_i^2}}$.

The proof of part (b) is exactly the same as that of part (a), and part (c) is an immediate consequence of parts (a) and (b). So only part (a) requires a proof. For this, we need a classic convexity lemma, originally used in Hoeffding (1963), and then a generalized version of it. Here is the first lemma.

(Hoeffding's Lemma). *Let X be a zero mean random variable such that $P(a \leq X \leq b) = 1$, where a, b are finite constants. Then, for any $s > 0$,*

$$E(e^{sX}) \leq e^{\frac{s^2(b-a)^2}{8}}.$$

Remark. It is important to note that the bound in this lemma depends only on $b - a$ and the mean zero assumption, but not on the individual values of a, b.

Proof of Hoeffding's Lemma. The proof uses convexity of the function $x \to e^{sx}$, and a calculus inequality on the function $\phi(u) = -pu + \log(1 - p + pe^u), u \geq 0$, where p is a fixed number in $(0, 1)$.

First, by the convexity of $x \to e^{sx}$, for $a \le x \le b$,

$$e^{sx} \le \frac{x-a}{b-a} e^{sb} + \frac{b-x}{b-a} e^{sa}.$$

Taking an expectation,

$$E(e^{sX}) \le p e^{sb} + (1-p) e^{sa}. \quad (*)$$

where $p = \frac{-a}{b-a}$; note that p belongs to $[0, 1]$. It now remains to show that $pe^{sb} + (1-p)e^{sa} \le e^{\frac{s^2(b-a)^2}{8}}$. Towards this, write

$$pe^{sb} + (1-p)e^{sa} = e^{sa}\left[1 - p + pe^{s(b-a)}\right] = e^{-sp(b-a)}\left[1 - p + pe^{s(b-a)}\right]$$
$$= e^{-sp(b-a) + \log(1 - p + pe^{s(b-a)})} = e^{-pu + \log(1-p+pe^u)},$$

writing u for $s(b-a)$.

A relatively simple calculus argument shows that the function $\phi(u) = -pu + \log(1 - p + pe^u)$ is bounded above by $\frac{u^2}{8}$ for all $u > 0$. Plugging this bound in $(*)$ results in the bound in the lemma.

(Generalized Hoeffding Lemma). *Let V, Z be two random variables such that*

$$E(V|Z) = 0, \text{ and } P(f(Z) \le V \le f(Z) + c) = 1$$

for some function $f(Z)$ of Z and some positive constant c. Then, for any $s > 0$,

$$E(e^{sV}|Z) \le e^{\frac{s^2 c^2}{8}}.$$

The generalized Hoeffding lemma has the same proof as Hoeffding's lemma itself. Refer to the remark that we made just before the proof of Hoeffding's lemma. It is the generalized Hoeffding lemma that gives us Azuma's inequality.

Proof of Azuma's Inequality. Still using the notation $V_i = X_i - X_{i-1}$, then, with $s > 0$,

$$P(X_n - X_0 \ge t) = P\left(e^{s(X_n - X_0)} \ge e^{st}\right) \le e^{-st} E\left(e^{s(X_n - X_0)}\right)$$
$$= e^{-st} E\left(e^{s \sum_{i=1}^n V_i}\right) = e^{-st} E\left(e^{s \sum_{i=1}^{n-1} V_i + sV_n}\right)$$
$$= e^{-st} E\left(e^{s \sum_{i=1}^{n-1} V_i} E\left[e^{sV_n} | X_0, \ldots, X_{n-1}\right]\right)$$
$$\le e^{-st} E\left(e^{s \sum_{i=1}^{n-1} V_i} e^{\frac{s^2(2c_n)^2}{8}}\right)$$

(because $E(V_n \mid X_0, \ldots, X_{n-1}) = 0$ by the martingale property of $\{X_n\}$, and then by applying the generalized Hoeffding lemma)

$$= e^{-st} e^{\frac{s^2 c_n^2}{2}} E\left(e^{s \sum_{i=1}^{n-1} V_i}\right) \leq e^{-st} e^{\frac{s^2 \sum_{i=1}^{n} c_i^2}{2}},$$

by repeating the same argument.

This latest inequality is true for any $s > 0$. Therefore, by minimizing the bound over $s > 0$,

$$P(X_n - X_0 \geq t) \leq \inf_{s>0} e^{-st} e^{\frac{s^2 \sum_{i=1}^{n} c_i^2}{2}} = e^{-\frac{t^2}{2 \sum_{i=1}^{n} c_i^2}},$$

where the infimum over s is easily established by a simple calculus argument. This proves Azuma's inequality. □

14.3.4 * Inequalities of McDiarmid and Devroye

McDiarmid (1989) and Devroye (1991) use novel martingale techniques to derive concentration inequalities and variance bounds for potentially complicated functions of independent random variables. The only requirement is that the function should not change by arbitrarily large amounts if all but one of the coordinates remain fixed. The first result below says that functions of certain types are concentrated near their mean value with a high probability.

Theorem 14.13. *Suppose $X_1, \ldots, X_n$ are independent random variables, and $f(x_1, \ldots, x_n)$ is a function such that for each $i, 1 \leq i \leq n$, there exist finite constant $c_i = c_{i,n}$ such that*

$$|f(x_1, \ldots, x_{i-1}, x_i, x_{i+1}, \ldots, x_n) - f(x_1, \ldots, x_{i-1}, x_i', x_{i+1}, \ldots, x_n)| \leq c_i$$

for all $x_1, \ldots, x_i, x_i', \ldots, x_n$. Let t be any positive number. Then,

(a) $P(f - E(f) \geq t) \leq e^{-\frac{2t^2}{\sum_{i=1}^{n} c_i^2}}$.

(b) $P(f - E(f) \leq -t) \leq e^{-\frac{2t^2}{\sum_{i=1}^{n} c_i^2}}$.

(c) $P(|f - E(f)| \geq t) \leq 2e^{-\frac{2t^2}{\sum_{i=1}^{n} c_i^2}}$.

Proof. Once again, only part (a) is proved, because (b) is proved exactly analogously, and (c) follows by adding the inequalities in (a) and (b). For notational convenience, we take $E(f)$ to be zero; this allows us to write f in place of $f - E(f)$ below.

The trick is to decompose f as $f = \sum_{k=1}^{n} V_k$, where $\{V_k\}$ is a martingale difference sequence such that it can be bounded in both directions, $Z_k \leq V_k \leq W_k$, in a manner so that $W_k - Z_k \leq c_k, k = 1, 2, \ldots, n$. Then, Azuma's inequality applies and the inequality of this theorem falls out. Construct the random variables V_k, Z_k, W_k as follows.

Define

$$\eta(x_1, \ldots, x_k) = E[f(X_1, \ldots, X_n) | X_1 = x_1, \ldots, X_k = x_k];$$

$$V_k = \eta(X_1, \ldots, X_k) - \eta(X_1, \ldots, X_{k-1}) \text{ for } k \geq 2, \text{ and } V_1 = \eta(X_1);$$

$$Z_k = \inf_{x_k} \eta(X_1, \ldots, X_{k-1}, x_k) - \eta(X_1, \ldots, X_{k-1}) \text{ for } k \geq 2,$$

and $Z_1 = \inf_{x_1} \eta(x_1)$;

$$W_k = \sup_{x_k} \eta(X_1, \ldots, X_{k-1}, x_k) - \eta(X_1, \ldots, X_{k-1}) \text{ for } k \geq 2,$$

and $W_1 = \sup_{x_1} \eta(x_1)$.

Now observe the following facts.

(a) By construction, $Z_k \leq V_k \leq W_k$ for each k.
(b) By hypothesis, $W_k - Z_k \leq c_k$ for each k.
(c) $f(X_1, \ldots, X_n) = \sum_{k=1}^{n} V_k$.
(d) $\{V_k\}_1^n$ forms a martingale difference sequence.

Therefore, we can once again apply the generalized Hoeffding lemma and simply repeat the proof of Azuma's inequality to obtain the inequality in part (a) of this theorem. □

An interesting feature of McDiarmid's inequality is that martingale methods were used to derive a probability inequality involving independent random variables. It turns out that martingale methods may also be used to derive variance bounds for functions of independent random variables. The following variance bound is taken from Devroye (1991).

Theorem 14.14. *Suppose $X_1, \ldots, X_n$ are independent random variables and $f(x_1, \ldots, x_n)$ is a function that satisfies the conditions of Theorem 14.13. Then,*

$$\operatorname{Var}(f(X_1, \ldots, X_n)) \leq \frac{\sum_{i=1}^{n} c_i^2}{4}.$$

Proof. We use the same notation as in the proof of Theorem 14.13. The proof consists of showing $\operatorname{Var}(f) = E(\sum_{i=1}^{n} V_i^2)$ and $E(V_i^2) \leq (c_i^2/4)$.

14.3 Martingale and Concentration Inequalities

To prove the first fact, we use the martingale decomposition as in Theorem 14.13 to get

$$\text{Var}(f) = \text{Var}\left(\sum_{i=1}^{n} V_i\right) = E\left[\left(\sum_{i=1}^{n} V_i\right)^2\right]$$

$$= \sum_{i=1}^{n} E[V_i^2] + 2\sum\sum_{i<j} E[V_i V_j]$$

$$= \sum_{i=1}^{n} E[V_i^2] + 2\sum\sum_{i<j} E(V_i E[V_j | X_1, \ldots, X_{j-1}])$$

$$= \sum_{i=1}^{n} E[V_i^2] + 2\sum\sum_{i<j} E(V_i \times 0) = \sum_{i=1}^{n} E[V_i^2].$$

To prove the second fact, we use an extremal property of two-point distributions, namely that the two-point distribution placing probability $\frac{1}{2}$ at each of a, b maximizes the variance among all distributions supported on $[a, b]$, and that this two-point distribution has variance $\frac{(b-a)^2}{4}$. From the proof of Theorem 14.13, $Z_i \leq V_i \leq W_i \leq Z_i + c_i$. Therefore, the conditional variance of V_i given $X_1, \ldots, X_{i-1}$ is at most $\frac{c_i^2}{4}$, and the conditional mean is zero. Putting these two facts together, we get our desired bound $E(V_i^2) \leq \frac{c_i^2}{4}$, which gives the variance bound stated in this theorem. □

The two theorems in this section give useful probability and variance bounds in many complicated problems in which direct evaluation would be essentially impossible.

Example 14.22 (The Kolmogorov–Smirnov Statistic). Suppose $X_1, X_2, \ldots, X_n$ are iid observations from some continuous CDF $F(x)$ on the real line. It is sometimes of interest in statistics to test the hypothesis that $F = F_0$, some specific CDF on the real line. By the Glivenko–Cantelli theorem (see Chapter 7), the empirical CDF F_n converges uniformly to the true CDF with probability one. So a measure of discrepancy of the observed data from the postulated CDF F_0 is $\Delta_n = \sup_x |F_n(x) - F_0(x)|$. The *Kolmogorov–Smirnov statistic* is $D_n = \sqrt{n}\Delta_n$. Exact calculations with D_n are very cumbersome, because of the complicated nature of its distribution for given n. The purpose of this example is to use the inequalities of McDiarmid and Devroye to get useful bounds on its tail probabilities and the variance.

The function f to which we would apply the inequalities of McDiarmid and Devroye is $f(X_1, \ldots, X_n) = \sup_x |F_n(x) - F_0(x)|$. We need to show that if just one data value changes, then the function f cannot change by too large an amount. Indeed, consider two datasets, $\{X_1, \ldots, X_i, \ldots, X_n\}$ and $\{X_1, \ldots, X'_i, \ldots, X_n\}$, where in the second set the X'_i value is different from X_i. Let the corresponding empirical CDFs be F_n, F'_n. Fix an x. The number of observations $\leq x$ in the two datasets can differ by at most one, and therefore $|F_n(x) - F'_n(x)| \leq \frac{1}{n}$. This holds for any x. By the triangular inequality,

$$|\sup_x |F_n(x) - F_0(x)| - \sup_x |F'_n(x) - F_0(x)|| \leq \sup_x |F_n(x) - F'_n(x)| \leq 1/n.$$

Thus, we may use $c_i = c_{i,n} = \frac{1}{n}$ in the inequalities of McDiarmid and Devroye. First, by simply plugging $c_i = \frac{1}{n}$ in Theorem 14.13, we get

$$P(|\Delta_n - E(\Delta_n)| \geq t) \leq 2e^{-2nt^2}$$
$$\Rightarrow P(|D_n - E(D_n)| \geq t) \leq 2e^{-2t^2}.$$

This concentration inequality holds for every fixed n and $t > 0$, and we do not need to deal with the exact distribution of D_n to arrive at this inequality.

Again plugging $c_i = \frac{1}{n}$ in Theorem 14.14, we get

$$\text{Var}(\Delta_n) \leq \frac{1}{4n} \Rightarrow \text{Var}(D_n) \leq \frac{1}{4},$$

for all $n \geq 1$. Once again, this is an attractive variance bound that is valid for every n, and we do not need to work with the exact distribution of D_n to arrive at this bound.

14.3.5 The Upcrossing Inequality

A final key inequality in martingale theory that we present is *Doob's upcrossing inequality*. The inequality is independently useful for studying fluctuations in the trajectory of a martingale (submartingale) sequence. It is also the result we need in the next section for establishing the fundamental convergence theorem for martingales (submartingales).

Given the discrete time process $\{X_n, n \geq 0\}$, fix an integer $N > 0$, and two numbers $a, b, a < b$. We now track the time instants at which this process crosses b from below, or a from above. Formally, let $T_0 = \inf\{k \geq 0 : X_k \leq a\}$. If $X_0 > a$, then this is the first *downcrossing* of a. If $X_0 \leq a$, then $T_0 = 0$. Now we wait for the first *upcrossing* of b after the time T_0. Formally, $T_1 = \inf\{k > T_0 : X_k \geq b\}$. We continue tracking the down and the upcrossings of the two levels a, b in this fashion. Here then is the formal definition for the entire sequence of stopping times T_n:

$$T_0 = \inf\{k \geq 0 : X_k \leq a\};$$
$$T_{2n+1} = \inf\{k > T_{2n} : X_k \geq b\}, \quad n \geq 0;$$
$$T_{2n+2} = \inf\{k > T_{2n+1} : X_k \leq a\}, \quad n \geq 0.$$

14.3 Martingale and Concentration Inequalities

The times $T_1, T_3, \ldots$ are then the instants of upcrossing, and the times $T_0, T_2, \ldots$ are the instants of downcrossing. The upcrossing inequality places a bound on the expected value of $U_{a,b,N}$, the number of upcrossings up to the time N. Note that this is simply the number of odd labels $2n + 1$ for which $T_{2n+1} \leq N$.

Theorem 14.15. *Let $\{X_n, n \geq 0\}$ be a submartingale. Then for any $a, b, N (a < b)$,*

$$E[U_{a,b,N}] \leq \frac{E[(X_N - a)^+] - E[(X_0 - a)^+]}{b - a} \leq \frac{E(|X_N|) + |a|}{b - a}.$$

Proof. The second inequality follows from the first inequality by the pointwise inequality $(x - a)^+ \leq x^+ + |a| \leq |x| + |a|$, and so, we prove only the first inequality.

First make the following reduction. Define a new nonnegative submartingale as $Y_n = (X_n - a)^+, n \geq 0$. This shifting by a is going to result in a useful reduction. There is a functional identity between the upcrossing variable that we are interested in, namely $U_{a,b,N}$ and the number of upcrossings $V_{0,b-a,N}$ of this new process $\{Y_n\}_0^N$ of the two new levels 0 and $b - a$. Indeed, $U_{a,b,N} = V_{0,b-a,N}$. So we need to show that $E[V_{0,b-a,N}] \leq \frac{E(Y_N - Y_0)}{b-a}$.

The key to proving this inequality is to write a clever decomposition

$$Y_N - Y_0 = \sum_{i=0}^{N}(Y_{\tau_i} - Y_{\tau_{i-1}}),$$

such that three things happen:

(i) The τ_i are increasing stopping times, so that the submartingale property is inherited by the Y_{τ_i} sequence.
(ii) The sum over the odd labels in this decomposition satisfy the pointwise inequality

$$\sum_{i:0 \leq i \leq N, i \text{ odd}} (Y_{\tau_i} - Y_{\tau_{i-1}}) \geq (b - a) V_{0,b-a,N}.$$

(iii) The sum over the even labels satisfy the inequality

$$E\left[\sum_{i:0 \leq i \leq N, i \text{ even}} (Y_{\tau_i} - Y_{\tau_{i-1}})\right] \geq 0.$$

If we put (ii) and (iii) together, we immediately get

$$E(Y_N - Y_0) \geq (b - a) E[V_{0,b-a,N}],$$

which is the needed result.

What are these stopping times τ_i, and why are (ii) and (iii) true? The stopping times $\tau_0 \leq \tau_1 \leq \ldots$ are defined in the following way. Analogous to the downcrossing and upcrossing times $T_0, T_1, \ldots$ of (a, b) for the original $\{X_n\}$ process, let

$T'_0, T'_1, \ldots$ be the downcrossing and upcrossing times of $(0, b-a)$ for the new $\{Y_n\}$ process. Now define $\tau_i = \min(T'_i, N)$. The τ_i are increasing, that is, $\tau_0 \leq \tau_1 \leq \ldots$, because the T'_i are. Note that these τ_i are stopping times adapted to $\{Y_n\}$.

Now look at the sum over the odd labels, namely $(Y_{\tau_1} - Y_{\tau_0}) + (Y_{\tau_3} - Y_{\tau_2}) + \cdots$. Break this sum further into two subsets of labels, $i \leq V = V_{0,b-a,N}$, and $i > V$. For each label i in the first subset, $(Y_{\tau_{2i+1}} - Y_{\tau_{2i}}) \geq b - a$, because $Y_{\tau_{2i+1}} \geq b$ and $Y_{\tau_{2i}} \leq a$. Adding over these labels, of which there are V many, we get the sum to be $\geq (b-a)V$. The labels in the other subset can be seen to give a sum ≥ 0 (just think of what V means, and a little thinking shows that the rest of the labels produce a sum ≥ 0). So, now adding over the two subsets of labels, we get our claimed inequality in (ii) above.

The claim in (iii) is automatic by the optional stopping theorem, because for each individual i, we will have $E(Y_{\tau_{i-1}}) \leq E(Y_{\tau_i})$ (actually, this is a slightly stronger demand than what the optional stopping theorem says; but it is true).

As was explained above, this completes the argument for the upcrossing inequality. □

14.4 Convergence of Martingales

14.4.1 The Basic Convergence Theorem

Paul Lévy initiated his study of martingales in his search for laws of large numbers beyond the case of means in the iid case. It turns out that martingales often converge to a limiting random variable, and even convergence of the means or higher moments can be arranged, provided that our martingale sequence is not allowed to fluctuate or grow out of control. To see why some such conditions would be needed, consider the case of the simple symmetric random walk $S_n = \sum_{i=1}^{n} X_i$, where the X_i are iid taking the values ± 1 with probability $\frac{1}{2}$ each. We know that the simple symmetric random walk is recurrent, and therefore it comes back infinitely often to every integer value x with probability one. So S_n, although a martingale, does not converge to some S_∞. The expected value of $|S_n|$ in the simple symmetric random walk case is of the order of $c\sqrt{n}$ for some constant c, and $c\sqrt{n}$ diverges as $n \to \infty$. A famous result in martingale theory says that if we can keep $E(|X_n|)$ in control (i.e., bounded away from ∞), then a martingale sequence $\{X_n\}$ will in fact converge to some suitable X_∞. Furthermore, some such condition is also essentially necessary for the martingale to converge. We start with an example.

Example 14.23 (*Convergence of the Likelihood Ratio*). Consider again the likelihood ratio $\Lambda_n = \prod_{i=1}^{n} \frac{f_1(X_i)}{f_0(X_i)}$, where f_0, f_1 are two densities and the sequence $X_1, X_2, \ldots$ is iid from the density f_0. We have seen that Λ_n is a martingale (see Example 14.8).

The likelihood ratio Λ_n gives a measure of the support in the first n data values for the density f_1. We know f_0 to be the true density from which the data values are coming, therefore we would like the support for f_1 to diminish as more data are accumulated. Mathematically, we would like Λ_n to converge to zero as $n \to \infty$. We recognize that this is therefore a question about convergence of a martingale sequence, because Λ_n, after all, is a martingale if the true density is f_0.

Does Λ_n indeed converge (almost surely) to zero? Indeed, it does, and we can verify it directly, without using any martingale convergence theorems that we have not yet encountered. Here is why we can verify the convergence directly.

Assume that f_0, f_1 are strictly positive for the same set of x values; that is, $\{x : f_1(x) > 0\} = \{x : f_0(x) > 0\}$. Since $u \to \log u$ is a strictly concave function on $(0, \infty)$, by Jensen's inequality,

$$m = E_{f_0}\left[\log \frac{f_1(X)}{f_0(X)}\right] < \log\left(E_{f_0}\left[\frac{f_1(X)}{f_0(X)}\right]\right) = \log 1 = 0.$$

Because $Z_i = \log \frac{f_1(X_i)}{f_0(X_i)}$ are iid with mean m, by the usual SLLN for iid random variables,

$$\frac{1}{n} \log \Lambda_n = \frac{1}{n} \sum_{i=1}^{n} Z_i \overset{a.s.}{\to} m < 0 \Rightarrow \log \Lambda_n \overset{a.s.}{\to} -\infty$$

$$\Rightarrow \Lambda_n \overset{a.s.}{\to} 0.$$

So, in this example, the martingale Λ_n does converge with probability one to a limiting random variable Λ_∞, and Λ_∞ happens to be a constant random variable, equal to zero. We remark that the martingale Λ_n satisfies $E(|\Lambda_n|) = E(\Lambda_n) = 1$ and so, a fortiori, $\sup_n E(|\Lambda_n|) < \infty$. This has something to do with the fact that Λ_n converges in this example, although the random walk, also a martingale, failed to converge. This is borne out by the next theorem, a famous result in martingale theory. The proof of this next theorem requires the use of two basic facts in measure theory, which we state below.

Theorem 14.16 (Fatou's Lemma). *Let $X_n, n \geq 1$ and X be random variables defined on a common sample space Ω. Suppose each X_n is nonnegative with probability one, and suppose $X_n \overset{a.s.}{\to} X$. Then, $\liminf_n E(X_n) \geq E(X)$.*

Theorem 14.17 (Monotone Convergence Theorem). *Let $X_n, n \geq 1$ and X be random variables defined on a common sample space Ω. Suppose each X_n is nonnegative with probability one, that $X_1 \leq X_2 \leq X_3 \leq \ldots$, and $X_n \overset{a.s.}{\to} X$. Then $E(X_n) \uparrow E(X)$.*

Theorem 14.18 (Submartingale Convergence Theorem). *(a) Let $\{X_n\}$ be a submartingale such that $\sup_n E(X_n^+) = c < \infty$. Then there exists a random variable $X = X_\infty$, almost surely finite, such that $X_n \overset{a.s.}{\to} X$.*

(b) Let $\{X_n\}$ be a nonnegative supermartingale, or a nonpositive submartingale. Then there exists a random variable $X = X_\infty$, almost surely finite, such that $X_n \overset{a.s.}{\to} X$.

Proof. The proof uses the upcrossing inequality, the monotone convergence theorem, and Fatou's lemma. The key idea is first to show that under the hypothesis of the theorem, the process $\{X_n\}$ cannot fluctuate indefinitely between two given numbers $a, b, a < b$. Then a standard analytical technique of approximation by rationals, and use of the monotone convergence theorem and Fatou's lemma produces the submartingale convergence theorem. Here are the steps of the proof. Define

$$U_{a,b,N} = \text{Number of upcrossings of (a,b) by } X_0, X_1, \ldots, X_N;$$
$$U_{a,b} = \text{Number of upcrossings of (a,b) by } X_0, X_1, \ldots;$$
$$\Theta_{a,b} = \{\omega \in \Omega : \liminf_n X_n \leq a < b \leq \limsup_n X_n\};$$
$$\Theta = \{\omega \in \Omega : \liminf_n X_n < \limsup_n X_n\}.$$

First, by the monotone convergence theorem, $E[U_{a,b,N}] \to E[U_{a,b}]$ as $N \to \infty$, because $U_{a,b,N}$ converges monotonically to $U_{a,b}$ as $N \to \infty$. Therefore, by the upcrossing inequality,

$$E[U_{a,b,N}] \leq \frac{E(|X_N|) + |a|}{b - a} \Rightarrow E[U_{a,b}] = \lim_N E[U_{a,b,N}]$$
$$\leq \frac{\limsup_N E(|X_N|) + |a|}{b - a} < \infty.$$

This means that $U_{a,b}$ must be finite with probability one (i.e., it cannot equal ∞ with a positive probability).

Next, note that $\Theta \subseteq \bigcup_{\{a<b, a, b \text{ rational}\}} \Theta_{a,b}$, and because we now have that $P(\Theta_{a,b}) = 0$ for any specific pair a, b, $P(\bigcup_{\{a<b, a, b \text{ rational}\}} \Theta_{a,b})$ must also be zero. This then implies that $P(\Theta) = 0$, which establishes the existence of an almost sure limit for the sequence X_n.

However, a subtle point still remains. The limit, X, could be ∞ or $-\infty$ with a positive probability. We use Fatou's lemma to rule out that possibility. Indeed, by Fatou's lemma,

$$E(|X|) \leq \liminf_n E(|X_n|) \leq \sup_n E(|X_n|) < \infty,$$

and so X must be finite with probability one. This finishes the proof of part (a) of the submartingale convergence theorem.

Part (b) is an easy consequence of part (a). For example, if $\{X_n\}$ is a nonpositive submartingale, then

$$\sup_n E(|X_n|) = \sup_n E(-X_n) = -\inf_n E(X_n) = -E(X_1) < \infty,$$

and so convergence of X_n to an almost surely finite X follows from part (a). □

14.4.2 Convergence in L_1 and L_2

The basic convergence theorem that we just proved says that an L_1 bounded submartingale converges to some random variable X. It is a bit disappointing that the apparently strong hypothesis that the submartingale is L_1 bounded is not strong enough to ensure convergence of the expectations: $E(X_n)$ need not converge to $E(X)$ in spite of the L_1 bounded assumption. A slightly stronger control on the growth of the submartingale sequence is needed to ensure convergence of expectations, in addition to the convergence of the submartingale itself. For example, $\sup_n E(|X_n|^p) < \infty$ for some $p > 1$ will suffice. A condition of this sort immediately reminds us of uniform integrability. Indeed, if $\sup_n E(|X_n|^p) < \infty$ for some $p > 1$, then $\{X_n\}$ will be uniformly integrable. It turns out that uniform integrability will be enough to assure us of convergence of the expectations in the basic convergence theorem, and it is almost the minimum that we can get away with. Statisticians are often interested in convergence of variances also. That is a stronger demand, and requires a stronger hypothesis. The next theorem records the conclusions on these issues. For reasons of space, this next theorem is not proved. One can see a proof in Fristedt and Gray (1997, p. 480).

Theorem 14.19. *Let $\{X_n, n \geq 0\}$ be a submartingale.*

(a) *Suppose $\{X_n\}$ is uniformly integrable. Then there exists an X such that $X_n \xrightarrow{a.s.} X$, and $E(|X_n - X|) \to 0$ as $n \to \infty$.*

(b) *Conversely, suppose there exists an X such that $E(|X_n - X|) \to 0$ as $n \to \infty$. Then $\{X_n\}$ must be uniformly integrable, and moreover, X_n necessarily converges almost surely to this X.*

(c) *If $\{X_n\}$ is a martingale, and is L_2 bounded (i.e., $\sup_n E(X_n^2) < \infty$), then there exists an X such that $X_n \xrightarrow{a.s.} X$, and $E(|X_n - X|^2) \to 0$ as $n \to \infty$.*

Example 14.24 (Pólya's Urn). We previously saw that the proportion of white balls in Pólya's urn, namely $R_n = \frac{a+S_n}{a+b+n}$ forms a martingale (see Example 14.6). This is an example in which the various convergences that we may want come easily. Because R_n is obviously a uniformly bounded sequence, by the theorem stated above, R_n converges almost surely and in L_2 (and therefore, in L_1) to a limiting random variable R, taking values in $[0, 1]$.

Neither the basic (sub)martingale convergence theorem nor the theorem in this section helps us in any way to identify the ditribution of R. In fact, in this case, R has a nondegenerate distribution, which is a Beta distribution with parameters a and b. As a consequence of this, $E(R_n) \to \frac{a}{a+b}$ and $\text{Var}(R_n) \to \frac{ab}{(a+b)^2(a+b+1)}$ as $n \to \infty$. A proof that R has a Beta distribution with parameters a, b is available in DasGupta (2010).

Example 14.25 (Bayes Estimates). We saw in Example 14.9 that the sequence of Bayes estimates (namely, the mean of the posterior distribution of the parameter) is a martingale adapted to the sequence of data values $\{X_n\}$. Continuing with the same notation as in Example 14.9, $Z_n = E(Y \mid X^{(n)})$ is our martingale sequence. Assume that the prior distribution for the parameter has a finite variance; that is, $E(Y^2) < \infty$. Then, by using Jensen's inequality for conditional expectations,

$$E(Z_n^2) = E[(E(Y \mid X^{(n)}))^2] \leq E[E(Y^2 \mid X^{(n)})] = E(Y^2).$$

Hence, by the theorem above in this section, the sequence of Bayes estimates Z_n converges to some Z almost surely, and moreover the mean and the variance of Z_n converge to the mean and the variance of Z.

A natural followup question is what exactly is this limiting random variable Z. We can only give partial answers in general. For example, for each n, $E(Z \mid X^{(n)}) = Z_n$ with probability one. It is tempting to conclude from here that Z is the same as Y with probability one. This will be the case if knowledge of the entire infinite data sequence $X_1, X_2, \ldots$ pins down Y completely, that is, if it is the case that someone who knows the infinite data sequence also knows Y with probability one.

14.5 * Reverse Martingales and Proof of SLLN

Partial sums of iid random variables are of basic interest in many problems in probability, such as the study of random walks, and as we know, the sequence of centered partial sums forms a martingale. On the other hand, the sequence of sample means is of fundamental interest in statistics; but the sequence of means does not form a martingale. Interestingly, if we measure time *backwards*, then the sequence of means does form a martingale, and then the rich martingale theory once again comes into play. This motivates the concept of a *reverse martingale*.

Definition 14.4. A sequence of random variables $\{X_n, n \geq 0\}$ defined on a common sample space Ω is called a *reverse submartingale* adapted to the sequence $\{Y_n, n \geq 0\}$, defined on the same sample space Ω, if $E(|X_n|) < \infty$ for all n and $E(X_n \mid Y_{n+1}, Y_{n+2}, \ldots) \geq X_{n+1}$ for each $n \geq 0$. The sequence $\{X_n\}$ is called a *reverse supermartingale* if $E(X_n \mid Y_{n+1}, Y_{n+2}, \ldots) \leq X_{n+1}$ for each n.

The sequence $\{X_n\}$ is called a *reverse martingale* if it is both a reverse submartingale and a reverse supermartingale with respect to the same sequence $\{Y_n\}$, that is, if $E(X_n \mid Y_{n+1}, Y_{n+2}, \ldots) = X_{n+1}$ for each n.

Example 14.26 (Sample Means). Let $X_1, X_2, \ldots$ be an infinite *exchangeable* sequence of random variables: for any $n \geq 2$ and any permutation π_n of $(1, 2, \ldots, n)$, $(X_1, X_2, \ldots, X_n)$ and $(X_{\pi_n(1)}, X_{\pi_n(2)}, \ldots, X_{\pi_n(n)})$ have the same joint distribution. For $n \geq 1$, let $\overline{X}_n = \frac{X_1 + \cdots + X_n}{n} = \frac{S_n}{n}$ be the sequence of sample means.

14.5 Reverse Martingales and Proof of SLLN

Then, by the exchanageability property of the $\{X_n\}$ sequence, for any given n, and any $k, 1 \leq k \leq n$,

$$\overline{X}_n = E(\overline{X}_n \mid S_n, S_{n+1}, \ldots) = \frac{1}{n} \sum_{i=1}^{n} E(X_i \mid S_n, S_{n+1}, \ldots)$$

$$= \frac{1}{n} n E(X_k \mid S_n, S_{n+1}, \ldots) = E(X_k \mid S_n, S_{n+1}, \ldots).$$

Consequently,

$$E(\overline{X}_{n-1} \mid S_n, S_{n+1}, \ldots) = \frac{1}{n-1} \sum_{k=1}^{n-1} E(X_k \mid S_n, S_{n+1}, \ldots)$$

$$= \frac{1}{n-1}(n-1)\overline{X}_n = \overline{X}_n,$$

which shows that the sequence of sample means is a reverse martingale (adapted to the sequence of partial sums).

There is a useful convex function theorem for reverse martingales as well, which is straightforward to prove.

Theorem 14.20 (Second Convex Function Theorem). *Let $\{X_n\}$ be a sequence of random variables defined on some sample space Ω, and f a convex function. Let $Z_n = f(X_n)$.*

(a) *If $\{X_n\}$ is a reverse martingale, then $\{Z_n\}$ is a reverse submartingale.*
(b) *If $\{X_n\}$ is a reverse submartingale, and f is also nondecreasing, then $\{Z_n\}$ is a reverse submartingale.*
(c) *If $\{X_{n,m}\}, m = 1, 2, \ldots$ is a countable family of reverse submartingales, defined on the same space Ω and all adapted to the same sequence, then $\{\sup_m X_{n,m}\}$ is also a reverse submartingale, adapted to the same sequence.*

Example 14.27 (A Paradoxical Statistical Consequence). Suppose Y is some real-valued random variable with mean η, and that we do not know the true value of η. Thus, we would like to estimate η. But, suppose that we cannot take any observations on the variable Y (for whatever reason). We can, however, take observations on a completely unrelated random variable X, where $E(|X|) < \infty$. Suppose we do take n iid observations on X. Call them $X_1, X_2, \ldots, X_n$ and let $\overline{X}_n$ be their mean. Then, by part (a) of the second convex function theorem, $|\overline{X}_n - \eta|$ forms a reverse submartingale, and hence $E(|\overline{X}_n - \eta|)$ is monotone nonincreasing in n. In other words, $E(|\overline{X}_{n+1} - \eta|) \leq E(|\overline{X}_n - \eta|)$ for all n, and so taking more observations on the useless variable X is going to be beneficial for estimating the mean of Y, a comical conclusion.

Note that there is really nothing special about using the absolute difference $|\overline{X}_n - \eta|$ as the criterion for the accuracy of estimation of η. The standard terminology in statistics for the criterion to be used is a *loss function*, and loss functions $L(\overline{X}_n, \eta)$ with a convexity property with respect to $\overline{X}_n$ for any fixed η will result in the same paradoxical conclusion. One needs to make sure that $E[L(\overline{X}_n, \eta)]$ is finite.

Reverse martingales possess a universal special property that is convenient in applications. The property is that a reverse martingale always converges almost surely to some finite random variable. The convergence property also holds for reverse submartingales, but the limiting random variable may equal $+\infty$ or $-\infty$ with a positive probability. An important and interesting consequence of this universal convergence property is a proof of the SLLN in the iid case by using martingale techniques. This is shown seen below as an example. The convergence property of reverse martingales is stated below.

Theorem 14.21 (Reverse Martingale Convergence Theorem). *(a) Let $\{X_n\}$ be a reverse martingale adapted to some sequence. Then it is necessarily uniformly integrable, and there exists a random variable X, almost surely finite, such that $X_n \xrightarrow{a.s.} X$, and $E|X_n - X| \to 0$ as $n \to \infty$.*
(b) Let $\{X_n\}$ be a reverse submartingale adapted to some sequence. Then there exists a random variable X taking values in $[-\infty, \infty]$ such that $X_n \xrightarrow{a.s.} X$.

See Fristedt and Gray (1997, pp. 483–484) for a proof using uniform integrability techniques. Here is an important application of this theorem.

Example 14.28 (Proof of Kolmogorov's SLLN). Let $X_1, X_2, \ldots$ be iid random variables, with $E(|X_1|) < \infty$, and let $E(X_1) = \mu$. The goal of this example is to show that the sequence of sample means, $\overline{X}_n$, converges almost surely to μ.

We use the reverse martingale convergence theorem to obtain a proof. Because we have already shown that $\{\overline{X}_n\}$ forms a reverse martingale sequence, by the reverse martingale convergence theorem we are assured of a finite random variable Y such that $\overline{X}_n$ converges almost surely to Y, and we are also assured that $E(Y) = \mu$. The only task that remains is to show that Y equals μ with probability one.

This is achieved by establishing that $P(Y \geq y) = [P(Y \geq y)]^2$ for all real y (i.e., $P(Y \geq y)$ is 0 or 1 for any y), which would force Y to be degenerate and therefore degenerate at μ. To prove that $P(Y \geq y) = [P(Y \geq y)]^2$ for all real y, define the double sequence

$$Y_{m,n} = \frac{X_{m+1} + X_{m+2} + \cdots + X_{m+n}}{n},$$

$m, n \geq 1$. Note that $\overline{X}_k$ and $Y_{m,n}$ are independent for any $m, k \leq m$, and any n, and that, furthermore, for any fixed m, $Y_{m,n}$ converges almost surely to Y (the same Y as above) as $n \to \infty$. These two facts together imply

$$P\left(Y \geq y, \max_{1 \leq k \leq m} \overline{X}_k \geq y\right) = P(Y \geq y) P\left(\max_{1 \leq k \leq m} \overline{X}_k \geq y\right)$$
$$\Rightarrow P(Y \geq y) = P(Y \geq y) P(Y \geq y) = [P(Y \geq y)]^2,$$

which is what we needed to complete the proof.

14.6 Martingale Central Limit Theorem

For an iid mean zero sequence of random variables $Z_1, Z_2, \ldots$ with variance one, the central limit theorem says that for large n, $\frac{Z_1 + \cdots + Z_n}{\sqrt{n}}$ is approximately standard normal. Suppose now that we consider a mean zero martingale (adapted to some sequence $\{Y_n\}$) $\{X_n, n \geq 0\}$ with $X_0 = 0$ and write $Z_i = X_i - X_{i-1}, i \geq 1$. Then, obviously we can write

$$X_n = X_n - X_0 = \sum_{i=1}^{n}(X_i - X_{i-1}) = \sum_{i=1}^{n} Z_i.$$

The summands Z_i are certainly no longer independent; however, they are uncorrelated (see the chapter exercises). The martingale central limit theorem says that under certain conditions on the growth of the conditional variances $\text{Var}(Z_n | Y_0, \ldots, Y_{n-1})$, $\frac{X_n}{\sqrt{n}}$ will still be approximately normally distributed for large n.

The area of martingale central limit theorems is a bit confusing due to an overwhelming variety of central limit theorems, each known as a martingale central limit theorem. In particular, the normalization of X_n can be deterministic or random. Also, there can be a double array of martingales and central limit theorems for them, analogous to Lyapounov's central limit theorem for the independent case. The best source and exposition of martingale central limit theorems is the classic book by Hall and Heyde (1980). We present two specific martingale central limit theorems in this section.

First, we need some notation. Let $\{X_n, n \geq 0\}$ be a zero mean martingale adapted to some sequence $\{Y_n\}$, with $X_0 = 0$. Let

$$Z_i = X_i - X_{i-1}, i \geq 1;$$
$$\sigma_j^2 = \text{Var}(Z_j | Y_0, \ldots, Y_{j-1}) = E(Z_j^2 | Y_0, \ldots, Y_{j-1});$$
$$V_n^2 = \sum_{j=1}^{n} \sigma_j^2;$$
$$s_n^2 = E(V_n^2) = E(X_n^2) = \text{Var}(X_n);$$

(see Section 14.3.2 for the fact that $E(V_n^2)$ and $E(X_n^2)$ are equal if $X_0 = 0$).

The desired result is that $\frac{X_n}{s_n}$ converges in distribution to $N(0, 1)$. The question is under what conditions can one prove such an asymptotic normality result. The conditions that we use are very similar to the corresponding Lindeberg–Lévy conditions in the independent case. Here are the two conditions we assume.

(A) Concentration Condition

$$\frac{V_n^2}{s_n^2} = \frac{V_n^2}{E(V_n^2)} \xrightarrow{P} 1.$$

(B) Martingale Lindeberg Condition

For any $\epsilon > 0$, $\quad \dfrac{\sum_{j=1}^{n} E(Z_j^2 I_{\{|Z_j| \geq \epsilon s_n\}})}{s_n^2} \xrightarrow{P} 0.$

Under condition (A), the Lindeberg condition (B) is nearly equivalent to the uniform asymptotic negligibility condition that $\dfrac{\max_{1 \leq j \leq n} \sigma_j^2}{s_n^2} \xrightarrow{P} 0$. We commonly see such uniform asymptotic negligibility conditions in the independent case central limit theorems. See Hall and Heyde (1980) and Brown (1971) for much additional discussion on the exact role of the Lindeberg condition in martingale central limit theorems. Here is our basic martingale CLT.

Theorem 14.22 (Basic Martingale CLT). *Suppose conditions (A) and (B) hold. Then $\dfrac{X_n}{s_n} \xRightarrow{\mathcal{L}} Z$, where $Z \sim N(0, 1)$.*

The proof of the Lindeberg–Lévy theorem for the independent case has to be suitably adapted to the martingale structure in order to prove this theorem. The two references above can be consulted for a proof. The Lindeberg condition can be difficult to verify. The following simpler version of martingale central limit theorems suffices for many applications. For this, we need the additional notation

$$\tau_t = \inf\left\{ n > 0 : \sum_{j=1}^{n} \sigma_j^2 \geq t \right\}.$$

Here is our simpler version of the martingale CLT.

Theorem 14.23. *Assume that*

$$|Z_i| \leq K < \infty \quad \text{for all } i \text{ and some } K;$$

$$\sum_{j=1}^{\infty} \sigma_j^2 = \infty \text{ almost surely};$$

$$\frac{t}{\tau_t} \xrightarrow{a.s.} \sigma^2 \quad \text{for some finite and positive constant } \sigma^2.$$

Then $\dfrac{X_n}{\sqrt{n}} \xRightarrow{\mathcal{L}} W$, where $W \sim N(0, \sigma^2)$.

Exercises

Exercise 14.1. Suppose $\{X_n, n \geq 1\}$ is a martingale adapted to some sequence $\{Y_n\}$. Show that $E(X_{n+m} | Y_1, \ldots, Y_n) = X_n$ for all $m, n \geq 1$.

Exercises

Exercise 14.2. Suppose $\{X_n, n \geq 1\}$ is a martingale adapted to some sequence $\{Y_n\}$. Fix $m \geq 1$ and define $Z_n = X_n - X_m, n \geq m + 1$. Is it true that $\{Z_n\}$ is also a martingale?

Exercise 14.3 (Product Martingale). Let $X_1, X_2, \ldots$ be iid nonnegative random variables with a finite positive mean μ. Identify a sequence of constants c_n such that $Z_n = c_n(\prod_{i=1}^n X_i), n \geq 1$ forms a martingale.

Exercise 14.4. Let $\{U_n\}, \{V_n\}$ be martingales, adapted to the same sequence $\{Y_n\}$. Identify, with proof, which of the following are also submartingales, and for those that are not necessarily submartingales, give a counterexample.

(a) $|U_n - V_n|$.
(b) $U_n^2 + V_n^2$.
(c) $U_n - V_n$.
(d) $\min(U_n, V_n)$.

Exercise 14.5 (Bayes Problem). Suppose given p, $X_1, X_2, \ldots$ are iid Bernoulli variables with a parameter p, and the marginal distribution of p is $U[0, 1]$. Let $S_n = X_1 + \cdots + X_n, n \geq 1$, and $Z_n = \frac{S_n+1}{n+2}$. Show that $\{Z_n\}$ is a martingale with respect to the sequence $\{X_n\}$.

Exercise 14.6 (Bayes Problem). Suppose given λ, $X_1, X_2, \ldots$ are iid Poisson variables with some mean λ, and the marginal density of λ is $\frac{\beta^\alpha e^{-\beta\lambda}\lambda^{\alpha-1}}{\Gamma(\alpha)}$, where $\alpha, \beta > 0$ are constants. Let $S_n = X_1 + \cdots + X_n, n \geq 1$, and $Z_n = \frac{S_n+\alpha}{n+\beta}$. Show that $\{Z_n\}$ is a martingale with respect to the sequence $\{X_n\}$.

Exercise 14.7 (Bayes Problem). Suppose given μ, $X_1, X_2, \ldots$ are iid $N(\mu, 1)$ variables, and that the marginal distribution of μ is standard normal. Let $S_n = X_1 + \cdots + X_n, n \geq 1$, and $Z_n = \frac{S_n}{n+1}$. Show that $\{Z_n\}$ is a martingale with respect to the sequence $\{X_n\}$.

Exercise 14.8. Suppose $\{X_n\}$ is known to be a submartingale with respect to some sequence $\{Y_n\}$. Show that $\{X_n\}$ is also a martingale if and only if $E(X_n) = E(X_m)$ for all m, n.

Exercise 14.9. Let $X_1, X_2, \ldots$ be a sequence of iid random variables such that $E(|X_1|) < \infty$. For $n \geq 1$, let $X_{n:n} = \max(X_1, \ldots, X_n)$. Show that $\{X_{n:n}\}$ is a submartingale adapted to itself.

Exercise 14.10 (Random Walk). Consider a simple asymmetric random walk with iid steps distributed as $P(X_i = 1) = p, P(X_i = -1) = 1 - p, p < \frac{1}{2}$. Let $S_n = X_1 + \cdots + X_n, n \geq 1$, Show that

(a) $V_n = (\frac{1-p}{p})^{S_n}$ is a martingale.
(b) Show that with probability one, $\sup_n S_n < \infty$.

Exercise 14.11 (Branching Process). Let $\{Z_{ij}\}$ be a double array of iid random variables with mean μ and variance $\sigma^2 < \infty$. Let $X_0 = 1$ and $X_{n+1} = \sum_{j=1}^{X_n} Z_{nj}$. Show that

(a) $W_n = \frac{X_n}{\mu^n}$ is a martingale.
(b) $\sup_n E(W_n) < \infty$.
(c) Is $\{W_n\}$ uniformly integrable? Prove or disprove it.

Remark. The process W_n is commonly called a branching process and is important in population studies.

Exercise 14.12 (A Time Series Model). Let $Z_0, Z_1, \ldots$ be iid standard normal variables. Let $X_0 = Z_0$, and for $n \geq 1$, $X_n = X_{n-1} + Z_n h_n(X_0, \ldots, X_{n-1})$, where for each n, $h_n(x_0, \ldots, x_{n-1})$ is an absolutely bounded function.
   Show that $\{X_n\}$ is a martingale adapted to some sequence $\{Y_n\}$, and explicitly identify such a sequence $\{Y_n\}$.

Exercise 14.13 (Another Time Series Model). Let $Z_0, Z_1, \ldots$ be a sequence of random variables such that $E(Z_{n+1}|Z_0, \ldots, Z_n) = cZ_n + (1-c)Z_{n-1}, n \geq 1$, where $0 < c < 1$. Let $X_0 = Z_0, X_n = \alpha Z_n + Z_{n-1}, n \geq 1$. Show that α may be chosen to make $\{X_n, n \geq 0\}$ a martingale with respect to $\{Z_n\}$.

Exercise 14.14 (Conditional Centering of a General Sequence). Let $Z_0, Z_1, \ldots$ be a general sequence of random variables, not necessarily independent, such that $E(|Z_k|) < \infty$ for all k. Let $V_n = \sum_{i=1}^{n}[Z_i - E(Z_i|Z_0, \ldots, Z_{i-1})], n \geq 1$. Show that $\{V_n\}$ is a martingale with respect to the sequence $\{Z_n\}$.

Exercise 14.15 (The Cross-Product Martingale). Let $X_1, X_2, \ldots$ be independent random variables, with $E(|X_i|) < \infty$ and $E(X_i) = 0$ for all i. For a fixed $k \geq 1$, let $V_{k,n} = \sum_{1 \leq i_1 < i_2 < \cdots < i_k \leq n} X_{i_1} \ldots X_{i_k}, n \geq k$. Show that $\{V_{k,n}\}$ is a martingale with respect to $\{X_n\}$.

Exercise 14.16 (The Wright–Fisher Markov Chain). Consider the Wright-Fisher Markov chain of Example 14.7. Let

$$V_n = \frac{X_n(N - X_n)}{(1 - \frac{1}{N})^n}, \quad n \geq 0.$$

Show that $\{V_n\}_0^N$ is a martingale.

Exercise 14.17 (An Example of Samuel Karlin). Let f be a continuous function defined on $[0, 1]$ and $U \sim U[0, 1]$. Let $X_n = \frac{\lfloor 2^n U \rfloor}{2^n}$, and $V_n = \frac{f(X_n + 2^{-n}) - f(X_n)}{2^{-n}}$. Show that $\{V_n\}$ is a martingale with respect to the sequence $\{X_n\}$.

Exercise 14.18. Let $X_1, X_2, \ldots$ be iid symmetric random variables with mean zero, and let $S_n = \sum_{i=1}^n X_i, n \geq 1$, and $S_0 = 0$. Let $\psi(t)$ be the characteristic function of X_1, and $V_n = [\psi(t)]^{-n} e^{itS_n}, n \geq 0$. Show that the real part as well as the imaginary part of $\{V_n\}$ is a martingale.

Exercises

Exercise 14.19 (Stopping Times). Consider the simple symmetric random walk S_n with $S_0 = 0$. Identify, with proof, which of the following are stopping times, and which among them have a finite expectation.

(a) $\inf\{n > 0 : |S_n| > 5\}$.
(b) $\inf\{n \geq 0 : S_n < S_{n+1}\}$.
(c) $\inf\{n > 0 : |S_n| = 1\}$.
(d) $\inf\{n > 0 : |S_n| > 1\}$.

Exercise 14.20. Let τ be a nonnegative integer-valued random variable, and $\{X_n, n \geq 0\}$ a sequence of random variables, all defined on a common sample space Ω. Prove or disprove that τ is a stopping time adapted to $\{X_n\}$ if and only if for every $n \geq 0$, $I_{\{\tau=n\}}$ is a function of only $X_0, \ldots, X_n$.

Exercise 14.21. Suppose τ_1, τ_2 are both stopping times with respect to some sequence $\{X_n\}$. Is $|\tau_1 - \tau_2|$ necessarily a stopping time with respect to $\{X_n\}$?

Exercise 14.22 (Condition for Optional Stopping Theorem). Suppose $\{X_n, n \geq 0\}$ is a martingale, and τ a stopping time, both adapted to a common sequence $\{Y_n\}$. Show that the equality $E(X_\tau) = E(X_0)$ holds if $E(|X_\tau|) < \infty$, and $E(X_{\min(\tau,n)} I_{\{\tau > n\}}) \to 0$ as $n \to \infty$.

Exercise 14.23 (The Random Walk). Consider the asymmetric random walk $S_n = \sum_{i=1}^n X_i$, where $P(X_i = 1) = p$, $P(X_i = -1) = q = 1 - p$, $p > \frac{1}{2}$, and $S_0 = 0$. Let x be a fixed positive integer, and $\tau = \inf\{n > 0 : S_n = x\}$. Show that for $0 < s < 1$, $E(s^\tau) = \left(\frac{1-\sqrt{1-4pqs^2}}{2qs}\right)^x$.

Exercise 14.24 (The Random Walk; continued). For the stopping time τ of the previous exercise, show that

$$E(\tau) = \frac{x}{p-q} \quad \text{and} \quad \text{Var}(\tau) = \frac{x[1-(p-q)^2]}{(p-q)^3}.$$

Exercise 14.25 (Gambler's Ruin). Consider the general random walk $S_n = \sum_{i=1}^n X_i$, where $P(X_i = 1) = p \neq \frac{1}{2}$, $P(X_i = -1) = q = 1 - p$, and $S_0 = 0$. Let a, b be fixed positive integers, and $\tau = \inf\{n > 0 : S_n = b \text{ or } S_n = -a\}$. Show that

$$E(\tau) = \frac{b}{p-q} - \frac{a+b}{p-q} \frac{[1-(\frac{p}{q})^b]}{[1-(\frac{p}{q})^{a+b}]},$$

and that by an application of L'Hospital's rule, this gives the correct formula for $E(\tau)$ even when $p = \frac{1}{2}$.

Exercise 14.26 (Martingales for Patterns). Consider the following martingale approach to a geometric distribution problem. Let $X_1, X_2, \ldots$ be iid Bernoulli variables, with $P(X_i = 1) = p$, $P(X_i = 0) = q = 1-p$. Let $\tau = \min\{k \geq 1 : X_k = 0\}$, and $\tau_n = \min(\tau, n), n \geq 1$.

Define $V_n = \frac{1}{q} \sum_{i=1}^{n} I_{\{X_i=0\}}, n \geq 1$.

(a) Show that $\{V_n - n\}$ is a martingale with respect to the sequence $\{X_n\}$.
(b) Show that $E(V_{\tau_n}) = E(\tau_n)$ for all n.
(c) Hence, show that $E(\tau) = E(V_\tau) = \frac{1}{q}$.

Exercise 14.27 (Martingales for Patterns). Let $X_1, X_2, \ldots$ be iid Bernoulli variables, with $P(X_i = 1) = p$, $P(X_i = 0) = q = 1 - p$. Let τ be the first k such that X_{k-2}, X_{k-1}, X_k are each equal to one (e.g., the number of tosses of a coin necessary to first obtain three consecutive heads), and $\tau_n = \min(\tau, n), n \geq 3$.
Define

$$V_n = \frac{1}{p^3} \sum_{i=1}^{n-2} I_{\{X_i = X_{i+1} = X_{i+2} = 1\}} + \frac{1}{p^2} I_{\{X_n = X_{n-1} = 1\}} + \frac{1}{p} I_{\{X_n = 1\}}, \quad n \geq 3.$$

(a) Show that $\{V_n - n\}$ is a martingale with respect to the sequence $\{X_n\}$.
(b) Show that $E(V_{\tau_n}) = E(\tau_n)$ for all n.
(c) Hence, show that

$$E(\tau) = E(V_\tau) = \frac{1}{p} + \frac{1}{p^2} + \frac{1}{p^3}.$$

(d) Generalize to the case of the expected waiting time for obtaining r consecutive 1s.

Exercise 14.28. Let $\{X_n, n \geq 0\}$ be a martingale.

(a) Show that $\lim_{n \to \infty} E(|X_n|)$ exists.
(b) Show that for any stopping time τ, $E(|X_\tau|) \leq \lim_{n \to \infty} E(|X_n|)$.
(c) Show that if $\sup_n E(|X_n|) < \infty$, then $E(|X_\tau|) < \infty$ for any stopping time τ.

Exercise 14.29 (Inequality for Stopped Martingales). Let $\{X_n, n \geq 0\}$ be a martingale, and τ a stopping time adapted to $\{X_n\}$. Show that $E(|X_\tau|) \leq 2 \sup_n E(X_n^+) - E(X_1) \leq 3 \sup_n E(|X_n|)$.

Exercise 14.30. Let $X_1, X_2, \ldots$ be iid random variables such that $E(|X_1|) < \infty$. Consider the random walk $S_n = \sum_{i=1}^{n} X_i, n \geq 1$ and $S_0 = 0$. Let τ be a stopping time adapted to $\{S_n\}$. Show that if $E(|S_\tau|) = \infty$, then $E(\tau)$ must also be infinite.

Exercise 14.31. Let $\{X_n, n \geq 0\}$ be a martingale, with $X_0 = 0$. Let $V_i = X_i - X_{i-1}, i \geq 1$. Show that for any $i \neq j$, V_i and V_j are uncorrelated.

Exercise 14.32. Let $\{X_n, n \geq 1\}$ be some sequence of random variables. Suppose $S_n = \sum_{i=1}^{n} X_i, n \geq 1$, and that $\{S_n, n \geq 1\}$ forms a martingale. Show that for any $i \neq j, E(X_i X_j) = 0$.

Exercise 14.33. Let $\{X_n, n \geq 0\}$ and $\{Y_n, n \geq 0\}$ both be square integrable martingales, adapted to some common sequence. Let $X_0 = Y_0 = 0$. Show that $E(X_n Y_n) = \sum_{i=1}^{n} E[(X_i - X_{i-1})(Y_i - Y_{i-1})]$ for any $n \geq 1$.

Exercise 14.34. Give an example of a submartingale $\{X_n\}$ and a convex function f such that $\{f(X_n)\}$ is not a submartingale.

Remark. Such a function f cannot be increasing.

Exercise 14.35 (Characterization of Uniformly Integrable Martingales). Let $\{X_n\}$ be uniformly integrable and a martingale with respect to some sequence $\{Y_n\}$. Show that there exists a random variable Z such that $E(|Z|) < \infty$ and such that for each n, $E(Z|Y_1,\ldots,Y_n) = X_n$ with probability one.

Exercise 14.36 (L_p-Convergence of a Martingale). Let $\{X_n, n \geq 0\}$ be a martingale, or a nonnegative submartingale. Suppose for some $p > 1$, $\sup_n E(|X_n|^p) < \infty$. Show that there exists a random variable X, almost surely finite, such that $E(|X_n - X|^p) \to 0$ and $X_n \xrightarrow{a.s.} X$ as $n \to \infty$.

Exercise 14.37. Let $\{X_n\}$ be a nonnegative martingale. Suppose $E(X_n) \to 0$ as $n \to \infty$. Show that $X_n \xrightarrow{a.s.} 0$.

Exercise 14.38. Let $X_1, X_2, \ldots$ be iid normal variables with mean zero and variance σ^2. Show that $\sum_{n=1}^{\infty} \frac{\sin(n\pi x)}{n} X_n$ converges with probability one for any given real number x.

Exercise 14.39 (Generalization of Maximal Inequality). Let $\{X_n, n \geq 0\}$ be a nonnegative submartingale, and $\{b_n, n \geq 0\}$ a nonnegative nonincreasing sequence of constants such that $b_n \to 0$ as $n \to \infty$, and $\sum_{n=0}^{\infty}[b_n - b_{n+1}]E(X_n)$ converges.

(a) Show that for any $x > 0$,

$$P(\sup_{n \geq 0} b_n X_n \geq x) \leq \frac{1}{x} \sum_{n=0}^{\infty}[b_n - b_{n+1}]E(X_n).$$

(b) Derive the Kolmogorov maximal inequality for nonnegative submartingales as a corollary to part (a).

Exercise 14.40 (Decomposition of an L_1-Bounded Martingale). Let $\{X_n\}$ be an L_1-bounded martingale adapted to some sequence $\{Y_n\}$, that is, $\sup_n E(|X_n|) < \infty$.

(a) Define $Z_{m,n} = E[|X_{m+1}||Y_1,\ldots,Y_n]$. Show that $Z_{m,n}$ is nondecreasing in m.
(b) Show that for fixed n, $Z_{m,n}$ converges almost surely.
(c) Let $U_n = \lim_m Z_{m,n}$. Show that $\{U_n\}$ is an L_1-bounded martingale.
(d) Show that X_n admits the decomposition $X_n = U_n - V_n$, where both U_n, V_n are nonnegative L_1-bounded martingales.

References

Azuma, K. (1967). Weighted sums of certain dependent random variables, *Tohoku Math. J.*, 19, 357–367.
Brown, B.M. (1971). Martingale central limit theorems, *Ann. Math. Statist.*, 42, 59–66.
Burkholder, D.L. (1973). Distribution function inequalities for martingales, *Ann. Prob.*, 1, 19–42.

Burkholder, D.L., Davis, B., and Gundy, R. F. (1972). Integral inequalities for convex functions of operators on martingales, *Proc. Sixth Berkeley Symp. Math. Statist. Prob.*, Vol. II, 223–240, University of California Press, Berkeley.

Chow, Y.S. and Teicher, H. (2003). *Probability Theory: Independence, Interchangeability, Martingales*, Springer, New York.

Chung, K.L. (1974). *A Course in Probability*, Academic Press, New York.

DasGupta, A. (2010). *Fundamentals of Probability: A First Course*, Springer, New York.

Davis, B. (1970). On the integrability of the martingale square function, *Israel J. Math.*, 8, 187–190.

Devroye, L. (1991). *Exponential Inequalities in Nonparametric Estimation, Nonparametric Functional Estimation and Related Topics*, 31–44, Kluwer Acad. Publ., Dordrecht.

Doob, J.L. (1971). What is a martingale?, *Amer. Math. Monthly*, 78, 451–463.

Fristedt, B. and Gray, L. (1997). *A Modern Approach to Probability Theory*, Birkhäuser, Boston.

Hall, P. and Heyde, C. (1980). *Martingale Limit Theory and Its Applications*, Academic Press, New York.

Heyde, C. (1972). Martingales: A case for a place in a statistician's repertoire, *Austr. J. Statist.*, 14, 1–9.

Hoeffding, W. (1963). Probability inequalities for sums of bounded random variables, *J. Amer. Statisto Assoc.*, 58, 13–30.

Karatzas, I. and Shreve, S. (1991). *Brownian Motion and Stochastic Calculus*, Springer, New York.

Karlin, S. and Taylor, H.M. (1975). *A First Course in Stochastic Processes*, Academic Press, New York.

McDiarmid, C. (1989). *On the Method of Bounded Differences, Surveys in Combinatorics*, London Math. Soc. Lecture Notes, 141, 148–188, Cambridge University Press, Cambridge, UK.

Williams, D. (1991). *Probability with Martingales*, Cambridge University Press, Cambridge, UK.

Chapter 15
Probability Metrics

The central limit theorem is a very good example of approximating a potentially complicated exact distribution by a simpler and easily computable approximate distribution. In mathematics, whenever we do an approximation, we like to quantify the error of the approximation. Common sense tells us that an error should be measured by some notion of distance between the exact and the approximate. Therefore, when we approximate one probability distribution (measure) by another, we need a notion of distances between probability measures. Fortunately, we have an abundant supply of distances between probability measures. Some of them are for probability measures on the real line, and others for probability measures on a general Euclidean space. Still others work in more general spaces. These distances on probability measures have other independent uses besides quantifying the error of an approximation. We provide a basic treatment of some common distances on probability measures in this chapter. Some of the distances have the so-called *metric property*, and they are called *probability metrics*, whereas some others satisfy only the weaker notion of being a *distance*. Our choice of which metrics and distances to include was necessarily subjective.

The main references for this chapter are Rachev (1991), Reiss (1989), Zolotarev (1983), Leise and Vajda (1987), Dudley (2002), DasGupta (2008), Rao (1987), and Gibbs and Su (2002). Diaconis and Saloff-Coste (2006) illustrate some concrete uses of probability metrics. Additional references are given in the sections.

15.1 Standard Probability Metrics Useful in Statistics

As we said above, there are numerous metrics and distances on probability measures. The choice of the metric depends on the need in a specific situation. No single metric or distance is the best or the most preferable. There is also the very important issue of analytic tractability and ease of computing. Some of the metrics are more easily bound, and some less so. Some of them are hard to compute. Our choice of metrics and distances to cover in this chapter is guided by all these factors, and also personal preferences. The definitions of the metrics and distances are given below. However, we must first precisely draw the distinction between metrics and distances.

Definition 15.1. Let $\mathcal{M}$ be a class of probability measures on a sample space Ω. A function $d : \mathcal{M} \otimes \mathcal{M} \to \mathcal{R}$ is called a *distance* on $\mathcal{M}$ if

(a) $d(P, Q) \geq 0 \; \forall \; P, Q, \in \mathcal{M}$ and $d(P, Q) = 0 \Leftrightarrow P = Q$;
(b) $d(P_1, P_3) \leq d(P_1, P_2) + d(P_2, P_3) \; \forall \; P_1, P_2, P_3 \in \mathcal{M}$ (Triangular inequality).
d is called a *metric* on $\mathcal{M}$ if, moreover.
(c) $d(P, Q) = d(Q, P) \; \forall \; P, Q, \in \mathcal{M}$ (Symmetry).

Here now are the probability metrics and distances that we mention in this chapter.

Definition 15.2 (Kolmogorov Metric). Let P, Q be probability measures on $\mathcal{R}^d$, $d \geq 1$, with corresponding CDFs F, G. The *Kolmogorov metric* is defined as

$$d(P, Q) = \sup_{x \in \mathcal{R}^d} |F(x) - G(x)|.$$

Wasserstein Metric. Let P, Q be probability measures on $\mathcal{R}$ with corresponding CDFs F, G. The *Wasserstein metric* is defined as

$$W(P, Q) = \int_{-\infty}^{\infty} |F(x) - G(x)| dx.$$

Total Variation Metric. Let P, Q be absolutely continuous probability measures on $\mathcal{R}^d, d \geq 1$, with corresponding densities f, g. The *total variation metric* is defined as

$$\rho(P, Q) = \rho(f, g) = \frac{1}{2} \int |f(x) - g(x)| dx.$$

If P, Q are discrete, with corresponding mass functions p, q on the set of values $\{x_1, x_2, \ldots, \}$, then the total variation metric is defined as

$$\rho(P, Q) = \rho(p, q) = \frac{1}{2} \sum_i |p(i) - q(i)|,$$

where $p(i), q(i)$ are the probabilities at x_i under P and Q, respectively.

Separation Distance. Let P, Q be discrete, with corresponding mass functions p, q, The *separation distance* is defined as

$$D(P, Q) = \sup_i \left(1 - \frac{p(i)}{q(i)}\right).$$

Note that the order of P, Q matters in defining $D(P, Q)$.

Hellinger Metric. Let P, Q be absolutely continuous probability measures on $\mathcal{R}^d, d \geq 1$, with corresponding densities f, g. The *Hellinger metric* is defined as

$$H(P, Q) = \left[\int \left(\sqrt{f(x)} - \sqrt{g(x)}\right)^2 dx\right]^{1/2}.$$

15.1 Standard Probability Metrics Useful in Statistics

If P, Q are discrete, with corresponding mass functions p, q, the Hellinger metric is defined as

$$H(P, Q) = \left[\sum_i (\sqrt{p_i} - \sqrt{q_i})^2\right]^{1/2}.$$

Kullback–Leibler Distance. Let P, Q be absolutely continuous probability measures on $\mathcal{R}^d, d \geq 1$, with corresponding densities f, g. The *Kullback–Leibler distance* is defined as

$$K(P, Q) = K(f, g) = \int f(x) \log \frac{f(x)}{g(x)} dx.$$

If P, Q are discrete, with corresponding mass functions p, q, then the Kullback–Leibler distance is defined as

$$K(P, Q) = K(p, q) = \sum_i p(i) \log \frac{p(i)}{q(i)}.$$

Note that the order of P, Q matters in defining $K(P, Q)$.

Lévy–Prokhorov Metric. Let P, Q be probability measures on $\mathcal{R}^d, d \geq 1$. The *Lévy–Prokhorov metric* is defined as

$$L(P, Q) = \inf\{\epsilon > 0 : \forall B \text{ Borel}, P(B) \leq Q(B^\epsilon) + \epsilon\},$$

where B^ϵ is, *the outer ϵ-parallel body of B*; that is,

$$B^\epsilon = \{x \in \mathcal{R}^d : \inf_{\{y \in B\}} ||x - y|| \leq \epsilon\}.$$

If $d = 1$, then $L(P, Q)$ equals

$$L(P, Q) = \inf\{\epsilon > 0 : \forall x, F(x) \leq G(x + \epsilon) + \epsilon\},$$

where F, G are the CDFs of P, Q.

f-Divergences. Let P, Q be absolutely continuous probability measures on $\mathcal{R}^d$, $d \geq 1$, with densities p, q, and f any real-valued convex function on $\mathcal{R}^+$, with $f(1) = 0$. The *f-divergence between P, Q* is defined as

$$d_f(P, Q) = \int q(x) f\left(\frac{p(x)}{q(x)}\right) dx.$$

If P, Q are discrete, with corresponding mass functions p, q, then the f-divergence is defined as

$$d_f(P, Q) = \sum_i q(i) f\left(\frac{p(i)}{q(i)}\right).$$

f-divergences have the *finite partition property* that $d_f(P, Q) = \sup_{\{A_j\}} \sum_j Q(A_j) f\left(\frac{P(A_j)}{Q(A_j)}\right)$, where the supremum is taken over all possible finite partitions $\{A_j\}$ of $\mathcal{R}^d$.

15.2 Basic Properties of the Metrics

Basic properties and interrelationships of these distances and metrics are now studied.

Theorem 15.1. *(a) Let P_n, P be probabilty measures on $\mathcal{R}$ such that $d(P_n, P) \to 0$ as $n \to \infty$. Then P_n converges weakly (i.e., in distribution) to P. If the CDF of P is continuous, then the converse is also true.*
(b) $\rho(P, Q) = \sup_B |P(B) - Q(B)|$, where the supremum is taken over all (Borel) sets B in $\mathcal{R}^d$.
(c) Given probability measures $P_1, P_2, \rho(P_1, P_2)$ satisfies the coupling identity

$$\rho(P_1, P_2) = \inf\{P(X \neq Y) : \text{ all jointly distributed } (X, Y) \text{ with } X \sim P_1, Y \sim P_2\}.$$

Furthermore,

$$\rho(P_1, P_2) = \frac{1}{2} \sup_{h: |h| \leq 1} \left[|E_{P_1}(h) - E_{P_2}(h)| \right].$$

(d) Let $\mathcal{H}$ denote the family of all functions $h : \mathcal{R} \to \mathcal{R}$ with the Lipschitz norm *bounded by one; that is,*

$$\mathcal{H} = \left\{ h : \sup_{\{x,y\}} \frac{|h(x) - h(y)|}{|x - y|} \leq 1 \right\}.$$

Then,

$$W(P, Q) = \sup\{|E_P(h) - E_Q(h)| : h \in \mathcal{H}\}.$$

(e) Let P_n, P be probabilty measures on $\mathcal{R}^d$. Then P_n converges weakly to P if and only if $L(P_n, P) \to 0$.
(f) If any of $\rho(P_n, P), W(P_n, P), D(P_n, P), H(P_n, P), K(P_n, P) \to 0$, then P_n converges weakly to P.
(g) The following converses of part (f) are true.

   (i) P_n converges weakly to $P \Rightarrow \rho(P_n, P) \to 0$ if P_n, P are absolutely continuous and unimodal probability measures on $\mathcal{R}$, or if P_n, P are discrete with mass functions p_n, p.
   (ii) P_n converges weakly to $P \Rightarrow W(P_n, P) \to 0$ if P_n, P are all supported on a bounded set Ω in $\mathcal{R}$.
   (iii) P_n converges weakly to $P \Rightarrow H(P_n, P) \to 0$ if P_n, P are absolutely continuous and unimodal probability measures on $\mathcal{R}$, or if P_n, P are discrete with mass functions p_n, p.

Proof. Due to the long nature of the proofs, we only refer to the proofs of parts (c)–(g). Parts (a) and (b) are proved below. We refer to Gibbs and Su (2002) for parts (c), (e), and (f). Part (d) is proved in Dudley (2002). Part (i) in (g) is a result in Ibragimov (1956). Part (ii) in (g) is a consequence of Theorem 2 in Gibbs and Su (2002), and part (e) in this theorem. The first statement in part (iii) in (g) is a

15.2 Basic Properties of the Metrics

consequence of inequality (8) in Gibbs and Su (2002) (which is a result in LeCam (1969)) and part (i) in (g) of this theorem. The second statement in part (iii) in (g) is a consequence of Scheffé's theorem (see Chapter 7), and once again inequality (8) in Gibbs and Su (2002).

As regards part (a), that P_n converges weakly to P if $d(P_n, P) \to 0$ is a consequence of the definition of convergence in distribution. That the converse is also true when P has a continuous CDF is a consequence of Pólya's theorem (see Chapter 7).

The proof of part (b) proceeds as follows. Consider the case when P, Q have densities f, g. Fix any (Borel) set B, and define $B_0 = \{x : f(x) \geq g(x)\}$. Then,

$$P(B) - Q(B) = \int_B [f(x) - g(x)]dx \leq \int_{B_0} [f(x) - g(x)]dx.$$

On the other hand,

$$\int |f(x) - g(x)|dx = \int_{B_0} [f(x) - g(x)]dx + \int_{B_0^c} [g(x) - f(x)]dx$$

$$= 2\int_{B_0} [f(x) - g(x)]dx.$$

Combining these two facts,

$$\frac{1}{2}\int |f(x) - g(x)|dx = \int_{B_0} [f(x) - g(x)]dx = P(B_0) - Q(B_0)$$

$$\leq \sup_B (P(B) - Q(B)) \leq \int_{B_0} [f(x) - g(x)]dx$$

$$\Rightarrow \sup_B (P(B) - Q(B)) = \frac{1}{2}\int |f(x) - g(x)|dx.$$

We can also switch P, Q in the above argument to show that $\sup_B (Q(B) - P(B)) = \frac{1}{2}\int |f(x) - g(x)|dx$, and therefore the statement in part (b) follows. □

Example 15.1 (Distances of Joint and Marginal Distributions). Suppose f_1, g_1 and f_2, g_2 are two pairs of univariate densities such that according to some metric (or distance), f_1 is close to g_1, and f_2 is close to g_2. Now make two bivariate random vectors, say (X_1, X_2) and (Y_1, Y_2), with $X_1 \sim f_1, X_2 \sim f_2, Y_1 \sim g_1, Y_2 \sim g_2$. Then we may expect that the joint distribution of (X_1, X_2) is close to the joint distribution of (Y_1, Y_2) in that same metric. Indeed, such concrete results hold if we assume that X_1, X_2 are independent, and that Y_1, Y_2 are independent. This example shows a selection of such results.

Consider first the total variation metric, with the independence assumptions stated in the above paragraph. Denote the distribution of X_i by $P_i, i = 1, 2$, and denote the distribution of Y_i by $Q_i, i = 1, 2$, Then, the total variation distance between the joint distribution P of (X_1, X_2) and the joint distribution Q of (Y_1, Y_2) is

$$2\rho(P,Q) = \int\int |f_1(x_1)f_2(x_2) - g_1(x_1)g_2(x_2)|dx_1 dx_2$$
$$= \int\int |f_1(x_1)f_2(x_2) - f_1(x_1)g_2(x_2) + f_1(x_1)g_2(x_2)$$
$$- g_1(x_1)g_2(x_2)|dx_1 dx_2$$
$$\leq \int\int f_1(x_1)|f_2(x_2) - g_2(x_2)|dx_1 dx_2 + \int\int g_2(x_2)|f_1(x_1)$$
$$- g_1(x_1)|dx_1 dx_2$$
$$= \int |f_2(x_2) - g_2(x_2)|dx_2 + \int |f_1(x_1) - g_1(x_1)|dx_1$$
$$= 2\rho(P_2, Q_2) + 2\rho(P_1, Q_1)$$
$$\Rightarrow \rho(P,Q) \leq \rho(P_1, Q_1) + \rho(P_2, Q_2).$$

In fact, quite evidently, the inequality generalizes to the case of k-variate joint distributions, $\rho(P,Q) \leq \sum_{i=1}^{k} \rho(P_i, Q_i)$, as long as we make the assumption of independence of the coordinates in the k-variate distributions.

Next consider the Kullback–Leibler distance between joint distributions, under the same assumption of independence of the coordinates. Consider the bivariate case for ease. Then,

$$K(P,Q) = \int\int f_1(x_1)f_2(x_2)\log\left(\frac{f_1(x_1)f_2(x_2)}{g_1(x_1)g_2(x_2)}\right)dx_1 dx_2$$
$$= \int\int f_1(x_1)f_2(x_2)[\log f_1(x_1) + \log f_2(x_2)$$
$$- \log g_1(x_1) - \log g_2(x_2)]dx_1 dx_2$$
$$= \int f_1(x_1)\log f_1(x_1)dx_1 + \int f_2(x_2)\log f_2(x_2)dx_2$$
$$- \int f_1(x_1)\log g_1(x_1)dx_1 - \int f_2(x_2)\log g_2(x_2)dx_2$$
$$= \int f_1(x_1)\log\frac{f_1(x_1)}{g_1(x_1)}dx_1 + \int f_2(x_2)\log\frac{f_2(x_2)}{g_2(x_2)}dx_2$$
$$= K(P_1, Q_1) + K(P_2, Q_2).$$

Once again, the result generalizes to the case of a general k; that is, under the assumption of independence of the coordinates of the k-variate joint distribution, $K(P,Q) = \sum_{i=1}^{k} K(P_i, Q_i)$.

A formula connecting $H(P,Q)$, the Hellinger distance, to $H(P_i, Q_i)$ is also possible, and is a chapter exercise.

Example 15.2 (Hellinger Distance Between Two Normal Distributions). The total variance distance between two general univariate normal distributions was worked out in Example 7.36. It was, in fact, somewhat involved. In comparison, the Hellinger and the Kullback–Leibler distance between two normal distributions is easy to find. We work out a formula for the Hellinger distance in this example. The Kullback–Leibler case is a chapter exercise.

15.2 Basic Properties of the Metrics

Let P be the measure corresponding to the d-variate $N_d(\mu_1, \Sigma)$ distribution, and Q the measure corresponding to the $N_d(\mu_2, \Sigma)$ distribution. We may assume without any loss of generality that $\mu_1 = \mathbf{0}$, and write simply μ for μ_2. Denote the corresponding densities by f, g. Then,

$$\int_{\mathcal{R}^d} \left(\sqrt{f} - \sqrt{g}\right)^2 dx$$

$$= 2\left[1 - \int_{\mathcal{R}^d} \sqrt{fg}\, dx\right]$$

$$= 2\left[1 - \frac{1}{(2\pi)^{d/2}|\Sigma|^{\frac{1}{2}}} \int_{\mathcal{R}^d} \sqrt{e^{-\frac{x'\Sigma^{-1}x}{2} - \frac{(x-\mu)'\Sigma^{-1}(x-\mu)}{2}}}\, dx\right]$$

$$= 2\left[1 - \frac{1}{(2\pi)^{d/2}|\Sigma|^{\frac{1}{2}}} \int_{\mathcal{R}^d} e^{-\frac{x'\Sigma^{-1}x}{4} - \frac{(x-\mu)'\Sigma^{-1}(x-\mu)}{4}}\, dx\right].$$

Now, writing $\Sigma^{-\frac{1}{2}}\mu = \eta$,

$$\frac{1}{(2\pi)^{d/2}|\Sigma|^{\frac{1}{2}}} \int_{\mathcal{R}^d} e^{-\frac{x'\Sigma^{-1}x}{4} - \frac{(x-\mu)'\Sigma^{-1}(x-\mu)}{4}}\, dx$$

$$= \frac{1}{(2\pi)^{d/2}} \int_{\mathcal{R}^d} e^{-\frac{(x'x)}{2} + \frac{\eta'x}{2}}\, dx \times e^{-\frac{\eta'\eta}{4}}$$

$$= \frac{1}{(2\pi)^{d/2}} \int_{\mathcal{R}^d} e^{-\frac{1}{2}x - \frac{\eta}{2}'(x-\frac{\eta}{2})}\, dx \times e^{-\frac{\eta'\eta}{8}} = e^{-\frac{\eta'\eta}{8}}.$$

This gives $H(P, Q) = \sqrt{2\left[1 - e^{-\frac{\mu'\Sigma^{-1}\mu}{8}}\right]}$. For the general case, when P corresponds to $N_d(\mu_1, \Sigma)$ and Q corresponds to $N_d(\mu_2, \Sigma)$, this implies that

$$H(P, Q) = \sqrt{2\left[1 - e^{-\frac{(\mu_2-\mu_1)'\Sigma^{-1}(\mu_2-\mu_1)}{8}}\right]}.$$

We recall from Chapter 7 that in the univariate case when the variances are equal, the total variation distance between $N(\mu_1, \sigma^2)$, and $N(\mu_2, \sigma^2)$ has the simple formula $2\Phi\left(\frac{|\mu_2-\mu_1|}{\sigma}\right) - 1$. The Hellinger distance and the total variation distance between $N(0, 1)$ and $N(\mu, 1)$ distributions in one dimension are plotted in Fig. 15.1 for visual comparison; the similarity of the two shapes is interesting.

Example 15.3 (Best Normal Approximation to a Cauchy Distribution). Consider the one-dimensional standard Cauchy distribution and denote it as P. Consider a general one-dimensional normal distribution $N(\mu, \sigma^2)$, and denote it as Q. We want to find the particular Q that minimizes some suitable distance between P and Q. We use the Kolmogorov metric $d(P, Q)$ in this example.

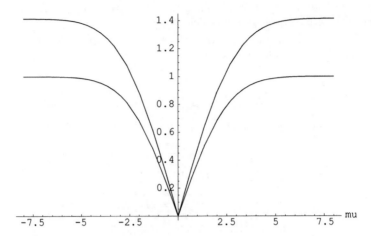

Fig. 15.1 Plot of Hellinger and variation distance between N(0, 1) and N(mu,1)

Because $d(P, Q) = \sup_x |F(x) - G(x)|$, where F, G are the CDFs of P, Q, respectively, it follows that there exists an x_0 such that $d(P, Q) = |F(x_0)-G(x_0)|$, and moreover, that at such an x_0, the two respective density functions f, g must cut. That is, we must have $f(x_0) = g(x_0)$ at such an x_0.

Now,

$$f(x_0) = g(x_0) \Leftrightarrow \left(1 + x_0^2\right) e^{-\frac{(x_0-\mu)^2}{2\sigma^2}} = \sigma\sqrt{\frac{2}{\pi}}$$

$$\Leftrightarrow \log\left(1 + x_0^2\right) - \frac{(x_0 - \mu)^2}{2\sigma^2} = c = \log\sigma + \frac{1}{2}\log\left(\frac{2}{\pi}\right).$$

The derivative with respect to x_0 of the lhs of the above equation equals

$$\frac{\mu + (2\sigma^2 - 1)x_0 + \mu x_0^2 - x_0^3}{\sigma^2(1 + x_0^2)}.$$

The numerator of this expression is a cubic, thus it follows that it can be zero at at most three values of x_0, and therefore the mean value theorem implies that the number of roots of the equation $f(x_0) = g(x_0)$ can be at most four. Let $-\infty < x_{1,\mu,\sigma} \leq x_{2,\mu,\sigma} \leq x_{3,\mu,\sigma} \leq x_{4,\mu,\sigma} < \infty$ be the roots of $f(x_0) = g(x_0)$. Then,

$$d(P, Q) = \max_{1 \leq i \leq 4} |F(x_{i,\mu,\sigma}) - G(x_{i,\mu,\sigma})| = d_{\mu,\sigma} \text{ say}.$$

and the best normal approximation to the standard Cauchy distribution is found by minimizing $d_{\mu,\sigma}$ over (μ, σ). This cannot be done in a closed-form analytical manner.

Numerical work gives that the minimum Kolmogorov distance is attained when $\mu = \pm.4749$ and $\sigma = 3.10$, resulting in the best normal approximation $N(\pm.4749, 9.61)$, and the corresponding minimum Kolmogorov distance of .0373.

15.2 Basic Properties of the Metrics

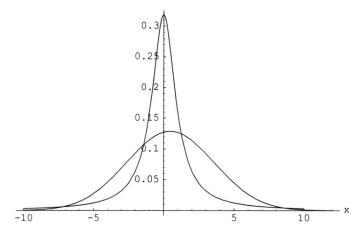

Fig. 15.2 Standard cauchy and the best Kolmogorov normal approximation

The best normal approximation is not symmetric around zero, although the standard Cauchy is. Moreover, the best normal approximation does not approximate the density well, although the Kolmogorov distance is only .0373. Both are plotted in Fig. 15.2 for a visual comparison.

Example 15.4 (f-Divergences). f-Divergences have a certain unifying character with respect to measuring distances between probability measures. A number of leading metrics and distances that we have defined above are f-divergences with special choices of the convex function f.

For example, if $f(x) = \frac{|x-1|}{2}$, then

$$d_f(P, Q) = \int q(x) f\left(\frac{p(x)}{q(x)}\right) dx = \frac{1}{2}\int q(x)\left|\frac{p(x)}{q(x)} - 1\right| dx$$

$$= \frac{1}{2}\int |p(x) - q(x)| dx = \rho(P, Q).$$

If we let $f(x) = -\log x$, then

$$d_f(P, Q) = \int q(x) f\left(\frac{p(x)}{q(x)}\right) dx = \int q(x)\left[-\log \frac{p(x)}{q(x)}\right] dx$$

$$= \int q(x) \log \frac{q(x)}{p(x)} dx = K(Q, P).$$

On the other hand, if we choose $f(x) = x \log x$, then

$$d_f(P, Q) = \int q(x)\left[\frac{p(x)}{q(x)} \log \frac{p(x)}{q(x)}\right] dx$$

$$= \int p(x) \log \frac{p(x)}{q(x)} dx = K(P, Q).$$

Note that two different choices of f were needed to produce $K(Q, P)$ and $K(P, Q)$; this is because the Kullback–Leibler distance is not symmetric between P and Q.

Next, if $f(x) = 2(1 - \sqrt{x})$, then

$$d_f(P, Q) = \int q(x) f\left(\frac{p(x)}{q(x)}\right) dx = 2 \int q(x) \left[1 - \sqrt{\frac{p(x)}{q(x)}}\right] dx$$

$$= 2 \left[1 - \int \sqrt{p(x)q(x)} dx\right] = H^2(P, Q).$$

Notice that we ended up with $H^2(P, Q)$ rather than $H(P, Q)$ itself. Interestingly, if we choose $f(x) = (1 - \sqrt{x})^2$, then

$$d_f(P, Q) = \int q(x) \left[1 - \frac{\sqrt{p}}{\sqrt{q}}\right]^2 dx$$

$$= \int q(x) \left[1 - 2\frac{\sqrt{p}}{\sqrt{q}} + \frac{p}{q}\right] dx = \int \left[q - 2\sqrt{pq} + p\right] dx$$

$$= 1 - 2\int \sqrt{pq} dx + 1 = 2\left[1 - \int \sqrt{pq} dx\right] = H^2(P, Q).$$

So, two different choices of the function f result in the f-divergence being the square of the Hellinger metric.

Example 15.5 (Distribution of Uniform Maximum). Suppose $X_1, X_2, \ldots$ are iid $U[0, 1]$ variables, and for $n \geq 1$, let $X_{(n)}$ be the maximum of $X_1, \ldots, X_n$. Then, for any given $x > 0$, and $n \geq x$,

$$P(n(1 - X_{(n)}) > x) = P\left(X_{(n)} < 1 - \frac{x}{n}\right) = \left(1 - \frac{x}{n}\right)^n \to e^{-x}$$

as $n \to \infty$. Thus, $n(1 - X_{(n)})$ converges in distribution to a standard exponential random variable, and moreover, the density of $n(1 - X_{(n)})$ at any fixed x, say $f_n(x)$, converges to the standard exponential density $f(x) = e^{-x}$. Therefore, by Scheffé's theorem (Chapter 7), $\rho(P_n, P) \to 0$, where P_n, P stand for the exact distribution of $n(1 - X_{(n)})$ and the standard exponential distribution, respectively. We obtain an approximation for $\rho(P_n, P)$ in this example.

Because $f_n(x) = (1 - \frac{x}{n})^{n-1} I_{\{0 < x < n\}}$, we have

$$2\rho(P_n, P) = \int_0^n \left|\left(1 - \frac{x}{n}\right)^{n-1} - e^{-x}\right| dx + \int_n^\infty e^{-x} dx$$

$$= \int_0^n \left|\left(1 - \frac{x}{n}\right)^{n-1} - e^{-x}\right| dx + e^{-n}.$$

15.3 Metric Inequalities

Now observe that the second derivative of $\log(1-\frac{x}{n})^{n-1} - \log(e^{-x})$ is $\frac{1-n}{n^2(1-\frac{x}{n})^2} < 0$ for all x in $(-\infty, n)$. Therefore, $\log(1 - \frac{x}{n})^{n-1}$ can cut $\log(e^{-x})$ at most twice on $(-\infty, n)$, which means that $(1 - \frac{x}{n})^{n-1}$ can cut e^{-x} at most twice on $(-\infty, n)$. Because there is obviously one cut at $x = 0$, there can be at most one more cut on $(-\infty, n)$, and hence at most one more cut on $(0, n)$. In fact, there is such a cut, as can be seen by observing that $(1 - \frac{x}{n})^{n-1} > e^{-x}$ for small positive x, whereas $(1 - \frac{x}{n})^{n-1} < e^{-x}$ at $x = n$. Denote this unique point of cut by x_n. Then,

$$2\rho(P_n, P) = \int_0^{x_n} \left[\left(1-\frac{x}{n}\right)^{n-1} - e^{-x}\right] dx + \int_{x_n}^n \left[e^{-x} - \left(1-\frac{x}{n}\right)^{n-1}\right] dx + e^{-n}.$$

The point of cut x_n can be analytically approximated as $x_n \approx 2$. This results, on doing the necessary integrations in the above line, to

$$2\rho(P_n, P) \approx \int_0^2 \left[\left(1-\frac{x}{n}\right)^{n-1} - e^{-x}\right] dx + \int_2^n \left[e^{-x} - \left(1-\frac{x}{n}\right)^{n-1}\right] dx + e^{-n}$$

$$= 2\left[e^{-2} - \left(1-\frac{2}{n}\right)^n\right] \approx \frac{4e^{-2}}{n}.$$

Thus, the total variation distance between the exact and the limiting distribution of $n(1 - X_{(n)})$ goes to zero at the rate $\frac{1}{n}$ and is asymptotic to $\frac{2e^{-2}}{n}$. Even for $n = 20$, the approximation $\frac{2e^{-2}}{n}$ is extremely accurate. The exact value for $n = 20$ is $\rho(P_n, P) = .01376$ and the approximation is $\frac{2e^{-2}}{n} = .01353$, the error being .00023.

15.3 Metric Inequalities

Due to an abundance of metrics and distances on probability measures, the question of which metric or distance to choose and what would happen if another metric or distance were chosen is very important. The interrelations between the various metrics are well illustrated by known inequalities that they share among themselves. These inequalities are also useful for the reason that some of the metrics are inherently easier to compute and some others are harder to compute. So, when an inequality is available, the computationally hard metric can be usefully approximated in terms of the computationally easier metric. The inequalities are also fundamentally interesting because of their theoretical elegance. A selection of metric inequalities is presented in this section.

Theorem 15.2.
(a) $d(P, Q) \leq \rho(P, Q) \leq H(P, Q) \leq \sqrt{K(P, Q)}$.
(b) $\rho(P, Q)$ satisfies the lower bound

$$\rho(P, Q) \geq \max\left\{L(P, Q), \frac{H^2(P, Q)}{2}, \frac{W(P, Q)}{diam(S)}\right\},$$

where S is the smallest set satisfying $P(S) = Q(S) = 1$, and $diam(S) = \sup\{\|x - y\| : x, y \in S\}$.

(c) If P, Q are discrete,

$$\rho(P, Q) \leq \min\left\{D(P, Q), \sqrt{\frac{K(P, Q)}{2}}\right\}.$$

(d) $[L(P, Q)]^2 \leq W(P, Q) \leq (1 + diam(S))L(P, Q)$.
(e) If P, Q are supported on the set of integers $\{0, \pm 1, \pm 2, \ldots\}$, then

$$W(P, Q) \geq \rho(P, Q).$$

Proof. We prove a few of these inequalities here; see Gibbs and Su (2002), and Reiss (1989) for the remaining parts.

Because $d(P, Q) = \sup_x |P((-\infty, x]) - Q((-\infty, x])|$, and $\rho(P, Q) = \sup_B |P(B) - Q(B)|$, it is obvious that $d(P, Q) \leq \rho(P, Q)$, because $\rho(P, Q)$ is a supremum over a larger collection of sets.

To prove that $\rho(P, Q) \leq H(P, Q)$, we use

$$(p - q) = (\sqrt{p} - \sqrt{q})(\sqrt{p} + \sqrt{q}),$$

and therefore, by the Cauchy–Schwarz inequality,

$$\left(\int |p - q|\right)^2 \leq \left(\int (\sqrt{p} - \sqrt{q})^2\right)\left(\int (\sqrt{p} + \sqrt{q})^2\right)$$

$$= H^2(P, Q)\left(\int (\sqrt{p} + \sqrt{q})^2\right)$$

$$\leq H^2(P, Q)\left(\int 2(p + q)\right) = 4H^2(P, Q)$$

$$\Rightarrow \rho^2(P, Q) = \frac{1}{4}\left(\int |p - q|\right)^2 \leq H^2(P, Q),$$

giving the inequality $\rho(P, Q) \leq H(P, Q)$.

We now prove $H(P, Q) \leq \sqrt{K(P, Q)}$, which is the same as $H^2(P, Q) \leq K(P, Q)$. For this, recall that $H^2(P, Q) = 2[1 - \int \sqrt{pq}]$. We now obtain a suitable lower bound on $\int \sqrt{pq}$, which leads to the desired upper bound on $H^2(P, Q)$. The lower bound on $\int \sqrt{pq}$ is

$$\int \sqrt{pq} = \int \sqrt{\frac{q}{p}}\, p = E_P\left(\sqrt{\frac{q}{p}}\right)$$

$$= E_P(e^{\frac{1}{2}\log \frac{q}{p}}) \geq e^{\frac{1}{2} E_P (\log \frac{q}{p})}$$

(by Jensen's inequality applied to the convex function $u \to e^u$)

$$= e^{\frac{1}{2}\int p \log \frac{q}{p}} \geq 1 + \frac{1}{2}\int p \log \frac{q}{p}$$

15.3 Metric Inequalities

(by using the inequality $e^u \geq 1 + u$)

$$= 1 - \frac{1}{2} \int p \log \frac{p}{q} = 1 - \frac{1}{2} K(P, Q)$$

$$\Rightarrow 1 - \int \sqrt{pq} \leq \frac{1}{2} K(P, Q) \Rightarrow H^2(P, Q) \leq K(P, Q).$$

We now sketch the proof of the particular inequality in part (c) that in the discrete case $\rho(P, Q) \leq \sqrt{\frac{K(P,Q)}{2}}$. For this, we show that $2\rho^2(P, Q) \leq K(P, Q)$. This bound is proved by reducing the general case to the case of P, Q being two-point distributions, supported on the same two points. Suppose then at first that P, Q are distributions given by $P(0) = 1 - P(1) = p$, $Q(0) = 1 - Q(1) = r$. Then,

$$2\rho^2(P, Q) = 2(p - r)^2 \leq p \log \frac{p}{r} + (1 - p) \log \frac{1-p}{1-r},$$

which is an entropy inequality. Thus, for such two-point distributions supported on the same two points, $2\rho^2(P, Q) \leq K(P, Q)$.

Now take general P, Q, and consider the set $B_0 = \{i : p(i) \geq q(i)\}$. Also, consider the special two-point distributions P^*, Q^* with supports on $\{0, 1\}$ and defined by $p = P^*(0) = P(B_0)$, and $Q^*(0) = Q(B_0)$. Then, from the proof of part (b) of Theorem 15.1, $\rho(P, Q) = \rho(P^*, Q^*)$. On the other hand, by the *finite partition property* of general f-divergences (see Section 15.1), $K(P, Q) \geq K(P^*, Q^*)$. Therefore,

$$2\rho^2(P, Q) = 2\rho^2(P^*, Q^*) \leq K(P^*, Q^*) \leq K(P, Q).$$

We finally give a proof of part (e) in a special case. The special case we consider is when the two mass functions $\{p_i\}$ and $\{r_i\}$ corresponding to the two distributions P, Q have one cut; that is, there exists i_0 such that $p_i \geq r_i$ for $i \leq i_0$ and $p_i \leq r_i$ for $i > i_0$. Then,

$$W(P, Q) = \int |F(x) - G(x)| dx = \sum_k \int_k^{k+1} |F(x) - G(x)| dx$$

$$= \sum_k \left| \sum_{i \leq k} (p_i - r_i) \right| \geq \left| \sum_{i \leq i_0} (p_i - r_i) \right| = \sum_{i \leq i_0} (p_i - r_i) = \rho(P, Q).$$

□

Example 15.6 (Total Variation versus Hellinger). The inequalities in parts (a) and (b) of the above theorem say that the planar point $(H(P, Q), \rho(P, Q))$ for any P and Q falls in the convex region bounded by the straight line $y \leq x$, the parabolic curve $y \geq \frac{x^2}{2}$, and the rectangle $[0, \sqrt{2}] \otimes [0, 1]$. In Fig. 15.3, we first provide a plot of the curve of the set of points $(H(P, Q), \rho(P, Q))$ as P, Q run through the family

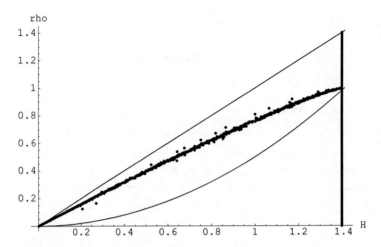

Fig. 15.3 Plot of total variation versus. Hellinger for P, Q in Poisson family

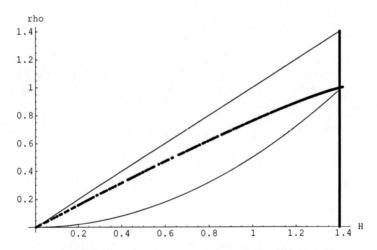

Fig. 15.4 Plot of total variation versus. Hellinger for P, Q univariate normal

of Poisson distributions. This curve is the dark curve inside the region bounded by $y = x, y = \frac{x^2}{2}, x \geq 0, x \leq \sqrt{2}$, which are also plotted to give a perspective for where the curve for the Poisson family lies within the admissible region. The curve for the Poisson family was obtained by computing $(H(P, Q), \rho(P, Q))$ at 1,000 randomly chosen pairs of Poisson distributions with means between 0 and 10. Then, the same plot is provided when P, Q run through the family of univariate normal distributions in Fig. 15.4. The curves for the Poisson and the normal case look very similar.

15.4 * Differential Metrics for Parametric Families

If the probability measures P, Q both belong to a common parametric family of distributions parametrized by some finite-dimensional parameter θ, then some of the distances and metrics discussed by us can be used to produce new simple distance measures between two distributions in that parametric family. The new distances are produced by consideration of geometric ideas and have some direct relevance to statistical estimation of unknown finite-dimensional Euclidean parameters. The approach was initiated in Rao (1945). Two subsequent references are Kass and Vos (1997) and Rao (1987). We begin with an illustrative example.

Example 15.7 (Curvature of the Kullback–Leibler Distance). Let $\{f(x|\theta)\}, \theta \in \Theta \subseteq \mathcal{R}$ be a family of densities indexed by a real parameter θ. Certain formal operations are done in this example, whose justifications require a set of regularity conditions on the densities $f(x|\theta)$. Let P_θ, P_ϕ be two distributions in this parametric family. Then, the Kullback–Leibler distance between them is $K(P_\theta, P_\phi) = \int f(x|\theta) \log \frac{f(x|\theta)}{f(x|\phi)} dx = J_\theta(\phi)$ (say). Formally expanding this as a function of ϕ around the point θ, we have

$$J_\theta(\phi) = J_\theta(\theta) + (\phi - \theta) \frac{d}{d\phi} J_\theta(\phi) \Big|_{\phi=\theta} + \frac{(\phi - \theta)^2}{2} \frac{d^2}{d\phi^2} J_\theta(\phi) \Big|_{\phi=\theta} + \cdots$$

$$= 0 + (\phi - \theta) \int f(x|\theta) \frac{d}{d\phi}(-\log f(x|\phi)) dx \Big|_{\phi=\theta}$$

$$+ \frac{(\phi - \theta)^2}{2} \int f(x|\theta) \frac{d^2}{d\phi^2}(-\log f(x|\phi)) dx \Big|_{\phi=\theta} + \cdots$$

$$= (\phi - \theta) \int f(x|\theta) \frac{-\frac{d}{d\theta} f(x|\theta)}{f(x|\theta)} dx$$

$$+ \frac{(\phi - \theta)^2}{2} \int f(x|\theta) \left[\frac{-\frac{d^2}{d\theta^2} f(x|\theta)}{f(x|\theta)} + \frac{\left(\frac{d}{d\theta} f(x|\theta)\right)^2}{f^2(x|\theta)} \right] dx + \cdots$$

$$= -(\phi - \theta) \int \frac{d}{d\theta} f(x|\theta) dx - \frac{(\phi - \theta)^2}{2} \int \frac{d^2}{d\theta^2} f(x|\theta) dx$$

$$+ \frac{(\phi - \theta)^2}{2} \int \frac{\left(\frac{d}{d\theta} f(x|\theta)\right)^2}{f(x|\theta)} dx + \cdots$$

$$= -(\phi - \theta) \frac{d}{d\theta} \int f(x|\theta) dx - \frac{(\phi - \theta)^2}{2} \frac{d^2}{d\theta^2} \int f(x|\theta) dx$$

$$+ \frac{(\phi - \theta)^2}{2} \int \frac{\left(\frac{d}{d\theta} f(x|\theta)\right)^2}{f(x|\theta)} dx + \cdots$$

$$= 0 + 0 + \frac{(\phi - \theta)^2}{2} \int \frac{\left(\frac{d}{d\theta} f(x|\theta)\right)^2}{f(x|\theta)} dx$$

$$= \frac{(\phi - \theta)^2}{2} \int \frac{\left(\frac{d}{d\theta} f(x|\theta)\right)^2}{f(x|\theta)} dx + \cdots.$$

The quantity

$$\int \frac{\left(\frac{d}{d\theta} f(x|\theta)\right)^2}{f(x|\theta)} dx = E_\theta \left[\frac{d}{d\theta} \log f(x|\theta)\right]^2$$

is called the *Fisher information function for the family* $\{f(x|\theta)\}$ and is usually denoted as $I_f(\theta)$. Thus, subject to the validity of the formal calculations done in the above lines, for $\phi \approx \theta$, $K(P_\theta, P_\phi) \approx \frac{(\phi-\theta)^2}{2} I_f(\theta)$. In other words, the Fisher information function $I_f(\theta)$ measures the curvature of the Kullback–Leibler distance at the particular distribution P_θ, and can be used as a measure for how quickly the measure P_θ is changing if we change θ slightly.

15.4.1 * Fisher Information and Differential Metrics

The calculations of this example can be generalized to the case of multivariate densities more general than the normal, and to the case of vector-valued parameters. Furthermore, the basic divergence to start with can be more general than the Kullback–Leibler distance. Such a general treatment is sketched below; we follow Rao (1987).

Let $F : \mathcal{R}^+ \otimes \mathcal{R}^+ \to \mathcal{R}$ satisfy the following properties.

(a) F is three times continuously partially differentiable with respect to each coordinate.
(b) $F(x, x) = 0 \, \forall x \in \mathcal{R}^+$.
(c) $F(x, y)$ is strictly convex in y for each given $x \in \mathcal{R}^+$.
(d) $\frac{\partial F}{\partial y} F(x, y)|_{y=x} = 0 \, \forall x \in \mathcal{R}^+$.

Let $\{f(x|\theta)\}$ be a family of densities (or mass functions) with common support $S \subseteq \mathcal{R}^k$ for some $k, 1 \le k < \infty$, and $\theta \in \Theta \subseteq \mathcal{R}^p$ for some $p, 1 \le p < \infty$. The functions $f(x|\theta)$ are assumed to satisfy the following properties.

(i) $\frac{\partial^3}{\partial \theta_i^3} f(x|\theta)$ exists for each $i = 1, 2, \ldots, p$ and for each x.
(ii) For each fixed θ, the divergence functional

$$J(\theta, \phi) = \int_S F(f(x|\theta), f(x|\phi)) dx$$

(the integral to be replaced by a sum if the distributions are discrete) can be partially differentiated twice with respect to each $\phi_i, i = 1, 2, \ldots, p$ inside the

15.4 Differential Metrics for Parametric Families

integral. Fix θ and consider another $\phi = \theta + d\theta \approx \theta$. Then the change in the divergence functional $J(\theta, \phi)$ from $0 = J(\theta, \theta)$ due to changing θ to ϕ is

$$\nabla J \approx \sum_{i=1}^{p} \frac{\partial}{\partial \phi_i} J(\theta, \phi)\bigg|_{\phi=\theta} d\theta_i + \sum_{i,j=1}^{p} \frac{\partial^2}{\partial \phi_i \partial \phi_j} J(\theta, \phi)\bigg|_{\phi=\theta} d\theta_i d\theta_j$$

$$= \sum_{i,j=1}^{p} \frac{\partial^2}{\partial \phi_i \partial \phi_j} J(\theta, \phi)|_{\phi=\theta} d\theta_i d\theta_j,$$

as the first-order partial derivatives will vanish under the assumptions made on F and f (in the above, when $i = j$, the iterated partial derivatives denote the second order partial derivative). Now, by a direct differentiation under the integral sign, we get

$$\frac{\partial^2}{\partial \phi_i \partial \phi_j} J(\theta, \phi)|_{\phi=\theta}$$

$$= \int F_{yy}(f(x|\theta), f(x|\theta)) \left(\frac{\partial}{\partial \theta_i} f(x|\theta) \right) \left(\frac{\partial}{\partial \theta_j} f(x|\theta) \right) dx = g^F_{i,j}(\theta)(\text{say}),$$

where the notation F_{yy} means the second partial derivative of $F(x, y)$ with respect to y. The quantity $\sum_{i,j=1}^{p} g^F_{i,j}(\theta) d\theta_i d\theta_j$ is called *the differential metric on Θ induced by F*.

Suppose now we specialize to the case of f-divergences, so that our functional $F(x, y)$ is now $x\phi \frac{y}{x}$. The notation ϕ is used so as not to cause confusion with the underlying densities, which have been denoted as f. In this case, $g^F_{i,j}(\theta)$ takes the form

$$g^F_{i,j}(\theta) = f''(1) \int \frac{1}{f(x|\theta)} \left(\frac{\partial f}{\partial \theta_i} \right) \left(\frac{\partial f}{\partial \theta_j} \right)$$

$$= f''(1) \int f(x|\theta) \left(\frac{\partial \log f}{\partial \theta_i} \right) \left(\frac{\partial \log f}{\partial \theta_j} \right).$$

The integral

$$\int f(x|\theta) \left(-\frac{\partial \log f}{\partial \theta_i} \right) \left(\frac{\partial \log f}{\partial \theta_j} \right) = E_\theta \left[-\left(\frac{\partial \log f}{\partial \theta_i} \right) \left(\frac{\partial \log f}{\partial \theta_j} \right) \right]$$

is the (i, j)th element in the *Fisher information matrix* corresponding to the regular family of densities $\{f(x|\theta)\}$. Thus, apart from the constant multiplier $f''(1)$, the differential metric arising from all of these divergences is the same. This is an interesting unifying phenomenon, and at the same time shows that the Fisher information function has a special role.

15.4.2 * Rao's Geodesic Distances on Distributions

The differential metric corresponding to the Fisher information function can be used to produce distances between two members P_θ, P_ϕ of a parametric family of distributions where the parameter belongs to a *manifold* in some Euclidean space. The distance between P_θ and P_ϕ is simply the geodesic distance between θ and ϕ arising from the differential metric on Θ, the parameter space. The geodesic distance is basically the distance between θ and ϕ along the shortest curve joining the two points along the manifold Θ, This approach was initiated in Rao (1945), and in special parametric families that we encounter in applications, the geodesic distance works out to neat and interesting distances. Specifically, the geodesic distance often has a connection to the form of a *variance-stabilizing transformation* (see Chapter 7). The geodesic distance is defined below.

Definition 15.3. Let Θ be a manifold in an Euclidean space $\mathcal{R}^p$ for some $p, 1 \leq p < \infty$, and d a metric on it. Let $\theta, \phi \in \Theta$, and let $\mathcal{C}$ be the family of curves $\eta(t)$ on $[0, 1]$ defined as

$$\mathcal{C} = \{\eta(t) : \eta(0) = \theta, \eta(1) = \phi, \eta(t) \text{ is piecewise continuously differentiable on}[0, 1]\}.$$

The *geodesic distance* between θ and ϕ is the infimum of the lengths of the curves $\eta(t) \in \mathcal{C}$, with length of a curve being defined with respect to the metric d.

Geodesic curves are usually hard to compute, and need not, in general, be unique. Using calculus of variation methods, it can be shown that a geodesic curve is a solution to the following boundary value problem. Find a curve $\eta(t)$ such that

$$\eta(0) = \theta; \quad \eta(1) = \phi;$$
$$\frac{d^2 \eta_k}{dt^2} + \Gamma_{ijk} \frac{d\eta_i}{dt} \frac{d\eta_j}{dt} = 0,$$

where

$$\Gamma_{ijk} = \frac{1}{2} \left[\frac{\partial}{\partial \theta_i} g_{jk}(\theta) + \frac{\partial}{\partial \theta_j} g_{ki}(\theta) - \frac{\partial}{\partial \theta_k} g_{ij}(\theta) \right].$$

The geodesic distances have been calculated in the literature for several well-known parametric families of distributions. They are always nontrivial to calculate. We provide a selection of these formulas; see Rao (1987) for these formulas.

Distribution	Density	Geodesic Distance
Binomial	$\binom{n}{x} p^x (1-p)^{n-x}$	$2\sqrt{n} \lvert \arcsin(\sqrt{p_1}) - \arcsin(\sqrt{p_2}) \rvert$
Poisson	$\dfrac{e^{-\lambda}\lambda^x}{x!}$	$2\lvert \sqrt{\lambda_1} - \sqrt{\lambda_2} \rvert$
Geometric	$(1-p)p^x$	$2\log \dfrac{1-\sqrt{p_1 p_2}+\lvert \sqrt{p_1}-\sqrt{p_2}\rvert}{\sqrt{(1-p_1)(1-p_2)}}$
Gamma	$\dfrac{e^{-\theta x}\theta^\alpha x^{\alpha-1}}{\Gamma(\alpha)}$	$\sqrt{\alpha}\,\lvert \log\theta_1 - \log\theta_2 \rvert$
Normal (fixed variance)	$\dfrac{1}{\sqrt{2\pi}\sigma} e^{-\frac{(x-\mu)^2}{2\sigma^2}}$	$\dfrac{\lvert \mu_1-\mu_2 \rvert}{\sigma}$
Normal (fixed mean)	$\dfrac{1}{\sqrt{2\pi}\sigma} e^{-\frac{(x-\mu)^2}{2\sigma^2}}$	$\sqrt{2}\,\lvert \log\sigma_1 - \log\sigma_2 \rvert$
General Normal	$\dfrac{1}{\sqrt{2\pi}\sigma} e^{-\frac{(x-\mu)^2}{2\sigma^2}}$	$2\sqrt{2}\tanh^{-1}\sqrt{\dfrac{(\mu_1-\mu_2)^2+2(\sigma_1-\sigma_2)^2}{(\mu_1-\mu_2)^2+2(\sigma_1+\sigma_2)^2}}$
p-Variate Normal (Σ fixed)	$\dfrac{e^{-\frac{1}{2}(x-\mu)'\Sigma^{-1}(x-\mu)}}{(2\pi)^{p/2}\lvert\Sigma\rvert^{1/2}}$	$(\mu_1-\mu_2)'\Sigma^{-1}(\mu_1-\mu_2)$
p-Variate Normal (μ fixed)	$\dfrac{e^{-\frac{1}{2}(x-\mu)'\Sigma^{-1}(x-\mu)}}{(2\pi)^{p/2}\lvert\Sigma\rvert^{1/2}}$ (here, $\{\lambda_i\}$ are the	$\dfrac{1}{\sqrt{2}}\sqrt{\sum_{i=1}^p \log \lambda_i^2}$ roots of $\lvert\Sigma_2 - \lambda\Sigma_1\rvert = 0$)
Multinomial	$\dfrac{n!}{\prod_{i=1}^k n_i!} p_i^{n_i}$	$2\sqrt{\pi}\arccos(\sum_{i=1}^k \sqrt{p_i\theta_i})$

Exercises

Exercise 15.1 (Skills Exercise). Compute the Hellinger and the Kullback–Leibler distance between P and Q when (a) $P = U[0,1]$, $Q = \text{Beta}(2,2)$; (b) $P = U[-a,a]$, $Q = N(0,1)$; (c) $P = \text{Exp}(\lambda)$, $Q = \text{Gamma}(\alpha, \lambda)$.

Exercise 15.2 (A Useful Formula). Compute the Kullback–Leibler distance between two general d-dimensional normal distributions.

Exercise 15.3. Suppose $X_1 \sim P_1, X_2 \sim P_2, Y_1 \sim Q_1, Y_2 \sim Q_2$, and that X_1, X_2 are independent and Y_1, Y_2 are independent. Let P, Q denote the joint distributions of (X_1, X_2) and (Y_1, Y_2). Show that $1 - \frac{1}{2}H^2(P,Q) = \left[1 - \frac{1}{2}H^2(P_1, Q_1)\right]\left[1 - \frac{1}{2}H^2(P_2, Q_2)\right]$.

Exercise 15.4. Suppose $X_1, X_2, \ldots, X_n$ are iid $U[0,1]$. Compute the Kullback–Leibler distance between $P = P_n$ and Q where P is the exact distribution of $n(1 - X_{(n)})$ and Q stands for an exponential distribution with mean one. Hence, find a sequence $c_n \to 0$ such that $\dfrac{K(P_n, Q)}{c_n} \to 1$ as $n \to \infty$.

Exercise 15.5 *(**How Large is the Class of t Distributions**). Consider the class $\mathcal{F}$ of all one-dimensional t densities symmetric about zero; that is,

$$\mathcal{F} = \left\{ f : f(x) = \dfrac{\Gamma\left(\frac{\alpha+1}{2}\right)}{\sqrt{\alpha\pi}\,\Gamma\left(\frac{\alpha}{2}\right)\left(1+\frac{x^2}{\alpha}\right)^{(\alpha+1)/2}}, \alpha > 0 \right\}.$$

Show that $\sup_{\{f,g \in \mathcal{F}\}} \rho(f,g) = 1$.

Exercise 15.6. Calculate the Kullback–Leibler distance between P and Q, where P, Q are the $\text{Bin}(n, p)$ and $\text{Bin}(n, \theta)$ distributions.

Exercise 15.7 * **(Binomial and Poisson).** Let P_n be the $\text{Bin}(n, \frac{\lambda}{n})$ distribution and Q a Poisson distribution with mean λ; here $0 < \lambda < \infty$ is a fixed number independent of n. Prove that the total variation distance between P_n and Q converges to zero as $n \to \infty$.

Exercise 15.8 * **(Convergence of Discrete Uniforms).** Let U_n have a discrete uniform distribution on $\{1, 2, \ldots, n\}$ and let $X_n = \frac{U_n}{n}$. Let P_n denote the distribution of X_n. Identify a distribution Q such that the Kolmogorov distance between P_n and Q goes to zero, and then compute this Kolmogorov distance.

Does there exist any distribution R such that the total variation distance between P_n and R converges to zero? Rigorously justify your answer.

Exercise 15.9 * **(Generalized Hellinger Distances).** For $\alpha > 0$, define

$$H_\alpha(P, Q) = \left[\int |f^\alpha(x) - g^\alpha(x)|^{\frac{1}{\alpha}} dx \right]^\alpha.$$

For what values of α, is H_α a distance?

Exercise 15.10 * **(An Interesting Plot).** Use the formula in the text (Example 7.36) for the total variation distance between two general univariate normal distributions to plot the following set.

$$S_\epsilon = \{(\mu, \sigma) : \rho(N(0, 1), N(\mu, \sigma^2)) \le \epsilon\}.$$

Use $\epsilon = .01, .05, .1$.

Exercise 15.11 (A Bayes Problem). Suppose $X_1, X_2, \ldots, X_n$ are iid $N(\mu, 1)$ and that μ has a $N(0, \tau^2)$ prior distribution. Compute the total variation distance between the posterior distribution of $\sqrt{n}(\mu - \bar{X})$ and the $N(0, 1)$ distribution, and show that it goes to zero as $n \to \infty$.

Exercise 15.12 * **(Variation Distance Between Multivariate Normals).** Let P be the $N_p(\mu_1, \Sigma)$ and Q the $N_p(\mu_2, \Sigma)$ distribution. Prove that the total variation distance between P and Q admits the formula

$$\rho(P, Q) = 2\Phi\left(\frac{\sqrt{(\mu_1 - \mu_2)' \Sigma^{-1} (\mu_1 - \mu_2)}}{2} \right) - 1,$$

where Φ denotes the standard normal CDF.

Exercise 15.13 (Metric Inequalities). Directly verify the series of inequalities $d(P, Q) \le \rho(P, Q) \le H(P, Q) \le \sqrt{K(P, Q)}$ when $P = N(0, 1)$ and $Q = N(\mu, \sigma^2)$. Are these inequalities all sharp in this special case?

Exercise 15.14 * (**Converse of Scheffé's Theorem Is False**). Give an example of a sequence of densities f_n and another density f such that f_n converges to f in total variation, but $f_n(x)$ does not converge pointwise to $f(x)$.

Exercise 15.15 (**Bhattacharya Affinity**). Given two densities f and g on a general Euclidean space $\mathcal{R}^k$, let $A(f, g) = 1 - \frac{2}{\pi} \arccos(\int_{\mathcal{R}^k} \sqrt{fg})$. Express the Hellinger distance between f and g in terms of $A(f, g)$.

Exercise 15.16 (**Chi-Square Distance**). If P, Q are probability measures on a general Euclidean space with densities f, g with a common support S, define $\chi^2(P, Q) = \sqrt{\int_S \frac{(f-g)^2}{g}}$.

(a) Show that $\chi^2(P, Q)$ is a distance, but not a metric.
(b) Show that $\rho(P, Q) \leq \frac{1}{2}\chi^2(P, Q)$.
(c) Show that $H(P, Q) \leq \chi^2(P, Q)$.

Exercise 15.17 (**Alternative Formula for Wasserstein Distance**). Let P, Q be distributions on the real line. with CDFs F, G. Show that the Wasserstein distance between P and Q satisfies $W(P, Q) = \int_0^1 |F^{-1}(\alpha) - G^{-1}(\alpha)| d\alpha$.

Exercise 15.18. * Prove that almost sure convergence is not metrized by any metric on probability measures.

Exercise 15.19. * Given probability measures P, Q on $\mathcal{R}$ with finite means, suppose $\mathcal{M}$ is the set of all joint distributions with marginals equal to P and Q. Show that the Wasserstein distance has the alternative representation $W(P, Q) = \inf\{E(|X - Y|) : (X, Y) \sim M, M \in \mathcal{M}\}$.

Exercise 15.20 * (**Generalized Wasserstein Distance**). Let $p \geq 1$. Given probability measures P, Q on a general Euclidean space $\mathcal{R}^k$ with finite pth moment, let $\mathcal{M}$ be the set of all joint distributions with marginals equal to P and Q. Consider $W_p(P, Q) = \inf\{E(||X - Y||_p) : (X, Y) \sim M, M \in \mathcal{M}\}$.

(a) Show that W_p is a metric.
(b) Suppose $W_p(P_n, P) \to 0$. For which values of p, does this imply that $P_n \overset{\mathcal{L}}{\Rightarrow} P$?

References

DasGupta, A. (2008). *Asymptotic Theory of Statistics and Probability*, Springer, New York.
Diaconis, P. and Saloff-Coste, L. (2006). Separation cut-offs for birth and death chains, *Ann. Appl. Prob.*, 16, 2098–2122.
Dudley, R. (2002). *Real Analysis and Probability*, Cambridge University Press, Cambridge, UK.
Gibbs, A. and Su, F. (2002). On choosing and bounding probability metrics, *Internat. Statist. Rev.*, 70, 419–435.
Ibragimov, I.A. (1956). On the composition of unimodal distributions, *Theor. Prob. Appl.*, 1, 283–288.

Kass, R. and Vos, p. (1997). *Geometrical Foundations of Asymptotic Inference*, Wiley, New York.
Lecam, L. (1969). Théorie Asymptotique de la Bécision statistique, Les presses de l'université de Montréal, Montréal.
Leise, F. and Vajda, I. (1987). *Convex Statistical Distances*, Teubner, Leipzig.
Rachev, S.T. (1991). *Probability Metrics and the Stability of Stochastic Models*, Wiley, New York.
Rao, C.R. (1945). Information and accuracy available in the estimation of statistical parameters, *Bull. Calcutta Math. Soc.*, 37, 81–91.
Rao, C.R. (1987). Differential metrics in probability spaces, in *Differential Geometry in Statistical Inference*, S.-I. Amari et al. Eds., IMS Lecture Notes and Monographs Series, Hayward, CA.
Reiss, R. (1989). *Approximation Theorems of Order Statistics*, Springer-Verlag, New York.
Zolotarev, V.M. (1983). Probability metrics, *Theor. Prob. Appl.*, 28, 2, 264–287.

Chapter 16
Empirical Processes and VC Theory

Like martingales, empirical processes also unify an incredibly large variety of problems in probability and statistics. Results in empirical processes theory are applicable to numerous classic and modern problems in probability and statistics; a few examples of applications are the study of central limit theorems in more general spaces than Euclidean spaces, the bootstrap, goodness of fit, density estimation, and machine learning. Familiarity with the basic theory of empirical processes is extremely useful across fields in probability and statistics.

Empirical process theory has seen a major revolution since the early 1970s with the emergence of the powerful *Vapnik–Chervonenkis theory*. The classic theory of empirical processes relied primarily on tools such as invariance principles and the Wiener process. With the advent of the VC (Vapnik–Chervonenkis) theory, combinatorics and geometry have increasingly become major tools in studies of empirical processes. In some sense, the theory of empirical processes has two quite different faces: the classic aspect and the new aspect. Both are useful. We provide an introduction to both aspects of the empirical process theory in this chapter. Among the enormous literature on empirical process theory, we recommend Shorack and Wellner, (2009) Csörgo and Révész (1981), Dudley (1984), van der Vaart and Wellner (1996), and Kosorok (2008) for comprehensive treatments, and Pollard (1989), Giné (1996), Csörgo (2002), and Del Barrio et al. (2007) for excellent reviews and overviews. A concise treatment is also available in DasGupta (2008). Other specific references are provided in the sections.

16.1 Basic Notation and Definitions

Given n real-valued random variables $X_1, \ldots, X_n$, the *empirical CDF* of $X_1, \ldots, X_n$ is defined as

$$F_n(t) = \frac{\#\{X_i : X_i \leq t\}}{n} = \frac{1}{n}\sum_{i=1}^{n} I_{\{X_i \leq t\}}, \quad t \in \mathcal{R}.$$

If $X_1, \ldots, X_n$ happen to be iid, with the common CDF F, then $nF_n(t) \sim \text{Bin}(n, F(t))$ for any fixed t. The mean of $F_n(t)$ is $F(t)$, and the variance is $\frac{F(t)(1-F(t))}{n}$.

So, a natural centering for $F_n(t)$ is $F(t)$, and a natural scaling is $\sqrt{n}$. Indeed, by the CLT for iid random variables with a finite variance, for any fixed t, $\sqrt{n}[F_n(t) - F(t)] \stackrel{\mathcal{L}}{\Rightarrow} N(0, F(t)(1 - F(t)))$. We can think of t as a running time parameter, and define a process $\beta_n(t) = \sqrt{n}[F_n(t) - F(t)], t \in \mathcal{R}$. This is the one-dimensional (normalized) *empirical process*. We can talk about it in more generality than the iid case; but we only consider the iid case in this chapter.

Note that the empirical CDF is a finitely supported CDF; it has the data values as its jump points. Any CDF gives rise to a probability distribution (measure), and so does the empirical CDF F_n. The *empirical measure* P_n is defined by $P_n(A) = \frac{\#\{X_i : X_i \in A\}}{n}$, for a general set $A \subseteq \mathcal{R}$ (we do not mention it later, but we need A to be a *Borel* set).

There is no reason why we have to restrict such a definition to the case of real-valued random variables. We can talk about an empirical measure in far more general spaces. In particular, given d-dimensional random vectors $\mathbf{X}_1, \ldots, \mathbf{X}_n$ from some distribution P in $\mathcal{R}^d$, $1 \leq d < \infty$, we can define the empirical measure in exactly the same way, namely, $P_n(A) = \frac{\#\{X_i : X_i \in A\}}{n}$, $A \subseteq \mathcal{R}^d$.

Why is a study of the empirical process (measure) so useful? A simple explanation is that all the information about the underlying P in our data is captured by the empirical measure. So, if we know how to manipulate the empirical measure properly, we should be able to analyze and understand the probabilistic aspects of a problem, and also analyze and understand particular methods to infer any unknown aspects of the underlying P.

A useful reduction for many problems in the theory of empirical processes is that we can often pretend as if the true F in the one-dimensional case is the CDF of the $U[0, 1]$ distribution. For this reason, we need a name for the empirical process in the special $U[0, 1]$ case. If $U_1, \ldots, U_n$ are iid $U[0, 1]$ variables, then the normalized *uniform empirical process* is $\alpha_n(t) = \sqrt{n}[G_n(t) - t], t \in [0, 1]$, where $G_n(t) = \frac{\#\{U_i : U_i \leq t\}}{n}$.

Just as the empirical CDF mimics the true underlying CDF well in large samples, the empirical percentiles mimic the true population percentiles well in large samples. Analogous to the empirical process, we can define a *quantile process*. Suppose $X_1, X_2, \ldots, X_n$ are iid with a common CDF F, which we assume to be continuous. The quantile function of F is defined in the usual way, namely,

$$Q(y) = F^{-1}(y) = \inf\{t : F(t) \geq y\}, \quad 0 < y \leq 1.$$

The empirical quantile function is defined in terms of the order statistics of the sample values $X_1, X_2, \ldots, X_n$, namely,

$$Q_n(y) = F_n^{-1}(y) = X_{k:n}, \quad \frac{k-1}{n} < y \leq \frac{k}{n},$$

$1 \leq k \leq n$, where $X_{1:n}, X_{2:n}, \ldots, X_{n:n}$ are the order statistics of $X_1, X_2, \ldots, X_n$. The *normalized quantile process* is simply $q_n(y) = \sqrt{n}[Q_n(y) - Q(y)]$, $0 < y \leq 1$. We often do not mention the normalized term, and just call it the quantile process. In the special case of $U[0, 1]$ variables, we of course have $Q(y) = y$. We use the notation $U_n(y)$ for the quantile function $Q_n(y)$ in the special $U[0, 1]$ case, which gives us the *uniform quantile process* $u_n(y) = \sqrt{n}[U_n(y) - y], 0 < y \leq 1$.

Given a real-valued function f on some interval $[a, b] \subseteq \mathcal{R}$, let $||f||_\infty = \sup_{a \leq x \leq b} |f(x)|$. We often call it the *uniform norm* or the L_∞ norm of f. The paths of the empirical process are only right continuous; indeed, $F_n(t)$ jumps at the data values $X_1, \ldots, X_n$. However, at any point t, $F_n(t)$ has a left limit. We need a name for functions of these types. Let $[a, b]$ be an interval with $-\infty \leq a < b \leq \infty$. Then, $C[a, b]$ denotes the class of all continuous real-valued functions on $[a, b]$, and $\ell^\infty[a, b]$ denotes the class of all real-valued functions f with $||f||_\infty < \infty$. The class of all real-valued functions that are right continuous on $[a, b]$ and have left limits everywhere is denoted by $D[a, b]$; they are commonly known as *cadlag functions*. An important inclusion property is $C[a, b] \subset D[a, b] \subset \ell^\infty[a, b]$.

As an estimate of the true CDF F, the empirical CDF F_n is uniformly accurate in large samples. Indeed, the Glivenko–Cantelli theorem says that (in the iid case) $||F_n - F||_\infty \stackrel{a.s.}{\to} 0$ as $n \to \infty$ (see Chapter 7). A common test statistic for goodness of fit in statistics is the Kolmogorov–Smirnov statistic $D_n = \sqrt{n}||F_n - F||_\infty$ (see Chapter 14). Empirical process theory is going to help us in pinning down finer properties of $||F_n - F||_\infty$ and the asymptotic distribution of D_n.

Note that $||F_n - F||_\infty$ is just the *Kolmogorov distance* (see Chater 15) between F_n and F. Hence, the Glivenko–Cantelli theorem may be rephrased as $d(F_n, F) \stackrel{a.s.}{\to} 0$. A simple point worthy of mention is that F_n need not be close to the true CDF F according to other common notions of distance. For example, consider the empirical measure P_n and the true underlying distribution (measure) P that corresponds to the CDF F. The *total variation distance* between P_n and P is $\rho(P_n, P) = \sup_A |P_n(A) - P(A)|$, the supremum being taken over arbitrary (Borel) sets of $\mathcal{R}$. Then, clearly, $\rho(P_n, P)$, the total variation distance between P_n and P, cannot converge to zero as $n \to \infty$ in general, because P_n is always supported on a finite set, and P may even give zero probability to all finite sets (e.g., if P had a density). In such a case, $\rho(P_n, P)$ would actually be equal to 1 for any n, and therefore would not converge to zero. This argument has nothing to do with the variables being real-valued. The same argument works in $\mathcal{R}^d$ for any $d < \infty$. Thus, the empirical measure need not estimate the true P accurately even in large samples, if our notion of accuracy is too strong. Empirical process theory is about the nature of deviation of P_n from P, especially in large samples, and to quantify the deviation in very precise terms, using highly powerful mathematical ideas and tools. Of course, the theory has numerous applications.

16.2 Classic Asymptotic Properties of the Empirical Process

In studying the classic results on empirical processes, it is helpful first to understand at a heuristic level that the one-dimensional empirical process and the standard Brownian bridge on $[0, 1]$ have some connections. The connections have exactly the same flavor as those between the partial sum process and Brownian motion, which was extensively discussed in Chapter 12. The connection between the empirical process and the Brownian bridge is the driving force behind many of the most valuable results on the asymptotic behavior of the one-dimensional empirical process, and it is useful to have a preview of it.

Preview. Consider first the uniform empirical process $\alpha_n(t), t \in [0, 1]$. As we remarked in the previous section, for any fixed t, $\alpha_n(t) \stackrel{\mathcal{L}}{\Rightarrow} Z_t$, where Z_t is distributed as $N(0, t(1-t))$. This is an immediate consequence of the one-dimensional central limit theorem in the iid case. We can quickly generalize this to the case of any finite number of times $0 \leq t_1 < t_2 < \ldots < t_k \leq 1$. First note that by a simple and direct calculation, for any fixed n, and a pair of times t_i, t_j,

$$\mathrm{Cov}(\alpha_n(t_i), \alpha_n(t_j)) = P(U_1 \leq t_i, U_1 \leq t_j) - P(U_1 \leq t_i)P(U_1 \leq t_j)$$
$$= \min(t_i, t_j) - t_i t_j.$$

Therefore, by the multivariate central limit theorem for the iid case (see Chapter 7), we have the convergence result that

$$(\alpha_n(t_1), \alpha_n(t_2), \ldots, \alpha_n(t_k)) \stackrel{\mathcal{L}}{\Rightarrow} (Z_{t_1}, Z_{t_2}, \ldots, Z_{t_k}),$$

where $(Z_{t_1}, Z_{t_2}, \ldots, Z_{t_k})$ has a k-dimensional normal distribution with means zero, and the covariance matrix with elements $\mathrm{Cov}(Z_{t_i}, Z_{t_j}) = \min(t_i, t_j) - t_i t_j$. Notice that the covariances $\min(t_i, t_j) - t_i t_j$ exactly coincide with the formula for the covariance between $B(t_i)$ and $B(t_j)$, where $B(t)$ is a Brownian bridge on $[0, 1]$ (see Chapter 12). In other words, the finite-dimensional distributions of the process $\alpha_n(t)$ converge to the corresponding finite-dimensional distributions of $B(t)$, a Brownian bridge on $[0, 1]$.

This would lead one to hope that perhaps $\alpha_n(t)$ converges to $B(t)$ as a process. If true, this would be stronger than convergence of just the finite-dimensional distributions, and would lead to better applications. It was a triumph of probability theory that in 1952, Monroe Donsker proved that indeed $\alpha_n(t)$ converges to $B(t)$ as a process (Donsker (1952)). There were certain technical problems of measurability in Donsker's proof. The problems were initially overcome by Anatoliy Skorohod and Andrei Kolmogorov. In order to avoid bringing in that part of the theory, we state Donsker's theorem in the form obtained in Dudley (1999).

The approximation of the one-dimensional uniform empirical process by a Brownian bridge easily carries over to the general one-dimensional empirical process, by essentially using the quantile transformation for real-valued random variables with a continuous CDF. This is seen below. However, the approximation through a Brownian bridge is actually much stronger than what convergence in distribution would imply. Starting in the 1970s, a much stronger form of approximation of the empirical process by a sequence of Brownian bridges was established. This is the analogue of what we called the *strong invariance principle* for the partial sum process in Chapter 12. These strong approximations for the one-dimensional empirical

16.2.1 Invariance Principle and Statistical Applications

Theorem 16.1. *(a) Glivenko–Cantelli Property* $||F_n - F||_\infty \overset{a.s.}{\Rightarrow} 0$ *as* $n \to \infty$.
(b) On a suitable probability space, it is possible to construct an iid sequence U_i of $U[0, 1]$ random variables, and a sequence of Brownian bridges $B_n(t)$, such that $||\alpha_n - B_n||_\infty \overset{a.s.}{\Rightarrow} 0$ *as* $n \to \infty$.
(c) **(Donsker's Theorem).** *The uniform empirical process $\alpha_n(t)$ converges in distribution to a Brownian bridge $B(t)$, as random elements of the class of functions $D[0, 1]$.*
(d) For a general continuous CDF F on the real line, if $X_1, X_2, \ldots$ are iid with the common CDF F, then the empirical process $\beta_n(t)$ converges in distribution to a Gaussian process $B_F(t)$ with $E[B_F(t)] = 0$ and $\text{Cov}(B_F(s), B_F(t)) = F(\min(s,t)) - F(s)F(t)$, as random elements of the class of functions $D(-\infty, \infty)$.
(e) **(Weak Invariance Principle).** *Given any real-valued functional defined on $D(-\infty, \infty)$ that is continuous with respect to the uniform norm, $h(\beta_n(t)) \overset{\mathcal{L}}{\Rightarrow} h(B_F(t))$.*

Due to the length and technical nature of the proof of this theorem, we refer to Billingsley (1968) and Dudley (1999) for its proof. Part (a) of this theorem was previously proved in Theorem 7.7. Part (b) is a special result in strong approximations of the empirical process. We treat the topic of strong approximations in more detail later. It is part (e) of the theorem that is of the greatest use in applications, as we demonstrate shortly in some examples. Part (e) is commonly known as the invariance principle. The spectacular aspect of the result in part (e) is that regardless of what the functional h is, as long as it is continuous with respect to the uniform norm, the distribution of $h(\beta_n(t))$ will be close to the distribution of $h(B_F(t))$ for large n. Therefore, if we know how to find the distribution of the functional $h(B_F(t))$ for our limiting Gaussian process $B_F(t)$, then we have bypassed the usually insurmountable problem of finding the finite sample distribution of $h(\beta_n(t))$, and we have done so all at one time for any continuous functional h. In applications, we can therefore pick and choose a suitable h, and simply apply the invariance principle. Here are some examples of the above theorem.

Example 16.1 (*The Kolmogorov–Smirnov Statistic*). The Kolmogorov–Smirnov statistic is commonly used for testing the hypothesis $F = F_0$, where F is an unknown true CDF on the real line from which we have n iid observations

$X_1, X_2, \ldots, X_n$, and F_0 is a specified continuous CDF (see Example 14.22). It is defined as $D_n = \sqrt{n}||F_n - F_0||_\infty = \sqrt{n}\sup_{-\infty<t<\infty}|F_n(t) - F_0(t)|$. The exact distribution of D_n is difficult to find except for quite small n; some calculations were done in Kolmogorov (1933). Usually, in applications of the Kolmogorov–Smirnov test, the exact distribution is replaced by its asymptotic distribution, and this can be obtained elegantly by using the invariance principle.

First note that under any continuous CDF F_0 on the real line, the quantile transform shows that for any n, the distribution of D_n is the same for all F_0. Thus, one may assume F_0 to be the $U[0, 1]$ distribution. Consider now the functional $h(f) = ||f||_\infty$ on $D[0, 1]$. It is obviously continuous with respect to the uniform norm, and therefore, by the invariance principle,

$$D_n = \sqrt{n}\sup_{0<t<1}|F_n(t) - t| = \sup_{0<t<1}|\sqrt{n}(F_n(t) - t)| = \sup_{0<t<1}|\alpha_n(t)|$$
$$\stackrel{\mathcal{L}}{\Rightarrow} \sup_{0<t<1}|B(t)|,$$

where $B(t)$ is a Brownian bridge on $[0, 1]$. From Chapter 12 (see Theorem 12.4), $\sup_{0<t<1}|B(t)|$ has the CDF $H(x) = 1 - 2\sum_{k=1}^\infty (-1)^{k-1}e^{-2k^2x^2}, x \geq 0$. Therefore, we have the result that for any $x > 0$, $P(D_n > x) \to 2\sum_{k=1}^\infty (-1)^{k-1}e^{-2k^2x^2}$ as $n \to \infty$. This is of tremendous practical utility in statistics.

Example 16.2 (Cramér–von Mises Statistic). It is conceivable that the empirical CDF F_n is a moderate distance away from the postulated CDF F_0 over large parts of the real line, although it is never too far away. In such a case, the Kolmogorov–Smirnov statistic may fail to detect the falsity of the postulated null hypothesis, but a statistic that measures an average deviation of F_n from F_0 may succeed. The Cramér–von Mises statistic $C_n^2 = n\int_{-\infty}^\infty (F_n(t) - F_0(t))^2 dF_0(t)$ is such a statistic, and is frequently used as an alternative to the Kolmogorov–Smirnov statistic, or as a complementary statistic.

Once again, the exact distribution of C_n^2 is difficult to find, but an application of the invariance principle leads to the asymptotic distribution, which is used as an approximation to the exact distribution. As long as F_0 is continuous, the distribution of C_n^2 is independent of F_0 for any n, and so once again, as in our previous example, we may take F_0 to be the $U[0, 1]$ CDF. Then, we have $C_n^2 = \int_0^1 [\sqrt{n}(F_n(t) - t)]^2 dt = \int_0^1 \alpha_n^2(t)dt$. The functional $h(f) = \int_0^1 f^2(t)dt$ is continuous on $D[0, 1]$ with respect to the uniform norm. To see this, use the sequence of inequalities

$$\left|\int_0^1 f^2(t)dt - \int_0^1 g^2(t)dt\right| \leq \int_0^1 |f^2(t) - g^2(t)|dt$$
$$= \int_0^1 |f(t) - g(t)||f(t) + g(t)|dt$$
$$\leq ||f - g||_\infty \int_0^1 (|f(t)| + |g(t)|)dt$$
$$\leq ||f - g||_\infty (||f||_\infty + ||g||_\infty).$$

16.2 Classic Asymptotic Properties of the Empirical Process

Therefore, by the invariance principle, $C_n^2 \stackrel{\mathcal{L}}{\Rightarrow} \int_0^1 B^2(t)dt$. It remains to characterize the distribution of $\int_0^1 B^2(t)dt$.

For this, we use the Karhunen–Loeve expansion of $B(t)$ (see Theorem 12.2) given by $B(t) = \sqrt{2} \sum_{m=1}^{\infty} \frac{\sin(m\pi t)}{m\pi} Z_m$, where $Z_1, Z_2, \ldots$ is an iid $N(0,1)$ sequence. By the orthogonality of the sequence of functions $\sin(m\pi t), m \geq 1$, we get

$$\int_0^1 B^2(t)dt = \int_0^1 \left[\sqrt{2} \sum_{m=1}^{\infty} \frac{\sin(m\pi t)}{m\pi} Z_m \right]^2 dt = \frac{2}{\pi^2} \sum_{m=1}^{\infty} \frac{Z_m^2 \int_0^1 \sin^2(m\pi t)dt}{m^2}$$

$$= \frac{2}{\pi^2} \sum_{m=1}^{\infty} \frac{Z_m^2}{2m^2} = \frac{1}{\pi^2} \sum_{m=1}^{\infty} \frac{Z_m^2}{m^2} = Y(\text{say}).$$

That is, $\int_0^1 B^2(t)dt$ is distributed as an infinite linear combination of iid chi-square random variables of one degree of freedom. At this point, the strategy is to find the cf (characteristic function) of Y, and to invert it to find a density for Y (see Theorem 8.1). The cf of a single chi-square random variable with one degree of freedom is $(1 - 2it)^{-1/2}$. It follows that the cf of Y is

$$\psi_Y(t) = \prod_{m=1}^{\infty} \left(1 - 2i \frac{t}{\pi^2 m^2} \right)^{-1/2} = \left(\prod_{m=1}^{\infty} (1 - 2i \frac{t}{\pi^2 m^2}) \right)^{-1/2}.$$

We now use the identity that for a complex number z, $\prod_{m=1}^{\infty}(1 - \frac{z}{m^2}) = \frac{\sin(\pi\sqrt{z})}{\pi\sqrt{z}}$. Using $z = \frac{2it}{\pi^2}$, we get, with a little algebra, that

$$\psi_Y(t) = \left(\frac{\sin(\sqrt{2it})}{\sqrt{2it}} \right)^{-1/2} = \sqrt{\frac{\sqrt{2it}}{\sin(\sqrt{2it})}}.$$

It is possible to invert this to write the CDF of Y as a convergent infinite series (see part (a) of Theorem 8.1). The CDF has the formula

$$F_Y(y) = 1 - \frac{1}{\pi} \sum_{j=0}^{\infty} (-1)^j \int_{(2j+1)^2\pi^2}^{(2j+2)^2\pi^2} \frac{1}{z} \sqrt{\frac{\sqrt{z}}{-\sin(\sqrt{z})}} e^{-\frac{yz}{2}} dz.$$

This is certainly complicated, even if partially closed-form. In practice, one would probably calculate a tail probability for this statistic by carefully simulating the null distribution of C_n^2 (which is independent of F_0). Nevertheless, the example demonstrates the power of the invariance principle and characteristic functions to solve important problems.

16.2.2 * Weighted Empirical Process

It is sometimes useful to consider normalized or otherwise weighted versions of the empirical process in order to accentuate its behavior at the tails, that is, for t near zero or one. Thus, given a nonnegative weight function w on $(0, 1)$, we may want to consider the weighted empirical process

$$\beta_{n,w}(t) = \frac{\sqrt{n}(F_n(t) - F(t))}{w(F(t))}.$$

Possible uses of such a weighted empirical process would include statistical tests of a null hypothesis $H_0 : F = F_0$ by using test statistics such as

$$D_{n,w} = \sup_{-\infty < t < \infty} \left| \frac{\sqrt{n}(F_n(t) - F_0(t))}{w(F_0(t))} \right|, \text{ or } C_{n,w}^2 = n \int_{-\infty}^{\infty} \frac{[F_n(t) - F_0(t)]^2}{w^2(F_0(t))} dF_0(t).$$

The limiting behavior of such statistics is not necessarily what one might expect intuitively. For instance, $D_{n,w}$ does not necessarily converge in law to the supremum of $\frac{|B(t)|}{w(t)}$. The tails of the function w must be such that $\frac{B(t)}{w(t)}$ does not blow up near $t = 0$ or 1. The specification of the weighting function w such that no such disasters occur is a very subtle and nontrivial problem. The following result (Chibisov (1964), O'Reilly (1974)) completely describes the properties required of w so that the weighted empirical process behaves well at the tails.

Theorem 16.2 (Chibisov–O'Reilly). *(a) Suppose the function w is nondecreasing in a neighborhood of zero and nonincreasing in a neighborhood of 1. The statistic $D_{n,w}$ has a nontrivial limiting distribution if and only if*

$$\int_0^1 \frac{e^{-\epsilon \frac{w^2(t)}{t(1-t)}}}{t(1-t)} dt < \infty \quad \text{for some } \epsilon > 0,$$

in which case $D_{n,w} \stackrel{\mathcal{L}}{\Rightarrow} \sup_{0 < t < 1} \frac{|B(t)|}{w(t)}$.
(b) The statistic $C_{n,w}^2$ has a nontrivial limiting distribution if and only if $\int_0^1 \frac{t(1-t)}{w^2(t)} dt < \infty$, in which case $C_{n,w}^2 \stackrel{\mathcal{L}}{\Rightarrow} \int_0^1 \left(\frac{B(t)}{w(t)} \right)^2 dt$.

Example 16.3. Consider the Anderson–Darling statistic

$$A_n^2 = n \int_{-\infty}^{\infty} \frac{(F_n(t) - F_0(t))^2}{F_0(t)(1 - F_0(t))} dF_0(t),$$

and the weighted Kolmogorov–Smirnov statistic

$$D_n^* = \sqrt{n} \sup_{-\infty < t < \infty} \frac{|F_n(t) - F_0(t)|}{\sqrt{F_0(t)(1 - F_0(t))}}.$$

16.2 Classic Asymptotic Properties of the Empirical Process

Note that A_n^2 and D_n^* have some formal similarity; D_n^* is the supremum (L_∞ norm) and A_n the L_2 norm of the weighted empirical process

$$\frac{\sqrt{n}(F_n(t) - F_0(t))}{\sqrt{F_0(t)(1 - F_0(t))}}.$$

However, the theorem of Chibishov and O'Reilly implies that D_n^* does not have a nontrivial limiting distribution, whereas A_n^2 does, and converges in distribution to $\int_0^1 \frac{B^2(t)}{t(1-t)} dt$. It is a chapter exercise to find the distribution of $\int_0^1 \frac{B^2(t)}{t(1-t)} dt$. The problem with D_n^* is that $\frac{1}{\sqrt{t(1-t)}}$ diverges to infinity at 0 and 1 too rapidly to balance the behavior of $B(t)$ near zero and one. If we weight the empirical process by a more modest weight function, the problems disappear and the weighted Kolmogorov–Smirnov statistic has a nontrivial limiting distribution.

As we just saw, the supremum of the standardized empirical process, namely D_n^*, does not have a nontrivial limiting distribution by itself. However, it can be centered and normed suitably to make it have a nontrivial limiting distribution. Very roughly speaking, D_n^* does not have a nontrivial limiting distribution because it blows up when $n \to \infty$, and it grows at the rate of $\sqrt{2 \log \log n}$. So, by centering it (more or less) at $\sqrt{2 \log \log n}$, and then norming it, we can obtain a limiting distribution. This can still be used to calculate large sample approximations to tail probabilities for D_n^* (i.e., P-values). The following results are due to Jaeschke (1979) and Eicker (1979), Csáki (1980), and Einmahl and Mason (1985).

Theorem 16.3.

(a) $\dfrac{D_n^*}{\sqrt{2 \log \log n}} \overset{P}{\Rightarrow} 1.$

(b) $\forall x,\ P\left(\sqrt{2 \log \log n}\left[D_n^* - \dfrac{2 \log \log n + \frac{1}{2} \log \log \log n + \frac{1}{2} \log \pi}{\sqrt{2 \log \log n}}\right] \leq x\right)$
$\to e^{-2e^{-x}}.$

(c) For any $\epsilon > 0$, with probability 1 for all large n, $D_n^* \leq (\log n)^{\frac{1}{2} + \epsilon}$.

Notice that part (a) of this theorem follows from part (b). It is to be noted that part (a) cannot be strengthened to almost sure convergence.

The preceding discussion and the results show that the weight function $\frac{1}{\sqrt{t(1-t)}}$ grows too rapidly near $t = 0$ and 1 for the supremum of the weighted empirical process to settle down. A particular special case of part (a) of Theorem 16.2 is sometimes useful in applications, and is given below.

Theorem 16.4. *For each $0 < v < \frac{1}{2}$,*

(a) *Almost surely,* $\sup\limits_{0 < t < 1} \dfrac{|B(t)|}{[t(1-t)]^v} < \infty.$

(b) $\sup\limits_{-\infty < t < \infty} \dfrac{|\sqrt{n}(F_n(t) - F_0(t))|}{[F_0(t)(1 - F_0(t))]^v} \overset{\mathcal{L}}{\Rightarrow} \sup\limits_{0 < t < 1} \dfrac{|B(t)|}{[t(1-t)]^v}.$

16.2.3 The Quantile Process

Just as the empirical CDF F_n approximates a true CDF F on the real line, the empirical quantile function $Q_n(y) = F_n^{-1}(y)$ approximates the true quantile function $Q(y) = F^{-1}(y)$. However, we can intuitively see that it is difficult to estimate quantiles of F beyond the range of the data, namely, $X_1, X_2, \ldots, X_n$. So, we have to be careful about the nature of uniform approximability of $Q(y)$ by $Q_n(y)$. In addition, to get convergence results to Brownian bridges akin to the case of the empirical process, we have to properly normalize the quantile process, so that the covariance structure matches that of a Brownian bridge. The theorem below collects the three most important asymptotic properties of the normalized quantile process; a proof for them can be seen in csörgo (1983). Quantile processes are useful in a number of statistical problems, such as *change point problems, goodness of fit, reliability, and survival analysis.*

Theorem 16.5. *Suppose F is an absolutely continuous CDF on the real line, with a density f. Let $q_n(y) = \sqrt{n}[Q_n(y) - Q(y)], 0 < y < 1, n \geq 1$, be the quantile process, and $\rho_n(y) = \sqrt{n} f(Q(y))[Q_n(y) - Q(y)], 0 < y < 1, n \geq 1$, be the standardized quantile process.*

(a) **(Restricted Glivenko–Cantelli Property).** *Suppose f has a bounded support, that f is differentiable, is bounded away from zero on its support, and that $|f'|$ is bounded on the support of f. Then*

$$\sup_{0<y<1} |Q_n(y) - Q(y)| \stackrel{a.s.}{\to} 0.$$

(b) If F has an unbounded support, then with probability one,

$$\lim_{n\to\infty} \sup_{0<y<1} |Q_n(y) - Q(y)| = \infty.$$

(c) Suppose F has a general support (not necessarily bounded) with a density f. Assume that f is strictly positive on its support, that f is differentiable, and that $F(x)(1 - F(x))\frac{|f'(x)|}{f^2(x)}$ is uniformly bounded on the support of f. Then, on a suitable probability space, one can construct an iid sequence $X_1, X_2, \ldots$ with the CDF F, and a sequence of Brownian bridges $B_n(y), 0 \leq y \leq 1$, such that

$$\sup_{\frac{1}{n+1} \leq y \leq \frac{n}{n+1}} |\rho_n(y) - B_n(y)| \stackrel{P}{\Rightarrow} 0.$$

(d) **(Weak Invariance Principle).** *Let h be a real-valued functional on $D[0, 1]$, continuous with respect to the uniform norm. Then, under the conditions of part (c), for any $c, 0 < c < 1, h(\rho_n(y))|_{c \leq y \leq 1-c} \stackrel{\mathcal{L}}{\Rightarrow} h(B(y))|_{c \leq y \leq 1-c}$, where $B(y)$ is a Brownian bridge on $[0, 1]$. In particular, $\sup_{c \leq y \leq 1-c} |\rho_n(y)| \stackrel{\mathcal{L}}{\Rightarrow} \sup_{c \leq y \leq 1-c} |B(y)|$.*

16.2.4 Strong Approximations of the Empirical Process

We proved in Chapter 12 that the partial sum process for an iid sequence with a finite variance converges in distribution to a Brownian motion. This is the invariance principle for the partial sum process (see Section 12.6). In Section 12.7, we showed that in fact for each fixed n, we can construct a Wiener process such that the partial sum process is uniformly close to the Wiener process with probability one. This was called the *strong invariance principle* for the partial sum process. A parallel strong approximation theory exists for the empirical process of an iid sequence. The strong approximation provides a handy tool in many situations for solving a particular problem. In addition, the error statements in the strong approximation results give a very precise idea about the accuracy of the Brownian bridge approximation of the empirical process. The results pin down what can and cannot be done. We remark in passing that part (a) in Theorem 16.1 and part (c) in Theorem 16.5 are in fact instances of strong approximations.

Theorem 16.6. *On a suitable probability space, one can construct an iid sequence $X_1, X_2, \ldots$ with the common CDF F, and a sequence of Brownian bridges $B_n(t)$, such that the empirical process $\beta_n(t) = \sqrt{n}(F_n(t) - F(t)), -\infty < t < \infty$ satisfies*

(a) *Almost surely,* $\sup_{-\infty < t < \infty} |\beta_n(t) - B_n(F(t))| = O\left(\frac{\log n}{\sqrt{n}}\right)$.

(b) $\forall n \geq 1$, *and* $\forall x \in \mathcal{R}$, $P\left(\sup_{-\infty < t < \infty} |\beta_n(t) - B_n(F(t))| > \frac{12 \log n + x}{\sqrt{n}}\right) \leq 2e^{-\frac{x}{6}}$.

(c) $\forall c > 0$, $\sup_{F^{-1}(\frac{c}{n}) \leq t \leq F^{-1}(1-\frac{c}{n})} \frac{|\beta_n(t) - B_n(F(t))|}{\sqrt{F(t)(1-F(t))}} = O_P(1)$.

(d) $\forall c > 0$, *and* $\forall 0 < \nu < \frac{1}{2}$, $\sup_{F^{-1}(\frac{c}{n}) \leq t \leq F^{-1}(1-\frac{c}{n})} \frac{|\beta_n(t) - B_n(F(t))|}{[F(t)(1-F(t))]^\nu} = O_P\left(n^{\nu - \frac{1}{2}}\right)$.

Part (a) of this theorem actually follows from part (b) by clever choices of x, and then an application of the Borel–Cantelli lemma. The choice of x is indicated in a chapter exercise. It is important to note that the inequality is valid for all n and all real x, and so in a specific application, n and x can be chosen to suit one's need. In particular, x can depend on n. Part (b) is called the Komlós, Major, Tusnady (KMT) theorem. See Komlós et al. (1975a, b), and Mason and van Zwet (1987) for a more detailed proof. Part (c) is an $O_P(1)$ result, rather than an $o_P(1)$ one, and reinforces our discussion in the previous section that if we weight the empirical process by $\sqrt{F(1-F)}$, then the invariance principle will fail. But if we weight it by a smaller power of $F(1-F)$, then not only can we recover the invariance principle, but even a strong approximation holds, as in part (d). The KMT rate $\frac{\log n}{\sqrt{n}}$ in part (a) cannot be improved.

16.3 Vapnik–Chervonenkis Theory

As we have seen, an important consequence of Donsker's invariance principle is the derivation of the limiting distribution of $D_n = \sqrt{n}\sup_{-\infty<t<\infty}|F_n(t) - F(t)|$, for a continuous one-dimensional CDF F. If we let $\mathcal{F} = \{I_{-\infty,t]} : t \in \mathcal{R}\}$, then the Kolmogorov–Smirnov result says that $\sqrt{n}\sup_{f\in\mathcal{F}}(E_{P_n}f - E_Pf) \overset{\mathcal{L}}{\Rightarrow}$ $\sup_{0\leq t\leq 1}|B(t)|$, P_n, P being the probability measures corresponding to F_n, F, and $B(t)$ being a Brownian bridge. Extensions of this involve study of the asymptotic behavior of $\sup_{f\in\mathcal{F}}(E_{P_n}f - E_Pf)$ for much more general classes of functions $\mathcal{F}$ and the range space of the random variables X_i; the range space need not be $\mathcal{R}$, or $\mathcal{R}^d$ for some finite d. It can be a much more general set $\mathcal{S}$. Examples of asymptotic behavior include derivation of laws of large numbers and central limit theorems. Because of the involvement of the supremum over $f \in \mathcal{F}$, this is much more difficult than consideration of $E_{P_n}f - E_Pf$ for a single or a finite number of functions f.

There are numerous applications of these extensions to classic statistical problems. The more modern applications are in areas of statistical classification of objects involving many variables, and other problems in machine learning. To give a simple example, suppose $X_1, X_2, \ldots, X_n$ are d-dimensional iid random vectors from some P and we want to test the null hypothesis that $P = P_0$ (specified). Then, a natural statistic to assess the truth of the hypothesis is $T_n = \sup_{C\in\mathcal{C}}|P_n(C) - P_0(C)|$ for a suitable class of sets $\mathcal{C}$. Now, if $\mathcal{C}$ is too rich, for example, if it is the class of all measurable sets, then clearly there cannot be any meaningful asymptotics if P_0 is absolutely continuous. On the other hand, if $\mathcal{C}$ is too small, then the statistic cannot be sensitive enough for detecting departures from the null hypothesis. So these extensions study the question of what kinds of families $\mathcal{C}$ or function classes $\mathcal{F}$ allow meaningful asymptotics and also result in good and common sense tests. *Vapnik–Chervonenkis theory* pins down exactly how rich such a class $\mathcal{C}$ can be, and even more, supplies the user with readymade classes that will work. It is important to understand that applications of the VC theory are far more wide ranging than testing problems in statistics.

The technical tools required for such generalizations are extremely sophisticated, and have led to striking new discoveries and mathematical advances in the theory of empirical processes. Along with these advances, have come numerous new and useful statistical and probabilistic applications. The literature is huge; we strongly recommend Wellner (1992), Giné (1996), Pollard (1989), and Giné and Zinn (1984) for comprehensive reviews, and sources for major theorems and additional references; specific references to some results are given later. We present a basic description of certain key results and tools in VC theory below.

16.3.1 Basic Theory

We first discuss the plausibility of strong laws more general than the well-known Glivenko–Cantelli theorem, which asserts that in the one-dimensional iid case

16.3 Vapnik–Chervonenkis Theory

$\sup_t |F_n(t) - F(t)| \overset{a.s.}{\to} 0$. We need a concept of combinatorial richness of a class of sets $\mathcal{C}$ that will allow us to make statements like such as $\sup_{C \in \mathcal{C}} |P_n(C) - P(C)| \overset{a.s.}{\to} 0$. A class of sets for which this property holds is called a *Glivenko–Cantelli Class*. A useful such concept of combinatorial richness is that of the *Vapnik–Chervonenkis dimension* of a class of sets. Meaningful asymptotics will exist for classes of sets that have a finite Vapnik–Chervonenkis dimension. It is therefore critical to know what it means and what are good examples of classes of sets with a finite Vapnik–Chervonenkis dimension. A basic treatment of this is given next.

We often use the notation below. Given a specific set of n elements $x_1, x_2, \ldots, x_n$ of a general set $\mathcal{S}$, which need not be distinct, and a specific class $\mathcal{C}$ of subsets of $\mathcal{S}$, we let

$$\Delta^{\mathcal{C}}(x_1, x_2, \ldots, x_n) = \text{Card}(\{x_1, x_2, \ldots, x_n\} \cap C : C \in \mathcal{C}),$$

where Card denotes cardinality. In words, $\Delta^{\mathcal{C}}(x_1, x_2, \ldots, x_n)$ counts how many distinct subsets of $x_1, x_2, \ldots, x_n$ can be generated by intersecting $\{x_1, x_2, \ldots, x_n\}$ with the members of $\mathcal{C}$.

Definition 16.1. Let $A \subset \mathcal{S}$ be a fixed set, and $\mathcal{C}$ a class of subsets of $\mathcal{S}$. A is said to be *shattered by* $\mathcal{C}$ if every subset U of A is the intersection of A with some member C of $\mathcal{C}$, that is, $\{A \cap C : C \in \mathcal{C}\} = \mathcal{P}(A)$, where $\mathcal{P}(A)$ denotes the power set of A.

Sometimes the phenomenon is colloquially described as *every subset of A is picked up by some member of $\mathcal{C}$*.

Definition 16.2. The Vapnik–Chervonenkis (VC) dimension of $\mathcal{C}$ is the size of the largest set A that can be shattered by $\mathcal{C}$.

Although this is already fine as a definition, a more formal definition is given by using the concept of *shattering coefficients*.

Definition 16.3. For $n \geq 1$, the nth *shattering coefficient* of $\mathcal{C}$ is defined to be

$$S(n, \mathcal{C}) = \max_{x_1, x_2, \ldots, x_n \in \mathcal{S}} \Delta^{\mathcal{C}}(x_1, x_2, \ldots, x_n).$$

That is, $S(n, \mathcal{C})$ is the largest possible number of subsets of some (wisely chosen) set of n points that can be formed by intersecting the set with members of $\mathcal{C}$. Clearly, for any n, $S(n, \mathcal{C}) \leq 2^n$.

Here is an algebraic definition of the VC dimension of a class of sets.

Definition 16.4. The VC dimension of $\mathcal{C}$ equals $VC(\mathcal{C}) = \min\{n : S(n, \mathcal{C}) < 2^n\} - 1 = \max\{n : S(n, \mathcal{C}) = 2^n\}$.

Definition 16.5. $\mathcal{C}$ is called a *Vapnik–Chervonenkis (VC) class* if $VC(\mathcal{C}) < \infty$.

The following remarkable result is known as *Sauer's lemma* (Sauer (1972)).

Proposition. *Either $S(n, \mathcal{C}) = 2^n \; \forall n$, or $\forall n, S(n, \mathcal{C}) \leq \sum_{i=0}^{VC(\mathcal{C})} \binom{n}{i}$.*

Remark. Sauer's lemma says that either a class of sets has infinite VC dimension, or its shattering coefficients grow polynomially. A few other important and useful properties of the shattering coefficients are listed below; most of them are derived easily. These properties are useful for generating new classes of VC sets from known ones by using various Boolean operations.

Theorem 16.7. *The shattering coefficients $S(n, C)$ of a class of sets C satisfy*

(a) $S(m, C) < 2^m$ for some $m \Rightarrow S(n, C) < 2^n \ \forall n > m$.
(b) $S(n, C) \leq (n+1)^{VC(C)} \forall n \geq 1$.
(c) $S(n, C^c) = S(n, C)$, where C^c is the class of complements of members of C.
(d) $S(n, B \cap C) \leq S(n, B)S(n, C)$, where the $\cap$ notation means the class of sets formed by intersecting members of B with those of C.
(e) $S(n, B \otimes C) \leq S(n, B)S(n, C)$, where the $\otimes$ notation means the class of sets formed by taking Cartesian products of members of B and those of C.
(f) $S(m+n, C) \leq S(m, C)S(n, C)$.

See Vapnik and Chervonenkis (1971) and Sauer (1972) for many of the parts in this theorem. Now we give examples for practical use.

16.3.2 Concrete Examples

Example 16.4. Let C be the class of all left unbounded closed intervals on the real line; that is, $C = \{(-\infty, x] : x \in \mathcal{R}\}$. To illustrate the general formula, suppose $n = 2$; what is $S(n, C)$? Clearly, if we pick up the larger one among x_1, x_2, we will pick up the smaller one too. Or, we may pick up none of them or just the smaller one. So we can pick up three distinct subsets from the power set of $\{x_1, x_2\}$. The same argument shows that the general formula for the shattering coefficients is $S(n, C) = n + 1$. Consequently, this is a VC class with VC dimension one.

Example 16.5. Although topologically, there are just as many left unbounded intervals on the real line as there are arbitrary intervals, in the VC index they act differently. This is interesting. Thus, let $C = \{(a, b) : a \leq b \in \mathcal{R}\}$. Then it is easy to establish the formula $S(n, C) = 1 + \binom{n+1}{2}$. For $n = 2$, this is equal to four, which is also 2^2. For $n = 3, 1 + \binom{n+1}{2} = 7 < 2^n$. Consequently, this is a VC class with VC dimension two.

Example 16.6. The previous example says that on $\mathcal{R}$, the class of all convex sets is a VC class. However, this is far from being true even in two dimensions. Indeed, if we let C be just the class of convex polygons in the plane, it is clear geometrically that for any n, C can shatter n points. So, convex polygons in $\mathcal{R}^2$ have an infinite VC dimension.

More examples of exact values of VC dimensions are given in the chapter exercises. For actual applications of these ideas to concrete extensions of Donsker's principles, it is extremely useful to know what other natural classes of sets in various spaces are VC classes. The various parts of the following result are available in Vapnik and Chervonenkis (1971) and Dudley (1978, 1979).

Theorem 16.8. *Each of the following classes of sets is a VC class.*

(a) *The class of all southwest quadrants of $\mathcal{R}^d$; that is, the class of all sets of the form $\prod_{i=1}^{d}(-\infty, x_i]$.*
(b) *The class of all closed half-spaces of $\mathcal{R}^d$.*
(c) *The class of all closed balls of $\mathcal{R}^d$.*
(d) *The class of all closed rectangles of $\mathcal{R}^d$.*
(e) *$\mathcal{C} = \{\{x \in \mathcal{R}^d : g(x) \geq 0\} : g \in G\}$, where G is a finite-dimensional vector space of real-valued functions defined on $\mathcal{R}^d$.*
(f) *Projections of a class of the form in part (e) onto a smaller number of coordinates.*

There are practically useful ways to generate new VC classes from known ones. By using the various parts of Theorem 16.7, one can obtain the following useful results.

Theorem 16.9. (a) *$\mathcal{C}$ is VC implies $\mathcal{C}^c = \{C^c : C \in \mathcal{C}\}$ is VC.*
(b) *$\mathcal{C}$ is VC implies $\phi(\mathcal{C})$ is VC for any one-to-one function ϕ, where $\phi(\mathcal{C})$ is the class of images of sets in $\mathcal{C}$ under ϕ.*
(c) *$\mathcal{C}, \mathcal{D}$ are VC implies $\mathcal{C} \cap \mathcal{D}$ and $\mathcal{C} \cup \mathcal{D}$ are VC, where the intersection and union classes are defined as classes of intersections $C \cap D$ and unions $C \cup D$, $C \in \mathcal{C}, D \in \mathcal{D}$.*
(d) *$\mathcal{C}$ is VC in some space $\mathcal{S}_1$, and $\mathcal{D}$ is VC in some space $\mathcal{S}_2$ implies that the Cartesian product $\mathcal{C} \otimes \mathcal{D}$ is VC in the product space $\mathcal{S}_1 \otimes \mathcal{S}_2$.*

We can now state a general version of the familiar Glivenko–Cantelli theorem. It say's that a VC class is a Glivenko–Cantelli class. The following famous theorem of Vapnik and Chervonenkis (1971) on Euclidean spaces is considered to be a penultimate result of the problem.

Theorem 16.10. *Let $X_1, X_2, \ldots \stackrel{iid}{\sim} P$, a probability measure on $\mathcal{R}^d$ for some finite d. Given any class of (measurable) sets $\mathcal{C}$, for $n \geq 1, \epsilon > 0$,*

$$P(\sup_{\mathcal{C}} |P_n(C) - P(C)| > \epsilon) \leq 8E[\Delta^{\mathcal{C}}(X_1, \ldots, X_n)]e^{-n\epsilon^2/32}$$

$$\leq 8S(n, \mathcal{C})e^{-n\epsilon^2/32}.$$

Remark. This theorem implies that for classes of sets that are of the right complexity as measured by the VC dimension, the empirical measure converges to the true at an essentially exponential rate. This is a sophisticated generalization of the one-dimensional DKW inequality. The first bound of this theorem is harder to implement because it involves computation of a hard expectation, namely $E[\Delta^{\mathcal{C}}(X_1, X_2, \ldots, X_n)]$. It would usually not be possible to find this expectation, although simulating the quantity $\Delta^{\mathcal{C}}(X_1, X_2, \ldots, X_n)$ would be an interesting exercise.

The general theorem is given next; see Giné (1996).

Theorem 16.11. *Let P be a probability measure on a general measurable space S and let $X_1, X_2, \ldots \stackrel{iid}{\sim} P$. Let P_n denote the sequence of empirical measures and let C be a class of sets in S. Then, under suitable measurability conditions,*

(a) $\sup_{C \in \mathcal{C}} |P_n(C) - P(C)| \stackrel{a.s.}{\to} 0$ iff $\sup_{C \in \mathcal{C}} |P_n(C) - P(C)| \stackrel{P}{\Rightarrow} 0$.
(b) $\sup_{C \in \mathcal{C}} |P_n(C) - P(C)| \stackrel{a.s.}{\to} 0$ iff $\frac{\log \Delta^{\mathcal{C}}(X_1, X_2, \ldots, X_n)}{n} \stackrel{P}{\to} 0$.
(c) Suppose $\mathcal{C}$ is a VC class of sets. Then $\sup_{C \in \mathcal{C}} |P_n(C) - P(C)| \stackrel{a.s.}{\to} 0$.

It is not easy to find $\Delta^{\mathcal{C}}(X_1, X_2, \ldots, X_n)$ for the classes of sets $\mathcal{C}$ that we will meet in practice. Note, however, that we always have the upper bound $\Delta^{\mathcal{C}}(X_1, X_2, \ldots, X_n) \leq S(n, \mathcal{C})$, and so, if we know the shattering coefficient $S(n, \mathcal{C})$, then application of the sufficiency part of this theorem becomes more practicable. The point is that one way or another, we need to find a bound on $\Delta^{\mathcal{C}}(X_1, X_2, \ldots, X_n)$ rather than working with it directly. Here is a very neat application of this result from Giné (1996).

Example 16.7 (Uniform Convergence of Relative Frequencies). This example could be characterized as the reason that statistics exists as a subject. Suppose we have a multinomial distribution with cell probabilities $p_1, p_2, \ldots$, and in a sample of n individuals, X_i are found to be of type i. Thus, the relative frequencies are $\frac{X_i}{n}, i = 1, 2, \ldots$. Of course, if the number of types (cells) is finite, then by just the SLLN, $\max_i |\frac{X_i}{n} - p_i|$ converges almost surely to zero. Suppose that the number of cells is infinite. Even then, the relative frequencies are uniformly close to the true cell probabilities in large samples statistical sampling will unveil the complete truth if large samples are taken. We give a proof of this by using part (a) of our theorem above.

To formulate the problem, suppose P is supported on a countable set, which without loss of generality we take to be $\mathcal{Z}_+ = \{1, 2, 3, \ldots\}$, and suppose $X_1, X_2, \ldots, X_n$ is an iid sample from P. Let P_n be the empirical measure. We want to prove that $||P_n - P||_{\mathcal{C}}$ converges almost surely to zero even if $\mathcal{C}$ is the power set of $\mathcal{Z}_+$. To derive this result, we use the binomial distribution tail inequality that if $W \sim \text{Bin}(n, p)$, then for any k, $P(W \geq k) \leq (\frac{enp}{k})^k$ (see, e.g., Giné (1996)). To verify part (a) of the above theorem, fix $\epsilon > 0$, and observe that

$$P\left(\frac{\log \Delta^{\mathcal{C}}(X_1, X_2, \ldots, X_n)}{n} > \epsilon \log 2\right) = P(\Delta^{\mathcal{C}}(X_1, X_2, \ldots, X_n) > 2^{\epsilon n})$$
$$\leq P(\text{Number of distinct values among } X_1, X_2, \ldots, X_n \geq \lfloor \epsilon n \rfloor)$$
$$\leq P(\text{At least half of } X_1, X_2, \ldots, X_n \text{ are } \geq \frac{1}{2}\lfloor \epsilon n \rfloor)$$
$$\leq \left(\frac{enP(X_1 \geq \frac{1}{2}\lfloor \epsilon n \rfloor)}{\frac{1}{2}\lfloor \epsilon n \rfloor}\right)^{\frac{1}{2}\lfloor \epsilon n \rfloor}$$

(this is the binomial distribution inequality referred to above)

$$\to 0,$$

because ϵ is a fixed number and $P(X_1 \geq \frac{1}{2}\lfloor \epsilon n \rfloor) \to 0$, making the quotient

$$enP\left(X_1 \geq \frac{1}{2}\lfloor \epsilon n \rfloor\right)\frac{1}{2}\lfloor \epsilon n \rfloor$$

eventually bounded away from 1. This proves that

$$\frac{\log \Delta^\mathcal{C}(X_1, X_2, \ldots, X_n)}{n} \overset{P}{\Rightarrow} 0,$$

as was desired.

16.4 CLTs for Empirical Measures and Applications

Theorem 16.8 and Theorem 16.9 give us hope for establishing CLTs for suitably normalized versions of $\sup_{C \in \mathcal{C}} |P_n(C) - P(C)|$ in general spaces and with general VC classes of sets. It is useful to think of this as an analogue of the one-dimensional Kolmogorov–Smirnov statistic for real-valued random variables, namely, $\sup_x |F_n(x) - F(x)|$. Invariance principles allowed us to conclude that the limiting distribution is related to a Brownian bridge, with real numbers in [0, 1] as the time parameter. Now, however, the setup is much more abstract. The space is not an Euclidean space, and the time parameter is a set or a function. So the formulation and description of the appropriate CLTs is more involved, and although suitable Gaussian processes will still emerge as the relevant processes that determine the asymptotics, they are not Brownian bridges, and they even depend on the underlying P from which we are sampling. Some of the most profound advances in the theory of statistics and probability in the twentieth century have taken place around this problem, resulting along the way in deep mathematical developments and completely new tools. A short description of this is provided next.

16.4.1 Notation and Formulation

First some notation and definitions. We recall that the notation $(P_n - P)(f)$ would mean $\int f dP_n - \int f dP$. Here, f is supposed to belong to some suitable class of functions $\mathcal{F}$. For example, $\mathcal{F}$ could be the class of indicator functions of the members C of a class of sets $\mathcal{C}$. In that case, $(P_n - P)(f)$ would simply mean

$P_n(C) - P(C)$; we have just talked about strong laws for their supremums as C varies over $\mathcal{C}$. That is a uniformity result. Likewise, we now need certain uniformity assumptions on the class of functions $\mathcal{F}$. We assume that

(a) $\sup_{f \in \mathcal{F}} |f(s)| := \overline{F(s)} < \infty \; \forall s \in \mathcal{S}$

(measurability of F is clearly not obvious, but is being ignored here);

(b) $\overline{F} \in L_2(P)$.

The function $\overline{F}$ is called the *envelope of* $\mathcal{F}$. In the case of real-valued random variables and for the problem of convergence of the process $F_n(t) - F(t)$, the corresponding functions, as we just noted, are indicator functions of $(-\infty, t)$, which are uniformly bounded functions. Now the time parameter has become a function itself, and we need to talk about uniformly bounded functionals of functions; we use the notation

$$l_\infty(\mathcal{F}) = \{h : \mathcal{F} \to \mathcal{R} : \sup_{f \in \mathcal{F}} |h(f)| < \infty\}.$$

Furthermore, we refer to $\sup_{f \in \mathcal{F}} |h(f)|$ as the uniform norm and denote it as $||h||_{\infty, \mathcal{F}}$.

The other two notions we need to formalize are those of convergence of the process $(P_n - P)(f)$ (on normalization) and of a limiting Gaussian process that plays the role of a Brownian bridge in these general circumstances.

The Gaussian process, which we denote as $B_P(f)$, continues to have continuous sample paths, as was the case for the ordinary Brownian bridge, but now the time parameter is a function, and continuity is with respect to $\rho_P(f, g) = \sqrt{E_P(f(X) - g(X))^2}$. B_P has mean zero, and the covariance kernel $\text{Cov}(B_P(f), B_P(g)) = P(fg) - P(f)P(g) := E_P(f(X)g(X)) - E_P(f(X))E_P(g(X))$. Note that due to our assumption that $\overline{F} \in L_2(P)$, the covariance kernel is well defined. Trajectories of our Gaussian process B_P are therefore members of $l_\infty(\mathcal{F})$, also (uniformly) continuous with respect to the norm ρ_P we have defined above.

Finally, as in the Portmanteau theorem in Chapter 7, convergence of the process $\sqrt{n}(P_n - P)(f)$ to $B_P(f)$ would mean that expectation of any functional H of $\sqrt{n}(P_n - P)(f)$ will converge to the expectation of $H(B_P(f))$, H being a bounded and continuous functional, defined on $l_\infty(\mathcal{F})$, and taking values in $\mathcal{R}$. We remind our reader that continuity on $l_\infty(\mathcal{F})$ is with respect to the uniform norm we have already defined there. A class of functions $\mathcal{F}$ for which this central limit property holds is called a *P-Donsker class*; if the property holds for every probability measure P on $\mathcal{S}$, it is called a *universal Donsker class*.

16.4.2 Entropy Bounds and Specific CLTs

We now discuss what sorts of assumptions on our class of functions $\mathcal{F}$ will ensure that weak convergence occurs (i.e., a CLT holds), and also what are some good applications of such CLTs. There are multiple sets of assumptions on the class of functions $\mathcal{F}$ that ensure a CLT. Here, we describe only two, one of which relates to the concept of VC classes. and the second related to *metric entropy* and *packing numbers*. Inasmuch as we are already familiar with the concept of VC classes, we first state a CLT based on a VC assumption of a suitable class of sets.

Definition 16.6. A family $\mathcal{F}$ of functions f on a (measurable) space $\mathcal{S}$ is called *VC-subgraph* if the class of subgraphs of $f \in \mathcal{F}$ is a VC class of sets, where the subgraph of f is defined to be $C_f = \{(x, y), x \in \mathcal{S}, y \in \mathcal{R} : 0 \le y \le f(x)$ or $f(x) \le y \le 0\}$.

Theorem 16.12. *Given* $X_1, X_2, \ldots \stackrel{iid}{\sim} P$, *a probability measure on a measurable space* $\mathcal{S}$, *and a family of functions* $\mathcal{F}$ *on* $\mathcal{S}$ *such that* $\overline{F}(s) := \sup_{f \in \mathcal{F}} |f(s)| \in L_2(P)$, $\sqrt{n}(P_n - P)(f) \stackrel{\mathcal{L}}{\Rightarrow} B_P(f)$ *if* $\mathcal{F}$ *is a VC-subgraph family of functions.*
An important application of this theorem is the following result.

Corollary 16.1. *Under the other assumptions made in the above theorem,* $\sqrt{n} (P_n - P)(f) \stackrel{\mathcal{L}}{\Rightarrow} B_P(f)$ *if* $\mathcal{F}$ *is a finite-dimensional space of functions or if* $\mathcal{F} = \{I_C : C \in \mathcal{C}\}$, *where* $\mathcal{C}$ *is any VC class of sets.*

This theorem beautifully connects the scope of a Glivenko–Cantelli theorem to that of a CLT via the same VC concept, modulo the extra qualification that $\overline{F} \in L_2(P)$. One can see more about this key theorem in Alexander (1987) and Giné (1996).

A pretty and useful statistical application of the above result is the following example on extension (due to Beran and Millar (1986)) of the familiar Kolmogorov–Smirnov test for goodness of fit to general spaces.

Example 16.8. Let $X_1, X_2, \ldots$ be iid observations from P on some space $\mathcal{S}$ and consider testing the null hypothesis $H_0 : P = P_0$ (specified). The natural Kolmogorov–Smirnov type test statistic for this problem is $T_n = \sqrt{n} \sup_{C \in \mathcal{C}} |P_n(C) - P_0(C)|$, for a judiciously chosen family of (measurable) sets $\mathcal{C}$. The above theorem implies that T_n converges under the null in distribution to the supremum of the absolute value of the Gaussian process $B_{P_0}(f)$, the sup being taken over all $f = I_C, C \in \mathcal{C}$, a VC class of subsets of $\mathcal{S}$. In principle, therefore, the null hypothesis can be tested by using this Kolmogorov–Smirnov type statistic. Note, however, that the limiting Gaussian process depends on P_0. Evaluation of the critical points of the limiting distribution of T_n under the null needs some type of numerical work. In particular, a powerful statistical tool known as the *bootstrap* can help in this specific situation; see Giné (1996) for more discussion and references on this computational issue.

The second CLT we present requires the concepts of metric entropy and bracketing numbers, which we introduce next.

Definition 16.7. Let $\mathcal{F}^*$ be a space of real-valued functions defined on some space $\mathcal{S}$, and suppose $\mathcal{F}^*$ is equipped with a norm $||.||$. Let $\mathcal{F}$ be a specific subcollection of $\mathcal{F}^*$. The *covering number* of $\mathcal{F}$ is defined to be the smallest number of balls $B(g, \epsilon) = \{h : ||h - g|| < \epsilon\}$ needed to cover $\mathcal{F}$, where $\epsilon > 0$ is arbitrary but fixed, $g \in \mathcal{F}^*$, and $||g|| < \infty$.

The covering number of $\mathcal{F}$ is denoted as $N(\epsilon, \mathcal{F}, ||.||)$. $\log N(\epsilon, \mathcal{F}, ||.||)$; is called the *entropy without bracketing* of $\mathcal{F}$.

Definition 16.8. In the same setup of the previous definition, a *bracket* is the set of functions sandwiched between two given functions l, u; that is, i.e., a bracket is the set $\{f : l(s) \leq f(s) \leq u(s) \,\forall s \in \mathcal{S}\}$. It is denoted as $[l, u]$.

Definition 16.9. The *bracketing number* of $\mathcal{F}$ is defined to be the smallest number of brackets $[l, u]$ needed to cover $\mathcal{F}$ under the restriction $||l - u|| < \epsilon, \epsilon > 0$ an arbitrary but fixed number.

The bracketing number of $\mathcal{F}$ is denoted as $N_{[]}(\epsilon, \mathcal{F}, ||.||)$; $\log N_{[]}(\epsilon, \mathcal{F}, ||.||)$ is called the *entropy with bracketing* of $\mathcal{F}$.

Discussion: Clearly, the smaller the radius of the balls or the width of the brackets, the greater is the number of balls or brackets necessary to cover the function class $\mathcal{F}$. The important thing is to pin down, qualitatively, the rate at which the entropy (with or without bracketing) is going to ∞ for a given $\mathcal{F}$, as $\epsilon \to 0$. It turns out, as we show, that for many interesting and useful classes of functions $\mathcal{F}$, this rate would be of the order of $(-\log \epsilon)$, and this will, by virtue of some theorems to be given below, ensure that the class $\mathcal{F}$ is P-Donsker.

Theorem 16.13. *Assume that* $\overline{F} \in L_2(P)$. *Then,* $\mathcal{F}$ *is* P-*Donsker if either*

(a) $\int_0^\infty \sqrt{\log N_{[]}(\epsilon, \mathcal{F}, ||.|| = L_2(P))} d\epsilon < \infty,$

or,

(b) $\int_0^\infty \sup_Q \left(\sqrt{\log N \left(\epsilon ||\overline{F}||_{2,Q}, \mathcal{F}, ||.|| = L_2(Q)\right)} \right) d\epsilon < \infty,$

where Q denotes a general probability measure on $\mathcal{S}$.

See van der Vaart and Wellner (1996, pp. 127–132) for a proof of this theorem, under the two separate sufficient conditions.

We have previously seen that if $\mathcal{F}$ is a VC-subgraph, then it is P-Donsker. It turns out that this result follows from our above theorem on the integrability of $\sup_Q \sqrt{\log N}$. What one needs is the following upper bound on the entropy without bracketing of a VC-subgraph class. See van der Vaart and Wellner (1996, p. 141) for its proof.

Proposition. *Given a VC-subgraph class $\mathcal{F}$, for any probability measure Q, and any $r \geq 1$, for all $0 < \epsilon < 1, N(\epsilon ||\overline{F}||_{r,Q}, \mathcal{F}, ||.|| = L_r(Q)) \leq C(\frac{1}{\epsilon})^{rVC(\mathcal{C})}$, where the constant C depends only on $VC(\mathcal{C}), \mathcal{C}$ being the subgraph class of $\mathcal{F}$.*

16.4.3 Concrete Examples

Here are some additional good applications of the entropy results.

Example 16.9. As mentioned above, the key to the applicability of the entropy theorems is a good upper bound on the rate of growth of the entropy numbers of the class. Such bounds have been worked out for many intuitively interesting classes. The bounds are sometimes sharp in the sense lower bounds can also be obtained that grow at the same rate as the upper bounds. In nearly every case mentioned in this example, the derivation of the upper bound is completely nontrivial. A very good reference is van der Vaart and Wellner (1996), particularly Chapter 2.7 there.

Uniformly Bounded Monotone Functions on $\mathcal{R}$ For this function class $\mathcal{F}$, $\log N_{[]}(\epsilon, \mathcal{F}, ||.|| = L_2(P)) \leq \frac{K}{\epsilon}$, where K is a universal constant independent of P, and so by part (a) of Theorem 16.11, this class is in fact universal P-Donsker.

Uniformly Bounded Lipschitz Functions on Bounded Intervals in $\mathcal{R}$ Let $\mathcal{F}$ be the class of real-valued functions on a bounded interval $\mathcal{I}$ in $\mathcal{R}$ that are uniformly bounded by a universal constant and are uniformly Lipschitz of some order $\alpha > \frac{1}{2}$; a that is, $|f(x) - f(y)| \leq M|x-y|^\alpha$, uniformly in x, y and for some finite universal constant M. For this class, $\log N_{[]}(\epsilon, \mathcal{F}, ||.|| = L_2(P)) \leq K\left(\frac{1}{\epsilon}\right)^{1/\alpha}$, where K depends only on the length of $\mathcal{I}, M$, and α, and so this class is also universal P-Donsker.

Compact Convex Subsets of a Fixed Compact Set in $\mathcal{R}^d$ Suppose S is a compact set in $\mathcal{R}^d$ for some finite d, and let $\mathcal{C}$ be the class of all compact convex subsets of S. For any absolutely continuous P, this class satisfies $\log N_{[]}(\epsilon, \mathcal{C}, ||.|| = L_2(P)) \leq K\left(\frac{1}{\epsilon}\right)^{d-1}$, where K depends on S, P, and d. Here it is meant that the function class is the set of indicators of the members of $\mathcal{C}$. Thus, for $d = 2, \mathcal{F}$ is P-Donsker for any absolutely continuous P.

A common implication of all of these applications of the entropy thorems is that in each of these cases, asymptotic goodness-of-fit tests can be constructed by using these function classes.

16.5 Maximal Inequalities and Symmetrization

Consider a general stochastic process $X(t), t \in T$, for some time-indexing set T. Inequalities on the tail probabilities or moments of the maxima of $X(t)$ are called *maximal inequalities* for the $X(t)$ process. More precisely, a maximal inequality places an upper bound on probabilities of the form $P(\sup_{t \in t} |X(t)| > \epsilon)$ or on $E[(\sup_{t \in t} |X(t)|)^p]$, often with $p = 2$. Constructing useful maximal inequalities is always a hard problem. Perhaps the maximum amount of concrete theory exists for suitable Gaussian processes (e.g., continuity of the sample paths is often required). In the context of developing maximal inequalities for $\sqrt{n}(P_n - P)(f), f \in \mathcal{F}$, this turns out to be a rather happy coincidence, because empirical processes are, after all, roughly Gaussian. A highly ingenious technique that has been widely used to turn this near Gaussianity into useful maximal inequalities for the empirical process is the *symmetrization method*. Ultimately, the inequalities on tail probabilities are aimed at establishing some sort of an exponential decay for the tail, and the moment inequalities often end up having ties to covering and entropy numbers of the class $\mathcal{F}$. In this way, maximal inequalities for Gaussian processes, the technique of symmetrization, and covering numbers come together very elegantly to produce a collection of sophisticated results on deviations of the empirical from the true. The treatment below gives a flavor of this part of the theory. We follow Pollard (1989) and Giné (1996).

Preview. It is helpful to have a summary of the main ideas behind the technical development of the ultimate inequalities. The path to the inequalities is roughly the following.

(a) Use existing Gaussian process theory to write down inequalities for the moments of the maxima of a Gaussian process in terms of its covariance function. This has connections to covering numbers.

(b) Invent a sequence of suitable new random variables $\xi_1, \xi_2, \ldots$, and from these, a sequence of new processes $Z_n, n \geq 1$, such that conditional on the X_i, Z_n is a Gaussian process, and such that the tails of Z_n are heavier than the tails of the empirical process. This is the symmetrization part of the story.

(c) Apply the Gaussian process moment inequalities to the Z_n process conditional on the X_i, and then uncondition them. Because Z_n has heavier tails than the empirical process, these unconditioned moment bounds will also be valid for the empirical process.

(d) Use the moment bounds to bound in turn tail probabilities for the empirical process.

These steps are sketched below. We need a new definition and a Gaussian process inequality.

Definition 16.10. Let T be an indexing set equipped with a pseudometric ρ. For a given $\epsilon > 0$, the *packing number* $D(\epsilon, T, \rho)$ is defined to be the largest possible n such that there exist $t_1, t_2, \ldots, t_n \in T$, with $\rho(t_i, t_j) > \epsilon$ for $1 \leq i \neq j \leq n$; $\log D(\epsilon, T, \rho)$ is called the ϵ-capacity of T.

The Gaussian process inequality that we need uses the ϵ-capacity of its time indexing set T. It is stated below.

Lemma. *Suppose $Z(t), t \in T$ is a Gaussian process whose paths are continuous with respect to some specific pseudometric ρ on T. Assume also that $Z(t)$ satisfies $E[Z(t) - Z(s)]^2 \leq \rho^2(t, s)$ for all $s, t \in T$. Fix $t_0 \in T$. Then, there exists a finite universal constant K such that*

$$\left(E[\sup_{t \in T} |Z(t)|^2]\right)^{\frac{1}{2}} \leq (E[Z^2(t_0)])^{\frac{1}{2}} + K \int_0^\delta (\log D(x, T, \rho))^{\frac{1}{2}} dx,$$

where $\delta = \sup_{t \in T} \rho(t, t_0)$.

16.5 Maximal Inequalities and Symmetrization

There are similar inequalities for other moments of $\sup_{t \in T} |Z(t)|$, and those too can be used to bound the tail probabilities.

Theorem 16.14. *Let $\mathcal{F}$ be a given class of functions and let, for $n \geq 1$, $\sup_{f \in \mathcal{F}} |\sqrt{n}(P_n - P)(f)| = \tau_n$. Suppose the family $\mathcal{F}$ has an envelope function $\overline{F}$ bounded by $M < \infty$. Then, for a suitable universal finite constant K,*

$$E[\tau_n^2] \leq 2\pi K^2 E \left[\int_0^M \sqrt{\log D(x, \mathcal{F}, P_n)} dx \right]^2,$$

where $D(x, \mathcal{F}, P_n)$ is the largest number N for which we can find functions $f_1, f_2, \ldots, f_N \in \mathcal{F}$ such that $\sqrt{P_n(f_i - f_j)^2} \geq x$ for any $i, j, i \neq j$.

Proof. We give a sketch of its proof. For this proof, we use the notation $E_{\{X\}}$ to denote conditional expectation given $X_1, X_2, \ldots, X_n$. The bulk of the proof consists of showing the following.

(a) Construct a new process $Z_n(X, f)$, $f \in \mathcal{F}$, based on a sequence of new random variables described below, such that for any $X_1, X_2, \ldots, X_n$,

$$\left(E_{\{X\}} \left[\sup_{f \in \mathcal{F}} |Z_n(X, f)|^2 \right] \right)^{1/2} \leq K \int_0^M \sqrt{\log D(x, \mathcal{F}, P_n)} dx.$$

(b) Show that $E[\tau_n^2] \leq 2\pi E[\sup_{f \in \mathcal{F}} |Z_n(X, f)|^2]$; note that here E means unconditional expectation.
(c) Square the bound in part (a) on the conditional expectation, and take another expectation to get a bound on the unconditional expectation. Then combine it with the bound in part (b), and the theorem falls out. Thus, part (a) and part (b) easily imply the result of the theorem.

To prove parts (a) and (b), we construct a duplicate iid sequence $X_1', X_2', \ldots, X_n'$ from our basic measure P, and iid $N(0, 1)$ variables $W_1, W_2, \ldots, W_n$, such that the entire collection of variables $\{X_1, X_2, \ldots, X_n\}$, $\{X_1', X_2', \ldots, X_n'\}$, $\{W_1, W_2, \ldots, W_n\}$ is mutually independent. Let $\sigma_i = \frac{W_i}{|W_i|}$, $1 \leq i \leq n$; note that σ_i are symmetric ± 1 valued random variables, and σ_i is independent of $|W_i|$. We now define our new process Z_n:

$$Z_n(X, f) = n^{-1/2} \sum_{i=1}^n W_i f(X_i) = n^{-1/2} \sum_{i=1}^n \sigma_i |W_i| f(X_i).$$

Note that conditionally, given the X_i, this is a Gaussian process, and the Gaussian process lemma stated above will apply to it. The duplicate sequence

$\{X'_1, X'_2, \ldots, X'_n\}$ has not so far been used, but is used very cleverly in the steps below:

$$E[\tau_n^2] = E\left[\sup_f \left|\sqrt{n}\left(\frac{1}{n}\sum_{i=1}^n f(X_i) - \int f\,dP\right)\right|^2\right]$$

$$= E\left[\sup_f \left|\sqrt{n}\left(\frac{1}{n}\sum_{i=1}^n f(X_i) - \frac{1}{n}\sum_{i=1}^n Ef(X'_i)\right)\right|^2\right]$$

$$= E\left[\frac{1}{n}\sup_f \left|\sum_{i=1}^n [f(X_i) - Ef(X'_i)]\right|^2\right]$$

$$= E\left[\frac{1}{n}\sup_f \left|\sum_{i=1}^n [f(X_i) - E_{\{X\}}f(X'_i)]\right|^2\right]$$

(because the $\{X'_i\}$ are independent of the $\{X_i\}$)

$$= E\left[\frac{1}{n}\sup_f \left|E_{\{X\}}\sum_{i=1}^n [f(X_i) - f(X'_i)]\right|^2\right]$$

(for the same reason)

$$\leq E\left[\frac{1}{n}\sup_f E_{\{X\}}\left|\sum_{i=1}^n [f(X_i) - f(X'_i)]\right|^2\right];$$

(inside the supremum, use the fact that for any U, $|E(U)|^2 \leq E(U^2)$);

$$\leq E\left[\frac{1}{n}E_{\{X\}}\sup_f \left|\sum_{i=1}^n [f(X_i) - f(X'_i)]\right|^2\right]$$

(use the fact that the supremum of $E_{\{X\}}$ of a collection is $\leq$ the $E_{\{X\}}$ of the supremum of the collection)

$$= E\left[\frac{1}{n}\sup_f \left|\sum_{i=1}^n [f(X_i) - f(X'_i)]\right|^2\right]$$

(because expectation of a conditional expectation is the marginal expectation)

$$= E\left[\frac{1}{n}\sup_f \left|\sum_{i=1}^n \sigma_i[f(X_i) - f(X'_i)]\right|^2\right]$$

(because the three sequences $\{\sigma_i\}, \{X_i\}, \{X'_i\}$ are mutually independent, and the inequality holds for each of the 2^n combinations for the vector $(\sigma_1, \sigma_2, \ldots, \sigma_n)$)

$$\leq 4E\left[\frac{1}{n} \sup_f |\sum_{i=1}^n \sigma_i f(X_i)|^2\right]$$

(use the triangular inequality $|\sum_{i=1}^n \sigma_i [f(X_i) - f(X'_i)]| \leq |\sum_{i=1}^n \sigma_i f(X_i)| + |\sum_{i=1}^n \sigma_i f(X'_i)|$, and then the fact that $(|a| + |b|)^2 \leq 2(a^2 + b^2)$).

This implies, by the definition of $Z_n(X, f) = n^{-1/2} \sum_{i=1}^n \sigma_i |W_i| f(X_i)$, and the fact that $E(|W_i|) = \sqrt{\frac{2}{\pi}}$,

$$E[\tau_n^2] \leq 4\frac{\pi}{2} E\left[\sup_f |Z_n(X, f)|^2\right] = 2\pi E\left[\sup_f |Z_n(X, f)|^2\right].$$

Finally, now, evaluate $E\left[\sup_f |Z_n(X, f)|^2\right]$ by first conditioning on $\{X_1, X_2, \ldots, X_n\}$, so that the Gaussian process lemma applies, and we get

$$E_{\{X\}}\left[\sup_f |Z_n(X, f)|^2\right] \leq K^2 \left[\int_0^\delta \sqrt{\log D(x, \mathcal{F}, P_n)} dx\right]^2$$

$$\leq K^2 \left[\int_0^M \sqrt{\log D(x, \mathcal{F}, P_n)} dx\right]^2.$$

This proves both part (a) and part (b) and that leads to the theorem. □

16.6 * Connection to the Poisson Process

In describing the *complete randomness property* of a homogeneous Poisson process on $\mathcal{R}_+$, we noted in Theorem 13.3 that the conditional distribution of the arrival times in the Poisson process given that there have been n events in $[0, 1]$ is the same as the joint distribution of the order statistics in a sample of size n from a $U[0, 1]$ distribution. Essentially the same proof leads to the result that for a sample of size n from a $U[0, 1]$ distribution, the counting process $\{nG_n(t), 0 \leq t \leq 1\}$, which counts the number of sample values that are $\leq t$, acts as a homogeneous Poisson process on $[0, 1]$ given that there have been n events in the time interval $[0, 1]$. This is useful, because the Poisson process has special properties, that can be used to manipulate the empirical process. In particular, this Poisson process connection leads to useful inequalities on the exceedance probabilities of the empirical process, that is, probabilities of the form $P(\sup_{0 \leq t \leq T} |\alpha_n(t)| \geq c)$ for $T < 1, c > 0$.

We first give the exact Poisson process connection, and then describe an exceedance inequality.

Theorem 16.15. *For given $n \geq 1$, let $U_1, U_2, \ldots, U_n$ be iid $U[0,1]$ variables, and let $G_n(t) = \frac{\#\{i : U_i \leq t\}}{n}, 0 \leq t \leq 1$. Let $X(t)$ be a homogeneous Poisson process with constant arrival rate n. Then,*

$$\{nG_n(t), 0 \leq t \leq 1\} \stackrel{\mathcal{L}}{=} \{X(t), 0 \leq t \leq 1 \mid X(1) = n\}.$$

This distributional equality allows us to essentially conclude that the probability that the path of a uniform empirical process has some property is comparable to the probability that the path of a certain homogeneous Poisson process has that same property. Once we have this, the machinery of Poisson processes can be brought to bear.

Theorem 16.16. *Let E be any set of nonnegative, nondecreasing, right-continuous functions on $[0, T]$, $T < 1$. Fix $n \geq 1$. Then, there exists a finite constant $C = C(T)$ such that*

$$P\left(\{nG_n(t), 0 \leq t \leq T\} \in E\right) \leq CP\left(\{X(t), 0 \leq t \leq T\} \in E\right),$$

where as above, $X(t)$ is a homogeneous Poisson process with constant arrival rate n.

Proof. Denote

$$M = X(T), \quad \bar{M} = X(1) - X(T),$$
$$\mathcal{A}_1 = \text{The event that } \{nG_n(t), 0 \leq t \leq T\} \in E;$$
$$\mathcal{A}_2 = \text{The event that } \{X(t), 0 \leq t \leq T\} \in E.$$

Because $X(t)$ is a Poisson process, $(\mathcal{A}_2, M)$ is (jointly) independent of $\bar{M}$. Using this, and the distributional equality fact in Theorem 16.15,

$$P(\mathcal{A}_1) = P(\mathcal{A}_2 \mid X(1) = n) = \frac{P(\mathcal{A}_2 \cap \{M + \bar{M} = n\})}{P(M + \bar{M} = n)}$$

$$= \frac{1}{P(M + \bar{M} = n)} \sum_{k=0}^{n} P(\mathcal{A}_2 \cap \{M = n - k\}) P(\bar{M} = k)$$

$$\leq \frac{\sup_{k \geq 0} P(\bar{M} = k)}{P(X(1) = n)} \sum_{k=0}^{n} P(\mathcal{A}_2 \cap \{M = n - k\})$$

$$\leq \frac{\sup_{k \geq 0} P(\bar{M} = k)}{P(X(1) = n)} P(\mathcal{A}_2).$$

Now we use the fact that $\lfloor \lambda \rfloor$ is a mode of a Poisson distribution with mean λ (see Theorem 1.25). This allows us to conclude

16.6 Connection to the Poisson Process

$$\sup_{k \geq 0} P(\bar{M} = k) \leq e^{-n(1-T)} \frac{[n(1-T)]^{\lfloor n(1-T) \rfloor}}{\lfloor n(1-T) \rfloor!}.$$

Plugging this into the previous line,

$$P(\mathcal{A}_1) \leq \frac{1}{e^{-n}\frac{n^n}{n!}} e^{-n(1-T)} \frac{[n(1-T)]^{\lfloor n(1-T) \rfloor}}{\lfloor n(1-T) \rfloor!} P(\mathcal{A}_2).$$

Now using Stirling's approximation for the factorials, it follows that

$$\frac{e^n n!}{n^n} e^{-n(1-T)} \frac{[n(1-T)]^{\lfloor n(1-T) \rfloor}}{\lfloor n(1-T) \rfloor!}$$

has a finite supremum over n, and we may take the constant C of the theorem to be this supremum. This completes the proof of this theorem. □

This theorem, together with the following Poisson process inequality ultimately leads to an exceedance probability for the empirical process. The function h in the theorem below is $h(x) = x \log x - x + 1, x > 0$.

Theorem 16.17. *Let $X(t), t \geq 0$ be a homogeneous Poisson process with constant arrival rate equal to 1. Fix $T, \alpha \geq 0$. Then,*

$$P\left(\sup_{0 \leq t \leq T} |X(t) - t| \geq \alpha T\right) \leq 2e^{-Th(1+\alpha)}.$$

See Del Barrio et al. (2007, p. 126) for this Poisson process inequality.

Here is our final exceedance probability for the uniform empirical process. By the usual means of a quantile transform, this can be translated into an exceedance probability for $\beta_n(t)$, the general one-dimensional empirical process. The theorem below establishes exponential bounds on the tail of the supremum of the empirical process. In spirit, this is similar to the DKW inequality (see Theorem 6.13). Processes with such an exponential rate of decay for the tail are called sub-Gaussian. Thus, the theorem below and the DKW inequality establish the sub-Gaussian nature of the one-dimensional empirical process.

Theorem 16.18 (Exceedance Probability for the Empirical Process). *Let $0 < T < 1$, and $\alpha > 0$. The uniform empirical process $\alpha_n(t), 0 \leq t \leq 1$ satisfies the inequality*

$$P\left(\sup_{0 \leq t \leq T} |\alpha_n(t)| \geq \alpha \sqrt{T}\right) \leq Ce^{-nTh\left(1+\frac{\alpha}{\sqrt{nT}}\right)}$$

$$\leq Ce^{-\frac{\alpha^2}{2} + \frac{\alpha^3}{6\sqrt{nT}}},$$

for some constant C depending only on T.

We refer to del Barrio, Deheuvels and van de Geer (2007, p. 147) for a detailed proof, using Theorem 16.16 and Theorem 16.17.

Exercises

Exercise 16.1. Simulate iid $U[0, 1]$ observations and plot the uniform empirical process $\alpha_n(t)$ for $n = 10, 25, 50$.

Exercise 16.2. Give examples of three functions that are members of $D[0, 1]$, but not of $C[0, 1]$.

Exercise 16.3. For a given n, what is the set of possible values of the Cramér–von Mises statistic C_n^2?

Exercise 16.4. Find an expression for $E(C_n^2)$.

Exercise 16.5. By using the Karhunen–Loeve expansion for a Brownian bridge, give a proof of the familiar calculus formula $\sum_{n=1}^{\infty} \frac{1}{n^2} = \frac{\pi^2}{6}$.

Exercise 16.6 * **(Anderson–Darling Statistic).** Find the characteristic function of the limiting distribution of the Anderson–Darling statistic A_n^2.

Exercise 16.7. Prove that $\frac{D_n^*}{\sqrt{2 \log \log n}} \stackrel{P}{\Rightarrow} 1$.

Exercise 16.8 * **(Deviations from the Three Sigma Rule).** Let $X_1, X_2, \ldots$ be iid $N(\mu, \sigma^2)$, and let $\overline{X}_n, s_n$ be the mean and the standard deviation of the first n observations. On suitable centering and normalization, find the limiting distribution of the number of observations among the first n that are between $\overline{X}_n \pm k s_n$, where k is a fixed general positive number.

Exercise 16.9 (Rate in Strong Approximation). Prove part (a) of Theorem 16.5 by using the KMT inequality in part (b), using $x = x_n = c \log n$, and then by invoking the Borel–Cantelli lemma.

Exercise 16.10 (Strong Approximation of Quantile Process). Show that the strong approximation result in part (c) of Theorem 16.5 holds for any normal and any Cauchy distribution.

Exercise 16.11 * **(Strong Approximation of Quantile Process).** Give an example of a distribution for which the conditions of part (c) of Theorem 16.5 do not hold.

Hint: Look at distributions with very slow tails.

Exercise 16.12 (Nearest Neighbors Are Close). Let $X_1, X_2, \ldots$ be iid from a distribution P supported on a compact set C in $\mathcal{R}^d$. Let $\mathcal{K}$ be a fixed compact set in $\mathcal{R}^d$. Prove that $\sup_{x \in \mathcal{K} \cap C} ||X_{n,NN,x} - x|| \stackrel{a.s.}{\to} 0$, where for a given x, $X_{n,NN,x}$ is a nearest neighbor of x among $X_1, X_2, \ldots, X_n$.

Is this result true without the assumption of compact support for P?

Exercise 16.13. Consider the weight function $w_\alpha(t) = t^{-\alpha}$ for $0 \leq t \leq \frac{1}{2}$, and $w_\alpha(t) = (1 - t)^{-\alpha}$ for $\frac{1}{2} \leq t \leq 1$. For what values of $\alpha, \alpha \geq 0$, does $w_\alpha(t)$ satisfy part (a) of the Chibisov–O'Reilly theorem? Part (b) of the Chibisov–O'Reilly theorem?

Exercises

Exercise 16.14. Let $B(t)$ be a Brownian bridge on $[0, 1]$. What can you say about $\limsup_{t \to 0} \frac{|B(t)|}{\sqrt{t(1-t)}}$?

Exercise 16.15 (Deviation of Sample Mean from Population Mean). Suppose $X_1, X_2, \ldots$ are iid from a distribution P on the real line with a finite variance σ^2. Does $\frac{|\bar{X} - \mu|}{E(|\bar{X} - \mu|)}$ converge almost surely to one? Prove your answer.

Exercise 16.16 (Evaluating Shattering Coefficients). Suppose $\mathcal{C}$ is the class of all closed rectangles in $\mathcal{R}$. Show that $S(n, \mathcal{C}) = 1 + \binom{n}{1} + \binom{n}{2}$, and hence show that the VC dimension of $\mathcal{C}$ is 2.

Exercise 16.17 * **(Evaluating VC Dimensions).** Find the VC dimension of the following classes of sets :

(a) Southwest quadrants of $\mathcal{R}^d$
(b) Closed half-spaces of $\mathcal{R}^d$
(c) Closed balls of $\mathcal{R}^d$
(d) Closed rectangles of $\mathcal{R}^d$

Exercise 16.18. Give examples of three nontrivial classes of sets in $\mathcal{R}^d$ that are not VC classes.

Exercise 16.19 * **(Evaluating VC Dimensions).** Find the VC dimension of the class of all polygons in the plane with four vertices.

Exercise 16.20. *Is the VC dimension of the class of all ellipsoids of $\mathcal{R}^d$ the same as that of the class of all closed balls of $\mathcal{R}^d$?

Exercise 16.21. Find the VC dimension of the class of all simplices in $\mathcal{R}^d$, where a simplex is a set of the form $C = \{x : x = \sum_{i=1}^{d+1} p_i x_i, p_i \geq 0, \sum_{i=1}^{d+1} p_i = 1\}$.

Exercise 16.22 (VC Dimension Under Translation). Suppose $\mathcal{F}$ is a given class of real valued functions, and g is some other fixed function. Let $\mathcal{G}$ be the family of functions $f + g$ with $f \in \mathcal{F}$. Show that the class of subgraphs of the functions in $\mathcal{F}$ and the class of subgraphs of the functions in $\mathcal{G}$ have the same VC dimension.

Exercise 16.23 * **(Huge Classes with Low VC Dimension).** Show that it is possible for a class of sets of integers to have uncountably many sets, and yet to have a VC dimension of just one.

Exercise 16.24 * **(A Concept Similar to VC Dimension).** Let $\mathcal{C}$ be a class of sets and define $\mathcal{D}(\mathcal{C}) = \inf\{\alpha > 0 : \sup_{n \geq 1} \frac{S(n, \mathcal{C})}{n^\alpha} < \infty\}$.
  Show that

(a) $\mathcal{D}(\mathcal{C}) \leq VC(\mathcal{C})$.
(b) $VC(\mathcal{C}) < \infty$ if and only if $\mathcal{D}(\mathcal{C}) < \infty$.

Exercise 16.25. Give a proof of the one dimensional Glivenko-Cantelli theorem by using part (b) of Theorem 16.11.

Exercise 16.26. *Design a test for testing that sample observations in $\mathcal{R}^2$ are iid from a uniform distribution in the unit square by using suitable VC classes and applying the CLT for empirical measures.

Exercise 16.27. Let P be a distribution on $\mathcal{R}^d$, and P_n the empirical measure. Let $\mathcal{C}$ be a VC class. Show that $E[\sup_{C \in \mathcal{C}} |P_n(C) - P(C)|] = O\left(\sqrt{\frac{\log n}{n}}\right)$.

Hint: Look at Theorem 16.8.

Exercise 16.28 *(Distinct Values in a Sample).* Suppose P is a countably supported distribution on the real line and $X_1, X_2, \ldots$ are iid with common distribution P. Show that

$$\frac{E\left[\mathrm{Card}(X_1, X_2, \ldots, X_n)\right]}{n} \to 0,$$

where as usual, $\mathrm{Card}(X_1, X_2, \ldots, X_n)$ denotes the number of distinct values among $X_1, X_2, \ldots, X_n$.

Exercise 16.29 *(Farthest Nearest Neighbor).* Suppose $X_1, X_2, \ldots$ are iid random variables from a continuous CDF F on the real line. For any $j, 1 \leq j \leq n$, let $X_{j,n,NN}$ denote the nearest neighbor of X_j among the rest of the observations $X_1, X_2, \ldots, X_n$. Let $d_{j,n} = |X_j - X_{j,n,NN}|$ and $M_n = \max_j d_{j,n}$. Show that $n M_n \stackrel{a.s.}{\to} \infty$.

Exercise 16.30 *(Orlicz Norms).* For a nondecreasing convex function ψ on $[0, \infty)$ with $\psi(0) = 0$, the Orlicz norm of a real-valued random variable X is $||X||_\psi = \inf\{c > 0 : E\left[\psi\left(\frac{|X|}{c}\right)\right] \leq 1\}$, and $||X||_\psi = \infty$ if no c exists for which $E\left[\psi\left(\frac{|X|}{c}\right)\right] \leq 1$.

(a) Show that the L_p norm of X, namely $||X||_p = (E[|X|^p])^{1/p}$ is an Orlicz norm for all $p \geq 1$.
(b) Consider $\psi_p(x) = e^{x^p} - 1$ for $p \geq 1$. Show that $||X||_p \leq p! ||X||_{\psi_1}$.
(c) Show that $||X||_2 \leq ||X||_{\psi_2}$.
(d) Obtain the tail bound $P(|X| \geq x) \leq \dfrac{1}{\psi\left(\frac{x}{||X||_\psi}\right)}$ for a general such ψ.
(e) For any constant $\alpha \geq 1$, $||X||_{\alpha\psi} \leq \alpha ||X||_\psi$.
(f) Show that the infimum in the definition of an Orlicz norm is attained when the Orlicz norm is finite.

Exercise 16.31 *(Orlicz Norms and Growth of the Maximum).* Let $X_1, X_2, \ldots, X_n$ be any n random variables and ψ an increasing convex function with $\psi(0) = 0$. Suppose also that $\limsup_{x,y \to \infty} \frac{\psi(x)\psi(y)}{\psi(cxy)} < \infty$ for some finite positive c.

(a) Then there exists a universal constant K depending only on ψ such that

$$\left|\left| \max_{1 \leq i \leq n} X_i \right|\right|_\psi \leq K \psi^{-1}(n) \max_{1 \leq i \leq n} ||X_i||_\psi.$$

(b) Use this inequality with the functions $\psi_p(x) = e^{x^p} - 1, p \geq 1$, for iid $N(0, 1)$ random variables $X_1, X_2, \ldots, X_n$.

Hint: See van der Vaart and Wellner (1996, p. 96).

Exercise 16.32 (**Maximum of Correlated Normals**). Suppose $X_1, X_2, \ldots, X_n$ are jointly normally distributed, with zero means, and $\text{Var}(X_i) = \sigma_i^2$. Show that
$$E\left[\max_{1 \leq i \leq n} |X_i|\right] \leq 2\sqrt{2 \log n} \max_{1 \leq i \leq n} \sigma_i.$$

Exercise 16.33 (**Metric Entropy and Packing Numbers**). Suppose $\mathcal{T}$ is a bounded set in an Euclidean space $\mathcal{R}^d$. Show that the metric entropy and the ϵ-capacity satisfy the inequality $D(2\epsilon, T, ||.||_2) \leq N(\epsilon, T, ||.||_2) \leq D(\epsilon, T, ||.||_2)$.

Exercise 16.34 (**Metric Entropy**). Suppose T is a compact subset of $[0, 1]$. Show that

(a) If T has positive Lebesgue measure, then $N(\epsilon, T, ||.||) \approx \epsilon^{-1}$, where the norm $||.||$ is the usual distance between two reals.
(b) If T has zero Lebesgue measure, then $N(\epsilon, T, ||.||) = o(\epsilon^{-1})$.

Exercise 16.35 * (**Packing Number of Spheres**). Let B be a ball of radius r in $\mathcal{R}^d$. Show that $D(\epsilon, B, ||.||_2) \leq \left(\frac{3r}{\epsilon}\right)^d$.

Hint: See p. 94 in van der Vaart and Wellner (1996).

Exercise 16.36. Prove the following generalization of the binomial tail inequality in Example 16.7: suppose $W_1, W_2, \ldots, W_n$ are independent Bernoulli variables with parameters $p_1, p_2, \ldots, p_n$. Then, for any k, $P(\sum_{i=1}^{n} W_i > k) \leq \left(\frac{en\bar{p}}{k}\right)^k$, where $\bar{p} = \frac{p_1 + p_2 + \ldots + p_n}{n}$.

Exercise 16.37 (**Exceedance Probability for Poisson Processes**). Suppose $X(t)$, $t \geq 0$ is a general Poisson process with intensity function $\lambda(t)$ and mean measure μ. By using Theorem 16.17, derive an upper bound on $P(\sup_{0 \leq t \leq T} |X(t) - \mu([0, t])| \geq \alpha T)$, where T, α are positive constants. State your assumptions on the intensity function $\lambda(t)$.

Exercise 16.38. By using Theorem 16.18, find an upper bound on $E[\sup_{0 \leq t \leq T} |\alpha_n(t)|]$ that is valid for every n and each $T < 1$.

References

Alexander, K. (1987). The central limit theorem for empirical processes on Vapnik-Chervonenkis classes, *Ann. Prob.*, 15, 178–203.
Beran, R. and Millar, P. (1986). Confidence sets for a multinomial distribution, *Ann. Statist.*, 14, 431–443.

Billingsley, P. (1968). *Convergence of Probability Measures*, John Wiley, New York.
Chibisov, D. (1964). Some theorems on the limiting behavior of an empirical distribution function, *Theor. Prob. Appl.* 71, 104–112.
Csáki, E. (1980). On the standardized empirical distribution function, *Nonparametric Statist. Infer. I*, Colloq. Math. Soc. János Bolyai, 32, North-Holland, Amsterdam. 123–138.
Csörgo, M. (1983). Quantile processes with statistical applications, *CBMS-NSF Regional conf. ser.*, 42, SIAM, Philadelphia.
Csörgo, M. (2002). A glimpse of the impact of Pal Erdös on probability and statistics, *Canad. J. Statist.*, 30, 4, 493–556.
Csörgo, M. and Révész, P. (1981). *Strong Approximations in Probability and Statistics*, Academic Press, New York.
DasGupta, A. (2008). *Asymptotic Theory of Statistics and Probability*, Springer, New York.
del Barrio, E., Deheuvels, P., and van de Geer, S. (2007). *Lectures on Empirical Processes*, European Mathematical Society, Zurich.
Donsker, M. (1952). Justification and extension to Doob's heuristic approach to Kolmogorov-Smirnov theorems, *Ann. Math. Statist.*, 23, 277–281.
Dudley, R. (1978). Central limit theorems for empirical measures, *Ann. Prob.*, 6, 899–929.
Dudley, R. (1979). Central limit theorems for empirical measures, *Ann. Prob.*, 7, 5, 909–911.
Dudley, R. (1984). *A Course on Empirical Processes*, Springer-Verlag, New York.
Dudley, R. (1999). *Uniform Central Limit Theorems*, Cambridge University Press, Cambridge,UK.
Eicker, F. (1979). The asymptotic distribution of the supremum of the standardized empirical process, *Ann. Statist.*, 7, 116-138.
Einmahl, J. and Mason, D. M. (1985). Bounds for weighted multivariate empirical distribution functions, *Z. Wahr. Verw. Gebiete*, 70, 563–571.
Giné, E. (1996). Empirical processes and applications: An overview, *Bernoulli*, 2, 1, 1–28.
Giné, E. and Zinn, J. (1984). Some limit theorems for empirical processes: With discussion, *Ann. Prob.*, 12, 4, 929–998.
Jaeschke, D. (1979). The asymptotic distribution of the supremum of the standardized empirical distribution function over subintervals, *Ann. Statist.*, 7, 108–115.
Komlós, J., Major, P., and Tusnady, G. (1975a). An approximation of partial sums of independent rvs and the sample df :I, *Zeit für Wahr. Verw. Geb.*, 32, 111–131.
Komlós, J., Major, P., and Tusnady, G. (1975b). An approximation of partial sums of independent rvs and the sample df :II, *Zeit für Wahr. Verw. Geb.*, 34, 33–58.
Kosorok, M. (2008). *Introduction to Empirical Processes and Semiparametric Inference*, Springer, New York.
Mason, D. and van Zwet, W.R. (1987). A refinement of the KMT inequality for the uniform empirical process, *Ann. Prob.*, 15, 871–884.
O'Reilly, N. (1974). On the weak convergence of empirical processes in sup-norm metrics, *Ann. Prob.*, 2, 642–651.
Pollard, D. (1989). Asymptotics via empirical processes, statist. Sci., 4, 4, 341–366.
Sauer, N. (1972). On the density of families of sets, *J. Comb. Theory, Ser A*, 13, 145–147.
Shorack, G. and Wellner, J. (2009). *Empirical Processes, with Applications to Statistics*, SIAM, Philadelphia.
van der Vaart, A. and Wellner, J. (1996). *Weak Convergence and Empirical Processes*, Springer-Verlag, New York.
Vapnik, V. and Chervonenkis, A. (1971). On the uniform convergence of relative frequencies of events to their probabilities, *Theory Prob. Appl.*, 16, 264–280.
Wellner, J. (1992). Empirical processes in action: A review, *Int. Statist. Rev.*, 60, 3, 247–269.

Chapter 17
Large Deviations

The mean μ of a random variable X is arguably the most common one number summary of the distribution of X. Although averaging is a primitive concept with some natural appeal, the mean μ is a useful summary only when the random variable X is concentrated around the mean μ, that is, probabilities of large deviations from the mean are small. The most basic large deviation inequality is Chebyshev's inequality, which says that if X has a finite variance σ^2, then $P(|X - \mu| > k\sigma) \leq \frac{1}{k^2}$. But, usually, this inequality is not strong enough in specific applications, in the sense that the assurance we seek is much stronger than what Chebyshev's inequality will give us. The theory of large deviations is a massive and powerful mathematical enterprise that gives bounds, usually of an exponential nature, on probabilities of the form $P(f(X_1, X_2, \ldots, X_n) - E[f(X_1, X_2, \ldots, X_n)] > t)$, where $X_1, X_2, \ldots, X_n$ is some set of n random variables, not necessarily independent, $f(X_1, X_2, \ldots, X_n)$ is a suitable function of them, and t is a given positive number. We expect that the probability $P(f(X_1, X_2, \ldots, X_n) - E[f(X_1, X_2, \ldots, X_n)] > t)$ is small for large n if we know from some result that $f(X_1, X_2, \ldots, X_n) - E[f(X_1, X_2, \ldots, X_n)] \stackrel{P}{\Rightarrow} 0$; the theory of large deviations attempts to quantify how small this probablity is.

The basic Chernoff–Bernstein inequality (see Theorem 1.34) is generally regarded to be the first and most fundamental large deviation inequality. The Chernoff–Bernstein inequality is for $f(X_1, X_2, \ldots, X_n) = \sum_{i=1}^{n} X_i$ in the one-dimensional iid case, assuming the existence of an mgf for the common distribution of the X_i. Since then, the theory of large deviations has grown by leaps and bounds in every possible manner. The theory covers dependent sequences, higher and even infinite dimensions, nonlinear functions f, much more abstract spaces than Euclidean spaces, and in certain cases the existence of an mgf is no longer assumed. Numerous excellent book-length treatments of the theory of large deviations are now available. We can only give a flavor of this elegant theory in a few selected cases. It should be emphasized that large deviations is one area of probability where beautiful mathematics has smoothly merged with outstanding concrete applications over a diverse set of problems and areas. It is now seeing applications in emerging areas of contemporary statistics, such as multiple testing in high-dimensional situations. The importance of large deviations is likely to increase even more than what it already is.

Among numerous excellent books on large deviations, we recommend Stroock (1984), Varadhan (1984), Dembo and Zeitouni (1993), den Hollander (2000), and Bucklew (2004). There is considerable overlap between the modern theory of concentration inequalities and the area of large deviations. Relevant references include Devroye et al. (1996), McDiarmid (1998), Lugosi (2004), Ledoux (2004), and Dubhashi and Panconesi (2009). Shao (1997), Hall and Wang (2004), and Dembo and Shao (2006), among many others, treat the modern problem of large deviations using self-normalization, applicable when an mgf or even sufficiently many moments fail to exist.

17.1 Large Deviations for Sample Means

Although large deviation theory has been worked out for statistics that are far more general than sample means, and without requiring that the underlying sequence $X_1, X_2, \ldots$ be iid or even independent, for historical importance we start with the case of sample means of iid random variables in one dimension.

17.1.1 The Cramér–Chernoff Theorem in $\mathcal{R}$

We start with the basic Cramér–Chernoff theorem. This result may be regarded as the starting point of large deviation theory.

Theorem 17.1. *Suppose $X_1, X_2, \ldots$ are iid zero mean random variables with an mgf $\psi(z) = E(e^{zX_1})$, assumed to exist for all real z. Let $k(z) = \log \psi(z)$ be the cumulant generating function of X_1. Then, for fixed $t > 0$,*

$$\lim_n \frac{1}{n} \log P(\overline{X} > t) = -I(t) = \inf_{z \in \mathcal{R}}(k(z) - tz) = -\sup_{z \in \mathcal{R}}(tz - k(z))$$

The function $I(t)$ is called the rate function *corresponding to F, the common distribution of the X_i. Because we assume the existence of a mean (in fact, the existence of an mgf), by the WLLN $\overline{X}$ converges in probability to zero. Therefore, we already know that $P(\overline{X} > t)$ is small for large enough n. According to the Cramér–Chernoff theorem, $\lim_n \frac{1}{n} \log P(\overline{X} > t) = -I(t)$, and so for large n, $\frac{1}{n} \log P(\overline{X} > t) \approx -I(t)$. Therefore, as a first approximation, $P(\overline{X} > t) \approx e^{-nI(t)}$. In other words, assuming the existence of an mgf, $P(\overline{X} > t)$ converges to zero at an exponential rate, and $I(t)$ exactly characterizes that exponential rate. This justifies the name* rate function *for $I(t)$.*

Actually, it would be natural to consider $P(\overline{X} > t)$ itself, rather than its logarithm. But the quantity $P(\bar{X} > t)$ is a sequence of the form $c_n(t)e^{-nI(t)}$, for some suitable sequence $c_n(t)$, which does not converge to zero at an exponential rate. Pinning down the exact asymptotics of the c_n sequence is a difficult problem;

17.1 Large Deviations for Sample Means

see Exercise 17.13. If we instead look at $\frac{1}{n} \log P(\overline{X} > t)$, then $I(t)$ becomes the dominant term, and analysis of the sequence c_n can be avoided.

Proof of the Cramér–Chernoff theorem. For $n \geq 1$, let $S_n = \sum_{i=1}^{n} X_i$. First note that we may, by a simple translation argument, take t to be zero, and correspondingly, the mean of X_1 to be < 0. This reduces the theorem to showing that $\lim_n \frac{1}{n} \log P(\overline{X} \geq 0) = -I(0)$. Because $\overline{X} \geq 0$ if and only if $S_n \geq 0$, we need to show that $\lim_n \frac{1}{n} \log P(S_n \geq 0) = -I(0)$.

If the common CDF F of the X_i is supported on $(-\infty, 0]$, that is, if $P(X_1 \leq 0) = 1$, then the theorem falls out easily. This case is left as an exercise. We therefore consider the case where $P(X_1 < 0) > 0$, $P(X_1 > 0) > 0$, $\mu = E(X_1) < 0$.

We now observe a few important facts that are used in the rest of the proof.

(a) $\psi(0) = 1, \psi'(0) = \mu < 0, \psi''(z) > 0 \, \forall \, z \in \mathcal{R}$,

implying that ψ is strictly convex. Let τ be the unique minima of ψ and let $\psi(\tau) = \inf_{z \in \mathcal{R}} \psi(z) = \rho$. Note that $\tau > 0$, and $\psi'(\tau) = \int x e^{\tau x} dF(x) = 0$.

(b) $I(0) = -\inf_{z \in \mathcal{R}} \log \psi(z) = -\log \inf_{z \in \mathcal{R}} \psi(z) = -\log \rho$

$\Rightarrow -I(0) = \log \rho$.

Therefore, we have to prove that

$$\lim_n \frac{1}{n} \log P(S_n \geq 0) = \log \rho.$$

This consists of showing that $\limsup_n \frac{1}{n} \log P(S_n \geq 0) \leq \log \rho$ and $\liminf_n \frac{1}{n} \log P(S_n \geq 0) \geq \log \rho$. Of these, the first inequality is nearly immediate. Indeed, for any $z > 0$,

$$P(S_n \geq 0) = P(zS_n \geq 0) = P(e^{zS_n} \geq 1)$$
$$\leq E[e^{zS_n}] = [\psi(z)]^n$$

(by Markov's inequality)

$$\Rightarrow \frac{1}{n} \log P(S_n \geq 0) \leq \inf_{z>0} \log \psi(z) = \log \inf_{z>0} \psi(z)$$
$$= \log \inf_{z \in \mathcal{R}} \psi(z) = \log \rho,$$

giving the first inequality $\limsup_n \frac{1}{n} \log P(S_n \geq 0) \leq \log \rho$.

The second inequality is more involved, and uses a common technique in large deviation theory known as *exponential tilting*. Starting from the CDF F of the X_i, define a new CDF

$$\tilde{F}(x) = \frac{\int_{(-\infty, x]} e^{\tau y} dF(y)}{\rho}.$$

Note that $d\tilde{F}(x) = \frac{e^{\tau x} dF(x)}{\rho} \Rightarrow dF(x) = \rho e^{-\tau x} d\tilde{F}(x)$. Let $\tilde{X}_1, \tilde{X}_2, \ldots, \tilde{X}_n$ be an iid sample from $\tilde{F}$. We have

$$E(\tilde{X}_i) = \int x \, d\tilde{F}(x) = \frac{\int x e^{\tau x} dF(x)}{\rho} = \frac{\psi'(\tau)}{\rho} = 0.$$

Also, $\tilde{X}_i$ have a finite variance. With all these, for any given $n \geq 1$,

$$P(S_n \geq 0) = \int_{(x_1, x_2, \ldots, x_n): x_1 + x_2 + \cdots + x_n \geq 0} dF(x_1) dF(x_2) \cdots dF(x_n)$$

$$= \rho^n \int_{(x_1, x_2, \ldots, x_n): x_1 + x_2 + \cdots + x_n \geq 0} e^{-\tau(x_1 + x_2 + \cdots + x_n)}$$

$$\times d\tilde{F}(x_1) d\tilde{F}(x_2) \cdots d\tilde{F}(x_n)$$

$$= \rho^n E\left[e^{-\tau \tilde{S}_n} I_{\tilde{S}_n \geq 0}\right],$$

where $\tilde{S}_n = \tilde{X}_1 + \tilde{X}_2 + \cdots + \tilde{X}_n$.
Therefore,

$$\frac{1}{n} \log P(S_n \geq 0) = \log \rho + \frac{1}{n} \log E\left[e^{-\tau \tilde{S}_n} I_{\tilde{S}_n \geq 0}\right].$$

By the CLT, $\frac{\tilde{S}_n}{\sqrt{n}}$ is approximately a mean zero normal random variable. This implies (with just a little manipulation) that

$$E\left[e^{-\tau \tilde{S}_n} I_{\tilde{S}_n \geq 0}\right]$$

is bounded below by $c e^{-a\tau \sqrt{n}}$ (with a lot of room to spare) for suitable $a, c > 0$, and so, by taking logarithms,

$$\liminf_n \frac{1}{n} \log E\left[e^{-\tau \tilde{S}_n} I_{\tilde{S}_n \geq 0}\right] \geq \liminf_n \left[\frac{\log c}{n} - \frac{a\tau \sqrt{n}}{n}\right] = 0,$$

and this gives us the second inequality $\liminf_n \frac{1}{n} \log P(S_n \geq 0) \geq \log \rho$. □

What is the practical value of a large deviation result? The practical value is that, in the absence of a large deviation result, we approximate a probability of the type $P(\bar{X} > t)$ by a CLT approximation, as that would be almost the only other approximation that we can think of. Indeed, it is common practice to use the CLT approximation in applied work. However, for *fixed* t, the CLT approximation is not

17.1 Large Deviations for Sample Means

going to give an accurate approximation to the true value of $P(\bar{X} > t)$. The CLT is supposed to be applied for those t that are typical values for $\bar{X}$, that is, t of the order of $\frac{1}{\sqrt{n}}$, but not for fixed t. The *exponential tilting technique* brilliantly reduces the problem to the case of a typical value, but for a new sequence $\tilde{X}_1, \tilde{X}_2, \ldots$. Thus, in comparison, an application of the Cramér–Chernoff theorem is going to produce a more accurate approximation than a straight CLT approximation. Whether it really is more accurate in a given case depends on the value of t. See Groeneboom and Oosterhoff (1977, 1980, 1981) for extensive finite sample numerics on the comparative accuracy of the CLT and the large deviation approximations to $P(\bar{X} > t)$.

We now work out some examples of the rate function $I(t)$ for some common choices of F.

Example 17.1 (Rate Function for Normal). Let $X_1, X_2, \ldots$ be iid $N(0, \sigma^2)$. Then $\psi(z) = e^{z^2\sigma^2/2}$ and $k(z) = \log \psi(z) = z^2\sigma^2/2$. Therefore, $k(z) - tz = z^2\sigma^2/2 - tz$ is a strictly convex quadratic in z with a unique minima at $z_0 = \frac{t}{\sigma^2}$, and a minimum value of $z_0^2\sigma^2/2 - tz_0 = -\frac{t^2}{2\sigma^2}$. Therefore, in this case, the rate function equals $I(t) = \frac{t^2}{2\sigma^2}$. This can be used to form the first approximation $P(\bar{X} > t) \approx e^{-nt^2/(2\sigma^2)}$, or, equivalently $P\left(\frac{\bar{X}}{\sigma} > t\right) \approx e^{-nt^2/2}$.

Note that in this example, the rate function $I(t)$ turns out to be convex and strictly positive unless $t = 0$. It is also smooth. These turn out to be true in general, as we later show.

Example 17.2 (Rate Function in Bernoulli Case). Here is an example where the rate function $I(t)$ is related to the *Kullback–Leibler distance* between two suitable Bernoulli distributions (see Chapter 15 for the definition and properties of the Kullback–Leibler distance). Suppose $X_1, X_2, \ldots$ is an iid sequence of Bernoulli variables with parameter p. Then the mgf $\psi(z) = pe^z + q$, where $q = 1 - p$. In the Bernoulli case, the question of deriving the rate function is interesting only for $0 < t < 1$, because $P(\bar{X} > t) = 1$ if $t < 0$ and $P(\bar{X} > t) = 0$ if $t \geq 1$. Also, if $t = 0$, then, trivially, $P(\bar{X} > t) = 1 - (1-p)^n$.

For $0 < t < 1$,

$$\frac{d}{dz}[tz - \log \psi(z)] = t - \frac{pe^z}{pe^z + q} = \frac{qt - pe^z(1-t)}{pe^z + q},$$

and,

$$\frac{d^2}{dz^2}[tz - \log \psi(z)] = \frac{-pqe^z}{(pe^z + q)^2} < 0 \; \forall \, z.$$

Therefore, $tz - \log \psi(z)$ has a unique maxima at z given by

$$qt = pe^z(1-t) \Leftrightarrow z = \log \frac{1-p}{p} + \log \frac{t}{1-t},$$

and the rate function equals

$$I(t) = \sup_{z \in \mathcal{R}}[tz - \log \psi(z)] = t\left[\log \frac{1-p}{p} + \log \frac{t}{1-t}\right] - \log\left(\frac{qt}{1-t} + q\right)$$

$$= t \log \frac{t}{p} - t \log \frac{1-t}{1-p} + \log \frac{1-t}{1-p} = t \log \frac{t}{p} + (1-t) \log \frac{1-t}{1-p}.$$

Interestingly, this exactly equals the Kullback–Leibler distance $K(P, Q)$ between the distributions P, Q with P as a Bernoulli distribution with parameter t and Q a Bernoulli distribution with parameter p. Indeed,

$$K(P, Q) = \sum_{x=0}^{1} t^x (1-t)^{1-x} \log \frac{t^x (1-t)^{1-x}}{p^x (1-p)^{1-x}} = E_P\left[X \log \frac{t}{p} + (1-X) \log \frac{1-t}{1-p}\right]$$

$$= t \log \frac{t}{p} + (1-t) \log \frac{1-t}{1-p}.$$

Example 17.3 (The Cauchy Case). The Cramér–Chernoff theorem requires the existence of the mgf for the underlying F. Suppose $X_1, X_2, \ldots$ are iid $C(0, 1)$. In this case, $\bar{X}$ is also distributed as $C(0, 1)$ for all n (see Chapter 8), and therefore, $\lim_n \frac{1}{n} \log P(\bar{X} > t) = \lim_n \frac{1}{n} \log P(X_1 > t) = 0$. As regards the mgf $\psi(z), \psi(0) = 1$ and at any $z \neq 0, \psi(z) = \infty$. Therefore, formally, $\sup_z[tz - \log \psi(z)] = 0$. That is, formally, the rate function $I(t) = 0$, which gives the correct answer for $\lim_n \frac{1}{n} \log P(\bar{X} > t)$.

17.1.2 Properties of the Rate Function

The rate function $I(t)$ satisfies a number of general shape and smoothness properties. We need a few definitions and some notation to describe these properties. Given a CDF F, we let μ be its mean, and let

$$\mathcal{D} = \{z \in \mathcal{R} : \psi(z) = E_F(e^{zX}) < \infty\}; \quad \mathcal{D}_* = \{t \in \mathcal{R} : I(t) < \infty\}.$$

Definition 17.1. A *level set* of a real-valued function f is a set of the form $\{t \in \mathcal{R} : f(t) \leq c\}$, for some constant c.

Definition 17.2. A real-valued function f is called *lower semicontinuous* if all its level sets are closed, or equivalently, for any t, and any sequence t_n converging to t, $\liminf_n f(t_n) \geq f(t)$.

Definition 17.3. A real-valued function f is called *real analytic* on an open set C if it is infinitely differentiable on C and can be expanded around any $t_0 \in C$ as a convergent Taylor series $f(t) = \sum_{n=0}^{\infty} a_n (t - t_0)^n$ for all t in a neighborhood of t_0.

Theorem 17.2. *The rate function $I(t)$ satisfies the following properties:*

(a) $I(t) \geq 0$ for all t and equals zero only when $t = \mu$.
(b) $I(t)$ is convex on $\mathcal{R}$. Moreover, it is strictly convex on the interior of $\mathcal{D}_*$.

17.1 Large Deviations for Sample Means

(c) $I(t)$ is lower semicontinuous on $\mathcal{R}$.
(d) $I(t)$ is a real analytic function on the interior of $\mathcal{D}_*$.
(e) If z solves $k'(z) = t$, then $I(t) = tz - k(z)$.

Proof. Part (a), the convexity part of part (b), and part (c) follow from simple facts in real analysis (such as the supremum of a collection of continuous functions is lower semicontinuous, and the supremum of a collection of linear functions is convex). Part (e) follows from the properties of strict convexity and differentiability, by simply setting the first derivative of $tz - k(z)$ to be zero. Part (d) is a deep property. See p. 121 in den Hollander (2000). □

Example 17.4 (Error Probabilities of Likelihood Ratio Tests). Suppose $X_1, X_2, \ldots$ are iid observations from a distribution P on some space $\mathcal{X}$ and suppose that we want to test a null hypothesis $H_0 : P = P_0$ against an alternative hypothesis $H_1 : P = P_1$. We assume, for simplicity, that P_0, P_1 have densities f_0, f_1, respectively. The likelihood function based on $X_1, X_2, \ldots, X_n$ is defined to be

$$\Lambda_n = \frac{\prod_{i=1}^n f_1(X_i)}{\prod_{i=1}^n f_0(X_i)} = \prod_{i=1}^n \frac{f_1(X_i)}{f_0(X_i)}.$$

A likelihood ratio test is a test that rejects H_0 when $\Lambda_n > \lambda_n$ for some sequence of numbers λ_n. The type I error probability of such a test is defined to be the probability of false rejection of H_0, namely,

$$\alpha_n = P_0(\Lambda_n > \lambda_n),$$

and the type II error probability is defined to be the probability of false acceptance of H_0, namely,

$$\beta_n = P_1(\Lambda_n \leq \lambda_n).$$

Again, for simplicity, we take $\lambda_n = \lambda^n$, where λ is a constant, and denote $\gamma = \log \lambda$. The purpose of this example is to show that the two error probabilities of the likelihood ratio test converge to zero at an exponential rate, when the number of observations $n \to \infty$, and that the rates of these exponential convergences can be paraphrased in large deviation terms. We denote $\log \frac{f_1(X_i)}{f_0(X_i)} = Y_i$ below. Note that the mgf of Y_1 under H_0 is

$$\psi_0(z) = E_{P_0}[e^{zY_1}] = \int e^{z \log \frac{f_1(x)}{f_0(x)}} f_0(x)dx = \int f_1^z f_0^{1-z} dx.$$

Then,

$$\alpha_n = P_0(\Lambda_n > \lambda_n) = P_0(\log \Lambda_n > \log \lambda_n)$$
$$= P_0\left(\sum_{i=1}^n \log \frac{f_1(X_i)}{f_0(X_i)} > \gamma n\right) = P_0\left(\sum_{i=1}^n Y_i > \gamma n\right) = P_0(\overline{Y} > \gamma).$$

It is already clear from this that by virtue of the Cramér–Chernoff theorem, α_n converges to zero exponentially. More specifically, suppose that we choose $\gamma = K(P_1, P_0)$, the Kullback–Leibler distance between P_1 and P_0. The motivation for choosing $\gamma = K(P_1, P_0)$ would be that $E_{P_1}(\overline{Y}) = E_{P_1}(Y_1) = \int \log \frac{f_1(x)}{f_0(x)} f_1(x)$, which is $K(P_1, P_0)$; that is, we accept P_1 to be the true P if $\overline{Y}$ exceeds its expected value under P_1. In that case,

$$\gamma z - \log \psi_0(z) = z K(P_1, P_0) - \log \int f_1^z f_0^{1-z} dx.$$

By direct differentiation,

$$\frac{d}{dz}\left[z K(P_1, P_0) - \log \int f_1^z f_0^{1-z} dx \right] = K(P_1, P_0) - \frac{\int \log \frac{f_1}{f_0} f_1^z f_0^{1-z} dx}{\int f_1^z f_0^{1-z} dx},$$

which equals zero at $z = 1$, which is therefore the maxima of $\gamma z - \log \psi_0(z)$, and the maximum value is $\gamma - \log \psi_0(1) = \gamma - \log 1 = \gamma$, giving eventually,

$$\frac{1}{n} \log \alpha_n \to -\gamma = -K(P_1, P_0).$$

So, as a first approximation, we may write $\alpha_n \approx e^{-nK(P_1, P_0)}$; the larger the distance between P_0 and P_1, the smaller is the chance that the test will make a type I error. The same analysis can also be done for the type II error probability β_n.

17.1.3 Cramér's Theorem for General Sets

The basic Cramér–Chernoff theorem in $\mathcal{R}$ deals with probabilities of the form $P(\overline{X} \in C)$, where C is a set of the form (t, ∞). The rate of exponential convergence of this probability to zero is determined by $I(t)$. In other words, the rate of exponential convergence is determined solely by the value of the rate function $I(.)$ at the *boundary point t* of our set $C = (t, \infty)$. The following generalization of the basic Cramér–Chernoff inequality says that this fundamental phenomenon is true in far greater generality. A large deviation probability is a probability of a collection of rare events. The generalization says that to the first order, the magnitude of the large deviation probability is determined by the least rare of the collection of rare events.

Theorem 17.3. *Suppose $X_1, X_2, \ldots$ are iid zero mean random variables, with an mgf $\psi(z) = E[e^{zX_1}]$, assumed to exist for all real z. Let F, C be general (measurable) closed and open sets in $\mathcal{R}$, respectively, and let $I(t) = \sup_{z \in \mathcal{R}}[tz - k(z)]$, where $k(z) = \log \psi(z)$. For any given set S, denote $I(S) = \inf_{t \in S} I(t)$. Then,*

(a) $\limsup_n \frac{1}{n} \log P(\overline{X} \in F) \leq -I(F)$.
(b) $\liminf_n \frac{1}{n} \log P(\overline{X} \in C) \geq -I(C)$.

See Dembo and Zeitouni (1998, p. 27) for a proof of this theorem. When we compare the basic Cramér–Chernoff theorem with this more general theorem, we see that in this greater generality, we can no longer assert the existence of a limit for $\frac{1}{n}\log P(\bar{X} \in F)$ (or for $\frac{1}{n}\log P(\bar{X} \in C)$). We can assert the existence of a limit for $\frac{1}{n}\log P(\bar{X} \in S)$ for a general set S, provided $\inf_{t \in S^0} I(t) = \inf_{t \in \bar{S}} I(t)$, where $S^0, \bar{S}$ denote the interior and the closure of S. If this holds, then $\frac{1}{n}\log P(\bar{X} \in S)$ has a limit, and

$$\lim_n \frac{1}{n}\log P(\bar{X} \in S) = -\inf_{t \in S^0} I(t) = -\inf_{t \in \bar{S}} I(t) = -\inf_{t \in S} I(t) = -I(S).$$

This property is aesthetically desirable, but we can ensure it only when $I(t)$ does not have discontinuities on the boundary of the set S.

For example, consider singleton sets $S = \{t\}$, for fixed $t \in \mathcal{R}$. Now $P(\bar{X} \in S) = P(\bar{X} = t)$ is actually zero if the summands X_i have a continuous distribution (i.e., if the X_i have a density). If we insist on the equality of $\lim_n \frac{1}{n}\log P(\bar{X} \in S)$ and $-I(S)$, this would force $I(t)$ to be identically equal to ∞. But if $I(t) \equiv \infty$, then the equality $\lim_n \frac{1}{n}\log P(\bar{X} \in S) = -I(S)$ will fail at other subsets S of $\mathcal{R}$. This explains why the general Cramér–Chernoff theorem is in the form of lower and upper bounds, rather than an exact inequality.

17.2 The Gärtner–Ellis Theorem and Markov Chain Large Deviations

For sample means of iid random variables, the mgf is determined from the mgf of the underlying distribution itself. And that underlying mgf determines the large deviation rate function $I(t)$. When we give up the iid assumption, there is no longer one underlying mgf that determines the final rate function, even if there are meaningful large deviation asymptotics in the new problem. Rather, one has a *sequence* of mgfs, one for each n, corresponding to whatever statistics T_n we want to consider; for example, T_n could still be a sequence of sample means, but when the X_i have some dependence among themselves. Or, T_n could be a more complicated statistic with some nonlinear structure.

It turns out that despite these complexities, under some conditions a large deviation rate can be established without imposing the restriction of an iid setup, or requiring that T_n be the sequence of sample means. This greatly expands the scope of applications of large deviation theory, but in exchange for considerably more subtlety in exactly what assumptions are needed for which result to hold. The Gärtner–Ellis theorem, a special case of which we present below, is regarded as a major advance in large deviation theory, due to its wide-reaching applications. Although the assumptions (namely, (a)–(d) below) can fail in simple-looking problems, the Gärtner–Ellis theorem has also been successfully used to find large deviation rates in important non-iid setups, for example, for functionals of the form $T_n = \sum_{i=1}^n f(X_i)$, where $X_1, X_2, \ldots$ forms a Markov chain; see the examples below.

Theorem 17.4. *Let $\{T_n\}$ be a sequence of random variables taking values in an Euclidean space $\mathcal{R}^d$. Let $M_n(z) = E[e^{z'T_n}]$ and $k_n(z) = \log M_n(z)$, defined for z at which $M_n(z)$ exists. Let $\psi(z) = \lim_{n\to\infty} \frac{1}{n} k_n(nz)$, assuming that the limit exists as an extended real-valued function.*

Assume the following conditions on the function ψ.

(a) $D_\psi = \{z : \psi(z) < \infty\}$ *has a nonempty interior.*
(b) ψ *is differentiable in the interior of D_ψ.*
(c) *If ξ_n is a sequence in the interior of D_ψ approaching a point ξ on the boundary of D_ψ, then the length of the gradient vector of ψ at ξ_n converges to ∞; that is, $\|\nabla\psi(\xi_n)\| \to \infty$ for any sequence $\xi_n \to \xi$, a point on the boundary of D_ψ.*
(d) *The origin belongs to the interior of D_ψ.*

Let $I(t) = \sup_{z \in \mathcal{R}^d} [t'z - \psi(z)]$. Then, for general closed and open sets F, C in $\mathcal{R}^d$ respectively,

(a) $\limsup_n \frac{1}{n} \log P(T_n \in F) \leq -\inf_{t \in F} I(t)$.
(b) $\liminf_n \frac{1}{n} \log P(T_n \in C) \geq -\inf_{t \in C} I(t)$.

Without requiring any of the assumptions (a)–(d) above, for compact sets S,

(c) $\limsup_n \frac{1}{n} \log P(T_n \in S) \leq -\inf_{t \in S} I(t)$.

A simple sufficient condition for both parts (a) and (b) of the theorem to hold is that $\psi(z) = \lim_{n\to\infty} \frac{1}{n} k_n(nz)$ exists, is finite, and is also differentiable at every $z \in \mathcal{R}^d$.
We refer to Dembo and Zeitouni (1998, pp. 44–51) for a proof.

Example 17.5 (Multivariate Normal Mean). Suppose $X_1, X_2, \ldots$ are iid $N_d(0, \Sigma)$ random vectors, and assume that Σ is positive definite. Let T_n be the sample mean vector $\frac{1}{n}\sum_{i=1}^n X_i$. Thus, T_n still has the sample mean structure in an iid setup, except it is not a real-valued random variable, but a vector. We use the Gärtner–Ellis theorem to derive the rate function for T_n.

Using the formula for the mgf of a multivariate normal distribution (see Chapter 5),

$$M_n(z) = E\left[e^{z'\bar{X}}\right] = e^{\frac{z'\Sigma z}{2n}}$$

$$\Rightarrow k_n(z) = \frac{z'\Sigma z}{2n} \Rightarrow \frac{1}{n} k_n(nz) = \frac{z'\Sigma z}{2},$$

and therefore $\psi(z) = \lim_n \frac{1}{n} k_n(nz) = \frac{z'\Sigma z}{2}$. Therefore, $I(t) = \sup_{z \in \mathcal{R}^d} [t'z - \frac{z'\Sigma z}{2}] = \frac{t'\Sigma^{-1}t}{2}$, because $t'z - \frac{z'\Sigma z}{2}$ is a strictly concave function of z due to the positive definiteness of Σ, and the unique maxima occurs at $z = \Sigma^{-1}t$. Note that $I(t)$ is smooth everywhere in $\mathcal{R}^d$.

17.2 The Gärtner–Ellis Theorem and Markov Chain Large Deviations

Suppose now S is a set separated from the origin. Then, by the Gärtner–Ellis theorem, $\frac{1}{n} \log P(\overline{X} \in S)$ is going to be approximately equal to the minimum of $I(t) = \frac{t'\Sigma^{-1}t}{2}$ as t ranges over S. Now the contours of the function $\frac{t'\Sigma^{-1}t}{2}$ are ellipsoids centered at the origin. So, to get the limiting value of $\frac{1}{n} \log P(\overline{X} \in S)$, we keep drawing ellipsoids centered at the origin and with orientation Σ, until for the first time the ellipsoid is just large enough to touch the set S. The point where the ellipsoid touches S will determine the large deviation rate. This is a very elegant geometric connection to the probability problem at hand.

Example 17.6 (Sample Mean of a Markov Chain). Let $X_1, X_2, \ldots$ be a stationary Markov chain with the finite state space $S = \{1, 2, \ldots, t\}$. We assume that the chain is irreducible (see Chapter 10), and denote the initial state to be i. Also let p_{ij} denote the one-step transition probabilities $p_{ij} = P(X_{n+1} = j \mid X_n = i)$. Although just irreducibility is enough, we assume in this example that p_{ij} is strictly positive for all i, j. We apply the Gärtner–Ellis theorem to derive the large deviation rate for the sequence of sample means $\overline{X}_n = \frac{1}{n}\sum_{i=1}^{n} X_i$.

Given a real z, define a matrix Π_z with entries $\Pi_{z,i,j} = p_{ij} e^{zj}$. Note that the entries of Π_z are all strictly positive. We now calculate $k_n(nz)$. We have,

$$k_n(nz) = \log E[e^{z\sum_{k=1}^n X_k}] = \log \sum_{x_1,x_2,\ldots,x_n} e^{z(x_1+x_2\cdots x_n)}$$
$$\times P(X_1 = x_1, X_2 = x_2, \ldots, X_n = x_n)$$
$$= \log \sum_{x_1,x_2,\ldots,x_n} e^{zx_1} p_{i,x_1} e^{zx_2} p_{x_1,x_2} \cdots e^{zx_n} p_{x_{n-1},x_n}$$

(recall that the notation i stands for the initial state of the chain)

$$= \log \sum_{x_n} [\Pi_z]_{i,x_n}^n,$$

where $[\Pi_z]^n$ denotes the nth power of the matrix Π_z and $[\Pi_z]_{i,x_n}^n$ means the (i, x_n) element of it. This formula for $k_n(nz)$ leads to the large deviation rate for the sample mean of our Markov chain.

By part (e) of the *Perron–Frobenius theorem* (see Chapter 10), we get

$$\lim_n (1/n) k_n(nz) = \lim_n \frac{1}{n} \log \sum_{x_n} [\Pi_z]_{i,x_n}^n$$
$$= \log \lambda_1(z),$$

where $\lambda_1(z)$ is the largest eigenvalue of Π_z (see part (a) of the Perron–Frobenius theorem). Therefore, $\psi(z) = \lim_n \frac{1}{n} k_n(nz)$ exists for every z, and it is differentiable by the *implicit function theorem* (i.e., basically because $\lambda_1(z)$ is the maximum root of the determinant of $\Pi_z - \lambda I$, and the determinant itself is a smooth function of z).

It now follows from the Gärtner–Ellis theorem that the conclusion of both parts (a) and (b) of the theorem hold, with the large deviation rate function being $I(t) = \sup_z [tz - \log \lambda_1(z)]$.

17.3 The t-Statistic

Given n random variables $X_1, X_2, \ldots, X_n$, the t-statistic is defined as $T_n = \frac{\sqrt{n}\bar{X}}{s}$, where $s^2 = \frac{1}{n-1} \sum_{i=1}^{n}(X_i - \bar{X})^2$ is the sample variance. If the sequence $X_1, X_2, \ldots$ is iid with zero mean and a finite variance, then $T_n \stackrel{\mathcal{L}}{\Rightarrow} N(0, 1)$ (see Chapter 7). The t-statistic is widely used in statistics in situations involving tests of hypotheses about a population mean. Effect of nonnormality on the t-statistic has been studied by several authors, among them, Efron (1969), Cressie (1980), Hall (1987), and Basu and DasGupta (1991). Logan et al. (1973) showed that if $X_1, X_2, \ldots$ are iid standard Cauchy, then T_n still does converge in distribution, but to a bimodal and unbounded nonstandard density. It was conjectured there that the t-statistic converges in distribution if and only if the underlying CDF, say F, is in the domain of attraction of a stable law. This conjecture has turned out to be correct. Even more, it is now known that the t-statistic converges in distribution to a normal distribution if and only if the underlying CDF is in the domain of attraction of the normal; see Giné et al. (1997). For example, the t-statistic converges in distribution to $N(0, 1)$ if F itself is a t-distribution with two degrees of freedom, which does not have a finite variance. Hall and Wang (2004) characterize the speed of convergence of the t-statistic to the standard normal under this weakest possible assumption.

Large deviations for the t-statistic are of interest on a few grounds. One interesting fact is that the large deviation rate function for the t-statistic is not the same as the rate function for the sample mean, even if we have a conventional finite mgf situation. Second, in certain multiple testing problems of recent interest in statistics, large deviations of the t-statistic have become important. The results on the large deviations of the t-statistic are complex, and the proofs are long. We refer to Shao (1997) for the derivation of the main result on the rate function of the t-statistic. The most striking part of this result is that large deviation rates are derived without making any moment conditions at all, in contrast to the Cramér–Chernoff theorem for the sample mean, which assumes the existence of the mgf itself. The multivariate case, usually known as Hotelling's T^2 (see Chapter 5), is treated in Dembo and Shao (2006).

We use the following notation. Given an iid sequence of random variables $X_1, X_2, \ldots$ from a CDF F on $\mathcal{R}$, we let, for $n \geq 2$,

$$S_n = \sum_{i=1}^{n} X_i, \quad V_n^2 = \sum_{i=1}^{n} X_i^2, \quad s_n^2 = \frac{1}{n-1} \sum_{i=1}^{n} (X_i - \bar{X})^2,$$

$$Z_n = \frac{S_n}{V_n}, \quad T_n = \frac{\sqrt{n}\bar{X}}{s_n}.$$

17.3 The t-Statistic

Then, the *self-normalized* statistic Z_n and the t-statistic T_n are related by the identity

$$T_n = Z_n \left[\frac{n-1}{n - Z_n^2} \right]^{1/2}.$$

The function $z \to \frac{z}{\sqrt{n-z^2}}$ is strictly increasing, as can be seen from its first derivative $\frac{n}{(n-z^2)^{3/2}}$, which is strictly positive on $(-\sqrt{n}, \sqrt{n})$. As a result,

$$P(T_n > t) = P\left(\frac{Z_n}{\sqrt{n - Z_n^2}} > \frac{t}{\sqrt{n-1}} \right) = P\left(Z_n > \frac{t\sqrt{n}}{\sqrt{n-1+t^2}} \right).$$

Therefore, the large deviation rate function of the t-statistic can be figured out from the large deviation rate function of the formally simpler statistic Z_n. This is the approach taken in Shao (1997); this convenient technique was also adopted in Efron (1969), and is generally a useful trick to remember in dealing with t-statistics.

Theorem 17.5. *Let $X_1, X_2, \ldots$ be an iid sequence of real-valued random variables.*

(a) Suppose $E(X_1)$ exists and equals zero. Then, for $x > 0$,

$$\lim_n \frac{1}{n} \log P\left(\frac{T_n}{\sqrt{n}} \geq x \right) = \sup_{c \geq 0} \inf_{t \geq 0} \log E\left[e^{ctX_1 - \frac{xt(c^2 + X_1^2)}{2\sqrt{1+x^2}}} \right].$$

(b) If $E(X_1)$ does not exist, then the result of part (a) still holds for all $x > 0$. Part (a) actually holds with a minor modification if $E(X_1)$ exists and is nonnegative, but we do not show the modification here; see Shao (1997).

Example 17.7 *(The Normal Case).* Suppose $X_1, X_2, \ldots$ are iid $N(0, 1)$. In this case, $E[e^{ctX_1 - \frac{xt(c^2 + X_1^2)}{(2\sqrt{1+x^2})}}]$ exists for all $t \geq 0$, and on calculation and algebra, we get

$$E\left[e^{ctX_1 - \frac{xt(c^2 + X_1^2)}{2\sqrt{1+x^2}}} \right] = \frac{(1+x^2)^{1/4} \left(\exp\left(\frac{c^2 t(t - x\sqrt{1+x^2})}{2(1+x^2+tx\sqrt{1+x^2})} \right) \right)}{\sqrt{tx + \sqrt{1+x^2}}}$$

$$= g(x, c, t) \text{ (say)}.$$

To find the limit of $\frac{1}{n} \log P\left(\frac{T_n}{\sqrt{n}} \geq x \right)$, we need to find $\sup_{c \geq 0} \inf_{t \geq 0} g(x, c, t)$. By taking the logarithm of $g(x, c, t)$, and differentiating it with respect to t, we find that the infimum over $t \geq 0$ occurs at the unique positive root of the following function of t.

$$-(1+c^2)x - 2(1+c^2)x^3 - (1+c^2)x^5 + \sqrt{1+x^2}t \left[2c^2 + (c^2-2)x^2 - (c^2+4)x^4 \right]$$
$$+ \left[3c^2 x + (3c^2-1)x^3 - x^5 \right] t^2 + c^2 x^2 \sqrt{1+x^2} t^3 = 0.$$

It turns out that the required root is $t = x\sqrt{1+x^2}$. Plugging this back into the expression for $g(x, c, t)$, we find that $\sup_{c \geq 0} \inf_{t \geq 0} g(x, c, t) = \frac{1}{\sqrt{1+x^2}}$. That is,

$$\lim_n \frac{1}{n} \log P\left(\frac{T_n}{\sqrt{n}} \geq x\right) = -\frac{1}{2}\log(1 + x^2).$$

We notice that the rate function $\frac{1}{2}\log(1 + x^2)$ is smaller than $\frac{x^2}{2}$, which was the rate function for the sample mean in the standard normal case. This is due to the fact that the t-statistic is more dispersed than the mean, and the rate functions of the t-statistic and the mean are demonstrating that.

17.4 Lipschitz Functions and Talagrand's Inequality

Suppose $Z \sim N(0, 1)$ and $f : \mathcal{R} \to \mathcal{R}$ is a function with a uniformly bounded first derivative, namely $|f'(z)| \leq M < \infty$ for all z. Then, by a first-order Taylor expansion,

$$f(Z) \approx f(0) + Zf'(0) \Rightarrow |f(Z) - f(0)| \approx |Z||f'(0)| \leq M|Z|.$$

Therefore, given $\epsilon > 0$,

$$P(|f(Z) - E[f(Z)]| > \epsilon) \approx P(|f(Z) - f(0)| > \epsilon) \approx P(|Z||f'(0)| > \epsilon)$$

$$\leq P(M|Z| > \epsilon) = P\left(|Z| > \frac{\epsilon}{M}\right) = 2\left[1 - \Phi\left(\frac{\epsilon}{M}\right)\right]$$

$$\leq e^{-\frac{\epsilon^2}{2M^2}},$$

because $1 - \Phi(z) \leq \frac{1}{2}e^{-z^2/2}$ for $z \geq 0$. These heuristics suggest that smooth functions of a normal random variable should be quite tightly concentrated around their mean, as long as they do not change by too much over small intervals. We present a major result to this effect. The result was originally proved in Borell (1975) and Sudakov and Cirelson (1974). The proof below is coined from Talagrand (1995), and uses the following lemma. We need the notion of *outer parallel body* for the proof, which is defined first.

Definition 17.4. Let $B \subset \mathcal{R}^d$, and let $\epsilon > 0$. Then the ϵ-outer parallel body of B is the set $B^\epsilon = \{x \in \mathcal{R}^d : \inf_{y \in B} ||x - y|| \leq \epsilon\}$.

Also, let $\mathcal{H}$ denote the collection of all half spaces in $\mathcal{R}^d$. The idea of the proof of the result in this section is that if a set B has probability at least .5 under a standard d-dimensional normal distribution, then with a large probability the normal random vector will lie within the ϵ-outer parallel body of B, and that the worst offenders to this rule are half-spaces. Here, the requirement that B have a probability at least .5 is indispensable.

17.4 Lipschitz Functions and Talagrand's Inequality

Lemma. Let $Z \sim N_d(0, I)$ and let $0 < \alpha < 1$. Then,

$$\sup_{B: P(Z \in B) = \alpha} P(Z \notin B^\epsilon) = \sup_{H \in \mathcal{H}: P(Z \in H) = \alpha} P(Z \notin H^\epsilon).$$

Theorem 17.6. *Let $Z \sim N_d(0, I)$, and let $f(z)$ be a function with Lipschitz norm $M < \infty$; that is,*

$$\sup_{x,y} \frac{|f(x) - f(y)|}{\|x - y\|} = M < \infty.$$

Let θ be either $E[f(Z)]$ or the median of $f(Z)$. Then, for any $\epsilon > 0$,

$$P(|f(Z) - \theta| > \epsilon) \le e^{-\frac{\epsilon^2}{2M^2}}.$$

Proof. We prove the theorem only for the case when θ is a median of $f(Z)$. Let $B = \{z : f(z) \le \theta\}$. Fix $\epsilon > 0$, and consider $y \in B^\epsilon$. Pick $z \in B$ such that $\|y - z\| \le \epsilon$. Then,

$$f(y) = f(y) - f(z) + f(z) \le M\|y - z\| + f(z) \le M\epsilon + \theta$$
$$\Rightarrow P(Z \in B^\epsilon) \le P(f(Z) \le M\epsilon + \theta) \Rightarrow P(f(Z) > M\epsilon + \theta) \le P(Z \notin B^\epsilon)$$
$$\le \sup_{A: P(A) \ge \frac{1}{2}} P(Z \notin A^\epsilon) = \sup_{H \in \mathcal{H}: P(Z \in H) \ge \frac{1}{2}} P(Z \notin H^\epsilon).$$

(by the lemma). This latest supremum can be proved to be equal to $1 - \Phi(\epsilon)$, leading to, for $x > 0$,

$$P(f(Z) > M\epsilon + \theta) \le 1 - \Phi(\epsilon) \Rightarrow P(f(Z) - \theta > x) \le 1 - \Phi\left(\frac{x}{M}\right).$$

Similarly,

$$P(f(Z) - \theta < -x) \le 1 - \Phi\left(\frac{x}{M}\right).$$

Adding the two inequalities together,

$$P(|f(Z) - \theta| > x) \le 2\left[1 - \Phi\left(\frac{x}{M}\right)\right] \le e^{-\frac{x^2}{2M^2}}.$$

$\square$

Example 17.8 (Maximum of Correlated Normals). We apply the above theorem to find a concentration inequality for the maximum of d jointly distributed normal random variables. Thus, let $X \sim N(0, \Sigma)$, where Σ is positive definite, and let $\sigma_i^2 = \text{Var}(X_i), i = 1, 2, \ldots, d$. Let $\Sigma = AA'$, where A is nonsingular.

Consider now the function $f : \mathcal{R}^d \to \mathcal{R}$ defined by $f(u) = \max\{(Au)_1, (Au)_2, \ldots, (Au)_d\}$, where $(Au)_1$ means the first coordinate of the vector Au, and so on. Denote $M = \max_i \sigma_i$. We claim that f is a Lipschitz function with Lipschitz norm bounded by M. To see this, for notational simplicity, consider the case when A is diagonal; the general case can be similarly handled. Consider two vectors u, w. Then,

$$(Au)_1 = a_{11}u_1 = a_{11}w_1 + a_{11}(u_1 - w_1) \leq a_{11}w_1 + M\|u - w\|$$
$$\leq \max\{(Aw)_1, (Aw)_2, \ldots, (Aw)_d\} + M\|u - w\|.$$

By the same argument, for each i, $(Au)_i \leq \max\{(Aw)_1, (Aw)_2, \ldots, (Aw)_d\} + M\|u - w\|$, and therefore,

$$f(u) = \max\{(Au)_1, (Au)_2, \ldots, (Au)_d\}$$
$$\leq \max\{(Aw)_1, (Aw)_2, \ldots, (Aw)_d\} + M\|u - w\|$$
$$= f(w) + M\|u - w\|.$$

By switching the roles of u and w, $f(w) \leq f(u) + M\|u - w\|$, and hence, $|f(u) - f(w)| \leq M\|u - w\|$.

Next observe that

$$\max\{X_1, X_2, \ldots, X_d\} \stackrel{\mathcal{L}}{=} \max\{(AZ)_1, (AZ)_2, \ldots, (AZ)_d\} = f(Z),$$

where $Z \sim N(0, I)$. Therefore, by our above theorem,

$$P(|\max\{X_1, X_2, \ldots, X_d\} - E[\max\{X_1, X_2, \ldots, X_d\}]| > x)$$
$$= P(|f(Z) - E[f(z)]| > x) \leq e^{-\frac{x^2}{2M^2}},$$

where $M = \max_i \sigma_i$. The striking parts of this are that although $X_1, X_2, \ldots, X_d$ are not assumed to be independent, we can prove an exponential concentration inequality by using only the coordinatewise variances, and that the inequality is valid for all d.

17.5 * Large Deviations in Continuous Time

Let $X(t), t \in T$, be a stochastic process in continuous time, where T is some indexing set. Random variables of the type $\sup_{t \in T} X(t)$, or $\sup_{t \in T} |X(t)|$, and so on are collectively known as extremes. Extreme statistics are widely used in assessing rarity of an event, or for disaster planning, and have become particularly important in recent years in genetics, finance, climate studies, environmental planning, and astronomy. Extremes for a discrete-time stochastic sequence were previously discussed in Chapters 6 and 9. The case of continuous time is treated in this section. For extreme statistics, we often want to know how rare was an observed value of

17.5 Large Deviations in Continuous Time

the extreme of some stochastic sequence or process. We are then content with good upper bounds on probabilities of the form $P(\sup_{t \in T} |X(t)| > x)$. This has a resemblance to a large deviation probability. A basic treatment of this topic when $X(t)$ is a Gaussian process is provided in this section. As usual, we first need some definitions and notation.

Throughout we take $X(t)$ to be a zero mean one-dimensional Gaussian process; it is referred to as a *centered Gaussian process*. We also take the indexing set T to be a set in an Euclidean space, usually the real line. The covariance kernel of $X(t)$ is denoted as $\rho(s, t)$, and when $X(t)$ is stationary, we drop one of the arguments, and simply use the notation $\rho(t)$. The covariance kernel induces a new metric on T, which we define below.

Definition 17.5. The *canonical metric* of T induced by the process $X(t)$ is defined as $\Delta(s, t) = \sqrt{E(X(t) - X(s))^2}, s, t \in T$.

It is meaningless to talk about $P(\sup_{t \in T} |X(t)| > x)$ unless $\sup_{t \in T} |X(t)|$ is finite with probability one. Thus, boundedness of $X(t)$ over T is going to be a key consideration for this section. On the other hand, boundedness of $X(t)$ over T will have close relations to continuity of the sample paths. These two properties are defined below.

Definition 17.6. Let $X(t), t \in T$ be a real-valued stochastic process. Let $t_1, t_2, \ldots, t_n \in T$. Then the joint distribution of $(X_{t_1}, X_{t_2}, \ldots, X_{t_n})$ is called a *finite-dimensional distribution* (fdd) of the process $X(t)$.

Definition 17.7. Let $X(t), Y(t), t \in T$ be two stochastic processes defined on T. They are called versions of each other if all of their finite-dimensional distributions are the same.

For the purpose of defining continuity of a process, we need to give a meaning to closeness of two different times s, t. For Gaussian processes, we can achieve this by simply using the canonical metric $\Delta(s, t)$. For more general processes, we achieve this by simply assuming that T is a subset in an Euclidean space. For defining the notion of boundedness of a process, we do not require definition of the closeness of times s, t. Thus, in defining boundedness, we can let the time set T be more general.

Definition 17.8. Let $X(t), t \in T$ be a stochastic process, where T is a subset of an Euclidean space $\mathcal{R}^d$. $X(t)$ is called continuous if there is a version of it, say $Y(t), t \in T$, such that with probability one, $Y(t)$ is continuous on T as a function of t.

Definition 17.9. Let $X(t), t \in T$ be a stochastic process. $X(t)$ is called bounded if there is a version of it, say $Y(t), t \in T$, such that with probability one, $\sup_{t \in T} |Y(t)| < \infty$.

In order to avoid mentioning the availability of a version that is continuous or bounded, we simply say that $X(t)$ is continuous, or bounded, on some relevant set T. This is not mentioned again.

17.5.1 * Continuity of a Gaussian Process

Continuity of $X(t)$ is a key factor in our final goal of writing bounds on probabilities of the form $P(\sup_{t \in T} X(t) > x)$ or $P(\sup_{t \in T} |X(t)| > x)$. This is because continuity and boundedness are connected, and also because a smooth process does not jump around too much, which helps in keeping the supremum of the process in control. We limit our treatment to Gaussian processes with the time set T a subset of an Euclidean space, and often just $[0, 1]$. Conditions for the continuity of a Gaussian process have evolved from simple sufficient conditions to more complex necessary and sufficient conditions. Obviously, if the time set T is compact, then continuity of the process implies its boundedness. If T is not compact, this need not be true. A simple example is that of the Wiener process $X(t)$ on $T = [0, \infty)$. We know from Chapter 12 that $X(t)$ is continuous. However, $P(\sup_{t \geq 0} X(t) = \infty, \inf_{t \geq 0} X(t) = -\infty) = 1$, so that $X(t)$ is not bounded, and is in fact almost surely unbounded. We first give a theorem with a set of classic sufficient conditions for the continuity of a Gaussian process. See Adler (1987) or Fernique (1974) for a proof.

Theorem 17.7. (a) *Suppose $X(t), t \in T = [0, 1]$ is a centered Gaussian process satisfying $\Delta(s, t) \leq \phi(|t - s|)$ for some function ϕ such that $\int_0^\infty \phi(e^{-x^2}) \, dx < \infty$. Then $X(t)$ is continuous on T.*
(b) *Suppose $X(t), t \in T$ is a centered Gaussian process on a compact set T in an Euclidean space. Suppose for some $C, \alpha,$ and $h > 0$,*

$$\Delta^2(s, t) = E[X(t) - X(s)]^2 \leq \frac{C}{|\log ||s - t|||^{1+\alpha}} \quad \forall \, s, t, ||s - t|| \leq h.$$

Then $X(t)$ is almost surely continuous on T.
(c) *Suppose we additionally assume that $X(t)$ is stationary on T. Suppose for some $A, \beta, \delta > 0$,*

$$\rho(0) - \rho(t) \geq \frac{A}{|\log ||t|||^{1-\beta}} \quad \forall \, t, ||t|| \leq \delta.$$

Then $X(t)$ is almost surely discontinuous on T.

Part (c) of the theorem says that the sufficient condition in part (b) is at the borderline of necessary and sufficient if the Gaussian process is stationary.

Example 17.9 (The Wiener Process). We prove by using the theorem above that paths of a Wiener process are continuous. If we can show that a Wiener process is continuous on $[0, 1]$, it follows that it is continuous on all of $[0, \infty)$. For the Wiener process, $\Delta^2(s, t) = E[X(t) - X(s)]^2 = t + s - 2 \min(s, t) = |t - s|$. Choose $\phi(u) = \sqrt{u}$. Then, we have $\Delta(s, t) \leq \phi(|t - s|)$. Furthermore, $\int_0^\infty \phi(e^{-x^2}) dx = \int_0^\infty e^{-x^2/2} dx < \infty$. Therefore, by part (a) of the above theorem, the Wiener process is continuous on $[0, \infty)$. Next, as regards part (b) of the theorem, because

17.5 Large Deviations in Continuous Time

$x(\log x)^2 \to 0$ as $x \downarrow 0$, the Wiener process also satisfies the sufficient condition in part (b) with, for example, $\alpha = 1, C = 1$, and $h = \frac{1}{2}$, and part (b) again shows that the Wiener process is continuous on $[0, 1]$, and hence, on $[0, \infty)$.

Example 17.10 (Logarithm Tail of the Maxima and the Landau–Shepp Theorem). This example illustrates a famous theorem of Landau and Shepp (1970). The Landau–Shepp theorem says that if $X(t)$ is a centered and almost surely bounded Gaussian process, then the tail of its supremum acts like the tail of a suitable single normal random variable. Precisely, let $X(t)$ be a centered and bounded Gaussian process on some set T. For instance, if T is a compact interval in $\mathcal{R}$ and if $X(t)$ satisfies one of the two sufficient conditions in our theorem above, then the Landau–Shepp theorem applies. Let $\sigma_T^2 = \sup_{t \in T} \text{Var}(X_t)$. The Landau–Shepp theorem says that $\lim_{u \to \infty} \frac{1}{u^2} \log P(M_T > u) = -\frac{1}{2\sigma_T^2}$, where $M_T = \sup_{t \in T} X(t)$.

Let us put this result in context. Take a single univariate normal random variable X with mean zero and variance σ^2. Then, the distribution of X satisfies the two-sided bounds

$$\left(\frac{\sigma}{\sqrt{2\pi}u} - \frac{\sigma^3}{\sqrt{2\pi}u^3}\right) e^{-\frac{u^2}{2\sigma^2}} \leq P(X > u) \leq \frac{\sigma}{\sqrt{2\pi}u} e^{-\frac{u^2}{2\sigma^2}},$$

for all $u > 0$. Taking logarithms, it follows on elementary manipulation that $\lim_{u \to \infty} \frac{1}{u^2} \log P(X > u) = -\frac{1}{2\sigma^2}$.

Suppose now that there exists $t_0 \in T$ such that $\text{Var}(X_{t_0}) = \sigma_T^2$; that is, the maximum variance value is attained. Then, by using the Landau–Shepp theorem, we see that M_T satisfies the same logarithmic tail rate as the single normal random variable X_{t_0}. That is, on a logarithmic scale, the tail of M_T is the same as that of a normal variable with the maximal variance over the set T.

To complete this example, we recall from Chapter 12 that for the Wiener process on $[0, T]$ for any finite T, the exact distribution of $M_T = \sup_{0 \leq t \leq T} X(t)$ is actually known, and $P(M_T > u) = 2\left[1 - \Phi\left(\frac{u}{\sqrt{T}}\right)\right]$, $u > 0$. Therefore, for the special case of the Wiener process, the Landau–Shepp theorem follows directly from the exact distribution of M_T.

17.5.2 * Metric Entropy of T and Tail of the Supremum

The classic sufficient conditions for the continuity of a Gaussian process have evolved into conditions on the size of the time set T that control continuity of the process and magnitude of the supremum at the same time. These conditions involve the metric entropy of the set T with respect to the canonical metric of T. The smoother the covariance kernel of the process is near zero, the smaller will be the metric entropy of T, and the easier it is to control the supremum of the process over T. Roughly, if the covering numbers $N(\epsilon, T, \Delta)$ of T grow at the rate of $\epsilon^{-\alpha}$ for some positive α, then the Gaussian process will be continuous, and various results on the tail of the supremum of the process can be derived. These metric entropy conditions and their ties to the magnitude of the supremum of the process, as in the theorem below, are due to Dudley (1967) and Talagrand (1996). The reader should compare the results below with the main theorem in Section 16.5, where the L_2 norm of the supremum of a continuous Gaussian process is bounded by the ϵ-capacity of T. We remark that stationarity of the process is not assumed in the theorem below.

Theorem 17.8. *Let $X(t), t \in T$ be a real-valued centered Gaussian process. Let $N(\epsilon, T, \Delta)$ be the covering number of T with respect to the canonical metric Δ of T. Assume:*

(a) *For each $\epsilon > 0$, $N(\epsilon, T, \Delta) < \infty$.*
(b) *The diameter of T with respect to Δ, that is, $L = \sup_{s,t \in T} \sqrt{E[X(s) - X(t)]^2}$, is finite.*
(c) $\int_0^L \sqrt{\log N(\epsilon, T, \Delta)} d\epsilon < \infty.$

Then $X(t)$ is (almost surely) continuous and bounded on T and

$$E[\sup_{t \in T} X(t)] \leq 12 \int_0^\infty \sqrt{\log N(\epsilon, T, \Delta)} d\epsilon = 12 \int_0^{\frac{L}{2}} \sqrt{\log N(\epsilon, T, \Delta)} d\epsilon.$$

If $N(\epsilon, T, \Delta)$ satisfies $N(\epsilon, T, \Delta) \leq K\epsilon^{-\alpha}$ for some finite $\alpha, K > 0$, then there exists a finite constant C such that for any $\delta > 0$,

$$P(\sup_{t \in T} X(t) > u) \leq C u^{\alpha + \delta} e^{-\frac{u^2}{2\sigma_T^2}},$$

for all large u, where $\sigma_T^2 = \sup_{t \in T} Var(X(t))$.

Example 17.11 (The Wiener Process). We examine various assumptions and conclusions of this theorem in the context of a Wiener process $X(t)$ on $[0, 1]$. Note that we know the exact distribution and the expectation of $M = \sup_{0 \leq t \leq 1} X(t)$, namely, $P(M > u) = 2[1 - \Phi(u)], u > 0$, and $E(M) = \sqrt{\frac{2}{\pi}} = .799$.

The canonical metric in this case is $\Delta(s, t) = \sqrt{|s - t|}$. Thus, the diameter of the time set $T = [0, 1]$ with respect to Δ is $L = 1$. Next, because $\Delta(s, t) = \sqrt{|s - t|}$, one ball of Δ-radius ϵ will cover an Euclidean length of $2\epsilon^2$. Therefore, if we consider the particular cover $[0, 2\epsilon^2], [2\epsilon^2, 4\epsilon^2], \ldots$, then we require at most $\frac{1}{2\epsilon^2} + 1$ balls to cover the time set $T = [0, 1]$. We therefore have

$$N(\epsilon, T, \Delta) \leq 1 + \frac{1}{2\epsilon^2}$$
$$\Rightarrow \log N(\epsilon, T, \Delta) \leq \log \frac{1}{2\epsilon^2} + 2\epsilon^2 \leq -2\log \epsilon + 2\epsilon^2 - \log 2.$$

We can evaluate the integral $\int_0^{.5} \sqrt{-2\log \epsilon + 2\epsilon^2 - \log 2} d\epsilon$. We can easily show analytically that the integral is finite. Our theorem above then implies that $X(t)$ is continuous and bounded on [0, 1] (something that we already knew). Moreover, the theorem also says that

$$E(M) \leq 12 \int_0^{.5} \sqrt{\log N(\epsilon, T, \Delta)} d\epsilon \leq 12 \int_0^{.5} \sqrt{-2\log \epsilon + 2\epsilon^2 - \log 2} d\epsilon$$
$$= 12 \times .811 = 9.732,$$

whereas the exact value is $E(M) = .799$. So the bounds are not yet practically very useful, but describe exactly how the covering numbers must behave for the tail of the supremum to go down at a Gaussian rate.

Exercises

Exercise 17.1. Derive the rate function $I(t)$ for the sample mean in the exponential case.

Exercise 17.2. Derive the rate function $I(t)$ for the sample mean in the Poisson case.

Exercise 17.3. Derive the rate function $I(t)$ for the sample mean for the three-point distribution giving probability $\frac{1}{3}$ to each of $0, \pm 1$.
Note: You will need a formula for the derivative of the cosine hyperbolic function.

Exercise 17.4 (SLLN from Cramér–Chernoff Theorem). Prove, by using the Cramér–Chernoff theorem, the SLLN for the mean of an iid sequence under the conditions of the Cramér–Chernoff theorem.

Exercise 17.5 * (Rate Function for Uncommon Distributions). For each of the following densities, derive the rate function $I(t)$ for the sample mean.

(a) $f(x) = \frac{1}{\text{Beta}(\mu,\alpha)} e^{-\mu x}(1 - e^{-x})^{\alpha-1}$, $x, \mu, \alpha > 0$.

(b) $f(x) = \frac{e^{-\mu x}}{\Gamma(\mu)\zeta(\mu)(e^{e^{-x}}-1)}$, $x > 0, \mu > 1$, where $\zeta(\mu) = \sum_{n=1}^{\infty} n^{-\mu}$.

(c) $f(x) = \frac{e^{-\theta^2 x^2}}{(1-\Phi(\theta))\pi e^{\theta^2}(1+x^2)}$, $-\infty < x < \infty, \theta > 0$.

Exercise 17.6 * (Multivariate Normal). Characterize the rate function $I(t)$ for the statistic $T_n = ||\overline{X}||$, when $X_1, X_2, \ldots$ are iid $N_d(0, I)$.

Exercise 17.7 * (Uniform Distribution in a Ball). Characterize the rate function $I(t)$ for the statistic $T_n = ||\overline{X}||$, when $X_1, X_2, \ldots$ are iid uniform in the d-dimensional unit ball.

Exercise 17.8 (Numerical Accuracy of Large Deviation Approximation). Let $W_n \sim \chi_n^2$. Do a straight CLT approximation for $P(W_n > (1 + t)n)$, and do an approximation using the Cramér–Chernoff theorem.
   Compare the numerical accuracy of these two approximations for $t = .5, 1, 2$ when $n = 30$.

Exercise 17.9. Prove that the rate function $I(t)$ for the sample mean of an iid sequence in the one-dimensional case cannot be zero at $t \neq \mu$, when the mgf exists for all z.

Exercise 17.10. Prove that the rate function $I(t)$ for the sample mean of an iid sequence satisfies $I''(\mu) = \frac{1}{\sigma^2}$.

Exercise 17.11 * **(Type II Error Rate of the Likelihood Ratio Test).** Find $\lim_n \frac{1}{n} \log \beta_n$ for the likelihood ratio test of Example 17.4.

Exercise 17.12 (Discontinuous Rate Function). Give an example of a distribution on $\mathcal{R}$ for which the rate function for the sample mean is not continuous.

Exercise 17.13 * **(Second-Order Term in the Cramér–Chernoff Theorem).** Suppose we consider $P(\overline{X} > t)$ itself, instead of its logarithm, in the one-dimensional iid case. Identify $c_n(t)$ in the representation $P(\overline{X} > t) = c_n(t) e^{-nI(t)}[1 + o(1)]$ when the X_i are iid $N(0, 1)$.

Exercise 17.14 * **(Rate Function When the Population Is t).** Suppose $X_1, X_2, \ldots$ are iid from a t-distribution with degrees of freedom $\alpha > 0$. Prove that whatever is α, $\lim_n \frac{1}{n} \log P(\overline{X} > t) = 0$ for $t > 0$.

Exercise 17.15 * **(Adjusted Large Deviation Rate When the Population Is t).** Suppose $X_1, X_2, \ldots$ are iid from a t-distribution with two degrees of freedom. In this case, $\overline{X}$ converges in probability to zero by the WLLN, but $P(\overline{X} > t)$ does not converge to zero exponentially. Find the exact rate at which $P(\overline{X} > t)$ converges to zero.

Exercise 17.16. Suppose $X_n \sim \text{Bin}(n, p)$. Show that for any $a > 0$, $P(X_n < np - a) \leq e^{-\frac{a^2}{2np}}$.

Hint: Use the technique to obtain the upper bound part in the Cramér–Chernoff theorem.

Exercise 17.17. Suppose $X \sim \text{Poi}(\lambda)$. Show that for any $\epsilon > 0$, $P(X < \lambda(1 - \epsilon)) \leq e^{-\frac{\lambda \epsilon^2}{2}}$.

Hint: Approximate a Poisson by a suitable binomial. Then use the binomial distribution inequality in the exercise above. Take a limit.

Exercise 17.18 * **(Verification of Conditions in the Gärtner–Ellis Theorem).** Suppose X_i are iid $N(\mu, 1)$ and Y is an independent standard exponential random variable. Consider the statistic $T_n = \overline{X} + \frac{Y}{\sqrt{n}}$.

(a) Is $\psi(z)$ finite and differentiable at every z?
(b) Which of the conditions (a)–(d) in the Gärtner–Ellis theorem hold?

Exercises

Exercise 17.19 (Gärtner–Ellis Theorem). For the two-state Markov chain of Example 10.1, derive the rate function for the sample mean.

Exercise 17.20 * **(Gärtner–Ellis Theorem).** For the transition matrix

$$P = \begin{pmatrix} 0 & 1 & 0 \\ 0 & 0 & 1 \\ p & 1-p & 0 \end{pmatrix},$$

(a) Is the Gärtner–Ellis theorem applicable?
(b) If it is, derive the rate function for the sample mean.

Hint: Verify if the chain is irreducible.

Exercise 17.21 (The t-statistic). Suppose $X_1, X_2, \ldots$ are iid $N(0, 1)$. Approximate $P\left(\frac{T_n}{\sqrt{n}} > x\right)$ by using the normal approximation, and the large deviation approximation (see the rate function worked out in Example 17.7). Compare them to the exact value of $P\left(\frac{T_n}{\sqrt{n}} > x\right)$, which is obtainable by using a t table. Use $n = 30$ and $x = 0.1, 0.25, 0.5$.

Exercise 17.22 (Mean Absolute Deviation of a Lipschitz Function). Suppose $Z \sim N_d(0, I)$ and $f(Z)$ is a Lipschitz function. Use Theorem 17.6 to find an upper bound on $E[|f(Z) - E(f(Z))|]$ in terms of the Lipschitz norm of f.

Exercise 17.23 * **(The Canonical Metric and the Euclidean Metric).** Suppose $X(t)$ is a centered Gaussian process on a compact interval T on $\mathcal{R}$. Show that $X(t)$ is continuous with respect to the canonical metric of T if and only if it is continuous with respect to the Euclidean metric on T.

Exercise 17.24 (Increments of a Wiener Process). Let $W(t)$ be a Wiener process on $[0, \infty)$, and let $X(t) = W(t+1) - W(t)$.

(a) Calculate the canonical metric on T.
(b) Without using the fact that $W(t)$ is continuous, prove the continuity of $X(t)$ by using the Fernique sufficient conditions in Theorem 17.7.

Hint: See Example 12.1.

Exercise 17.25 (Ornstein–Uhlenbeck Process). Consider the Ornstein–Uhlenbeck process $X(t) = \frac{\sigma}{\sqrt{\alpha}} e^{-\frac{\alpha t}{2}} W(e^{\alpha t}), t \geq 0$, where $W(t)$ is a Wiener process. Verify the Fernique conditions for continuity of the $X(t)$ process.

Exercise 17.26 * **(Metric Entropy of the Ornstein–Uhlenbeck Process).** Consider the Ornstein–Uhlenbeck process of the above exercise.

(a) Calculate the canonical metric for this process.
(b) Show that the metric entropy condition in part (c) of Theorem 17.8 holds for $0 \leq t \leq T$ for any $T < \infty$.
(c) Find a bound on $E[\sup_{0 \leq t \leq T} X(t)]$ by using part (c) of Theorem 17.8. Evaluate the bound when $\sigma = \alpha = T = 1$.

References

Adler, R.J. (1987). *An Introduction to Continuity, Extrema, and Related Topics*, IMS, Lecture Notes and Monograph Series, Hayward, CA.

Basu, S. and DasGupta, A. (1991). Robustness of standard confidence intervals for location parameters under departure from normality, *Ann. Stat.*, 23, 1433–1442.

Borell, C. (1975). Convex functions in d-space, *Period. Math. Hungar.*, 6, 111–136.

Bucklew, J. (2004). *Introduction to Rare Event Simulation*, Springer, New York.

Cressie, N. (1980). Relaxing assumptions in the one sample t-test, *Austr. J. Statist.*, 22, 143–153.

Dembo, A. and Shao, Q.M. (2006). Large and moderate deviations for Hotelling's T^2 statistic, *Electron. Comm. Prob.*, 11, 149–159.

Dembo, A. and Zeitouni, O. (1998). *Large Deviations, Techniques and Applications*, Jones and Bartlett, Boston.

den Hollander, F. (2000). *Large Deviations*, AMS, Providence, RI.

Devroye, L., Gyorfi, L., and Lugosi, G. (1996). *A Probabilistic Theory of Pattern Recognition*, Springer, New York.

Dubhashi, D. and Panconesi, A. (2009). *Concentration of Measure for the Analysis of Randomized Algorithms*, Cambridge University Press, Cambridge, UK.

Dudley, R.M. (1967). The sizes of compact subsets of Hilbert space and continuity of Gaussian processes, *J. Funct. Anal.*, 1, 290–330.

Efron, B. (1969). Student's t-test under symmetry conditions, *J. Amer. Statist. Assoc.*, 64, 1278–1302.

Fernique, X. (1974). Des resultats nouveaux sur les processus gaussiens, *C. R. Acad. Sci. Paris, Ser. A*, 278, 363–365.

Giné, E., Götze, F., and Mason, D.M. (1997). When is the Student's t-statistic asymptotically standard normal?, *Ann. Prob.*, 25, 1514–1531.

Groeneboom, P. and Oosterhoff, J. (1977). Bahadur efficiency and probabilities of large deviations, *Statist. Neerlandica*, 31, 1, 1–24.

Groeneboom, P. and Oosterhoff, J. (1980). Bahadur Efficiency and Small Sample Efficiency : A Numerical Study, *Mathematisch Centrum*, Amsterdam.

Groeneboom, P. and Oosterhoff, J. (1981). Bahadur efficiency and small sample efficiency, *Internat. Statist. Rev.*, 49, 2, 127–141.

Hall, P. (1987). Edgeworth expansions for Student's t-statistic under minimal moment conditions, *Ann. Prob.*, 15, 920–931.

Hall, P. and Wang, Q. (2004). Exact convergence rate and leading term in central limit theorem for Student's t-statistic, *Ann. Prob.*, 32, 1419–1437.

Landau, H.J. and Shepp, L. (1970). On the supremum of a Gaussian process, *Sankhyā, Ser. A*, 32, 369–378.

Ledoux, M. (2004). Spectral gap, logarithmic Sobolev constant, and geometric bounds, *Surveys in Differential Geometry, IX*, 219–240, International. Press, Somerville, MA.

Logan, B.F., Mallows, C.L., Rice, S.O., and Shepp, L. (1973). Limit distributions of self-normalized sums, *Ann. Prob.*, 1, 788–809.

Lugosi, G. (2004). Concentration Inequalities, Preprint.

McDiarmid, C. (1998). Concentration, Prob. Methods for Algorithmic Discrete Math., 195–248, Algorithm. Combin., 16, Springer, Berlin.

Shao, Q.M. (1997). Self-normalized large deviations, *Ann. Prob.*, 25, 285–328.

Stroock, D. (1984). *An Introduction to the Theory of Large Deviations*, Springer, New York.

Sudakov, V.N. and Cirelson, B.S. (1974). Extremal properties of half spaces for spherically invariant measures, *Zap. Nauchn. Sem. Leningrad Otdel. Mat. Inst. Steklov*, 41, 14–24.

Talagrand, M. (1995). Concentration of measures and isoperimetric inequalities, *Inst. Hautes Etudes Sci. Publ. Math.*, 81, 73–205.

Talagrand, M. (1996). Majorizing measures: the generic chaining, *Ann. Prob.*, 24, 1049–1103.

Varadhan, S.R.S. (1984). *Large Deviations and Applications*, SIAM, Philadelphia.

Chapter 18
The Exponential Family and Statistical Applications

The exponential family is a practically convenient and widely used unified family of distributions on finite-dimensional Euclidean spaces parametrized by a finite-dimensional parameter vector. Specialized to the case of the real line, the exponential family contains as special cases most of the standard discrete and continuous distributions that we use for practical modeling, such as the normal, Poisson, binomial, exponential, Gamma, multivariate normal, and so on. The reason for the special status of the exponential family is that a number of important and useful calculations in statistics can be done all at one stroke within the framework of the exponential family. This generality contributes to both convenience and larger-scale understanding. The exponential family is the usual testing ground for the large spectrum of results in parametric statistical theory that require notions of *regularity* or *Cramér–Rao regularity*. In addition, the unified calculations in the exponential family have an element of mathematical neatness. Distributions in the exponential family have been used in classical statistics for decades. However, it has recently obtained additional importance due to its use and appeal to the machine learning community. A fundamental treatment of the general exponential family is provided in this chapter. Classic expositions are available in Barndorff-Nielsen (1978), Brown (1986), and Lehmann and Casella (1998). An excellent recent treatment is available in Bickel and Doksum (2006).

18.1 One-Parameter Exponential Family

Exponential families can have any finite number of parameters. For instance, as we should a normal distribution with a known mean is in the one-parameter exponential family, where as a normal distribution with both parameters unknown is in the two-parameter exponential family. A bivariate normal distribution with all parameters unknown is in the five-parameter exponential family. As another example, if we take a normal distribution in which the mean and the variance are functionally related (e.g., the $N(\mu, \mu^2)$ distribution), then the distribution is neither in the one-parameter nor in the two-parameter exponential family, but in a family called a *curved exponential family*. We start with the one-parameter regular exponential family.

18.1.1 Definition and First Examples

We start with an illustrative example that brings out some of the most important properties of distributions in an exponential family.

Example 18.1 (Normal Distribution with a Known Mean). Suppose $X \sim N(0, \sigma^2)$. Then the density of X is

$$f(x \mid \sigma) = \frac{1}{\sigma\sqrt{2\pi}} e^{-\frac{x^2}{2\sigma^2}} I_{x \in \mathcal{R}}.$$

This density is parametrized by a single parameter σ. Writing

$$\eta(\sigma) = -\frac{1}{2\sigma^2}, \quad T(x) = x^2, \quad \psi(\sigma) = \log \sigma, \quad h(x) = \frac{1}{\sqrt{2\pi}} I_{x \in \mathcal{R}},$$

we can represent the density in the form

$$f(x \mid \sigma) = e^{\eta(\sigma)T(x) - \psi(\sigma)} h(x),$$

for any $\sigma \in \mathcal{R}_+$.

Next, suppose that we have an iid sample $X_1, X_2, \ldots, X_n \sim N(0, \sigma^2)$. Then the joint density of $X_1, X_2, \ldots, X_n$ is

$$f(x_1, x_2, \ldots, x_n \mid \sigma) = \frac{1}{\sigma^n (2\pi)^{n/2}} e^{-\frac{\sum_{i=1}^n x_i^2}{2\sigma^2}} I_{x_1, x_2, \ldots, x_n \in \mathcal{R}}.$$

Now writing

$$\eta(\sigma) = -\frac{1}{2\sigma^2}, \quad T(x_1, x_2, \ldots, x_n) = \sum_{i=1}^n x_i^2, \quad \psi(\sigma) = n \log \sigma,$$

and

$$h(x_1, x_2, \ldots, x_n) = \frac{1}{(2\pi)^{n/2}} I_{x_1, x_2, \ldots, x_n \in \mathcal{R}},$$

once again we can represent the joint density in the same general form

$$f(x_1, x_2, \ldots, x_n \mid \sigma) = e^{\eta(\sigma)T(x_1, x_2, \ldots, x_n) - \psi(\sigma)} h(x_1, x_2, \ldots, x_n).$$

We notice that in this representation of the joint density $f(x_1, x_2, \ldots, x_n \mid \sigma)$, the statistic $T(X_1, X_2, \ldots, X_n)$ is still a one-dimensional statistic, namely, $T(X_1, X_2, \ldots, X_n) = \sum_{i=1}^n X_i^2$. Using the fact that the sum of squares of n independent standard normal variables is a chi-square variable with n degrees of freedom, we have that the density of $T(X_1, X_2, \ldots, X_n)$ is

18.1 One-Parameter Exponential Family

$$f_T(t\,|\sigma) = \frac{e^{-\frac{t}{2\sigma^2}} t^{\frac{n}{2}-1}}{\sigma^n 2^{n/2}\Gamma(\frac{n}{2})} I_{t>0}.$$

This time, writing

$$\eta(\sigma) = -\frac{1}{2\sigma^2}, \quad S(t) = t, \psi(\sigma) = n\log\sigma, \quad h(t) = \frac{1}{2^{n/2}\Gamma(\frac{n}{2})} I_{t>0},$$

once again we are able to write even the density of $T(X_1, X_2, \ldots, X_n) = \sum_{i=1}^{n} X_i^2$ in that same general form

$$f_T(t\,|\sigma) = e^{\eta(\sigma)S(t) - \psi(\sigma)} h(t).$$

Clearly, something very interesting is going on. We started with a basic density in a specific form, namely, $f(x\,|\sigma) = e^{\eta(\sigma)T(x) - \psi(\sigma)} h(x)$, and then we found that the joint density and the density of the relevant one-dimensional statistic $\sum_{i=1}^{n} X_i^2$ in that joint density, are once again densities of exactly that same general form. It turns out that all of these phenomena are true of the entire family of densities that can be written in that general form, which is the one-parameter exponential family. Let us formally define it and we then extend the definition to distributions with more than one parameter.

Definition 18.1. Let $X = (X_1, \ldots, X_d)$ be a d-dimensional random vector with a distribution $P_\theta, \theta \in \Theta \subseteq \mathcal{R}$.

Suppose $X_1, \ldots, X_d$ are jointly continuous. The family of distributions $\{P_\theta, \theta \in \Theta\}$ is said to belong to the *one-parameter exponential family* if the density of $X = (X_1, \ldots, X_d)$ may be represented in the form

$$f(x\,|\theta) = e^{\eta(\theta)T(x) - \psi(\theta)} h(x),$$

for some real-valued functions $T(x), \psi(\theta)$ and $h(x) \geq 0$.

If $X_1, \ldots, X_d$ are jointly discrete, then $\{P_\theta, \theta \in \Theta\}$ is said to belong to the one-parameter exponential family if the joint pmf $p(x\,|\theta) = P_\theta(X_1 = x_1, \ldots, X_d = x_d)$ may be written in the form

$$p(x\,|\theta) = e^{\eta(\theta)T(x) - \psi(\theta)} h(x),$$

for some real-valued functions $T(x), \psi(\theta)$ and $h(x) \geq 0$. Note that the functions η, T, and h are not unique. For example, in the product ηT, we can multiply T by some constant c and divide η by it. Similarly, we can play with constants in the function h.

Definition 18.2. Suppose $X = (X_1, \ldots, X_d)$ has a distribution $P_\theta, \theta \in \Theta$, belonging to the one parameter exponential family. Then the statistic $T(X)$ is called *the natural sufficient statistic* for the family $\{P_\theta\}$.

The notion of a sufficient statistic is a fundamental one in statistical theory and its applications. Sufficiency was introduced into the statistical literature by Sir Ronald A. Fisher (Fisher (1922)). Sufficiency attempts to formalize the notion of *no loss of information*. A sufficient statistic is supposed to contain by itself all of the information about the unknown parameters of the underlying distribution that the entire sample could have provided. In that sense, there is nothing to lose by restricting attention to just a sufficient statistic in one's inference process. However, the form of a sufficient statistic is very much dependent on the choice of a particular distribution P_θ for modeling the observable X. Still, reduction to sufficiency in widely used models usually makes just simple common sense. We come back to the issue of sufficiency once again later in this chapter.

We now show examples of a few more common distributions that belong to the one-parameter exponential family.

Example 18.2 (Binomial Distribution). Let $X \sim \text{Bin}(n, p)$, with $n \geq 1$ considered as known, and $0 < p < 1$ a parameter. We represent the pmf of X in the one-parameter exponential family form.

$$f(x|p) = \binom{n}{x} p^x (1-p)^{n-x} I_{\{x \in \{0,1,\ldots,n\}\}} = \binom{n}{x} \left(\frac{p}{1-p}\right)^x (1-p)^n I_{\{x \in \{0,1,\ldots,n\}\}}$$

$$= \binom{n}{x} e^{x \log p/(1-p) + n \log(1-p)} I_{\{x \in \{0,1,\ldots,n\}\}}.$$

Writing $\eta(p) = \log \frac{p}{1-p}$, $T(x) = x$, $\psi(p) = -n\log(1-p)$, and $h(x) = \binom{n}{x} I_{\{x \in \{0,1,\ldots,n\}\}}$, we have represented the pmf $f(x|p)$ in the one-parameter exponential family form, as long as $p \in (0, 1)$. For $p = 0$ or 1, the distribution becomes a one-point distribution. Consequently, the family of distributions $\{f(x|p), 0 < p < 1\}$ forms a one-parameter exponential family, but if either of the boundary values $p = 0, 1$ is included, the family is not in the exponential family.

Example 18.3 (Normal Distribution with a Known Variance). Suppose $X \sim N(\mu, \sigma^2)$, where σ is considered known, and $\mu \in \mathcal{R}$ a parameter. Then,

$$f(x|\mu) = \frac{1}{\sqrt{2\pi}} e^{-\frac{x^2}{2} + \mu x - \frac{\mu^2}{2}} I_{x \in \mathcal{R}},$$

which can be written in the one-parameter exponential family form by witing $\eta(\mu) = \mu$, $T(x) = x$, $\psi(\mu) = \frac{\mu^2}{2}$, and $h(x) = e^{-\frac{x^2}{2}} I_{x \in \mathcal{R}}$. So, the family of distributions $\{f(x|\mu), \mu \in \mathcal{R}\}$ forms a one-parameter exponential family.

Example 18.4 (Errors in Variables). Suppose U, V, W are independent normal variables, with U and V being $N(\mu, 1)$ and W being $N(0, 1)$. Let $X_1 = U + W$ and $X_2 = V + W$. In other words, a common error of measurement W contaminates both U and V.

18.1 One-Parameter Exponential Family

Let $X = (X_1, X_2)$. Then X has a bivariate normal distribution with means μ, μ, variances $2, 2$, and a correlation parameter $\rho = \frac{1}{2}$. Thus, the density of X is

$$f(x \mid \mu) = \frac{1}{2\sqrt{3}\pi} e^{-\frac{2}{3}\left[\frac{(x_1-\mu)^2}{2} + \frac{(x_2-\mu)^2}{2} - 2(x_1-\mu)(x_2-\mu)\right]} I_{x_1,x_2 \in \mathcal{R}}$$

$$= \frac{1}{2\sqrt{3}\pi} e^{[\frac{2}{3}\mu(x_1+x_2) - \frac{2}{3}\mu^2]} e^{-\frac{x_1^2 + x_2^2 - 4x_1 x_2}{3}} I_{x_1,x_2 \in \mathcal{R}}.$$

This is in the form of a one-parameter exponential family with the natural sufficient statistic $T(X) = T(X_1, X_2) = X_1 + X_2$.

Example 18.5 (Gamma Distribution). Suppose X has the Gamma density

$$\frac{e^{-\frac{x}{\lambda}} x^{\alpha-1}}{\lambda^\alpha \Gamma(\alpha)} I_{x>0}.$$

As such, it has two parameters λ, α. If we assume that α is known, then we may write the density in the one-parameter exponential family form:

$$f(x \mid \lambda) = e^{-\frac{x}{\lambda} - \alpha \log \lambda} \frac{x^{\alpha-1}}{\Gamma(\alpha)} I_{x>0},$$

and recognize it as a density in the exponential family with $\eta(\lambda) = -\frac{1}{\lambda}, T(x) = x, \psi(\lambda) = \alpha \log \lambda, h(x) = \frac{x^{\alpha-1}}{\Gamma(\alpha)} I_{x>0}.$

If we assume that λ is known, once again, by writing the density as

$$f(x \mid \alpha) = e^{\alpha \log x - \alpha(\log \lambda) - \log \Gamma(\alpha)} \frac{e^{-\frac{x}{\lambda}}}{x} I_{x>0},$$

we recognize it as a density in the exponential family with $\eta(\alpha) = \alpha, T(x) = \log x, \psi(\alpha) = \alpha(\log \lambda) + \log \Gamma(\alpha), h(x) = \frac{e^{-\frac{x}{\lambda}}}{x} I_{x>0}.$

Example 18.6 (An Unusual Gamma Distribution). Suppose we have a Gamma density in which the mean is known, say, $E(X) = 1$. This means that $\alpha\lambda = 1 \Rightarrow \lambda = 1/\alpha$. Parametrizing the density with α, we have

$$f(x \mid \alpha) = e^{-\alpha x + \alpha \log x} \frac{\alpha^\alpha}{\Gamma(\alpha)} \frac{1}{x} I_{x>0}$$

$$= e^{\alpha[\log x - x] - [\log \Gamma(\alpha) - \alpha \log \alpha]} \frac{1}{x} I_{x>0},$$

which is once again in the one-parameter exponential family form with $\eta(\alpha) = \alpha$, $T(x) = \log x - x, \psi(\alpha) = \log \Gamma(\alpha) - \alpha \log \alpha, h(x) = \frac{1}{x} I_{x>0}.$

Example 18.7 (A Normal Distribution Truncated to a Set). Suppose a certain random variable W has a normal distribution with mean μ and variance one. We saw in Example 18.3 that this is in the one-parameter exponential family. Suppose now that the variable W can be physically observed only when its value is inside some set A. For instance, if $W > 2$, then our measuring instruments cannot tell what the value of W is. In such a case, the variable X that is truly observed has a normal distribution truncated to the set A. For simplicity, take A to be $A = [a, b]$, an interval. Then, the density of X is

$$f(x \mid \mu) = \frac{e^{-\frac{(x-\mu)^2}{2}}}{\sqrt{2\pi}[\Phi(b-\mu) - \Phi(a-\mu)]} I_{a \leq x \leq b}.$$

This can be written as

$$f(x \mid \mu) = \frac{1}{\sqrt{2\pi}} e^{\mu x - \frac{\mu^2}{2} - \log[\Phi(b-\mu) - \Phi(a-\mu)]} e^{-\frac{x^2}{2}} I_{a \leq x \leq b},$$

and we recognize this to be in the exponential family form with $\eta(\mu) = \mu$, $T(x) = x$, $\psi(\mu) = \frac{\mu^2}{2} + \log[\Phi(b-\mu) - \Phi(a-\mu)]$, and $h(x) = e^{-\frac{x^2}{2}} I_{a \leq x \leq b}$. Thus, the distribution of W truncated to $A = [a, b]$ is still in the one-parameter exponential family. This phenomenon is in fact more general.

Example 18.8 (Some Distributions Not in the Exponential Family). It is clear from the definition of a one-parameter exponential family that if a certain family of distributions $\{P_\theta, \theta \in \Theta\}$ belongs to the one-parameter exponential family, then each P_θ has exactly the same support. Precisely, for any fixed θ, $P_\theta(A) > 0$ if and only if $\int_A h(x)dx > 0$, and in the discrete case, $P_\theta(A) > 0$ if and only if $A \cap \mathcal{X} \neq \emptyset$, where $\mathcal{X}$ is the countable set $\mathcal{X} = \{x : h(x) > 0\}$. As a consequence of this common support fact, the so-called *irregular distributions* whose support depends on the parameter cannot be members of the exponential family. Examples would be the family of $U[0, \theta]$, $U[-\theta, \theta]$ distributions, and so on. Likewise, the *shifted exponential density* $f(x \mid \theta) = e^{\theta - x} I_{x > \theta}$ cannot be in the exponential family.

Some other common distributions are also not in the exponential family, but for other reasons. An important example is the family of Cauchy distributions given by the location parameter form $f(x \mid \mu) = \frac{1}{\pi[1+(x-\mu)^2]} I_{x \in \mathcal{R}}$. Suppose that it is. Then, we can find functions $\eta(\mu), T(x)$ such that for all x, μ,

$$e^{\eta(\mu)T(x)} = \frac{1}{1 + (x-\mu)^2} \Rightarrow \eta(\mu)T(x) = -\log(1 + (x-\mu)^2)$$

$$\Rightarrow \eta(0)T(x) = -\log(1 + x^2) \Rightarrow T(x) = -c\log(1 + x^2)$$

for some constant c.

Plugging this back, we get, for all x, μ,

$$-c\eta(\mu)\log(1+x^2) = -\log(1+(x-\mu)^2) \Rightarrow \eta(\mu) = \frac{1}{c}\frac{\log(1+(x-\mu)^2)}{\log(1+x^2)}.$$

This means that $\frac{\log(1+(x-\mu)^2)}{\log(1+x^2)}$ must be a constant function of x, which is a contradiction. The choice of $\mu = 0$ as the special value of μ is not important.

18.2 The Canonical Form and Basic Properties

Suppose $\{P_\theta, \theta \in \Theta\}$ is a family belonging to the one-parameter exponential family, with density (or pmf) of the form $f(x\,|\,\theta) = e^{\eta(\theta)T(x) - \psi(\theta)} h(x)$. If $\eta(\theta)$ is a one-to-one function of θ, then we can drop θ altogether, and parametrize the distribution in terms of η itself. If we do that, we get a reparametrized density g in the form $e^{\eta T(x) - \psi^*(\eta)} h(x)$. By a slight abuse of notation, we again use the notation f for g and ψ for ψ^*.

Definition 18.3. Let $X = (X_1, \ldots, X_d)$ have a distribution $P_\eta, \eta \in \mathcal{T} \subseteq \mathcal{R}$. The family of distributions $\{P_\eta, \eta \in \mathcal{T}\}$ is said to belong to the *canonical one-parameter exponential family* if the density (pmf) of P_η may be written in the form

$$f(x\,|\,\eta) = e^{\eta T(x) - \psi(\eta)} h(x),$$

where

$$\eta \in \mathcal{T} = \left\{ \eta : e^{\psi(\eta)} = \int_{\mathcal{R}^d} e^{\eta T(x)} h(x) dx < \infty \right\},$$

in the continuous case, and

$$\mathcal{T} = \left\{ \eta : e^{\psi(\eta)} = \sum_{x \in \mathcal{X}} e^{\eta T(x)} h(x) < \infty \right\},$$

in the discrete case, with $\mathcal{X}$ being the countable set on which $h(x) > 0$.

For a distribution in the canonical one-parameter exponential family, the parameter η is called the *natural parameter*, and $\mathcal{T}$ is called *the natural parameter space*. Note that $\mathcal{T}$ describes the largest set of values of η for which the density (pmf) can be defined. In a particular application, we may have extraneous knowledge that η belongs to some proper subset of $\mathcal{T}$. Thus, $\{P_\eta\}$ with $\eta \in \mathcal{T}$ is called the *full canonical one-parameter exponential family*. We generally refer to the full family, unless otherwise stated.

The canonical exponential family is called *regular* if $\mathcal{T}$ is an open set in $\mathcal{R}$, and it is called *nonsingular* if $\text{Var}_\eta(T(X)) > 0$ for all $\eta \in \mathcal{T}^0$, the interior of the natural parameter space $\mathcal{T}$.

It is analytically convenient to work with an exponential family distribution in its canonical form. Once a result has been derived for the canonical form, if desired we can rewrite the answer in terms of the original parameter θ. Doing this retransformation at the end is algebraically and notationally simpler than carrying the original function $\eta(\theta)$ and often its higher derivatives with us throughout a calculation. Most of our formulas and theorems below are given for the canonical form.

Example 18.9 (Binomial Distribution in Canonical Form). Let $X \sim \text{Bin}(n, p)$ with the pmf $\binom{n}{x} p^x (1-p)^{n-x} I_{x \in \{0,1,\ldots,n\}}$. In Example 18.2, we represented this pmf in the exponential family form

$$f(x \mid p) = e^{x \log \frac{p}{(1-p)} - n \log(1-p)} \binom{n}{x} I_{x \in \{0,1,\ldots,n\}}.$$

If we write $\log \frac{p}{1-p} = \eta$, then $\frac{p}{1-p} = e^\eta$, and hence, $p = \frac{e^\eta}{1+e^\eta}$, and $1 - p = \frac{1}{1+e^\eta}$. Therefore, the canonical exponential family form of the binomial distribution is

$$f(x \mid \eta) = e^{\eta x - n \log(1+e^\eta)} \binom{n}{x} I_{x \in \{0,1,\ldots,n\}},$$

and the natural parameter space is $\mathcal{T} = \mathcal{R}$.

18.2.1 Convexity Properties

Written in its canonical form, a density (pmf) in an exponential family has some convexity properties. These convexity properties are useful in manipulating with moments and other functionals of $T(X)$, the natural sufficient statistic appearing in the expression for the density of the distribution.

Theorem 18.1. *The natural parameter space $\mathcal{T}$ is convex, and $\psi(\eta)$ is a convex function on $\mathcal{T}$.*

Proof. We consider the continuous case only, as the discrete case admits basically the same proof. Let η_1, η_2 be two members of $\mathcal{T}$, and let $0 < \alpha < 1$. We need to show that $\alpha\eta_1 + (1-\alpha)\eta_2$ belongs to $\mathcal{T}$; that is,

$$\int_{\mathcal{R}^d} e^{(\alpha\eta_1 + (1-\alpha)\eta_2)T(x)} h(x) dx < \infty.$$

18.2 The Canonical Form and Basic Properties

But

$$\int_{\mathcal{R}^d} e^{(\alpha\eta_1+(1-\alpha)\eta_2)T(x)} h(x)dx = \int_{\mathcal{R}^d} e^{\alpha\eta_1 T(x)} \times e^{(1-\alpha)\eta_2 T(x)} h(x)dx$$

$$= \int_{\mathcal{R}^d} \left(e^{\eta_1 T(x)}\right)^\alpha \left(e^{\eta_2 T(x)}\right)^{1-\alpha} h(x)dx$$

$$\leq \left(\int_{\mathcal{R}^d} e^{\eta_1 T(x)} h(x)dx\right)^\alpha \left(\int_{\mathcal{R}^d} e^{\eta_2 T(x)} h(x)dx\right)^{1-\alpha}$$

(by Holder's inequality)

$$< \infty,$$

because, by hypothesis, $\eta_1, \eta_2 \in \mathcal{T}$, and hence, $\int_{\mathcal{R}^d} e^{\eta_1 T(x)} h(x)dx$, and $\int_{\mathcal{R}^d} e^{\eta_2 T(x)} h(x)dx$ are both finite.

Note that in this argument, we have actually proved the inequality

$$e^{\psi(\alpha\eta_1+(1-\alpha)\eta_2)} \leq e^{\alpha\psi(\eta_1)+(1-\alpha)\psi(\eta_2)}.$$

But this is the same as saying

$$\psi(\alpha\eta_1 + (1-\alpha)\eta_2) \leq \alpha\psi(\eta_1) + (1-\alpha)\psi(\eta_2);$$

that is, $\psi(\eta)$ is a convex function on $\mathcal{T}$. □

18.2.2 Moments and Moment Generating Function

The next result is a very special fact about the canonical exponential family, and is the source of a large number of closed-form formulas valid for the entire canonical exponential family. The fact itself is actually a fact in mathematical analysis. Due to the special form of exponential family densities, the fact in analysis translates to results for the exponential family, an instance of interplay between mathematics and statistics and probability.

Theorem 18.2. *(a) The function $e^{\psi(\eta)}$ is infinitely differentiable at every $\eta \in \mathcal{T}^0$. Furthermore, in the continuous case, $e^{\psi(\eta)} = \int_{\mathcal{R}^d} e^{\eta T(x)} h(x)dx$ can be differentiated any number of times inside the integral sign, and in the discrete case, $e^{\psi(\eta)} = \sum_{x \in \mathcal{X}} e^{\eta T(x)} h(x)$ can be differentiated any number of times inside the sum.*
(b) In the continuous case, for any $k \geq 1$,

$$\frac{d^k}{d\eta^k} e^{\psi(\eta)} = \int_{\mathcal{R}^d} [T(x)]^k e^{\eta T(x)} h(x)dx,$$

and in the discrete case,

$$\frac{d^k}{d\eta^k}e^{\psi(\eta)} = \sum_{x \in \mathcal{X}}[T(x)]^k e^{\eta T(x)}h(x).$$

Proof. Take $k = 1$. Then, by the definition of derivative of a function, $\frac{d}{d\eta}e^{\psi(\eta)}$ exists if and only if $\lim_{\delta \to 0}[\frac{e^{\psi(\eta+\delta)}-e^{\psi(\eta)}}{\delta}]$ exists. But

$$\frac{e^{\psi(\eta+\delta)} - e^{\psi(\eta)}}{\delta} = \int_{\mathcal{R}^d} \frac{e^{(\eta+\delta)T(x)} - e^{\eta T(x)}}{\delta}h(x)dx,$$

and by an application of the dominated convergence theorem (see Chapter 7),

$$\lim_{\delta \to 0}\int_{\mathcal{R}^d} \frac{e^{(\eta+\delta)T(x)} - e^{\eta T(x)}}{\delta}h(x)dx$$

exists, and the limit can be carried inside the integral, to give

$$\lim_{\delta \to 0}\int_{\mathcal{R}^d} \frac{e^{(\eta+\delta)T(x)} - e^{\eta T(x)}}{\delta}h(x)dx = \int_{\mathcal{R}^d} \lim_{\delta \to 0} \frac{e^{(\eta+\delta)T(x)} - e^{\eta T(x)}}{\delta}h(x)dx$$

$$= \int_{\mathcal{R}^d} \frac{d}{d\eta}e^{\eta T(x)}h(x)dx$$

$$= \int_{\mathcal{R}^d} T(x)e^{\eta T(x)}h(x)dx.$$

Now use induction on k by using the dominated convergence theorem again. □

This compact formula for an arbitrary derivative of $e^{\psi(\eta)}$ leads to the following important moment formulas.

Theorem 18.3. *At any $\eta \in T^0$,*

(a) $E_\eta[T(X)] = \psi'(\eta);\quad Var_\eta[T(X)] = \psi''(\eta);$
(b) *The coefficients of skewness and kurtosis of $T(X)$ equal*

$$\beta(\eta) = \frac{\psi^{(3)}(\eta)}{[\psi''(\eta)]^{3/2}}; \text{ and } \gamma(\eta) = \frac{\psi^{(4)}(\eta)}{[\psi''(\eta)]^2};$$

(c) *At any t such that $\eta + t \in T$, the mgf of $T(X)$ exists and equals*

$$M_\eta(t) = e^{\psi(\eta+t)-\psi(\eta)}.$$

18.2 The Canonical Form and Basic Properties

Proof. Again, we take just the continuous case. Consider the result of the previous theorem that for any $k \geq 1$, $\frac{d^k}{d\eta^k} e^{\psi(\eta)} = \int_{\mathcal{R}^d} [T(x)]^k e^{\eta T(x)} h(x) dx$. Using this for $k = 1$, we get

$$\psi'(\eta) e^{\psi(\eta)} = \int_{\mathcal{R}^d} T(x) e^{\eta T(x)} h(x) dx \Rightarrow \int_{\mathcal{R}^d} T(x) e^{\eta T(x) - \psi(\eta)} h(x) dx = \psi'(\eta),$$

which gives the result $E_\eta[T(X)] = \psi'(\eta)$.

Similarly,

$$\frac{d^2}{d\eta^2} e^{\psi(\eta)} = \int_{\mathcal{R}^d} [T(x)]^2 e^{\eta T(x)} h(x) dx \Rightarrow [\psi''(\eta) + \{\psi'(\eta)\}^2] e^{\psi(\eta)}$$

$$= \int_{\mathcal{R}^d} [T(x)]^2 e^{\eta T(x)} h(x) dx$$

$$\Rightarrow \psi''(\eta) + \{\psi'(\eta)\}^2 = \int_{\mathcal{R}^d} [T(x)]^2 e^{\eta T(x) - \psi(\eta)} h(x) dx,$$

which gives $E_\eta[T(X)]^2 = \psi''(\eta) + \{\psi'(\eta)\}^2$. Combine this with the already obtained result that $E_\eta[T(X)] = \psi'(\eta)$, and we get $\text{Var}_\eta[T(X)] = E_\eta[T(X)]^2 - (E_\eta[T(X)])^2 = \psi''(\eta)$.

The coefficient of skewness is defined as $\beta_\eta = \frac{E[T(X) - ET(X)]^3}{(\text{Var} T(X))^{3/2}}$. To obtain $E[T(X) - ET(X)]^3 = E[T(X)]^3 - 3E[T(X)]^2 E[T(X)] + 2[ET(X)]^3$, use the identity $\frac{d^3}{d\eta^3} e^{\psi(\eta)} = \int_{\mathcal{R}^d} [T(x)]^3 e^{\eta T(x)} h(x) dx$. Then use the fact that the third derivative of $e^{\psi(\eta)}$ is $e^{\psi(\eta)} [\psi^{(3)}(\eta) + 3\psi'(\eta)\psi''(\eta) + \{\psi'(\eta)\}^3]$. As we did in our proofs for the mean and the variance above, transfer $e^{\psi(\eta)}$ into the integral on the right-hand side and then simplify. This gives $E[T(X) - ET(X)]^3 = \psi^{(3)}(\eta)$, and the skewness formula follows. The formula for kurtosis is proved by the same argument, using $k = 4$ in the derivative identity $\frac{d^k}{d\eta^k} e^{\psi(\eta)} = \int_{\mathcal{R}^d} [T(x)]^k e^{\eta T(x)} h(x) dx$.

Finally, for the mgf formula,

$$M_\eta(t) = E_\eta[e^{tT(X)}] = \int_{\mathcal{R}^d} e^{tT(X)} e^{\eta T(x) - \psi(\eta)} h(x) dx$$

$$= e^{-\psi(\eta)} \int_{\mathcal{R}^d} e^{(t+\eta)T(x)} h(x) dx$$

$$= e^{-\psi(\eta)} e^{\psi(t+\eta)} \int_{\mathcal{R}^d} e^{(t+\eta)T(x) - \psi(t+\eta)} h(x) dx = e^{-\psi(\eta)} e^{\psi(t+\eta)} \times 1$$

$$= e^{\psi(t+\eta) - \psi(\eta)}.$$

An important consequence of the mean and the variance formulas is the following monotonicity result. □

Corollary 18.1. *For a nonsingular canonical exponential family, $E_\eta[T(X)]$ is strictly increasing in η on T^0.*

Proof. From part (a) of Theorem 18.3, the variance of $T(X)$ is the derivative of the expectation of $T(X)$, and by nonsingularity, the variance is strictly positive. This implies that the expectation is strictly increasing.

As a consequence of this strict monotonicity of the mean of $T(X)$ in the natural parameter, nonsingular canonical exponential families may be reparametrized by using the mean of T itself as the parameter. This is useful for some purposes.

Example 18.10 (Binomial Distribution). From Example 18.9, in the canonical representation of the binomial distribution, $\psi(\eta) = n\log(1 + e^\eta)$. By direct differentiation,

$$\psi'(\eta) = \frac{ne^\eta}{1+e^\eta}; \qquad \psi''(\eta) = \frac{ne^\eta}{(1+e^\eta)^2};$$

$$\psi^{(3)}(\eta) = \frac{-ne^\eta(e^\eta - 1)}{(1+e^\eta)^3}; \qquad \psi^{(4)}(\eta) = \frac{ne^\eta(e^{2\eta} - 4e^\eta + 1)}{(1+e^\eta)^4}.$$

Now recall from Example 18.9 that the success probability p and the natural parameter η are related as $p = \frac{e^\eta}{1+e^\eta}$. Using this, and our general formulas from Theorem 18.3, we can rewrite the mean, variance, skewness, and kurtosis of X as

$$E(X) = np; \quad \mathrm{Var}(X) = np(1-p); \quad \beta_p = \frac{1-2p}{\sqrt{np(1-p)}}; \quad \gamma_p = \frac{\frac{1}{p(1-p)} - 6}{n}.$$

For completeness, it is useful to have the mean and the variance formula in an original parametrization, and they are stated below. The proof follows from an application of Theorem 18.3 and the chain rule.

Theorem 18.4. *Let $\{P_\theta, \theta \in \Theta\}$ be a family of distributions in the one-parameter exponential family with density (pmf)*

$$f(x|\theta) = e^{\eta(\theta)T(x) - \psi(\theta)} h(x).$$

Then, at any θ at which $\eta'(\theta) \neq 0$,

$$E_\theta[T(X)] = \frac{\psi'(\theta)}{\eta'(\theta)}; \quad \mathrm{Var}_\theta(T(X)) = \frac{\psi''(\theta)}{[\eta'(\theta)]^2} - \frac{\psi'(\theta)\eta''(\theta)}{[\eta'(\theta)]^3}.$$

18.2.3 Closure Properties

The exponential family satisfies a number of important closure properties. For instance, if a d-dimensional random vector $X = (X_1, \ldots, X_d)$ has a distribution in the exponential family, then the conditional distribution of any subvector given the rest is also in the exponential family. There are a number of such closure properties, of which we discuss only four.

18.2 The Canonical Form and Basic Properties

First, if $X = (X_1, \ldots, X_d)$ has a distribution in the exponential family, then the natural sufficient statistic $T(X)$ also has a distribution in the exponential family. Verification of this in the greatest generality cannot be done without using measure theory. However, we can easily demonstrate this in some particular cases. Consider the continuous case with $d = 1$ and suppose $T(X)$ is a differentiable one-to-one function of X. Then, by the Jacobian formula (see Chapter 1), $T(X)$ has the density

$$f_T(t \mid \eta) = e^{\eta t - \psi(\eta)} \frac{h(T^{-1}(t))}{|T'(T^{-1}(t))|}.$$

This is once again in the one-parameter exponential family form, with the natural sufficient statistic as T itself, and the ψ function unchanged. The h function has changed to a new function $h^*(t) = \frac{h(T^{-1}(t))}{|T'(T^{-1}(t))|}$.

Similarly, in the discrete case, the pmf of $T(X)$ is given by

$$P_\eta(T(X) = t) = \sum_{x:T(x)=t} e^{\eta T(x) - \psi(\eta)} h(x) = e^{\eta t - \psi(\eta)} h^*(t),$$

where $h^*(t) = \sum_{x:T(x)=t} h(x)$.

Next, suppose $X = (X_1, \ldots, X_d)$ has a density (pmf) $f(x \mid \eta)$ in the exponential family and $Y_1, Y_2, \ldots, Y_n$ are n iid observations from this density $f(x \mid \eta)$. Note that each individual Y_i is a d-dimensional vector. The joint density of $Y = (Y_1, Y_2, \ldots, Y_n)$ is

$$f(y \mid \eta) = \prod_{i=1}^{n} f(y_i \mid \eta) = \prod_{i=1}^{n} e^{\eta T(y_i) - \psi(\eta)} h(y_i)$$

$$= e^{\eta \sum_{i=1}^{n} T(y_i) - n\psi(\eta)} \prod_{i=1}^{n} h(y_i).$$

We recognize this to be in the one-parameter exponential family form again, with the natural sufficient statistic as $\sum_{i=1}^{n} T(Y_i)$, the new ψ function as $n\psi$, and the new h function as $\prod_{i=1}^{n} h(y_i)$. The joint density $\prod_{i=1}^{n} f(y_i \mid \eta)$ is known as the *likelihood function* in statistics (see Chapter 3). So, likelihood functions obtained from an iid sample from a distribution in the one-parameter exponential family are also members of the one-parameter exponential family.

The closure properties outlined in the above are formally stated in the next theorem.

Theorem 18.5. *Suppose $X = (X_1, \ldots, X_d)$ has a distribution belonging to the one-parameter exponential family with the natural sufficient statistic $T(X)$.*

(a) *$T = T(X)$ also has a distribution belonging to the one-parameter exponential family.*

(b) *Let $Y = AX + u$ be a nonsingular linear transformation of X. Then Y also has a distribution belonging to the one-parameter exponential family.*

(c) Let $\mathcal{I}_0$ be any proper subset of $\mathcal{I} = \{1, 2, \ldots, d\}$. Then the joint conditional distribution of $X_i, i \in \mathcal{I}_0$ given $X_j, j \in \mathcal{I} - \mathcal{I}_0$ also belongs to the one-parameter exponential family.
(d) For given $n \geq 1$, suppose $Y_1, \ldots, Y_n$ are iid with the same distribution as X. Then the joint distribution of $(Y_1, \ldots, Y_n)$ also belongs to the one-parameter exponential family.

18.3 Multiparameter Exponential Family

Similar to the case of distributions with only one-parameter, several common distributions with multiple parameters also belong to a general multiparameter exponential family. An example is the normal distribution on $\mathcal{R}$ with both parameters unknown. Another example is a multivariate normal distribution. Analytic techniques and properties of multiparameter exponential families are very similar to those of the one-parameter exponential family. For that reason, most of our presentation in this section dwells on examples.

Definition 18.4. Let $X = (X_1, \ldots, X_d)$ have a distribution $P_\theta, \theta \in \Theta \subseteq \mathcal{R}^k$. The family of distributions $\{P_\theta, \theta \in \Theta\}$ is said to belong to the k-parameter exponential family if its density (pmf) may be represented in the form

$$f(x|\theta) = e^{\sum_{i=1}^{k} \eta_i(\theta) T_i(x) - \psi(\theta)} h(x).$$

Again, obviously, the choice of the relevant functions η_i, T_i, h is not unique. As in the one-parameter case, the vector of statistics $(T_1, \ldots, T_k)$ is called the natural sufficient statistic, and if we reparametrize by using $\eta_i = \eta_i(\theta), i = 1, 2, \ldots, k$, the family is called the k-parameter canonical exponential family.

There is an implicit assumption in this definition that the number of *freely varying* θ is the same as the number of freely varying η, and that these are both equal to the specific k in the context. The formal way to say this is to assume the following.

Assumption. The dimension of Θ as well as the dimension of the image of Θ under the map $(\theta_1, \theta_2, \ldots, \theta_k) \longrightarrow (\eta_1(\theta_1, \theta_2, \ldots, \theta_k), \eta_2(\theta_1, \theta_2, \ldots, \theta_k), \ldots, \eta_k(\theta_1, \theta_2, \ldots, \theta_k))$ is equal to k.

There are some important examples where this assumption does not hold. They are not counted as members of a k-parameter exponential family. The name curved exponential family is commonly used for them, and this is discussed in the last section.

The terms *canonical form, natural parameter,* and *natural parameter space* mean the same things as in the one-parameter case. Thus, if we parametrize the distributions by using $\eta_1, \eta_2, \ldots, \eta_k$ as the k parameters, then the vector $\eta = (\eta_1, \eta_2, \ldots, \eta_k)$ is called the natural parameter vector, the parametrization $f(x|\eta) = e^{\sum_{i=1}^{k} \eta_i T_i(x) - \psi(\eta)} h(x)$ is called the canonical form, and the set of all

18.3 Multiparameter Exponential Family

vectors η for which $f(x\,|\,\eta)$ is a valid density (pmf) is called the natural parameter space. The main theorems for the case $k = 1$ hold for a general k.

Theorem 18.6. *The results of Theorem 18.1 and 18.5 hold for the k-parameter exponential family.*

The proofs are almost verbatim the same. The moment formulas differ somewhat due to the presence of more than one-parameter in the current context.

Theorem 18.7. *Suppose $X = (X_1, \ldots, X_d)$ has a distribution $P\eta, \eta \in \mathcal{T}$, belonging to the canonical k-parameter exponential family, with a density (pmf)*

$$f(x\,|\,\eta) = e^{\sum_{i=1}^{k} \eta_i T_i(x) - \psi(\eta)} h(x),$$

where

$$\mathcal{T} = \left\{\eta \in \mathcal{R}^k : \int_{\mathcal{R}^d} e^{\sum_{i=1}^{k} \eta_i T_i(x)} h(x)\, dx < \infty\right\}$$

(and the integral being replaced by a sum in the discrete case).

(a) At any $\eta \in \mathcal{T}^0$,

$$e^{\psi(\eta)} = \int_{\mathcal{R}^d} e^{\sum_{i=1}^{k} \eta_i T_i(x)} h(x)\, dx$$

is infinitely partially differentiable with respect to each η_i, and the partial derivatives of any order can be obtained by differentiating inside the integral sign.

(b) $E_\eta[T_i(X)] = \frac{\partial}{\partial \eta_i} \psi(\eta)$; $\mathrm{Cov}_\eta(T_i(X), T_j(X)) = \frac{\partial^2}{\partial \eta_i \partial \eta_j} \psi(\eta)$, $1 \le i, j \le k$.

(c) If η, t are such that $\eta, \eta + t \in \mathcal{T}$, then the joint mgf of $(T_1(X), \ldots, T_k(X))$ exists and equals

$$M_\eta(t) = e^{\psi(\eta+t) - \psi(\eta)}.$$

An important new terminology is that of a full rank.

Definition 18.5. A family of distributions $\{P\eta, \eta \in \mathcal{T}\}$ belonging to the canonical k-parameter exponential family is called *full rank* if at every $\eta \in \mathcal{T}^0$, the $k \times k$ covariance matrix $\left(\left(\frac{\partial^2}{\partial \eta_i \partial \eta_j} \psi(\eta)\right)\right)$ is nonsingular.

Definition 18.6 (Fisher Information Matrix). Suppose a family of distributions in the canonical k-parameter exponential family is nonsingular. Then, for $\eta \in \mathcal{T}^0$, the matrix $\left(\left(\frac{\partial^2}{\partial \eta_i \partial \eta_j} \psi(\eta)\right)\right)$ is called the Fisher information matrix (at η).

The Fisher information matrix is of paramount importance in parametric statistical theory and lies at the heart of finite and large sample optimality theory in statistical inference problems for general regular parametric families.

We now show some examples of distributions in k-parameter exponential families where $k > 1$.

Example 18.11 (Two-Parameter Normal Distribution). Suppose $X \sim N(\mu, \sigma^2)$, and we consider both μ, σ to be parameters. If we denote $(\mu, \sigma) = (\theta_1, \theta_2) = \theta$, then parametrized by θ, the density of X is

$$f(x\mid\theta)=\frac{1}{\sqrt{2\pi}\theta_2}e^{-\frac{(x-\theta_1)^2}{2\theta_2^2}}I_{x\in\mathcal{R}}=\frac{1}{\sqrt{2\pi}\theta_2}e^{-\frac{x^2}{2\theta_2^2}+\frac{\theta_1 x}{\theta_2^2}-\frac{\theta_1^2}{2\theta_2^2}}I_{x\in\mathcal{R}}.$$

This is in the two-parameter exponential family with

$$\eta_1(\theta)=-\frac{1}{2\theta_2^2},\quad \eta_2(\theta)=\frac{\theta_1}{\theta_2^2},\quad T_1(x)=x^2,\quad T_2(x)=x,$$

$$\psi(\theta)=\frac{\theta_1^2}{2\theta_2^2}+\log\theta_2,\quad h(x)=\frac{1}{\sqrt{2\pi}}I_{x\in\mathcal{R}}.$$

The parameter space in the θ parametrization is

$$\Theta=(-\infty,\infty)\otimes(0,\infty).$$

If we want the canonical form, we let

$$\eta_1=-\frac{1}{2\theta_2^2},\quad \eta_2=\frac{\theta_1}{\theta_2^2},\quad\text{and}\quad \psi(\eta)=-\frac{\eta_2^2}{4\eta_1}-\frac{1}{2}\log(-\eta_1).$$

The natural parameter space for (η_1,η_2) is $(-\infty,0)\otimes(-\infty,\infty)$.

Example 18.12 (Two-Parameter Gamma). It was seen in Example 18.5 that if we fix one of the two-parameters of a Gamma distribution, then it becomes a member of the one-parameter exponential family. We show in this example that the general Gamma distribution is a member of the two-parameter exponential family. To show this, just observe that with $\theta=(\alpha,\lambda)=(\theta_1,\theta_2)$,

$$f(x\mid\theta)=e^{-\frac{x}{\theta_2}+\theta_1\log x-\theta_1\log\theta_2-\log\Gamma(\theta_1)}\frac{1}{x}I_{x>0}.$$

This is in the two-parameter exponential family with $\eta_1(\theta)=-\frac{1}{\theta_2}, \eta_2(\theta)=\theta_1, T_1(x)=x, T_2(x)=\log x, \psi(\theta)=\theta_1\log\theta_2+\log\Gamma(\theta_1)$, and $h(x)=\frac{1}{x}I_{x>0}$. The parameter space in the θ-parametrization is $(0,\infty)\otimes(0,\infty)$. For the canonical form, use $\eta_1=-\frac{1}{\theta_2}, \eta_2=\theta_1$, and so, the natural parameter space is $(-\infty,0)\otimes(0,\infty)$. The natural sufficient statistic is $(X,\log X)$.

Example 18.13 (The General Multivariate Normal Distribution). Suppose $X\sim N_d(\mu,\Sigma)$, where μ is arbitrary and Σ is positive-definite (and of course, symmetric). Writing $\theta=(\mu,\Sigma)$, we can think of θ as a subset in an Euclidean space of dimension

$$k=d+d+\frac{d^2-d}{2}=d+\frac{d(d+1)}{2}=\frac{d(d+3)}{2}.$$

18.3 Multiparameter Exponential Family

The density of X is

$$f(x \mid \theta) = \frac{1}{(2\pi)^{d/2}|\Sigma|^{1/2}} e^{-\frac{1}{2}(x-\mu)'\Sigma^{-1}(x-\mu)} I_{x \in \mathcal{R}^d}$$

$$= \frac{1}{(2\pi)^{d/2}|\Sigma|^{1/2}} e^{-\frac{1}{2}x'\Sigma^{-1}x + \mu'\Sigma^{-1}x - \frac{1}{2}\mu'\Sigma^{-1}\mu} I_{x \in \mathcal{R}^d}$$

$$= \frac{1}{(2\pi)^{d/2}|\Sigma|^{1/2}} e^{-\frac{1}{2}\sum\sum_{i,j}\sigma^{ij}x_i x_j + \sum_i (\sum_k \sigma^{ki}\mu_k)x_i - \frac{1}{2}\mu'\Sigma^{-1}\mu} I_{x \in \mathcal{R}^d}$$

$$= \frac{1}{(2\pi)^{d/2}|\Sigma|^{1/2}} e^{-\frac{1}{2}\sum_i \sigma^{ii}x_i^2 - \sum\sum_{i<j}\sigma^{ij}x_i x_j + \sum_i (\sum_k \sigma^{ki}\mu_k)x_i}$$
$$-\frac{1}{2}\mu'\Sigma^{-1}\mu I_{x \in \mathcal{R}^d}.$$

We have thus represented the density of X in the k-parameter exponential family form with the k-dimensional natural sufficient statistic

$$T(X) = (X_1, \ldots, X_d, X_1^2, \ldots, X_d^2, X_1 X_2, \ldots, X_{d-1} X_d),$$

and the natural parameters defined by

$$\sum_k \sigma^{k1}\mu_k, \ldots, \sum_k \sigma^{kd}\mu_k, -\frac{1}{2}\sigma^{11}, \ldots, -\frac{1}{2}\sigma^{dd}, -\sigma^{12}, \ldots, -\sigma^{d-1,d}.$$

Example 18.14 (Multinomial Distribution). Consider the $k+1$ cell multinomial distribution with cell probabilities $p_1, p_2, \ldots, p_k, p_{k+1} = 1 - \sum_{i=1}^{k} p_i$. Writing $\theta = (p_1, p_2, \ldots, p_k)$, the joint pmf of $X = (X_1, X_2, \ldots, X_k)$, the cell frequencies of the first k cells, is

$$f(x \mid \theta) = \frac{n!}{(\prod_{i=1}^{k} x_i!)(n - \sum_{i=1}^{k} x_i)!} \prod_{i=1}^{k} p_i^{x_i} \left(1 - \sum_{i=1}^{k} p_i\right)^{n - \sum_{i=1}^{k} x_i}$$
$$\times I_{x_1, \ldots, x_k \geq 0, \sum_{i=1}^{k} x_i \leq n}$$

$$= \frac{n!}{(\prod_{i=1}^{k} x_i!)(n - \sum_{i=1}^{k} x_i)!} e^{\sum_{i=1}^{k} (\log p_i) x_i - \log(1 - \sum_{i=1}^{k} p_i)(\sum_{i=1}^{k} x_i)}$$
$$+ n \log(1 - \sum_{i=1}^{k} p_i) I_{x_1, \ldots, x_k \geq 0, \sum_{i=1}^{k} x_i \leq n}$$

$$= \frac{n!}{(\prod_{i=1}^{k} x_i!)(n - \sum_{i=1}^{k} x_i)!} e^{\sum_{i=1}^{k} \left(\log \frac{p_i}{1 - \sum_{i=1}^{k} p_i}\right) x_i + n \log\left(1 - \sum_{i=1}^{k} p_i\right)}$$
$$\times I_{x_1, \ldots, x_k \geq 0, \sum_{i=1}^{k} x_i \leq n}.$$

This is in the k-parameter exponential family form with the natural sufficient statistic and natural parameters

$$T(X) = (X_1, X_2, \ldots, X_k), \quad \eta_i = \log \frac{p_i}{1 - \sum_{i=1}^{k} p_i}, \quad 1 \leq i \leq k.$$

Example 18.15 (Two-parameter Inverse Gaussian Distribution). It was shown in Theorem 11.5 that for the simple symmetric random walk on $\mathcal{R}$, the time of the rth return to zero ν_r satisfies the weak convergence result

$$P\left(\frac{\nu_r}{r^2} \leq x\right) \to 2\left[1 - \Phi\left(\frac{1}{\sqrt{x}}\right)\right], \quad x > 0,$$

as $r \to \infty$. The density of this limiting CDF is $f(x) = \frac{e^{-\frac{1}{2x}} x^{-\frac{3}{2}}}{\sqrt{2\pi}} I_{x>0}$. This is a special *inverse Gaussian distribution*. The general inverse Gaussian distribution has the density

$$f(x \mid \theta_1, \theta_2) = \left(\frac{\theta_2}{\pi x^3}\right)^{1/2} e^{-\theta_1 x - \frac{\theta_2}{x} + 2\sqrt{\theta_1 \theta_2}} I_{x>0};$$

the parameter space for $\theta = (\theta_1, \theta_2)$ is $[0, \infty) \otimes (0, \infty)$. Note that the special inverse Gaussian density ascribed to the above corresponds to $\theta_1 = 0, \theta_2 = \frac{1}{2}$. The general inverse Gaussian density $f(x \mid \theta_1, \theta_2)$ is the density of the first time that a Wiener proces (starting at zero) hits the straight line with the equation $y = \sqrt{2\theta_2} - \sqrt{2\theta_1} t, t > 0$.

It is clear from the formula for $f(x \mid \theta_1, \theta_2)$ that it is a member of the two-parameter exponential family with the natural sufficient statistic $T(X) = (X, \frac{1}{X})$ and the natural parameter space $\mathcal{T} = (-\infty, 0] \otimes (-\infty, 0)$. Note that the natural parameter space is not open.

18.4 * Sufficiency and Completeness

Exponential families under mild conditions on the parameter space Θ have the property that if a function $g(T)$ of the natural sufficient statistic $T = T(X)$ has zero expected value under each $\theta \in \Theta$, then $g(T)$ itself must be essentially identically equal to zero. A family of distributions that has this property is called a *complete family*. The completeness property, particularly in conjunction with the property of sufficiency, has had an historically important role in statistical inference. Lehmann (1959), Lehmann and Casella (1998), and Brown (1986) give many applications. However, our motivation for studying the completeness of a full rank exponential family is primarily for presenting a well-known theorem in statistics, which actually is also a very effective and efficient tool for probabilists. This theorem, known as *Basu's theorem* (Basu (1955)), is an efficient tool for probabilists in minimizing clumsy distributional calculations. Completeness is required in order to state Basu's theorem.

18.4 Sufficiency and Completeness

Definition 18.7. A family of distributions $\{P_\theta, \theta \in \Theta\}$ on some sample space $\mathcal{X}$ is called *complete* if $E_{P_\theta}[g(X)] = 0$ for all $\theta \in \Theta$ implies that $P_\theta(g(X) = 0) = 1$ for all $\theta \in \Theta$.

It is useful to first see an example of a family that is not complete.

Example 18.16. Suppose $X \sim \text{Bin}(2, p)$, and the parameter p is $\frac{1}{4}$ or $\frac{3}{4}$. In the notation of the definition of completeness, Θ is the two-point set $\{\frac{1}{4}, \frac{3}{4}\}$. Consider the function g defined by

$$g(0) = g(2) = 3, \quad g(1) = -5.$$

Then,

$$E_p[g(X)] = g(0)(1-p)^2 + 2g(1)p(1-p) + g(2)p^2$$
$$= 16p^2 - 16p + 3 = 0, \quad \text{if } p = \frac{1}{4} \text{ or } \frac{3}{4}.$$

Therefore, we have exhibited a function g that violates the condition for completeness of this family of distributions.

Thus, completeness of a family of distributions is not universally true. The problem with the two-point parameter set in the above example is that it is too small. If the parameter space is richer, the family of binomial distributions for any fixed n is in fact complete. In fact, any distribution in the general k-parameter exponential family as a whole is a complete family, provided the set of parameter values is not too thin. Here is a general theorem.

Theorem 18.8. *Suppose a family of distributions $\mathcal{F} = \{P_\theta, \theta \in \Theta\}$ belongs to a k-parameter exponential family, and that the set Θ to which the parameter θ is known to belong has a nonempty interior. Then the family $\mathcal{F}$ is complete.*

The proof of this requires the use of properties of functions that are analytic on a domain in C^k, where C is the complex plane. We do not prove the theorem here; see Brown (1986, p. 43) for a proof. The nonempty interior assumption is protecting us from the set Θ being too small.

Example 18.17. Suppose $X \sim \text{Bin}(n, p)$, where n is fixed, and the set of possible values for p contains an interval (however small). Then, in the terminology of the theorem above, Θ has a nonempty interior. Therefore, such a family of binomial distributions is indeed complete. The only function $g(X)$ that satisfies $E_p[g(X)] = 0$, for all p in a set Θ that contains in it an interval, is the zero function $g(x) = 0$ for all $x = 0, 1, \ldots, n$. Contrast this with Example 18.16.

We require one more definition before we can state Basu's theorem.

Definition 18.8. Suppose X has a distribution P_θ belonging to a family $\mathcal{F} = \{P_\theta, \theta \in \Theta\}$. A statistic $S(X)$ is called $\mathcal{F}$-*ancillary* (or, simply *ancillary*), if for any set A, $P_\theta(S(X) \in A)$ does not depend on $\theta \in \Theta$, that is, if $S(X)$ has the same distribution under each $P_\theta \in \mathcal{F}$.

Example 18.18. Suppose X_1, X_2 are iid $N(\mu, 1)$, and μ belongs to some subset Θ of the real line. Let $S(X_1, X_2) = X_1 - X_2$. then, under any P_μ, $S(X_1, X_2) \sim N(0, 2)$, a fixed distribution that does not depend on μ. Thus, $S(X_1, X_2) = X_1 - X_2$ is ancillary, whatever the set of values of μ is.

Example 18.19. Suppose X_1, X_2 are iid $U[0, \theta]$, and θ belongs to some subset Θ of $(0, \infty)$. Let $S(X_1, X_2) = \frac{X_1}{X_2}$. We can write $S(X_1, X_2)$ as

$$S(X_1, X_2) \stackrel{\mathcal{L}}{=} \frac{\theta U_1}{\theta U_2} = \frac{U_1}{U_2},$$

where U_1, U_2 are iid $U[0, 1]$. Thus, under any P_θ, $S(X_1, X_2)$ is distributed as the ratio of two independent $U[0, 1]$ variables. This is a fixed distribution that does not depend on θ. Thus, $S(X_1, X_2) = \frac{X_1}{X_2}$ is ancillary, whatever the set of values of θ is.

Example 18.20. Suppose $X_1, X_2, \ldots, X_n$ are iid $N(\mu, 1)$, and μ belongs to some subset Θ of the real line. Let $S(X_1, \ldots, X_n) = \sum_{i=1}^{n}(X_i - \overline{X})^2$. We can write $S(X_1, \ldots, X_n)$ as

$$S(X_1, \ldots, X_n) \stackrel{\mathcal{L}}{=} \sum_{i=1}^{n}(\mu + Z_i - [\mu + \overline{Z}])^2 = \sum_{i=1}^{n}(Z_i - \overline{Z})^2,$$

where $Z_1, \ldots, Z_n$ are iid $N(0, 1)$. Thus, under any P_μ, $S(X_1, \ldots, X_n)$ has a fixed distribution, namely the distribution of $\sum_{i=1}^{n}(Z_i - \overline{Z})^2$ (actually, this is a χ^2_{n-1} distribution). Thus, $S(X_1, \ldots, X_n) = \sum_{i=1}^{n}(X_i - \overline{X})^2$ is ancillary, whatever the set of values of μ is.

Theorem 18.9 (Basu's Theorem for the Exponential Family). *In any k-parameter exponential family $\mathcal{F}$, with a parameter space Θ that has a nonempty interior, the natural sufficient statistic of the family $T(X)$ and any $\mathcal{F}$-ancillary statistic $S(X)$ are independently distributed under each $\theta \in \Theta$.*

We show applications of this result following the next section.

18.4.1 ∗ Neyman–Fisher Factorization and Basu's Theorem

There is a more general version of Basu's theorem that applies to arbitrary parametric families of distributions. The intuition is the same as it was in the case of an exponential family, namely, a *sufficient statistic*, which contains all the information, and an *ancillary statistic*, which contains no information, must be independent. For this, we need to define what a sufficient statistic means for a general parametric family. Here is Fisher's original definition (Fisher (1922)).

Definition 18.9. Let $n \geq 1$ be given, and suppose $X = (X_1, \ldots, X_n)$ has a joint distribution $P_{\theta, n}$ belonging to some family

$$\mathcal{F}_n = \{P_{\theta, n} : \theta \in \Theta\}.$$

18.4 Sufficiency and Completeness

A statistic $T(X) = T(X_1, \ldots, X_n)$ taking values in some Euclidean space is called *sufficient* for the family $\mathcal{F}_n$ if the joint conditional distribution of $X_1, \ldots, X_n$ given $T(X_1, \ldots, X_n)$ is the same under each $\theta \in \Theta$.

Thus, we can interpret the sufficient statistic $T(X_1, \ldots, X_n)$ in the following way: once we know the value of T, the set of individual data values $X_1, \ldots, X_n$ has nothing more to convey about θ. We can think of sufficiency as data reduction at no cost; we can save only T and discard the individual data values without losing any information. However, what is sufficient depends, often crucially, on the functional form of the distributions $P_{\theta,n}$. Thus, sufficiency is useful for data reduction subject to loyalty to the chosen functional form of $P_{\theta,n}$.

Fortunately, there is an easily applicable universal recipe for automatically identifying a sufficient statistic for a given family $\mathcal{F}_n$. This is the *factorization theorem*.

Theorem 18.10 (**Neyman–Fisher Factorization Theorem**). *Let $f(x_1, \ldots, x_n \mid \theta)$ be the joint density function (joint pmf) corresponding to the distribution $P_{\theta,n}$. Then, a statistic $T = T(X_1, \ldots, X_n)$ is sufficient for the family $\mathcal{F}_n$ if and only if for any $\theta \in \Theta$, $f(x_1, \ldots, x_n \mid \theta)$ can be factorized in the form*

$$f(x_1, \ldots, x_n \mid \theta) = g(\theta, T(X_1, \ldots, X_n)) h(x_1, \ldots, x_n).$$

See Bickel and Doksum (2006) for a proof.

The intuition of the factorization theorem is that the only way that the parameter θ is tied to the data values $X_1, \ldots, X_n$ in the likelihood function $f(x_1, \ldots, x_n \mid \theta)$ is via the statistic $T(X_1, \ldots, X_n)$, because there is no θ in the function $h(x_1, \ldots, x_n)$. Therefore, we should only care to know what T is, but not the individual values $X_1, \ldots, X_n$.

Here is one example on using the factorization theorem.

Example 18.21 (*Sufficient Statistic for a Uniform Distribution*). Suppose $X_1, \ldots, X_n$ are iid and distributed as $U[0, \theta]$ for some $\theta > 0$. Then, the likelihood function is

$$f(x_1, \ldots, x_n \mid \theta) = \prod_{i=1}^{n} \frac{1}{\theta} I_{\theta \geq x_i} = \left(\frac{1}{\theta}\right)^n \prod_{i=1}^{n} I_{\theta \geq x_i}$$

$$= \left(\frac{1}{\theta}\right)^n I_{\theta \geq x_{(n)}},$$

where $x_{(n)} = \max(x_1, \ldots, x_n)$. If we let

$$T(X_1, \ldots, X_n) = X_{(n)}, \quad g(\theta, t) = \left(\frac{1}{\theta}\right)^n I_{\theta \geq t}, \, h(x_1, \ldots, x_n) \equiv 1,$$

then, by the factorization theorem, the sample maximum $X_{(n)}$ is sufficient for the $U[0, \theta]$ family. The result does make some intuitive sense.

Here is now the general version of Basu's theorem.

Theorem 18.11 (**General Basu Theorem**). *Let $\mathcal{F}_n = \{P_{\theta,n} : \theta \in \Theta\}$ be a family of distributions. Suppose $T(X_1, \ldots, X_n)$ is sufficient for $\mathcal{F}_n$, and $S(X_1, \ldots, X_n)$ is ancillary under $\mathcal{F}_n$. Then T and S are independently distributed under each $P_{\theta,n} \in \mathcal{F}_n$.*

See Basu (1955) for a proof.

18.4.2 * Applications of Basu's Theorem to Probability

We had previously commented that the sufficient statistic by itself captures all of the information about θ that the full knowledge of X could have provided. On the other hand, an ancillary statistic cannot provide any information about θ, because its distribution does not even involve θ. Basu's theorem says that a statistic which provides all the information, and another that provides no information, must be independent, provided the additional nonempty interior condition holds, in order to ensure completeness of the family $\mathcal{F}$. Thus, the concepts of information, sufficiency, ancillarity, completeness, and independence come together in Basu's theorem. However, our main interest is simply to use Basu's theorem as a convenient tool to arrive quickly at some results that are purely results in the domain of probability. Here are a few such examples.

Example 18.22 (*Independence of Mean and Variance for a Normal Sample*). Suppose $X_1, X_2, \ldots, X_n$ are iid $N(\eta, \tau^2)$ for some η, τ. It was stated in Chapter 4 that the sample mean $\overline{X}$ and the sample variance s^2 are independently distributed for any n, and whatever η and τ are. We now prove it. For this, first we establish the claim that if the result holds for $\eta = 0, \tau = 1$, then it holds for all η, τ. Indeed, fix any η, τ, and write $X_i = \eta + \tau Z_i, 1 \leq i \leq n$, where $Z_1, \ldots, Z_n$ are iid $N(0, 1)$. Now,

$$\left(\overline{X}, \sum_{i=1}^n (X_i - \overline{X})^2\right) \stackrel{\mathcal{L}}{=} \left(\eta + \tau \overline{Z}, \tau^2 \sum_{i=1}^n (Z_i - \overline{Z})^2\right).$$

Therefore, $\overline{X}$ and $\sum_{i=1}^n (X_i - \overline{X})^2$ are independently distributed under (η, τ) if and only if $\overline{Z}$ and $\sum_{i=1}^n (Z_i - \overline{Z})^2$ are independently distributed. This is a step in getting rid of the parameters η, τ from consideration.

But, now, we import a parameter! Embed the $N(0, 1)$ distribution into a larger family of $\{N(\mu, 1), \mu \in \mathcal{R}\}$ distributions. Consider now a fictitious sample $Y_1, Y_2, \ldots, Y_n$ from $P_\mu = N(\mu, 1)$. The joint density of $Y = (Y_1, Y_2, \ldots, Y_n)$ is a one-parameter exponential family density with the natural sufficient statistic $T(Y) = \sum_{i=1}^n Y_i$. By Example 18.20, $\sum_{i=1}^n (Y_i - \overline{Y})^2$ is ancillary. The parameter space for μ obviously has a nonempty interior, thus all the conditions of Basu's theorem are satisfied, and therefore, under each μ, $\sum_{i=1}^n Y_i$ and $\sum_{i=1}^n (Y_i - \overline{Y})^2$ are independently distributed. In particular, they are independently distributed under $\mu = 0$, that is, when the samples are iid $N(0, 1)$, which is what we needed to prove.

18.4 Sufficiency and Completeness

Example 18.23 (An Exponential Distribution Result). Suppose $X_1, X_2, \ldots, X_n$ are iid exponential random variables with mean λ. Then, by transforming $(X_1, X_2, \ldots, X_n)$ to

$$\left(\frac{X_1}{X_1 + \cdots + X_n}, \ldots, \frac{X_{n-1}}{X_1 + \cdots + X_n}, X_1 + \cdots + X_n\right),$$

one can show by carrying out the necessary Jacobian calculation (see Chapter 4), that

$$\left(\frac{X_1}{X_1 + \cdots + X_n}, \ldots, \frac{X_{n-1}}{X_1 + \cdots + X_n}\right)$$

is independent of $X_1 + \cdots + X_n$. We can show this without doing any calculations by using Basu's theorem.

For this, once again, by writing $X_i = \lambda Z_i, 1 \leq i \leq n$, where the Z_i are iid standard exponentials, first observe that

$$\left(\frac{X_1}{X_1 + \cdots + X_n}, \ldots, \frac{X_{n-1}}{X_1 + \cdots + X_n}\right)$$

is a (vector) ancillary statistic. Next observe that the joint density of $X = (X_1, X_2, \ldots, X_n)$ is a one-parameter exponential family, with the natural sufficient statistic $T(X) = X_1 + \cdots + X_n$. Because the parameter space $(0, \infty)$ obviously contains a nonempty interior, by Basu's theorem, under each λ,

$$\left(\frac{X_1}{X_1 + \cdots + X_n}, \ldots, \frac{X_{n-1}}{X_1 + \cdots + X_n}\right) \quad \text{and} \quad X_1 + \cdots + X_n$$

is independently distributed.

Example 18.24 (A Covariance Calculation). Suppose $X_1, \ldots, X_n$ are iid $N(0, 1)$, and let $\overline{X}$ and M_n denote the mean and the median of the sample set $X_1, \ldots, X_n$. By using our old trick of importing a mean parameter μ, we first observe that the difference statistic $\overline{X} - M_n$ is ancillary. On the other hand, the joint density of $X = (X_1, \ldots, X_n)$ is of course a one-parameter exponential family with the natural sufficient statistic $T(X) = X_1 + \cdots + X_n$. By Basu's theorem, $X_1 + \cdots + X_n$ and $\overline{X} - M_n$ are independent under each μ, which implies

$$\text{Cov}(X_1 + \cdots + X_n, \overline{X} - M_n) = 0 \Rightarrow \text{Cov}(\overline{X}, \overline{X} - M_n) = 0$$
$$\Rightarrow \text{Cov}(\overline{X}, M_n) = \text{Cov}(\overline{X}, \overline{X}) = \text{Var}(\overline{X}) = \frac{1}{n}.$$

We have achieved this result without doing any calculations at all. A direct attack on this problem requires handling the joint distribution of $(\overline{X}, M_n)$.

Example 18.25 (An Expectation Calculation). Suppose $X_1, \ldots, X_n$ are iid $U[0, 1]$, and let $X_{(1)}, X_{(n)}$ denote the smallest and the largest order statistic of $X_1, \ldots, X_n$. Import a parameter $\theta > 0$, and consider the family of $U[0, \theta]$ distributions. We have shown that the largest order statistic $X_{(n)}$ is sufficient; it is also complete. On the other hand, the quotient $\frac{X_{(1)}}{X_{(n)}}$ is ancillary. To see this, again, write $(X_1, \ldots, X_n) \stackrel{\mathcal{L}}{=} (\theta U_1, \ldots, \theta U_n)$, where $U_1, \ldots, U_n$ are iid $U[0, 1]$. As a consequence, $\frac{X_{(1)}}{X_{(n)}} \stackrel{\mathcal{L}}{=} \frac{U_{(1)}}{U_{(n)}}$. So, $\frac{X_{(1)}}{X_{(n)}}$ is ancillary. By the general version of Basu's theorem which works for any family of distributions (not just an exponential family), it follows that $X_{(n)}$ and $\frac{X_{(1)}}{X_{(n)}}$ are independently distributed under each θ. Hence,

$$E[X_{(1)}] = E\left[\frac{X_{(1)}}{X_{(n)}} X_{(n)}\right] = E\left[\frac{X_{(1)}}{X_{(n)}}\right] E[X_{(n)}]$$

$$\Rightarrow E\left[\frac{X_{(1)}}{X_{(n)}}\right] = \frac{E[X_{(1)}]}{E[X_{(n)}]} = \frac{\frac{\theta}{n+1}}{\frac{n\theta}{n+1}} = \frac{1}{n}.$$

Once again, we can get this result by using Basu's theorem without doing any integrations or calculations at all.

Example 18.26 (A Weak Convergence Result Using Basu's Theorem). Suppose $X_1, X_2, \ldots$ are iid random vectors distributed as a uniform in the d-dimensional unit ball. For $n \geq 1$, let $d_n = \min_{1 \leq i \leq n} ||X_i||$, and $D_n = \max_{1 \leq i \leq n} ||X_i||$. Thus, d_n measures the distance to the closest data point from the center of the ball, and D_n measures the distance to the farthest data point. We find the limiting distribution of $\rho_n = d_n/D_n$. Although this can be done by using other means, we do so by an application of Basu's theorem.

Toward this, note that for $0 \leq u \leq 1$,

$$P(d_n > u) = (1 - u^d)^n; \quad P(D_n > u) = 1 - u^{nd}.$$

As a consequence, for any $k \geq 1$,

$$E[D_n]^k = \int_0^1 k u^{k-1}(1 - u^{nd}) du = \frac{nd}{nd + k},$$

and,

$$E[d_n]^k = \int_0^1 k u^{k-1}(1 - u^d)^n du = \frac{n!\Gamma(\frac{k}{d} + 1)}{\Gamma(n + \frac{k}{d} + 1)}.$$

Now, embed the uniform distribution in the unit ball into the family of uniform distributions in balls of radius θ and centered at the origin. Then, D_n is complete and sufficient (akin to Example 18.24), and ρ_n is ancillary. Therefore, once again, by the general version of Basu's theorem, D_n and ρ_n are independently distributed under each $\theta > 0$, and so, in particular under $\theta = 1$. Thus, for any $k \geq 1$,

$$E[d_n]^k = E[D_n \rho_n]^k = E[D_n]^k E[\rho_n]^k$$

$$\Rightarrow E[\rho_n]^k = \frac{E[d_n]^k}{E[D_n]^k} = \frac{n!\Gamma(\frac{k}{d}+1)}{\Gamma(n+\frac{k}{d}+1)} \frac{nd+k}{nd}$$

$$\sim \frac{\Gamma(\frac{k}{d}+1) e^{-n} n^{n+1/2}}{e^{-n-k/d}(n+\frac{k}{d})^{n+\frac{k}{d}+1/2}}$$

(by using Stirling's approximation)

$$\sim \frac{\Gamma\left(\frac{k}{d}+1\right)}{n^{\frac{k}{d}}}.$$

Thus, for each $k \geq 1$,

$$E\left[n^{1/d} \rho_n\right]^k \to \Gamma\left(\frac{k}{d}+1\right) = E[V]^{k/d} = E[V^{1/d}]^k,$$

where V is a standard exponential random variable. This implies, because $V^{1/d}$ is uniquely determined by its moment sequence, that

$$n^{1/d} \rho_n \overset{\mathcal{L}}{\Rightarrow} V^{1/d},$$

as $n \to \infty$.

18.5 Curved Exponential Family

There are some important examples in which the density (pmf) has the basic exponential family form $f(x|\theta) = e^{\sum_{i=1}^{k} \eta_i(\theta) T_i(X) - \psi(\theta)} h(x)$, but the assumption that the dimensions of Θ, and that of the range space of $(\eta_1(\theta), \ldots, \eta_k(\theta))$ are the same is violated, more precisely, the dimension of Θ is some positive integer q strictly less than k. Let us start with an example.

Example 18.27. Suppose $X \sim N(\mu, \mu^2)$, $\mu \neq 0$. Writing $\mu = \theta$, the density of X is

$$f(x|\theta) = \frac{1}{\sqrt{2\pi}|\theta|} e^{-\frac{1}{2\theta^2}(x-\theta)^2} I_{x \in \mathcal{R}}$$

$$= \frac{1}{\sqrt{2\pi}} e^{-\frac{x^2}{2\theta^2} + \frac{x}{\theta} - \frac{1}{2} - \log|\theta|} I_{x \in \mathcal{R}}.$$

Writing $\eta_1(\theta) = -\frac{1}{2\theta^2}$, $\eta_2(\theta) = \frac{1}{\theta}$, $T_1(x) = x^2$, $T_2(x) = x$, $\psi(\theta) = \frac{1}{2} + \log|\theta|$, and $h(x) = \frac{1}{\sqrt{2\pi}} I_{x \in \mathcal{R}}$, this is in the form $f(x|\theta) = e^{\sum_{i=1}^{k} \eta_i(\theta) T_i(x) - \psi(\theta)} h(x)$, with $k = 2$, although $\theta \in \mathcal{R}$, which is only one-dimensional. The two functions

$\eta_1(\theta) = -\frac{1}{2\theta^2}$ and $\eta_2(\theta) = \frac{1}{\theta}$ are related to each other by the identity $\eta_1 = -\frac{\eta_2^2}{2}$, so that a plot of (η_1, η_2) in the plane would be a curve, not a straight line. Distributions of this kind go by the name of *curved exponential family*. The dimension of the natural sufficient statistic is more than the dimension of Θ for such distributions.

Definition 18.10. Let $X = (X_1, \ldots, X_d)$ have a distribution $P_\theta, \theta \in \Theta \subseteq \mathcal{R}^q$. Suppose P_θ has a density (pmf) of the form

$$f(x \mid \theta) = e^{\sum_{i=1}^{k} \eta_i(\theta) T_i(x) - \psi(\theta)} h(x),$$

where $k > q$. Then, the family $\{P_\theta, \theta \in \Theta\}$ is called a *curved exponential family*.

Example 18.28 (A Specific Bivariate Normal). Suppose $X = (X_1, X_2)$ has a bivariate normal distribution with zero means, standard deviations equal to one, and a correlation parameter $\rho, -1 < \rho < 1$. The density of X is

$$f(x \mid \rho) = \frac{1}{2\pi\sqrt{1-\rho^2}} e^{-\frac{1}{2(1-\rho^2)}[x_1^2 + x_2^2 - 2\rho x_1 x_2]} I_{x_1, x_2 \in \mathcal{R}}$$

$$= \frac{1}{2\pi\sqrt{1-\rho^2}} e^{-\frac{x_1^2 + x_2^2}{2(1-\rho^2)} + \frac{\rho}{1-\rho^2} x_1 x_2} I_{x_1, x_2 \in \mathcal{R}}.$$

Therefore, here we have a curved exponential family with $q = 1, k = 2, \eta_1(\rho) = -\frac{1}{2(1-\rho^2)}, \eta_2(\rho) = \frac{\rho}{1-\rho^2}, T_1(x) = x_1^2 + x_2^2, T_2(x) = x_1 x_2, \psi(\rho) = \frac{1}{2}\log(1-\rho^2)$, and $h(x) = \frac{1}{2\pi} I_{x_1, x_2 \in \mathcal{R}}$.

Example 18.29 (Poissons with Random Covariates). Suppose given $Z_i = z_i, i = 1, 2, \ldots, n, X_i$ are independent Poi(λz_i) variables, and $Z_1, Z_2, \ldots, Z_n$ have some joint pmf $p(z_1, z_2, \ldots, z_n)$. It is implicitly assumed that each $Z_i > 0$ with probability one. Then, the joint pmf of $(X_1, X_2, \ldots, X_n, Z_1, Z_2, \ldots, Z_n)$ is

$$f(x_1, \ldots, x_n, z_1, \ldots, z_n \mid \lambda) = \prod_{i=1}^{n} \frac{e^{-\lambda z_i}(\lambda z_i)^{x_i}}{x_i!} p(z_1, z_2, \ldots, z_n) I_{x_1, \ldots, x_n \in \mathcal{N}_0}$$

$$I_{z_1, z_2, \ldots, z_n \in \mathcal{N}_1}$$

$$= e^{-\lambda \sum_{i=1}^{n} z_i + (\sum_{i=1}^{n} x_i) \log \lambda} \prod_{i=1}^{n} \frac{z_i^{x_i}}{x_i!} p(z_1, z_2, \ldots, z_n)$$

$$I_{x_1, \ldots, x_n \in \mathcal{N}_0} I_{z_1, z_2, \ldots, z_n \in \mathcal{N}_1},$$

where $\mathcal{N}_0$ is the set of nonnegative integers, and $\mathcal{N}_1$ is the set of positive integers. This is in the curved exponential family with

$$q = 1, \quad k = 2, \quad \eta_1(\lambda) = -\lambda, \quad \eta_2(\lambda) = \log \lambda, \quad T_1(x, z) = \sum_{i=1}^{n} z_i,$$

$$T_2(x, z) = \sum_{i=1}^{n} x_i,$$

and
$$h(x,z) = \prod_{i=1}^{n} \frac{z_i^{x_i}}{x_i!} p(z_1, z_2, \ldots, z_n) I_{x_1,\ldots,x_n \in \mathcal{N}_0} I_{z_1,z_2,\ldots,z_n \in \mathcal{N}_1}.$$

If we consider the covariates as fixed, the joint distribution of $(X_1, X_2, \ldots, X_n)$ becomes a regular one-parameter exponential family.

Exercises

Exercise 18.1. Show that the geometric distribution belongs to the one-parameter exponential family if $0 < p < 1$, and write it in the canonical form and by using the mean parametrization.

Exercise 18.2 (Poisson Distribution). Show that the Poisson distribution belongs to the one-parameter exponential family if $\lambda > 0$. Write it in the canonical form and by using the mean parametrization.

Exercise 18.3 (Negative Binomial Distribution). Show that the negative binomial distribution with parameters r and p belongs to the one-parameter exponential family if r is considered fixed and $0 < p < 1$. Write it in the canonical form and by using the mean parametrization.

Exercise 18.4 * (Generalized Negative Binomial Distribution). Show that the generalized negative binomial distribution with the pmf $f(x|p) = \frac{\Gamma(\alpha+x)}{\Gamma(\alpha)x!} p^\alpha (1-p)^x, x = 0, 1, 2, \ldots$ belongs to the one-parameter exponential family if $\alpha > 0$ is considered fixed and $0 < p < 1$.

Show that the two-parameter generalized negative binomial distribution with the pmf $f(x|\alpha, p) = \frac{\Gamma(\alpha+x)}{\Gamma(\alpha)x!} p^\alpha (1-p)^x, x = 0, 1, 2, \ldots$ does not belong to the two-parameter exponential family.

Exercise 18.5 * (Normal with Equal Mean and Variance). Show that the $N(\mu, \mu)$ distribution belongs to the one-parameter exponential family if $\mu > 0$. Write it in the canonical form and by using the mean parametrization.

Exercise 18.6 * (Hardy–Weinberg Law). Suppose genotypes at a single locus with two alleles are present in a population according to the relative frequencies $p^2, 2pq$, and q^2, where $q = 1 - p$, and p is the relative frequency of the dominant allele. Show that the joint distribution of the frequencies of the three genotypes in a random sample of n individuals from this population belongs to a one-parameter exponential family if $0 < p < 1$. Write it in the canonical form and by using the mean parametrization.

Exercise 18.7 (Beta Distribution). Show that the two-parameter Beta distribution belongs to the two-parameter exponential family if the parameters $\alpha, \beta > 0$. Write it in the canonical form and by using the mean parametrization.

Show that symmetric Beta distributions belong to the one-parameter exponential family if the single parameter $\alpha > 0$.

Exercise 18.8 * **(Poisson Skewness and Kurtosis).** Find the skewness and kurtosis of a Poisson distribution by using Theorem 18.3.

Exercise 18.9 * **(Gamma Skewness and Kurtosis).** Find the skewness and kurtosis of a Gamma distribution, considering α as fixed, by using Theorem 18.3.

Exercise 18.10 * **(Distributions with Zero Skewness).** Show that the only distributions in a canonical one-parameter exponential family such that the natural sufficient statistic has a zero skewness are the normal distributions with a fixed variance.

Exercise 18.11 * **(Identifiability of the Distribution).** Show that distributions in the nonsingular canonical one-parameter exponential family are identifiable; that is, $P_{\eta_1} = P_{\eta_2}$ only if $\eta_1 = \eta_2$.

Exercise 18.12 * **(Infinite Differentiability of Mean Functionals).** Suppose P_θ, $\theta \in \Theta$ is a one-parameter exponential family and $\phi(x)$ is a general function. Show that at any $\theta \in \Theta^0$ at which $E_\theta[|\phi(X)|] < \infty$, $\mu_\phi(\theta) = E_\theta[\phi(X)]$ is infinitely differentiable, and can be differentiated any number of times inside the integral (sum).

Exercise 18.13 * **(Normalizing Constant Determines the Distribution).** Consider a canonical one-parameter exponential family density (pmf) $f(x|\theta) = e^{\eta x - \psi(\eta)} h(x)$. Assume that the natural parameter space $\mathcal{T}$ has a nonempty interior. Show that $\psi(\eta)$ determines $h(x)$.

Exercise 18.14. Calculate the mgf of a $(k+1)$ cell multinomial distribution by using Theorem 18.7.

Exercise 18.15 * **(Multinomial Covariances).** Calculate the covariances in a multinomial distribution by using Theorem 18.7.

Exercise 18.16 * **(Dirichlet Distribution).** Show that the Dirichlet distribution defined in Chapter 4, with parameter vector $\alpha = (\alpha_1, \ldots, \alpha_{n+1})$, $\alpha_i > 0$ for all i, is an $(n+1)$-parameter exponential family.

Exercise 18.17 * **(Normal Linear Model).** Suppose given an $n \times p$ nonrandom matrix X, a parameter vector $\beta \in \mathcal{R}^p$, and a variance parameter $\sigma^2 > 0$, $Y = (Y_1, Y_2, \ldots, Y_n) \sim N_n(X\beta, \sigma^2 I_n)$, where I_n is the $n \times n$ identity matrix. Show that the distribution of Y belongs to a full rank multiparameter exponential family.

Exercise 18.18 (Fisher Information Matrix). For each of the following distributions, calculate the Fisher information matrix:

(a) Two-parameter Beta distribution.
(b) Two-parameter Gamma distribution.

(c) Two-parameter inverse Gaussian distribution.

(d) Two-parameter normal distribution.

Exercise 18.19 *(Normal with an Integer Mean). Suppose $X \sim N(\mu, 1)$, where $\mu \in \{1, 2, 3, \ldots\}$. Is this a regular one-parameter exponential family?

Exercise 18.20 *(Normal with an Irrational Mean). Suppose $X \sim N(\mu, 1)$, where μ is known to be an irrational number. Is this a regular one-parameter exponential family?

Exercise 18.21 *(Normal with an Integer Mean). Suppose $X \sim N(\mu, 1)$, where $\mu \in \{1, 2, 3, \ldots\}$. Exhibit a function $g(X) \not\equiv 0$ such that $E_\mu[g(X)] = 0$ for all μ.

Exercise 18.22 (Application of Basu's Theorem). Suppose $X_1, \ldots, X_n$ is an iid sample from a standard normal distribution, and suppose $X_{(1)}, X_{(n)}$ are the smallest and the largest order statistics of $X_1, \ldots, X_n$, and s^2 is the sample variance. Prove, by applying Basu's theorem to a suitable two-parameter exponential family, that

$$E\left[\frac{X_{(n)} - X_{(1)}}{s}\right] = 2\frac{E[X_{(n)}]}{E(s)}.$$

Exercise 18.23 (Mahalanobis's D^2 and Basu's Theorem). Suppose $X_1, \ldots, X_n$ is an iid sample from a d-dimensional normal distribution $N_d(0, \Sigma)$, where Σ is positive definite. Suppose S is the sample covariance matrix (see Chapter 5) and $\overline{X}$ the sample mean vector. The statistic $D_n^2 = n\overline{X}'S^{-1}\overline{X}$ is called the *Mahalanobis D^2-statistic*. Find $E(D_n^2)$ by using Basu's theorem.

Hint: Look at Example 18.13, and Theorem 5.10.

Exercise 18.24 (Application of Basu's Theorem). Suppose $X_i, 1 \leq i \leq n$ are iid $N(\mu_1, \sigma_1^2)$, $Y_i, 1 \leq i \leq n$ are iid $N(\mu_2, \sigma_2^2)$, where $\mu_1, \mu_2 \in \mathcal{R}$, and $\sigma_1^2, \sigma_2^2 > 0$. Let $\overline{X}, s_1^2$ denote the mean and the variance of $X_1, \ldots, X_n$, and $\overline{Y}, s_2^2$ denote the mean and the variance of $Y_1, \ldots, Y_n$. Also let r denote the sample correlation coefficient based on the pairs $(X_i, Y_i), 1 \leq i \leq n$. Prove that $\overline{X}, \overline{Y}, s_1^2, s_2^2, r$ are mutually independent under all $\mu_1, \mu_2, \sigma_1, \sigma_2$.

Exercise 18.25 (Mixtures of Normal). Show that the mixture distribution $.5N(\mu, 1) + .5N(\mu, 2)$ does not belong to the one-parameter exponential family. Generalize this result to more general mixtures of normal distributions.

Exercise 18.26 (Double Exponential Distribution). (a) Show that the double exponential distribution with a known σ value and an unknown mean does not belong to the one-parameter exponential family, but the double exponential distribution with a known mean and an unknown σ belongs to the one-parameter exponential family.

(b) Show that the two-parameter double exponential distribution does not belong to the two-parameter exponential family.

Exercise 18.27 *(A Curved Exponential Family).** Suppose $X \sim \text{Bin}(n, p), Y \sim \text{Bin}(m, p^2)$, and that X, Y are independent. Show that the distribution of (X, Y) is a curved exponential family.

Exercise 18.28 (Equicorrelation Multivariate Normal). Suppose $(X_1, X_2, \ldots, X_n)$ are jointly multivariate normal with general means μ_i, variances all one, and a common pairwise correlation ρ. Show that the distribution of $(X_1, X_2, \ldots, X_n)$ is a curved exponential family.

Exercise 18.29 (Poissons with Covariates). Suppose $X_1, X_2, \ldots, X_n$ are independent Poissons with $E(X_i) = \lambda e^{\beta z_i}, \lambda > 0, -\infty < \beta < \infty$. The covariates $z_1, z_2, \ldots, z_n$ are considered fixed. Show that the distribution of $(X_1, X_2, \ldots, X_n)$ is a curved exponential family.

Exercise 18.30 (Incomplete Sufficient Statistic). Suppose $X_1, \ldots, X_n$ are iid $N(\mu, \mu^2), \mu \neq 0$. Let $T(X_1, \ldots, X_n) = (\sum_{i=1}^n X_i, \sum_{i=1}^n X_i^2)$. Find a function $g(T)$ such that $E_\mu[g(T)] = 0$ for all μ, but $P_\mu(g(T) = 0) < 1$ for any μ.

Exercise 18.31 *(Quadratic Exponential Family).** Suppose the natural sufficient statistic $T(X)$ in some canonical one-parameter exponential family is X itself. By using the formula in Theorem 18.3 for the mean and the variance of the natural sufficient statistic in a canonical one-parameter exponential family, characterize all the functions $\psi(\eta)$ for which the variance of $T(X) = X$ is a quadratic function of the mean of $T(X)$, that is, $\text{Var}_\eta(X) \equiv a[E_\eta(X)]^2 + bE_\eta(X) + c$ for some constants a, b, c.

Exercise 18.32 (Quadratic Exponential Family). Exhibit explicit examples of canonical one-parameter exponential families which are quadratic exponential families.

Hint: There are six of them, and some of them are common distributions, but not all. See Morris (1982), Brown (1986).

References

Barndorff-Nielsen, O. (1978). *Information and Exponential Families in Statistical Theory*, Wiley, New York.
Basu, D. (1955). On statistics independent of a complete sufficient statistic, *Sankhyā*, 15, 377–380.
Bickel, P.J. and Doksum, K. (2006). *Mathematical Statistics, Basic Ideas and Selected Topics*, Vol I, Prentice Hall, UpperSaddle River, NJ.
Brown, L.D. (1986). *Fundamentals of Statistical Exponential Families*, IMS, Lecture Notes and Monographs Series, Hayward, CA.
Fisher, R.A. (1922). On the mathematical foundations of theoretical statistics, *Philos. Trans. Royal Soc. London, Ser. A*, 222, 309–368.
Lehmann, E.L. (1959). *Testing Statistical Hypotheses*, Wiley, New York.
Lehmann, E. L. and Casella, G. (1998). *Theory of Point Estimation*, Springer, New York.
Morris, C. (1982). Natural Exponential families with quadratic variance functions, *Ann. Statist.* 10, 65–80.

Chapter 19
Simulation and Markov Chain Monte Carlo

Simulation is a computer-based exploratory exercise that aids in understanding how the behavior of a random or even a deterministic process changes in response to changes in input or the environment. It is essentially the only option left when exact mathematical calculations are impossible, or require an amount of effort that the user is not willing to invest. Even when the mathematical calculations are quite doable, a preliminary simulation can be very helpful in guiding the researcher to theorems that were not a priori obvious or conjectured, and also to identify the more productive corners of a particular problem. Although simulation in itself is a machine-based exercise, credible simulation must be based on appropriate theory. A simulation algorithm must be theoretically justified before we use it. This chapter gives a fairly broad introduction to the classic theory and techniques of probabilistic simulation, and also to some of the modern advents in simulation, particularly Markov chain Monte Carlo (MCMC) methods based on ergodic Markov chain theory.

The classic theory of simulation includes such time-tested methods as the original Monte Carlo, and textbook techniques of simulation from standard distributions in common use. They involve a varied degree of sophistication. *Markov chain Monte Carlo* is the name for a collection of simulation algorithms for simulating from the distribution of very general types of random variables taking values in quite general spaces. MCMC methods have truly revolutionized simulation because of an inherent simplicity in applying them, the generality of their scopes, and the diversity of applied problems in which some suitable form of MCMC has helped in making useful practical advances. MCMC methods are the most useful when conventional Monte Carlo is difficult or impossible to use. This happens to be the case when one has a complicated distribution in a high-dimensional space, or the state space of the underlying random variable is exotic or huge, or the distribution to simulate from, called the target distribution, is known only up to a normalizing constant, and this normalizing constant cannot be computed. The normalizing constant issue is common when the target distribution is the posterior distribution of an unknown parameter when the prior is not a simple one. The problem of huge state spaces arises in numerous problems in image processing and statistical physics. Indeed, MCMC methods originated in statistical physics, and only gradually made their way into probability and statistics.

The principle underlying all MCMC methods is extremely simple. Given a target distribution π on some state space S from which one wants to simulate, one simply constructs a Markov chain $X_n, n = 0, 1, 2, \ldots$, such that X_n has a unique stationary distribution, and that stationary distribution is π. So, if we simply generate successive states $X_0, X_1, \ldots, X_B, X_{B+1}, \ldots, X_n$ for some suitable large values of B, n, then we can act as if $X_{B+1}, \ldots, X_n$ is a dependent sample of size $n - B$ from π. We can then use these $n - B$ values to approximate probabilities, to approximate expectations, or for whatever reason we want to use them. The practical part is to devise a Markov chain that has π as its stationary distribution, and it must be reasonably easy to run this chain as a matter of convenient implementation.

There are a number of popular MCMC methods in use. These include the Hastings algorithm, the Metropolis algorithm, and the Gibbs sampler. Implementing these algorithms is usually a simple matter, and computer codes for many of them are now publicly available. The difficult part in justifying MCMC methods is answering the question of how long to let the chain run so as to allow it to get close to stationarity. This is critical, because the stationary distribution of the chain is the target distribution. In principle, the speed of convergence of the chain to stationarity is exponential under mild conditions on the chain. General Markov chain theory implies such exponential convergence. However, concrete statements on the exact exponential speed, or asserting theoretical bounds on the closeness of the chain to stationarity can usually be done only on a case-by-case basis, and typically each new case needs a new creative technique.

We start with a treatment of conventional Monte Carlo and textbook simulation techniques, and then provide the most popular MCMC algorithms with a selection of applications. We then provide some of the available general theory on the question of speed of convergence of a particular chain to stationarity. This is linked to choosing the number B, the burn-in period, and the total run length n. These convergence theorems form the most appealing theoretical aspect of MCMC methods. They are also practically useful, because without a theorem ensuring that the chain has come very close to its equilibrium state, the user can never be confident that the simulation output is reliable.

For conventional Monte Carlo and techniques of simulation, some excellent references are Fishman (1995), Ripley (1987), Robert and Casella (2004), and Ross (2006). Markov Chain Monte Carlo started off with the two path-breaking articles: Metropolis et al. (1953), and Hastings (1970). Geman and Geman (1984), Tanner and Wong (1987), Smith and Roberts (1993), and Gelfand and Smith (1987) are among the pioneering articles in the statistical literature. Excellent book-length treatments of MCMC include Robert and Casella (2004), Gamerman (1997), and Liu (2008). Geyer (1992), Gilks et al. (1995), and Gelman et al. (2003) are excellent readings on MCMC with an applied focus. Diaconis (2009) is an up-to-date review. Literature on convergence of an MCMC algorithm has been growing steadily. Brémaud (1999) is one of the best sources to read about general theory of convergence of MCMC algorithms. Diaconis et al. (2008) is a tour de force on convergence of the Gibbs sampler, and Diaconis and Saloff-Coste (1998) on the Metropolis algorithm. Other important references on the difficult issue of convergence are Diaconis and Stroock (1991), Tierney (1994), Athreya, Doss and

Sethuraman (1996), Propp and Wilson (1998), and Rosenthal (2002); Dimakos (2001) is a useful survey. Various useful modifications of the basic MCMC have been suggested to address specific important applications. Notable among them are Green (1995), Besag and Clifford (1991), Fill (1998), Diaconis and Sturmfels (1998), Higdon (1998), and Kendall and Thönnes (1999), among many others. Various other specific references are given in the sections.

19.1 The Ordinary Monte Carlo

The ordinary Monte Carlo is a simulation technique for approximating the expectation of a function $\phi(X)$ for a general random variable X, when the exact expectation cannot be found analytically, or by other numerical means, such as quadrature. The idea of Monte Carlo simulation originated around 1940 in connection with secret nuclear weapon projects in which physicists wanted to understand how the physical properties of neutrons would be affected by various possible scenarios following a collison with a nucleus. The *Monte Carlo* term was picked by Stanislaw Ulam and von Neumann during that time.

19.1.1 Basic Theory and Examples

The basis for the ordinary Monte Carlo is Kolmogorov's SLLN (see Chapter 7), which says that if $X_1, X_2, \ldots$ are iid copies of X, the basic underlying random variable, then $\hat{\mu}_\phi = \frac{1}{n}\sum_{i=1}^{n} \phi(X_i)$ converges almost surely to $\mu_\phi = E[\phi(X)]$, as long as we know that the expectation exists. This is because, if $X_1, X_2, \ldots$ are iid copies of X, then $Z_i = \phi(X_i), i = 1, 2, \ldots$ are iid copies of $Z = \phi(X)$ and therefore, by the canonical SLLN, $\frac{1}{n}\sum_{i=1}^{n} Z_i$ converges almost surely to $E[Z]$. Therefore, provided that we can actually do the simulation in practice, we can simply simulate a large number of iid copies of X and approximate the true value of $E[\phi(X)]$ by the sample mean $\frac{1}{n}\sum_{i=1}^{n} \phi(X_i)$. Of course, there will be a Monte Carlo error in this approximation, and if we run the simulation again, we get a different approximated value for $E[\phi(X)]$. Thus, some reliability measure for the Monte Carlo estimate is necessary. A quick one at hand is the variance of the Monte Carlo estimate

$$\text{Var}(\hat{\mu}_\phi) = \frac{\text{Var}[\phi(X)]}{n} = \frac{E[\phi^2(X)] - (E[\phi(X)])^2}{n}.$$

However, this involves quantities that we do not know, in particular $E[\phi(X)]$, the very quantity we set out to approximate! However, we can fall back on the sample variance

$$s_z^2 = \frac{1}{n-1}\sum_{i=1}^{n}(Z_i - \overline{Z})^2$$

to estimate this uncomputable exact variance, that is, a reliability measure could be

$$\widehat{\text{Var}(\hat{\mu}_\phi)} = \frac{1}{n(n-1)} \sum_{i=1}^{n} (Z_i - \overline{Z})^2,$$

where $Z_i = \phi(X_i), i = 1, 2, \ldots, n$. We can also compute confidence intervals for the true value of μ_ϕ. By the central limit theorem, for a large Monte Carlo size $n, \hat{\mu} \approx N\left(\mu, \frac{\sigma^2}{n}\right)$, where $\sigma^2 = \text{Var}(\phi(X))$. As we commented above, in a practical problem, we would not know the true value of σ^2, and therefore the usual confidence interval for μ_ϕ, namely, $\overline{Z} \pm z_{\alpha/2} \frac{\sigma}{\sqrt{n}}$ would not be usable. The standard proxy in such a case is the t confidence interval $\overline{Z} \pm t_{\frac{\alpha}{2}, n-1} \frac{s_z}{\sqrt{n}}$, where $t_{\frac{\alpha}{2}, n-1}$ stands for the $100\left(1 - \frac{\alpha}{2}\right)$ percentile of a t-distribution with $n-1$ degrees of freedom (see Chapter 4).

In the special case that we want to find a Monte Carlo estimate of a probability $P(X \in A)$, the function of interest becomes an indicator function $\phi(X) = I_{X \in A}$, and $Z_i = I_{X_i \in A}, i = 1, 2, \ldots, n$. Then, the Monte Carlo estimate of $p = P(X \in A)$ is

$$\hat{p} = \frac{1}{n} \sum_{i=1}^{n} Z_i = \frac{\#\{i : X_i \in A\}}{n}.$$

We can also construct confidence intervals on p as in Chapter 1. The score confidence interval based on the Monte Carlo samples is

$$\frac{\hat{p} + \frac{z_{\frac{\alpha}{2}}^2}{2n}}{1 + \frac{z_{\frac{\alpha}{2}}^2}{n}} \pm \frac{z_{\frac{\alpha}{2}} \sqrt{n}}{n + z_{\frac{\alpha}{2}}^2} \sqrt{\hat{p}(1-\hat{p}) + \frac{z_{\frac{\alpha}{2}}^2}{4n}}.$$

In principle, this seems to give an acceptable numerical solution to our problem; we can give an approximate value for $E[\phi(X)]$, an accuracy measure, and also a confidence interval. It is important to understand that although the Monte Carlo principle is applicable, in principle, very widely and to any number of dimensions, the Monte Carlo sample size n must be quite large to provide reliable estimates for the true value of the expectation. This is evidenced in some of the examples below.

Example 19.1 (A Cauchy Distribution Expectation). Suppose we want to evaluate $\mu = E(\log|X|)$, where $X \sim C(0, 1)$, the standard Cauchy distribution. Actually, the value of μ can be analytically calculated, and $\mu = 0$ (this is a chapter exercise). We use the Monte Carlo simulation method to approximate μ, and then we investigate its accuracy. For this, we calculate the Monte Carlo estimate for μ itself, and then a 95% t confidence interval for μ. We use four different values of the Monte Carlo sample size $n, n = 100, 250, 500, 1000$, to inspect the increase in accuracy obtainable with an increase in the Monte Carlo sample size.

19.1 The Ordinary Monte Carlo

n	Monte Carlo Estimate of $\mu = 0$	95% Confidence Interval
100	.0714	.0714 ± .2921
250	−.0232	−.0232 ± .2185
500	.0712	.0712 ± .1435
1000	.0116	.0116 ± .1005

The Monte Carlo estimate itself oscillates. But the confidence interval gets tighter as the Monte Carlo sample size n increases. Only for $n = 1000$, the results of the Monte Carlo simulation approach even barely acceptable accuracy. It is common practice to use n in several thousands when applying ordinary Monte Carlo for estimating one single μ_ϕ. If we have to estimate μ_ϕ for several different choices of ϕ, and if the functions ϕ have awkward behavior at regions of low density of X, then the Monte Carlo sample size has to be increased. Formal recommendations can be given by using a pilot or guessed value of σ^2, the true variance of $\phi(X)$.

Example 19.2 (Estimating the Volume of a Set). Let S be a set in an Euclidean space $\mathcal{R}^d$, and suppose we want to find the volume of S. Unless S has a very specialized shape, an exact volume formula will be difficult or impossible to write (and especially so, when d is large). However, Monte Carlo can assist us in this difficult problem. We have to make some assumptions. We assume that S is a bounded set, and that there is an explicit d-dimensional rectangle R such that $S \subseteq R$. Without loss of generality, we may take R to be $[0, 1]^d$. Denote the volume of S by $\lambda = \text{Vol}(S)$. Note that $\lambda = \int_S dx$.

There is nothing probabilistic in the problem so far. It appears to be a purely a mathematical problem. But, we can think of λ probabilistically by writing it as $\lambda = \int_S f(x)dx$, where $f(x) = I_{x \in R}$ is the density of the d-dimensional uniform distribution on $R = [0, 1]^d$. Therefore, $\lambda = P(X \in S)$, where X is distributed uniformly in R. We now realize the potential of Monte Carlo in estimating λ.

Indeed, let $X_1, X_2, \ldots, X_n$ be independent uniformly distributed points in R. Then, from our general discussion above, a Monte Carlo estimate of λ is $\hat{\lambda} = \frac{\#\{i: X_i \in S\}}{n}$. We can also construct confidence intervals for λ by following the score interval's formula given above. Thus, a potentially very difficult mathematical problem is reduced to simply simulating n uniform random vectors from the rectangle R, which is the same as simulating nd iid $U[0, 1]$ random variables, an extremely simple task. Of course, it is necessary to remember that Monte Carlo is not going to give us the exact value of the volume of S. With luck, it will give a good approximation, which is useful.

Example 19.3 (Monte Carlo Estimate of the Volume of a Cone). This is a specific example of the application of the Monte Carlo method to estimate the volume of a set. Take the case of a right circular cone S of base radius $r = 1$ and height $h = 1$. The true volume of S is $\lambda = \frac{1}{3}\pi r^2 h = \frac{\pi}{3} = 1.047$. The apex aperture of S is given by $\theta = 2\arctan(\frac{r}{h}) = 2\frac{\pi}{4} = \frac{\pi}{2}$. Therefore, S is described in the Cartesian coordinates as

$$S = \{(x, y, z) : 0 \leq z \leq 1, x^2 + y^2 \leq z^2\}.$$

S is contained in the rectangle $R = [-1, 1] \otimes [-1, 1] \otimes [0, 1]$. To approximate the volume λ of S, we simulate n random vectors $X_1, X_2, \ldots, X_n$ from R and form the equation

$$\frac{\lambda}{\text{Vol}(R)} = \frac{\#\{i : X_i \in S\}}{n},$$

which gives the estimate of λ

$$\hat{\lambda} = \text{Vol}(R) \frac{\#\{i : X_i \in S\}}{n} = 4 \frac{\#\{i : X_i \in S\}}{n}.$$

The table below reports Monte Carlo estimates and 95% score confidence intervals for λ for a range of values of n.

n	Monte Carlo Estimate of $\lambda = 1.047$	95% Confidence Interval
100	1.08	$1.08 \pm .343$
250	1.168	$1.168 \pm .224$
500	1.016	$1.016 \pm .152$
1000	1.100	$1.100 \pm .1105$
10,000	1.064	$1.064 \pm .035$

Once again, as in Example 19.1, we see that the Monte Carlo estimate itself oscillates, and the Monte Carlo error is not monotone decreasing in n. However, the width of the confidence interval consistently decreases as the Monte Carlo sample size n increases. To get a really good estimate and a tight confidence interval, we appear to need a Monte Carlo sample size of about 10,000.

Example 19.4 (A Bayesian Example). Monte Carlo methods can be used not only to approximate expectations, but also to approximate percentiles of a distribution. Precisely, if $X_1, X_2, \ldots$ are iid sample observations from a continuous distribution F with a strictly positive density f, then for any $\alpha, 0 < \alpha < 1$, the sample quantile $F_n^{-1}(\alpha)$ converges almost surely to the corresponding quantile $F^{-1}(\alpha)$ of F. So, as in the case of Monte Carlo estimates for expectations, we can generate a Monte Carlo sample from F and estimate $F^{-1}(\alpha)$ by $F_n^{-1}(\alpha)$.

Suppose $X \sim \text{Bin}(m, p)$, and the unknown parameter p is assigned a beta prior density, with parameters α, β. Then the *posterior density* of p is another Beta, namely $\text{Be}(x + \alpha, m - x + \beta)$ (see Chapter 3). To give a specific example, suppose $m = 100, x = 45, \alpha = \beta = 1$ (i.e., p has a $U[0, 1]$ prior). The percentiles of a Beta density do not have closed-form formulas. So, one has to resort to numerical methods to evaluate them.

We first estimate the posterior median by using a Monte Carlo sample for various values of the Monte Carlo sample size n. For comparison, a true value for the posterior median in this case is reported by *Mathematica* as 0.4507.

19.1 The Ordinary Monte Carlo

n	Monte Carlo Estimate of Posterior Median = 0.4507
50	.4584
100	.4463
250	.4486
500	.4553
1000	.4498

The estimates oscillate slightly. But even for a Monte Carlo sample size as small as $n = 50$, the estimate is impressively accurate. But change the problem to Monte Carlo estimation of an extreme quantile, say the 99.9th percentile of the posterior distribution of p. The true value is reported by Mathematica to be 0.6028. The Monte Carlo estimates for various n are reported below. We can see that the quality of the estimation has deteriorated.

n	Monte Carlo Estimate of 99.9th Posterior Percentile = 0.6028
50	.5662
100	.5618
250	.5714
500	.5735
1000	.5889
5000	.5956
10,000	.5987

The need for a much larger Monte Carlo sample size is clearly seen from the table. Although we could estimate the posterior median accurately with a Monte Carlo sample size of 100, for the extreme quantile, we need five to ten thousand Monte Carlo samples.

Example 19.5 (Computing a Nasty Integral). Monte Carlo methods can be extremely useful in computing the value of a complicated integral that cannot be evaluated in closed form. Monte Carlo is especially useful for this purpose in high dimensions, because methods of numerical integration are hard to implement and quite unreliable when the number of dimensions is even moderately high (as few as four). The basic idea is very simple. Suppose we wish to know the value of the definite integral $I = \int_{\mathcal{R}^d} f(x)dx$ for some (possibly complicated) function f. We are of course assuming we know that the integral exists. This must be verified mathematically before we start on the Monte Carlo journey. Monte Carlo cannot verify the existence of the integral.

The idea now is to use a suitable density function $g(x)$ on $\mathcal{R}^d$ and write I as

$$I = \int_{\mathcal{R}^d} f(x)dx = \int_{\mathcal{R}^d} \frac{f(x)}{g(x)} g(x)dx = E_g \left[\frac{f(X)}{g(X)} \right].$$

Now the Monte Carlo method is usable. We simulate iid random vectors $X_1, \ldots, X_n$ from g, and approximate I by $\hat{I} = \frac{1}{n} \sum_{i=1}^{n} \frac{f(X_i)}{g(X_i)}$. The choice of g is obviously not unique. It is usually chosen such that it is easy to simulate from g, and such that $\frac{f(x)}{g(x)}$ can be reliably computed. In other words, the function $\frac{f(x)}{g(x)}$ preferably should not have singularities, or cusps, or too many local maximas or minimas. Some preliminary investigation on the choice of g is needed, and in fact, this choice issue is related to a topic known as *importance sampling*, which we discuss later.

As a concrete example, suppose we want to find the value of

$$I = \int_{-\infty}^{\infty} \int_{-\infty}^{\infty} \int_{-\infty}^{\infty} \int_{-\infty}^{\infty} \frac{e^{-x^2-y^2-z^2-w^2}}{\sqrt{1+x^2+y^2+z^2+w^2}} dx dy dz dw.$$

Suppose now we let

$$g(x,y,z,w) = \frac{1}{\pi^2} e^{-x^2-y^2-z^2-w^2},$$

and

$$f(x,y,z,w) = \frac{e^{-x^2-y^2-z^2-w^2}}{\sqrt{1+x^2+y^2+z^2+w^2}}.$$

Then,

$$I = \int_{-\infty}^{\infty} \int_{-\infty}^{\infty} \int_{-\infty}^{\infty} \int_{-\infty}^{\infty} f(x,y,z,w) dx dy dz dw$$

$$= E_g \left[\frac{f(X)}{g(X)} \right] = \pi^2 E_g \left[\frac{1}{\sqrt{1+X^2+Y^2+Z^2+W^2}} \right].$$

Therefore, we can find a Monte Carlo estimate of I by simulating (X_i, Y_i, Z_i, W_i), $i = 1, \ldots, n$, where X_i, Y_i, Z_i, W_i are all iid $N(0, \frac{1}{2})$, and by using the estimate

$$\hat{I} = \pi^2 \frac{1}{n} \sum_{i=1}^{n} \frac{1}{\sqrt{1+X_i^2+Y_i^2+Z_i^2+W_i^2}}.$$

The table below reports Monte Carlo estimates of I for various values of n. The exact value of I is useful for comparison and can be obtained after transformation to polar coordinates (see Chapter 4) to be

$$I = \pi^2 \left[1 + e\sqrt{\pi} \left(\Phi\left(\sqrt{2}\right) - 1 \right) \right] = 6.1297.$$

19.1 The Ordinary Monte Carlo

n	Monte Carlo Estimate of $I = 6.1297$
50	6.3503
100	6.2117
250	6.2264
500	6.1854
1000	6.1946
5000	6.0806
10,000	6.1044

By the time the Monte Carlo sample size is 10,000, we get fairly accurate estimates for the value of I.

Example 19.6 (Monte Carlo Evaluation of π). In numerous probability examples, the number π arises in the formula for the probability of some suitable event A in some random experiment, say $p = P(A) = h(\pi)$. Then, we can form a Monte Carlo estimate for p, say $\hat{p}$, and estimate π as $h^{-1}(\hat{p})$, assuming that the function h is one to one. This is not a very effective method, as it turns out. But the idea has an inherent appeal and we describe such a method in this example.

Suppose we fix a positive integer N, and let X, Y be independent discrete uniforms on $\{1, 2, \ldots, N\}$. Then, it is known that

$$\lim_{N \to \infty} P(X, Y \text{ are coprime}) = \frac{6}{\pi^2}.$$

Here, coprime means that X, Y do not have any common factors >1. So, in principle, we may choose a large N, choose n pairs (X_i, Y_i) independently at random from the discrete set $\{1, 2, \ldots, N\}$, and find a Monte Carlo estimate $\hat{p}$ for $p = P(X, Y \text{ are coprime})$, and invert it to form an estimate for the value of π as

$$\hat{\pi} = \sqrt{\frac{6}{\hat{p}}}.$$

The table below reports the results of such a Monte Carlo experiment.

N	n	Monte Carlo Estimate of $\pi = 3.14159$
500	100	3.0252
1000	100	3.0817
1000	250	3.1308
5000	250	3.2225
5000	500	3.2233
10000	1000	3.1629
10,000	5000	3.1395

This is an interesting example of the application of Monte Carlo where two indices N, n have to take large values simultaneously. Only when $N = 10{,}000$ and $n = 5{,}000$, do we come close to matching the second digit after the decimal.

19.1.2 Monte Carlo P-Values

Suppose we have a statistical hypothesis-testing problem in which we wish to test a particular null hypothesis H_0 against a particular alternative hypothesis H_1. We have at our disposal a data vector $X^{(m)} = (X_1, \ldots, X_m)$. Often, the testing is done by a judicious choice of a statistic $T_m = T_m(X_1, \ldots, X_m)$ such that large values of T_m cast doubt on the truth of the null hypothesis H_0. Suppose that for the dataset $(x_1, \ldots, x_m)$ that is actually observed, the statistic T_m takes the value $T_m(x_1, \ldots, x_m) = t_m$. In that case, it is common statistical practice to calculate the P-value defined as

$$p_m = p_m(X_1, \ldots, X_m) = P_{H_0}(T_m > t_m),$$

and reject the null hypothesis if p_m is small (say, smaller than 0.01). The program assumes that $p_m = P_{H_0}(T_m > t_m)$ can be computed (and that it does not depend on any unknown parameters not specified by H_0). However, except in some special cases, we do not know the exact null distribution of our test statistic T_m, and so, computing the P-value will require some suitable approximation method. The Monte Carlo P-value method simply simulates a lot of datasets $(X_1, \ldots, X_m)$ under the null, computes $T_m(X_1, \ldots, X_m)$ for each generated dataset, and then computes the percentile rank of our value t_m among these synthetic sets of values of T_m. For example, if we generated $n = 99$ additional datasets, and we find our value t_m to be the 98th order statistic among these 100 values of T_m, we declare the Monte Carlo P-value to be 0.02. If $T_m(X_1, \ldots, X_m)$ has a continuous distribution under H_0, then the rank of t_m among the $n+1$ values of T_m, say $R_{m,n}$, is simply a discrete uniform random variable on the set $\{1, 2, \ldots, n+1\}$, and therefore, we can estimate the true value of p_m by using $\frac{n+1-R_{m,n}}{n+1}$. This lets us avoid a potentially impossible distributional calculation for the exact calculation of p_m.

The idea of a Monte Carlo P-value is attractive. But it should be noted that if we repeat the simulation, we get a different P-value by this method for the same original dataset. Second, one can work out at best a case-by-case Monte Carlo sample size necessary for accurate approximation of the true value of p_m. Such calculations are very likely to need the same difficult calculations that we are trying to avoid. Also, if the test statistic T_m is hard to compute, then evaluation of a Monte Carlo P-value may require a prohibitive amount of computing. There will often be theoretically grounded alternative methods for approximating the true value of p_m, such as central limit theorems or Edgeworth expansions. So, use of Monte Carlo P-values need not be the only method available to us, or the best method available to us in a given problem. Monte Carlo P-values were originally suggested in

19.1 The Ordinary Monte Carlo

Barnard (1963) and Besag and Clifford (1989). They have become quite popular in certain applied sciences, notably biology. For deeper theoretical studies of Monte Carlo P-values, one can see Hall and Titterington (1989).

Example 19.7 (Testing for the Center of Cauchy). This is an example where the calculation of the exact P-value using any reasonable test statistic is not easy. Suppose $X_1, \ldots, X_m$ are iid $C(\mu, 1)$ and suppose we wish to test $H_0 : \mu = 0$ against $\mu > 0$. As a test statistic, the sample mean $\overline{X}$ is an extremely poor choice in this case. The sample median is reasonable, and the unique maximum likelihood estimate of μ is asymptotically the best, but it is not easy to compute the maximum likelihood estimate in this case. Although the sample median and the maximum likelihood estimate are both asymptotically normal, neither of them has a tractable CDF for given $m > 2$. Therefore, an exact calculation of the P-value $p_m = P_{\mu=0}(T_m > t_m)$ is essentially impossible using either of these two statistics.

With $m = 25$ and various values of the Monte Carlo sample size n, the following table reports the Monte Carlo P-values with T_m as the sample median. An approximation to the P-value obtained from the asymptotic normality result $\sqrt{m}(T_m - \mu) \stackrel{\mathcal{L}}{\Rightarrow} N(0, \frac{\pi^2}{4})$ is $p_m \approx 0.386$. The original dataset of size $m = 25$ was also generated under the null, that is, from $C(0, 1)$.

n	Monte Carlo P-Value
50	0.32
100	0.38
250	0.388
500	0.374
1000	0.371

The Monte Carlo P-values stabilize when the Monte Carlo sample size n is about 100. They closely match the normal approximation $p_m \approx 0.386$. However, it is impossible to say which one is more accurate, the Monte Carlo P-value or the normal approximation.

19.1.3 Rao–Blackwellization

Let X, Y be any two random variables such that $\text{Var}(X)$ and $\text{Var}(X \mid Y)$ exist. Then, the iterated variance formula says that $\text{Var}(X) = E[\text{Var}(X \mid Y)] + \text{Var}(E(X \mid Y))$ (see Chapter 2). As a consequence, $\text{Var}(E(X \mid Y)) \leq \text{Var}(X)$. Therefore, if we define $h(Y) = E(X \mid Y)$, then $E(h(Y)) = E[E(X \mid Y)] = E(X)$, and $\text{Var}(h(Y)) \leq \text{Var}(X)$. This says that as an estimate of $\mu = E(X)$, the conditional expectation $h(Y) = E(X \mid Y)$ is at least as good an unbiased estimate as X itself. The technique is similar to the well-known technique in statistics of conditioning with respect to a sufficient statistic, which is due to David Blackwell and C. R. Rao. So, $h(Y)$ is called a Rao–Blackwellized Monte Carlo estimate for $\mu = E(X)$. For the method to be applicable, the function $h(y) = E[E(X \mid Y = y)$ must be explicitly computable for any y. To apply this method, we need to choose a random variable Y, simulate iid values $Y_1, \ldots, Y_n$ from the distribution of Y, calculate $h(Y_1), \ldots, h(Y_n)$, and estimate μ by using $\hat{\mu} = \frac{1}{n} \sum_{i=1}^{n} h(Y_i)$.

Example 19.8. Suppose X, Y are iid standard normal variables and that we wish to estimate $\mu = E(e^{tXY})$. The ordinary Monte Carlo estimate of it requires a simulation $(X_i, Y_i), i = 1, \ldots, n$, all mutually independent standard normal variables, and estimates μ as $\hat{\mu} = \frac{1}{n} \sum_{i=1}^{n} e^{tX_i Y_i}$. For the Rao–Blackwellized estimate, we condition on Y, so that we have

$$h(y) = E(e^{tXY} \mid Y = y) = E(e^{tyX} \mid Y = y) = E(e^{tyX}) = e^{\frac{t^2 y^2}{2}}.$$

The Rao–Blackwellized Monte Carlo estimate is $\hat{\mu} = \frac{1}{n} \sum_{i=1}^{n} h(Y_i) = \frac{1}{n} \sum_{i=1}^{n} e^{\frac{t^2 Y_i^2}{2}}$. Note that this requires simulation of only $Y_1, \ldots, Y_n$, and not the pairs $(X_i, Y_i), i = 1, \ldots, n$.

19.2 Textbook Simulation Techniques

As was commented before, the entire Monte Carlo method is based on the assumption that we can in fact simulate the Monte Carlo sample observations $X_1, \ldots, X_n$ from whatever distribution is the relevant one for the given problem. Widely available commercial software exists for simulating from nearly every common distribution in one dimension, and many common distributions in higher dimensions. With the modern high-speed computers that most researchers now generally use, the efficiency issue of these commercial algorithms has become less important than before. Still, the fact is that often one has to customize one's own algorithm to a given problem, either because the problem is unusual and commercial software is not available, or that commercial software is unacceptably slow. We do not intend to delve into customized simulation algorithms for special problems in this text. We give a basic description of a few widely used simulation methods, and some easily applied methods for 25 special distributions, for the purpose of quick reference. Textbook and more detailed treatments of simulation from standard distributions are available in Fishman (1995), Robert and Casella (2004), and Ross (2006), among others. Schmeiser (1994, 2001) provides lucidly written summary accounts.

19.2.1 *Quantile Transformation and Accept–Reject*

Quantile Transformation. We are actually already familiar with this method (see Chapter 1 and Chapter 6). Suppose F is a continuous CDF on the real line with the quantile function F^{-1}. Suppose $X \sim F$. Then $U = F(X) \sim U[0, 1]$. Therefore,

19.2 Textbook Simulation Techniques

to simulate a value of $X \sim F$, we can simulate a value of $U \sim U[0, 1]$ and use $X = F^{-1}(U)$. As long as the quantile function F^{-1} has a formula, this method will work for any one-dimensional random variable X with a continuous CDF.

Example 19.9. Suppose we want to simulate a value for $X \sim \text{Exp}(1)$. The quantile function of the standard exponential distribution is $F^{-1}(p) = -\log(1 - p), 0 < p < 1$. Therefore, to simulate X, we can generate $U \sim U[0, 1]$, and use $X = -\log(1 - U)$ ($-\log U$ will work too).

As another example, suppose we want to simulate $X \sim \text{Be}(\frac{1}{2}, \frac{1}{2})$, the Beta distribution with parameters $\frac{1}{2}$ each. The density of the $\text{Be}(\frac{1}{2}, \frac{1}{2})$ distribution is $\frac{1}{\pi\sqrt{x(1-x)}}, 0 < x < 1$. By direct integration, we get that the CDF is $F(x) = \frac{2}{\pi} \arcsin(\sqrt{x})$. Therefore, the quantile function is $F^{-1}(p) = \sin^2(\frac{\pi}{2} p)$, and so, to simulate $X \sim \text{Be}(\frac{1}{2}, \frac{1}{2})$, we generate $U \sim U[0, 1]$, and use $X = \sin^2(\frac{\pi}{2} U)$.

The quantile function F^{-1} does not have closed-form formulas if F is a normal, or a Beta, or a Gamma, and so on. In such cases, the use of the quantile transform method by numerically evaluating $F^{-1}(U)$ will cause a slight error. The error may be practically negligible. But, for these distributions, simulated values are usually obtained in practice by using special techniques, rather than the quantile transform technique. For example, simulations for a standard normal variable are obtainable by using the *Box–Muller method* (see the chapter exercises in Chapter 4).

Accept–Reject. The *accept–reject* method is useful when it is difficult to directly simulate from a target density $f(x)$ on the real line, but we can construct another density $g(x)$ such that $\frac{f(x)}{g(x)}$ is uniformly bounded, and it is much easier to simulate from g. Then we do simulate X from g, and retain it or toss it according to some specific rule. The set of X values that are retained may be treated as independent simulations from the original target density f. Because an X value is either retained or discarded, depending on whether it passes the admission rule, the method is called the *accept–reject method*. The density g is called the *envelope density*.

The method proceeds as follows.

(a) Find a density function g and a finite constant c such that $\frac{f(x)}{g(x)} \leq c$ for all x.
(b) Generate $X \sim g$.
(c) Generate $U \sim U[0, 1]$, independently of X.
(d) Retain this generated X value if $U \leq \frac{f(X)}{cg(X)}$.
(e) Repeat the steps until the required number of n values of X has been obtained.

The following result justifies this indirect algorithm for generating values from the actual target density f.

Theorem 19.1. *Let $X \sim g$, and U, independent of X, be distributed as $U[0, 1]$. Then the conditional density of X given that $U \leq \frac{f(X)}{cg(X)}$ is f.*

Proof. Denote the CDF of f by F. Then,

$$P\left(X \le x \mid U \le \frac{f(X)}{cg(X)}\right)$$
$$= \frac{P\left(X \le x, U \le \frac{f(X)}{cg(X)}\right)}{P\left(U \le \frac{f(X)}{cg(X)}\right)} = \frac{\int_{-\infty}^{x}\int_{0}^{\frac{f(t)}{cg(t)}} g(t)dudt}{\int_{-\infty}^{\infty}\int_{0}^{\frac{f(t)}{cg(t)}} g(t)dudt}$$
$$= \frac{\int_{-\infty}^{x} \frac{f(t)}{cg(t)} g(t)dt}{\int_{-\infty}^{\infty} \frac{f(t)}{cg(t)} g(t)dt} = \frac{\int_{-\infty}^{x} f(t)dt}{\int_{-\infty}^{\infty} f(t)dt} = \frac{F(x)}{1} = F(x).$$

□

Example 19.10 (Generating from Normal via Accept–Reject). Suppose we want to generate $X \sim N(0, 1)$. Thus our target density f is just the standard normal density. Since there is no formula for the quantile function of the standard normal, the quantile transform method is usually not used to generate from a standard normal distribution. We can, however, use the accept–reject method to generate standard normal values. For this, we need an envelope density g such that $\frac{f(x)}{g(x)}$ is uniformly bounded, and furthermore, it should be easier to sample from this g.

One possibility is to use the standard double exponential density $g(x) = \frac{1}{2}e^{-|x|}$. Then,

$$\frac{f(x)}{g(x)} = \frac{\frac{1}{\sqrt{2\pi}}e^{-x^2/2}}{\frac{1}{2}e^{-|x|}}$$
$$= \sqrt{\frac{2}{\pi}}e^{|x|-x^2/2} \le \sqrt{\frac{2e}{\pi}}$$

for all real x, by elementary differential calculus.

We take $c = \sqrt{\frac{2e}{\pi}}$ in the accept–reject scheme and, of course, g as the standard double exponential density. The scheme works out to the following: generate $U \sim U[0, 1]$, and a double exponential value X, and retain X if $U \le e^{|X|-X^2/2-\frac{1}{2}}$. Note that a double exponential value can be generated easily by several means:

(a) Generate a standard exponential value Y and assign it a random sign (+ or − with an equal probability).
(b) Generate two independent standard exponential values Y_1, Y_2 and set $X = Y_1 - Y_2$.
(c) Use the quantile transform method, as there is a closed-form formula for the quantile function of the standard double exponential.

It is helpful to understand this example by means of a plot. The generated X value is retained if and only if the pair (X, U) is below the graph of the function in Fig. 19.1 namely the function $u = e^{|x|-x^2/2-\frac{1}{2}}$. We can see that one of the two generated values will be accepted, and the other rejected.

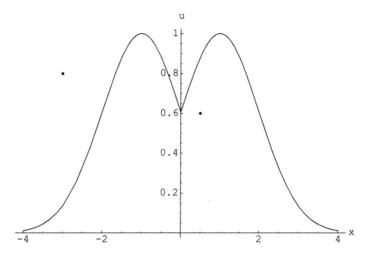

Fig. 19.1 Acceept–reject plot for generating N(0, 1) values

Example 19.11 (Generating a Beta). Special-purpose customized algorithms are the most efficient for generating simulations from general Beta distributions. For values of the parameters α, β of the Beta distribution in various ranges, separate treatments are necessary for producing the most efficient schemes. If α and β are both larger than one, then a Beta density is strictly unimodal with an interior mode at $\frac{\alpha-1}{\alpha+\beta-2}$. As a result, for $\alpha, \beta > 1$, a Beta density is uniformly bounded, and hence the $U[0, 1]$ density can serve as an envelope density for generating from such a Beta distribution by using the accept-reject scheme. Precisely, generate $U, X \sim U[0, 1]$ (independently), and retain the X value if $U \leq \frac{f(X)}{\sup_x f(x)}$, where

$$f(x) = \frac{\Gamma(\alpha+\beta)}{\Gamma(\alpha)\Gamma(\beta)} x^{\alpha-1}(1-x)^{\beta-1}, 0 < x < 1.$$

Because

$$\sup_x f(x) = f\left(\frac{\alpha-1}{\alpha+\beta-2}\right) = \frac{\Gamma(\alpha+\beta)}{\Gamma(\alpha)\Gamma(\beta)} \frac{(\alpha-1)^{\alpha-1}(\beta-1)^{\beta-1}}{(\alpha+\beta-2)^{\alpha+\beta-2}},$$

the scheme finally works out to the following.

Generate independent $U, X \sim U[0, 1]$ and retain the X value if

$$U \leq \frac{X^{\alpha-1}(1-X)^{\beta-1}(\alpha+\beta-2)^{\alpha+\beta-2}}{(\alpha-1)^{\alpha-1}(\beta-1)^{\beta-1}}.$$

It is to be noted that this accept–reject scheme with g as the $U[0, 1]$ density would not be very efficient if α, β are large. If α, β are large, the beta density tapers off very rapidly from its mode, and the uniform envelope density would be a poor choice for g. For α, β not too far from 1, the scheme of this example would be reasonably efficient.

An important practical issue about an accept–reject scheme is the acceptance percentage. One must strive to make this as large as possible in order to increase the efficiency of the method. This is achieved by choosing c to be the smallest possible number that one can, as the following result shows.

Proposition. *For an accept–reject scheme, the probability that an $X \sim g$ is accepted is $\frac{1}{c}$, and is maximized when c is chosen to be $c = \sup_x \frac{f(x)}{g(x)}$.*

We have essentially already proved it, because the acceptance probability is

$$P\left(U \leq \frac{f(X)}{cg(X)}\right) = \int_{-\infty}^{\infty} \int_0^{\frac{f(t)}{cg(t)}} g(t) du\, dt$$

$$= \int_{-\infty}^{\infty} \frac{f(t)}{cg(t)} g(t) dt = \int_{-\infty}^{\infty} \frac{f(t)}{c} dt = \frac{1}{c}.$$

Because any c that can be chosen must be at least as large as $\sup_x \frac{f(x)}{g(x)}$, obviously $\frac{1}{c}$ is maximized by choosing $c = \sup_x \frac{f(x)}{g(x)}$.

Example 19.12 (*Efficiency of Accept–Reject Scheme*). In Example 19.8, we used an accept–reject scheme with g as the standard double exponential density to simulate from the standard normal distribution. Because

$$\sup_x \frac{f(x)}{g(x)} = \sqrt{\frac{2e}{\pi}},$$

the acceptance rate, by our previous theorem, would be $\sqrt{\frac{\pi}{2e}} = .7602$. If we generate 100 X-values from g, we can expect that about 75 of them would be retained, and the others discarded.

Suppose now we use $g(x) = \frac{1}{\pi(1+x^2)}$, the standard Cauchy density. This density also satisfies the requirement that $\sup_x \frac{f(x)}{g(x)} < \infty$, and in fact with this choice of g, $\sup_x \frac{f(x)}{g(x)} = \sqrt{\frac{2\pi}{e}}$. Therefore, with g as the standard Cauchy density, the acceptance rate would be $\sqrt{\frac{e}{2\pi}} = .6577$, which is lower than the acceptance rate when g is the standard double exponential. In general, when using an accept–reject scheme, one should choose the envelope density g to be a density that matches the target density f as closely as possible, while being easier to simulate from. The standard Cauchy density does not match the standard normal density well, and so we get a lower acceptance rate. Rubin (1976) gives some ideas on improving the efficiency of acceptance–rejection schemes.

19.2.2 Importance Sampling and Its Asymptotic Properties

There are two different ways to think about importance sampling. The more traditional one is to go back to the primary problem that Monte Carlo wants to solve, namely to approximate the value of an expectation $\mu = \int \phi_0(x) dF_0(x)$ for some function ϕ_0 and some CDF F_0. However, (ϕ_0, F_0) is not the only pair (ϕ, F) for which $\int \phi(x) dF(x)$ equals the specific number μ. Indeed, given any other CDF F_1,

$$\mu = \int \phi_0(x) dF_0(x) = \int \phi_0(x) \frac{dF_0}{dF_1}(x) dF_1(x) = \int \lambda(x) \phi_0(x) dF_1(x),$$

where $\lambda(x) = \frac{dF_0}{dF_1}(x)$. If F_0, F_1 have densities f_0, f_1, then $\lambda(x) = \frac{f_0(x)}{f_1(x)}$; if F_0, F_1 have respective pmfs f_0, f_1, then also $\lambda(x) = \frac{f_0(x)}{f_1(x)}$ (if one is continuous and the other discrete, $\frac{dF_0}{dF_1}(x)$ need not be defined; the general treatment needs the use of Radon–Nikodym derivatives).

This raises the interesting possibility that we can sample from a general F_1, and subsequently use the usual Monte Carlo estimate

$$\hat{\mu} = \frac{1}{n} \sum_{i=1}^{n} \lambda(X_i) \phi_0(X_i),$$

where $X_1, X_2, \ldots, X_n$ is a Monte Carlo sample from F_1. Importance sampling poses the problem of finding an optimal choice of F_1 from which to sample, so that $\hat{\mu}$ has the smallest possible variance. The distribution F_1 that ultimately gets chosen is called the *importance sampling distribution*.

A more contemporary view of importance sampling is that we do not approach importance sampling as an optimization problem, but because the circumstances force us to consider different sampling distributions F, and possibly even different functions ϕ within the boundaries of the same general problem that we are trying to solve. For example, in studies of Bayesian robustness, it is imperative that one consider different prior distributions on the parameter of the problem, which would force us to look at different posterior distributions for the parameter. If we want to compute the mean and the variance of all these posterior distributions, then we automatically have multiple pairs (ϕ, F) to consider simultaneously. The traditional view of importance sampling as a technique to decrease Monte Carlo variance was perhaps more relevant when simulation was not as cheap as it is today. The modern view that we have to consider importance sampling out of necessity is arguably the more prevalent viewpoint now.

Coming back to the question of the choice of the importance sampling distribution, we present the proposed solution in a format that allows us to confront a recurring problem in Bayesian calculations, which is that the basic underlying CDF (density) F_0 as well as a candidate importance sampling distribution F_1 are known only up to uncomputable normalizing constants (Section 19.3 discusses this dilemma in greater detail). We also assume that F_0, F_1 both have densities, say f_0, f_1. The presentation given below carries over with only notational change if F_0, F_1 are both discrete with the same support. Suppose then $f_i(x) = \frac{h_i(x)}{c_i}, i = 0, 1$, where the assumption is that h_0, h_1 are completely known and also computable, but c_0, c_1 are unknown and are not even computable. Then, as we showed above, for any function ϕ for which the expectation $E_{F_0}[\phi(X)]$ exists,

$$\mu = E_{F_0}[\phi(X)] == \int \lambda(x)\phi(x)f_1(x)dx$$
$$= \frac{c_1}{c_0} \int \frac{\phi(x)h_0(x)}{h_1(x)} f_1(x)dx$$
$$= \frac{c_1}{c_0} E_{F_1}\left[\frac{\phi(X)h_0(X)}{h_1(X)}\right].$$

This is a useful reduction, but we still have to deal with the fact that the ratio $\frac{c_1}{c_0}$ is not known to us. Fortunately, if we use the special function $\phi(x) \equiv 1$, the same representation above gives us

$$1 = \frac{c_1}{c_0} E_{F_1}\left[\frac{h_0(X)}{h_1(X)}\right] \Rightarrow \frac{c_1}{c_0} = \frac{1}{E_{F_1}\left[\frac{h_0(X)}{h_1(X)}\right]},$$

and because h_0, h_1 are explicitly known to us, we have a way to get rid of the quotient $\frac{c_1}{c_0}$ and write the final *importance sampling identity*

$$E_{F_0}[\phi(X)] = \frac{E_{F_1}\left[\frac{\phi(X)h_0(X)}{h_1(X)}\right]}{E_{F_1}\left[\frac{h_0(X)}{h_1(X)}\right]}.$$

We can now use an available Monte Carlo sample $X_1, \ldots, X_n$ from F_1 to find Monte Carlo estimates for $\mu = E_{F_0}[\phi(X)]$.

The basic plug-in estimate for μ is the so-called *ratio estimate*

$$\hat{\mu} = \frac{\sum_{i=1}^{n} \frac{\phi(X_i)h_0(X_i)}{h_1(X_i)}}{\sum_{i=1}^{n} \frac{h_0(X_i)}{h_1(X_i)}}.$$

If the Monte Carlo sample size is small, this estimate will probably have quite a bit of bias, and some bias correction would be desirable. In any case, we have at hand at least a first-order approximation for $E_{F_0}[\phi(X)]$ based on a Monte Carlo sample from a general candidate importance sampling distribution F_1. The issue of which F_1 to actually choose has not been addressed yet. All we have done so far is some algebraic manipulation and then a proposal for using the ratio estimate of survey theory. The choice issue is addressed below. But first let us see an example.

19.2 Textbook Simulation Techniques

Example 19.13 (Binomial Bayes Problem with an Atypical Prior). Suppose $X \sim$ Bin(m, p) for some fixed m, and p has the prior density $c \sin^2(\pi p)$, where c is a normalizing constant. Throughout the example, c denotes a generic constant, and is not intended to mean the same constant at every use.

The posterior density of p given $X = x$ is

$$\pi(p \mid X = x) = cp^x(1-p)^{m-x} \sin^2(\pi p), \quad 0 < p < 1.$$

The problem is to find the posterior mean

$$\mu = c \int_0^1 p[p^x(1-p)^{m-x} \sin^2(\pi p)] dp.$$

We use importance sampling to approximate the value of μ. Towards this, choose

$$\phi(p) = p, \quad h_0(p) = p^x(1-p)^{m-x} \sin^2(\pi p), \quad h_1(p) = p^x(1-p)^{m-x},$$

so that if $p_1, p_2, \ldots, p_n$ are samples from F_1, (i.e., a Beta distribution with parameters $x + 1$ and $m - x + 1$), then the importance sampling estimate of the posterior mean μ is

$$\hat{\mu} = \frac{\sum_{i=1}^n \frac{\phi(p_i)h_0(p_i)}{h_1(p_i)}}{\sum_{i=1}^n \frac{h_0(p_i)}{h_1(p_i)}}$$

$$= \frac{\sum_{i=1}^n p_i \sin^2(\pi p_i)}{\sum_{i=1}^n \sin^2(\pi p_i)}.$$

Note that we did not need to calculate the normalizing constant in the posterior density. We take $m = 100, x = 45$ for specificity. The following table gives values of this importance sampling estimate and also the value of μ computed by using a numerical integration routine, so that we can assess the accuracy of the importance sampling estimate.

n	Importance Sampling Estimate of $\mu = .4532$
20	.4444
50	.4476
100	.4558
250	.4537
500	.4529

Even with an importance sampling size of $n = 20$, the estimation error is less than 2%. This has partly to do with the choice of the importance sampling distribution. In this example, the importance sampling distribution was chosen to match the shape of the posterior distribution well. This generally enhances the accuracy of importance sampling.

Let us now study two basic theoretical properties of the importance sampling estimate $\hat{\mu}$ in general. The first issue is whether asymptotically it estimates μ correctly, and the second issue is what can we say about the amount of the error $\hat{\mu} - \mu$ in general. Fortunately, we can handle both questions by using our asymptotic toolbox from Chapter 7.

Theorem 19.2. *Suppose*

$$\mathrm{Var}_{F_1}\left(\phi(X)\frac{h_0(X)}{h_1(X)}\right) \text{ and } \mathrm{Var}_{F_1}\left(\frac{h_0(X)}{h_1(X)}\right)$$

both exist. Denote

$$\mu_1 = E_{F_1}\left(\phi(X)\frac{h_0(X)}{h_1(X)}\right), \quad \mu_2 = E_{F_1}\left(\frac{h_0(X)}{h_1(X)}\right),$$

$$\sigma_1^2 = \mathrm{Var}_{F_1}\left(\phi(X)\frac{h_0(X)}{h_1(X)}\right), \quad \sigma_2^2 = \mathrm{Var}_{F_1}\left(\frac{h_0(X)}{h_1(X)}\right),$$

$$\sigma_{12} = \mathrm{Cov}_{F_1}\left(\phi(X)\frac{h_0(X)}{h_1(X)}, \frac{h_0(X)}{h_1(X)}\right).$$

Then, as $n \to \infty$,

(a) $\hat{\mu} \stackrel{a.s.}{\to} \mu = \frac{\mu_1}{\mu_2}$.

(b) $\sqrt{n}(\hat{\mu} - \mu) \stackrel{\mathcal{L}}{\Rightarrow} N(0, \tau^2)$, *where*

$$\tau^2 = \frac{\sigma_2^2 \mu_1^2}{\mu_2^4} + \frac{\sigma_1^2}{\mu_2^2} - \frac{2\sigma_{12}\mu_1}{\mu_2^3}.$$

Proof. For part (a), note that by Kolmogorov's SLLN (see Chapter 7) $\frac{1}{n}\sum_{i=1}^{n} \frac{\phi(X_i)h_0(X_i)}{h_1(X_i)}$ converges almost surely to μ_1, and $\frac{1}{n}\sum_{i=1}^{n} \frac{h_0(X_i)}{h_1(X_i)}$ converges almost surely to $\mu_2 > 0$. Therefore,

$$\hat{\mu} = \frac{\frac{1}{n}\sum_{i=1}^{n} \frac{\phi(X_i)h_0(X_i)}{h_1(X_i)}}{\frac{1}{n}\sum_{i=1}^{n} \frac{h_0(X_i)}{h_1(X_i)}}$$

converges almost surely to $\frac{\mu_1}{\mu_2} = \mu$.

For part (b), denote

$$U_i = \frac{\phi(X_i)h_0(X_i)}{h_1(X_i)}, \quad V_i = \frac{h_0(X_i)}{h_1(X_i)},$$

and note that by the multivariate central limit theorem (see Chapter 7),

$$\sqrt{n}(\overline{U} - \mu_1, \overline{V} - \mu_2) \stackrel{\mathcal{L}}{\Rightarrow} N_2(0, \Sigma),$$

where
$$\Sigma = \begin{pmatrix} \sigma_1^2 & \sigma_{12} \\ \sigma_{12} & \sigma_2^2 \end{pmatrix}.$$

Now define the transformation $g(u, v) = \frac{u}{v}$, so that $\hat{\mu} = g(\overline{U}, \overline{V})$. By the delta theorem (see Chapter 7), one has

$$\sqrt{n}(g(\overline{U}, \overline{V}) - g(\mu_1, \mu_2)) \xrightarrow{\mathcal{L}} N(0, [\nabla g(\mu_1, \mu_2)]' \Sigma [\nabla g(\mu_1, \mu_2)]).$$

But $g(\mu_1, \mu_2) = \mu$, and $[\nabla g(\mu_1, \mu_2)]' \Sigma [\nabla g(\mu_1, \mu_2)] = \tau^2$, on algebra, proving part (b) of the theorem. □

19.2.3 Optimal Importance Sampling Distribution

We now address the question of the optimal choice of the importance sampling distribution. There is no unique way to define what an optimal choice means. We formulate one definition of optimality and provide an optimal importance sampling distribution. The optimal choice would not be practically usable, as we shown. However, the solution still gives useful insight.

Theorem 19.3. *Consider the importance sampling estimator $\hat{\mu} = \frac{1}{n} \sum_{i=1}^{n} \lambda(X_i) \phi(X_i)$ for $\mu = \int \phi(x) f_0(x) dx$, where $\lambda(x) = \frac{f_0(x)}{f_1(x)}$, and $X_1, \ldots, X_n$ are iid observations from F_1. Assume that $\phi(x) \geq 0$, and $\mu > 0$. Then, $\mathrm{Var}_{F_1}(\hat{\mu})$ is minimized when $f_1(x) = \frac{\phi(x) f_0(x)}{\mu}$.*

Proof. Because $X_1, \ldots, X_n$ are iid, so are $\lambda(X_1)\phi(X_1), \ldots, \lambda(X_n)\phi(X_n)$, and hence,

$$\mathrm{Var}_{F_1}(\hat{\mu}) = \frac{1}{n} \mathrm{Var}_{F_1}(\lambda(X_1)\phi(X_1)).$$

Clearly, this is minimized when with probability one under F_1, $\lambda(X_1)\phi(X_1)$ is a constant, say k. The constant k must be equal to the mean of $\lambda(X_1)\phi(X_1)$, that is,

$$k = \int \lambda(x) \phi(x) f_1(x) dx = \int \frac{\phi(x) f_0(x)}{f_1(x)} f_1(x) dx$$
$$= \int \phi(x) f_0(x) dx = \mu.$$

Therefore, the optimal importance sampling density satisfies $\lambda(x)\phi(x) = \mu \Rightarrow f_1(x) = \frac{\phi(x) f_0(x)}{\mu}$. □

This is not usable in practice, because it involves μ, which is precisely the unknown number we want to approximate. However, the theoretically optimal solution suggests that the importance sampling density should follow key properties of the unnormalized function $\phi(x) f_0(x)$. For example, f_1 should have the same shape and tail behavior as $\phi(x) f_0(x)$. Do and Hall (1989) show the advantages of using importance sampling and choosing the correct shape in distribution estimation problems that arise in the bootstrap. This reduces the variance of $\hat{\mu}$, and increases its accuracy.

19.2.4 Algorithms for Simulating from Common Distributions

Standard software for simulating from common univariate and multivariate distributions is now widely available. *Mathematica* permits simulation from essentially all common distributions. However, for the sake of quick simulations when efficiency is not of primary concern, a few simple rules for simulating from 25 common distributions are listed below. Their justification comes from various well-known results in standard distribution theory, many of which have been previously derived in this text itself.

Standard Exponential. To generate $X \sim \text{Exp}(1)$, generate $U \sim U[0, 1]$ and use $X = -\log U$.

Gamma with Parameters n and λ. To generate $X \sim G(n, \lambda)$, generate n independent values $X_1, \ldots, X_n$ from a standard exponential, and use $X = \lambda(X_1 + \cdots + X_n)$.

Beta with Integer Parameters m, n. To generate $X \sim \text{Be}(m, n)$, generate $U \sim G(m, 1)$, $V \sim G(n, 1)$ independently, and use $X = \frac{U}{U+V}$.

Weibull with General Parameters. To generate X from a Weibull distribution with parameters β, λ, generate $Y \sim \text{Exp}(1)$ and use $X = \lambda Y^{\frac{1}{\beta}}$.

Standard Double Exponential. To generate X from a standard double exponential, generate $X_1, X_2 \sim \text{Exp}(1)$ independently, and use $X = X_1 - X_2$.

Standard Normal. To generate $X \sim N(0, 1)$, use either of the following methods.
(i) Generate $U, V \sim U[0, 1]$ independently, and use $X = \sqrt{-2 \log U} \cos(2\pi V)$, $Y = \sqrt{-2 \log U} \sin(2\pi V)$. Then X, Y are independent $N(0, 1)$.
(ii) Use the accept–reject method with $g(x) = \frac{1}{2} e^{-|x|}$ and $c = \sqrt{\frac{2e}{\pi}}$.

Standard Cauchy. To generate $X \sim C(0, 1)$, generate $U \sim U[0, 1]$ and use $X = \tan[\pi(U - \frac{1}{2})]$.

t with Integer Degree of Freedom. To generate $X \sim t(n)$, generate $Z_1, Z_2, \cdots Z_{n+1} \sim N(0, 1)$ independently, and use

$$X = \frac{Z_1}{\sqrt{\frac{Z_2^2 + \cdots + Z_{n+1}^2}{n}}}.$$

lognormal with General Parameters. To generate X from a lognormal distribution with parameters μ, σ^2, generate $Z \sim N(0, 1)$ and use $X = e^{\mu + \sigma Z}$.

19.2 Textbook Simulation Techniques

Pareto with General Parameters. To generate X from a Pareto distribution with parameters α, θ, generate $U \sim U[0, 1]$ and use $X = \theta(1 - U)^{-\frac{1}{\alpha}}$.

Gumbel with General Parameters. To generate X from a Gumbel law with parameters μ, σ, generate $U \sim U[0, 1]$ and use $X = \mu - \sigma \log \log \frac{1}{U}$.

Multivariate Normal with General Parameters. To generate $X \sim N_d(\mu, \Sigma)$,

(i) Find the square root of Σ, that is, a $d \times d$ matrix A such that $\Sigma = AA'$.
(ii) Generate independent standard normals $Z_1, \ldots, Z_d$.
(iii) Use $X = \mu + AZ$, where $Z' = (Z_1, \ldots, Z_d)$.

Wishart with General Parameters. To generate $S \sim W_p(n, \Sigma)$, generate $X_1, X_2, \ldots X_n \sim N_p(0, \Sigma)$ independently, and use $S = \sum_{i=1}^{n} X_i X_i'$.

Dirichlet with Integer Parameters. To generate $(p_1, \ldots, p_n)$ from a Dirichlet distribution with integer parameters $m_1, \ldots, m_{n+1}$, generate $X_i \sim G(m_i, 1), i = 1, 2, \ldots, n + 1$ independently, and set

$$p_i = \frac{X_i}{\sum_{j=1}^{n+1} X_j}, \quad i = 1, 2, \ldots, n, \quad \text{and} \quad p_{n+1} = 1 - \sum_{j=1}^{n} p_j.$$

Uniform on the Boundary of the Unit Ball. To generate X from the uniform distribution on the boundary of the unit ball, generate $Z_1, \ldots, Z_d \sim N(0, 1)$ independently, and use $X_i = \frac{Z_i}{\sqrt{Z_1^2 + \cdots + Z_d^2}}, i = 1, \ldots, d$.

Uniform Inside the Unit Ball. To generate X from the uniform distribution in the d-dimensional unit ball, use either of the following methods.

(i) Generate Z according to a uniform distribution on the boundary of the d-dimensional unit ball, and independently generate $U \sim U[0, 1]$, and use $X = U^{\frac{1}{d}} Z$.
(ii) Generate $U_1, \ldots, U_d \sim U[-1, 1]$ independently, and use $X = (U_1, \ldots, U_d)'$, if $\sum_{i=1}^{d} U_i^2 \leq 1$. Otherwise, discard it and repeat.

Bernoulli with a General Parameter. To generate $X \sim \text{Ber}(p)$, generate $U \sim U[0, 1]$ and set $X = I_{U > 1-p}$.

Binomial with General Parameters. To generate $X \sim \text{Bin}(n, p)$, generate $X_1, \ldots, X_n \sim \text{Ber}(p)$ independently, and use $X = \sum_{i=1}^{n} X_i$.

Geometric with a General Parameter. To generate $X \sim \text{Geo}(p)$, generate $Y \sim \text{Exp}(1)$ and use $X = 1 + \left\lfloor \frac{Y}{-\log p} \right\rfloor$, where $\lfloor \ \rfloor$ denotes the integer part.

Negative Binomial with General Parameters. To generate $X \sim NB(k, p)$, generate $X_1, \ldots, X_k \sim \text{Geo}(p)$ independently, and use $X = \sum_{i=1}^{k} X_i$.

Poisson with a General Parameter. To generate $X \sim \text{Poi}(\lambda)$, generate $Y \sim \text{Bin}(n, \frac{\lambda}{n})$ with $n > 100\lambda$, and use $X = Y$.

Multinomial with General Parameters. To generate $(X_1, X_2, \ldots, X_k)$ from a multinomial distribution with general parameters $n, p_1, p_2, \ldots, p_k$,

(i) Generate $X_1 \sim \text{Bin}(n, p_1)$.
(ii) Generate $X_2 \sim \text{Bin}(n - X_1, \frac{p_2}{p_2 + \cdots + p_k})$.
(iii) Generate $X_3 \sim \text{Bin}(n - X_1 - X_2, \frac{p_3}{p_3 + \cdots + p_k})$, and so on.

Random Permutation. To generate $(\pi(1), \ldots, \pi(n))$ according to a uniform distribution on the set of $n!$ permutations of $(1, 2, \ldots, n)$,

(i) Choose one of the numbers $1, 2, \ldots, n$ at random and set it equal to $\pi(1)$.
(ii) Choose one of the remaining $n - 1$ numbers at random and set it equal to $\pi(2)$, and so on.

Brownian Motion on [0, 1]. To generate a path of a standard Brownian motion $W(t)$ on $[0, 1]$,

(i) Generate Bernoulli random variables with parameter $\frac{1}{2}$ $B_1, B_2, \ldots, B_n$ for $n \geq 100{,}000$.
(ii) Set $X_i = 2B_i - 1, i = 1, 2, \ldots, n$.
(iii) With $S_0 = 0$, $S_k = \sum_{i=1}^{k} X_i, k \geq 1$, set $W(t) = \frac{S_k}{\sqrt{n}}$ for $k \leq nt < k+1, k = 0, 1, \ldots, n$.

Homogeneous Poisson Process. To generate a path of a homogeneous Poisson process $X(t)$ with rate λ for $t \geq 0$,

(i) Generate standard exponential random variables $X_1, X_2, \ldots$.
(ii) Set $T_i = \frac{X_i}{\lambda}, i = 1, 2, \ldots$.
(iii) Set $X(t) = k$ if $T_1 + \cdots + T_k \leq t < T_1 + \cdots + T_{k+1}$.

Example 19.14 (Random points on the surface of a circle). We generate 20 points according to a uniform distribution on the surface of the unit circle in the two-dimensional plane by using the algorithm that was presented above (for a general dimension). To generate one point, we first generate iid standard normals Z_1, Z_2, and then the uniform random point X on the surface of the circle is taken to be

$$X = \left(\frac{Z_1}{\sqrt{Z_1^2 + Z_2^2}}, \frac{Z_2}{\sqrt{Z_1^2 + Z_2^2}} \right)'.$$

This is then repeated 20 times.

Two of the four quadrants have 7 of the 20 points each, and the other two have 3 points each. The expected number per quadrant is of course 5. The observed discrepancy from the expected number is not at all unusual. One can calculate the P value by using the method of Section 6.7. The second picture shows 200 random points inside the unit circle. Although the points are chosen according to the uniform distribution inside the circle, visually there are signs of some clustering and some void regions in the picture. Once again, this is not unusual.

19.3 Markov Chain Monte Carlo

The standard simulation techniques are difficult to apply, or even do not apply, when the target distribution is an unconventional one, or even worse, it is known only up to a normalizing constant: that is, $f(x) = \frac{h(x)}{c}$ for some explicit function h, but only an implicit normalizing constant c, because c cannot be computed exactly, or even to a high degree of accuracy. This problem often arises in simulating from posterior densities of a parameter (perhaps a vector)

$$\pi(\theta \mid x) = \frac{f(x \mid \theta)\pi(\theta)}{m(x)},$$

where $f(x \mid \theta)$ is the *likelihood function*, $\pi(\theta)$ is the prior density, and $m(x)$ is the marginal density of the observable X induced by (f, π) (see Chapter 3). Thus, $m(x) = \int_\Theta f(x \mid \theta)\pi(\theta)d\theta$, and it serves as the normalizing constant to the function $h(\theta) = f(x \mid \theta)\pi(\theta)$. But, if the parameter θ is high-dimensional, and the prior density $\pi(\theta)$ is not a very conveniently chosen one, then $m(x)$ usually cannot be calculated in closed-form, or even to a high degree of numerical approximation. All the simulation methods discussed in the previous section are useless in such a

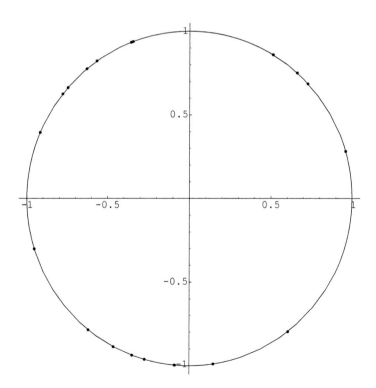

Fig. 19.2 Twenty random points on the unit circle

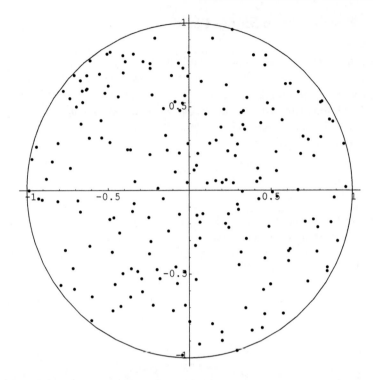

Fig. 19.3 Two hundred random points inside the unit circle

situation. It is a remarkable marriage of mathematical theory and dire practical need that Markov chain Monte Carlo methods allow one to simulate from the target density (distribution) π in such a situation. The simulated values can then be used to approximate probabilities $P_\pi(X \in A)$, or to approximate expectations, $E_\pi[\phi(X)]$, etc. Strikingly, the actual algorithms are remarkably simple and are very general. MCMC methods have certainly made statistical computing a whole lot easier, and the main bulk of the development in statistics took place in a short span of about 20 years. However, important questions on the speed of convergence of MCMC algorithms await additional concrete results, and these are difficult to obtain. MCMC schemes need not work in every problem; if the target distribution is multimodal, or has gaps in its support, then MCMC methods usually give misleading results and estimates. We present the basic MCMC principle, some popular special algorithms, and fundamental convergence theory for discrete state spaces alone in this section. The case of more general state spaces (e.g., $S = \mathcal{R}$), are treated in a separate section. We chose to present the discrete state space first, because the ideas are easier to assimilate in that case.

19.3.1 Reversible Markov Chains

As was mentioned in the introduction, MCMC methods follow this route.

(a) Identify the target distribution (or density) π and the state space S (i.e., the support of π) from which one wants to simulate.
(b) Construct a stationary Markov chain $X_n, n \geq 0$ on S with suitable additional properties such that X_n has a unique stationary distribution and the stationary distribution coincides with the target distribution π.
(c) Run the chain X_n for a sufficiently long time, and use the values of the chain $X_{B+1}, \cdots, X_n$ as correlated samples from π for some suitably large B.
(d) Approximate $E_\pi[\phi(X)]$ by $\frac{1}{(n-B)} \sum_{k=B+1}^{n} \phi(X_k)$.
(e) Form confidence intervals for $E_\pi[\phi(X)]$ based on $X_{B+1}, \ldots, X_n$. This is much more complicated than the previous steps.

The second step is the main algorithmic state, and choice of B is the main theoretical step. The construction of the chain is not unique. Given a target distribution π on the state space S, there are many stationary Markov chains with π as the stationary distribution. We have to answer two questions.

Question 1. How do we isolate an initial class of chains that definitely have π as the unique stationary distribution?

Question 2. How do we choose a specific chain from this class for implementation in a given problem?

Reversibility of a Markov chain plays an important role in answering the first question. The choice issue has been answered in classic research on MCMC, and general easy-to-apply recipes for choosing particular chains to run have been worked out. A few popular recipes are the Metropolis–Hastings algorithm, the Barker algorithm, the independent Metropolis algorithm, and the Gibbs sampler. We start with a description of reversible Markov chains and an explanation of why reversibility is a convenient way to address our *Question 1*.

Definition 19.1. A stationary Markov chain $X_n, n \geq 0$ on a countable state space S with the transition matrix $P = ((p_{ij})), i, j \in S$, is called *reversible* if there exists a nonnegative function $\pi(x)$ on S such that the *detailed balance equation* $p_{ij} \pi(i) = p_{ji} \pi(j)$ holds for all $i, j \in S, i \neq j$.

The justification for the name *reversible* stems from the following simple calculation.

Suppose $X_n, n \geq 0$ is reversible, and that the nonnegative function $\pi(x)$ is a probability function. Assume also that the initial distribution of the chain is π (i.e., $P(X_0 = k) = \pi(k), k \in S$). Then the marginal distribution of the state of the chain at all subsequent times remains equal to this initial distribution, that is, for all $n \geq 1, P(X_n = k) = \pi(k), k \in S$.

Now fix n. Then, from the defining equation for reversibility and Bayes' theorem,

$$P(X_n = j \mid X_{n+1} = i) = \frac{P(X_{n+1} = i \mid X_n = j) P(X_n = j)}{P(X_{n+1} = i)}$$

$$= \frac{p_{ji}\pi(j)}{\pi(i)} = \frac{p_{ij}\pi(i)}{\pi(i)} = p_{ij}$$

$$= P(X_{n+1} = j \mid X_n = i).$$

In other words, in the statement, "The probability that the next state is j given that the current state is i is p_{ij}" we have the liberty to take either n as the current time and $n+1$ as the next, or, $n+1$ as the current time and n as the next. That is, if we run our clock backwards, then it would seem to us that the chain is evolving the same way as it did when the clock ran forward; hence the name a reversible chain.

But we get even more. Assume that the chain is regular (see Chapter 10) and that the state space S is finite. By summing the defining equation for reversibility over j, we get, for every $i \in S$,

$$\sum_{j \in S} p_{ji}\pi(j) = \sum_{j \in S} p_{ij}\pi(i) = \pi(i) \sum_{j \in S} p_{ij} = \pi(i).$$

Therefore, by Theorem 10.5, π must be the unique stationary distribution of the Markov chain $X_n, n \geq 0$. In fact we do not require the finiteness of the state space or the assumption of regularity of the chain. The fact that a reversible chain has π as its stationary distribution holds under weaker conditions. Here is a standard version of that result.

Theorem 19.4. *Suppose $X_n, n \geq 0$ is an irreducible and aperiodic stationary Markov chain on a discrete state space S such that for some probability distribution π on S with $\pi(k) > 0$ for all k, the reversibility equation $p_{ij}\pi(i) = p_{ji}\pi(j)$ holds for all $i, j \in S$. Then, π is a unique stationary distribution for the chain $X_n, n \geq 0$.*

As a consequence, in order to devise an MCMC algorithm to draw simulations from a target distribution π on a countable set S, we only have to invent a stationary Markov chain (or equivalently, a transition matrix P), which is irreducible and aperiodic and is such that the reversibility equation $p_{ij}\pi(i) = p_{ji}\pi(j)$ holds. There are infinitely many such transition matrices P for a given target π. But a few special recipes for choosing P have earned a special status over the years. These algorithms go by the collective name of Metropolis algorithms, *and are discussed in the next section.*

On the justifiability of approximating the true average $E_\pi[\phi(X)]$ by the sample average $\frac{1}{n}\sum_{k=1}^{n} \phi(X_k)$ (or $\frac{1}{(n-B)}\sum_{k=B+1}^{n} \phi(X_k)$ after throwing out an initial segment), we have the Markov chain SLLN, generally known as the strong ergodic theorem. *Compare it with the* weak ergodic theorem, *Theorem 10.7.*

Theorem 19.5. *Suppose $X_n, n \geq 0$ is an irreducible stationary Markov chain on a discrete state space S. Suppose X_n is known to possess a stationary distribution π.*

19.3 Markov Chain Monte Carlo

Let $\phi : S \to \mathcal{R}$ be such that $E_\pi[\phi(X)]$ exists, that is, $\sum_{i \in S} |\phi(i)| \pi(i) < \infty$. Then, for any initial distribution μ, $\frac{1}{n} \sum_{k=1}^{n} \phi(X_k) \overset{a.s.}{\to} E_\pi[\phi(X)]$; that is,
$$P_\mu\left(\frac{1}{n} \sum_{k=1}^{n} \phi(X_k) \overset{a.s.}{\to} E_\pi[\phi(X)]\right) = 1.$$

Proof. The proof uses the fact that for a positive recurrent Markov chain on a discrete state space, the proportion of times that the chain is at a particular state i converges almost surely to the stationary probability of that state i. That is, given $n \geq 1$ and $i \in S$, suppose $V_i(n) = \sum_{k=1}^{n} I_{X_k = i}$. Then, $\frac{V_i(n)}{n} \overset{a.s.}{\to} \pi(i)$. We can now see intuitively why the strong ergodic theorem is true. We have:

$$\frac{1}{n} \sum_{k=1}^{n} \phi(X_k) = \frac{1}{n} \sum_{i \in S} \sum_{k: X_k = i} \phi(X_k)$$
$$= \frac{1}{n} \sum_{i \in S} \phi(i) \sum_{k: X_k = i} 1 = \frac{1}{n} \sum_{i \in S} \phi(i) V_i(n)$$
$$= \sum_{i \in S} \phi(i) \frac{V_i(n)}{n} \approx \sum_{i \in S} \phi(i) \pi(i) = E_\pi[\phi(X)].$$

Formally, fix an $\epsilon > 0$, and find a finite subset of states $E \subseteq S$ such that $\pi(S - E) < \epsilon$. Let T denote the number of elements in the set E, and suppose n is large enough that for each $i \in E$, $\left|\frac{V_i(n)}{n} - \pi(i)\right| < \epsilon$. Such an n can be chosen because E is a finite set, and because we have the almost sure convergence property that $\frac{V_i(n)}{n}$ converges to $\pi(i)$ for any fixed i, as was mentioned above. Also, note for the sake of the proof below that $\sum_{i \in S} V_i(n) = n \Rightarrow \sum_{i \in S} \left[\frac{V_i(n)}{n} - \pi(i)\right] = 0$. We prove the theorem for functions ϕ that are bounded, and in that case we may assume that $|\phi| \leq 1$.

Hence,

$$\left|\frac{1}{n} \sum_{k=1}^{n} \phi(X_k) - \sum_{i \in S} \phi(i) \pi(i)\right|$$
$$= \left|\sum_{i \in S} \phi(i) \frac{V_i(n)}{n} - \sum_{i \in S} \phi(i) \pi(i)\right| = \left|\sum_{i \in S} \phi(i) \left[\frac{V_i(n)}{n} - \pi(i)\right]\right|$$
$$\leq \sum_{i \in S} |\phi(i)| \left|\frac{V_i(n)}{n} - \pi(i)\right|$$
$$= \sum_{i \in E} |\phi(i)| \left|\frac{V_i(n)}{n} - \pi(i)\right| + \sum_{i \in S-E} |\phi(i)| \left|\frac{V_i(n)}{n} - \pi(i)\right|$$
$$\leq \sum_{i \in S} \left|\frac{V_i(n)}{n} - \pi(i)\right| + \sum_{i \in S-E} \left[\frac{V_i(n)}{n} + \pi(i)\right]$$

$$\leq \sum_{i \in S} \left| \frac{V_i(n)}{n} - \pi(i) \right| + \sum_{i \in S-E} \left[\frac{V_i(n)}{n} - \pi(i) \right] + 2 \sum_{i \in S-E} \pi(i)$$

$$= \sum_{i \in S} \left| \frac{V_i(n)}{n} - \pi(i) \right| + \sum_{i \in E} \left[\pi(i) - \frac{V_i(n)}{n} \right] + 2 \sum_{i \in S-E} \pi(i)$$

$$\leq T\epsilon + T\epsilon + 2\epsilon = 2(T+1)\epsilon.$$

Because ϵ is arbitrary, the theorem follows. $\square$

19.3.2 Metropolis Algorithms

The general principle of any Metropolis algorithm is the following. Suppose the chain is at some state $i \in S$ at the current instant. Then, as a first step, one of the states, say j, from the states in S is picked according to some probability distribution for possibly moving to that state j. The state j is commonly called *a candidate state*. The distribution used to pick this candidate state is called a *proposal distribution*. Then, as a second step, if j happens to be different from i, then we either move to the candidate step j with some designated probability, or we decline to move, and therefore remain at the current state i. Thus, the entries of the overall transition matrix P have the multiplicative structure

$$p_{ij} = \theta_{ij}\gamma_{ij}, i, j \in S, \quad j \neq i,$$
$$p_{ii} = 1 - \sum_{j \neq i} p_{ij},$$

where

$\theta_{ij} = P(\text{State j is picked as the candidate state});$

$\gamma_{ij} = P(\text{The chain actually moves to the candidate state j}).$

The matrix $((\theta_{ij}))$ is chosen to be irreducible, in order that the ultimate transition matrix $P = ((p_{ij}))$ is irreducible. Note that the Metropolis algorithm has a formal similarity to the accept–reject scheme. A candidate state j is picked according to the *proposal probabilities* θ_{ij} and once picked, is accepted according to the *acceptance probabilities* γ_{ij}. Special choices for θ_{ij}, γ_{ij} lead to the following well-known algorithms.

Independent Sampling. Choose

$$\theta_{ij} = \pi(j) \,\forall\, i, \text{ and } \gamma_{ij} \equiv 1.$$

19.3 Markov Chain Monte Carlo

Metropolis–Hastings Algorithm. Choose

$$\theta_{ij} = c = \text{constant}, \text{ and } \gamma_{ij} = \min\left\{1, \frac{\pi(j)}{\pi(i)}\right\}.$$

Barker's Algorithm. Choose

$$\theta_{ij} = c = \text{constant}, \text{ and } \gamma_{ij} = \frac{\pi(j)}{\pi(i) + \pi(j)}.$$

Independent Metropolis Algorithm. For all i, choose

$$\theta_{ij} = p_j, \text{ and } \gamma_{ij} = \min\left\{1, \frac{\frac{\pi(j)}{p_j}}{\frac{\pi(i)}{p_i}}\right\}.$$

It is implicitly assumed that the state space S is finite for the Metropolis–Hastings and the Barker algorithm to apply. Note that the Metropolis–Hastings, independent Metropolis, and the Barker algorithm only require the specification of the target distribution π up to normalizing constants. That is, if we only knew that $\pi(k) = \frac{h(k)}{c}$, where $h(k)$ is explicit, but the normalizing constant c is not, we can still execute all of these algorithms. It is also worth noting that in the Metropolis–Hastings algorithm, if a state j gets picked as the candidate state, and j happens to be more likely than the current state i, then the chain surely moves to j. This is not the case for Barker's algorithm.

A fourth MCMC algorithm that is especially popular in statistics, and particularly in Bayesian statistics, is the Gibbs sampler, which is treated separately. Note that for each proposed algorithm, one needs to verify two things; that the chain is irreducible and aperiodic and that the time-reversibility equation indeed holds.

Example 19.15 (Verifying the Assumptions). First consider the case of purely independent sampling. Then, $p_{ij} = \pi(j)$ and $p_{ji} = \pi(i)$. So irreducibility holds, because P itself has all entries strictly positive. Also, $\pi(i)p_{ij} = \pi(i)\pi(j) = p_{ji}\pi(j)$, and so time-reversibility also holds.

Next, consider the Metropolis–Hastings algorithm. Fix i, j. Consider the case that $\pi(j) \geq \pi(i)$. Then, $p_{ij} = c$, and $p_{ji} = c\frac{\pi(i)}{\pi(j)}$. Therefore, every entry in P is again strictly positive, and so the chain is irreducible. Furthermore, $p_{ji}\pi(j) = c\pi(i) = p_{ij}\pi(i)$, and therefore, time-reversibility holds. For the other case, namely $\pi(i) \geq \pi(j)$, the proof of reversibility is the same. The verification of the assumptions for the Barker algorithm and the independence Metropolis sampler is a chapter exercise.

Example 19.16 (Simulation from a Beta–Binomial Distribution). If X given p is distributed as $\text{Bin}(n, p)$ for some fixed n, and p has a Beta distribution with parameters α and β, then the marginal distribution of X on the state space $S = \{0, 1, \ldots, n\}$ is called a Beta–Binomial distribution. It has the pmf

$$m(x) = \frac{\Gamma(\alpha+\beta)}{\Gamma(\alpha)\Gamma(\beta)}\binom{n}{x}\int_0^1 p^{x+\alpha-1}(1-p)^{n-x+\beta-1}dp$$

$$= \frac{\Gamma(\alpha+\beta)}{\Gamma(\alpha)\Gamma(\beta)}\binom{n}{x}\frac{\Gamma(x+\alpha)\Gamma(n-x+\beta)}{\Gamma(n+\alpha+\beta)}, \quad x = 0, 1, \ldots, n.$$

To simulate from the Beta–Binomial distribution by using the Metropolis–Hastings algorithm, if we are currently at some state $i, 0 \leq i \leq n$, then one of the states j is chosen at random as a candidate state for the chain to move, and the move is actually made with a probability of $\min\{1, \frac{m(j)}{m(i)}\}$. By using the formula for $m(x)$ from the above,

$$\frac{m(j)}{m(i)} = \frac{i!(n-i)!\Gamma(j+\alpha)\Gamma(n-j+\beta)}{j!(n-j)!\Gamma(i+\alpha)\Gamma(n-i+\beta)}.$$

Thus, the overall transition matrix of the Metropolis–Hastings chain is

$$p_{ij} = \frac{1}{n+1}\min\left\{1, \frac{i!(n-i)!\Gamma(j+\alpha)\Gamma(n-j+\beta)}{j!(n-j)!\Gamma(i+\alpha)\Gamma(n-i+\beta)}\right\}, j \neq i,$$

$$p_{ii} = 1 - \sum_{j \neq i} p_{ij}.$$

Note that the chain is in fact regular. Interestingly, if $\alpha = \beta = 1$, so that $p \sim U[0, 1]$, then this works out to $p_{ij} \equiv \frac{1}{n+1}$. That is, if the chain moves, it moves to one of the noncurrent states with an equal probability.

Example 19.17 (Simulation from a Truncated Geometric Distribution). Suppose $Y \sim \text{Geo}(p)$, and we want to simulate from the conditional distribution of Y given that $Y \leq n$, where $n > 1$ is some specified integer. The pmf of this conditional distribution is

$$\pi(x) = \frac{P(Y=x)}{P(Y \leq n)} I_{1 \leq x \leq n}$$

$$= \frac{p(1-p)^{x-1}}{1-\sum_{x=n+1}^{\infty} p(1-p)^{x-1}} I_{1 \leq x \leq n} = \frac{p(1-p)^{x-1}}{1-(1-p)^n} I_{1 \leq x \leq n}.$$

Therefore, for any $i, j \leq n$,

$$\frac{\pi(j)}{\pi(i)} = (1-p)^{j-i} \leq 1 \text{ iff } j \geq i.$$

Thus, the overall transition matrix of the Metropolis–Hastings chain would be

$$p_{ij} = \frac{1}{n} \text{ if } j < i;$$

$$p_{ij} = \frac{1}{n}(1-p)^{j-i} \text{ if } j > i;$$

$$p_{ii} = 1 - \sum_{j \neq i} p_{ij}.$$

The general Metropolis algorithm uses more general forms for the probabilities θ_{ij}, γ_{ij}, and we provide this general form next.

Definition 19.2. The *general Metropolis algorithm* corresponds to the transition probabilities $p_{ij} = \theta_{ij}\gamma_{ij}$, where $((\theta_{ij}))$ is a general irreducible transition probability matrix, and γ_{ij} are defined as $\gamma_{ij} = \min\left\{1, \frac{\pi(j)\theta_{ji}}{\pi(i)\theta_{ij}}\right\}$.

If the proposal distribution is symmetric (i.e., if $\theta_{ji} = \theta_{ij}$ for all pairs i, j), then the general Metropolis algorithm reduces to the previously described basic Metropolis algorithm. Thus, more generality is attained only by choosing the matrix $((\theta_{ij}))$ to be not symmetric. The general Metropolis algorithm has the desired convergence property and it is a consequence of the general convergence theorem for irreducible reversible chains (Theorem 19.4).

Theorem 19.6. *The general Metropolis algorithm has π as its stationary distribution.*

19.4 The Gibbs Sampler

The Metropolis algorithms of the previous section can be difficult to apply when the dimension of the state space is high, and the likelihood ratios $\frac{\pi(y)}{\pi(x)}, x, y \in S$ depend on all the coordinates of x, y. The generation of the chain becomes too much of a multidimensional problem and becomes at least unwieldy, and possibly undoable.

A collection of special Metropolis algorithms very cleverly reduces the multidimensional nature of the Markov chain generation problem to a sequence of *one-dimensional* problems. To explain it more precisely, suppose a state x in the state space S is a vector in some m-dimensional space with coordinates $(x_1, x_2, \ldots, x_m)$. Suppose we are currently in state x. To make a transition to a new state $y \in S$, we change coordinates one at a time, such as $(x_1, x_2, \ldots, x_m) \longrightarrow (y_1, x_2, \ldots, x_m) \longrightarrow (y_1, y_2, \ldots, x_m) \longrightarrow \cdots (y_1, y_2, \ldots, y_m)$, and each coordinate change is made by using the conditional distribution of that coordinate given the rest of the coordinates. For example, the transition $(x_1, x_2, \ldots, x_m) \longrightarrow (y_1, x_2, \ldots, x_m)$ is made by simulating from the distribution $\pi(x_1 \mid x_2, \ldots, x_m)$. These conditional distributions of one coordinate given all the rest are called *full conditionals*. Therefore, as long as we can calculate and also simulate from all the full conditionals, a complicated multidimensional problem is broken into m one-dimensional problems, which is a productive simplification in many applications. It can be shown that the sequence of observations thus generated forms a stationary Markov chain and under suitable conditions, the chain has the target distribution π as its stationary distribution. It should be noted that in some applications, the full conditionals do not depend on all the $m-1$ coordinates that one conditions on, but only on a smaller number of them. This happy coincidence makes the algebraic aspect of simulating runs of the Markov chain even simpler.

This special Metropolis algorithm, that uses only the full conditionals to form the proposal probabilities and uses acceptance probabilities equal to one, is called the *Gibbs sampler*. It arose in mathematical physics in the works of Glauber (1963) in a more general setup than the finite-dimensional Euclidean setup described here. The Gibbs sampler is a hugely popular tool in several areas of statistics, notably simulation of posterior distributions in Bayesian statistics, image processing, generation of random contingency tables, random walks on gigantic state spaces, such as those that arise in models of random graphs and Bayesian networks, molecular biology, and many others.

The Gibbs sampler, by definition, makes its transitions by changing one coordinate of the m-dimensional vector at a time. At any particular step of this transition, the coordinate to be changed may be chosen at random, or according to some deterministic order. If the selection is random, the corresponding algorithm is called a *random scan Gibbs sampler*, and if the selection is deterministic, the algorithm is called a *systematic scan Gibbs sampler*. There are some important optimality issues regarding which type of scan is preferable, and we discuss them briefly later in this section.

Suppose, for specificity, that we use a random scan Gibbs sampler. Denote the full conditionals by using the notation $\pi(x_i \mid x_{-i})$. Denote the current state by $x = (x_1, \ldots, x_m)$. Pick the coordinate to be changed at random from the m available coordinates. If the coordinate picked is i, then make a transition from x to a new vector $y = (x_1, \ldots, x_{i-1}, y_i, x_{i+1}, \ldots, x_m)$ by using the full conditional $\pi(y_i \mid x_{-i})$. Then, the transition probability matrix of the chain is $P = \frac{1}{m}(P_1 + \cdots + P_m)$, where P_i has entries given by

$$p_{i,xy} = \pi(y_i \mid x_{-i}) I_{y_j = x_j \, \forall \, j \neq i}.$$

Each P_i is reversible, and hence so is P. To show that a given P_i is reversible, we verify directly

$$\pi(x) p_{i,xy} = \pi(x_1, \ldots, x_{i-1}, x_i, x_{i+1}, \ldots, x_m)$$
$$\times \frac{\pi(x_1, \ldots, x_{i-1}, y_i, x_{i+1}, \ldots, x_m)}{\sum_{y_i} \pi(x_1, \ldots, x_{i-1}, y_i, x_{i+1}, \ldots, x_m)} I_{y_j = x_j \, \forall \, j \neq i}$$
$$= \frac{\pi(y_1, \ldots, y_{i-1}, x_i, y_{i+1}, \ldots, y_m) \pi(y_1, \ldots, y_{i-1}, y_i, y_{i+1}, \ldots, y_m)}{\sum_{y_i} \pi(y_1, \ldots, y_{i-1}, y_i, y_{i+1}, \ldots, y_m)}$$
$$\times I_{y_j = x_j \, \forall \, j \neq i}$$

19.4 The Gibbs Sampler

$$= \pi(y_1, \ldots, y_{i-1}, y_i, y_{i+1}, \ldots, y_m)$$
$$\times \frac{\pi(y_1, \ldots, y_{i-1}, x_i, y_{i+1}, \ldots, y_m)}{\sum_{y_i} \pi(y_1, \ldots, y_{i-1}, y_i, y_{i+1}, \ldots, y_m)} I_{y_j = x_j \,\forall\, j \neq i}$$
$$= \pi(y) p_{i, yx}.$$

Likewise, if the P_i are irreducible, so will be P. A simple sufficient condition for irreducibility is that $\pi(x) > 0$ on the state space. This is a simple sufficient condition, and weaker conditions for irreducibility are available.

For the systematic scan Gibbs sampler, we change the coordinates in a predetermined sequence. For example, if we change the coordinates in the natural sequence $1, 2, \ldots m$, then the transition probability matrix P of the chain is $P = P_1 P_2 \cdots P_m$. So, provided that the P_i are irreducible, P is still irreducible. However, reversibility will typically be lost. Reversibility is not absolutely essential, because the chain may have π as the unique stationary distribution even if reversibility fails. But it is useful to have the reversibility property, because then Theorem 19.4 guarantees the desired convergence property, without requiring further verification. To ensure reversibility, we may use a hybrid of the systematic scan and the random scan Gibbs sampler, by first choosing the ordering of the change of coordinates at random from the $m!$ possible orderings of $1, 2, \ldots, m$. We call this the *order randomized systematic scan Gibbs sampler*. The transition probability matrix P now is

$$P = \frac{1}{m!} \sum_{S(m)} P_{i_1} P_{i_2} \cdots P_{i_m},$$

where $\sum_{S(m)}$ denotes the sum over all $m!$ permutations of $1, 2, \ldots, m$. Once we have both reversibility and irreducibility, we get the desired convergence result (Theorem 19.4). We state this formally.

Theorem 19.7. *Suppose $\pi(x) > 0$ for all $x \in S$. Then the random scan and the order randomized systematic scan Gibbs sampler both have π as the unique stationary distribution of the chain.*

Example 19.18 (The Beta–Binomial Pair). To continue with Example 19.14, suppose that X given p is distributed as $\text{Bin}(m, p)$ and p has a Beta prior distribution with parameters α and β. Then, the posterior distribution of p given $X = x$ is $\text{Be}(x + \alpha, m - x + \beta)$. The systematic scan Gibbs sampler runs the bivariate Markov chain $(X_k, p_k), k = 0, 1, 2, \ldots$ in the following steps:

(a) Choose an initial value $p_0 \sim \text{Be}(\alpha, \beta)$.
(b) Choose an initial value $X_0 = x_0$ from $\text{Bin}(m, p_0), 0 \le x_0 \le m$.
(c) Generate p_1 from $\text{Be}(x_0 + \alpha, m - x_0 + \beta)$.
(d) Generate $X_1 = x_1$ from $\text{Bin}(m, p_1)$.
(e) Repeat.

This produces a bivariate chain that has the bivariate distribution

$$g(x,p) = \frac{\Gamma(\alpha+\beta)}{\Gamma(\alpha)\Gamma(\beta)}\binom{m}{x}p^{x+\alpha-1}(1-p)^{m-x+\beta-1}, \quad x = 0,1,\ldots,m, 0 < p < 1$$

as its stationary distribution. We can reverse the order of the systematic scan, that is, start with a fixed x_0, get p_0 from $\text{Be}(x_0+\alpha, m-x_0+\beta)$, then x_1 from $\text{Bin}(m, p_0)$, and so on, and the convergence result will still hold.

Example 19.19 (A Gaussian Bayes Example). Posterior densities were introduced in this text in Chapter 3 as conditional distributions of the parameter when there is a bona fide joint distribution on the data value and the parameter, induced by the likelihood function and a prior probability distribution on the parameter. Sometimes, a nonnegative function $\pi(\theta)$ of θ is used as a *formal prior* on θ, although $\pi(\theta)$ is not a valid probability density; that is, $\int_\Theta \pi(\theta)d\theta$ is actually infinite. Such formal priors are called *improper priors*. It can happen, however, that if we forget the fact that $\pi(\theta)$ is not a probability density, but plug it into the formula for the conditional density of θ given $X = x$, that is, in the formula

$$\pi(\theta \mid x) = cf(x \mid \theta)\pi(\theta),$$

then $\pi(\theta \mid x)$ is indeed a probability density with a suitable finite normalizing constant c. In that case, one sometimes proceeds as if $\pi(\theta \mid x)$ is a bona fide posterior density and does inference with it. This is such an example of the application of the Gibbs sampler to a posterior density arising from an improper prior.

Suppose given $\theta = (\mu, \sigma^2)$, $X_1, X_2, \ldots, X_n$ are iid $N(\mu, \sigma^2)$, and that for $\theta = (\mu, \sigma^2)$ we use the nonnegative function $\pi(\theta) = \frac{1}{\sigma^2}d\mu d\sigma^2$ as an improper prior. It is important to note that σ^2 is being treated as the second coordinate of θ, not σ. One may informally interpret such a prior as treating μ and σ^2 to be independent, and then putting, respectively, the improper prior densities $\pi_1(\mu) \equiv 1$ and $\pi_2(\sigma^2) = \frac{1}{\sigma^2}(\sigma^2 > 0)$.

Our final goal is to simulate observations from the posterior distribution of $\theta = (\mu, \sigma^2)$ given the data $X = (X_1, \ldots, X_n)$. To apply the Gibbs sampler, we need the *full conditionals*. Because θ is two-dimensional, there are only two full conditionals to find, namely, the posterior distribution of μ given σ^2, and the posterior distribution of σ^2 given μ. Note that when we consider posteriors, the data values are considered as fixed. So X does not enter into this Gibbs picture as another variable.

It turns out that even though the prior is an improper prior, the formal posterior is a bona fide probability density on $\theta = (\mu, \sigma^2)$ for any $n \geq 1$. Furthermore, these two full conditionals are very easy to find. In the following lines, c denotes a generic normalizing constant, and is not intended to be the same number at each occurrence. Consider the conditional of μ given σ^2. This is formed as

19.4 The Gibbs Sampler

$$\pi(\mu \mid \sigma^2, X) = c \frac{1}{\sigma^n} e^{-\frac{1}{2\sigma^2} \sum_{i=1}^n (X_i - \mu)^2}$$

$$= c \frac{1}{\sigma^n} e^{-\frac{1}{2\sigma^2} \sum_{i=1}^n (X_i - \overline{X})^2} e^{-\frac{n}{2\sigma^2}(\overline{X} - \mu)^2}$$

$$= c e^{-\frac{n}{2\sigma^2}(\overline{X} - \mu)^2},$$

and therefore, $\pi(\mu \mid \sigma^2, X)$ is the density of a $N(\overline{X}, \frac{\sigma^2}{n})$ distribution.

Next, the conditional of σ^2 given μ is

$$\pi(\sigma^2 \mid \mu, X) = c \frac{1}{\sigma^n} e^{-\frac{1}{2\sigma^2} \sum_{i=1}^n (X_i - \mu)^2} \frac{1}{\sigma^2}$$

$$= c \frac{1}{(\sigma^2)^{\frac{n}{2}+1}} e^{-\frac{1}{(2\sigma^2)} \sum_{i=1}^n (X_i - \mu)^2}.$$

Therefore, by transforming to $w = \frac{1}{\sigma^2}$, from the Jacobian formula, the conditional of w given μ is

$$\pi(w \mid \mu, X) = c w^{\frac{n}{2}-1} e^{-\frac{w}{2} \sum_{i=1}^n (X_i - \mu)^2}.$$

Making one final transformation to $v = w \sum_{i=1}^n (X_i - \mu)^2$, the conditional of v given μ is

$$\pi(v \mid \mu, X) = c v^{\frac{n}{2}-1} e^{-\frac{v}{2}}.$$

We recognize this to be the density of a chi-square distribution with $\frac{n}{2}$ degrees of freedom. To summarize the steps of the systematic scan Gibbs sampler for this problem,

Step 1. Start with an initial value σ_0^2.
Step 2. Generate μ_0 from $N(\overline{X}, \frac{\sigma_0^2}{n})$.
Step 3. Generate σ_1^2 by first simulating v_1 from $\chi_{\frac{n}{2}}^2$, the chi-square distribution with $\frac{n}{2}$ degrees of freedom, and then set $\sigma_1^2 = \frac{\sum_{i=1}^n (X_i - \mu_0)^2}{v_1}$.
Step 4. Generate μ_1 from $N\left(\overline{X}, \frac{\sigma_1^2}{n}\right)$, and so on and repeat.

Example 19.20 (Gibbs Sampling from Dirichlet Distributions). The success of the Gibbs sampler hinges on one's ability to write and easily simulate from the one-dimensional full conditionals. One important case where this works out very nicely is that of the n-dimensional Dirichlet distribution (see Chapter 4). The Dirichlet distribution arises naturally as models for data on a simplex. They also arise prominently in Bayesian statistics in numerous situations, for example, as priors for the parameters of a multinomial distribution, and also in dealing with infinite-dimensional Bayesian problems.

A canonical representation of a Dirichlet random vector with parameter $\alpha = (\alpha_1, \alpha_2, \ldots, \alpha_{n+1})$ is given by

$$(p_1, p_2, \ldots, p_n) \stackrel{\mathcal{L}}{=} \left(\frac{X_1}{X_1 + X_2 + \cdots + X_{n+1}}, \frac{X_2}{X_1 + X_2 + \cdots + X_{n+1}}, \ldots, \frac{X_n}{X_1 + X_2 + \cdots + X_{n+1}} \right),$$

where X_i are independent Gamma variables with parameters α_i and 1. Therefore, it is possible to simulate a Dirichlet random vector by simulating $n+1$ independent Gamma variables. However, the Gibbs sampler is an alternative and relatively simple method for this case, because by Theorem 4.9, the full conditionals are simply Beta distributions. Precisely,

$$\frac{p_i}{1 - \sum_{j \neq i=1}^{n} p_j} \mid \{p_j, j \neq i\} \sim \mathrm{Be}(\alpha_i, \alpha_{n+1}).$$

So, to simulate from a Dirichlet distribution by using the systematic scan Gibbs sampler, we can

Step 1. Fix initial values $p_{20}, \ldots, p_{n0}$ for $p_2, \ldots, p_n$.
Step 2. Simulate a Beta variable Y_1 from $\mathrm{Be}(\alpha_1, \alpha_{n+1})$ and set $p_{10} = (1 - \sum_{j=2}^{n} p_{j0}) Y_1$.
Step 3. Simulate a Beta variable Y_2 from $\mathrm{Be}(\alpha_2, \alpha_{n+1})$ and set $p_{21} = (1 - \sum_{j \neq 2} p_{j0}) Y_2$, and so on, by updating the coordinates one at a time. The resulting chain will not be reversible, but reversibility can be retrieved by using a somewhat more involved updating scheme; see the general principles presented at the beginning of this section.

Example 19.21 (Gibbs Sampling in a Change Point Problem). Suppose for some given $n \geq 2$, and $1 \leq k \leq n$, we have $X_1, \ldots, X_k \sim \mathrm{Poi}(\theta)$, and $X_{k+1}, \ldots, X_n \sim \mathrm{Poi}(\nu)$, where all n observations are mutually independent, and k, θ, ν are unknown parameters. This is an example of a *change point problem*, where an underlying sequence of Poisson distributed random variables has a level change at some unknown time within the interval of the study. This unknown change point is the parameter k. The problem is unusual in one parameter being integer-valued, and the others continuous. Change point problems are rather difficult from a theoretical perspective.

Suppose the problem is attacked by assuming a prior distribution on (k, θ, ν), with

$$k \sim \mathrm{Unif}\{1, 2, \ldots, n\};$$
$$\theta \sim G(\alpha, \lambda_1), \quad \nu \sim G(\beta, \lambda_2),$$

where we assume that k, θ, ν are independent.

19.5 Convergence of MCMC and Bounds on Errors

To implement the systematic scan Gibbs sampler in the order of updating $k \to \theta \to \nu$, we only need to know what the full conditionals are. By direct calculations,

$$k \mid \theta, \nu \sim \pi(k \mid \theta, \nu) = \frac{e^{(\nu-\theta)k} \left(\frac{\nu}{\theta}\right)^{\sum_{i=1}^{k} x_i}}{\sum_{k=1}^{n} e^{(\nu-\theta)k} \left(\frac{\nu}{\theta}\right)^{\sum_{i=1}^{k} x_i}} I_{1 \leq k \leq n};$$

$$\theta \mid (k, \nu) \sim G\left(\alpha + \sum_{i=1}^{k} x_i, \left(\frac{1}{\lambda_1} + k\right)^{-1}\right);$$

$$\nu \mid (k, \theta) \sim G\left(\beta + \sum_{i=k+1}^{n} x_i, \left(\frac{1}{\lambda_2} + n - k\right)^{-1}\right).$$

Each of these full conditionals is a standard distribution. The full conditional corresponding to k is just a multinomial distribution, and the other two full conditionals are Gamma distributions. It is easy to simulate from each of them, and so the Gibbs chain can be generated easily. Special interest lies in estimating the change point k. Once one has the three-component Gibbs chain, the k-component of it can be used to approximate the posterior mean of k; there are also methods to estimate the posterior distribution itself from the output of the Gibbs chain.

19.5 Convergence of MCMC and Bounds on Errors

Theorem 19.4 does ensure that a stationary Markov chain with suitable additional properties has the target distribution π as its unique stationary distribution. In particular, whatever the initial distribution μ is, for any state $i \in S$, $P_\mu(X_n = i) \to \pi(i)$ as $n \to \infty$. However, in practice we can only run the chain for a finite amount of time, and hence only for finitely many steps. So, a question of obvious practical importance is how large the value of n should be in a given application for the true distribution of X_n to closely approximate the target distribution π, according to some specified measure of distance between distributions. Several probability metrics introduced in Chapter 15 now become useful. A related question is what can we say about the distance between the true distribution of X_n and the target distribution π for a given n.

Without any doubt, a study of these questions forms the most challenging and the most sophisticated part of the MCMC theme. A broad general rule is that the second largest eigenvalue in modulus (SLEM) of our transition probability matrix P is going to determine the rapidity of the convergence of the chain to stationarity. The smaller the SLEM is, the faster will be the convergence. However, it is usually extremely difficult to go further, and find the SLEM explicitly. We can instead try to give concrete bounds on the SLEM, or directly the distance between the true distribution of X_n and the target distribution, and each new problem usually requires ingenious new methods. The area is still flourishing, and a fascinating interplay of powerful tools from pure mathematics and probability is producing increasingly sophisticated results.

There are some general theorems on the SLEM and hence speed of convergence, and these are useful to know. They are mostly for reversible chains on finite-state spaces, although there are some exceptions. Some of these are presented with illustrative examples in this section. These results are taken primarily from Dobrushin (1956), Diaconis and Stroock (1991), Diaconis and Saloff-Coste (1996), and Brémaud (1999). References on convergence of the Gibbs sampler are given later.

First, we need to choose the specific distances that we use to determine the rapidity of convergence of the chain. Use of the *total variation distance* is the most common; the *separation* and the *chi-square distances* are also used. We redefine these distances here for easy reference; they were introduced and treated in detail in Chapter 15.

Suppose $P^n = \big((p_{ij}(n))\big)_{t \times t}$ is the n-step transition probability matrix for our stationary chain on the finite state space $S = \{1, 2, \ldots, t\}$. Suppose that enough conditions on the chain have been assumed so that we know that a unique stationary distribution π exists. If the initial distribution of the chain is μ, then the distribution of X_n for fixed n is

$$P_\mu(X_n = i) = \sum_{k \in S} P(X_0 = k) p_{ki}(n) = \sum_{k \in S} \mu_k p_{ki}(n),$$

which is the ith element in the vector μP^n. If the initial distribution μ is a one-point distribution at some $x \in S$ (i.e., if $P(X_0 = x) = 1$), then we follow the notation $P^n(x, .)$ to denote the probabilities

$$P^n(x, A) = P(X_n \in A \mid X_0 = x).$$

The total variation distance between the true distribution of X_n and the stationary distribution π under a general initial distribution μ is

$$\rho(\mu P^n, \pi) = \sup_A \big|P_{\mu P^n}(A) - P_\pi(A)\big| = \sup_A \left| \sum_{i \in A} \sum_{k \in S} \mu_k p_{ki}(n) - \sum_{i \in A} \pi(i) \right|.$$

If we know that for a particular n_0, $\rho(\mu P^{n_0}, \pi)$ is small, and if we also know that $\rho(\mu P^n, \pi)$ is monotone decreasing in n, then we can be assured that MCMC output after step n_0 will produce sufficiently accurate estimates of probabilities of arbitrary sets under the target distribution π. This is the motivation for wanting a small value for $\rho(\mu P^n, \pi)$.

19.5 Convergence of MCMC and Bounds on Errors

Two other distances that are also used in this context are the separation and the chi-square distances. The separation distance is defined as

$$D(\mu P^n, \pi) = \sup_{i \in S}\left[1 - \frac{P_\mu(X_n = i)}{\pi(i)}\right],$$

and the chi-square distance is defined as

$$\chi^2(\mu P^n, \pi) = \sum_{i \in S} \frac{(P_\mu(X_n = i) - \pi(i))^2}{\pi(i)}.$$

Here are the principal notions of convergence that are used.

Definition 19.3. Suppose $X_n, n \geq 0$ is a stationary Markov chain on a discrete state space S with a unique stationary distribution π. Assume that $\pi(x) > 0$ for all $x \in S$. The chain is said to be *ergodic in variation norm* if for any $x \in S$,

$$\rho(P^n(x,.), \pi) = \sup_{A} |P(X_n \in A \mid X_0 = x) - \pi(A)| \to 0 \text{ as } n \to \infty.$$

The chain is said to be *geometrically ergodic* if for any $x \in S$, there exist a finite constant $M(x)$ and $0 < r < 1$ such that

$$\rho(P^n(x,.), \pi) \leq M(x) r^n \text{ for all } n \geq 1.$$

The chain is said to be *uniformly ergodic* if there exist a fixed finite constant M and $0 < r < 1$ such that for any $x \in S$,

$$\rho(P^n(x,.), \pi) \leq M r^n \text{ for all } n \geq 1.$$

If the chain has a finite state space and is geometrically ergodic, then it is obviously uniformly ergodic. Otherwise, and especially so for MCMC chains, uniform ergodicity usually does not hold, although geometric ergodicity does.

19.5.1 Spectral Bounds

In the case of an irreducible and aperiodic chain with a finite state space, convergence to stationarity occurs geometrically fast. That is,

$$P^n = \begin{pmatrix} \pi \\ \pi \\ \vdots \\ \pi \end{pmatrix} + O(n^{\alpha_2 - 1} c^n),$$

where c is the second largest value among the moduli of the eigenvalues of P (called the *SLEM*), and α_2 is the algebraic multiplicity of that particular eigenvalue. This follows from general spectral decomposition of stochastic matrices, and is also known as *the Perron–Frobenius theorem*; see Chapter 6 in Brémaud (1999).

The important point is that irreducibility and aperiodicity guarantee that all eigenvalues except one are strictly less than one in modulus (although they can be complex), and so, in particular $c < 1$, which causes each row of P^n to converge to the stationary vector π at a geometric (i.e., exponentially fast) rate, and therefore, the chain is geometrically ergodic. If, in addition, the chain is reversible, then the eigenvalues are real, say

$$1 = \lambda_1 > \lambda_2 \geq \lambda_3 \geq \cdots \geq \lambda_t > -1,$$

and so in that case, $c = \max\{\lambda_2, |\lambda_t|\}$.

Example 19.22 (Two-State Chain). This is the simplest possible example of a stationary Markov chain. Consider the two-state chain with the transition probability matrix

$$P = \begin{pmatrix} \alpha & 1-\alpha \\ 1-\beta & \beta \end{pmatrix},$$

where $0 < \alpha, \beta < 1$. The eigenvalues of P can be analytically computed, and they are

$$\lambda_1 = 1, \quad \lambda_2 = \alpha + \beta - 1.$$

Thus, the SLEM equals $c = |\alpha + \beta - 1|$.

Example 19.23. Consider the three-state stationary Markov chain with the transition probability matrix

$$P = \begin{pmatrix} 0 & 1 & 0 \\ 0 & .5 & .5 \\ .5 & 0 & .5 \end{pmatrix}.$$

Note that the chain is irreducible and aperiodic (see Example 10.13). But it is not reversible. Thus, there can be complex eigenvalues. Indeed, the eigenvalues are

$$\lambda_1 = 1; \quad \lambda_2 = -\frac{i}{2}; \quad \lambda_3 = \frac{i}{2}.$$

The moduli of the eigenvalues are therefore $1, \frac{1}{2},$ and $\frac{1}{2}$, and so, the SLEM equals $c = \frac{1}{2}$.

The practical difficulty is that it is usually difficult or impossible to derive expressions for λ_2 and λ_t in closed-form, and especially so when the number of possible states t is large. Hence, much of the research effort in this field has been directed towards finding effective and explicit upper bounds on c. The following theorem gives some of the most significant bounds currently available on the distance between

19.5 Convergence of MCMC and Bounds on Errors

μP^n and π for some of the distances that we defined above. The bounds in the theorem below all depend on our ability to evaluate or bound the SLEM c, and they are all for the reversible finite-state case.

Theorem 19.8. *Let $X_n, n \geq 0$ be an irreducible, stationary, reversible Markov chain on the finite-state space $S = \{1, 2, \ldots, t\}$. Let π be the stationary distribution of the chain, and c the SLEM of its transition probability matrix P. Then,*

(a) For all $n \geq 1$ and any $i \in S$,

$$\sup_A |P^n(i, A) - \pi(A)| \leq \sqrt{\frac{1 - \pi(i)}{\pi(i)}} \frac{c^n}{2}.$$

(b) For all $n \geq 1$ and any $i \in S$,

$$\sup_A |P^n(i, A) - \pi(A)| \leq \sqrt{\frac{p_{ii}(2)}{\pi(i)}} c^{n-1},$$

where $p_{ii}(2)$ is the ith diagonal element in P^2.
(c) For all $n \geq 1$ and any initial distribution μ,

$$\chi^2(\mu P^n, \pi) \leq c^{2n} \chi^2(\mu, \pi).$$

(d) For all $n \geq 1$ and any initial distribution μ,

$$\sup_A |P_\mu(X_n \in A) - \pi(A)| \leq \frac{c^n}{2} \sqrt{\chi^2(\mu, \pi)}.$$

Proof. See pp. 208–210 in Brémaud (1999).

Example 19.24 (Two-State Chain Revisited). Consider again the chain of Example 19.19. We have the eigenvalues of P as $\lambda_1 = 1, \lambda_2 = \alpha + \beta - 1$. We can calculate P^n exactly by spectral decomposition, and we get

$$P^n = \frac{1}{2 - \alpha - \beta} \begin{pmatrix} 1 - \beta & 1 - \alpha \\ 1 - \beta & 1 - \alpha \end{pmatrix} + \frac{(\alpha + \beta - 1)^n}{2 - \alpha - \beta} \begin{pmatrix} 1 - \alpha & \alpha - 1 \\ \beta - 1 & 1 - \beta \end{pmatrix}.$$

The first matrix in this formula represents the matrix $\begin{pmatrix} \pi \\ \pi \end{pmatrix}$ (the stationary distribution π was derived in Example 10.18). Therefore, we can also calculate $\sup_A |P^n(i, A) - \pi(A)|$ exactly. Indeed, if we take the initial state to be $i = 1$, then we have

$$\sup_A |P(X_n \in A \mid X_0 = i) - \pi(A)| = \frac{1}{2} \sum_{j=1}^{2} |p_{1j}(n) - \pi(j)|$$

(this is a standard formula for the total variation distance; see Section 15.1)

$$= \frac{1}{2} \frac{|\alpha + \beta - 1|^n}{2 - \alpha - \beta} (|1 - \alpha| + |\alpha - 1|) = \frac{1 - \alpha}{2 - \alpha - \beta} |\alpha + \beta - 1|^n.$$

On the other hand, the first bound in part (a) of Theorem 19.8 equals

$$\sqrt{\frac{1 - \pi(1)}{\pi(1)}} \frac{c^n}{2} = \sqrt{\frac{\pi(2)}{\pi(1)}} \frac{c^n}{2}$$

$$= \sqrt{\frac{1 - \alpha}{1 - \beta}} \frac{|\alpha + \beta - 1|^n}{2}.$$

Therefore, the relative error of the upper bound is

$$\frac{\sqrt{\frac{1-\alpha}{1-\beta}}}{\frac{1-\alpha}{2-\alpha-\beta}} - 1 = \frac{2 - \alpha - \beta}{2\sqrt{(1-\alpha)(1-\beta)}} - 1 = \frac{\frac{(1-\alpha)+(1-\beta)}{2}}{\sqrt{(1-\alpha)(1-\beta)}} - 1.$$

That is, we get the very interesting conclusion that the relative error of the bound in part (a) of Theorem 19.8 is the quotient of the arithmetic and the geometric mean of $1 - \alpha$ and $1 - \beta$ minus one. If $\alpha = \beta$, then the error is zero, and the bound becomes exact. If α and β are very different, the bound becomes inefficient.

Example 19.25 (SLEM for Metropolis–Hastings Algorithm). As a rule, the eigenvalues of P are hard to evaluate in symbolic form, except when the size of the state space is very small. An important example where the eigenvalues can, in fact, be found in closed-form is the Metropolis–Hastings algorithm. The availability of general formulas for the eigenvalues makes the SLEM bounds applicable when using the Metropolis–Hastings algorithm.

In this case, the transition probabilities are given by

$$p_{ij} = \frac{1}{t} \min\left\{1, \frac{\pi(j)}{\pi(i)}\right\}, \quad j \neq i;$$

$$p_{ii} = 1 - \sum_{j \neq i} p_{ij}.$$

We label the states in decreasing order of the values of their stationary probabilities. That is, we label the states such that $\pi(1) \geq \pi(2) \geq \ldots \geq \pi(k)$. The first eigenvalue, as always, is $\lambda_1 = 1$, and the remaining eigenvalues are

$$\lambda_k = \frac{1}{t} \left[\sum_{j=k-1}^{t} \frac{\pi(k-1) - \pi(j)}{\pi(k-1)} \right], \quad k \geq 2.$$

19.5 Convergence of MCMC and Bounds on Errors

For example, take $t = 3$. Then, denoting $\pi(1) = a \geq \pi(2) = b \geq \pi(3) = c = 1 - a - b$, the transition probability matrix works out to

$$P = \frac{1}{3} \begin{pmatrix} 4 - \frac{1}{a} & \frac{b}{a} & \frac{c}{a} \\ 1 & 2 - \frac{c}{b} & \frac{c}{b} \\ 1 & 1 & 1 \end{pmatrix}.$$

By direct verification, the eigenvalues of P are

$$1, 1 - \frac{1}{3a}, \frac{2}{3} - \frac{1-a}{3b}.$$

To see that these correspond to the eigenvalue formulas,

$$\frac{1}{3}\left[\sum_{j=1}^{3} \frac{\pi(1) - \pi(j)}{\pi(1)}\right]$$

$$= \frac{1}{3}\left[3 - \frac{1}{\pi(1)}\sum_{j=1}^{3}\pi(j)\right] = 1 - \frac{1}{3\pi(1)} = 1 - \frac{1}{3a},$$

because $\sum_{j=1}^{3} \pi(j) = 1$. Likewise,

$$\frac{1}{3}\left[\sum_{j=2}^{3} \frac{\pi(2) - \pi(j)}{\pi(2)}\right]$$

$$= \frac{1}{3}\left[2 - \frac{1}{\pi(2)}\sum_{j=2}^{3}\pi(j)\right] = \frac{2}{3} - \frac{1 - \pi(1)}{3\pi(2)} = \frac{2}{3} - \frac{1-a}{3b}.$$

The general case is proved by direct verification in Liu (1995).

19.5.2 ∗ *Dobrushin's Inequality and Diaconis–Fill–Stroock Bound*

The bounds in the preceding section are only for the reversible case. There are some examples in which convergence occurs, although the chain is not reversible. There are two principal ways to deal with the nonreversible situation. A classic inequality in Dobrushin (1956) provides an upper bound on the SLEM of the transition probability matrix P by using a very simple computable index, known as *Dobrushin's coefficient*, even if the chain is not reversible. A second method introduced in Diaconis and Stroock (1991) and Fill (1991) uses a technique of reversal of a nonreversible transition probability matrix, and derives the needed bounds in terms of the SLEM of a reversible matrix M, constructed from the original transition probability matrix P.

First, we need some notation and definitions. Given a $t \times t$ transition probability matrix (not necessarily reversible), let

$$\vec{p}_i = (p_{i1}, \ldots, p_{it}),$$
$$\tilde{p}_{ij} = \frac{p_{ji}\pi(j)}{\pi(i)}; \quad \tilde{P} = ((\tilde{p}_{ij})); \quad M = P\tilde{P},$$

where π is the stationary distribution corresponding to P. Note that M is reversible.

Definition 19.4. Let P be the transition probability matrix of a stationary Markov chain on a finite-state space $S = \{1, 2, \ldots, t\}$. The *Dobrushin coefficient* of P is defined as

$$\Delta(P) = \max_{i,j \in S} \rho(\vec{p}_i, \vec{p}_j) = \frac{1}{2} \max_{i,j \in S} \sum_{k \in S} |p_{ik} - p_{jk}|.$$

In the above, recall that we are using the notation ρ to denote total variation distance. Note that $0 \leq \Delta(P) \leq 1$, and usually, $0 < \Delta(P) < 1$.

Here is our main theorem on handling rates of convergence to stationarity when the chain is not reversible. These are not the only bounds available, but the specific bounds in the theorem below have some simplicity about them.

Theorem 19.9. *Let P be the transition probability matrix of a stationary Markov chain on a finite-state space, with unique stationary distribution π. Let c be the SLEM of P and λ the SLEM of M. Then,*

(a) $c \leq \Delta(P)$.
(b) *For any two initial distributions μ, ν, and for all $n \geq 1$,*

$$\rho(\mu P^n, \nu P^n) \leq \rho(\mu, \nu)[\Delta(P)]^n.$$

(c) *For all $n \geq 1$ and any $i \in S$,*

$$\left[\sup_A |P^n(i, A) - \pi(A)|\right]^2 \leq \frac{\lambda^n}{4\pi(i)}.$$

See Brémaud (1999, pp. 237–238) for a proof of this theorem.

Example 19.26 (Two-State Chain). Consider our two-state case with the transition probability matrix P

$$P = \begin{pmatrix} \alpha & 1 - \alpha \\ 1 - \beta & \beta \end{pmatrix}.$$

Then, the Dobrushin coefficient equals

$$\Delta(P) = \frac{1}{2}\left[|\alpha - 1 + \beta| + |1 - \alpha - \beta|\right] = 2|\alpha + \beta - 1|.$$

19.5 Convergence of MCMC and Bounds on Errors

We have previously seen that the eigenvalues of P are 1 and $\alpha + \beta - 1$ (see Example 19.21), so that the SLEM equals $c = |\alpha + \beta - 1|$. Hence, in this case, $c = \Delta(P)$.

Example 19.27 (Metropolis–Hastings for Truncated Geometric). It was shown in Example 19.15 that the Metropolis–Hastings algorithm for simulating from the distribution of Y given $Y \leq m$, where $Y \sim \text{Geo}(p)$, has the transition matrix with entries

$$p_{ij} = \frac{1}{m} \text{ if } j < i; \quad p_{ij} = \frac{1}{m}(1-p)^{j-i} \text{ if } j > i; \quad p_{ii} = 1 - \sum_{j \neq i} p_{ij}.$$

The eigenvalues of P are remarkably structured, and are

$$\lambda_1 = 1, \quad \lambda_{i+1} = \frac{i - \sum_{j=1}^{i} q^j}{m}, \quad i = 1, \ldots, m-1,$$

where $q = 1 - p$. The eigenvalues are all in the unit interval $[0, 1]$, and therefore the second largest eigenvalue is also the SLEM. The second largest eigenvalue is the last one, namely,

$$\lambda_m = \frac{m - 1 - \sum_{j=1}^{m-1} q^j}{m} = 1 - \frac{1 - q^m}{m(1 - q)} = 1 - \frac{1 - q^m}{mp}.$$

There is no such general closed-form formula for the Dobrushin coefficient in this case, but for any given m and p, it is easily computable.

Example 19.28 (Dobrushin Bound May Not Be Useful). Consider again the transition matrix of Example 19.20

$$P = \begin{pmatrix} 0 & 1 & 0 \\ 0 & .5 & .5 \\ .5 & 0 & .5 \end{pmatrix}.$$

In this case, $\rho(\vec{p}_1, \vec{p}_2) = .5, \rho(\vec{p}_2, \vec{p}_3) = .5$, and $\rho(\vec{p}_1, \vec{p}_3) = 1$. Therefore, $\Delta(P) = 1$. Therefore, the results of Theorem 19.9 involving the Dobrushin coefficient are not useful in this example. A closer examination reveals that the Dobrushin coefficient is rendered equal to one because $\vec{p}_1$ and $\vec{p}_3$ are orthogonal. In any such case, the bounds involving $\Delta(P)$ would not be informative.

19.5.3 * Drift and Minorization Methods

The bounds of this section on the distance between the exact distribution of the chain at some fixed time n and the stationary distribution π have the appealing feature that they apply to even nonreversible chains. As we have seen, certain Gibbs chains are not reversible. However, in principle, the *drift* and *minorization* methods of this section can apply to them, and geometric convergence to stationarity may be established by using the methods below.

We consider the case of discrete state spaces below. For more general state spaces (e.g., if $S = \mathcal{R}_+$), the methods of this section need an added assurance of a form of recurrence, known as *Harris recurrence*. For the special case of a Gibbs chain on a discrete state space, the Harris recurrence condition will automatically hold under the conditions that we assume. However, it should be emphasized that Harris recurrence must be verified before using drift methods if the underlying chain has a more general state space. We first need the appropriate definitions.

Definition 19.5. Let $X_n, n \geq 0$ be an irreducible and aperiodic stationary Markov chain on a discrete state space S. The chain is said to satisfy the *drift condition* with the *energy function* $V : S \to \mathcal{R}_+$ if there exist $\epsilon > 0$, a real number b, and a finite subset of states $R \subseteq S$ such that

$$E[V(X_{k+1}) \mid X_k = x] = \sum_{y \in S} V(y) p_{xy} \leq (1-\epsilon)V(x) + b$$

for all $x \notin R$.

Definition 19.6. Let $X_n, n \geq 0$, be an irreducible and aperiodic stationary Markov chain on a discrete state space S. Suppose also that the chain satisfies the drift condition with the energy function V, and the associated constants ϵ, b. Then, the chain is said to satisfy the *minorization condition* with respect to q, a probability distribution on S, if there exist $\delta > 0$ and $\alpha > \frac{2b}{\epsilon}$ such that

$$p_{xy} \geq \delta q(y) \,\forall\, y \in S, \quad x \in C,$$

where $C = \{x \in S : V(x) \leq \alpha\}$.

Remark. The set C in the definition of the minorization condition is generally called a *small set*. The drift condition says that if the chain is sitting at some $x \notin R$, then it will tend to drift towards some state with a smaller energy than x. However, the energy function is bounded from below, and so it cannot get lower at a steady rate ad infinitum. So, eventually, the chain will seek shelter in R, and stay there. Convergence will occur geometrically fast; that is, the total variation distance will satisfy a bound of the form $\sup_A |P^n(x, A) - \pi(A)| \leq K\gamma^n$ for some $\gamma \in (0, 1)$ and some K. If we have the minorization condition, we can furthermore place concrete bounds on γ, typically of the form $\gamma \leq 1 - \delta$.

It is an unfortunate aspect of the approach via drift and minorization conditions, that it is usually very difficult to manufacture the energy function V and the probability distribution q even in simple problems. In addition, if an energy function V satisfying the drift condition exists, it is not unique, and different choices produce different constants K, γ. However, the approach has met some success, and it is a

19.5 Convergence of MCMC and Bounds on Errors

main theoretical approach in the convergence area. In particular, the drift approach has been successfully used for some Gibbs chains for simulating from the posterior distribution of a vector of means in so-called *hierarchical Bayes linear models*.

At first glance, the drift and the minorization conditions appear to be rather obscure, and it is not clear how they lead to bounds on the total variation distance between the exact distribution of the chain and the stationary distribution. For the case that the minorization condition $p_{xy} \geq \delta q(y)$ holds for all x and y in the state space S, we show the proof of the following theorem.

Theorem 19.10. *Let $X_n, n \geq 0$, be an irreducible and aperiodic stationary Markov chain on a discrete state space S. Suppose also that the chain satisfies the minorization condition $p_{xy} \geq \delta q(y)$ for all $x, y \in S$. Then, for all $n \geq 1$, and any $i \in S$,*

$$\sup_A |P^n(i, A) - \pi(A)| \leq (1 - \delta)^n.$$

Proof. The proof uses a famous technique in probability theory, known as *coupling*. We manufacture (only for the purpose of the proof) two new chains, $Z_n, n \geq 0$, and $Y_n, n \geq 0$, such that the two chains will eventually *couple*. That is, from some random time $T = t$ onwards, we will have $Z_n = Y_n$. It turns out that the chains are so constructed that this coupling time $T \sim \text{Geo}(\delta)$, and it also turns out that $|P(X_n \in A \mid X_0 = i) - \pi(A)| \leq P(T > n)$ for any A, i, and n, where X_n is our original chain (with π as the stationary distribution). Because $P(T > n) = (1-\delta)^n$, the theorem follows.

Here is how the chains $Z_n, Y_n, n \geq 0$ are constructed. Define the probability density $r_x(y)$ on the state space S by the defining equation

$$p_{xy} = \delta q(y) + (1 - \delta) r_x(y);$$

we can define such a density because the minorization condition has been assumed to hold for all $x, y \in S$. Set $Z_0 = i \in S$, $Y_0 \sim \pi$, and choose a Bernoulli variable $B_0 \sim \text{Ber}(\delta)$. If $B_0 = 0$, choose $Y_1 \sim r_{y_0}(y)$ and choose $Z_1 \sim r_{z_0}(y)$, independently of each other. If $B_0 = 1$, choose a single $z \sim q(y)$ and set $Z_1 = Y_1 = z$. This is one sweep. Repeat this sweep, mutually independently, until the Z-chain and the Y-chain couple and call the coupling time T. Once the chains couple, at all subsequent steps, the two chains remain identical. Because the sweeps are independent and at each sweep the same Bernoulli experiment with success probability δ is performed, we have $T \sim \text{Geo}(\delta)$.

To complete the proof, we now show that $|P(X_n \in A \mid X_0 = i) - \pi(A)| \leq P(T > n)$ for any A, i, and n. For this, note the important fact that because Y_0 was chosen from the stationary distribution π, and because the transition probabilities satisfy the mixture representation $p_{xy} = \delta q(y) + (1 - \delta) r_x(y)$, the marginal distribution of Y_n coincides with π for every n. Furthermore, because the Z-chain started at the state i, its marginal distribution coincides with the distribution of X_n given that $X_0 = i$. Hence,

$$|P(X_n \in A \mid X_0 = i) - \pi(A)|$$
$$= |P(X_n \in A \mid X_0 = i) - P(Y_n \in A)|$$
$$= |P(Z_n \in A) - P(Y_n \in A)|$$
$$= |P(Z_n \in A, Z_n = Y_n) + P(Z_n \in A, Z_n \neq Y_n)$$
$$\quad - P(Y_n \in A, Y_n = Z_n) - P(Y_n \in A, Y_n \neq Z_n)|$$
$$= |P(Z_n \in A, Z_n = Y_n) + P(Z_n \in A, Z_n \neq Y_n)$$
$$\quad - P(Z_n \in A, Z_n = Y_n) - P(Y_n \in A, Y_n \neq Z_n)|$$
$$= |P(Z_n \in A, Z_n \neq Y_n) - P(Y_n \in A, Z_n \neq Y_n)|$$
$$\leq \max\left(P(Z_n \in A, Z_n \neq Y_n), P(Y_n \in A, Z_n \neq Y_n)\right)$$
$$\leq P(Z_n \neq Y_n) = P(T > n). \qquad \square$$

19.6 MCMC on General Spaces

Markov chain Monte Carlo methods are not limited to target distributions π that have a discrete set S as their support. They are routinely used for simulating from continuous distributions with general supports. The basic definitions, results, and phenomena in these general cases closely match the corresponding results in the discrete case. For example, we still have Metropolis schemes and the Gibbs sampler for these general cases, and convergence theory and convergence conditions are very similar to those for the discrete case. We describe the continuous versions of the MCMC algorithms and the basic associated theory with examples in this section. First, we set up the necessary notation and provide the necessary definitions.

19.6.1 General Theory and Metropolis Schemes

Definition 19.7. A discrete-time stochastic process $X_n, n \geq 0$, is called a stationary or a homogeneous Markov chain on a general state space $S \subseteq \mathcal{R}^d$ if there exists a *Markov transition kernel* $P(x, .)$, which is a function of two arguments, $x \in S$ and $A \subseteq S$, such that for fixed x, $P(x, .)$ is a probability distribution (measure) on S, and for any $n \geq 0$ and any $x \in S$,

$$P(X_{n+1} \in A \mid X_n = x) = P(x, A).$$

The Markov transition kernel $P(x, A)$ is the direct generalization of the probability $P(X_{n+1} \in A \mid X_n = i) = \sum_{j \in A} p_{ij}$ in the discrete state space case. In the discrete case, the diagonal elements p_{ii} can be strictly positive. Likewise, in the general state space case, $P(x, \{x\})$ can be strictly positive. Therefore, in general, a Markov transition kernel $P(x, A)$ need not be a continuous distribution; that is, $P(x, A)$

19.6 MCMC on General Spaces

need not have a density. If it does, the density corresponding to the measure $P(x, .)$ is called a *transition density*, and we denote it by $p(x, y)$. So, a transition density, if it exists, satisfies the two usual properties:

$$p(x, y) \geq 0; \quad \int_S p(x, y)dy = 1.$$

Caution. In the next definition, a probability measure and its density function have been denoted by using the same notation π. This is an abuse of notation, but is very common in the literature in the present context.

Definition 19.8. Suppose $X_n, n \geq 0$, is a stationary Markov chain with Markov transition kernel $P(x, .)$. A probability measure π with density π on S is called a stationary or an invariant measure for the chain if for all $A \subseteq S$,

$$\pi(A) = \int P(x, A)\pi(x)dx.$$

This is a direct extension of the corresponding equation $\pi(i) = \sum_{j \in S} p_{ji}\pi(j)$ in the discrete case. The distribution of the chain at time n, having started from state x at time zero, is still denoted as

$$P^n(x, A) = P(X_n \in A \mid X_0 = x).$$

As in the discrete case, we must remember that

(a) A stationary distribution π need not exist.
(b) If it exists, it need not be unique.
(c) Even if a certain distribution π is a stationary distribution for a chain X_n, we need not have the convergence property $\sup_A |P^n(x, A) - \pi(A)| \to 0$ as $n \to \infty$.

And, as in the discrete case, the three conditions that help us eliminate these obstacles are *reversibility, irreducibility, and aperiodicity*. We need to redefine them in the new notation and the new general state space setup.

Definition 19.9. A stationary Markov chain on a general state space S with a transition density $p(x, y)$ is called *reversible* with respect to π if the *detailed balance equation* $\pi(x)p(x, y) = \pi(y)p(y, x)$ holds for all $x, y, \in S, y \neq x$.

Definition 19.10. A stationary Markov chain $X_n, n \geq 0$ on a general state space S is called *irreducible* with respect to a given π if for any $x \in S$ and any $A \subseteq S$ such that $\pi(A) > 0$, $P(\exists n \text{ such that } X_n \in A \mid X_0 = x) > 0$.

Definition 19.11. A stationary Markov chain $X_n, n \geq 0$ on a general state space S is called *periodic of period* $d > 1$ if there exists a partition $A_1, \ldots, A_d$ of the state space S such that the one-step transitions of the chain must be of the form $A_1 \to A_2 \to A_3 \to \cdots \to A_d \to A_1$. The chain is called *aperiodic* if there is no $d > 1$ such that the chain is periodic of period d.

Metropolis chains on general state spaces will usually (but not always) have these three helpful properties of reversibility with respect to the target distribution π that is in our mind, and irreducibility with respect to that same π, and aperiodicity. This enables us to conclude that the chain indeed has the targeted π as its unique stationary distribution, and that convergence takes place as well. Those issues are picked up later in this section.

The *Metropolis chain* in this general case is defined in essentially the same manner as in the discrete case. If the chain is currently at state x, a *candidate state* y is generated according to a *proposal distribution* and once a candidate state has been picked, it is accepted or declined according to an *acceptance distribution*. We need notations for these two things. We assume below that the proposal distribution is continuous; that is, it has a density, and this proposal density is denoted as $\theta(x, y)$. So, the properties of a proposal density are $\theta(x, y) \geq 0$; $\int_S \theta(x, y) dy = 1$. The acceptance probabilities are denoted by $\gamma(x, y)$. So, for each fixed x and each fixed y, $0 \leq \gamma(x, y) \leq 1$. The overall transition kernel of our Metropolis chain is then

$$p(x, y) = \theta(x, y)\gamma(x, y), \quad x, y \in S, y \neq x;$$

$$P(x, \{x\}) = \int_S \theta(x, y)[1 - \gamma(x, y)] dy.$$

Throughout we make the assumption that the stationary density $\pi(x)$ and the proposal densities $\theta(x, y)$ are strictly positive for all x, y. This is not essential, but it saves us from a lot of tedious fixing if we do not allow them to take the value zero. Now we proceed to define the special MCMC schemes exactly as we did in the discrete case.

Independent Sampling.

$$\theta(x, y) = \pi(y) \; \forall \; y; \quad \gamma(x, y) \equiv 1.$$

General Metropolis Scheme. Choose a general $\theta(x, y)$ and

$$\gamma(x, y) = \min\left\{\frac{\pi(y)\theta(y, x)}{\pi(x)\theta(x, y)}, 1\right\}.$$

Random Walk Metropolis Scheme. In the general Metropolis scheme, choose $\theta(x, y) = \theta(y, x) = q(|y - x|)$ for some fixed density q, so that

$$\gamma(x, y) = \min\left\{\frac{\pi(y)}{\pi(x)}, 1\right\}.$$

19.6 MCMC on General Spaces

Independent Metropolis Scheme. Choose $\theta(x, y) = p(y)$, that is, a fixed strictly positive density $p(y)$ independent of x, and

$$\gamma(x, y) = \min\left\{1, \frac{\frac{\pi(y)}{p(y)}}{\frac{\pi(x)}{p(x)}}\right\}.$$

Example 19.29 (Simulating from a t-Distribution). Suppose the target distribution is a t-distribution with α degrees of freedom, where we take $\alpha > 1$. Thus,

$$\pi(x) = \frac{c}{\left(1 + \frac{x^2}{\alpha}\right)^{\left(\frac{\alpha+1}{2}\right)}}, \quad -\infty < x < \infty.$$

The normalizing constant c is unimportant for the purpose of implementing an MCMC scheme. Note that in this example, the state space is $S = \mathcal{R}$. We illustrate the use of a random walk Metropolis scheme with a proposal distribution that is Cauchy. Precisely,

$$\theta(x, y) = \frac{c}{1 + (y - x)^2}, \quad -\infty < y < \infty;$$

once again, c is a normalizing constant (not the same one as in the t density), and is not important for our purpose. Therefore,

$$\frac{\pi(y)\theta(y, x)}{\pi(x)\theta(x, y)}$$

$$= \frac{\left(1 + \frac{x^2}{\alpha}\right)^{\frac{\alpha+1}{2}}(1 + (y - x)^2)}{(1 + \frac{y^2}{\alpha})^{\frac{\alpha+1}{2}}(1 + (x - y)^2)} = \left(\frac{\alpha + x^2}{\alpha + y^2}\right)^{\frac{\alpha+1}{2}}$$

$$\geq 1 \quad \text{iff } x^2 \geq y^2 \text{ iff } |x| \geq |y|.$$

Hence the acceptance probabilities of the Metropolis scheme are given by

$$\gamma(x, y) = 1 \text{ if } |x| \geq |y|, \text{ and } \gamma(x, y) = \left(\frac{\alpha + x^2}{\alpha + y^2}\right)^{\frac{\alpha+1}{2}} \text{ if } |x| < |y|.$$

So, finally, the Metropolis scheme for simulating from the t-distribution proceeds as follows.

Step 1. If the current value of the chain is x, generate a candidate state $y \sim C(x, 1)$.
Step 2. If $|x| \geq |y|$, accept the candidate value y; if $|x| < |y|$, perform a Bernoulli experiment with success probability $\left(\frac{\alpha+x^2}{\alpha+y^2}\right)^{\frac{\alpha+1}{2}}$, and accept the candidate value y if the Bernoulli experiment results in a success.

Example 19.30 (Simulating from a Nonconventional Multivariate Distribution). Suppose we want to simulate from a trivariate distribution with the density

$$\pi(x_1, x_2, x_3) = \frac{c}{(1 + x_1 + x_2 + x_3)^4}, \quad x_1, x_2, x_3 > 0.$$

Thus, in this example, the state space is $S = \mathcal{R}_+^3$. We use a suitable independent Metropolis scheme to simulate from π. Towards this, choose the proposal density

$$p(y_1, y_2, y_3) = \frac{c}{(1 + y_1^2 + y_2^2 + y_3^2)^2}, \quad y_1, y_2, y_3 > 0$$

Note that the proposal density is a trivariate spherically symmetric Cauchy density, restricted to the first quadrant. We can simulate quite easily from this proposal density. For instance, we can first simulate a value for $r = \sqrt{y_1^2 + y_2^2 + y_3^2}$ according to the density

$$g(r) = \frac{4}{\pi} \frac{r^2}{(1 + r^2)^2}, \quad 0 < r < \infty,$$

and then generate (y_1, y_2, y_3) as $(y_1, y_2, y_3) = r(u_1, u_2, u_3)$ where (u_1, u_2, u_3) is a point on the boundary of the unit sphere (see Section 19.2.4 on how to do these simulations). When the point (u_1, u_2, u_3) does not fall in the first quadrant, we have to fix the signs.

We only have to work out the acceptance probabilities of our Metropolis scheme now. We have

$$\frac{\frac{\pi(y)}{p(y)}}{\frac{\pi(x)}{p(x)}} = \frac{(1 + x_1 + x_2 + x_3)^4 (1 + y_1^2 + y_2^2 + y_3^2)^2}{(1 + y_1 + y_2 + y_3)^4 (1 + x_1^2 + x_2^2 + x_3^2)^2}$$

$$= \alpha(x, y) \text{ say.}$$

Then, the acceptance probability is $\gamma(x, y) = \min\{1, \alpha(x, y)\}$. For example, if the current state is $(x_1, x_2, x_3) = (1, 1, 1)$, and the candidate state is $(y_1, y_2, y_3) = (2, 2, 2)$, then $\alpha(x, y) = 1.126$, and therefore $\gamma(x, y) = \min\{1, 1.126\} = 1$; that is, the candidate state is definitely accepted.

19.6.2 Convergence

The convergence issues, conditions needed, proofs of the convergence theorems, and verification of the conditions needed for convergence of Metropolis schemes on general state spaces are all substantially more complicated than in the discrete state space case. The questions of importance are the following.

19.6 MCMC on General Spaces

Question 1. Does a particular Metropolis scheme in a specific problem have a stationary distribution at all, and if so, is it our target distribution π, and if so, is π a unique stationary distribution?

Question 2. For a particular Metropolis scheme in a specific problem, does $\sup_A |P^n(x, A) - \pi(A)| \to 0$?

Question 3. If it does, is the chain geometrically ergodic?

Question 4. Is the chain uniformly ergodic?

Of these, the answer to the first question is usually straightforward. By construction, our Metropolis chain will usually be reversible with respect to the given π that we have in mind, and exactly as in the discrete state space case, this ensures that π is a stationary distribution. Uniqueness follows if we also have irreducibility and aperiodicity of our Metropolis chain. So, what we strive for is the construction of a Metropolis chain that is reversible, and irreducible, and aperioodic. The nice thing about Metropolis chains is that we can usually (but not always) achieve all three without imposing stringent conditions on π or the chain.

An added complexity in the case of general state spaces, which does not arise in the discrete case, is that one may have $\sup_A |P^n(x, A) - \pi(A)| \to 0$ for almost all initial states x, but not for all x. Because we do not know for exactly which initial states x, $\sup_A |P^n(x, A) - \pi(A)| \to 0$, we cannot be sure that convergence occurs for the particular initial state with which we started.

The best theorems on convergence and the speed of convergence for the case of general state spaces are generally awkward, either in the statement of the conditions, or in their verification. We present results and techniques that are simpler to state, understand, and verify, rather than focus on the weakest possible conditions. The principal references for the results that we present here are Tierney (1994), Mengersen and Tweedie (1996), and Roberts and Rosenthal (2004). Some of the best theorems on convergence of general stationary Markov chains on general state spaces under essentially the weakest conditions can be seen in Athreya, Doss and Sethuraman (1996).

We start with the most basic question of whether Metropolis chains on general state spaces at all converge to the desired target distribution π, and if so, under what conditions. To answer this question, we need to define *Harris recurrence*; this is defined below.

Definition 19.12. A stationary Markov chain $X_n, n \geq 0$, on a general state space S is called *Harris recurrent with respect to π* if given any initial state $x \in S$, and any subset $B \subseteq S$ such that $\pi(B) > 0$,

$$P(\exists n \text{ such that } X_n \in B \mid X_0 = x) = 1.$$

This is the same as saying that regardless of where the chain started, any set B with $\pi(B) > 0$ will be visited infinitely often by the chain with probability one. On comparing this with the definition of irreducibility, we find that Harris recurrence is stronger than irreducibility.

Harris recurrence is a difficult object to verify in general. However, for Metropolis chains, if we make the proposal density nice enough, and if the target density is also nice enough, then Harris recurrence obtains. So, one does not have to verify Harris recurrence on a case-by-case basis, under some general conditions on the Metropolis chain. Once we have Harris recurrence, and aperiodicity, we get the desired convergence property, although rates of convergence still need a separate study. Here is a very concrete theorem from Tierney (1994).

Theorem 19.11. *Suppose the state space of the chain is $S \subseteq \mathcal{R}^d$.*

(a) The random walk Metropolis chain is aperiodic and Harris recurrent if either of the following holds.
  (A) q is positive on all of $\mathcal{R}^d$.
  (B) q is positive in a neighborhood of zero, and $\Omega = \{x : \pi(x) > 0\}$ is open and connected.
(b) The independent Metropolis chain is aperiodic and Harris recurrent if $\pi(x) > 0 \Rightarrow p(x) > 0$.
(c) Provided that the sufficient conditions stated above hold, the random walk Metropolis chain and the independent Metropolis chain satisfy the convergence property

$$\text{For any } x \in S, \sup_A |P^n(x, A) - \pi(A)| \to 0 \quad \text{as } n \to \infty.$$

(d) In addition, under the sufficient conditions stated above, the random walk Metropolis chain and the independent Metropolis chain are strongly ergodic; that is, for any $\phi : S \to \mathcal{R}$ such that $\int_S |\phi(x)|\pi(x)dx < \infty$, $\frac{1}{n}\sum_{k=1}^n \phi(X_k) \xrightarrow{a.s.} \int_S \phi(x)\pi(x)dx$, for any initial distribution μ.

Example 19.31 (Simulating from a t-Distribution, Continued). Consider again the setup of Example 19.26, with $\pi(x)$ as the density of a t-distribution with α degrees of freedom and $\theta(x, y)$ as the density of $C(x, 1)$. Thus, in our notation, $q(x) = \frac{c}{1+x^2} I_{-\infty < x < \infty}$. Because q satisfies condition (A) in part (a) of Theorem 19.11, the desired convergence property $\sup_A |P^n(x, A) - \pi(A)| \to 0$ is assured.

Suppose we change the proposal density to $\theta(x, y) = \frac{1}{2} I_{x-1 \leq y \leq x+1}$. Then $q(x) = \frac{1}{2} I_{-1 \leq x \leq 1}$. So, q does not satisfy condition (A) in part (a), but it still satisfies condition (B). Furthermore, π being a t-density, $\Omega = \mathcal{R}$, which is open and connected. Hence, we are again assured of $\sup_A |P^n(x, A) - \pi(A)| \to 0$.

Suppose to force quick jumps, we choose $\theta(x, y) = \frac{1}{2} I_{x-2 \leq y \leq x-1} + \frac{1}{2} I_{x+1 \leq y \leq x+2}$. Then $q(x) = \frac{1}{2} I_{-2 \leq x \leq -1} + \frac{1}{2} I_{1 \leq x \leq 2}$. Now, q is not positive in any neighborhood of zero, and so it satisfies neither condition (A) nor condition (B) of part (a) in Theorem 19.11, and the theorem fails to guarantee $\sup_A |P^n(x, A) - \pi(A)| \to 0$.

The above theorem suggests that if all we care for is simple convergence, then Metropolis chains will usually deliver it. However, for practical use, what we really need are concrete bounds on $\sup_A |P^n(x, A) - \pi(A)|$ for given n. For the

19.6 MCMC on General Spaces

discrete state space case, we obtained such bounds by spectral methods, drift and minorization, and by using Dobrushin's coefficient. In the general state space case, we provide a result based on drift and minorization, and another based on the tails of the target and the proposal density. We need another definition before the next theorem.

Definition 19.13. A density $\pi(x)$ on the real line is called *logconcave in the tails* with the associated exponential constant $\alpha > 0$ if for some x_0,

$$\frac{\log \pi(x) - \log \pi(y)}{|x - y|} \geq \alpha$$

for $y > x \geq x_0$ and $y < x \leq -x_0$.

Here is a rather explicit theorem on convergence and bounds on approximation error due to Mengersen and Tweedie (1996).

Theorem 19.12. *(a) The independent Metropolis sampler is geometrically and uniformly ergodic if the proposal density $p(y)$ satisfies the condition that for some finite constant c, $\frac{\pi(y)}{p(y)} \leq c$ for all y. Furthermore, in this case, for any $x \in S$, and any $n \geq 1$,*

$$\rho(P^n(x,.), \pi) = \sup_A |P^n(x, A) - \pi(A)| \leq \left(1 - \frac{1}{c}\right)^n.$$

(b) Suppose the state space $S \subseteq \mathcal{R}$. Assume the following conditions on π and the proposal densities $\theta(x, y)$:

(i) $\pi(x)$ is logconcave in the tails with the associated exponential constant α.
(ii) $\theta(x, y) = q(|y - x|)$ for some fixed density function q, which satisfies the fat tail condition $q(x) \leq Ke^{-\alpha|x|}$ with α as in (i) above.

Then, with $V(x)$ defined as $V(x) = e^{a|x|}$ with $a < \alpha$, the random walk Metropolis chain with the proposal density $\theta(x, y)$ is geometrically, but not uniformly, ergodic in the following sense:

$$\sup_{f: |f(y)| \leq V(y)} |E(f(X_n) | X_0 = x) - \int_S f(y)\pi(y)dy| \leq MV(x)r^n$$

for some finite constant M, and $0 < r < 1$. This holds for any initial state x, and any $n \geq 1$.

Example 19.32 (A Geometrically Ergodic Metropolis Chain). This is an example where we can verify the conditions in part (b) of the above theorem easily. Suppose the target density is $\pi(x) = ce^{-|x|}, -\infty < x < \infty$. We use a Metropolis sampling scheme with the proposal density $\theta(x, y) = q(|y - x|)$, where q is the uniform density $q(y) = \frac{1}{2b}I_{|y|\leq b}, 0 < b < \infty$. Then, $\pi(x)$ is certainly logconcave in the tails. Indeed, taking x_0 to be any positive number, we have for $y > x > x_0$, or for $y < x < -x_0$,

$$\frac{\log \pi(x) - \log \pi(y)}{|x-y|} = 1,$$

and hence we can choose $\alpha = 1$ in the tail logconcavity property. Also, clearly, q satisfies the fat tail condition $q(x) \leq Ke^{-|x|}$ for all x (note that $q = 0$ outside $[-b, b]$). Therefore, by part (b) of the above theorem, the Metropolis sampler of this example is geometrically ergodic. It is, however, not uniformly ergodic.

Example 19.33 (Proposal Densities Should Be Heavier Tailed Than Target). Suppose the target density is a standard exponential, and suppose we consider an independent Metropolis sampler with a proposal density that is exponential with mean λ. Thus, for $y > 0$,

$$\frac{\pi(y)}{p(y)} = \frac{e^{-y}}{\frac{1}{\lambda}e^{-y/\lambda}} = \lambda e^{\frac{1-\lambda}{\lambda y}}.$$

This is uniformly bounded by a finite constant c if $\lambda \geq 1$, and therefore, from part (a) of Theorem 19.12, we can conclude that the independence Metropolis chain is geometrically ergodic.

For $\lambda < 1$, $\frac{\pi(y)}{p(y)}$ is unbounded. How does the independent Metropolis chain behave in that case? Indeed, the chain behaves poorly when $\lambda < 1$. For $\lambda < 1$, the acceptance probabilities are

$$\gamma(x, y) = \min \left\{ 1, \frac{\frac{\pi(y)}{p(y)}}{\frac{\pi(x)}{p(x)}} \right\} = \min \left\{ 1, e^{\frac{1-\lambda}{\lambda}(y-x)} \right\}.$$

Suppose now that the chain is started at $X_0 = x_0 = 1$. Then, the candidate value $y = y_0$, it being from an exponential distribution with a small mean, will likely be smaller than x_0, and therefore, from our expression above, the probability $\gamma(x_0, y_0)$ that the candidate value will be accepted is small. This cycle will persist, and the chain will mix very slowly. When a candidate value gets accepted, it will tend to be still lower than the initial value x_0. Even after many steps, averages such as $\frac{1}{n}\sum_{k=1}^{n} \phi(X_k)$ will produce poor estimates of $E_\pi(\phi(X))$. This example illustrates that it is important that the proposal density not be thinner tailed than the target density when using an independent Metropolis sampler.

19.6.3 Convergence of the Gibbs Sampler

It is important to understand that the Gibbs sampler is not guaranteed to always converge; it can fail. Indeed, it can fail in extremely simple-looking problems. When it fails, it is usually due to a lack of communicability problem. Here is an well-known example.

19.6 MCMC on General Spaces

Example 19.34 (Failure of the Gibbs Sampler to Converge). Suppose U, V are iid $N(0, 1)$ (normality is not important for this example), and suppose that we wish to simulate from the joint distribution of (X, Y) where $Y = U, X = |V|\text{sgn}(U) = |V|\text{sgn}(Y)$. Therefore, the full conditionals, that of $X \mid Y$ and $Y \mid X$, are such that $P(Y > 0 \mid X > 0) = P(X > 0 \mid Y > 0) = 1$. Suppose the starting values in a Gibbs chain are $x_0 = 1, y_0 = 1$. Then, with probability one, the Gibbs update produces x_1, y_1 such that first $x_1 > 0$ (because x_1 is simulated from $f(x \mid Y = y_0)$ and so x_1 has the same sign as y_0), and then, $y_1 > 0$. Then, with probability one, the Gibbs update produces x_2, y_2 such that $x_2, y_2 > 0$, and so on. Thus, if we let A be the event that $X > 0, Y > 0$, the estimate of $P(A)$ from a Gibbs sampler, regardless of the length of the chain, is 1. But the true value of $P(A)$ is $P(A) = P(X > 0, Y > 0) = \frac{1}{2}$.

The reason that the Gibbs sampler fails to converge to the correct joint distribution of (X, Y) in the above example is that the support of the joint distribution is confined to two disjoint subsets, the first quadrant, and the third quadrant. Whenever the target distribution has such a *topologically disconnected* support, the full conditionals of course preserve it, and so the Markov chain corresponding to Gibbs sampling fails to be irreducible. Without the irreducibility property, the chain fails to mix, and gets lost.

Convergence issues for the Gibbs sampler are usually treated separately for the cases of discrete and continuous state spaces. The Gibbs sampler is a special Metropolis scheme, in which the acceptance probabilities $\gamma(x, y)$ are equal to one. So, any general theorem on the convergence of a Metropolis scheme, and any general bound on the error of the approximation works, in principle, for the Gibbs sampler. For example, in the finite-state space case, if a particular Gibbs chain is reversible (recall that Gibbs chains are not necessarily reversible and the exact updating scheme matters; see Section 19.4), then all the results in Theorem 19.8 apply. However, their practical use depends on whether the constant c, that is, the SLEM, can be evaluated or at least bounded for that specific Gibbs chain. Unfortunately, this has to be done on a case-by-case basis. Diaconis et al. (2008) have succeeded in some exponential family problems, where π is the posterior distribution of a scalar parameter, and the prior has a conjugate character (see Chapter 3 for an explanation of conjugacy).

The case of a continuous-state space is more complicated. There are some general theorems with conditions on π that at least guarantee convergence. However, if we seek geometric ergodicity over and above just convergence, then typically one has to fall back on drift and minorization methods, and for Gibbs chains, this approach is essentially problem specific, and not easy. Construction of the *energy function* $V(x)$ corresponding to the drift condition (see Section 19.5.3, and also Theorem 19.12 part (b)) is often too difficult for Gibbs chains. Rosenthal (1995, 1996), Chan (1993), and Jones and Hobert (2001) are some specific success stories. Uniform ergodicity usually does not hold for Gibbs chains unless the state space is bounded.

No very general satisfactory error bounds are available in the Gibbs context. The following concrete theorem is based on Athreya et al. (1996) and Tierney (1994).

Theorem 19.13. *(a) Suppose the target distribution has the joint density $\pi(x_1, x_2, \ldots, x_m)$ and that it satisfies the condition*

(A) $\pi(x_1, x_2, \ldots, x_i \mid x_{i+1}, \ldots, x_m) > 0 \,\forall\, i < m$ and $\forall\, (x_1, x_2, \ldots, x_m)$.

Then the systematic scan Gibbs sampler that updates the coordinates in the order $1 \to 2 \to \cdots \to m$ has the convergence property

$$\sup_A |P^n(x, A) - \pi(A)| \to 0 \quad \text{as } n \to \infty,$$

for almost all starting values x (with respect to the target distribution π).
(b) Suppose that the target distribution has the joint density $\pi(x_1, x_2, \ldots, x_m)$ and that $\Omega = \{x : \pi(x) > 0\}$ is an open connected set in an Euclidean space. Then the random scan Gibbs sampler is reversible, Harris recurrent, has π as its unique stationary distribution, and has the convergence property

$$\sup_A |P^n(x, A) - \pi(A)| \to 0 \quad \text{as } n \to \infty$$

for all initial states x.

Example 19.35 *(The Drift Method in a Bayes Example).* Here is a simple enough example where the drift method can be made to work with ease. Suppose $X_i \mid (\theta, \lambda) \stackrel{iid}{\sim} N(\theta, \lambda), i = 1, 2, \ldots, m$, and (θ, λ) are given the joint improper formal prior density $\frac{1}{\sqrt{\lambda}}, -\infty < \theta < \infty, \lambda > 0$. Then the formal joint posterior density is

$$\pi(\theta, \lambda \mid x_1, \ldots, x_m) = c\lambda^{-\frac{m+1}{2}} e^{-\frac{1}{2\lambda}\sum_{i=1}^m (x_i - \theta)^2}.$$

Even though the formal prior is not a probability density, the formal posterior is integrable for $m \geq 3$. We assume that $m \geq 5$ in order that the drift method fully works out. The full conditionals are derived easily from the joint posterior, and they are:

$$\frac{1}{\lambda} \mid (\theta, x_1, \ldots, x_m) \sim G\left(\frac{m-1}{2}, \frac{2}{\sum_{i=1}^m (x_i - \theta)^2}\right);$$

$$\theta \mid (\lambda, x_1, \ldots, x_m) \sim N\left(\bar{x}, \frac{\lambda}{m}\right).$$

Consider now the systematic scan Gibbs sampler that updates in the order $(\lambda, \theta) \to (\lambda', \theta) \to (\lambda', \theta')$. The transition density $k\left((\theta, \lambda), (\theta', \lambda')\right)$ of this Gibbs chain is

$$k\left((\theta, \lambda), (\theta', \lambda')\right) = \pi(\lambda' \mid \theta, x_1, \ldots, x_m)\pi(\theta' \mid \lambda', x_1, \ldots, x_m),$$

and these two full conditionals have been described above.

Now notice, from this, that given θ, (θ', λ') is conditionally independent of λ, and given λ', θ' is conditionally independent of θ. Therefore, by iterated expectation, for any function $V(\theta, \lambda)$, as long as the expectations below exist,

$$E[V(\theta', \lambda') \mid (\theta, \lambda)] = E[V(\theta', \lambda') \mid \theta]$$
$$= E_{\lambda' \mid \theta} \Big(E[V(\theta', \lambda') \mid (\theta, \lambda')] \Big).$$

Now choose the energy function to be $V(\theta, \lambda) = (\theta - \bar{x})^2$. Then, from the above,

$$E[V(\theta', \lambda') \mid (\theta, \lambda)] = E_{\lambda' \mid \theta} \Big(\frac{\lambda'}{m}\Big)$$
$$= \frac{\sum_{i=1}^{m}(x_i - \theta)^2}{m(m-3)}.$$

Now decompose $\sum_{i=1}^{m}(x_i - \theta)^2$ as $\sum_{i=1}^{m}(x_i - \theta)^2 = m(\theta - \bar{x})^2 + \sum_{i=1}^{m}(x_i - \bar{x})^2$. Therefore,

$$E[V(\theta', \lambda') \mid (\theta, \lambda)] = \frac{mV(\theta, \lambda) + \sum_{i=1}^{m}(x_i - \bar{x})^2}{m(m-3)} = \frac{V(\theta, \lambda)}{m-3} + \frac{\sum_{i=1}^{m}(x_i - \bar{x})^2}{m(m-3)}$$
$$\leq (1 - \epsilon)V(\theta, \lambda) + b,$$

where $\epsilon \leq 1 - \frac{1}{m-3} = \frac{m-4}{m-3}$, and $b \geq \frac{\sum_{i=1}^{m}(x_i - \bar{x})^2}{m(m-3)}$. This establishes the drift condition in this example (see Definition 19.5).

19.7 Practical Convergence Diagnostics

The spectral methods as well as the drift methods for studying convergence rates of MCMC algorithms may sometimes give very conservative answers to the all-important practical question: how long should the chain be run to make $\rho(\mu P^n, \pi) \leq \epsilon$? The spectral bounds can be difficult to apply, if the eigenvalues of P are intractable, and the drift methods are usually difficult to apply in any case. In addition the spectral and the drift methods require each new problem to be treated separately. Therefore, effective and simple omnibus convergence diagnostics are useful. They can be graphical, or numerical. Fueled by this practical need, a bewildering array of MCMC convergence diagnostics have been proposed over the last 25 years. A major review of methods up to that time is Cowles and Carlin (1996). A more recent literature summary is Mengersen et al. (2004).

These convergence diagnostics have often been received with skepticism by actual users, because none of them has been found to be generally superior to the others, and more importantly, because in a given application, they often give contradictory answers. Although one diagnostic may lead us to conclude that approximate convergence has already been achieved, another may suggest that we are still very far from convergence. However, in conjunction with the use of spectral or drift methods, and collectively, they have some practical utility. We provide a brief overview of a few of the diagnostics that have become popular in the area.

Gelman-Rubin Multiple Chain Method. This is based on Gelman and Rubin (1992). In this method, m parallel chains, say $X_{kj}, k = 1, 2, \ldots, 2n, j = 1, 2, \ldots, m$ are generated by using an MCMC algorithm. The method applies to any MCMC algorithm for which convergence to a stationary distribution is theoretically known.

Suppose the target distribution π is d-dimensional for some $d \geq 1$, and let $g : \mathcal{R}^d \to \mathcal{R}$ be a function of interest. Let

$$g_{kj} = g(X_{kj}), \quad \overline{g}_j = \frac{1}{n} \sum_{k=n+1}^{2n} g_{kj}, \quad \bar{g} = \frac{1}{mn} \sum_{j=1}^{m} \sum_{k=n}^{2n} g_{kj},$$

$$s_j^2 = \frac{1}{n} \sum_{k=n+1}^{2n} (g_{kj} - \overline{g}_j)^2.$$

Thus, g_{kj} is the value of g based on the kth sweep of the jth run, $\overline{g}_j$ is their mean, s_j^2 is their variance, and $\bar{g}$ is the overall mean. Now define the *between* and *within* variances by

$$B = \frac{1}{m} \sum_{j=1}^{m} (\overline{g}_j - \bar{g})^2, \quad W = \frac{1}{m} \sum_{j=1}^{m} s_j^2.$$

The method also computes an estimate $\hat{\pi}_n$ of the target distribution π by pooling the last n values from the m parallel chains. Convergence is monitored by calculating

$$\text{GR} = \text{GR}_{g,n} = \left[\frac{n-1}{n} + \frac{m+n}{mn} \frac{B}{W} \right] c_n,$$

where c_n is a certain concrete parameter of the estimating distribution $\hat{\pi}_n$. GR has the interpretation of the additional possible gain in efficiency for estimating $E_\pi(g(X))$ if the run size is increased indefinitely. Thus, convergence is concluded when GR ≈ 1. If there are several different functions g of interest, then each corresponding GR has to be monitored and sampling continues until each is close to 1.

Yu–Mykland Single Chain Graphic. This is based om Yu and Mykland (1994). In this method, a single chain $X_j, j = 0, 1, \ldots, n$ is generated using any convergent MCMC chain. Once again, let $g : \mathcal{R}^d \to \mathcal{R}$ be a function of interest, and let $\bar{g} = (1/(n-B)) \sum_{j=B+1}^{n} g(X_j)$ be the mean of the value of g after discarding an initial B value, B has to be chosen by the user. Define the *cumulative sum values* $S_t = \sum_{j=B+1}^{t} [g(X_j) - \bar{g}], t = B, B+1, \ldots, n$. The method plots the pairs $(t, S_t), t = B, B+1, \ldots, n$. The plot is called a CUSUM plot. This plot starts and ends at $S_t = 0$.

19.7 Practical Convergence Diagnostics

A necessary condition for accurate estimation of means and other functionals of the target distribution is that the MCMC chain has had enough time to traverse through the entire state space. This is called *good mixing*. Chains that mix slowly are likely to produce well-behaved CUSUM plots, such as large segments of monotone curves. Chains that mix rapidly are likely to produce oscillatory plots. The CUSUM plot is visually examined for its smoothness, and if it appears nonsmooth, then rapid mixing is concluded. Rapid mixing is indicative of rapid convergence. So the plot assesses convergence indirectly. The CUSUM plot could be nonsmooth for some g, and smooth for another g, however. Apart from the need to select the burn-in period B, the method has a simplistic appeal.

Garren–Smith Multiple Chain Gibbs Method. This applies only to reversible Gibbs chains $X_n, n \geq 0$, because of the theoretical need that the eigenvalues of the chain's transition matrix be such that the second largest eigenvalue is unique. Here is the precise meaning of that. Suppose P is reversible, and let its eigenvalues (which are real) be arranged in the order of decreasing absolute values, $1 > |\lambda_2| \geq |\lambda_3| \geq \ldots$. Let A be any subset of the state space S. Provided that $|\lambda_2| > |\lambda_3|$ (strictly greater), one gets

$$P(X_k \in A \mid X_0 = i) = \pi(A) + c\lambda_2^k + o(|\lambda_2|^k).$$

Suppose that m parallel chains $X_{kj}, k = 1, 2, \ldots, n, j = 1, 2, \ldots, m$, all with the same starting value, say i, are generated, and select a value for the burn-in period B. Denote $\pi(A) = p$, and form estimates of p by pooling simulated values from the multiple chains:

$$\hat{p}_k = \frac{1}{m} \sum_{j=1}^{m} I_{X_{kj} \in A}, \quad k = B+1, \ldots, n.$$

Denote $\theta = (p, c, \lambda_2)$, and estimate θ as

$$\hat{\theta} = \hat{\theta}_B = \operatorname{argmin} \sum_{k=B+1}^{n} (\hat{p}_k - p - c\lambda_2^k)^2.$$

As a preliminary diagnostic, one plots each coordinate of $\hat{\theta}_B$ against B, as B ranges over $1, 2, \ldots$ (or some smaller subset of it). Initially, the plot would look unstable. The burn-in period is chosen as that B where the plot clearly begins to stabilize, if such a value of B can be identified. This may be difficult and ambiguous. As regards convergence, if the plot stabilizes at some $B << n$, then convergence may be suspected. Garren and Smith also give more complex second stage diagnostics in their article (Garren and Smith (1993)).

Exercises

Exercise 19.1. Find a Monte Carlo estimate for the value of $\mu = \int_0^{10} e^{\sqrt{x}} dx$, and construct a 95% confidence interval for μ. Use a selection of values for the Monte Carlo sample size n. Compare the Monte Carlo estimate to the true value of μ.

Exercise 19.2. * Find a Monte Carlo estimate for the value of $\mu = \int_{-\infty}^{\infty} e^{-x^2} \sin^5(x) dx$, and construct a 95% confidence interval for μ. Use a selection of values for the Monte Carlo sample size n. Compare the Monte Carlo estimate to the true value of μ.

Exercise 19.3 * **(Monte Carlo Estimate of a Three-Dimensional Integral).** Find a Monte Carlo estimate for the value of

$$\mu = \int_{-\infty}^{\infty} \int_{-\infty}^{\infty} \int_{-\infty}^{\infty} \frac{e^{-(1/2)(x^2+y^2+z^2)}}{1+|x|+|y|+|z|} dxdydz,$$

and construct a 95% confidence interval for μ. Use a selection of values for the Monte Carlo sample size n.

Exercise 19.4. With reference to Example 19.1, give a rigorous proof that $E(\log|X|) = 0$ if $X \sim C(0, 1)$.

Exercise 19.5 * **(Monte Carlo Estimate of Area of a Triangle).** Find a Monte Carlo estimate for the area μ of a triangle with sides $a = 3, b = 2, c = 2$, and construct a 95% confidence interval for μ. Use a selection of values for the Monte Carlo sample size n. Compare the Monte Carlo estimate to the true value of μ.

Exercise 19.6 * **(Monte Carlo Estimate of Volume of a Cylinder).** Find a Monte Carlo estimate for the volume μ of a right circular cylinder with height $h = 2$ and radius $r = 1$. Construct a 95% confidence interval for μ. Use a selection of values for the Monte Carlo sample size n. Compare the Monte Carlo estimate to the true value of μ.

Exercise 19.7 * **(Monte Carlo Estimate of Volume of a Pyramid).** Find a Monte Carlo estimate for the volume μ of a square-based pyramid with base sides equal to $b = 1$ and height $h = 5$. Construct a 95% confidence interval for μ. Use a selection of values for the Monte Carlo sample size n. Compare the Monte Carlo estimate to the true value of μ.

Exercise 19.8 * **(Monte Carlo Estimate of Surface Area).** Devise a Monte Carlo scheme for finding the surface area of the d-dimensional unit ball. Use it to find estimates for the surface area when $d = 2, 3, 5, 10, 50$. Use a selection of values for the Monte Carlo sample size n. Compare the Monte Carlo estimate to the true value of the surface area of the unit ball (which is $\frac{d\pi^{d/2}}{\Gamma(\frac{d}{2}+1)}$).

Exercises

Exercise 19.9 (Using the Monte Carlo on a Divergent Integral). The integral $\mu = \int_0^1 \frac{\cos x}{x}$ diverges. Investigate what happens if you try to find a Monte Carlo estimate for the integral anyway. Use a selection of values for the Monte Carlo sample size n.

Exercise 19.10 * (Using the Monte Carlo on an Infinite Expectation). For any $m \geq 2$, if $X_1, \ldots, X_m$ are iid standard Cauchy, then $E(X_{(m)} - X_{(m-1)}) = \infty$. Investigate what happens if you try to find a Monte Carlo estimate for this expectation anyway; use $m = 2, 5, 20$, and a selection of values for the Monte Carlo sample size n.

Exercise 19.11. Suppose $X_i \sim \text{Exp}(i), i = 1, 2, 3, 4$, and suppose that they are independent. Find a Monte Carlo estimate for $P(X_1 < X_2 < X_3 < X_4)$. Use a selection of values for the Monte Carlo sample size n.

Exercise 19.12. Suppose $X_i \stackrel{iid}{\sim} N(0, 1), i = 1, 2, \ldots, m$. Find a Monte Carlo estimate for $E(X_{(m)})$, the expectation of the maximum of $X_1, \ldots, X_m$. Use $m = 2, 5, 10, 25, 50$, and a selection of values for the Monte Carlo sample size n. Compare the Monte Carlo estimate with the true value of $E(X_{(m)})$ (see Chapter 6).

Exercise 19.13 * (Monte Carlo in a Bayes Problem). Suppose $X \sim \text{Bin}(100, p)$ and p has the prior density $cp^p(1-p)^{1-p}, 0 < p < 1$, where c is a normalizing constant. Find a Monte Carlo estimate for the posterior mean of p if $X = 45$. Use a selection of values for the Monte Carlo sample size n.

Exercise 19.14 * (Monte Carlo in a Bayes Problem). Suppose $X \sim \text{Poi}(\lambda)$ and λ has the discrete prior $\pi(n) = \frac{c}{n^2}, n = 1, 2, \ldots$, where c is a normalizing constant. Find a Monte Carlo estimate for the posterior mean of λ if $X = 1$. Use a selection of values for the Monte Carlo sample size n.

Exercise 19.15. * Suppose $X \sim N_3(\mu, I)$, and that the mean vector μ has a prior density

$$\pi(\mu) = \frac{c}{1 + (|\mu_1| + |\mu_2| + |\mu_3|)^{3.5}}.$$

Find a Monte Carlo estimate for the posterior mean of $\mu_i, i = 1, 2, 3$ if $X = (1, 0, -1)$. Use a selection of values for the Monte Carlo sample size n.

Exercise 19.16 (Monte Carlo Estimate of e). Find a Monte Carlo estimate for e by using the identity $P(X > 1) = \frac{1}{e}$ if $X \sim \text{Exp}(1)$. Use a selection of values for the Monte Carlo sample size n, and plot the Monte Carlo estimates against n.

Exercise 19.17 (Monte Carlo P-value). With reference to Example 19.7, calculate the Monte Carlo P-value for the median-based test for the center of a Cauchy distribution when the original dataset is generated under the alternative $C(\mu, 1)$ with $\mu = 2.5$. Compare with the P-value approximated by using the normal approximation, as in Example 19.7.

Exercise 19.18 * **(Monte Carlo P-value).** Calculate the Monte Carlo P-value for the Kolmogorov–Smirnov statistic for testing $H_0 : F = N(0, 1)$, when the original dataset of size $m = 50$ is generated under the null; under $F = U[-2, 2]$. Compare with the P-value approximated by using the Brownian bridge approximation, as in Chapter 16, Section 16.2.1.

Exercise 19.19 * **(Outlier Detection and Monte Carlo P-value).** Suppose $X_1, \ldots, X_m$ are iid observations from a continuous CDF F_0 on the real line. In outlier detection studies, one often declares the largest observation i.e., the mth order statistic, an outlier if

$$\frac{X_{(m)} - X_{(m-1)}}{X_{(m)} - M}$$

is large, where M is the median of $X_1, \ldots, X_m$, $X_{(m)}$ is the mth order statistic, and $X_{(m-1)}$ the $(m-1)$th order statistic.

Calculate the Monte Carlo P-value for this test when the original dataset is generated from $F_0 = C(0, 1)$; from $F_0 = N(0, 1)$. Use $m = 50$, and a selection of Monte Carlo sample sizes n.

Exercise 19.20 (Testing for Randomness). Suppose a computer has produced n numbers $U_1, \ldots, U_n$, that are supposed to be iid $U[0, 1]$. Devise a formal test of it by using the Kolmogorov–Smirnov statistic (see Chapter 16). Answer the following question. Does this method test whether there are possible dependencies between the obtained values $U_1, \ldots, U_n$?

Exercise 19.21 * **(Testing for Randomness).** Suppose a computer has produced n numbers $U_1, \ldots, U_n$, that are supposed to be iid $U[0, 1]$. Let $W_0 = U_{(1)}, W_i = U_{(i+1)} - U_{(i)}, 1 \leq i \leq n-1$, where $U_{(1)}, U_{(n)}$ are the order statistics of $U_1, \ldots, U_n$. Devise formal tests of whether $U_1, \ldots, U_n$ are indeed independent uniform by using the following facts from Chapter 6 (Theorem 6.6):

(i) For any r, $U_{(r)} \sim Be(r, n - r + 1)$.
(ii) Jointly, $(W_0, W_1, \ldots, W_{n-1})$ is uniformly distributed in the n-dimensional simplex.

Exercise 19.22 (Accept–Reject for a General Beta). Devise an accept–reject algorithm to simulate from a $Be(\alpha, \beta)$ distribution, by using an envelope density $g(x) = cx^{\alpha-1}(1 - x^\alpha)^{\frac{\beta}{\alpha}-1}$. Find the acceptance probability of your accept–reject algorithm.

Exercise 19.23 (Accept–Reject for a Truncated Exponential). Devise an accept-reject algorithm to simulate from the density $f(x) = e^{\mu-x}, x \geq \mu$, by using an $\text{Exp}(\lambda)$ envelope density. Identify the value of λ for which the accept–reject scheme is the most efficient.

Exercise 19.24. Devise an accept–reject scheme to simulate from the density $f(x) = \frac{16}{3}x, 0 \leq x \leq \frac{1}{4}, f(x) = \frac{4}{3}, \frac{1}{4} \leq x \leq \frac{3}{4}, f(x) = \frac{16}{3}(1 - x), \frac{3}{4} \leq x \leq 1$. Find the efficiency value of your accept–reject scheme.

Exercises

Exercise 19.25 * (**Accept–Reject for Multidimensional Truncated Normal**). Devise an accept–reject scheme to simulate from the density $f(x) = ce^{-\sum_{j=1}^{d} x_j^2}$, $\sum_{j=1}^{d} x_j^2 \leq 1$, where c is a normalizing constant, and find the efficiency value of your accept–reject scheme.

Exercise 19.26 (**Accept–Reject for Truncated Gamma**). Devise an accept–reject scheme with exponential envelope densities to simulate from the density $f(x) = cx^n e^{-x}, 0 \leq x \leq 1$, where $n \geq 1$, and $c = c_n$ is a normalizing constant. Find the best envelope density from the exponential class.

Exercise 19.27 * (**Asymptotic Zero Efficiency of Accept–Reject**). Consider the scheme of generating a uniformly distributed random vector X in the d-dimensional unit ball by generating iid variables $U_i \sim U[0, 1], i = 1, 2, \ldots, d$, and by setting $X = (U_1, \ldots, U_d)$ if $\sum_{i=1}^{d} U_i^2 \leq 1$. Show that the acceptance rate of this scheme converges to zero as $d \to \infty$. Explicitly evaluate the acceptance rate for $2 \leq d \leq 6$.

Exercise 19.28 * (**Generating Points in the Unit Circle**). Generate n points uniformly in the unit circle and visually identify the radius of the largest empty circle. Use $n = 50, 100, 500, 1000$.

Exercise 19.29 * (**Generating Cauchy from Two Uniforms**).

(a) Suppose U, V are iid $U[0, 1]$. Show that the conditional distribution of $\frac{U}{V}$ given that $U^2 + V^2 \leq 1$ is standard Cauchy.
(b) Hence, devise an algorithm for generating a standard Cauchy random variable.

Exercise 19.30 * (**General Ratio of Uniforms Method**). Suppose $f(x)$ is a density function on the real line, and suppose (U, V) is uniformly distributed in the set in the plane defined as $C = \{(u, v) : 0 \leq u \leq \sqrt{f(\frac{v}{u})}\}$. Show that the marginal density of $\frac{V}{U}$ is f.

Use this result to generate a standard Cauchy random variable.

Exercise 19.31 (**Quantile Transform Method**). Show how to simulate from the density

$$f(x) = c \frac{x^{\alpha-1}}{\sqrt{1 - x^{2\alpha}}}, \quad 0 < x < 1,$$

by using the quantile transform method; here, c is a normalizing constant.

Exercise 19.32 (**Quantile Transform Method**). Show how to simulate a standard double exponential random variable by using the quantile transform method.

Exercise 19.33 * (**Quantile Transform Method to Simulate from t**). Suppose X has a t-distribution with two degrees of freedom (you can write the CDF in a simple closed-form). Show how to simulate X by using the quantile transform method.

Exercise 19.34 (**Quantile Transform Method to Simulate from Pareto**). Suppose $X \sim Pa(\alpha, \theta)$. Show how to simulate X by using the quantile transform method.

Exercise 19.35 (Quantile Transform Method to Simulate from the Gumbel Distribution). Show how to simulate a standard Gumbel random variable by using the quantile transform method.

Exercise 19.36 (Importance Sampling in a Bayes Problem). Suppose $X \sim$ Poi(λ) and that λ has the prior density $c \frac{1}{1+\lambda^2}, \lambda > 0$, where c is a normalizing constant. By following the methods of Example 19.12, approximate the posterior mean of λ when $X = 5$. Use a selection of values of the importance sample size n.

Exercise 19.37 * (Asymptotic Bias of the Importance Sampling Estimate). Derive an expression of the form $E(\hat{\mu}) = \mu + \frac{\alpha(h_0, h_1)}{n} + o(n^{-1})$ for the importance sampling estimate $\hat{\mu}$, in the notation of Section 19.2.2.

Exercise 19.38 (Importance Sampling Estimate of Normal Tail Probability). Suppose $Z \sim N(0, 1)$. Find an importance sampling estimate of $P(Z > z)$ by using a truncated exponential importance sampling density $g(x) = e^{z-x}, x > z$. Use it for $z = 1, 2, 3, 4$. Compare with a standard normal table.

Exercise 19.39 * (Simulating from the Tail of a Normal Density). Suppose $Z \sim N(0, 1)$. Use the family of truncated exponential densities $\frac{1}{\lambda} e^{-\frac{x-z}{\lambda}}, x > z$ to devise an importance sampling scheme for simulating from the distribution of Z given $Z > z$.

Use your scheme to generate ten observations from the distribution of Z given $Z > 3.5$.

Exercise 19.40 (Optimal Importance Sampling Density). Suppose $X \sim $ Exp(1) and that we want to find Monte Carlo estimates for $\mu = E[\phi(X)]$.

(a) Find the optimal importance sampling density from the family of all possible truncated exponential densities $g(x \mid \mu, \lambda) = \frac{1}{\lambda} e^{-\frac{x-\mu}{\lambda}}, x \geq \mu$, if $\phi(x) = I_{x>a}$.
(b) Find the optimal importance sampling density from the same family if $\phi(x) = x^k$.

Exercise 19.41 * (Importance Sampling and Delta Theorem). Suppose $X \sim N(0, 1)$ and you want to approximate $\mu = E(|X|)$ by using importance sampling with a standard double exponential importance sampling density. Find an expression for the asymptotic variance of the importance sampling estimate of μ, by using Theorem 19.2.

Exercise 19.42 * (Error of the Importance Sampling Estimate). Suppose $X \sim $ Be(5, 5) and that you want to estimate $\mu = E(X^6)$ by using importance sampling with a $U[0, 1]$ importance sampling density. Find an approximation to the probability $P(|\hat{\mu} - \mu| > .02)$ by using the asymptotic normality result of Theorem 19.2. Evaluate this approximation when the importance sample size is $n = 100, 250$.

Exercises

Exercise 19.43 *(Hit-or-Miss Method).** Let $0 \leq \phi(x) \leq M < \infty, 0 \leq x \leq c$. Consider iid bivariate simulations $(X_i, Y_i), i = 1, 2, \ldots, n$ from the bivariate uniform distribution on $[0, c] \otimes [0, M]$. Consider the following two estimates for the definite integral $\mu = \int_0^c \phi(x)dx$:

$$\hat{\mu}_1 = \frac{c}{n} \sum_{i=1}^n \phi(X_i);$$

$$\hat{\mu}_2 = \frac{Mc}{n} \sum_{i=1}^n I_{Y_i \leq \phi(X_i)}.$$

(a) Show that each estimate is unbiased; that is, $E[\hat{\mu}_i] = \mu, i = 1, 2$.
(b) One of the two estimates always has a smaller variance than the other one. Identify it with a proof.

Exercise 19.44 (Rao–Blackwellization). A chicken lays a Poisson number N of eggs per week. The eggs hatch, mutually independently, with probability p. Suppose X is the number of eggs actually hatched in a week. We wish to estimate $E(X)$.

(a) Describe an ordinary Monte Carlo estimate for $E(X)$.
(b) Describe a Rao–Blackwellized estimate, by inventing an appropriate variable on which you condition.
(c) How much is the percentage reduction in variance achieved by your Rao–Blackwellized estimate?

Exercise 19.45 (Rao–Blackwellization). Suppose X, Y are iid iid continuous random variables with CDF F. We wish to estimate $P(X < 3Y - 3Y^2 + Y^3)$.

(a) Describe an ordinary Monte Carlo estimate.
(b) Describe a Rao–Blackwellized estimate, by inventing an appropriate variable on which you condition.

Exercise 19.46 *(Quantile Transform for Discrete Distributions).** Suppose X takes values $x_1 < x_2 < \cdots$ with probabilities $p_i = P(X = x_i), i = 1, 2, \ldots$. Let $U \sim U[0, 1]$ and set $Y = x_i$ if $p_1 + \cdots + p_{i-1} < U \leq p_1 + \cdots + p_i$. Show that Y has the distribution of X.

Exercise 19.47 (Quantile Transform for Poisson). Use the general quantile transform method of the above exercise to simulate n independent values from $\text{Poi}(\lambda)$; use a selection of values of λ and n.

Exercise 19.48 (Quantile Transform for Geometric). By using the fact that $P(X > i)$ has a closed-form formula for any geometric random variable, simulate from a $\text{Geo}(p)$ distribution by using the general quantile transform method of Exercise 19.46.

Exercise 19.49 *(Generating Poisson from Exponential).** Suppose $X_1, X_2, \ldots$ are iid exponential random variables with mean $\frac{1}{\lambda}$. Define $Y = i$ if $X_1 + \cdots + X_i \leq 1 < X_1 + \cdots + X_{i+1}$. Show that $Y \sim \text{Poi}(\lambda)$.

Exercise 19.50 (Generating Exponential-Order Statistics). By using Reyni's representation (see Chapter 6), devise an algorithm for simulating the set of order statistics of n iid standard exponentials.

Exercise 19.51 *(Generating Random Permutations).** Suppose $U_1, \ldots, U_n$ are iid $U[0, 1]$, and let $U_{(1)} < U_{(2)} < \cdots < U_{(n)}$ be their order statistics. Set $\pi(i) = j$ if $U_i = U_{(j)}$. Show that $(\pi(1), \pi(2), \ldots, \pi(n))$ is uniformly distributed on the set of $n!$ permutations of $(1, 2, \ldots, n)$.

Exercise 19.52 (Generating Points from a Homogeneous Poisson Process). Use Theorem 13.3 to generate points from a homogeneous Poisson process with constant rate 1 on the interval $[0, 10]$, and then plot the path of the process.

Exercise 19.53 *(Generating Points from a Nonhomogeneous Poisson Process).** Use the transformation method of Section 13.6 to generate points from a Poisson process with the intensity function $\lambda(x) = \min(x, 5)$ on $[0, 10]$, and then plot the path of the process.

Exercise 19.54 (Metropolis–Hastings Chain). For the special case that $n = 4$ and $\alpha = \beta = 2$, find explicitly the transition probabilities of the Metropolis–Hastings chain for the Beta–Binomial case (Example 19.15), and verify that the chain is irreducible and aperiodic.

Exercise 19.55 (Barker's Algorithm). Suppose $Y \sim \text{Geo}(p)$. Find explicitly the transition probabilities for Barker's algorithm for simulating from the conditional distribution of Y given that $Y \leq n$, where $n > 1$ is a fixed integer.

Exercise 19.56. Write the following scheme formally as a Metropolis chain: generate a candidate state j according to the target distribution π, regardless of the current state i, and accept the candidate state only if the target distribution assigns at least ϵ probability to it.

Exercise 19.57 *(Autoregressive Metropolis Chain).** Suppose the state space $S = \mathcal{R}^d$ for some $d \geq 1$. Formally write the proposal probabilities and the acceptance probabilities of the following Metropolis scheme. If we are currently at state x, the candidate state j is generated as $y = \mu + A(x - \mu) + z$, where μ is a fixed element of the state space, A is a fixed matrix, and z is a random element generated from a density $q(z)$.

Exercise 19.58 (Practical MCMC). Use the independent Metropolis sampler when the target distribution is $N(0, 1)$ and the proposal density is a standard double exponential. Run the chain to length $n = 2500$. Try to determine what is a good burn-in value B by using some graphical method (e.g., you may use the Yu–Mykland CUSUM plot).

Now repeat it when the proposal density is $N(0, 4)$. For which proposal density, did you settle for a smaller burn-in time? Is that what you would have expected? Why?

Exercises

Exercise 19.59 (Practical MCMC). Generate a Metropolis–Hastings chain of length $n = 50$ when the target distribution is Bin(20, .5).

Exercise 19.60 (Practical MCMC). Generate a Metropolis–Hastings chain of length $n = 50$ when the target distribution is a Poisson with mean 1, truncated to $\{0, 1, \ldots, 5\}$.

Exercise 19.61 (Practical MCMC). Generate an independent Metropolis chain of length $n = 50$ when the target distribution is a Poisson with mean 1, and the proposal distribution is a failure geometric with the pmf $p_j = (\frac{1}{2})^{j+1}, j = 0, 1, 2, \ldots$.

Exercise 19.62 (Visual Comparison of MCMC Chains). For a target distribution of Poisson with mean 1, truncated at 10, generate a Metropolis–Hastings chain of length $n = 50$, plot a histogram, and compare it visually to the histogram of the independent Metropolis chain of the previous exercise.

Exercise 19.63 (Practical MCMC in a Bayes Problem). Suppose $X \sim \text{Bin}(m, p)$ and $p \sim \text{Beta}(\text{alpha}, \beta)$. Use the systematic scan Gibbs sampler to simulate from the joint distribution of (X, p), and take the marginal p-output to estimate the mean and the variance of the posterior. Compare these answers to the known exact values (see Chapter 3). Use $m = 25, \alpha = \beta = 2$, and $x = 10$.

Exercise 19.64 (Practical MCMC in a Bayes Problem). Suppose $X \sim \text{Poi}(\lambda), \lambda \sim G(\alpha, \beta)$. Use the systematic scan Gibbs sampler to simulate from the joint distribution of (X, λ) and take the λ-output to estimate the mean and the variance of the posterior. Compare these answers to the known exact values (see Chapter 3). Use $\alpha = 4, \beta = 1, x = 8$.

Exercise 19.65 * (Where to Start). Suppose you wish to use some MCMC scheme for simulating from a posterior distribution of a parameter. What are reasonable starting values?

Exercise 19.66 (Gibbs Chain in a Normal Problem). Suppose $X_1, \ldots, X_n$ are iid $N(\mu, \theta)$, and (μ, θ) has the improper prior density $\pi(\mu, \theta) = \frac{1}{\sqrt{\theta}}$. Find the full conditionals $\pi(\mu \mid \theta)$ and $\pi(\theta \mid \mu)$.

Exercise 19.67 * (Gibbs Chain for Bivariate Poisson). Suppose X, Y, Z are independent Poissons with means λ, μ, θ. Let $U = X + Z, V = Y + Z$. The joint distribution of (U, V) is called a bivariate Poisson. Find the full conditionals $\pi(u \mid v)$ and (by symmetry) $\pi(v \mid u)$.

Exercise 19.68 * (Gibbs Chain for a Bimodal Joint Density). Suppose (X, Y) has the joint density

$$\pi(x, y) = ce^{-\frac{1}{2}[(x-4)^2 + (y-4)^2 + x^2 y^2]}, \quad -\infty < x, y < \infty,$$

where c is a normalizing constant. Devise a Gibbs sampler to simulate from this target density π.

Exercise 19.69 * (**Monotonicity of Gibbs Error**). Show that the systematic scan Gibbs chain for simulating an m-dimensional random vector has the property that the total variation distance $\sup_A |P^n(x, A) - \pi(A)|$ is monotonically nonincreasing in n.

Exercise 19.70 (**A Nonreversible Chain**). Consider a stationary Markov chain with the transition probability matrix

$$P = \begin{pmatrix} \frac{1}{2} & \frac{1}{2} & 0 \\ 0 & \frac{1}{2} & \frac{1}{2} \\ \frac{1}{2} & 0 & \frac{1}{2} \end{pmatrix}.$$

Show that the chain has a stationary distribution π, but that the chain is not reversible.

Exercise 19.71 * (**Nonreversibility of Gibbs Chains**). Give a counterexample to show that the systematic scan Gibbs sampler in the fixed update order $1 \to 2 \to \cdots \to m$ need not be reversible.

Exercise 19.72 (**Reversibility of Gibbs Chains**). Show that the Gibbs chain with the transition amtrix $P = \frac{1}{m!} \sum_{S(m)} P_{i_1} P_{i_2} \cdots P_{i_m}$ introduced in Section 19.4 is reversible.

Exercise 19.73 (**Necessary and Sufficient Condition for Reversibility**). Given discrete vectors x, y, and a probability distribution π, suppose $(x, y)_\pi = \sum_{i \in S} x_i y_i \pi(i)$. Suppose P is the transition probability matrix of a stationary Markov chain on the state space S. Prove that (P, π) satisfy the equation of detailed balance if and only if $(Px, y)_\pi = (x, y)_\pi$ for all vectors x, y such that $\sum_{i \in S} x_i^2 \pi(i) < \infty$ and $\sum_{i \in S} y_i^2 \pi(i) < \infty$.

Exercise 19.74 * (**Necessary and Sufficient Condition for Reversibility**). Suppose P is the transition probability matrix of a stationary Markov chain on the finite-state space S, and π is a probability distribution on S. Let Δ be the diagonal matrix with diagonal elements $\pi(i), i = 1, 2, \ldots, t$. Prove that (P, π) satisfy the equation of detailed balance if and only if the matrix $\Delta^{\frac{1}{2}} P \Delta^{-\frac{1}{2}}$ is symmetric.

Exercise 19.75. Generate a Gibbs chain of length $n = 50$ using the random scan Gibbs chain, when the target distribution is a bivariate normal with means 0, standard deviations 1, and correlation 0.5.

Exercise 19.76 * (**Gibbs Chain for Poissons with Covariates**). Suppose $X_i \mid \lambda_i \sim \text{Poi}(x_i \lambda_i), i = 1, 2, \ldots, m$, and that $X_1, \ldots, X_m$ are independent. Next, $\lambda_i \mid \beta \sim G(\alpha_i, y_i \beta)$, where $\lambda_1, \ldots, \lambda_m$ are independent, and finally $\frac{1}{\beta} \sim G(\gamma, \delta)$. The numbers $x_i, y_i, \alpha_i, \gamma, \delta$ are taken to be given constants. Find the full conditionals $\pi(\lambda_1, \ldots, \lambda_m \mid \beta, X_1, \ldots, X_m)$ and $\pi(\beta \mid \lambda_1, \ldots, \lambda_m, X_1, \ldots, X_m)$. How are these useful in simulating from the posterior distribution of the first-stage parameters $(\lambda_1, \ldots, \lambda_m)$?

Exercises

Exercise 19.77 * (**Gibbs Chain for the ZIP Model**). Suppose $X_i, B_i, i = 1, 2, \ldots, m$ are mutually independent, with $X_i \mid \lambda \sim \text{Poi}(\lambda)$, $B_i \mid p \sim \text{Ber}(p)$, $\lambda \sim G(\alpha, \beta)$, $p \sim U[0, 1]$, and that λ, p are independent. Find the full conditionals $\pi(\lambda \mid p)$ and $\pi(p \mid \lambda)$.

Exercise 19.78. Find the eigenvalues in analytical form for the three-state stationary Markov chain with the transition matrix

$$P = \begin{pmatrix} \alpha & 1-\alpha & 0 \\ 0 & \beta & 1-\beta \\ 1-\gamma & 0 & \gamma \end{pmatrix}.$$

Hence find an expression for the SLEM. Is this an irreducible chain?

Exercise 19.79 * (**Nonconvergent Gibbs Sampler**). Give an example, different from the one in the text, for which the systematic scan Gibbs chain does not converge to the target distribution.

Exercise 19.80. Give a proof that for a reversible chain, the eigenvalues of the transition matrix are all in the interval $[0, 1]$.

Exercise 19.81 * (**Ehrenfest Chain**). Consider the symmetric Ehrenfest chain of Examples 10.20 and 10.4. For $m = 7$, calculate the SLEM of the transition probability matrix P. Plug it into the bounds in parts (a) and (b) of Theorem 19.8, and find the best answer to the following question.

How large an n is needed to make

$$\sup_i \sup_A |P^n(i, A) - \pi(A)| \leq .01?$$

Note that the stationary distribution of the Ehrenfest chain was worked out in Example 10.20.

Exercise 19.82 * (**SLEM Calculation for Metropolis–Hastings Chain**). Use Example 19.24 to evaluate the SLEM of the Metropolis–Hastings chain when $t = 4$ and the target distribution is $\pi(1) = \frac{1}{3}, \pi(2) = \pi(3) = \frac{1}{4}, \pi(4) = \frac{1}{6}$.

Exercise 19.83 (**Extreme Values of the Dobrushin Coefficient**). Give necessary and sufficient conditions for $\Delta(P)$ to take the values $0, 1$.

Exercise 19.84. Calculate the Dobrushin coefficient as well as the SLEM for the nonreversible transition matrix of Example 19.22, and verify that $c \leq \Delta(P)$.

Exercise 19.85. * Construct an example in which the SLEM and the Dobrushin coefficient coincide.

Exercise 19.86 *(SLEM versus Dobrushin Coefficients). Prove the better inequality $c \leq \min_n \left(\Delta(P^n) \right)^{\frac{1}{n}}$.

Exercise 19.87 *(A Way to Numerically Approximate the SLEM). Show that $\log \Delta(P) = \lim_{n \to \infty} \frac{1}{n} \log \left(\Delta(P^n) \right)$.

References

Athreya, K., Doss, H., and Sethuraman, J. (1996). On the convergence of the Markov chain simulation method, *Ann. Statist.*, 24, 89–100.

Barnard, G. (1963). Discussion of paper by M.S. Bartlett, *JRSS Ser. B*, 25, 294.

Besag, J. and Clifford, P. (1989). Generalized Monte Carlo significance tests, *Biometrika*, 76, 633–642.

Besag, J. and Clifford, P. (1991). Sequential Monte Carlo p-values, *Biometrika*, 78, 301–304.

Brémaud, P. (1999). *Markov Chains*, Springer, New York.

Chan, K. (1993). Asymptotic behavior of the Gibbs samples, *J. Amer. Statist. Assoc.*, 88, 320–326.

Cowles, M. and Carlin, B. (1996). Markov chain Monte Carlo convergence diagnostics: A comparative review, *J. Amer. Statist. Assoc.*, 91, 883–904.

Diaconis, P. (2009). The MCMC revolution, *Bull. Amer. Math. Soc.*, 46, 179–205.

Diaconis, P. and Saloff-Coste, L. (1996). Logarithmic Sobolev inequalities for finite Markov chains, *Ann. Appl. Prob.*, 6, 695–750.

Diaconis, P. and Saloff-Coste, L. (1998). What do we know about the Metropolis algorithm, *J. Comput. System Sci.*, 57, 20–36.

Diaconis, P. and Stroock, D. (1991). Geometric bounds for eigenvalues of Markov chains, *Ann. Appl. Prob.*, 1, 36–61.

Diaconis, P. and Sturmfels, B. (1998). Algebraic algorithms for sampling from conditional distributions, *Ann. Statist.*, 26, 363–398.

Diaconis, P., Khare, K., and Saloff-Coste, L. (2008). Gibbs sampling, exponential families, and orthogonal polynomials, with discussion, *Statist. Sci.*, 23, 2, 151–200.

Dimakos, X.K. (2001). A guide to exact simulation, *Internat. Statist. Rev.*, 69, 27–48.

Do, K.-A. and Hall, P. (1989). On importance resampling for the bootstrap, *Biometrika*, 78, 161–167.

Dobrushin, R.L. (1956). Central limit theorems for non-stationary Markov chains II, *Ther. Prob. Appl.*, 1, 329–383.

Fill, J. (1991). Eigenvalue bounds on convergence to stationarity for non-reversible Markov chains, with an application to the exclusion process, *Ann. Appl. Prob.*, 1, 62–87.

Fill, J. (1998). An interruptible algorithm for perfect sampling via Markov chains, *Ann. App. Prob.*, 8, 131–162.

Fishman, G. S. (1995). *Monte Carlo, Concepts, Algorithms, and Applications*, Springer, New York.

Gamerman, D. (1997). *Markov Chain Monte Carlo: Stochastic Simulation for Bayesian Inference*, Chapman and Hall, London.

Garren, S. and Smith, R.L. (1993). Convergence diagnostics for Markov chain samplers, Manuscript.

Gelfand, A. and Smith, A.F.M. (1987). Sampling based approaches to calculating marginal densities, *J. Amer. Stat. Assoc.*, 85, 398–409.

Gelman, A. and Rubin, D. (1992). Inference from iterative simulation using multiple sequences, with discussion, *Statist. Sci.*, 7, 457–511.

Gelman, A., Carlin, B., Stern, H., and Rubin, D. (2003). *Bayesian Data Analysis*, Chapman and Hall/CRC, Boca Raton.

References

Geman, S. and Geman, D. (1984). Stochastic relaxation, Gibbs distributions, and the Bayesian restoration of images, *IEEE Trans. Pattern Anal. Mach. Intele.*, 721–740.
Geyer, C. (1992). Practical Markov chain Monte Carlo, with discussion, *Statist. Sci.*, 7, 473–511.
Gilks, W., Richardson, S., and Spiegelhalter, D. (Eds.), (1995). *Markov Chain Monte Carlo in Practice*, Chapman and Hall, London.
Glauber, R. (1963). Time dependent statistics of the Ising Model, *J. Math. Phys.*, 4, 294–307.
Green, P.J. (1995). Reversible jump Markov Chain Monte Carlo computation and Bayesian model determination, *Biometrika*, 82, 711–732.
Hall, P. and Titterington, D.M. (1989). The effect of simulation order on level accuracy and power of Monte Carlo tests, *JRSS Ser. B*, 51, 459–467.
Hastings, W. (1970). Monte Carlo sampling methods using Markov chains and their applications, *Biometrika*, 57, 92–109.
Higdon, D. (1998). Auxiliary variables methods for Markov chain Monte Carlo applications, *J. Amer. Statist. Assoc.*, 93, 585–595.
Jones, G. and Hobert, J. (2001). Honest exploration of intractable probability distributions via Markov Chain Monte Carlo, *Statist. Sci.*, 16, 312–334.
Kendall, W. and Thönnes, E. (1999). Perfect simulation in stochastic geometry, *Patt. Recogn.*, 32, 1569–1586.
Liu, J. (1995). Eigenanalysis for a Metropolis sampling scheme with comparisons to rejection sampling and importance sampling, Manuscript.
Liu, J. (2008). *Monte Carlo Strategies in Scientific Computing*, Springer, New York.
Mengersen, K. and Tweedie, R. (1996). Rates of convergence of Hastings and Metropolis algorithms, *Ann. Statist.*, 24, 101–121.
Mengersen, K., Knight, S., and Robert, C. (2004). MCMC: How do we know when to stop?, Manuscript.
Metropolis, N., Rosenbluth, A., Rosenbluth, M., Teller, A., and Teller, E. (1953). Equations of state calculations by fast computing machines, *J. Chem. Phys.*, 21, 1087–1092.
Propp, J. and Wilson, B. (1998). How to get a perfectly random sample from a generic Markov chain and generate a random spanning tree to a directed graph, *J. Alg.*, 27, 170–217.
Ripley, B. D. (1987). *Stochastic Simulation*, Wiley, New York.
Robert, C. and Casella, G. (2004). *Monte Carlo Statistical Methods*, Springer, New York.
Roberts, G. and Rosenthal, J.S. (2004). General state space Markov chains and MCMC algorithms, *Prob. Surveys*, 1, 20–71.
Rosenthal, J. (1995). Minorization conditions and convergence rates for Markov chain Monte Carlo, *J. Amer. Statist. Assoc.*, 90, 558–566.
Rosenthal, J. (1996). Analysis of the Gibbs sampler for a model related to the James–Stein estimations, *Statist. Comput.*, 6, 269–275.
Rosenthal, J. (2002). Quantitative convergence rates of Markov chains: A simple account, *Electr. Comm. Prob.*, 7, 123–128.
Ross, S. (2006). *Simulation*, Academic Press, New York.
Rubin, H. (1976). Some fast methods of generating random variables with pre-assigned distributions: General acceptance-rejection procedures, Manuscript.
Schmeiser, B. (1994). Modern simulation environments: Statistical issues, *Proceedings of the First IE Research Conference*, 139–144.
Schmeiser, B. (2001). Some myths and common errors in simulation experiments, B. Peters et al. Eds., *Proceedings of the Winter Simulation Conference*, 39–46.
Smith, A.F.M. and Roberts, G. (1993). Bayesian computation via the Gibbs sampler, with discussion, *JRSS Ser. B*, 55, 3–23.
Tanner, M. and Wong, W. (1987). The calculation of posterior distributions, with discussions, *J. Amer. Statist. Assoc.*, 82, 528–550.
Tierney, L. (1994). Markov chains for exploring posterior distributions, with discussion, *Ann. Statist.*, 22, 1701–1762.
Yu, B. and Mykland, P. (1994). Looking at Markov samplers through CUSUM path plots: A simple diagnostic idea, Manuscript.

Chapter 20
Useful Tools for Statistics and Machine Learning

As much as we would like to have analytical solutions to important problems, it is a fact that many of them are simply too difficult to admit closed-form solutions. Common examples of this phenomenon are finding exact distributions of estimators and statistics, computing the value of an exact optimum procedure, such as a maximum likelihood estimate, and numerous combinatorial algorithms of importance in computer science and applied probability. Unprecedented advances in computing powers and availability have inspired creative new methods and algorithms for solving old problems; often, these new methods are better than what we had in our toolbox before. This chapter provides a glimpse into a few selected computing tools and algorithms that have had a significant impact on the practice of probability and statistics, specifically, the bootstrap, the EM algorithm, and the use of kernels for smoothing and modern statistical classification. The treatment is supposed to be introductory, with references to more advanced parts of the literature.

20.1 The Bootstrap

The bootstrap is a resampling mechanism designed to provide approximations to the sampling distribution of a functional $T(X_1, X_2, \ldots, X_n, F)$, where F is a CDF, typically on some Euclidean space, and $X_1, X_2, \ldots, X_n$ are independent sample observations from F. For example, F could be some continuous CDF on the real line, and $T(X_1, X_2, \ldots, X_n, F)$ could be $\sqrt{n}(\overline{X} - \mu)$, where $\mu = E_F(X_1)$. The problem of approximating the distribution is important, because even when the statistic $T(X_1, X_2, \ldots, X_n, F)$ is a simple one, such as the sample mean, we usually cannot find the distribution of $T(X_1, X_2, \ldots, X_n, F)$ exactly for given n. Sometimes, there may be a suitable asymptotic normality result known about the statistic T, which may be used to form an approximation to the distribution of T. A remarkable fact about the bootstrap is that even if such an asymptotic normality result is available, the bootstrap often provides a better approximation to the true distribution of T than does the normal approximation.

The bootstrap is not limited to the iid situation. It has been studied for various kinds of dependent data and highly complex situations. In fact, this versatility of the bootstrap is the principal reason for its huge popularity and the impact that it has had on practice. We recommend Hall (1992) and Shao and Tu (1995) for detailed theoretical developments of the bootstrap, and Efron and Tibshirani (1993) for an application-oriented readable exposition. Modern reviews include Hall (2003), Bickel (2003), Efron (2003), and Lahiri (2006). Lahiri (2003) is a rigorous treatment of the bootstrap for various kinds of dependent data, including problems that arise in time series and spatial statistics.

Suppose $X_1, X_2, \ldots, X_n \stackrel{iid}{\sim} F$, and $T(X_1, X_2, \ldots, X_n, F)$ is a functional, for example, $T(X_1, X_2, \ldots, X_n, F) = \frac{\sqrt{n}(\bar{X}-\mu)}{\sigma}$, where $\mu = \mu(F) = E_F(X_1)$, and $\sigma^2 = \sigma^2(F) = \text{Var}_F(X_1)$, assumed to be finite. In statistical problems, we frequently need to know something about the sampling distribution of T, for example, $P_F(T(X_1, X_2, \ldots, X_n, F) \leq t)$. If we had replicated samples from the population, resulting in a series of values for the statistic T, then we could form estimates of $P_F(T \leq t)$ by counting how many of the T_is are $\leq t$. But statistical sampling is not done that way. Usually, we do not obtain replicated samples; we obtain just one set of data values of some size n. The intuition of the canonical bootstrap is that by the Glivenko–Cantelli theorem (see Chapter 7), the empirical CDF F_n should be very close to the true underlying CDF F, and so, sampling from F_n, which amounts to simply resampling n values with replacement from the already available data $(X_1, X_2, \ldots, X_n)$, should produce new sets of values that act like samples from F itself. So, although we did not have replicated datasets to start with, it is as if by resampling from the available dataset we now have the desired replications. There is a certain element of faith in this idea, unless we have demonstrable proofs that this simple idea will in fact work, that is, that these resamples lead us to accurate approximations to the true distribution of T. It turns out that such theorems are available, and have led to the credibility and popularity of the bootstrap as a distribution approximation tool. To implement the bootstrap, we only need to be able to generate enough resamples from the original dataset. So, in a sense, the bootstrap replaces a hard mathematical calculation in probability theory by an omnibus and almost automated computing exercise. It is the automatic nature of the bootstrap that makes it so appealing. However, it is also frequently misused in situations where it should not be used, because it is theoretically unjustifiable in those problems, and will in fact give incorrect and inaccurate answers.

Suppose for some number B, we draw B resamples of size n from the original sample. Denoting the resamples from the original sample as

$$(X_{11}^*, X_{12}^*, \ldots, X_{1n}^*), (X_{21}^*, X_{22}^*, \ldots, X_{2n}^*), \ldots, (X_{B1}^*, X_{B2}^*, \ldots, X_{Bn}^*),$$

with corresponding values $T_1^*, T_2^*, \ldots, T_B^*$ for the functional T, one can use simple frequency-based estimates such as $\frac{\#\{j : T_j^* \leq t\}}{B}$ to estimate $P_F(T \leq t)$. This is the basic idea of the bootstrap.

The formal definition of the *bootstrap distribution of a functional* is as follows.

Definition 20.1. Suppose $X_1, \ldots, X_n$ are iid observations from a CDF F on some space $\mathcal{X}$, and $T(X_1, X_2, \ldots, X_n, F)$ is a given real-valued functional. The *ordinary bootstrap distribution* of T is defined as

20.1 The Bootstrap

$$H_{\text{boot}}(x) = P_{F_n}(T(X_1^*, \ldots, X_n^*, F_n) \leq x),$$

where $(X_1^*, \ldots, X_n^*)$ refers to an iid sample of size n from the empirical CDF F_n.

It is common to use the notation P_* to denote probabilities under the bootstrap distribution. $P_{F_n}(\cdot)$ corresponds to probability statements corresponding to all the n^n possible with replacement resamples from the original sample $X_1, \ldots, X_n$. Recalculating T from all n^n resamples is basically impossible unless n is very small, therefore one uses a smaller number of B resamples and recalculates T only B times. Thus $H_{\text{boot}}(x)$ is itself estimated by a Monte Carlo, known as *bootstrap Monte Carlo*. So the final estimate for $P_F(T(X_1, X_2, \ldots, X_n, F) \leq x)$ absorbs errors from two sources: (i) pretending that $(X_{i1}^*, X_{i2}^*, \ldots, X_{in}^*)$ are bona fide samples from F; (ii) estimating the true $H_{\text{boot}}(x)$ by a Monte Carlo. By choosing B adequately large, the issue of the Monte Carlo error is generally ignored. The choice of which B would let one ignore the Monte Carlo error is a hard mathematical problem; Hall (1986, 1989) are two key references. It is customary to choose $B \approx 500-1000$ for variance estimation and a somewhat larger value for estimating quantiles. It is hard to give any general reliable prescriptions for B.

At first glance, the bootstrap idea of resampling from the original sample appears to be a bit too simple to actually work. One has to have a definition for what one means by the bootstrap working in a given situation. For estimating the CDF of a statistic, one should want $H_{\text{boot}}(x)$ to be numerically close to the true CDF, say $H_n(x)$, of T. This would require consideration of metrics on CDFs, a topic we covered in Chapter 15. For a general metric ρ, the definition of the bootstrap working in a given problem is the following.

Definition 20.2. Let $\rho(F, G)$ be a metric on the space of CDFs on $\mathcal{X}$. For a given functional $T(X_1, X_2, \ldots, X_n, F)$, let

$$H_n(x) = P_F(T(X_1, X_2, \ldots, X_n, F) \leq x),$$
$$H_{\text{boot}}(x) = P_*(T(X_1^*, X_2^*, \ldots, X_n^*, F_n) \leq x).$$

We say that the bootstrap is *weakly consistent* under ρ for T if $\rho(H_n, H_{\text{boot}}) \xrightarrow{P} 0$ as $n \to \infty$. We say that the bootstrap is *strongly consistent* under ρ for T if $\rho(H_n, H_{\text{boot}}) \xrightarrow{a.s.} 0$.

Note that the need for mentioning convergence to zero in probability or a.s. in this definition is due to the fact that the bootstrap distribution H_{boot} is a random CDF. It is a random CDF because as a function it depends on the original sample $(X_1, X_2, \ldots, X_n)$. Thus, the bootstrap uses a random CDF to approximate a deterministic but unknown CDF, namely the true CDF H_n of the functional T. In principle, a sequence of random CDFs could very well converge to another random CDF, or not converge at all! It is remarkable that under certain minimal conditions, those disasters do not happen, and H_{boot} and H_n get close as $n \to \infty$.

Example 20.1 (Applying the Bootstrap). How does one apply the bootstrap in practice? Suppose for example, $T(X_1, X_2, \ldots, X_n, F) = \frac{\sqrt{n}(\bar{X}-\mu)}{\sigma}$. In the canonical bootstrap scheme, we take iid samples from F_n. By a simple calculation, the mean and the variance of the empirical distribution F_n are $\bar{X}$ and $s^2 = \frac{1}{n}\sum_{i=1}^{n}(X_i - \bar{X})^2$ (note the n rather than $n-1$ in the denominator). The bootstrap is a device for estimating

$$P_F\left(\frac{\sqrt{n}(\bar{X} - \mu(F))}{\sigma}\right) \leq x) \text{ by } P_{F_n}\left(\frac{\sqrt{n}(\bar{X}_n^* - \bar{X})}{s} \leq x\right).$$

We further approximate $P_{F_n}(\frac{\sqrt{n}(\bar{X}_n^* - \bar{X})}{s}) \leq x)$ by resampling only B times from the original sample set $\{X_1, X_2, \ldots, X_n\}$. In other words, we finally report as our estimate for $P_F(\frac{\sqrt{n}(\bar{X} - \mu)}{\sigma} \leq x)$ the number

$$\frac{\#\left\{j : \frac{\sqrt{n}(\bar{X}_{n,j}^* - \bar{X})}{s} \leq x\right\}}{B}.$$

This number depends on the original sample set $\{X_1, X_2, \ldots, X_n\}$, the particular resampled sets $(X_{i1}^*, X_{i2}^*, \cdots, X_{in}^*)$, and the bootstrap Monte Carlo sample size B. If the bootstrap Monte Carlo is repeated, then for the same B, and of course, the same original sample set $\{X_1, X_2, \ldots, X_n\}$, the bootstrap estimate will be a different number. We would like the bootstrap estimate to be close to the true value of $P_F(\frac{\sqrt{n}(\bar{X} - \mu)}{\sigma} \leq x)$; consistency of the bootstrap is about our ability to guarantee that for large n, and an implicit unspoken assumption of a large B.

20.1.1 Consistency of the Bootstrap

We start with the case of the sample mean of iid random riables. If $X_1, X_2, \ldots, X_n \stackrel{iid}{\sim} F$ and if $\text{Var}_F(X_i) < \infty$, then $\sqrt{n}(\bar{X} - \mu)$ has a limiting normal distribution, by the CLT. So a probability such as $P_F(\sqrt{n}(\bar{X} - \mu) \leq x)$ could be approximated by, for example, $\Phi(\frac{x}{s})$, where s is the sample standard deviation. An interesting property of the bootstrap approximation is that even when the CLT approximation $\Phi(\frac{x}{s})$ is available, the bootstrap approximation may be more accurate. Such results generally go by the name of *higher-order accuracy* of the bootstrap.

But first we present two consistency results corresponding to two specific metrics that have earned a special status in this literature. The two metrics are

(i) Kolmogorov metric

$$K(F, G) = \sup_{-\infty < x < \infty} |F(x) - G(x)|;$$

20.1 The Bootstrap

(ii) Mallows–Wasserstein metric

$$\ell_2(F, G) = \inf_{\Gamma_{2,F,G}} (E|Y - X|^2)^{\frac{1}{2}},$$

where $X \sim F$, $Y \sim G$ and $\Gamma_{2,F,G}$ is the class of all joint distributions of (X, Y) with marginals F and G, each with a finite second moment. See Chapter 15 for detailed treatment of these two metrics. We recall from Chapter 15 that the Kolmogorov metric is universally regarded as a natural one. The metric ℓ_2 is a natural metric for many statistical problems because of its interesting property that $\ell_2(F_n, F) \to 0$ iff $F_n \overset{\mathcal{L}}{\Rightarrow} F$ and $E_{F_n}(X^i) \to E_F(X^i)$ for $i = 1, 2$. One might want to use the bootstrap primarily for estimating the CDF, and the mean and the variance of a statistic, thus consistency in ℓ_2 is just the right result for that purpose.

Theorem 20.1. *Suppose* $X_1, X_2, \ldots, X_n \overset{iid}{\sim} F$ *and suppose* $E_F(X_1^2) < \infty$. *Let* $T(X_1, X_2, \ldots, X_n, F) = \sqrt{n}(\bar{X} - \mu)$. *Then* $K(H_n, H_{\text{boot}})$ *and* $\ell_2(H_n, H_{\text{boot}}) \overset{a.s}{\to} 0$ *as* $n \to \infty$.

Strong consistency in K is proved in Singh (1981) and that for ℓ_2 is proved in Bickel and Freedman (1981). Notice that $E_F(X_1^2) < \infty$ guarantees that $\sqrt{n}(\bar{X}-\mu)$ admits a CLT. And the theorem above says that the bootstrap is strongly consistent (wrt K and ℓ_2) under that very assumption. This is in fact a very good rule of thumb: if a functional $T(X_1, X_2, \ldots, X_n, F)$ admits a CLT, then the bootstrap would be at least weakly consistent for T. Strong consistency might require more assumptions.

Proof. We sketch a proof of the strong consistency in K. The proof requires use of the Berry–Esseen inequality, Polya's theorem (see Chapter 7 and Chapter 8), and a strong law known as the Zygmund–Marcinkiewicz strong law, which we state below without a proof.

Proposition (Zygmund–Marcinkiewicz SLLN). *Let* $Y_1, Y_2, \ldots$ *be iid random variables with CDF* F *and suppose, for some* $0 < \delta < 1$, $E_F|Y_1|^\delta < \infty$. *Then* $n^{-1/\delta} \sum_{i=1}^n Y_i \overset{a.s.}{\Rightarrow} 0$.

We are now ready to sketch the proof of strong consistency of H_{boot} under the Kolmogorov metric K.

Proof of Theorem 20.1. Using the definition of K, we can write

$$K(H_n, H_{\text{boot}}) = \sup_x |P_F\{T_n \leq x\} - P_*\{T_n^* \leq x\}|$$

$$= \sup_x \left| P_F\left\{\frac{T_n}{\sigma} \leq \frac{x}{\sigma}\right\} - P_*\left\{\frac{T_n^*}{s} \leq \frac{x}{s}\right\} \right|$$

$$= \sup_x \left| P_F\left\{\frac{T_n}{\sigma} \leq \frac{x}{\sigma}\right\} - \Phi\left(\frac{x}{\sigma}\right) + \Phi\left(\frac{x}{\sigma}\right) - \Phi\left(\frac{x}{s}\right) \right.$$

$$\left. + \Phi\left(\frac{x}{s}\right) - P_*\left\{\frac{T_n^*}{s} \leq \frac{x}{s}\right\} \right|$$

$$\le \sup_x \left| P_F \left\{ \frac{T_n}{\sigma} \le \frac{x}{\sigma} \right\} - \Phi\left(\frac{x}{\sigma}\right) \right| + \sup_x \left| \Phi\left(\frac{x}{\sigma}\right) - \Phi\left(\frac{x}{s}\right) \right|$$
$$+ \sup_x \left| \Phi\left(\frac{x}{s}\right) - P_* \left\{ \frac{T_n^*}{s} \le \frac{x}{s} \right\} \right|$$
$$= A_n + B_n + C_n, \quad \text{say}.$$

That $A_n \to 0$ is a direct consequence of Polya's theorem (see Chapter 7). Also, s^2 converges almost surely to σ^2 and so, by the continuous mapping theorem, s converges almost surely to σ. Then $B_n \Rightarrow 0$ almost surely by the fact that $\Phi(\cdot)$ is a uniformly continuous function. Finally, we can apply the Berry–Esseen theorem (see Chapter 8) to show that C_n goes to zero:

$$C_n \le \frac{4}{5\sqrt{n}} \cdot \frac{E_{F_n}|X_1^* - \bar{X}|^3}{[\mathrm{Var}_{F_n}(X_1^*)]^{3/2}}$$
$$= \frac{4}{5\sqrt{n}} \cdot \frac{\sum_{i=1}^n |X_i - \bar{X}|^3}{n s^3}$$
$$\le \frac{4}{5 n^{3/2} s^3} \cdot 2^3 \left[\sum_{i=1}^n |X_i - \mu|^3 + n|\mu - \bar{X}|^3 \right]$$
$$= \frac{M}{s^3} \left[\frac{1}{n^{3/2}} \sum_{i=1}^n |X_i - \mu|^3 + \frac{|\bar{X} - \mu|^3}{\sqrt{n}} \right],$$

where $M = \frac{32}{5}$.

Because $s \Rightarrow \sigma > 0$ and $\bar{X} \Rightarrow \mu$, it is clear that $|\bar{X} - \mu|^3/(\sqrt{n} s^3) \Rightarrow 0$ almost surely. As regards the first term, let $Y_i = |X_i - \mu|^3$ and $\delta = 2/3$. Then the $\{Y_i\}$ are iid and

$$E|Y_i|^\delta = E_F|X_i - \mu|^{3 \cdot 2/3} = \mathrm{Var}_F(X_1) < \infty.$$

It now follows from the Zygmund–Marcinkiewicz SLLN that

$$\frac{1}{n^{3/2}} \sum_{i=1}^n |X_i - \mu|^3 = n^{-1/\delta} \sum_{i=1}^n Y_i \Rightarrow 0, \text{ a.s.,} \quad \text{as } n \to \infty.$$

Thus, $A_n + B_n + C_n \to 0$ almost surely, and hence $K(H_n, H_{\text{boot}}) \to 0$. □

It is natural to ask if the bootstrap is consistent for $\sqrt{n}(\bar{X} - \mu)$ even when $E_F(X_1^2) = \infty$. If we insist on strong consistency, then the answer is negative. The point is that the sequence of bootstrap distributions is a sequence of random CDFs and so it can converge to a random CDF, depending on the particular realization $X_1, X_2, \ldots$, if $E_F(X_1^2) = \infty$. See Athreya (1987), Giné and Zinn (1989), and Hall (1990) for proofs and additional detail.

20.1 The Bootstrap

Example 20.2 (Practical Accuracy of Bootstrap). How does the bootstrap compare with the CLT approximation in actual applications? The question can only be answered by case-by-case simulation. The results are mixed in the following numerical table. The X_i are iid Exp(1) in this example and $T = \sqrt{n}(\bar{X} - 1)$, with $n = 20$. For the bootstrap approximation, $B = 250$ was used.

t	$H_n(t)$	CLT Approximation	$H_{\text{boot}}(t)$
-2	0.0098	0.0228	0.0080
-1	0.1563	0.1587	0.1160
0	0.5297	0.5000	0.4840
1	0.8431	0.8413	0.8760
2	0.9667	0.9772	0.9700

The bootstrap approximation is more accurate than the normal approximation in the tails in this specific example. This greater accuracy is related to the higher-order accuracy of the bootstrap; see Section 20.1.3.

The ordinary bootstrap, which resamples with replacement from the empirical CDF F_n, is consistent for many other natural statistics in addition to the sample mean and even higher-order accurate for some, but under additional conditions. Examples of such statistics are sample percentiles, and smooth functions of a sample mean vector. The consistency of the bootstrap for the sample mean under finite second moments is also true for the multivariate case. The basic theorems are stated below; see DasGupta (2008) for further details.

Theorem 20.2. *Let $\vec{X}_1, \ldots, \vec{X}_n, \ldots$ be iid d-dimensional vectors with common CDF F, with $\text{Cov}_F(\vec{X}_1) = \Sigma$, Σ finite. Let $T(\vec{X}_1, \vec{X}_2, \ldots, \vec{X}_n, F) = \sqrt{n}(\vec{\bar{X}} - \vec{\mu})$. Then $K(H_{\text{boot}}, H_n) \xrightarrow{a.s.} 0$ as $n \to \infty$.*

We know from the ordinary delta theorem (see Chapter 7) that if a sequence of statistics T_n admits a CLT and if $g(\cdot)$ is a smooth transformation, then $g(T_n)$ also admits a CLT. If we were to believe in our rule of thumb, then this would suggest that the bootstrap should be consistent also for $g(T_n)$ if it is already consistent for T_n. For the case when T_n is a sample mean vector, the following result holds. This theorem has numerous applications to practically important statistics whose exact distributions are very difficult to find, and so the bootstrap is a very natural and effective tool for approximating their distributions. The mathematical assurance of consistency supplied by the theorem below gives us some confidence that the bootstrap approximation will be reasonable.

Theorem 20.3. *Let $\vec{X}_1, \vec{X}_2, \ldots, \vec{X}_n \overset{iid}{\sim} F$, and let $\Sigma_{d \times d} = \text{Cov}_F(\vec{X}_1)$ be finite. Let $T(\vec{X}_1, \vec{X}_2, \ldots, \vec{X}_n, F) = \sqrt{n}(\vec{\bar{X}} - \vec{\mu})$ and for some $m \geq 1$, let $g : \mathbb{R}^d \to \mathbb{R}^m$. If $\nabla g(\cdot)$ exists in a neighborhood of $\vec{\mu}$, $\nabla g(\vec{\mu}) \neq \vec{0}$, and if $\nabla g(\cdot)$ is continuous at $\vec{\mu}$, then the bootstrap is strongly consistent with respect to the Kolmogorov metric K for $\sqrt{n}[g(\vec{\bar{X}}) - g(\vec{\mu})]$. See Shao and Tu (1995, p. 80) for a sketch of a proof of this theorem.*

20.1.2 Further Examples

The bootstrap is used in practice for a variety of purposes. It is used to estimate a CDF, or a percentile, or the bias or variance of a statistic T_n. For example, if T_n is an estimate for some parameter θ, and if $E_F(T_n - \theta)$ is the bias of T_n, the bootstrap estimate $E_{F_n}(T_n^* - T_n)$ can be used to estimate the bias. Likewise, variance estimates can be formed by estimating $\text{Var}_F(T_n)$ by $\text{Var}_{F_n}(T_n^*)$. In other words, to estimate $\text{Var}_F(T_n)$, we sample B sets of samples of size n with replacement from the original sample set, say

$$(X_{11}^*, \ldots, X_{1n}^*), \ldots, (X_{B1}^*, \ldots, X_{Bn}^*).$$

We compute $T_i^* = T(X_{i1}^*, \ldots, X_{in}^*)$, $1 = 1, 2, \ldots, B$, and their mean $\bar{T}^*$, and estimate $\text{Var}_F(T_n)$ by $\frac{1}{B}\sum_{i=1}^{B}(T_i^* - \bar{T}^*)^2$. This is the basic bootstrap variance estimate. One wants to know how accurate the bootstrap-based estimates are in reality.

This can only be answered on the basis of case-by-case investigation. Some overall qualitative phenomena have emerged from these investigations. For instance,

(a) The bootstrap distribution estimate captures information about skewness that the CLT will miss.
(b) But the bootstrap tends to underestimate the variance of a statistic T.

Here are a few more illustrative examples.

Example 20.3 (Bootstrapping the Sample Variance). Let $X_1, X_2, \ldots$ be iid one-dimensional random variables with the CDF F, and suppose $E_F(X_1^4) < \infty$. Let $\vec{Y}_i = \begin{pmatrix} X_i \\ X_i^2 \end{pmatrix}$. Then with $d = 2$, $\vec{Y}_1, \vec{Y}_2, \ldots, \vec{Y}_n$ are iid d-dimensional vectors with $\text{Cov}(\vec{Y}_1)$ finite. Note that

$$\bar{\vec{Y}} = \begin{pmatrix} \bar{X} \\ \frac{1}{n}\sum_{i=1}^{n} X_i^2 \end{pmatrix}.$$

Consider the transformation $g : \mathbb{R}^2 \to \mathbb{R}^1$ defined as $g(u, v) = v - u^2$. Then

$$\frac{1}{n}\sum_{i=1}^{n}(X_i - \bar{X}])^2 = \frac{1}{n}\sum_{i=1}^{n} X_i^2 - (\bar{X})^2 = g(\bar{\vec{Y}}).$$

If we let $\vec{\mu} = E(\vec{Y}_1)$, then $g(\vec{\mu}) = \sigma^2 = \text{Var}(X_1)$. Because $g(\cdot)$ satisfies the conditions of Theorem 20.3, it follows that the bootstrap is strongly consistent with respect to the Kolmogorov metric K for

$$\sqrt{n}\left(\frac{1}{n}\sum_{i=1}^{n}(X_i - \bar{X})^2 - \sigma^2\right).$$

Example 20.4 (Bootstrapping the Sample Skewness Coefficient). Suppose X_1, $X_2, \ldots$ are iid observations from a CDF F on the real line with $E_F(X_1^6) < \infty$. Then, it follows from the delta theorem (see Chapter 7 and Exercise 7.36) that the standardized sample skewness $\sqrt{n}[b_1 - \beta]$ is asymptotically normally distributed with zero mean and a variance that will depend on F. Here,

$$\beta = \frac{E_F(X_1 - \mu)^3}{\sigma^3} \quad \text{and} \quad b = \frac{\frac{1}{n}\sum_{i=1}^n (X_i - \overline{X})^3}{s^3}.$$

In fixed-size samples when n is not too large, the true distribution of $\sqrt{n}[b_1 - \beta]$ will of course not be normal, and will depart from normal in various ways, depending on F. We sampled $n = 30$ samples from a standard normal distribution, and then bootstrapped the functional $\sqrt{n}[b_1 - \beta]$ using a bootstrap Monte Carlo size of $B = 600$. A histogram of the bootstrapped values in Fig. 20.1 shows the long right tail, which the normal approximation would not have shown. The histogram also shows a heavier central part in the distribution of $\sqrt{n}[b_1 - \beta]$ than a normal approximation would have shown. The true distribution of $\sqrt{n}[b_1 - \beta]$ would be impossible to find, and so, this is fertile ground for use of the bootstrap.

Example 20.5 (Bootstrapping the Correlation Coefficient). Suppose (X_i, Y_i), $i = 1, 2, \ldots, n$ are iid $BVN(0, 0, 1, 1, \rho)$ and let r be the sample correlation coefficient. Let $T_n = \sqrt{n}(r - \rho)$. We know that $T_n \overset{\mathcal{L}}{\Rightarrow} N(0, (1-\rho^2)^2)$; see Chapter 7. Convergence to normality is very slow. There is also an exact formula for the density of r. For $n \geq 4$, the exact density is,

$$f(r|\rho) = \frac{2^{n-3}(1-\rho^2)^{(n-1)/2}}{\pi(n-3)!}(1-r^2)^{(n-4)/2}\sum_{k=0}^{\infty}\Gamma\left(\frac{n+k-1}{2}\right)^2\frac{(2\rho r)^k}{k!};$$

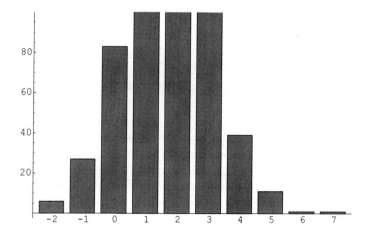

Fig. 20.1 Histogram of bootstrapped values

see Tong (1990). In the table below, we give simulation averages of the estimated standard deviation of r by using the bootstrap. We used $n = 20$, and $B = 200$. The bootstrap estimate was calculated for 1000 independent simulations; the table reports the average of the standard deviation estimates over the 1000 simulations.

n	True ρ	True s.d. of r	CLT Estimate	Bootstrap Estimate
20	0.0	0.230	0.232	0.217
	0.5	0.182	0.175	0.160
	0.9	0.053	0.046	0.046

Except when ρ is large the bootstrap underestimates the variance and the CLT estimate is better.

Example 20.6 (The t-Statistic for Poisson Data). Suppose $X_1, \ldots, X_n$ are iid Poi(μ) and let T_n be the t-statistic $T_n = \sqrt{n}(\bar{X} - \mu)/s$. In this example $n = 20$ and $B = 200$ and for the actual data, μ was chosen to be 1. Apart from the bias and the variance of T_n, in this example we also report percentile estimates for T_n. The bootstrap percentile estimates are found by calculating T_n^* for the B resamples and calculating the corresponding percentile value of the B values of T_n^*. The bias and the variance are estimated to be -0.18 and 1.614, respectively. The estimated percentiles are reported in the table. Note that the 5th and the 95th percentiles are not equal in absolute value in this table; neither are the 10th and the 90th. Such potential strong skewness would have remained undetected, if we had simply used a normal approximation.

α	Estimated 100α Percentile
0.05	-2.45
0.10	-1.73
0.25	-0.76
0.50	-0.17
0.75	0.49
0.90	1.25
0.95	1.58

Example 20.7 (Bootstrap Failure). In spite of the many consistency theorems in the previous section, there are instances where the ordinary bootstrap with replacement sampling from F_n actually does not work. Typically, these are instances where the functional T_n fails to admit a CLT. Here is a simple example where the ordinary bootstrap fails to consistently estimate the true distribution of a statistic.

Let $X_1, X_2, \ldots$ be iid $U(0, \theta)$ and let $T_n = n(\theta - X_{(n)})$, $T_n^* = n(X_{(n)} - X_{(n)}^*)$. The ordinary bootstrap will fail in this example in the sense that the conditional distribution of T_n^* given $X_{(n)}$ does not converge to the $Exp(\theta)$ distribution a.s. Let us take, for notational simplicity, $\theta = 1$. Then for $t \geq 0$,

20.1 The Bootstrap

$$\begin{aligned}
P_{F_n}(T_n^* \leq t) &\geq P_{F_n}\left(T_n^* = 0\right) \\
&= P_{F_n}\left(X_{(n)}^* = X_{(n)}\right) \\
&= 1 - P_{F_n}\left(X_{(n)}^* < X_{(n)}\right) \\
&= 1 - \left(\frac{n-1}{n}\right)^n \\
&\xrightarrow{n\to\infty} 1 - e^{-1}
\end{aligned}$$

For example, take $t = 0.0001$; then $\lim_n P_{F_n}(T_n^* \leq t) \geq 1 - e^{-1}$, whereas $\lim_n P_F(T_n \leq t) = 1 - e^{-0.0001} \approx 0$. So $P_{F_n}(T_n^* \leq t) \not\to P_F(T_n \leq t)$. The phenomenon of this example can be generalized essentially to any CDF F with a compact support $[\underline{\omega}(F), \overline{\omega}(F)]$ with some conditions on F, such as existence of a smooth and positive density. There are variants of the ordinary bootstrap that correct the bootstrap failure in this example; see DasGupta (2008).

20.1.3 * Higher-Order Accuracy of the Bootstrap

One question about the use of the bootstrap is whether the bootstrap has any advantages at all when a CLT is already available. To be specific, suppose $T(X_1, \ldots, X_n, F) = \sqrt{n}(\bar{X} - \mu)$. If $\sigma^2 = \text{Var}_F(X) < \infty$, then $\sqrt{n}(\bar{X} - \mu) \xrightarrow{\mathcal{L}} N(0, \sigma^2)$ and $K(H_{\text{boot}}, H_n) \xrightarrow{a.s.} 0$. So two competitive approximations to $P_F(T(X_1, \ldots, X_n, F) \leq x)$ are $\Phi(\frac{x}{\sigma})$ and $P_{F_n}(\sqrt{n}(\bar{X}^* - \bar{X}) \leq x)$. It turns out that for certain types of statistics, the bootstrap approximation is (theoretically) more accurate than the approximation provided by the CLT. The CLT, because any normal distribution is symmetric, cannot capture information about the skewness in the finite sample distribution of T. The bootstrap approximation does so. So the bootstrap succeeds in correcting for skewness, just as an Edgeworth expansion would do (see Chapter 1). This is called Edgeworth correction by the bootstrap and the property is called *second-order accuracy* of the bootstrap. It is important to remember that second-order accuracy is not automatic; it holds for certain types of T but not for others. It is also important to understand that practical accuracy and theoretical higher-order accuracy can be different things. The following heuristic calculation illustrates when second-order accuracy can be anticipated. The first result on higher-order accuracy of the bootstrap is due to Singh (1981). In addition to the references we provided in the beginning, Lehmann (1999) gives a very readable treatment of higher-order accuracy of the bootstrap.

Suppose $X_1, \ldots, X_n \stackrel{iid}{\sim} F$ and $T(X_1, \ldots, X_n, F) = \frac{\sqrt{n}(\bar{X}-\mu)}{\sigma}$; here $\sigma^2 = \text{Var}_F(X_1) < \infty$. We know that T admits the Edgeworth expansion (see Chapter 1):

$$P_F(T \leq x) = \Phi(x) + \frac{p_1(x|F)}{\sqrt{n}}\phi(x) + \frac{p_2(x|F)}{n}\phi(x)$$
$$+ \text{ smaller order terms,}$$

$$P_*(T^* \leq x) = \Phi(x) + \frac{p_1(x|F_n)}{\sqrt{n}} \phi(x) + \frac{p_2(x|F_n)}{n} \phi(x)$$
$$+ \text{ smaller order terms,}$$
$$H_n(x) - H_{\text{boot}}(x) = \frac{p_1(x|F) - p_1(x|F_n)}{\sqrt{n}} + \frac{p_2(x|F) - p_2(x|F_n)}{n}$$
$$+ \text{ smaller order terms.}$$

Recall now that the polynomials p_1, p_2 are given as

$$p_1(x|F) = \frac{\gamma}{6}(1-x^2),$$
$$p_2(x|F) = x\left[\frac{\kappa-3}{24}(3-x^2) - \frac{\kappa^2}{72}(x^4 - 10x^2 + 15)\right],$$

where $\gamma = \frac{E_F(X_1-\mu)^3}{\sigma^3}$ and $\kappa = \frac{E_F(X_1-\mu)^4}{\sigma^4}$. Because $\gamma_{F_n} - \gamma = O_p(\frac{1}{\sqrt{n}})$ and $\kappa_{F_n} - \kappa = O_p(\frac{1}{\sqrt{n}})$, just from the CLT for γ_{F_n} and κ_{F_n} (under finiteness of four moments), one obtains $H_n(x) - H_{\text{boot}}(x) = O_p(\frac{1}{n})$. If we contrast this to the CLT approximation, in general, the error in the CLT is $O(\frac{1}{\sqrt{n}})$, as is known from the Berry–Esseen theorem (Chapter 8). The $\frac{1}{\sqrt{n}}$ rate of the error of the CLT cannot be improved in general even if there are four moments. Thus, by looking at the standardized statistic $\frac{\sqrt{n}(\bar{X}-\mu)}{\sigma}$, we have succeeded in making the bootstrap one order more accurate than the CLT. This is called second-order accuracy of the bootstrap. If one does not standardize, then

$$P_F\left(\sqrt{n}(\bar{X}-\mu) \leq x\right) = P_F\left(\frac{\sqrt{n}(\bar{X}-\mu)}{\sigma} \leq \frac{x}{\sigma}\right) \to \Phi\left(\frac{x}{\sigma}\right)$$

and the leading term in the bootstrap approximation in this unstandardized case would be $\Phi(\frac{x}{s})$. So in the unstandardized case, the bootstrap approximates the true CDF $H_n(x)$ also at the rate $\frac{1}{\sqrt{n}}$, that is, if one does not standardize, then $H_n(x) - H_{\text{boot}}(x) = O_p(\frac{1}{\sqrt{n}})$. We have now lost the second-order accuracy. The following second rule of thumb often applies.

Rule of Thumb. Let $X_1, \ldots, X_n \stackrel{\text{iid}}{\sim}$ and $T(X_1, \ldots, X_n, F)$ a functional. If $T(X_1, \ldots, X_n, F) \stackrel{\mathcal{L}}{\Rightarrow} N(0, \tau^2)$ where τ is independent of F, then second-order accuracy is likely. Proving it depends on the availability of an Edgeworth expansion for T. If τ depends on F (i.e., $\tau = \tau(F)$), then the bootstrap should be just first-order accurate.

Thus, as we now show, canonical bootstrap is second-order accurate for the standardized mean $\frac{\sqrt{n}(\bar{X}-\mu)}{\sigma}$. From an inferential point of view, it is not particularly useful to have an accurate approximation to the distribution of $\frac{\sqrt{n}(\bar{X}-\mu)}{\sigma}$, because

20.1 The Bootstrap

σ would usually be unknown, and the accurate approximation could not really be used to construct a confidence interval for μ. Still, the second-order accuracy result is theoretically insightful.

We state a specific result below for the case of standardized and nonstandardized sample means. Let

$$H_n(x) = P_F\left(\sqrt{n}\left(\bar{X} - \mu\right) \leq x\right), \quad H_{n,0}(x) = P_F\left(\frac{\sqrt{n}\left(\bar{X} - \mu\right)}{\sigma} \leq x\right),$$

$$H_{\text{boot}}(x) = P_*\left(\sqrt{n}\left(\bar{X}^* - \bar{X}\right) \leq x\right), \quad H_{\text{boot},0}(x) = P_{F_n}\left(\frac{\sqrt{n}\left(\bar{X}^* - \bar{X}\right)}{s} \leq x\right).$$

Theorem 20.4. *Let* $X_1, \ldots, X_n \overset{\text{iid}}{\sim} F$.

(a) *If* $E_F|X_1|^3 < \infty$, *and* F *is nonlattice, then* $K(H_{n,0}, H_{\text{boot},0}) = o_p(\frac{1}{\sqrt{n}})$.

(b) *If* $E_F|X_1|^3 < \infty$, *and* F *is lattice, then* $\sqrt{n} K(H_{n,0}, H_{\text{boot},0}) \overset{P}{\to} c, 0 < c < \infty$.

See Shao and Tu (1995 p. 92–94) for a proof. The constant c in the lattice case equals $\frac{h}{\sigma\sqrt{2\pi}}$, *where h is the span of the lattice* $\{a + kh, k = 0, \pm 1, \pm 2, \ldots\}$ *on which the* X_i *are supported. Note also that part (a) says that higher-order accuracy for the standardized case obtains with three moments; Hall (1988) showed that finiteness of three absolute moments is in fact necessary and sufficient for higher-order accuracy of the bootstrap in the standardized case.*

20.1.4 Bootstrap for Dependent Data

The ordinary bootstrap that resamples observations with replacement from the original dataset does not work when the sample observations are dependent. This was already pointed out in Singh (1981). It took some time before consistent bootstrap schemes were offered for dependent data. There are consistent schemes that are meant for specific dependence structures (e.g., stationary autoregression of a known order) and there are also general bootstrap schemes that work for large classes of stationary time series without requiring any particular dependence structure. The model-based schemes are better for the specific models, but can completely fail if some assumption about the specific model does not hold. *Block bootstrap methods are regarded as the bread and butter of resampling for dependent sequences.* These are general and mostly all-purpose resampling schemes that provide at least consistency for a wide selection of dependent data models.

The basic idea of the block bootstrap method is that if the underlying series is a stationary process with short-range dependence, then blocks of observations of suitable lengths should be approximately independent. Also, the joint distribution of the variables in different blocks would be about the same, due to stationarity.

So, if we resample blocks of observations, rather than observations one at a time, then that should bring us back to the nearly iid situation, a situation in which the bootstrap is known to succeed. Block bootstrap was first suggested in Carlstein (1986) and Künsch (1989). Various block bootstrap schemes are now available. We only present three such schemes, for which the block length is nonrandom. A small problem with some of the blocking schemes is that the "starred" time series is not stationary, although the original series is, by hypothesis, stationary. A version of the block bootstrap that resamples blocks of random length allows the "starred" series to be provably stationary. This is called the *stationary bootstrap*, proposed in Politis and Romano (1994) and Politis et al. (1999). However, later theoretical studies have established that the auxilliary randomization to determine the block lengths can make the stationary bootstrap less accurate. For this reason, we only discuss three blocking methods with nonrandom block lengths.

(a) **(Nonoverlapping Block Bootstrap (NBB))**. In this scheme, one splits the observed series $\{y_1, \ldots, y_n\}$ into nonoverlapping blocks

$$B_1 = \{y_1, \ldots, y_h\}, \; B_2 = \{y_{h+1}, \ldots, y_{2h}\}, \ldots, \; B_m = \{y_{(m-1)h+1}, \ldots, y_{mh}\},$$

where it is assumed that $n = mh$. The common block length is h. One then resamples $B_1^*, B_2^*, \ldots, B_m^*$ at random, with replacement, from $\{B_1, \ldots, B_m\}$. Finally, the B_i^*s are pasted together to obtain the "starred" series $y_1^*, \ldots, y_n^*$.

(b) **(Moving Block Bootstrap (MBB))**. In this scheme, the blocks are

$$B_1 = \{y_1, \ldots, y_h\}, \quad B_2 = \{y_2, \ldots, y_{h+1}\}, \ldots, \quad B_N = \{y_{n-h+1}, \ldots, y_n\},$$

where $N = n - h + 1$. One then resamples $B_1^*, \ldots, B_m^*$ from $B_1, \ldots, B_N$, where still $n = mh$.

(c) **(Circular Block Bootstrap (CBB))**. In this scheme, one periodically extends the observed series as $y_1, y_2, \ldots, y_n, y_1, y_2, \ldots, y_n, \ldots$. Suppose we let z_i be the members of this new series, $i = 1, 2, \ldots$. The blocks are defined as

$$B_1 = \{z_1, \ldots, z_h\}, \quad B_2 = \{z_{h+1}, \ldots, z_{2h}\}, \ldots, \quad B_n = \{z_n, \ldots, z_{n+h-1}\}.$$

One then resamples $B_1^*, \ldots, B_m^*$ from $B_1, \ldots, B_n$.

We now give some theoretical properties of the three block bootstrap methods described above. The results below are due to Lahiri (1999). We need a definition for the result below.

Definition 20.3. Let $Y_n, n = 0, \pm 1, \pm 2, \ldots$ be a stationary time series with covariance function $\gamma(k) = \mathrm{Cov}(Y_t, Y_{t+k}), k = 0, \pm 1, \pm 2, \ldots$. The *spectral density* of the series is the function

$$f(\omega) = \frac{1}{2\pi} \sum_{k=-\infty}^{\infty} \gamma(k) e^{-ik\omega}, \quad -\pi < \omega \leq \pi,$$

where $i = \sqrt{-1}$.

20.1 The Bootstrap

Just as the covariance function characterizes second-order properties of the stationary time series in the time domain, the spectral density does the same working with the frequency domain. Both approaches are useful, and they complement each other.

Suppose $\{y_i : -\infty < i < \infty\}$ is a d-dimensional stationary process with a finite mean μ and spectral density f. Let $h : \mathcal{R}^d \to \mathcal{R}$ be a sufficiently smooth function. Let $\theta = h(\mu)$ and $\hat{\theta}_n = h(\bar{y}_n)$, where $\bar{y}_n$ is the mean of the realized series. We propose to use the block bootstrap schemes to estimate the bias and variance of $\hat{\theta}_n$. Precisely, let $b_n = E(\hat{\theta}_n - \theta)$ be the bias and let $\sigma_n^2 = \text{Var}(\hat{\theta}_n)$ be the variance. We use the block bootstrap based estimates of b_n and σ_n^2, denoted by $\hat{b}_n$ and $\hat{\sigma}_n^2$, respectively.

Next, let $T_n = \hat{\theta}_n - \theta = h(\bar{y}_n) - h(\mu)$, and let $T_n^* = h(\bar{y}_n^*) - h(E_* \bar{y}_n^*)$. The estimates $\hat{b}_n$ and $\hat{\sigma}_n^2$ are defined as $\hat{b}_n = E_* T_n^*$ and $\hat{\sigma}_n^2 = \text{Var}_*(T_n^*)$. Then the following asymptotic expansions hold; see Lahiri (1999).

Theorem 20.5. *Let $h : \mathcal{R}^d \to \mathcal{R}$ be a sufficiently smooth function.*

(a) For each of the NBB, MBB, and CBB, there exists $c_1 = c_1(f)$ such that

$$E\hat{b}_n = b_n + \frac{c_1}{nh} + o((nh)^{-1}), \quad n \to \infty.$$

(b) For the NBB, there exists $c_2 = c_2(f)$ such that

$$\text{Var}(\hat{b}_n) = \frac{2\pi^2 c_2 h}{n^3} + o(hn^{-3}), \quad n \to \infty,$$

and for the MBB and CBB,

$$\text{Var}(\hat{b}_n) = \frac{4\pi^2 c_2 h}{3n^3} + o(hn^{-3}), \quad n \to \infty.$$

(c) For each of NBB, MBB, and CBB, there exists $c_3 = c_3(f)$ such that $E(\hat{\sigma}_n^2) = \sigma_n^2 + \frac{c_3}{nh} + o((nh)^{-1})$, $n \to \infty$.

(d) For NBB, there exists $c_4 = c_4(f)$ such that

$$\text{Var}(\hat{\sigma}_n^2) = \frac{2\pi^2 c_4 h}{n^3} + o(hn^{-3}), \quad n \to \infty,$$

and for the MBB and CBB,

$$\text{Var}(\hat{\sigma}_n^2) = \frac{4\pi^2 c_4 h}{3n^3} + o(hn^{-3}), \quad n \to \infty.$$

We now use these expansions to derive optimal block sizes. The asymptotic expansions for the bias and the variance are combined to derive mean-squared error optimal block sizes. For example, for estimating b_n by $\hat{b}_n$, the leading term in the expansion for the mean-squared error is

$$m(h) = \frac{4\pi^2 c_2 h}{3n^3} + \frac{c_1^2}{n^2 h^2}.$$

To minimize $m(\cdot)$, we solve $m'(h) = 0$ to get

$$h_{opt} = \left(\frac{3c_1^2}{2\pi^2 c_2}\right)^{1/3} n^{1/3}.$$

Similarly, an optimal block length can be derived for estimating σ_n^2 by $\hat{\sigma}_n^2$. We state the following optimal block length result of Lahiri (1999) below.

Theorem 20.6. *For the MBB and the CBB, the optimal block length for estimating b_n by $\hat{b}_n$ satisfies*

$$h_{opt} = \left(\frac{3c_1^2}{2\pi^2 c_2}\right)^{1/3} n^{1/3}(1 + o(1)),$$

and the optimal block length for estimating σ_n^2 by $\hat{\sigma}_n^2$ satisfies

$$h_{opt} = \left(\frac{3c_3^2}{2\pi^2 c_4}\right)^{1/3} n^{1/3}(1 + o(1)).$$

The constants c_i depend on the spectral density f of the process, which would be unknown in a statistical context. So, the optimal block lengths cannot be directly used. Plug-in estimates for the c_i may be substituted. Or, the formulas can be used to try block lengths proportional to $n^{1/3}$, with flexible proportionality constants. There are also other methods in the literature on selection of block lengths; see Hall et al. (1995) and Politis and White (2004).

20.2 The EM Algorithm

Maximum likelihood is a mainstay of parametric statistical inference. The idea of maximizing the likelihood function globally over the parameter space Θ has an intuitive appeal. In addition, well-known theorems exist that show the asymptotic optimaility of the MLE for finite-dimensional and suitably smooth parametric problems. Lehmann and Casella (1998), Bickel and Doksum (2006), Le Cam and Yang (1990), and DasGupta (2008) are a few sources where the maximum likelihood estimate is carefully studied. A long history of success, the intrinsic intuition, and the theoretical support have all led to the mostly deserved reputation of maximum likelihood estimates as the estimate to be preferred in problems that do not have too many parameters.

20.2 The EM Algorithm

However, closed-form formulas for maximum likelihood estimates are rare outside of the exponential family structure. In such cases, one must understand the shape and boundedness properties of the likelihood function, and carefully compute the maximum likelihood estimate numerically for the observed data. Driven by this need, Fisher gave the well-known *scoring method*, the first iterative method for numerical calculation of the maximum likelihood estimate. In problems with a small number of parameters, the scoring method is known to work well, under some conditions. It is awkward to use when the number of parameters is even moderately large.

The EM algorithm, formally introduced in Dempster et al. (1977) as a general-purpose iterative numerical method for approximating the maximum likelihood estimate can be applicable, and even successful, when the scoring method is difficult to apply. The EM algorithm has become a mainstay of the numerical approximation of the maximum likelihood estimate, with widespread applications, quite like maximum likelihood itself is the mainstay of the estimation paradigm in parametric inference. The reputation of one seems to fuel the popularity of the other, although one of them is a principle, and the other a numerical scheme. The standard reference on the EM algorithm, its various mutations, and practical applications and properties is McLachlan and Krishnan (2008). Algorithms very similar to the EM algorithm were previously described in several places, notably Sundberg (1974) for the case of exponential families. The basic general algorithm is presented in this section with a description of some of its known properties and known weaknesses.

Underlying each application of the EM algorithm, there is an implicit element of *missing data*, say Z, and some observed data, say Y. If the missing data Z did become available, one would have the *complete data* $X = (Y, Z)$. Truly, the likelihood function, say $l(\theta, Y)$, depends on only the data Y that we have, and not the data that we might have had. However, the EM algorithm effectively fills in the missing data Z using the observed data Y, and a current value for the parameter θ, thereby producing a fictitious complete data likelihood $l(\theta, X)$. One finds the projection of this fictitious complete data likelihood $l(\theta, X)$ onto the class of functions that depend only on the actual observed data Y, which is then maximized over $\theta \in \Theta$ to produce a candidate maximum likelihood estimate $\hat{\theta}$. This $\hat{\theta}$ is used as the next current value for θ, and the process is let run until convergence within tolerable fluctuation appears to have been achieved. The filling in part corresponds to the E-part of the algorithm, and the maximization corresponds to the M-part of the algorithm. Because statistical models are often such that the logarithm of the likelihood function is a more manipulable function than the likelihood itself, the algorithm works with the log-likelihood, for which we use the notation $L = \log l$ below.

It is important to note that the so-called missing data Z may be really physically missing in some problems, whereas in other problems the missing data are imaginary, deviously thought of so that the complete data likelihood $l(\theta, X)$ becomes particularly pleasant and receptive to easy global maximization. In those problems where the missing data are an artificial construct, there would be a choice as to how to embed the problem into a missing data structure, and part of the art of the method is to pick a wise embedding.

20.2.1 The Algorithm and Examples

The EM algorithm runs as follows.
(a) Start with an initial value θ_0.
(b) At the kth stage of the iterative algorithm, find the best guess for what the idealized complete data likelihood would have been, by finding the conditional expectation
$$\hat{L}_k(\theta, Y) = E_{\theta_k}[L(\theta, X) | Y].$$
(c) Maximize this predicted complete data log-likelihood $\hat{L}_k(\theta, Y)$ over θ.
(d) Set the next stage current value of θ to be
$$\theta_{k+1} = \text{argmax}_{\theta \in \Theta} \hat{L}_k(\theta, Y).$$

In general, $\text{argmax}_{\theta \in \Theta} \hat{L}_k(\theta, Y)$ may be a set, rather than a single point. In that case, choose any member of that set to be the next stage current value.
(e) Iterate until convergence to satisfaction appears to have been achieved.

Thus, implementation of the EM algorithm is substantially more straightforward when the calculation of the conditional expectation $E_{\theta_k}[L(\theta, X) | Y]$ and the maximization of $\hat{L}_k(\theta, Y)$ can be done in closed-form. These two closed-form calculations may not be possible unless one has an exponential family structure in the complete data X. There have been newer versions of the EM algorithm that try to bypass closed-form calculations of these two quantities; for example, the self-evident idea of using Monte Carlo to calculate $E_{\theta_k}[L(\theta, X) | Y]$ when it cannot be done in closed-form is one of the newer versions of the EM algorithm, and is often called the *Monte Carlo EM algorithm*. We now give some illustrative examples.

Example 20.8 (Poisson with Missing Values). Suppose for some fixed $n \geq 1$, complete data $X_1, \ldots, X_n$ are iid Poi(λ), but the data value is actually reported only if $X_i \geq 2$. This sort of missing data can occur if, for example, the X_i are supposed to be counts of minor accidents per week in n locations, but the values do not get reported if there are too few incidents. If the number of recorded values is $m \leq n$, then denoting the recorded values as $Y_1, \ldots, Y_m$, the number of unreported zero values as Z_0, and the number of unreported values that equal one as Z_1, the complete data X can be represented as $(Y_1, \ldots, Y_m, m, Z_0, Z_1)$; the reported values $Y_1, \ldots, Y_m$ are iid from the conditional distribution of a Poisson variable with mean λ given that the variable is larger than 1. Therefore, writing $S_y = \sum_{i=1}^m y_i$, the likelihood based on the complete data is

$$l(\lambda, X) = \frac{e^{-m\lambda} \lambda^{S_y}}{\prod_{i=1}^m [y_i! (1 - e^{-\lambda} - \lambda e^{-\lambda})]} \left(1 - e^{-\lambda} - \lambda e^{-\lambda}\right)^m e^{-\lambda z_0} \lambda^{z_1} e^{-\lambda z_1}$$

$$= \frac{e^{-n\lambda} \lambda^{S_y + z_1}}{\prod_{i=1}^m y_i!}$$

$$\Rightarrow L(\lambda, X) = \log l(\lambda, X) = -n\lambda + (S_y + z_1) \log \lambda - \sum_{i=1}^m \log y_i!.$$

20.2 The EM Algorithm

Therefore,

$$E_{\lambda_k}[L(\lambda, X) \mid (Y_1, \ldots, Y_m, m)]$$

$$= -n\lambda + S_y \log \lambda + \log \lambda \, E_{\lambda_k}[Z_1 \mid (Y_1, \ldots, Y_m, m)] - \sum_{i=1}^{m} \log y_i!$$

$$= -n\lambda + S_y \log \lambda + (\log \lambda)(n - m)\frac{\lambda_k}{1 + \lambda_k} - \sum_{i=1}^{m} \log y_i!,$$

Because $Z_1 \mid (Y_1, \ldots, Y_m, m) \sim \text{Bin}(n - m, \frac{\lambda_k}{1+\lambda_k})$. This is the E-step of the problem. For the M-step, we have to maximize over $\lambda > 0$ the function

$$-n\lambda + \log \lambda \left[S_y + (n - m)\frac{\lambda_k}{1 + \lambda_k} \right].$$

By an easy calculus argument, this is maximized at

$$\lambda = \frac{S_y + (n - m)\frac{\lambda_k}{1+\lambda_k}}{n}.$$

This takes the position of λ_{k+1}, and the process is iterated until apparent convergence.

Example 20.9 (Bivariate Normal with Missing Coordinates). Suppose for some $n \geq 1$, complete data are iid bivariate normal vectors (X_{1j}, X_{2j}), $j = 1, 2, \ldots, n \sim N_2(\mu, \Sigma)$. However, for n_1 of the n units, only the X_1 coordinate is available, and for another n_2 distinct units, only the X_2 coordinate is available. For the rest of the $m = n - n_1 - n_2$ units, the data on both coordinates are available. We can therefore write the complete data in the canonical form

$$X = \big(Y_1, \ldots, Y_{n_1}, Y_{n_1+1}, \ldots, Y_{n_1+n_2}, (Y_{11}, Y_{21}), \ldots, (Y_{1m}, Y_{2m}),$$
$$Z_1, \ldots, Z_{n_1}, Z_{n_1+1}, \ldots, Z_{n_1+n_2}\big),$$

where

$(Y_i, Z_i), 1 \leq i \leq n_1, (Z_i, Y_i), n_1 + 1 \leq i \leq n_1 + n_2, (Y_{1j}, Y_{2j}), 1 \leq j \leq m, \overset{\text{iid}}{\sim} N_2(\mu, \Sigma).$

As usual, the notation Z is supposed to stand for the missing data. The parameter vector θ is $\theta = (\mu, \Sigma) = (\mu_1, \mu_2, \sigma_{11}, \sigma_{12}, \sigma_{22})$. The corresponding values in the kth iteration of the algorithm are denoted as $\theta_k = (\mu_1^{(k)}, \mu_2^{(k)}, \sigma_{11}^{(k)}, \sigma_{12}^{(k)}, \sigma_{22}^{(k)})$. We also use the following notation in the rest of this example, because they naturally arise in the calculations of the E-step:

$$\Sigma^{-1} = R = \begin{pmatrix} r_{11} & r_{12} \\ r_{12} & r_{22} \end{pmatrix};$$

$$\rho^{(k)} = \frac{\sigma_{12}^{(k)}}{\sqrt{\sigma_{11}^{(k)}\sigma_{22}^{(k)}}}; \quad \alpha_i^{(k)} = \mu_2^{(k)} + \frac{\sigma_{12}^{(k)}}{\sigma_{11}^{(k)}}\left(y_i - \mu_1^{(k)}\right);$$

$$\beta_i^{(k)} = \mu_1^{(k)} + \frac{\sigma_{12}^{(k)}}{\sigma_{22}^{(k)}}\left(y_i - \mu_2^{(k)}\right);$$

$$v_1^{(k)} = \sigma_{11}^{(k)}\left(1 - \left[\rho^{(k)}\right]^2\right); \quad v_2^{(k)} = \sigma_{22}^{(k)}\left(1 - \left[\rho^{(k)}\right]^2\right).$$

The complete data likelihood function is $l(\theta, X)$

$$= \frac{1}{|\Sigma|^{n/2}} e^{-\frac{1}{2}\sum_{i=1}^{n_1}(y_i-\mu_1,\, z_i-\mu_2)\Sigma^{-1}\begin{pmatrix} y_i - \mu_1 \\ z_i - \mu_2 \end{pmatrix} -\frac{1}{2}\sum_{i=n_1+1}^{n_1+n_2}(z_i-\mu_1,\, y_i-\mu_2)\Sigma^{-1}\begin{pmatrix} z_i - \mu_1 \\ y_i - \mu_2 \end{pmatrix}}$$

$$\times\, e^{-\frac{1}{2}\sum_{j=1}^{m}(y_{1j}-\mu_1,\, y_{2j}-\mu_2)\Sigma^{-1}\begin{pmatrix} y_{1j} - \mu_1 \\ y_{2j} - \mu_2 \end{pmatrix}}.$$

Therefore, $L(\theta, X) = \log l(\theta, X)$

$$= \frac{n}{2}\log|R| - \frac{1}{2}\left[\sum_{i=1}^{n_1}\{(y_i-\mu_1)^2 r_{11} + (z_i-\mu_2)^2 r_{22} + 2(y_i-\mu_1)(z_i-\mu_2)r_{12}\}\right.$$

$$+ \sum_{i=n_1+1}^{n_1+n_2}\{(z_i-\mu_1)^2 r_{11} + (y_i-\mu_2)^2 r_{22} + 2(z_i-\mu_1)(y_i-\mu_2)r_{12}\}$$

$$\left.+ \sum_{j=1}^{m}\{(y_{1j}-\mu_1)^2 r_{11} + (y_{2j}-\mu_2)^2 r_{22} + 2(y_{1j}-\mu_1)(y_{2j}-\mu_2)r_{12}\}\right].$$

To complete the E-step, we now need to evaluate the following conditional expectations:

$$E_{\theta_k}\left[(Z_i - \mu_2)^2 \mid y_i\right], \quad E_{\theta_k}\left[Z_i - \mu_2 \mid y_i\right], \quad 1 \leq i \leq n_1;$$

$$E_{\theta_k}\left[(Z_i - \mu_1)^2 \mid y_i\right], \quad E_{\theta_k}\left[Z_i - \mu_1 \mid y_i\right], \quad n_1 + 1 \leq i \leq n_1 + n_2.$$

These follow from standard bivariate normal conditional expectation formulas (see Chapter 3). Indeed, in our notation introduced above,

$$\text{For } 1 \leq i \leq n_1, \quad E_{\theta_k}[Z_i - \mu_2 \mid y_i] = \alpha_i^{(k)} - \mu_2;$$

$$E_{\theta_k}\left[(Z_i - \mu_2)^2 \mid y_i\right] = \left(\alpha_i^{(k)} - \mu_2\right)^2 + v_2^{(k)};$$

$$\text{For } n_1 + 1 \leq i \leq n_1 + n_2, \quad E_{\theta_k}[Z_i - \mu_1 \mid y_i] = \beta_i^{(k)} - \mu_1;$$

$$E_{\theta_k}\left[(Z_i - \mu_1)^2 \mid y_i\right] = \left(\beta_i^{(k)} - \mu_1\right)^2 + v_1^{(k)}.$$

Plugging these in, $E_{\theta_k}[L(\theta, X) \mid Y]$

20.2 The EM Algorithm

$$= \frac{n}{2}\log|R| - \frac{1}{2}\left[\sum_{i=1}^{n_1}\left\{(y_i-\mu_1)^2 r_{11} + \left(\left(\alpha_i^{(k)}-\mu_2\right)^2 + v_2^{(k)}\right)r_{22}\right.\right.$$

$$+ 2(y_i-\mu_1)\left(\alpha_i^{(k)}-\mu_2\right)r_{12}\bigg\}$$

$$+ \sum_{i=n_1+1}^{n_1+n_2}\left\{\left(\left(\beta_i^{(k)}-\mu_1\right)^2 + v_1^{(k)}\right)r_{11}\right.$$

$$+ (y_i-\mu_2)^2 r_{22} + 2(y_i-\mu_2)\left(\beta_i^{(k)}-\mu_1\right)r_{12}\bigg\}$$

$$+ \sum_{j=1}^{m}\left\{(y_{1j}-\mu_1)^2 r_{11} + (y_{2j}-\mu_2)^2 r_{22} + 2(y_{1j}-\mu_1)(y_{2j}-\mu_2)r_{12}\right\}\Bigg].$$

So, the E-step can indeed be done in closed-form. The M-step now requires maximization of this expression over $(\mu_1, \mu_2, r_{11}, r_{12}, r_{22})$. Although the expression above for $E_{\theta_k}[L(\theta, X) \mid Y]$ is long, on inspection we can see that it has the same structure as the logarithm of the density of a general bivariate normal distribution. Therefore, even the M-step can be done in closed-form by using standard formulas for maximum likelihood estimates of the mean vector and the covariance matrix in the general multivariate normal case. Alternatively, the M-step can be done from first principles by simply taking the partial derivative of the expression above with respect to $\mu_1, \mu_2, r_{11}, r_{12}, r_{22}$ and by solving the five equations obtained from setting these partial derivatives equal to zero. We do not show that calculation here.

Example 20.10 (EM in Estimating ABO Allele Proportions). The ABO blood classification system is perhaps the most clinically important blood typing system for humans. All humans can be classified into one of the four *phenotypes* A, B, AB, and O. ABO blood typing is essential before blood transfusions, because infusion of an incompatible blood type has fatal consequences. In fact, it was these observed fatalities during blood transfusions that led to the discovery of the ABO blood types.

The specific blood type is governed by a single gene with three alleles, which are also usually denoted as A, B, and O. Each individual receives one of these three alleles from the father, and one from the mother. Alleles A and B dominate over allele O. Thus, individuals who have one A allele and one O allele will show as phenotype A, and so on, although the true genotype is AO. Because A and B dominate over allele O, an individual can have phenotype O only if she or he receives an O allele from each parent.

EM is a natural tool for estimating the allele frequencies, that is, the respective proportions of A, B, and O alleles among all individuals in a sampling population. We think of the EM algorithm naturally, because although there are only four phenotypes A, B, AB, and O, there are six *genotypes*, AA, AO, BB, BO, AB, and OO. Because of the dominance property of the A and the B alleles, we cannot phenotypically distinguish between AA and AO, or between BB and BO. So, we have some missing data, and EM fits in very naturally.

The mathematical formulation needs some notation. The parameter vector is simply the vector of the three allele proportions in the population, namely $\theta = (p_A, p_B, p_O)$. The complete data would correspond to the six genotypical frequencies in a sample of n individuals, namely $X_{AA}, X_{AO}, \ldots, X_{OO}$. To reduce clutter, we call them $X_1, X_2, \ldots, X_6$. The observed data consist of the phenotypical frequencies Y_A, Y_B, Y_{AB}, Y_O; once again, to reduce clutter, we call them Y_1, Y_2, Y_3, Y_4. Note that $p_A + p_B + p_O = 1$, and also, $\sum_{i=1}^{6} X_i = \sum_{i=1}^{4} Y_i = n$. Also note the important relationships $X_1 + X_2 = Y_1, X_3 + X_4 = Y_2, X_5 = Y_3$, and $X_6 = Y_4$. From the usual formula for a multinomial pmf, the complete data likelihood function is

$$l(\theta, X) = c \; (p_A)^{2X_1} (2p_A p_O)^{X_2} (p_B)^{2X_3} (2p_B p_O)^{X_4} (2p_A p_B)^{X_5} (p_O)^{2X_6},$$

where c is a constant not involving θ. Therefore,

$$L(\theta, X) = \log l(\theta, X) = \text{constant} + (2X_1 + X_2 + X_5) \log p_A$$
$$+ (2X_3 + X_4 + X_5) \log p_B + (2X_6 + X_2 + X_4) \log p_O$$
$$= X_1(\log p_A - \log p_O) + X_3(\log p_B - \log p_O) + Y_1(\log p_A + \log p_O)$$
$$+ Y_2(\log p_B + \log p_O) + Y_3(\log p_A + \log p_B) + 2Y_4 \log p_O.$$

Because

$$\hat{X}_{1,k} := E_{\theta_k}(X_1 \mid Y) = Y_1 \frac{p_{A,k}^2}{p_{A,k}^2 + 2p_{A,k} p_{O,k}};$$

$$\hat{X}_{3,k} := E_{\theta_k}(X_3 \mid Y) = Y_2 \frac{p_{B,k}^2}{p_{B,k}^2 + 2p_{B,k} p_{O,k}},$$

we get $E_{\theta_k}[L(\theta, X) \mid Y]$

$$= \hat{X}_{1,k}(\log p_A - \log p_O) + \hat{X}_{3,k}(\log p_B - \log p_O)$$
$$+ Y_1(\log p_A + \log p_O) + Y_2(\log p_B + \log p_O)$$
$$+ Y_3(\log p_A + \log p_B) + 2Y_4 \log p_O.$$

This finishes the E-step, and once again, we are fortunate that we can do it in closed-form.

For the M-step, we have to maximize this with respect to (p_A, p_B, p_O) over the simplex

$$S = \{(p_A, p_B, p_O) : p_A, p_B, p_O \geq 0, p_A + p_B + p_O = 1\}.$$

Standard calculus methods using Lagrange multipliers lead to the closed-form maximas

$$p_A = \frac{Y_1 + Y_3 + \hat{X}_{1,k}}{2n};$$

$$p_B = \frac{Y_2 + Y_3 + \hat{X}_{3,k}}{2n};$$

$$p_O = \frac{Y_1 + Y_2 + 2Y_4 - \hat{X}_{1,k} - \hat{X}_{3,k}}{2n}.$$

These serve as the values of θ_{k+1}, the next iterate.

20.2.2 Monotone Ascent and Convergence of EM

In its search for the global maximum of the likelihood function, the EM algorithm has some positive properties and some murky properties. The EM algorithm does not behave erratically. In fact, in each iteration the EM algorithm produces a value of the likelihood function that is at least as large as the value at the previous iteration. This is known as *monotonicity of the EM algorithm*. This property is mathematically demonstrable, and is proved below. The ideal goal of the EM algorithm is to ultimately arrive at or very very close to the global maximum value of the likelihood function, and the MLE. In this, the EM algorithm has mixed success. There are no all-at-one-time theorems which show that iterates of the EM algorithm are guaranteed to lead eventually to the correct global maximum. Indeed, there cannot be such a theorem, because there are widely available counterexamples to it. What is true is that under frequently satisfied conditions, iterates of the EM algorithm will converge to a point of stationarity of the likelihood function. The starting value θ_0 determines to which stationary point the EM iterates converge. If we are willing to assume quite a bit more structure, such as that of a multiparameter exponential family, or a strongly unimodal likelihood function, then convergence to a global maximum can be assured. However, the EM algorithm has the reputation of converging very slowly to the global maximum, if it does at all. Although it has the monotonicity property, the ascent to the peak value can be slow. The main reference for this topic is Wu (1983). The work is nicely summarized in several other places, specifically McLachlan and Krishnan (2008). The main facts are described with a classic example below.

Theorem 20.7. *Let $l(\theta, y)$ denote the likelihood function on the basis of the observed data $Y = y$, and let $\theta_k, k \geq 0$ be the sequence of EM iterates. Then $l(\theta_{k+1}, y) \geq l(\theta_k, y)$ for all $k \geq 0$.*

Proof. We recall the notation. The complete data $X = (Y, Z)$, where Y are the actually observed data. The joint density of (Y, Z) under θ is $f_\theta(y, z)$, and the marginal density of Y under θ is $g_\theta(y)$. Thus, $l(\theta, y) = g_\theta(y)$ and $L(\theta, y) = \log l(\theta, y) = \log g_\theta(y)$. We also need the Kullback–Leibler distance inequality $E_p(\log p) \geq E_p(\log q)$, where p, q are two densities on some common space (see Chapter 15).

The key to the theorem is to show that $\hat{L}_k(\theta, y) - L(\theta, y)$ is maximized at $\theta = \theta_k$. For, if we can show this, then we will have

$$\hat{L}_k(\theta_{k+1}, y) - L(\theta_{k+1}, y) \leq \hat{L}_k(\theta_k, y) - L(\theta_k, y)$$
$$\leq \hat{L}_k(\theta_{k+1}, y) - L(\theta_k, y)$$

(because the function $\hat{L}_k(\theta, y)$ is maximized at $\theta = \theta_{k+1}$)

$$\Rightarrow L(\theta_{k+1}, y) \geq L(\theta_k, y),$$

which is the claim of the theorem. To prove that $\hat{L}_k(\theta, y) - L(\theta, y)$ is maximized at $\theta = \theta_k$, we observe that

$$\hat{L}_k(\theta, y) - L(\theta, y) = E_{\theta_k}(\log f_\theta(Y, Z) \mid Y = y) - \log g_\theta(y)$$
$$= E_{\theta_k}(\log f_\theta(y, Z) \mid Y = y) - \log g_\theta(y)$$
$$= E_{Z \mid Y = y, \theta_k}(\log f_\theta(y, Z)) - \log g_\theta(y)$$
$$= E_{Z \mid Y = y, \theta_k}\left(\log \frac{f_\theta(y, Z)}{g_\theta(y)}\right)$$
$$\leq E_{Z \mid Y = y, \theta_k}\left(\log \frac{f_{\theta_k}(y, Z)}{g_{\theta_k}(y)}\right)$$

(because if we identify $\frac{f_\theta(y,Z)}{g_\theta(y)}$ as q, and $\frac{f_{\theta_k}(y,Z)}{g_{\theta_k}(y)}$ as p, then the Kullback–Leibler distance inequality $E_p(\log p) \geq E_p(\log q)$ is exactly the inequality in the last line)

$$= \hat{L}_k(\theta_k, y) - L(\theta_k, y).$$

But that precisely means that $\hat{L}_k(\theta, y) - L(\theta, y)$ is maximized at $\theta = \theta_k$, and so this proves the theorem. □

The above theorem implies that if $L(\theta, y) = \log g_\theta(y)$ is bounded in θ for each fixed y, then the sequence of iterates $L(\theta_k, y)$ has a limit, say L^*. So, under the boundedness condition, the chain of EM values of the log-likelihood function ultimately converges to something. The question is: to what? We would like L^* to be the global maximum of $L(\theta, y)$, if such a global maximum exists. Unfortunately, it cannot be assured in general. The following example is taken from Wu (1983), who cites Murray (1977).

Example 20.11 (Failure of the EM Algorithm). Suppose the complete data consist of $n = 12$ iid observations from a bivariate normal distribution with means known to be zero, and the other three parameters $\sigma_1^2, \sigma_2^2, \rho$ unknown. The observed data correspond to our Example 20.9: for some sampling units, data on one coordinate are missing. The observed data are:

20.2 The EM Algorithm

$$(1, 1), (1, -1), (-1, 1), (-1, -1), (-2, *), (-2, *),$$

$$(2, *), (2, *), (*, -2), (*, -2), (*, 2), (*, 2).$$

The data have been deliberately so constructed as to produce a log-likelihood function $L(\theta, y)$ to have several local maximas, and two global maximas. The global maximas are $\sigma_1^2 = \sigma_2^2 = \frac{8}{3}, \rho = \pm\frac{1}{2}$. There are other stationary points of the log-likelihood function $L(\theta, y)$, one of which has $\sigma_1^2 = \sigma_2^2 = \frac{5}{2}, \rho = 0$. Actually, this point is moreover a saddle point of $L(\theta, y)$; that is, the Hessian matrix of the function $L(\theta, y)$ is indefinite at this particular stationary point. So, the point is not a local maximum, or a local minimum. Coming to what the EM algorithm does in this case, if the initial choice θ_0 has $\rho_0 = 0$, then for all $k \geq 1, \rho_k = 0$, and the sequence of EM iterates converges to the saddle point given above. The problem with the application of the EM algorithm in this example is exactly the fact that once the EM reaches $\rho = 0$ at any iteration, it fails to move out of there at any subsequent iteration. It is indefinitely trapped at $\rho = 0$, and can only maximize $L(\theta, y)$ over the submanifold $\{(\sigma_1^2, \sigma_2^2, \rho) : \rho = 0\}$. On that submanifold, the saddle point is the unique maxima; but it is not a global, or even a local maxima.

Wu (1983) gives the following theorem on the convergence of the EM algorithm, and this theorem is essentially the best possible that can be said.

Theorem 20.8. *Define the map*

$$\hat{L}(\theta, \phi) = E_\phi(L(\theta, X) | Y).$$

Assume that

(a) Θ *is a subset of some Euclidean space* $\mathcal{R}^d$.
(b) *For any* $y, L(\theta, y)$ *is continuous on* Θ *and once partially differentiable with respect to each coordinate of* θ *in the interior of* Θ.
(c) $L(\theta_0, y) > -\infty$.
(d) *The sets* $\{\theta : L(\theta, y) \geq c\}$ *are compact for all real c.*
(e) $\hat{L}(\theta, \phi)$ *is jointly continuous on* $\Theta \otimes \Theta$.

Then,

(i) *The sequence of iterates* $\{L(\theta_k, y)\}$ *converges to* $L(\theta^*, y)$ *for some stationary point* θ^* *of* $L(\theta, y)$.
(ii) *Any sub-sequence of the iterates* $\{\theta_k\}$ *converges to some stationary point of* $L(\theta, y)$.

Suppose, in addition, we assume that

(f) $L(\theta, y)$ *is concave on* Θ *with a unique stationary point* $\hat{\theta}$.
(g) *The gradient vector* $\nabla_\theta \hat{L}(\theta, \phi)$ *is jointly continuous on* $\Theta \otimes \Theta$.

Then the sequence of iterates $\{\theta_k\}$ *has only one limit point, and the limit coincides with* $\hat{\theta}$, *which is the unique MLE of* θ.

Wu (1983) uses general theorems on limit points of iterated point-to-set maps for proving the above theorem.

20.2.3 * Modifications of EM

The basic EM algorithm may not be exactly applicable, or inefficient due to slow convergence in some important problems. The basic EM algorithm rests on two assumptions: that the E-step can be done in closed-form, and that the M-step can also be done in closed form. Of these, failure of the second assumption may not be computationally too damaging in low dimensions, because numerical maximization algorithms on the complete data likelihood may be easily implementable. For example, numerical maximizations that use Newton–Raphson methods in the M-step are known as the *EM gradient algorithm*; see McLachlan and Krishnan (2008) and Lange (1999).

Failure to perform the E-step in closed-form almost certainly necessitates Monte Carlo evaluation of the expectation $E_{\theta_k}(L(\theta, X) \mid Y)$. This is called the *Monte Carlo EM algorithm*. However, this has to be done for a fine grid of θ values, because the subsequent M-step requires the full function $\hat{L}_k(\theta, y)$. This may be time consuming, depending on d, the dimension of θ. It is also very important to note that substitution of Monte Carlo for analytic calculations in the E-step calls for simulation from the conditional distribution of Z (the missing data) given Y (the observed data). This can be cumbersome, and even very difficult. In such a case, the Gibbs sampler may be useful, because the Gibbs sampler is specially designed for this (see Chapter 19). An unfortunate consequence of using Monte Carlo to accomplish the E-step is that the monotone ascent property germane to the basic EM algorithm is now usually lost. These ideas are described in McLachlan and Krishnan (2008), Wei and Tanner (1990), Chan and Ledolter (1995), and Levine and Casella (2001).

The EM is an optimization scheme for wanting to find the maximum value of a function. As such, the idea can also be applied, verbatim, to approximate the maximum of a posterior density, that is, to approximate a posterior mode. This is known as *Bayesian EM*. The posterior density of the parameter based on the complete data likelihood is $\pi(\theta \mid X) = c\, l(\theta, X) \pi(\theta)$, $\pi(\theta)$ being the prior density. Consequently, the E-step has only the trivial modification that we now have the extra term $\log \pi(\theta)$ added to the usual term $\hat{L}_k(\theta, y)$. The M-step should not be much more complex, unless the prior density is multimodal or somehow badly behaved. To an error of $O(n^{-1})$, the approximation to the mode will also provide an approximation to the mean of the posterior under enough regularity conditions; see DasGupta (2008) and Bickel and Doksum (2006) for such Bayesian asymptotia.

Accumulated user experience shows that the monotone ascent of the EM algorithm is often very slow. Certain modifications of the basic EM algorithm have been suggested to enhance the speed of practical convergence. Typically, these blend time-tested purely numerical analysis tools, such as one-or two-term Taylor expansions, with the EM algorithm itself. Collectively, these schemes are known as *accelerated EM algorithms*. Once again, accelerated EM algorithms do not have the monotone ascent property, and are also more difficult to code. The methods are described in detail with many references in McLachlan and Krishnan (2008).

20.3 Kernels and Classification

Smoothing noisy data or a rough function in order to extract the main features out of the noisy data is a long-standing and time-tested principle in quantitative science. For example, consider the CDF of a Poisson random variable $Y \sim \text{Poi}(5)$. The CDF of Y of course is not smooth; it is not even continuous. Suppose now we add to Y a small independent Gaussian random variable $Z \sim N(0, .01^2)$. The sum $X = Y + Z$ has a continuous distribution with a density, and the CDF of X is not only continuous, but even infinitely differentiable. In this example, we used convolution to smooth a nonsmooth CDF. Convolution is a special case of *kernel smoothing*, a particular type of smoothing that has found wide applications in statistics, machine learning, and approximation theory. This section provides a basic treatment of the theory and applications of kernels, with examples.

Definition 20.4. A function $K : \mathcal{R}^d \to \mathcal{R}$ is called a *kernel* if $\int_{\mathcal{R}^d} |K(x)| dx < \infty$, and $\int_{\mathcal{R}^d} K(x) dx = 1$.

In applications, we often take $K \geq 0$, in which case a kernel is just a probability density function on $\mathcal{R}^d$. Moreover, we also often take K to be symmetric, in the sense that $K(x) = K(-x)$ for all $x \in \mathcal{R}^d$.

20.3.1 Smoothing by Kernels

Kernels are often used as a smoothing device via the operation of convolution, as defined below.

Definition 20.5. Let $f : \mathcal{R}^d \to \mathcal{R}$ be an L_1 function: $\int_{\mathcal{R}^d} |f(x)| dx < \infty$. Let K be any kernel on $\mathcal{R}^d$. The *convolution* of f and K is defined as

$$(f * K)(x) = \int_{\mathcal{R}^d} K(x-y) f(y) dy.$$

Convolutions with kernels generally have two important properties that lead to the wide acceptance of the principle of kernel smoothing:

(a) The convolution $f * K$ will generally have some extra smoothness in comparison to f. The exact nature of the extra smoothness will depend on both f and K.
(b) If we smooth f by a sequence of increasingly spiky kernels K_n, then the sequence of convolutions $f * K_n$ will converge in some meaningful sense, and often in a very strong sense, to f.

These two properties together give us what we want: close approximation of a noisy function by a nicer smooth function. The following basic theorem gives a smoothing and an approximation property of convolutions with kernels. For the rest of this chapter we use the following standard notation for L_p norms and L_p spaces of functions:

$$\|f\|_p = \left(\int_{\mathcal{R}^d} |f|^p dx\right)^{1/p}, \quad 0 < p < \infty; \quad \|f\|_\infty = \sup_{x \in \mathcal{R}^d} |f(x)|;$$

$$L_p\left(\mathcal{R}^d\right) = \{f : \|f\|_p < \infty\}, \quad 0 < p \leq \infty.$$

We remark that in fact strictly rigorous definitions of L_p norms and L_p spaces require measure-theoretic considerations because of difficulties presented by null sets. We do not mention these again in the subsequent development.

Theorem 20.9. *Given a kernel K, define $K_n(x) = n^d K(nx), n \geq 1$.*

(a) *For any L_1 function f, $f * K_n \in L_1$ for all $n \geq 1$,*
(b)
$$\int_{\mathcal{R}^d} |f * K_n - f| dx \to 0 \text{ as } n \to \infty.$$

(c) *If K is r times differentiable with bounded partial derivatives $\frac{\partial^j}{\partial x_i^j} K, i = 1, 2, \ldots, d, j = 1, 2, \ldots, r$, then for any $f \in L_1$, $f * K_n$ is r times differentiable for all $n \geq 1$.*

(d) *If K is simply uniformly bounded, then $f * K_n$ is bounded and uniformly continuous for all $n \geq 1$.*

Proof. For part (a), simply note that

$$\int_{\mathcal{R}^d} |(f * K)(x)| dx = \int_{\mathcal{R}^d} \left|\int_{\mathcal{R}^d} K(x-y) f(y) dy\right| dx$$

$$\leq \int_{\mathcal{R}^d} \int_{\mathcal{R}^d} |K(x-y) f(y)| dy dx$$

$$= \int_{\mathcal{R}^d} |K(x-y)| dx \int_{\mathcal{R}^d} |f(y)| dy$$

$$= \|K\|_1 \|f\|_1 < \infty.$$

For part (b), the key step is to use the fact that K_n is a kernel for any $n \geq 1$, and hence each K_n integrates to 1 on $\mathcal{R}^d$. Therefore, writing $g(z) = \int_{\mathcal{R}^d} |f(x-z) - f(x)| dx$, we get

$$\int_{\mathcal{R}^d} |f * K_n - f| dx = \int_{\mathcal{R}^d} \left|\int_{\mathcal{R}^d} K_n(x-y)[f(y) - f(x)] dy\right| dx$$

$$= \int_{\mathcal{R}^d} \left|\int_{\mathcal{R}^d} K_n(z)[f(x-z) - f(x)] dz\right| dx$$

$$\leq \int_{\mathcal{R}^d} |K_n(z)| g(z) dz.$$

20.3 Kernels and Classification

Break the integral in the last line into two sets, a ball $B(0, \delta)$ of a suitable radius δ and the complement $\mathcal{R}^d - B(0, \delta)$. The first integral $\int_{B(0,\delta)} |K_n(z)| g(z) dz$ becomes small for large n, because g is continuous at $z = 0$ and $g(0) = 0$, and because K_n is integrable and indeed, $||K_n||_1$ is just a fixed constant. The second integral $\int_{\mathcal{R}^d - B(0,\delta)}$ can also be made small for large n, because f is integrable, which forces g to be integrable, and because as n gets large, the integrals $\int_{\mathcal{R}^d - B(0,\delta)} |K_n|$ become small. Putting these together, we get the convergence result of part (b).

Part (c) is an easy consequence of the dominated convergence theorem. For example, by the assumed boundedness of the partial derivatives $\frac{\partial}{\partial x_i} K$,

$$\int_{\mathcal{R}^d} ||\nabla K(x-y)|| \, |f(y)| dy \leq c \int_{\mathcal{R}^d} |f(y)| dy < \infty,$$

and therefore, by the dominated convergence theorem, at any $x \in \mathcal{R}^d$, $(f * K)(x)$ is partially differentiable with respect to each coordinate of x. The same argument works for any $n \geq 1$, and for the higher-order partial derivatives.

Part (d) is a sophisticated result, and requires use of a deep theorem in analysis called *Lusin's theorem*. We do not prove part (d) here. □

With added smoothness conditions on f, we can get convergence of $f * K_n$ to f in some suitable uniform sense. Here is such a result; see pp. 149 and 156 in Cheney and Light (2000).

Theorem 20.10. *(a) Suppose f is bounded and continuous on $\mathcal{R}^d$. Then $f * K_n$ converges to f uniformly on any given compact set $C \subset \mathcal{R}^d$.*
(b) Suppose f is uniformly continuous on $\mathcal{R}^d$ and K has compact support. Then $f * K_n$ converges to f uniformly on all of $\mathcal{R}^d$. In particular, if f is Lipschitz of order α for some $\alpha > 0$, then $f * K_n$ converges to f uniformly on $\mathcal{R}^d$, whenever K has compact support.

20.3.2 Some Common Kernels in Use

The only mathematical requirement of a kernel is that it be integrable, and that $\int_{\mathcal{R}^d} K(x) dx$ should equal 1. In practice, we choose our kernels to have several additional properties from the following list.

(a) **Nonnegativity.** $K(x) \geq 0$ for all x.
(b) **Isotropic.** $K(x) = h(||x||)$ for some function $h : \mathcal{R}_+ \to \mathcal{R}$.
(c) **Fourier or Positive Definiteness Property.** K is isotropic, and $h(t)$ is the characteristic function of some symmetric random variable X; that is, $h(t)$ has the representation $h(t) = \int_{\mathcal{R}} e^{itx} dF(x) = \int_{\mathcal{R}} \cos(tx) dF(x)$ for some CDF F that has the symmetry property $P_F(X \leq x) = P_F(X \geq -x)$ for all x.

(d) **Rapid Decay.** $K(x)$ converges rapidly to zero as $||x|| \to \infty$. One type of rapid decay is that $\int_{\mathcal{R}^d} (|x_1|^{\alpha_1} \ldots |x_d|^{\alpha_d}) K(x_1, \ldots, x_d) dx_1 \ldots dx_d$ should be finite for suitably large $\alpha_1, \ldots, \alpha_d > 0$.
(e) **Compact Support.** For some compact set C, $K(x) = 0$ if $x \notin C$.
(f) **Smoothness.** $K(x)$ has sufficient smoothness, for examples, that it be continuous, or uniformly continuous, or have some derivatives, or that it belongs to a suitable *Sobolev space*.

The choice of the kernel depends on the nature of the problem for which it will be used. Kernel methods are widely used in statistical density estimation, in classification, in simply smoothing an erratic function, and in various other approximate reconstruction problems. Good kernels for density estimation need not be good for classification. The following table lists some common kernels in use in statistics, computer science, and machine learning.

Kernel Name	Unnormalized $K(x)$
Uniform	$I_{||x|| \leq a}$
Weierstrass	$(1 - ||x||^2)^k I_{||x|| \leq 1}, k \geq 1$
Epanechnikov	$(1 - ||x||^2) I_{||x|| \leq 1}$
Gaussian	$e^{-a^2 ||x||^2}$
Cauchy	$\frac{1}{(1 + a^2 ||x||^2)^{\frac{d+1}{2}}}$
Exponential	$e^{-a||x||}$
Laplace	$e^{-a||x||_1}$
Spherical	$\left(1 - \frac{3}{2}||x|| + \frac{1}{2}||x||^3\right) I_{||x|| \leq 1}$
Polynomial	$\left(\sum_{i=1}^{m} c_i ||x||^{k_i}\right) I_{||x|| \leq 1}$
Hermite	$(1 - ||x||^2) e^{-a^2 ||x||^2}$
Wave	$\frac{\sin(c||x||)}{c||x||}$
Product	$\prod_{i=1}^{d} (1 - x_i^2) I_{|x_i| \leq 1}$
Product	$\prod_{i=1}^{d} \frac{1}{1 + x_i^2}$
Fejér	$\frac{1}{k} \left(\frac{\sin \frac{kx}{2}}{\sin \frac{x}{2}}\right)^2 \; (d = 1)$
de la Vallée-Poussin	$\left(\frac{\sin x}{x}\right)^2 \; (d = 1)$

We plot some of these kernels in one and two dimensions in Fig. 20.2. Note how some are quite flat, and others spiky, and some unimodal and the Fejer kernel wavy. The choice would depend on exactly what one wants to achieve in a particular problem. Generally, a flatter kernel would lead to more smoothing.

20.3 Kernels and Classification

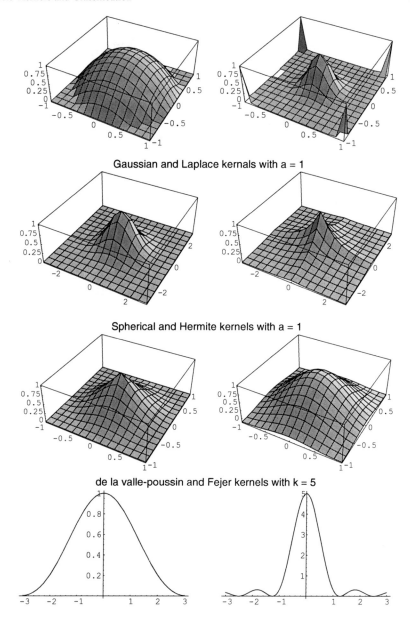

Fig. 20.2 Weierstrass kernels with k = 1, 8 when d = 2

20.3.3 Kernel Density Estimation

The basic problem of density estimation is the following. Suppose we have n iid observations $X_1, \ldots, X_n$ from a density $f(x)$ on an Euclidean space $\mathcal{R}^d$. We do not know the density function $f(x)$ and we want to estimate it. In some problems, a parametric assumption will be fine; for example, when $d = 1$, we may assume that f is a normal density with some mean μ and some variance σ^2. Then, we may use the *plug-in estiamte* $N(\overline{X}, s^2)$ as our estimate for f. In some other problems, a specific parametric form will be too restrictive. Then, we may dispose of any specific parametric form, and estimate f nonparametrically. Here, nonparametric is supposed to simply mean lack of a parametric functional form. Nonparametric density estimation is one of the most researched and still active areas in statistical theory, and the techniques and theory are highly sophisticated and elegant. A lot of development in statistics has taken place around the themes, methods, and mathematics of density estimation. For example, research in nonparametric regression and nonparametric function estimation has been heavily influenced by the density estimation literature.

Several standard types of density estimators are briefly discussed below.

(a) **Plug-In Estimates**. Assume $f \in \mathcal{F}_\Theta = \{f_\theta : \theta \in \Theta \subseteq \mathcal{R}^k, k < \infty\}$. Let $\hat{\theta}$ be any reasonable estimator of θ. Then the plug-in estimate of f is $\hat{f} = f_{\hat{\theta}}$.

(b) **Histograms**. Let the support of f be $[a, b]$ and consider the partition $a = t_0 < t_1 < \cdots < t_m = b$. Let $c_0, \ldots, c_{m-1}$ be constants. The histogram density estimate is

$$\hat{f}(x) = \begin{array}{ll} c_i, & t_i \leq x < t_{i+1} \\ c_{m-1}, & x = b \\ 0, & x \notin [a, b] \end{array}.$$

The restriction $c_i \geq 0$ and $\sum_i c_i(t_{i+1} - t_i) = 1$ makes $\hat{f}$ a true density; that is, $\hat{f} \geq 0$ and $\int \hat{f}(x)\,dx = 1$. The canonical histogram estimator is

$$\hat{f}_0(x) = \frac{n_i}{n(t_{i+1} - t_i)}, \quad t_i \leq x < t_{i+1},$$

where $n_i = \#\{i : t_i \leq X_i < t_{i+1}\}$. Then $\hat{f}_0$ is, in fact, the nonparametric maximum likelihood estimator of f. For a small interval $[t_i, t_{i+1})$ containing a given point x,

$$E_f(n_i) = n\left[P_f(t_i \leq X < t_{i+1})\right] \approx n(t_{i+1} - t_i)f(x).$$

Then it follows that

$$E_f[\hat{f}_0(x)] = E_f\left(\frac{n_i}{n(t_{i+1} - t_i)}\right) \approx f(x).$$

(c) **Series Estimates**. Suppose that $f \in L_2(\mathcal{R})$; that is, $\int_\mathcal{R} f^2(x)dx < \infty$. Let $\{\phi_k : k \geq 0\}$ be a collection of orthonormal basis functions (see Section 20.3.4 for details on orthonormal bases). Then f admits a Fourier expansion

$$f(x) = \sum_{k=0}^{\infty} c_k \phi_k(x), \quad \text{where } c_k = \int \phi_k(x) f(x)\,dx.$$

20.3 Kernels and Classification

Because $c_k = E_f[\phi_k(X)]$, as an estimate one may use

$$\hat{f}(x) = \sum_{k=0}^{l_n} \left[\frac{1}{n}\sum_{i=1}^{n}\phi_k(X_i)\right]\phi_k(x),$$

for a suitably large cutoff l_n. This is the orthonormal series estimator.

(d) **Kernel Estimates.** This is a natural and massive extension of the idea of histograms. Let $K(z) \geq 0$ be a general nonnegative function, integrating to 1, and let

$$\hat{f}_K(x) = \frac{1}{nh_n}\sum_{i=1}^{n} K\left(\frac{x - X_i}{h_n}\right).$$

The function $K(\cdot)$ is a *kernel* and $\hat{f}_K$ is called the *kernel density estimator*. The scaling factor $h = h_n$ is called the *bandwidth*; it must be suitably small so the density estimate does not suffer from large biases. Suppose the kernel function is a Gaussian kernel. In such a case, $\hat{f}_K$ is a mixture of $N(X_i, h_n^2)$ densities. Thus, the kernel density estimate will take normal densities with small widths centered at the data values and then blend them together. This fundamental and seminal idea was initiated in Rosenblatt (1956). Kernel density estimates are by far the most popular nonparametric density estimates. They generally provide the best rates of convergence, as well as provide a great deal of flexibility through the choice of the kernel. However, in an asymptotic sense, the choice of the kernel is of less importance than the choice of bandwidth.

As in most statistical inference problems, there are two issues: systematic error (in the form of bias) and random error (in the form of variance). To include both of these aspects, we consider the mean squared error (MSE). If $\hat{f}_n$ is some estimate of the unknown density f, we want to consider

$$\text{MSE}[\hat{f}_n] \equiv E[\hat{f}_n(x) - f(x)]^2 = \text{Var}[\hat{f}_n(x)] + E^2[\hat{f}_n(x) - f(x)].$$

As usual, there is a bias–variance trade-off. The standard density estimates require specification of a suitable tuning parameter. For instance, for kernel density estimates, the tuning parameter would be the bandwidth h. The tuning parameter is optimally chosen after weighing in the bias–variance trade-off.

To study the performance of kernel estimates, we consider both the variance and the bias, starting with the bias. Under various sets of assumptions on f and K, asymptotic unbiasedness indeed holds. One such result is given below; see Rosenblatt (1956).

Theorem 20.11. *Assume that K is nonnegative and integrable. Also assume that f is uniformly bounded by some $M < \infty$. Suppose $h = h_n \to 0$ as $n \to \infty$. Then, for any x that is a continuity point of f,*

$$E[\hat{f}_n(x)] \longrightarrow f(x) \int K(z)\,dz, \quad \text{as } n \to \infty.$$

Proof. Since the X_is are iid and $\hat{f}_n(x) = \frac{1}{nh}\sum K\left(\frac{x-X_i}{h}\right)$, we get,

$$E[\hat{f}_n(x)] = \frac{1}{h}\int_{-\infty}^{\infty} K\left(\frac{x-z}{h}\right) f(z)\,dz = \int_{-\infty}^{\infty} K(z) f(x - hz)\,dz.$$

But f is continuous at x and uniformly bounded; so $K(z) f(x - hz) \leq M K(z)$ and $f(x - hz) \to f(x)$ as $h \to 0$. Because $K(\cdot) \in L_1$ (i.e., $\int |K(z)|\,dz < \infty$), we can apply the dominated convergence theorem to interchange the limit and the integral. That is, if we let $h = h_n$ and $h_n \to 0$ as $n \to \infty$, then

$$\lim_{n\to\infty} E[\hat{f}_n(x)] = \int \lim_{n\to\infty} K(z) f(x - hz)\,dz$$

$$= \int K(z) f(x)\,dz$$

$$= f(x) \int K(z)\,dz$$

In particular, if $\int K(z)\,dz = 1$, then $\hat{f}_n(x)$ is asymptotically unbiased at all continuity points of f, provided that $h = h_n \to 0$. □

Next, we consider the variance of $\hat{f}_n$. Consistency of the kernel estimate does not follow from asymptotic unbiasedness alone; we need something more. To get a stronger result, we need to assume more than simply $h_n \to 0$. Loosely stated, we want to drive the variance of $\hat{f}_n$ to zero, and for this we must have more than $h_n \to 0$. Obviously, because $\hat{f}_n(x)$ is essentially a sample mean,

$$\text{Var}[\hat{f}_n(x)] = \frac{1}{n}\text{Var}\left[\frac{1}{h}K\left(\frac{x-X}{h}\right)\right],$$

which implies

$$nh\,\text{Var}[\hat{f}_n(x)] = h\,\text{Var}\left[\frac{1}{h}K\left(\frac{x-X}{h}\right)\right].$$

By an application of the dominated convergence theorem, as in the proof above, $E[\hat{f}_n(x)]^2$ converges, at continuity points x, to $f(x)\|K\|_2^2$, which is finite if

20.3 Kernels and Classification

$K \in \mathcal{L}_2$. We already know that for all continuity points x, $E[\hat{f}_n(x)]$ converges to $f(x)\|K\|_1$; so it follows that

$$h\left(E[\hat{f}_n(x)]\right)^2 \longrightarrow 0, \quad h \to 0.$$

Combining these results, we get, for continuity points x,

$$nh\text{Var}[\hat{f}_n(x)] = h\text{Var}\left[\frac{1}{h}K\left(\frac{x-X}{h}\right)\right] \longrightarrow f(x)\|K\|_2^2 < \infty,$$

provided $K \in \mathcal{L}_2$. Consequently, if $h \to 0$, $nh \to \infty$, $f \leq M$ and $K \in \mathcal{L}_2$, then, at continuity points x,

$$\text{Var}[\hat{f}_n(x)] \longrightarrow 0.$$

We summarize the above derivation in the following theorem. □

Theorem 20.12. *Suppose f is uniformly bounded by $M < \infty$, that $K \in \mathcal{L}_2$, and $\|K\|_1 = 1$. At any continuity point x of f,*

$$\hat{f}_n(x) \stackrel{P}{\Rightarrow} f(x), \quad \text{provided} \quad h \to 0 \quad \text{and} \quad nh \to \infty.$$

Let $k_r = \frac{1}{r!}\int z^r K(z)dz$. By decomposing the MSE into variance and the squared bias, we get

$$E[\hat{f}_n(x) - f(x)]^2 \asymp \frac{f(x)}{nh}\|K\|_2^2 + h^{2r}|k_r f^{(r)}(x)|^2,$$

provided $K \in \mathcal{L}_2$ and f has $(r+1)$ continuous derivatives at x. Minimizing the right-hand side of the above expansion with respect to h, we have the asymptotically optimal local bandwidth as

$$h_{loc,opt} = \left[f(x)\|K\|_2^2\right]^{\frac{1}{2r+1}}\left[2nr|k_r f^{(r)}(x)|^2\right]^{-\frac{1}{2r+1}}$$

and on plugging this in, the convergence rate of the MSE is

$$MSE \sim n^{-2r/(2r+1)}.$$

We often assume that $r = 2$, in which case the pointwise convergence rate is $MSE = O(n^{-4/5})$, slower than the parametric n^{-1} rate. Note that it is expected that in a nonparametric setup, we cannot get the parametric $n^{-1/2}$ convergence rate. See DasGupta (2008) for the proofs.

20.3.4 Kernels for Statistical Classification

One of the main applications of kernels is in the problem of *classification*. Here, based on *training data* $(x_i, y_i), i = 1, 2, \ldots, n$, where x_i are observed values of a d-dimensional relevant variable X (often called *covariates*), and y_i denotes the *group membership* of the ith sampled unit, one needs to classify a future individual with a known x value, but an unknown group membership. That is, the future individual has to be *classified* as belonging to one of p possible groups, based on the past data $(x_i, y_i), i = 1, 2, \ldots, n$, and the present value $X = x$. This is done by designing a classification function or *classification rule* $\hat{y} = \phi_n(x_1, \ldots, x_n, y_1, \ldots, y_n, x)$, with ϕ_n taking values in the set $\{1, 2, \ldots, p\}$. For example, when there are only two possible groups so that $p = 2$, an intuitive rule is the *majority rule*, which takes a suitably small neighborhood U of the present value $X = x$, counts how many of the training values $x_1, \ldots, x_n$ that fall inside U correspond to members of group 1, and how many correspond to members of group 2, and assigns the new x value to whichever group has the majority. For example, if we choose our neighborhood U to be a ball of radius h centered at x, then the mathematical definition of the majority rule works out to

$$\phi_n(x_1, \ldots, x_n, y_1, \ldots, y_n, x) = 1 \text{ iff } \sum_{i=1}^{n} I_{\|x_i - x\| \leq h, y_i = 1} \geq \sum_{i=1}^{n} I_{\|x_i - x\| \leq h, y_i = 2}.$$

A natural extension is a *kernel classification rule*:

$$\phi_{n,K}(x_1, \ldots, x_n, y_1, \ldots, y_n, x) = 1 \text{ iff } \sum_{i=1}^{n} I_{y_i = 1} K\left(\frac{x_i - x}{h}\right)$$

$$\geq \sum_{i=1}^{n} I_{y_i = 2} K\left(\frac{x_i - x}{h}\right),$$

where $K : \mathcal{R}^d \to \mathcal{R}$ is a kernel. We usually choose K to be nonnegative, and perhaps to have some other properties, as described in our list of properties in the previous section. The kernel K essentially quantifies the similarity of a new point x to a data point x_i. The more similar x and x_i are, the larger will be the numerical value of $K(\frac{x_i - x}{h})$. The scaling constant h is usually called a *bandwidth*, and should be chosen to be appropriately small. Thus, kernels arise very naturally in classification problems. We want to address the question of how good are kernel classification rules and under what conditions.

Suppose F denotes the joint distribution of (X, Y), namely the covariate vector and the group membership variable. If we knew F, we could try to build classification rules based on our knowledge of F. Precisely, suppose $g : \mathcal{R}^d \to \{1, 2\}$ is a classification function that classifies an individual by using just the X value of that individual, and suppose $\alpha(F, g)$ is its error probability

$$\alpha(F, g) = P_F(g(X) \neq Y).$$

20.3 Kernels and Classification

The *oracular error probability* is

$$\underline{\alpha}(F) = \inf_g \alpha(F, g),$$

where the infimum is taken over all possible functions $g : \mathcal{R}^d \to \{1, 2\}$. The word *oracular* is supposed to convey the concept that only a person with oracular access to knowledge of F can find the rule g_0 that makes $\alpha(F, g_0) = \underline{\alpha}(F)$.

On the other hand, we have our real field classification rules $\phi_n(x_1, \ldots, x_n, y_1, \ldots, y_n, x)$, and they have their conditional error probabilities

$$\alpha_n(F, \phi_n) = P_F\Big(\phi_n \neq Y \mid (x_1, \ldots, x_n, y_1, \ldots, y_n)\Big).$$

Note that $\alpha_n(F, \phi_n)$ is a sequence of random variables. We would like $\alpha_n(F, \phi_n)$ to be close to $\underline{\alpha}(F)$ for large n; i.e., with a lot of training data, we would like to perform at the oracular level.

The following theorem (see pp 150-159 in Devroye, Györfi, and Lugosi (1996)) shows such an *oracle inequality*.

Theorem 20.13. *Suppose K is a nonnegative, bounded, and uniformly continuous kernel. Suppose*

$$h = h_n \to 0; \; nh^d \to \infty.$$

Then for any joint distribution F of (X, Y), the kernel classification rule $\phi_{n,K}$ satisfies

$$P_F(\alpha_n(F, \phi_n) - \underline{\alpha}(F) > \epsilon) \leq 2e^{-cn\epsilon^2},$$

where $c > 0$ depends only on the kernel K and the dimension d, but not on F.

An easy consequence (see the Chapter exercises) of this result is that $\alpha_n(F, \phi_n)$ converges almost surely to $\underline{\alpha}(F)$ whatever be F. In the classification literature, this is called **strong universal consistency**. The result illustrates the advantages of using kernels that have a certain amount of smoothness, and do not take negative values. The Gaussian kernel and several other hernels in our illustrative list satisfy the conditions of the above theorem, and assure oracular performance universally for all F.

20.3.4.1 Reproducing Kernel Hilbert Spaces

Reproducing kernels allow a potentially very difficult optimization problem on some suitable infinite dimensional function space into an ordinary finite dimensional optimization problem on an Euclidean space. For example, suppose a set of n data points $(x_i, y_i), i = 1, 2, \ldots, n$ are given to us, and we want to fit a suitable function $f(x)$ to these points. We can find exact fits by using rough or oscillatory functions f, such as polynomials of high degrees, or broken line segnents. If the fit is an exact fit, then the *prediction error* $\sum_{i=1}^n (y_i - f(x_i))^2$ will be zero. But the zero error is achieved by using an erratic curve $f(x)$. We could trade an inexact, but reasonable, fit for some smoothness in the form of the function $f(x)$. A mathematical formulation of such a constrained minimization problem is to minimize

$$\sum_{i=1}^{n} L(y_i, f(x_i)) + \lambda \Theta(f)$$

over some specified function space $\mathcal{F}$, where $L(y, f(x))$ is a *loss function* that measures the goodness of our fit, $\Theta(f)$ is a real valued functional that measures the roughness of f, and λ is a tuning constant that reflects the importance that we place on using a smooth function. For example, if we use $\lambda = 0$, then that means that all we care for is a good fit, and smoothness is of no importance to us. The loss function L is typically a function such as $(y - f(x))^2$ or $|y - f(x)|$, but could be more general. The roughness penalty functional $\Theta(f)$ is often something like $\int (f'(x))^2 dx$, although it too can be more general.

On the face of it, this is an infinite dimensional optimization problem, because the function space $\mathcal{F}$ would usually be infinite dimensional, unless we make the choice of functions too restrictive. Reproducing kernels allow us to transform such an infinite dimensional problem into finding a function of the form $\sum_{i=1}^{n} c_i K(x_i, x)$, where $K(x, x')$ is a *kernel*, associated in a unique way with the function space $\mathcal{F}$. So, as long as we can identify what this *reproducing kernel* $K(x, x')$ is, all we have to do to solve our original infinite dimensional optimization problem is to find the n optimal constants $c_1, \cdots, c_n$. Such a kernel $K(x, x')$ uniquely associated with the function space $\mathcal{F}$ can be found, as long as $\mathcal{F}$ has a nice amount of structure. The structure needed is that of a special kind of *Hilbert space*. Aronszajn (1950) is the original reference on the theory of reproducing kernels. We first provide a basic treatment of Hilbert spaces themselves; this is essential for studying reproducing kernels. Rudin (1986) is an excellent first exposition on Hilbert spaces.

Definition 20.6. A real vector space $\mathcal{H}$ is called an *inner product space* if there is a function $(x, y) : \mathcal{H} \otimes \mathcal{H} \to \mathcal{R}$ such that

(a) $(x, x) \geq 0$ for all $x \in \mathcal{H}$, with $(x, x) = 0$ if and only if $x = 0$, the null element of $\mathcal{H}$.

(b) $(a_1 x_1 + a_2 x_2, y) = a_1 (x_1, y) + a_2 (x_2, y)$ for all $x_1, x_2, y \in \mathcal{H}$, and all real numbers a_1, a_2.

The function (x, y) is called the *inner product* of x and y, and $\sqrt{(x, x)} = ||x||_\mathcal{H}$ the *norm* of x. The function $d(x, y) = ||x - y||_\mathcal{H}$ is called the *distance* between x and y.

Inner product spaces are the most geometrically natural generalizations of Euclidean spaces, because we can talk about length, distance, and angle on inner product spaces by considering $||x||_\mathcal{H}$, (x, y), and $d(x, y)$.

Definition 20.7. Let $\mathcal{H}$ be an inner product space. The *angle* between x and y is defined as $\theta = \arccos \frac{(x, y)}{||x||_\mathcal{H} ||y||_\mathcal{H}}$. x, y are called *orthogonal* if $(x, y) = 0$, so that $\theta = \frac{\pi}{2}$. For n given vectors $x_1, \cdots, x_n$, the linear span of $x_1, \cdots, x_n$ is defined to be

20.3 Kernels and Classification

the set of all $z \in \mathcal{H}$ of the form $z = \sum_{i=1}^{n} c_i x_i$, where $c_1, \cdots, c_n$ are arbitrary real constants. The *projection* of a given y onto the linear span of $x_1, \cdots, x_n$ is defined to be $P_{x_1, \cdots, x_n} y = \sum_{i=1}^{n} c_i^* x_i$, where $(c_1^*, \cdots, c_n^*) = \mathrm{argmin}_{c_1, \cdots, c_n} d(y, \sum_{i=1}^{n} c_i x_i)$.

Example 20.12. Some elementary examples of inner product spaces are

1. The n-dimensional Euclidean space $\mathcal{R}^n$ with (x, y) defined as $(x, y) = x'y = \sum_{i=1}^{n} x_i y_i$.
2. A more general inner product on $\mathcal{R}^n$ is $(x, y) = x'Ay$ where A is an $n \times n$ symmetric positive definite matrix.
3. Let $\mathcal{H} = L_2[a, b]$ be the class of square integrable functions on a bounded interval $[a, b]$,

$$L_2[a, b] = \left\{ f : [a, b] \to \mathcal{R}, \int_a^b f^2(x) dx < \infty \right\}.$$

Then $\mathcal{H}$ is an inner product space, with the inner product

$$(f, g) = \int_a^b f(x) g(x) dx.$$

In this, we identify any two functions f, g such that $f = g$ almost everywhere on $[a, b]$ as being equivalent. Thus, any function f such that $f = 0$ almost everywhere on $[a, b]$ will be called a zero function.

4. Consider again a bounded interval $[a, b]$, but consider $\mathcal{H} = C[a, b]$, the class of all continuous functions on $[a, b]$. This is an inner product space with the same inner product $(f, g) = \int_a^b f(x) g(x) dx$.

Some elementary facts about inner product spaces are summarized for reference in the next result.

Proposition. *Let $\mathcal{H}$ be an inner product space. Then,*

1. **(Joint Continuity of Inner Products).** *Suppose $x_n, y_n, x, y \in \mathcal{H}$, and $d(x_n, x), d(y_n, y)$ converge to 0 as $n \to \infty$. Then, $(x_n, y_n) \to (x, y)$ as $n \to \infty$.*
2. **(Cauchy-Schwarz Inequality)** $|(x, y)| \leq ||x||_{\mathcal{H}} ||y||_{\mathcal{H}}$ *for all $x, y \in \mathcal{H}$, with equality if and only if $y = cx$ for some constant c.*
3. **(Triangular Inequality)** $||x + y||_{\mathcal{H}} \leq ||x||_{\mathcal{H}} + ||y||_{\mathcal{H}}$ *for all $x, y \in \mathcal{H}$.*
4. $d(x, z) \leq d(x, y) + d(y, z)$ *for all $x, y, z \in \mathcal{H}$.*
5. **(Pythagorean Identity)** *If x, y are orthogonal, then $||x + y||_{\mathcal{H}}^2 = ||x||_{\mathcal{H}}^2 + ||y||_{\mathcal{H}}^2$.*
6. **(Parallelogram Law)** *If x, y are orthogonal, then $||x + y||_{\mathcal{H}}^2 + ||x - y||_{\mathcal{H}}^2 = 2 \left(||x||_{\mathcal{H}}^2 + ||y||_{\mathcal{H}}^2 \right)$.*
7. *The projection of $y \in \mathcal{H}$ onto the linear span of some fixed $x \neq 0 \in \mathcal{H}$ equals $P_x y = \frac{(x, y)}{||x||_{\mathcal{H}}^2} x$.*

A generalization of part (g) is given in Exercise 20.27. Inner product spaces that have the property of completeness are Hilbert spaces. The following example demonstrates what completeness is all about, and why it need not hold for arbitrary inner product spaces.

Example 20.13 (*An Incomplete Inner Product Space*). Consider $\mathcal{H} = C[0, 1]$ equipped with the inner product $(f, g) = \int_0^1 f(x)g(x)dx$. Consider the sequence of functions $f_n \in \mathcal{H}$ defined as follows:

$$f_n(x) = 0 \text{ for } x \in \left[0, \frac{1}{2}\right], \quad f_n(x) = 1 \text{ for } x \in \left[\frac{1}{2} + \frac{1}{n}, 1\right],$$

$$\text{and } f_n(x) = n\left(x - \frac{1}{2}\right) \text{ for } x \in \left(\frac{1}{2}, \frac{1}{2} + \frac{1}{n}\right).$$

Now choose $m, n \to \infty$, and suppose $m < n$. Then, it is clear that the graphs of f_m and f_n coincide except on $[\frac{1}{2}, \frac{1}{2} + \frac{1}{m}]$, and it follows easily that $d(f_m, f_n) = \left(\int_0^1 (f_m - f_n)^2\right)^{\frac{1}{2}} \to 0$. This means that the sequence $f_n \in \mathcal{H}$ is a Cauchy sequence. However, the sequence does not have a continuous limit, i.e., there is no function f, continuous on $[0, 1]$ such that $\int_0^1 (f_n - f)^2 \to 0$. So, here we have an inner product space which is not *complete*, in the sense that we can have sequences which are Cauchy with respect to the distance induced by the inner product, which nevertheless do not have a limit within the same inner product space. Spaces on which such Cauchy sequences cannot be found form Hilbert spaces.

Definition 20.8. Let $\mathcal{H}$ be an inner product space such that every Cauchy sequence $x_n \in \mathcal{H}$ converges to some $x \in \mathcal{H}$. Then $\mathcal{H}$ is called a *Hilbert space* with the *Hilbert norm* $||x||_\mathcal{H} = \sqrt{(x, x)}$.

Among the L_p spaces, $L_2[a, b]$ is a Hilbert space; it has an inner product, and the inner product is complete. Besides L_2, the other L_p spaces are not Hilbert spaces, because their norms $||f||_\mathcal{H} = ||f||_p = \left(\int |f|^p\right)^{1/p}$ are not induced by an inner product. We already saw that $C[a, b]$ is not a Hilbert space, because the inner product is not complete. The real line with the usual inner product, however, is a Hilbert space, because a standard theorem in real analysis says that all Cauchy sequences must converge to some real number. In fact, essentially the same proof shows that all finite dimensional Euclidean spaces are Hilbert spaces.

In the finite dimensional Euclidean space $\mathcal{R}^n$, the standard unit vectors $e_k = (0, \cdots, 0, 1, 0, \cdots, 0), k = 1, 2, \cdots, n$ form an orthonomal basis, in the sense that the set $\{e_k\}$ is an orthonormal set of n-vectors, and any $x \in \mathcal{R}^n$ may be represented as $x = \sum_{k=1}^n c_k e_k$, where c_k are real constants. Furthermore, $||x||^2 = \sum_{k=1}^n c_k^2$. Hilbert spaces, in general, are infinite-dimensional, and an orthonormal basis would not be a finite set in general. In spite of this, a representation akin to the finite-dimensional $\mathcal{R}^n$ case exists, but there are a few subtle elements of differences. The exact result is the following.

Proposition (*Orthonormal Bases and Parseval's Identity*). *Let $\mathcal{H}$ be a Hilbert space with the inner product (x, y). Then,*

20.3 Kernels and Classification

(a) $\mathcal{H}$ has an orthonormal basis B, that is, a set of vectors $\{e_\alpha\}$ of $\mathcal{H}$ such that $||e_\alpha||_\mathcal{H} = 1$, $(e_\alpha, e_\beta) = 0$ for all $\alpha, \beta, \alpha \neq \beta$, and the linear span of the vectors in B is dense in $\mathcal{H}$ with respect to the Hilbert norm on $\mathcal{H}$.
(b) Given any $x \in \mathcal{H}$, at most countably many among (x, e_α) are not equal to zero, and x may be represented in the form $x = \sum_\alpha (x, e_\alpha) e_\alpha$.
(c) $||x||_\mathcal{H}^2 = \sum_\alpha |(x, e_\alpha)|^2$.

See Rudin (1986) for a proof. It may be shown that all orthonormal bases of a Hilbert space $\mathcal{H}$ have the same cardinality, which is called the dimension of $\mathcal{H}$.

With these preliminaries, we can now proceed to the topic of reproducing kernels. We need a key theorem about Hilbert spaces that plays a central role in the entire concept of a reproducing kernel Hilbert space. This theorem, a classic in analysis, gives a representation of *continuous linear functionals* on a Hilbert space. For completeness, we first define what is meant by a continuous linear functional.

Definition 20.9. Let $\mathcal{H}$ be a Hilbert space and $\delta : \mathcal{H} \to \mathcal{R}$ a real-valued *linear functional or operator* on $\mathcal{H}$; that is, $\delta(ax + by) = a\delta(x) + b\delta(y)$ for all $x, y \in \mathcal{H}$ and all real constants a, b. The *norm or operator norm* of δ is defined to be $||\delta|| = \sup_{x \in \mathcal{H}} \frac{|\delta(x)|}{||x||_\mathcal{H}}$.

Definition 20.10. Let $\mathcal{H}$ be a Hilbert space and $\delta : \mathcal{H} \to \mathcal{R}$ a linear functional on $\mathcal{H}$. The functional δ is called continuous if $x_n, x \in \mathcal{H}, d(x_n, x) \to 0 \Rightarrow \delta(x_n) \to \delta(x)$.

In general, linear functionals need not be continuous. But they are if they have a *finite operator norm*. The following extremely important result says that these two properties of continuity and *boundedness* are really the same.

Theorem 20.14. *Let $\mathcal{H}$ be a Hilbert space and $\delta : \mathcal{H} \to \mathcal{R}$ a linear functional. Then δ is continuous if and only if it is bounded; that is, $||\delta|| < \infty$. Equivalently, a linear operator is continuous if and only if there exists a finite real constant c such that $|\delta(x)| \leq c ||x||_\mathcal{H}$ for all $x \in \mathcal{H}$.*

Proof. For the "if" part, suppose δ is a bounded operator. Take $x_n \in \mathcal{H} \to 0$ (the null element). Then, by definition of operator norm,

$$|\delta(x_n)| \leq ||\delta|| \, ||x_n||_\mathcal{H} \to 0,$$

because $||\delta|| < \infty$. This proves that δ is continuous at 0, and therefore continuous everywhere by linearity. Conversely, if $||\delta|| = \infty$, find a sequence $x_n \in \mathcal{H}$ such that $||x_n||_\mathcal{H} \leq 1$, but $|\delta(x_n)| \to \infty$. This is possible, because $||\delta||$ is easily shown to be equal to $\sup\{|\delta(x)| : ||||x||_\mathcal{H} \leq 1\}$. Now define $z_n = \frac{x_n}{|\delta(x_n)|}, n \geq 1$. Then $z_n \to 0$, but $\delta(z_n)$ does not go to zero, because $|\delta(z_n)|$ is equal to 1 for all n. This proves the "only if" part of the theorem.

Here is the classic representation theorem for continuous linear functionals on a Hilbert space that we promised. The theorem says that any continuous linear functional on a Hilbert space $\mathcal{H}$ can be recovered as an inner product with a fixed element of $\mathcal{H}$, associated in a one-to-one way with the particular continuous functional.

Theorem 20.15 (Riesz Representation Theorem). *Let $\mathcal{H}$ be a Hilbert space and $\delta : \mathcal{H} \to \mathcal{R}$ a continuous linear functional on $\mathcal{H}$. Then there exists a unique $v \in \mathcal{H}$ such that $\delta(u) = (u, v)$ for all $u \in \mathcal{H}$.*

See Rudin (1986) for a proof.

We use a special name for Hilbert spaces whose elements are functions on some space. This terminology is useful when discussing reproducing kernel Hilbert spaces below.

Definition 20.11. Let $\mathcal{X}$ be a set and $\mathcal{H}$ a class of real-valued functions f on $\mathcal{X}$. If $\mathcal{H}$ is a Hilbert space, it is called a *Hilbert Function Space*.

Here is the result that leads to the entire topic of reproducing kernel Hilbert spaces.

Theorem 20.16. *Let $\mathcal{H}$ be a Hilbert function space. Consider the* point evaluation operators *defined by*

$$\delta_x(f) = f(x), \quad f \in \mathcal{H}, x \in \mathcal{X}.$$

If the operators $\delta_x(f) : \mathcal{H} \to \mathcal{R}$ are all continuous, then for each $x \in \mathcal{X}$, there is a unique element K_x of $\mathcal{H}$ such that $f(x) = \delta_x(f) = (f(.), K_x(.))$.

The proof is a direct consequence of the Riesz representation theorem, because the operators δ_x are linear operators. We can colloquially characterize this theorem as saying that the original functions $f(x)$ can be recovered by taking the inner product of f itself with a unique kernel function K_x. For example, if the inner product on our relevant Hilbert space $\mathcal{H}$ was an integral, namely, $(f, g) = \int_{\mathcal{X}} f(y)g(y)dy$, then our theorem above says that we can recover each function f in our function space $\mathcal{H}$ in the very special form

$$f(x) = \int_{\mathcal{X}} f(y) K_x(y) dy.$$

Writing $K_x(y) = K(x, y)$, we get the more conventional notation

$$f(x) = \int_{\mathcal{X}} f(y) K(x, y) dy.$$

It is common to call $K(x, y)$ the *reproducing kernel* of the function space $\mathcal{H}$. Note that in a deviation with how kernels were defined in the context of smoothing, the reproducing kernel is a function on $\mathcal{X} \otimes \mathcal{X}$. It is also important to understand that not all Hilbert function spaces possess a reproducing kernel. The point evaluation operators must be continuous for the function space to possess a reproducing kernel. In what follows, more is said of this, and about which kernels can at all be a reproducing kernel of some Hilbert function space.

Some basic properties of a reproducing kernel are given below.

Proposition. *Let $K(x, y)$ be the reproducing kernel of a Hilbert function space $\mathcal{H}$. Then,*

(a) $K(x, y) = (K_x, K_y)$ *for all $x, y \in \mathcal{X}$.*
(b) K *is symmetric; that is, $K(x, y) = K(y, x)$ for all $x, y \in \mathcal{X}$.*

20.3 Kernels and Classification

(c) $K(x, x) \geq 0$ for all $x \in \mathcal{X}$.
(d) $(K(x, y))^2 \leq K(x, x)K(y, y)$ for all $x, y \in \mathcal{X}$.

Proof of each part of this result is simple. Part (a) follows from the fact that $K_x \in \mathcal{H}$, and hence, $K_x(y) = K(x, y) = (K_x, K_y)$ by definition of a reproducing kernel. Part (b) follows from part (a) because (real) inner products are symmetric. Part (c) follows on noting that $K(x, x) = (K_x, K_x) = \|K_x\|_\mathcal{H}^2 \geq 0$, and part (d) follows from the Cauchy–Schwarz inequality for inner product spaces. □

It now turns out that the question of which real-valued functions $K(x, y)$ on $\mathcal{X} \otimes \mathcal{X}$ can act as a reproducing kernel of some Hilbert function space has an intimate connection to the question of which functions can be the covariance kernel of a one-dimensional Gaussian process $X(t)$, where the time parameter t runs through the set $\mathcal{X}$. In the special case where $\mathcal{X}$ is a subset of some Euclidean space $\mathcal{R}^n$, and the function $K(x, y)$ in question is of the form $K(x, y) = \psi(\|x - y\|)$, it moreover turns out that characterizing which functions can be reproducing kernels is intimately connected to characterizing which functions $\psi(t)$ on $\mathcal{R}$ can be the characteristic function (see Chapter 8) of a probability distribution on $\mathcal{R}$. We are now beginning to see that a question purely in the domain of analysis is connected to classic questions in probability theory. To describe the main characterization theorem, we need a definition.

Definition 20.12. A symmetric real-valued function $K : \mathcal{X} \otimes \mathcal{X} \to \mathcal{R}$ is called *positive-definite* if for each $n \geq 1, x_1, \ldots x_n \in \mathcal{X}$, and real constants $a_1, \ldots, a_n$, we have $\sum_{i=1}^n \sum_{j=1}^n a_i a_j K(x_i, x_j) \geq 0$. The function K is called *strictly positive-definite* if $\sum_{i=1}^n \sum_{j=1}^n a_i a_j K(x_i, x_j) > 0$ unless $a_i = 0$ for all i or $x_1, \cdots, x_n$ are identical.

So, K is positive-definite if matrices of the form

$$\left(\left(K(x_i, x_j)\right)\right)\Big|_{i,j=1}^n$$

are nonnegative-definite. Before giving examples of positive-definite functions, we point out what exactly the connection is between positive-definite functions and reproducing kernels of Hilbert function spaces.

Theorem 20.17. *Let $\mathcal{X}$ be a set and $K(x, y)$ a positive-definite function on $\mathcal{X} \otimes \mathcal{X}$. Then $K(x, y)$ is the unique reproducing kernel of some Hilbert function space $\mathcal{H}$ on $\mathcal{X}$. Conversely, suppose $\mathcal{H}$ is a Hilbert function space on a set $\mathcal{X}$ that has some kernel $K(x, y)$ as its reproducing kernel. Then $K(x, y)$ must be a positive-definite function on $\mathcal{X} \otimes \mathcal{X}$.*

A proof can be seen in Cheney and Light (2000, pp. 233–234), or in Berlinet and Thomas-Agnan (2004, p. 22).

20.3.5 Mercer's Theorem and Feature Maps

Theorem 20.15 says that there is a one-to-one correspondence between positive-definite functions on a set $\mathcal{X}$ and Hilbert function spaces on $\mathcal{X}$ that possess a reproducing kernel. The question arises as to how one finds positive-definite functions on some set $\mathcal{X}$, or verifies that a given function is indeed positive-definite. The definition of a positive-definite function is not the most efficient or practical method for verifying positive-definiteness of a function. Here is a clean and practical result.

Theorem 20.18. *A symmetric function $K(x, y)$ on $\mathcal{X} \otimes \mathcal{X}$ is the reproducing kernel of a Hilbert function space if and only if there is a family of maps $\phi(x), x \in \mathcal{X}$ with range space $\mathcal{F}$, where $\mathcal{F}$ is an inner product space, such that $K(x, y) = (\phi(x), \phi(y))_\mathcal{F}$, where the notation $(u, v)_\mathcal{F}$ denotes the inner product of the inner product space $\mathcal{F}$.*

The "if" part of the theorem is trivial, simply using the characterization of reproducing kernels as positive-definite functions. The "only if" part is nontrivial; see Aizerman, Braverman and Rozonoer (1964). A version of this result especially suited for L_2 spaces of functions on an Euclidean space is known as Mercer's theorem. It gives a constructive method for finding these maps $\phi(x)$ that is akin to finding the Gramian matrix of a finite-dimensional nonnegative-definite matrix. The maps $\phi(x)$ corresponding to a given kernel are called the feature maps, and the space $\mathcal{F}$ is called the feature space. The feature space $\mathcal{F}$ may, in applications, turn out to be much higher-dimensional than $\mathcal{X}$, in the case that $\mathcal{X}$ is a finite-dimensional space, for example, some Euclidean space. Still, this theorem gives us the flexibility to play with various feature maps and thereby choose a suitable kernel function, which would then correspond to a suitable reproducing kernel Hilbert space.

We now present two illustrative examples.

Example 20.14. Suppose $\mathcal{X} = [0, 1]$, $\phi_x(t) = \cos(xt)$, and $\mathcal{F} = L_2[-1, 1]$. Then, we get the kernel $K(x, y)$ on $\mathcal{X} \otimes \mathcal{X} = [0, 1] \otimes [0, 1]$ given by

$$K(x, y) = (\phi_x, \phi_y) = \int_{-1}^{1} \cos(xt) \cos(yt) dt$$

$$= \frac{\sin(x-y)}{x-y} + \frac{\sin(x+y)}{x+y},$$

where $\frac{\sin 0}{0}$ is interpreted as the limit $\lim_{z \to 0} \frac{\sin z}{z} = 1$. Thus, $K(x, x) = 1 + \frac{\sin(2x)}{2x}$ if $x \neq 0$, and $K(0, 0) = 2$. By the characterization theorem, Theorem 20.17, this is a reproducing kernel of a Hilbert space of functions on $\mathcal{X} = [0, 1]$.

Example 20.15. Suppose $\mathcal{X} = \mathcal{R}^n$ for some $n \geq 1$. Consider the kernel $K(x, y) = (x'y)^2$, where $x'y$ denotes the usual Euclidean inner product $x_1 y_1 + x_2 y_2 + \cdots + x_n y_n$. We find the feature maps corresponding to the positive-definite kernel K. Define the maps

$$\phi(x) = \left(x_1^2, x_1 x_2, \ldots, x_1 x_n, \ldots, x_n x_1, x_n x_2, \ldots, x_n^2\right),$$

20.3 Kernels and Classification

that is, the n^2-dimensional Euclidean vector with coordinates $x_i x_j$, $1 \leq i, j \leq n$. We are going to look at $\phi(x)$ as an element of $\mathcal{F} = \mathcal{R}^{n^2}$. Then,

$$(\phi(x), \phi(y))_{\mathcal{F}} = \sum_{k=1}^{n^2} (\phi(x))_k (\phi(y))_k = \sum_{i=1}^{n} \sum_{j=1}^{n} (x_i x_j)(y_i y_j)$$

$$= \sum_{i=1}^{n} \sum_{j=1}^{n} (x_i y_i)(x_j y_j) = \left(\sum_{i=1}^{n} x_i y_i\right)\left(\sum_{j=1}^{n} x_j y_j\right)$$

$$= (x'y)^2.$$

Therefore, the maps $\phi(x)$ are the feature maps corresponding to K.

We end this section with a statement of Mercer's theorem (Mercer (1909)). Extensions of Mercer's theorem to much more abstract spaces are now available; see Berlinet and Thomas-Agnan (2004).

Theorem 20.19. *Suppose $\mathcal{X}$ is a closed subset of a finite-dimensional Euclidean space, and μ a σ-finite measure on $\mathcal{X}$. Let $K(x, y)$ be a symmetric positive-definite function (i.e., a reproducing kernel) on $\mathcal{X} \otimes \mathcal{X}$, and suppose that K is square integrable in the sense $\int_{\mathcal{X}} \int_{\mathcal{X}} K^2(x, y) d\mu(x) d\mu(y) < \infty$. Define the linear operator $A_K(f) : L_2(X, \mu) \to L_2(X, \mu)$*

$$A_K(f)(y) = \int_{\mathcal{X}} K(x, y) f(x) d\mu(x), \quad y \in \mathcal{X}.$$

Then,

(a) *The operator A_K has a countable number of nonnegative eigenvalues $\lambda_i, i \geq 1$, and a corresponding sequence of mutually orthonormal eigenfunctions ψ_i, $i \geq 1$, satisfying $A_K(\psi_i) = \lambda_i \psi_i, i \geq 1$.*
(b) *The kernel K admits the representation*

$$K(x, y) = \sum_{i=1}^{\infty} \phi_i(x) \phi_i(y),$$

where $\phi_i(x) = \sqrt{\lambda_i} \psi_i(x), i \geq 1$.

The theorem covers the two most practically important cases of $\mathcal{X}$ being a rectangle (possibly unbounded) in a finite-dimensional Euclidean space with μ as Lebesgue measure, and $\mathcal{X}$ being a finite set in a finite-dimensional Euclidean space with μ as the counting measure. The success of Mercer's theorem in explicitly identifying the feature map

$$\phi(x) = \Big(\phi_i(x)\Big)_{i=1}^{\infty},$$

depends on our ability to find the eigenfunctions and the eigenvalues of the linear operator A_K. In some cases, we can find them explicitly, and in some cases, we are out of luck. It is worth noting that the linear operator A_K has certain additional properties (compactness in particular,) which allows the representation as in Mercer's theorem to hold; see, for example, Theorem 1 on p. 93 in Cheney (2001) for conditions needed for the spectral decomposition of an operator on a Hilbert space. Minh, Niyogi and Yao (2006) give some very nice examples of the calculation of λ_i and ψ_i in Mercer's theorem when the input space is the surface of a sphere in some finite-dimensional Euclidean space, or a discrete set in a finite-dimensional Euclidean space.

Theorem 20.15 says that symmetric positive-definite functions and reproducing kernel Hilbert spaces are in a one-to-one relationship. If we produce a symmetric positive-definite function, it will correspond to a suitable Hilbert space of functions with an inner product, and an induced norm. For the sake of applications, it is useful to know this correspondence for some special kernels. A list of these correspondences is given below for practical use.

$\mathcal{X}$	$K(x, y)$	$\mathcal{H}$	$\|f\|_H$
$[a, b]$	$e^{-\alpha\|x-y\|}$	$\{u : u' \in L_2[a,b]\}$	$\frac{1}{2}\left[u^2(a) + u^2(b) + \alpha^{-1}\int_a^b \left[(u')^2 + \alpha^2 u^2\right]\right]$
$[0, b]$	$e^{-x}\sinh y$	$\{u : u(0) = 0, u' \in L_2[0,b]\}$	$\int_0^b (u' + u)^2$
$[0, 1]$	$(1-x)(1-y) + xy + (x-y)_+^3 - \frac{x}{6}(1-y)(x^2 - 2y + y^2)$	$\{u : u'' \in L_2[0,1]\}$	$u^2(0) + u^2(1) + \int_0^1 (u'')^2$
$\mathcal{R}$	$\frac{\sin(M(x-y))}{M(x-y)}$	functions $u \in L_2(\mathcal{R})$ with $\int_{\mathcal{R}} e^{itx} u(x) dx = 0$ for $\|t\| > M$	usual L_2 norm
$[0, 1]$	$1 + \frac{(-1)^{m-1}}{(2m)!} B_{2m}(\|x-y\|)$	$\{u : u^{(m)} \in L_2[0,1], u^{(j)}(0) = u^{(j)}(1) = 0 \; \forall \, j < m\}$	$(\int_0^1 u)^2 + \int_0^1 (u^{(m)})^2$

Note: Above, B_j denotes the jth Bernoulli polynomial. For example, $B_2(x) = x^2 - x + \frac{1}{6}$, $B_4(x) = x^4 - 2x^3 + x^2 - \frac{1}{30}$.

20.3.5.1 Support Vector Machines

Let us now return to the two group statistical classification problem. Suppose the covariate vector X is a d-dimensional multivariate normal under each group, namely, $X \mid Y = 1 \sim N_d(\mu_1, \Sigma), X \mid Y = 2 \sim N_d(\mu_2, \Sigma)$. Suppose also that $P(Y = 1) = p, P(Y = 2) = 1 - p$. The marginal distribution of Y and the conditional distribution of X given Y determine the joint distribution F of (X, Y). A classic result is that the misclassification probability $P_F(g(X) \neq Y)$ is minimized by a linear classification rule that classifies a given X value into group 1 (i.e., sets $g(X) = 1$) if $c'x \geq b$ for a suitable vector c and a suitable real

20.3 Kernels and Classification

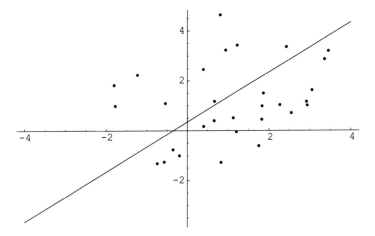

Fig. 20.3 Linearly separable data cloud

constant b. In the case that μ_1, μ_2 and Σ are known to the user, the vector c has the formula $c = (\mu_1 - \mu_2)'\Sigma^{-1}$. Usually, these mean vectors and the covariance matrix are unknown to the user, in which case c is estimated by using training data $(x_i, y_i), i = 1, 2, \ldots, n$.

This is the historically famous *Fisher linear classification rule*. Geometrically, the Fisher classification rule takes a suitable hyperplane in $\mathcal{R}^d$, and classifies X values on one side of the hyperplane to the first group and X values on the other side of the hyperplane to the second group. For Gaussian data with identical or nearly identical covariance structures, this idea of *linear separation* works quite well. See Fig. 20.3. However, linear separability is too optimistic for many kinds of data; for example, even in the Gaussian case itself, if the covariance structure of X is different under the two groups, then linear separation is not going to work well. On the other hand, linear separation has some advantages.

(a) A linear classification rule is easier to compute.
(b) A linear rule has geometric appeal.
(c) It may be easier to study operating characteristics, such as misclassification probabilities, of linear rules.

An appealing idea is to map the original input vector X into a feature space, say some Euclidean space $\mathcal{R}^D$, by using a feature map $\phi(X)$, and use a *linear rule* in the feature space. That is, use classification rules that classify an X value by using a classification function of the form $\sum_{i=1}^{D} c_i (\Phi(x_i))' (\Phi(X))$. As we remarked, the dimension of the feature space may be much higher, and in fact the feature space may even be infinite dimensional, in which case computing these inner products $(\Phi(x_i))' (\Phi(X))$ is going to be time consuming. However, now, our previous discussion of the theory of reproducing kernel Hilbert spaces is going to help us in avoiding computation of these very high-dimensional inner products. As we saw in Theorem 20.17, every symmetric positive-definite function $K(x, y)$ on the product space $\mathcal{X} \otimes \mathcal{X}$ arises as such an inner product, and vice versa. Hence, we can just directly choose a kernel function $K(x, y)$, and use a classification rule of the form $\sum_{i=1}^{n} c_i K(x, x_i)$. The feature maps are not directly used, but provide the necessary motivation and intuition for using a rule of the form $\sum_{i=1}^{n} c_i K(x, x_i)$. To choose the particular kernel, we can use a known collection of kernels, such as $K(x, y) = K_0(x - y)$, where K_0 is as in our table of kernels in Section 20.3.2, or generate new kernels.

The kernel function should act as a similarity measure. If two points x, y in the input space are *similar*, then $K(x, y)$ should be relatively large. The more dissimilar x, y are, the smaller should be $K(x, y)$. For example, if $\mathcal{X}$ is an Euclidean space, and we select a kernel of the form $K(x, y) = K_0(||x - y||)$, then $K_0(t)$ should be a nonincreasing function of t. A few common kernels in the classification and machine learning literature are the following.

(a) **Gaussian Radial Kernel.** $K(x, y) = e^{-c^2 ||x-y||^2}$.
(b) **Polynomial Kernel.** $K(x, y) = (x'y + c)^d$.
(c) **Exponential Kernel.** $K(x, y) = e^{-\alpha ||x-y||}$, $\alpha > 0$.
(d) **Sigmoid Kernel.** $K(x, y) = \tanh(cx'y + d)$.
(e) **Inverse Quadratic Kernel.** $K(x, y) = \frac{1}{\alpha^2 ||x-y||^2 + 1}$.

We close with a remark on why support vector classification appears to be more promising than other existing methods. Strictly linear separation on the input space $\mathcal{X}$ itself is clearly untenable from practical experience. But, figuring out an appropriate nonlinear classifier from the data is first of all difficult, secondly ad hoc, and thirdly will have small generalizability. Support vector machines, indirectly, map the data into another space, and then select an optimal linear classifier in that space. This is a well-posed problem. Furthermore, the support vector approach takes away the overriding emphasis on good fitting for the obtained data to place some emphasis on generalizability. More precisely, one could pose the classification problem in terms of minimizing the *empirical risk* $\frac{1}{n} \sum_{i=1}^{n} L(y_i, g(x_i))$ in some class $\mathcal{G}$ of classifiers g. However, this tends to put the boundary of the selected classifier too close to one of the two groups, and on a future data set, the good risk performance would not generalize. Support vector machines, in contrast, also place some emphasis on placing the boundary at a good geographical margin from the data cloud in each group. The risk, therefore, has an element of empirical risk minimization, and also an element of regularization. The mathematics of the support vector machines approach also shows that the optimal classifier is ultimately chosen by a few influential data points, which are the so-called *support vectors*. Vapnik and Chervonenkis (1964), Vapnik (1995), and Cristianini and Shawe-Taylor (2000) are major references on the support vector machine approach to learning from data for classification.

Exercises

Exercise 20.1 (Simple Practical Bootstrap). For $n = 10, 30, 50$ take a random sample from an $N(0, 1)$ distribution, and bootstrap the sample mean $\bar{X}$ using a bootstrap Monte Carlo size $B = 500$. Construct a histogram and superimpose on it the exact density of $\bar{X}$. Compare.

Exercise 20.2. For $n = 5, 20, 50$, take a random sample from an Exp(1) density, and bootstrap the sample mean $\bar{X}$ using a bootstrap Monte Carlo size $B = 500$. Construct the corresponding histogram and superimpose it on the exact density. Compare.

Exercise 20.3 * **(Bootstrapping a Complicated Statistic).** For $n = 15, 30, 60$, take a random sample from an $N(0, 1)$ distribution, and bootstrap the sample kurtosis coefficient using a bootstrap Monte Carlo size $B = 500$. Next, find the approximate normal distribution obtained from the delta theorem, and superimpose the bootstrap histogram on this approximate normal density. Compare.

Exercise 20.4. For $n = 20, 40, 75$, take a random sample from the standard Cauchy distribution, and bootstrap the sample median using a bootstrap Monte Carlo size $B = 500$. Next, find the approximate normal distribution obtained from Theorem 9.1 and superimpose the bootstrap histogram on this approximate normal density. Compare.

Exercise 20.5 * **(Bootstrapping in an Unusual Situation).** For $n = 20, 40, 75$, take a random sample from the standard Cauchy distribution, and bootstrap the t-statistic $\frac{\sqrt{n}\bar{X}}{s}$ using a bootstrap Monte Carlo size $B = 500$. Plot the bootstrap histogram. What special features do you notice in this histogram? In particular, comment on whether the true density appears to be unimodal, or bounded.

Exercise 20.6 * **(Bootstrap Variance Estimate).** For $n = 15, 30, 50$, find a bootstrap estimate of the variance of the sample median for a sample from a Beta(3, 3) density. Use a bootstrap Monte Carlo size $B = 500$.

Exercise 20.7. For $n = 10, 20, 40$, find a bootstrap estimate of the variance of the sample mean for a sample from a Beta(3, 3) density. Use a bootstrap Monte Carlo size $B = 500$. Compare with the known exact value of the variance of the sample mean.

Exercise 20.8 (Comparing Bootstrap with an Exact Answer). For $n = 15, 30, 50$, find a bootstrap estimate of the probability $P(\frac{\sqrt{n}\bar{X}}{s} \leq 1)$ for samples from a standard normal distribution. Use a bootstrap Monte Carlo size $B = 500$. Compare the bootstrap estimate with the exact value of this probability. (Why is the exact value easily computable?)

Exercise 20.9. * Prove that under appropriate moment conditions, the bootstrap is consistent for the sample correlation coefficient r between two jointly distributed variables X, Y.

Exercise 20.10 * (Conceptual). Give an example of
(a) A density such that the bootstrap is not consistent for the mean.
(b) A density such that the bootstrap is consistent, but not second-order accurate for the mean.

Exercise 20.11 * (Conceptual). In which of the following cases, can you use the canonical bootstrap justifiably?

(a) Approximating the distribution of the largest order statistic of a sample from a Beta distribution;
(b) Approximating the distribution of the median of a sample from a Beta distribution;
(c) Approximating the distribution of the maximum likelihood estimate of $P(X_1 \leq \lambda)$ for an exponential density with mean λ;
(d) Approximating the distribution of $\frac{\frac{1}{n}\sum_{i=1}^{n}(X_i - M_X)(Y_i - M_Y)}{s_X s_Y}$, where (X_i, Y_i) are independent samples from a bivariate normal distribution, M_X and s_X are the median and the standard deviation of the X_i values, and M_Y and s_Y are the median and the standard deviation of the Y_i values;
(e) Approximating the distribution of the sample mean for a sample from a t-distribution with two degrees of freedom.

Exercise 20.12. * Suppose $\bar{X}_n$ is the sample mean of an iid sample from a CDF F with a finite variance, and $\bar{X}_n^*$ is the mean of a bootstrap sample. Consistency of the bootstrap is a statement about the bootstrap distribution, conditional on the observed data. What can you say about the unconditional limit distribution of $\sqrt{n}(\bar{X}_n^* - \mu)$, where μ is the mean of F?

Exercise 20.13 (EM for Truncated Geometric). Suppose $X_1, \ldots, X_n$ are iid Geo(p), but the value of X_i reported only if it is ≤ 4. Explicitly derive the E-step of the EM algorithm for finding the MLE of p.

Exercise 20.14 (EM in a Background plus Signal Model).

(a) Suppose $Y = U + V$, where U, V are independent Poisson variables with mean λ and c, respectively, where c is known and λ is unknown. Only Y is observed. Design an EM algotithm for estimating λ, and describe the E-step and the M-step.
(b) Generalize part (a) to the case of independent replications $Y_i = U_i + V_i, 1 \leq i \leq n, U_i$ with common mean λ and V_i with common mean c.

Exercise 20.15 (EM with Data). The following are observations from a bivariate normal distribution, with $*$ indicating a missing value. Use the derivations in Example 20.9 to find the first six EM iterates for the MLE of the vector of five parameters. Comment on how close to convergence you are.

Data: $(0, .1), (0, -1), (1, *), (2, *), (.5, .75), (*, -3), (*, 2), (.2, -.2)$.

Exercises

Exercise 20.16 (Conceptual). Consider again the bivariate normal problem with missing values, except now there are no complete observations:

Data: $(0, *), (2, *), (*, -3), (*, 2), (.5, *), (-.6, *), (*, 0), (*, 1)$.
Is the EM algorithm useful now? Explain your answer.

Exercise 20.17 (EM for t-Distributions).

(a) Suppose $Y \sim t(\mu, m)$ with the density

$$\frac{c_m}{\left(1 + \frac{(x-\mu)^2}{m}\right)^{\frac{m+1}{2}}},$$

where c_m is the normalizing constant. Show that Y can be represented as

$$Y \mid Z = z \sim N\left(\mu, \frac{m}{z}\right), \sim Z \sim \chi^2(m).$$

(See Chapter 4).

(b) Show that under a general μ', $Z \mid Y = y \sim \frac{W}{m+(y-\mu')^2}$, where $W \sim \chi^2(m+1)$.

(c) Use this to derive an expression for $E_{\mu'}(Z \mid Y = y)$ under a general μ'.

(d) Design an EM algorithm for estimating μ by writing the complete data as $X_i = (Y_i, Z_i), i = 1, \ldots, n$.

(e) Write the complete data likelihood, and derive the E-step explicitly, by using the result in parts (a) and (b).

Exercise 20.18 (EM in a Genetics Problem). Consider the ABO blood group problem worked out in Example 20.10. For the data values $Y_A = 182, Y_B = 60, Y_{AB} = 17, Y_O = 176$ (McLachlan and Krishnan (2008)), find the first four EM iterates, using the starting values $(p_A, p_B, p_O) = (.264, .093, .643)$.

Exercise 20.19 (EM in a Mixture Problem). Suppose for some given $n \geq 1$, $X_i = (Y_i, Z_i)$, where Z_i takes values $1, 2, \ldots, p$ with probabilities $\pi_1, \pi_2, \ldots, \pi_p$. Conditional on $Z_i = j$, $Y_i \sim f_j(y \mid \theta)$. As usual, $X_1, \ldots, X_n$ are assumed to be independent. We only get to observe $Y_1, \ldots, Y_n$, but not the group memberships $Z_1, \ldots, Z_n$.

(a) Write with justification the complete data likelihood in the form $\prod_{j=1}^{p} \left(\pi_j^{n_j} \prod_{i:Z_i=j} f_j(y_i \mid \theta)\right)$, where n_j is the number of Z_i equal to j.

(b) By using Bayes' theorem, derive an expression for $P_{\theta'}(Z_i = j \mid y_i)$, for a general θ'.

(c) Use this to find $\hat{L}_k(\theta, y)$, where $y = (y_1, \ldots, y_n)$.

(d) Hence, complete the E-step.

(e) Suppose that the component densities f_j are $N(\mu_j, \sigma_j^2)$. Show how to complete the M-step, first for $\pi_1, \ldots, \pi_p$, and then, for $\mu_j, \sigma_j^2, j = 1, 2, \ldots, p$.

Exercise 20.20 (EM in a Censoring Situation). Let $X_i, i = 1, 2, \ldots, n$ be iid from a density $f_\theta(x)$. Let T be a fixed censoring time, and let $U_i = \min(X_i, T)$, $V_i = I_{X_i \leq T}$, $Y_i = (U_i, V_i)$. Suppose that we only get to observe Y_i, but not the X_i.

(a) Write a general expression for the conditional distribution of X_i given Y_i under a general θ'.
(b) Now suppose that $f_\theta(x)$ is the exponential density with mean θ. Use your representation of part (a) to derive an expression for $E_{\theta'}(X_i \mid U_i, V_i = v)$, where $v = 0, 1$.
(c) Hence complete the E-step of an EM algorithm for estimating θ.

Exercise 20.21 (Plug-In Density Estimate). Suppose the true model is a $N(\theta, \sigma^2)$ and a parametric plug-in estimate using MLEs of the parameters is used. Derive an expression for the global error index $E_f \int [f(x) - \hat{f}(x)]^2 dx$. At what rate does this converge to zero?

Exercise 20.22 (Choosing the Wrong Model). Suppose the true model is a double exponential location parameter density, but you thought it was a $N(\theta, 1)$ and used a parametric plug-in estimate with an MLE. Does $E_f \int |f(x) - \hat{f}(x)| dx$ converge to zero?

Exercise 20.23 (Trying out Density Estimates). Consider the $.5N(-2, 1) + .5N(2, 1)$ density, which is bimodal. Simulate a sample of size $n = 40$ from this density and compute and plot each the following density estimates, and write a comparative report:

(a) a histogram that uses $m = 11$ equiwidth cells;
(b) a kernel estimate that uses a Gaussian kernel and a bandwidth equal to .05;
(c) a kernel estimate that uses a Gaussian kernel and a bandwidth equal to .005;
(d) a kernel estimate that uses a Gaussian kernel and a bandwidth equal to .15.

Exercise 20.24 (Practical Effect of the Kernel). For the simulated data in the previous exercise, repeat part (b) with the Epanechnikov and Exponential kernels. Plot the three density estimates, and write a report.

Exercise 20.25. Suppose $\mathcal{H}$ is an inner product space. Prove the parallelogram law.

Exercise 20.26. Suppose $\mathcal{H}$ is an inner product space, and let $x, y \in \mathcal{H}$. Show that $x = y$ if and only if $(x, z) = (y, z)$ for all $z \in \mathcal{H}$.

Exercise 20.27. Suppose $\mathcal{H}$ is an inner product space, and let $x \in \mathcal{H}$. Show that $||x||_\mathcal{H} = \sup\{(x, v) : ||v||_\mathcal{H} = 1\}$.

Exercise 20.28 (Sufficient Condition for Convergence). Suppose $\mathcal{H}$ is an inner product space, and $x_n, y \in \mathcal{H}, n \geq 1$. Show that $d(x_n, y) \to 0$ if $||x_n||_\mathcal{H} \to ||y||_\mathcal{H}$, and $(x_n, y) \to ||y||^2_\mathcal{H}$.

Exercises

Exercise 20.29 *(Normalization Helps). Suppose $\mathcal{H}$ is an inner product space, and $x, y \in \mathcal{H}$. Suppose that $||x||_\mathcal{H} = 1$, but $||y||_\mathcal{H} > 1$. Show that $||x - \frac{y}{||y||_\mathcal{H}}||_\mathcal{H} \leq ||x - y||_\mathcal{H}$. Is this inequality always a strict inequality?

Exercise 20.30 (An Orthonormal System). Consider $\mathcal{H} = L_2[-\pi, \pi]$ with the usual inner product $(f, g) = \int_{-\pi}^{\pi} f(t)g(t)dt$. Define the sequence of functions

$$f_0(t) \equiv \frac{1}{\sqrt{2}}, \quad f_n(t) = \cos(nt), \quad n = -1, -2, -3, \ldots,$$

$$f_n(t) = \sin(nt), \quad n = 1, 2, 3, \ldots.$$

Show that $\{f_n\}_{n=-\infty}^{\infty}$ forms an orthonormal set on $\mathcal{H}$; that is, $||f_n||_\mathcal{H} = 1$ and $(f_m, f_n) = 0$, for all $m, n, m \neq n$.

Exercise 20.31 (Legendre Polynomials). Show that the polynomials $P_n(t) = \frac{d^n}{dt^n}(t^2 - 1)^n$, $n = 0, 1, 2, \ldots$ form an orthogonal set on $L_2[-1, 1]$.
Find, explicitly, $P_i(t), i = 0, 1, 2, 3, 4$.

Exercise 20.32 (Hermite Polynomials). Show that the polynomials

$$H_n(t) = (-1)^n e^{\frac{t^2}{2}} \frac{d^n}{dt^n}\left(e^{-\frac{t^2}{2}}\right), n = 0, 1, 2, \ldots$$

form an orthogonal set on $L_2(-\infty, \infty)$ with the inner product $(f, g) = \int_{-\infty}^{\infty} f(t)g(t)e^{-\frac{t^2}{2}} dt$.
Find, explicitly, $H_i(t), i = 0, 1, 2, 3, 4$.

Exercise 20.33 (Laguerre Polynomials). Show that for $\alpha > -1$, the polynomials

$$L_{n,\alpha}(t) = \frac{e^t t^{-\alpha}}{n!} \frac{d^n}{dt^n}\left(e^{-t} t^{n+\alpha}\right), n = 0, 1, 2, \ldots$$

form an orthogonal set on $L_2(0, \infty)$ with the inner product $(f, g) = \int_0^{\infty} f(t)g(t) e^{-t} t^\alpha dt$.
Find, explicitly, $L_{i,\alpha}(t), i = 0, 1, 2, 3, 4$.

Exercise 20.34 (Jacobi Polynomials). Show that for $\alpha, \beta > -1$, the polynomials $P_n^{(\alpha,\beta)}(t) = \frac{(-1)^n}{2^n n!}(1-t)^{-\alpha}(1+t)^{-\beta} \frac{d^n}{dt^n}\left((1-t)^{\alpha+n}(1+t)^{\beta+n}\right), n = 0, 1, 2, \ldots$ form an orthogonal set on $L_2(-1, 1)$ with the inner product $(f, g) = \int_{-1}^{1} f(t)g(t)(1-t)^\alpha (1+t)^\beta dt$.
Find, explicitly, $P_i^{(\alpha,\beta)}(t), i = 0, 1, 2, 3, 4$.

Exercise 20.35 (Gegenbauer Polynomials). Find, explicitly, the first five Gegenbauer polynomials, defined as $P_i^{(\alpha,\beta)}(t), i = 0, 1, 2, 3, 4$, when $\alpha = \beta$.

Exercise 20.36 (Chebyshev Polynomials). Find, explicitly, the first five Chebyshev polynomials $T_n(x), n = 0, 1, 2, 3, 4$, which are the Jacobi polynomials in the special case $\alpha = \beta = -\frac{1}{2}$. The polynomials $T_n, n = 0, 1, 2, \ldots$ form an orthogonal set on $L_2[-1, 1]$ with the inner product $(f, g) = \int_{-1}^{1} f(t)g(t)(1-t^2)^{-\frac{1}{2}} dt$.

Exercise 20.37 (Projection Formula). Suppose $\{x_1, x_2, \ldots, x_n\}$ is an orthonormal system in an inner product space $\mathcal{H}$. Show that the projection of an $x \in \mathcal{H}$ to the linear span of $\{x_1, x_2, \ldots, x_n\}$ is given by $\sum_{j=1}^{n}(x, x_j)x_j$.

Exercise 20.38 (Bessel's Inequality). Use the formula of the previous exercise to show that for a general countable orthonormal set $B = \{x_1, x_2, \ldots\}$ in an inner product space $\mathcal{H}$, and any $x \in \mathcal{H}$, one has $||x||_\mathcal{H}^2 \geq \sum_{i=1}^{\infty} |(x, x_i)|^2$.

Exercise 20.39 * **(When Is a Normed Space an Inner Product Space).** Suppose $\mathcal{H}$ is a normed space, that is, $\mathcal{H}$ has associated with it a norm $||x||_\mathcal{H}$. Show that this norm is induced by an inner product (x, y) on $\mathcal{H}$ if and only if the norm satisfies the parallelogram law $||x + y||_\mathcal{H}^2 + ||x - y||_\mathcal{H}^2 = 2\left(||x||_\mathcal{H}^2 + ||y||_\mathcal{H}^2\right)$ for all $x, y \in \mathcal{H}$.

Exercise 20.40 * **(Fact About L_p Spaces).** Let X be any subset of an Euclidean space. Show that for any $p \geq 1$, $L_p(X)$ is not an inner product space if $p \neq 2$.

Exercise 20.41. Let $\mathcal{I}$ be an interval in the real line. Show that the family of all real-valued continuous functions on $\mathcal{I}$, with the norm $||f||_\mathcal{H} = \sup_{x \in \mathcal{I}} |f(x)|$ is not an inner product space.

Exercise 20.42. Suppose K is the Epanechnikov kernel in $\mathcal{R}^d$, and f is an isotropic function $f(x) = g(||x||)$, where g is Lipschitz of some order $\alpha > 0$.

(a) Show that f is Lipschitz on $\mathcal{R}^d$.
(b) Show that $f * K_n$ converges uniformly to f where K_n means $K_n(x) = n^d K(nx)$.
(c) Prove this directly when $f(x) = ||x||$.

Exercise 20.43 (Kernel Plots). Plot the Cauchy and the exponential kernel, as defined in Section 20.3.2, for some selected values of a. What is the effect of increasing a?

Exercise 20.44 * **(Kernel Proof of Weierstrass's Theorem).** Use Weierstrass's kernels and Theorem 20.10 to show that every continuous function on a closed bounded interval in $\mathcal{R}$ can be uniformly well approximated by a polynomial of a suitable degree.

Exercise 20.45. Suppose $f, g \in L_1(\mathcal{R}^d)$. Prove that $|||f| * |g|||_1 = ||f||_1 ||g||_1$.

Exercise 20.46. Show that, pointwise, $|f * (gh)| \leq ||g||_\infty (|f| * |h|)$.

Exercise 20.47. Consider the linear operator $\delta : L_2[a, b] \to L_2[a, b]$, where a, b are finite real constants, defined as $\delta(f)(x) = \int_a^x f(y) dy$. Show that $||\delta|| \leq b - a$.

Exercises

Exercise 20.48. Suppose $K(x, y)$ is the reproducing kernel of a Hilbert function space $\mathcal{H}$. Suppose for some given $x_0 \in \mathcal{X}$, $K(x_0, x_0) = 0$. Show that $f(x_0) = 0$ for all $f \in \mathcal{H}$.

Exercise 20.49 * **(Reproducing Kernel of an Annhilator).** Suppose $K(x, y)$ is the reproducing kernel of a Hilbert function space $\mathcal{H}$ on the domain set $\mathcal{X}$. Fix $z \in \mathcal{H}$, and define $\mathcal{H}_z = \{f \in \mathcal{H} : f(z) = 0\}$.

(a) Show that $\mathcal{H}_z$ is a reproducing kernel Hilbert space.
(b) Show that the reproducing kernel of $\mathcal{H}_z$ is $K_z(x, y) = K(x, y) - \frac{K(x,z)K(z,y)}{K(z,z)}$.

Exercise 20.50 * **(Riesz Representation for an RKHS).** Consider a function space $\mathcal{H}$ that is known to be a reproducing kernel Hilbert space, and let δ be a continuous linear functional on it. Show that the unique $v \in \mathcal{H}$ that satisfies $\delta(u) = (u, v)$ for all $u \in \mathcal{H}$ is given by the function $v(x) = \delta(K(x, .))$.

Exercise 20.51 (Generating New Reproducing Kernels). Let $K_1(x, y)$, $K_2(x, y)$ be reproducing kernels on $\mathcal{X} \otimes \mathcal{X}$. Show that $K = K_1 + K_2$ is also a reproducing kernel. Can you generalize this result?

Exercise 20.52. Show that for any $p \leq 2, \alpha > 0, e^{-\alpha\|x-y\|^p}$, $x, y \in \mathcal{R}^d$ is a reproducing kernel.

Exercise 20.53. Show that $\min(x, y), x, y \in [0, 1]$ is a reproducing kernel.

Exercise 20.54. Show that under the conditions of Theorem 20.11, $\alpha_n(F, \phi_n) \xrightarrow{a.s.} \alpha(F)$ for any F.

Exercise 20.55 * **(Classification Using Kernels).** Suppose $K(x)$ is a kernel in $\mathcal{R}^d$ in the usual sense, that is, as defined in Definition 20.4. Suppose K has a verifiable isotropic upper bound $K(x) \leq g(\|x\|)$ for some g. Give sufficient conditions on g so that the kernel classification rule based on K is strongly universally consistent in the sense of Theorem 20.11.

Exercise 20.56 * **(Consistency of Kernel Classification).** For which of the Gaussian, Cauchy, exponential, Laplace, spherical, and polynomial kernels, defined as in Section 20.3.2, does strong universal consistency hold? Use Theorem 20.11.

Exercise 20.57 * **(Consistency of Product Kernels).** Suppose $K_j(x_j), j = 1, 2, \ldots, d$ are kernels on $\mathcal{R}$. Consider the classification rule that classifies a new x value into group 1 if

$$\sum_{i=1}^{n} I_{y_i=1} \prod_{j=1}^{d} K_j\left(\frac{x_j - x_{i,j}}{h_j}\right) \geq \sum_{i=1}^{n} I_{y_i=2} \prod_{j=1}^{d} K_j\left(\frac{x_j - x_{i,j}}{h_j}\right),$$

where $h_j = h_{j,n}$ are the coordinatewise bandwidths, $j = 1, 2, \ldots, d$, and $x_j, x_{i,j}$ stand for the jth coordinate of x and the ith data value x_i.

Show that universal strong consistency of this rule holds if each K_j satisfies the conditions of Theorem 20.11, and if $h_{j,n} \to 0$ for all j and $n h_{1,n} h_{2,n} \ldots h_{d,n} \to \infty$ as $n \to \infty$.

Exercise 20.58 * **(Classification with Fisher's Iris Data).** For *Fisher's Iris dataset* (e.g., wikipedia.org), form a kernel classification rule for the pairwise cases, *setosa versus. versicolor; setosa versus. virginica; versicolor versus. virginica*, by using

(a) Fisher's linear classification rule
(b) A kernel classification rule with an exponential kernel, where the parameter of the kernel is to be chosen by you
(c) A kernel classification rule with the inverse quadratic kernel, where the parameter of the kernel is to be chosen by you

Find the empirical error rate for each rule, and write a report.

Exercise 20.59 * **(Classification with Simulated Data).**

(a) Simulate $n = 50$ observations from a d-dimensional normal distribution, $N_d(0, I)$; do this for $d = 3, 5, 10$.
(b) Simulate $n = 50$ observations from a d-dimensional radial (i.e., spherically symmetric) Cauchy distribution, which has the density

$$\frac{c}{(1+||x||^2)^{\frac{d+1}{2}}}$$

c being the normalizing constant. Do this for $d = 3, 5, 10$.
(c) Form a kernel classification rule by using the exponential kernel and the inverse quadratic kernel, where the parameters of the kernel are to be chosen by you.

Find the empirical error rate for each rule, and write a report.

References

Aizerman, M., Braverman, E., and Rozonoer, L. (1964). Theoretical foundations of the potential function method in pattern recognition learning, *Autom. Remote Control*, 25, 821–837.

Aronszajn, N. (1950). Theory of reproducing kernels, *Trans. Amer. Math. Soc.*, 68, 307–404.

Athreya, K. (1987). Bootstrap of the mean in the infinite variance case, *Ann. Statist.*, 15, 724–731.

Berlinet, A. and Thomas-Agnan, C. (2004). *Reproducing Kernel Hilbert Spaces in Probability and Statistics*, Kluwer, Boston.

Bickel, P.J. (2003). Unorthodox bootstraps, Invited paper, *J. Korean Statist. Soc.*, 32, 213–224.

Bickel, P.J. and Doksum, K. (2006). *Mathematical Statistics, Basic Ideas and Selected Topics*, Prentice Hall, upper Saddle River, NJ.

Bickel, P.J. and Freedman, D. (1981). Some asymptotic theory for the bootrap, *Ann. Statist.*, 9, 1196–1217.

Carlstein, E. (1986). The use of subseries values for estimating the variance of a general statistic from a stationary sequence, *Ann. Statist.*, 14, 1171–1179.

Chan, K. and Ledolter, J. (1995). Monte Carlo estimation for time series models involving counts, *J. Amer. Statist. Assoc.*, 90, 242–252.

Cheney, W. (2001). *Analysis for Applied Mathematics*, Springer, New York.

Cheney, W. and Light, W. (2000). *A Course in Approximation Theory*, Pacific Grove, Brooks/Cole, CA.

Cristianini, N. and Shawe-Taylor, J. (2000). *An Introduction to Support Vector Machines and other Kernel Based Learning Methods,* Cambridge Univ. Press, Cambridge, UK.
DasGupta, A. (2008). *Asymptotic Theory of Statistics and Probability,* Springer, New York.
Dempster, A., Laird, N., and Rubin, D. (1977). Maximum likelihood from incomplete data via the EM algorithm, *JRSS, Ser. B,* 39, 1–38.
Devroye, L., Györfi, L., and Lugosi, G. (1996). *A Probabilistic Theory of Pattern Recognition,* Springer, New York.
Efron, B. (2003). Second thoughts on the bootstrap, *Statist. Sci.,* 18, 135–140.
Efron, B. and Tibshirani, R. (1993). *An Introduction to the Bootstrap,* Chapman and Hall, London.
Giné, E. and Zinn, J. (1989).Necessary conditions for bootstrap of the mean, *Ann. Statist.,* 17, 684–691.
Hall, P. (1986). On the number of bootstrap simulations required to construct a confidence interval, *Ann. Statist.,* 14, 1453–1462.
Hall, P. (1988). Rate of convergence in bootstrap approximations, *Ann. prob,* 16,4, 1665–1684.
Hall, P. (1989). On efficient bootstrap simulation, *Biometrika,* 76, 613–617.
Hall, P. (1990). Asymptotic properties of the bootstrap for heavy-tailed distributions, *Ann. Prob.,* 18, 1342–1360.
Hall, P. (1992). *The Bootstrap and Edgeworth Expansion,* Springer, New York.
Hall, P., Horowitz, J. and Jing, B. (1995). On blocking rules for the bootstrap with dependent data, *Biometrika,* 82, 561–574.
Hall, P. (2003). A short prehistory of the bootstrap, *Statist. Sci.,* 18, 158–167.
Künsch, H.R. (1989). The Jackknife and the bootstrap for general stationary observations, *Ann. Statist.,* 17, 1217–1241.
Lahiri, S.N. (1999). Theoretical comparisons of block bootstrap methods, *Ann. Statist.,* 27, 386–404.
Lahiri, S.N. (2003). *Resampling Methods for Dependent Data,* Springer-Verlag, New York.
Lahiri, S.N. (2006). Bootstrap methods, a review, in *Frontiers in Statistics,* J. Fan and H. Koul Eds., 231–256, Imperial College Press, London.
Lange, K. (1999). *Numerical Analysis for Statisticians,* Springer, New York.
Le Cam, L. and Yang, G. (1990). *Asymptotics in Statistics, Some Basic Concepts,* Springer, New York.
Lehmann, E.L. (1999). *Elements of Large Sample Theory,* Springer, New York.
Lehmann, E.L. and Casella, G. (1998). *Theory of Point Estimation,* Springer, New York.
Levine, R. and Casella, G. (2001). Implementation of the Monte Carlo EM algorithm, *J. Comput. Graph. Statist.,* 10, 422–439.
McLachlan, G. and Krishnan, T. (2008). *The EM Algorithm and Extensions,* Wiley, New York.
Mercer, J. (1909). Functions of positive and negative type and their connection with the theory of integral equations, *Philos. Trans. Royal Soc. London, A,* 415–416.
Minh, H., Niyogi, P., and Yao, Y. (2006). Mercer's theorem, feature maps, and smoothing, *Proc. Comput. Learning Theory, COLT,* 154–168.
Murray, G.D. (1977). Discussion of paper by Dempster, Laird, and Rubin (1977), *JRSS Ser. B,* 39, 27–28.
Politis, D. and Romano, J. (1994). The stationary bootstrap, *JASA,* 89, 1303–1313.
Politis, D. and White, A. (2004). Automatic block length selection for the dependent bootstrap, *Econ. Rev.,* 23, 53–70.
Politis, D., Romano, J. and Wolf, M. (1999). *Subsampling,* Springer, New York.
Rudin, W. (1986). *Real and Complex Analysis,* 3rd edition, McGraw-Hill, Columbus, OH.
Rosenblatt, M. (1956). Remarks on some nonparametric estimates of a density function, *Ann. Math, Statist.,* 27, 832–835. 3rd Edition, McGraw-Hill, Columbus, OH.
Shao, J. and Tu, D. (1995). *The Jackknife and Bootstrap,* Springer, New York.
Singh, K. (1981). On the asymptotic accuracy of Efron's bootstrap, *Ann. Statist.,* 9, 1187–1195.
Sundberg, R. (1974). Maximum likelihood theory for incomplete data from exponential family, *Scand. J. Statist.,* 1, 49–58.
Tong, Y. (1990). *The Multivariate Normal Distribution,* Springer, New York.

Vapnik, V. and Chervonenkis, A. (1964). A note on one class of perceptrons, *Autom. Remote Control*, 25.

Vapnik, V. (1995). *The Nature of Statistical Learning Theory*, Springer, New York.

Wei, G. and Tanner, M. (1990). A Monte Carlo implementation of the EM algorithm, *J. Amer. Statist. Assoc.*, 85, 699–704.

Wu, C.F.J. (1983). On the convergence properties of the EM algorithm, *Ann. Statist.*, 11, 95–103.

Appendix A
Symbols, Useful Formulas, and Normal Table

A.1 Glossary of Symbols

Ω	sample space
$P(B \mid A)$	conditional probability
$\{\pi(1), \ldots, \pi(n)\}$	permutation of $\{1, \ldots, n\}$
F	CDF
$\bar{F}$	$1 - F$
F^{-1}, Q	quantile function
iid, IID	independent and identically distributed
$p(x, y), f(x_1, \ldots, x_n)$	joint pmf; joint density
$F(x_1, \ldots, x_n)$	joint CDF
$f(y \mid x), E(Y \mid X = x)$	conditional density and expectation
var, Var	variance
Var$(Y \mid X = x)$	conditional variance
Cov	covariance
$\rho_{X,Y}$	correlation
$G(s), \psi(t)$	generating function; mgf or characteristic function
$\psi(t_1, \ldots, t_n)$	joint mgf
β, γ	skewness and kurtosis
μ_k	$E(X - \mu)^k$
m_k	sample kth central moment
κ_r	rth cumulant
ρ_r	rth standardized cumulant; correlation of lag r
r, θ	polar coordinates
J	Jacobian
F_n	empirical CDF
F_n^{-1}	sample quantile function
$X^{(n)}$	sample observation vector $(X_1, \ldots, X_n)$
M_n	sample median
$X_{(k)}, X_{k:n}$	kth-order statistic

W_n	sample range
IQR	interquartile range
$\overset{P}{\Rightarrow}, \overset{P}{\to}$	convergence in probability
$o_p(1)$	convergence in probability to zero
$O_p(1)$	bounded in probability
$a_n \sim b_n$	$0 < \liminf \frac{a_n}{b_n} \leq \limsup \frac{a_n}{b_n} < \infty$
$a_n \asymp b_n, a_n \approx b_n$	$\lim \frac{a_n}{b_n} = 1$
$\overset{a.s.}{\Rightarrow}, \overset{a.s.}{\to}$	almost sure convergence
w.p. 1	with probability 1
a.e.	almost everywhere
i.o.	infinitely often
$\overset{\mathcal{L}}{\Rightarrow}, \overset{\mathcal{L}}{\to}$	convergence in distribution
$\overset{r}{\Rightarrow}, \overset{r}{\to}$	convergence in rth mean
u. i.	uniformly integrable
LIL	law of iterated logarithm
VST	variance stabilizing transformation
δ_x	point mass at x
$P(\{x\})$	probability of the point x
λ	Lebesgue measure
$*$	convolution
$\ll$	absolutely continuous
$\frac{dP}{d\mu}$	Radon–Nikodym derivative
$\otimes$	product measure; Kronecker product
$I(\theta)$	Fisher information function or matrix
$\mathcal{T}$	natural parameter space
$f(\theta \mid x), \pi(\theta \mid X^{(n)})$	posterior density of θ
LRT	likelihood ratio test
Λ_n	likelihood ratio
S	sample covariance matrix
T^2	Hotelling's T^2 statistic
MLE	maximum likelihood estimate
$l(\theta, X)$	complete data likelihood in EM
$L(\theta, X)$	$\log l(\theta, X)$
$l(\theta, Y)$	likelihood for observed data
$L(\theta, Y)$	$\log l(\theta, Y)$
$\hat{L}_k(\theta, Y)$	function to be maximized in M-step
H_{Boot}	bootstrap distribution of a statistic
P_*	bootstrap measure
p_{ij}	transition probabilities in a Markov chain

A.1 Glossary of Symbols

$p_{ij}(n)$	n-step transition probabilities
T_i	first passage time
π	stationary distribution of a Markov chain
S_n	random walk; partial sums
$\xi(x,n), \eta(x,T)$	local time
$W(t), B(t)$	Brownian motion and Brownian bridge
$W^d(t)$	d-dimensional Brownian motion
$\rho(s,t)$	covariance kernel
$X(t), N(t)$	Poisson process
Π	Poisson point process
$\alpha_n(t), u_n(y)$	uniform empirical and quantile process
$F_n(t), \beta_n(t)$	general empirical process
P_n	empirical measure
D_n	Kolmogorov–Smirnov statistic
MCMC	Markov chain Monte Carlo
θ_{ij}	proposal probabilities
γ_{ij}	acceptance probabilities
$S(n,\mathcal{C})$	shattering coefficient
SLEM	second largest eigenvalue in modulus
$\Delta(P)$	Dobrushin's coefficient
$V(x)$	energy or drift function
$VC(\mathcal{C})$	VC dimension
$N(\epsilon, \mathcal{F}, \|\cdot\|)$	covering number
$N_{\sqcup}(\epsilon, \mathcal{F}, \|\cdot\|)$	bracketing number
$D(x, \epsilon, P)$	packing number
$\lambda(x)$	intensity function of a Poisson process
$\mathcal{R}$	real line
$\mathcal{R}^d$	d-dimensional Euclidean space
$C(X)$	real-valued continuous functions on X
$C_k(\mathcal{R})$	k times continuously differentiable functions
$C_0(\mathcal{R})$	real continuous functions f on $\mathcal{R}$ such that $f(x) \to 0$ as $|x| \to \infty$
$\mathcal{F}$	family of functions
∇, Δ	gradient vector and Laplacian
$f^{(m)}$	mth derivative
$D^k f(x_1, \ldots, x_n)$	$\sum_{m_1, m_2, \ldots, m_n \geq 0, m_1 + \ldots m_n = k} \frac{\partial^{m_1} f}{\partial x_1^{m_1}} \cdots \frac{\partial^{m_n} f}{\partial x_n^{m_n}}$
D_+, D^+	Dini derivatives
$\|\cdot\|$	Euclidean norm
$\|\cdot\|_\infty$	supnorm
tr	trace of a matrix
$|A|$	determinant of a matrix
$K(x), K(x,y)$	kernels

$\|x\|_1, \|x\|$	L_1, L_2 norm in $\mathcal{R}^n$
$\mathcal{H}$	Hilbert space
$\|x\|_{\mathcal{H}}$	Hilbert norm
(x, y)	inner product
$g(X), \phi_n$	classification rules
$\phi_x, \phi(x)$	Mercer's feature maps
$B(x, r)$	sphere with center at x and radius r
$U, U^0, \bar{U}$	domain, interior, and closure
I_A	indicator function of A
$I(t)$	large deviation rate function
$\{\}$	fractional part
$\lfloor \cdot \rfloor$	integer part
sgn, sign	signum function
x_+, x^+	$\max\{x, 0\}$
max, min	maximum, minimum
sup, inf	supremum, infimum
K_ν, I_ν, J_ν	Bessel functions
$L_p(\mu), L^p(\mu)$	set of functions such that $\int |f|^p d\mu < \infty$
d, L, ρ	Kolmogorov, Levy, and total variation distance
H, K	Hellinger and Kullback–Leibler distance
W, ℓ_2	Wasserstein distance
D	separation distance
d_f	f-divergence
$N(\mu, \sigma^2)$	normal distribution
ϕ, Φ	standard normal density and CDF
$N_p(\mu, \Sigma), MVN(\mu, \Sigma)$	multivariate normal distribution
BVN	bivariate normal distribution
$MN(n, p_1, \ldots, p_k)$	multinomial distribution with these parameters
$t_n, t(n)$	t-distribution with n degrees of freedom
Ber(p), Bin(n, p)	Bernoulli and binomial distribution
Poi(λ)	Poisson distribution
Geo(p)	geometric distribution
NB(r, p)	negative binomial distribution
$Exp(\lambda)$	exponential distribution with mean λ
Gamma(α, λ)	Gamma density with shape parameter α, and scale parameter λ
$\chi_n^2, \chi^2(n)$	chi-square distribution with n degrees of freedom
$C(\mu, \sigma)$	Cauchy distribution
$\mathcal{D}_n(\alpha)$	Dirichlet distribution
$W_p(k, \Sigma)$	Wishart distribution
DoubleExp(μ, λ)	double exponential with parameters μ, λ

A.2 Moments and MGFs of Common Distributions

Discrete Distributions

Distribution	$p(x)$	Mean	Variance	Skewness	Kurtosis	MGF
Uniform	$\frac{1}{n}, x = 1,\ldots,n$	$\frac{n+1}{2}$	$\frac{n^2-1}{12}$	0	$-\frac{6(n^2+1)}{5(n^2-1)}$	$\frac{e^{(n+1)t}-e^t}{n(e^t-1)}$
Binomial	$\binom{n}{x}p^x(1-p)^{n-x}, x=0,\ldots,n$	np	$np(1-p)$	$\frac{1-2p}{\sqrt{np(1-p)}}$	$\frac{1-6p(1-p)}{np(1-p)}$	$(pe^t + 1 - p)^n$
Poisson	$\frac{e^{-\lambda}\lambda^x}{x!}, x=0,1,\ldots$	λ	λ	$\frac{1}{\sqrt{\lambda}}$	$\frac{1}{\lambda}$	$e^{\lambda(e^t-1)}$
Geometric	$p(1-p)^{x-1}, x=1,2,\ldots$	$\frac{1}{p}$	$\frac{1-p}{p^2}$	$\frac{2-p}{\sqrt{1-p}}$	$6 + \frac{p^2}{1-p}$	$\frac{pe^t}{1-(1-p)e^t}$
Neg. Bin.	$\binom{x-1}{r-1}p^r(1-p)^{x-r}, x \geq r$	$\frac{r}{p}$	$\frac{r(1-p)}{p^2}$	$\frac{2-p}{\sqrt{r(1-p)}}$	$\frac{6}{r} + \frac{p^2}{r(1-p)}$	$\left(\frac{pe^t}{1-(1-p)e^t}\right)^r$
Hypergeom	$\frac{\binom{D}{x}\binom{N-D}{n-x}}{\binom{N}{n}}$	$n\frac{D}{N}$	$n\frac{D}{N}(1-\frac{D}{N})\frac{N-n}{N-1}$	Complex	Complex	Complex
Benford	$\frac{\log(1+\frac{1}{x})}{\log 10}, x = 1,\ldots,9$	3.44	6.057	.796	2.45	$\sum_{x=1}^{9} e^{tx} p(x)$

Continuous Distributions

Distribution	$f(x)$	Mean	Variance	Skewness	Kurtosis
Uniform	$\frac{1}{b-a}, a \leq x \leq b$	$\frac{a+b}{2}$	$\frac{(b-a)^2}{12}$	0	$-\frac{6}{5}$
Exponential	$\frac{e^{-x/\lambda}}{\lambda}, x \geq 0$	λ	λ^2	2	6
Gamma	$\frac{e^{-x/\lambda}x^{\alpha-1}}{\lambda^\alpha \Gamma(\alpha)}, x \geq 0$	$\alpha\lambda$	$\alpha\lambda^2$	$\frac{2}{\sqrt{\alpha}}$	$\frac{6}{\alpha}$
χ_m^2	$\frac{e^{-x/2}x^{m/2-1}}{2^{m/2}\Gamma(\frac{m}{2})}, x \geq 0$	m	$2m$	$\sqrt{\frac{8}{m}}$	$\frac{12}{m}$
Weibull	$\frac{\beta}{\lambda}\left(\frac{x}{\lambda}\right)^{\beta-1}e^{-(\frac{x}{\lambda})^\beta}, x > 0$	$\lambda\Gamma(1+\frac{1}{\beta})$	$\lambda^2\Gamma(1+\frac{2}{\beta})-\mu^2$	$\frac{\lambda^3\Gamma(1+\frac{3}{\beta})-3\mu\sigma^2-\mu^3}{\sigma^3}$	Complex
Beta	$\frac{x^{\alpha-1}(1-x)^{\beta-1}}{B(\alpha,\beta)}, 0 \leq x \leq 1$	$\frac{\alpha}{\alpha+\beta}$	$\frac{\alpha\beta}{(\alpha+\beta)^2(\alpha+\beta+1)}$	$\frac{2(\beta-\alpha)\sqrt{\alpha+\beta+1}}{\sqrt{\alpha\beta}(\alpha+\beta+2)}$	Complex
Normal	$\frac{1}{\sigma\sqrt{2\pi}}e^{-(x-\mu)^2/(2\sigma^2)}, x \in \mathcal{R}$	μ	σ^2	0	0
lognormal	$\frac{1}{\sigma\sqrt{2\pi}x}e^{-\frac{(\log x - \mu)^2}{2\sigma^2}}, x > 0$	$e^{\mu+\sigma^2/2}$	$(e^{\sigma^2}-1)e^{2\mu+\sigma^2}$	$e^{\sigma^2}+2\sqrt{e^{\sigma^2}-1}$	Complex

(continued)

A.2 Moments and MGFs of Common Distributions

Distribution	$f(x)$	Mean	Variance	Skewness	Kurtosis
Cauchy	$\frac{1}{\sigma\pi(1+(x-\mu)^2/\sigma^2)}, x \in \mathcal{R}$	None	None	None	None
t_m	$\frac{\Gamma(\frac{m+1}{2})}{\sqrt{m\pi}\Gamma(\frac{m}{2})}\frac{1}{(1+x^2/m)^{(m+1)/2}}, x \in \mathcal{R}$	$0 (m > 1)$	$\frac{m}{m-2}(m > 2)$	$0(m > 3)$	$\frac{6}{m-4}(m > 4)$
F	$\frac{(\frac{\beta}{\alpha})^3 x^{\alpha-1}}{B(\alpha,\beta)(x+\frac{\beta}{\alpha}y+\beta)}, x > 0$	$\frac{\beta}{\beta-1}(\beta > 1)$	$\frac{\beta^2(\alpha+\beta-1)}{\alpha(\beta-2)(\beta-1)^2}(\beta > 2)$	Complex	Complex
Double Exp.	$\frac{e^{-|x-\mu|/\sigma}}{2\sigma}, x \in \mathcal{R}$	μ	$2\sigma^2$	0	3
Pareto	$\frac{\alpha\theta^\alpha}{x^{\alpha+1}}, x \geq \theta > 0$	$\frac{\alpha\theta}{\alpha-1}(\alpha > 1)$	$\frac{\alpha\theta^2}{(\alpha-1)^2(\alpha-2)}(\alpha > 2)$	$\frac{2(\alpha+1)}{\alpha-3}\sqrt{\frac{\alpha-2}{\alpha}}(\alpha > 3)$	Complex
Gumbel	$\frac{1}{\sigma}(-e^{-\frac{x-\mu}{\sigma}})e^{-\frac{x-\mu}{\sigma}}, x \in \mathcal{R}$	$\mu + \gamma\sigma$	$\frac{\pi^2}{6}\sigma^2$	$\frac{12\sqrt{6}\zeta(3)}{\pi^3}$	$\frac{12}{5}$

Note: For the Gumbel distribution, $\gamma \approx .577216$ is the Euler constant, and $\zeta(3)$ is Riemann's zeta function $\zeta(3) = \sum_{n=1}^{\infty}\frac{1}{n^3} \approx 1.20206$.

Table of MGFs of Continuous Distributions

Distribution	$f(x)$	MGF
Uniform	$\frac{1}{b-a}, a \leq x \leq b$	$\frac{e^{bt}-e^{at}}{(b-a)t}$
Exponential	$\frac{e^{-x/\lambda}}{\lambda}, x \geq 0$	$(1-\lambda t)^{-1} (t < 1/\lambda)$
Gamma	$\frac{e^{-x/\lambda} x^{\alpha-1}}{\lambda^{\alpha} \Gamma(\alpha)}, x \geq 0$	$(1-\lambda t)^{-\alpha} (t < 1/\lambda)$
χ_m^2	$\frac{e^{-x/2} x^{m/2-1}}{2^{m/2} \Gamma(\frac{m}{2})}, x \geq 0$	$(1-2t)^{-m/2} (t < \frac{1}{2})$
Weibull	$\frac{\beta}{\lambda}(\frac{x}{\lambda})^{\beta-1} e^{-(\frac{x}{\lambda})^{\beta}}, x > 0$	$\sum_{n=0}^{\infty} \frac{(\lambda t)^n}{n!} \Gamma(1+\frac{n}{\beta})$
Beta	$\frac{x^{\alpha-1}(1-x)^{\beta-1}}{B(\alpha,\beta)}, 0 \leq x \leq 1$	$1F_1(\alpha, \alpha+\beta, t)$
Normal	$\frac{1}{\sigma \sqrt{2\pi}} e^{-(x-\mu)^2/(2\sigma^2)}, x \in \mathcal{R}$	$e^{t\mu + t^2 \sigma^2/2}$
lognormal	$\frac{1}{\sigma \sqrt{2\pi} x} e^{-\frac{(\log x - \mu)^2}{2\sigma^2}}, x > 0$	None
Cauchy	$\frac{1}{\sigma \pi (1+(x-\mu)^2/\sigma^2)}, x \in \mathcal{R}$	None
t_m	$\frac{\Gamma(\frac{m+1}{2})}{\sqrt{m\pi}\Gamma(\frac{m}{2})} \frac{1}{(1+x^2/m)^{(m+1)/2}}, x \in \mathcal{R}$	None
F	$\frac{(\frac{\beta}{\alpha})^\beta x^{\alpha-1}}{B(\alpha,\beta)(x+\frac{\beta}{\alpha})^{\alpha+\beta}}, x > 0$	None
Double Exp.	$\frac{e^{-|x-\mu|/\sigma}}{2\sigma}, x \in \mathcal{R}$	$\frac{e^{t\mu}}{1-\sigma^2 t^2} (|t| < 1/\sigma)$
Pareto	$\frac{\alpha \theta^\alpha}{x^{\alpha+1}}, x \geq \theta > 0$	None
Gumbel	$\frac{1}{\sigma} e^{(-e^{-\frac{x-\mu}{\sigma}})} e^{-\frac{x-\mu}{\sigma}}, x \in \mathcal{R}$	$e^{t\mu} \Gamma(1-t\sigma)(t < 1/\sigma)$

A.3 Normal Table

Standard Normal Probabilities $P(Z \leq t)$ and Standard Normal Percentiles

Quantity tabulated in the next page is $\Phi(t) = P(Z \leq t)$ for given $t \geq 0$, where $Z \sim N(0, 1)$. For example, from the table, $P(Z \leq 1.52) = .9357$.

For any positive t, $P(-t \leq Z \leq t) = 2\Phi(t) - 1$, and $P(Z > -t) = P(Z > t) = 1 - \Phi(t)$.

Selected standard normal percentiles z_α are given below. Here, the meaning of z_α is $P(Z > z_\alpha) = \alpha$.

α	z_α
.25	.675
.2	.84
.1	1.28
.05	1.645
.025	1.96
.02	2.055
.01	2.33
.005	2.575
.001	3.08
.0001	3.72

	0	1	2	3	4	5	6	7	8	9
0.0	0.5000	0.5040	0.5080	0.5120	0.5160	0.5199	0.5239	0.5279	0.5319	0.5359
0.1	0.5398	0.5438	0.5478	0.5517	0.5557	0.5596	0.5636	0.5675	0.5714	0.5753
0.2	0.5793	0.5832	0.5871	0.5910	0.5948	0.5987	0.6026	0.6064	0.6103	0.6141
0.3	0.6179	0.6217	0.6255	0.6293	0.6331	0.6368	0.6406	0.6443	0.6480	0.6517
0.4	0.6554	0.6591	0.6628	0.6664	0.6700	0.6736	0.6772	0.6808	0.6844	0.6879
0.5	0.6915	0.6950	0.6985	0.7019	0.7054	0.7088	0.7123	0.7157	0.7190	0.7224
0.6	0.7257	0.7291	0.7324	0.7357	0.7389	0.7422	0.7454	0.7486	0.7517	0.7549
0.7	0.7580	0.7611	0.7642	0.7673	0.7704	0.7734	0.7764	0.7794	0.7823	0.7852
0.8	0.7881	0.7910	0.7939	0.7967	0.7995	0.8023	0.8051	0.8078	0.8106	0.8133
0.9	0.8159	0.8186	0.8212	0.8238	0.8264	0.8289	0.8315	0.8340	0.8365	0.8389
1.0	0.8413	0.8438	0.8461	0.8485	0.8508	0.8531	0.8554	0.8577	0.8599	0.8621
1.1	0.8643	0.8665	0.8686	0.8708	0.8729	0.8749	0.8770	0.8790	0.8810	0.8830
1.2	0.8849	0.8869	0.8888	0.8907	0.8925	0.8944	0.8962	0.8980	0.8997	0.9015
1.3	0.9032	0.9049	0.9066	0.9082	0.9099	0.9115	0.9131	0.9147	0.9162	0.9177
1.4	0.9192	0.9207	0.9222	0.9236	0.9251	0.9265	0.9279	0.9292	0.9306	0.9319
1.5	0.9332	0.9345	0.9357	0.9370	0.9382	0.9394	0.9406	0.9418	0.9429	0.9441
1.6	0.9452	0.9463	0.9474	0.9484	0.9495	0.9505	0.9515	0.9525	0.9535	0.9545
1.7	0.9554	0.9564	0.9573	0.9582	0.9591	0.9599	0.9608	0.9616	0.9625	0.9633
1.8	0.9641	0.9649	0.9656	0.9664	0.9671	0.9678	0.9686	0.9693	0.9699	0.9706
1.9	0.9713	0.9719	0.9726	0.9732	0.9738	0.9744	0.9750	0.9756	0.9761	0.9767
2.0	0.9772	0.9778	0.9783	0.9788	0.9793	0.9798	0.9803	0.9808	0.9812	0.9817
2.1	0.9821	0.9826	0.9830	0.9834	0.9838	0.9842	0.9846	0.9850	0.9854	0.9857
2.2	0.9861	0.9864	0.9868	0.9871	0.9875	0.9878	0.9881	0.9884	0.9887	0.9890
2.3	0.9893	0.9896	0.9898	0.9901	0.9904	0.9906	0.9909	0.9911	0.9913	0.9916
2.4	0.9918	0.9920	0.9922	0.9925	0.9927	0.9929	0.9931	0.9932	0.9934	0.9936
2.5	0.9938	0.9940	0.9941	0.9943	0.9945	0.9946	0.9948	0.9949	0.9951	0.9952
2.6	0.9953	0.9955	0.9956	0.9957	0.9959	0.9960	0.9961	0.9962	0.9963	0.9964
2.7	0.9965	0.9966	0.9967	0.9968	0.9969	0.9970	0.9971	0.9972	0.9973	0.9974
2.8	0.9974	0.9975	0.9976	0.9977	0.9977	0.9978	0.9979	0.9979	0.9980	0.9981
2.9	0.9981	0.9982	0.9982	0.9983	0.9984	0.9984	0.9985	0.9985	0.9986	0.9986
3.0	0.9987	0.9987	0.9987	0.9988	0.9988	0.9989	0.9989	0.9989	0.9990	0.9990
3.1	0.9990	0.9991	0.9991	0.9991	0.9992	0.9992	0.9992	0.9992	0.9993	0.9993
3.2	0.9993	0.9993	0.9994	0.9994	0.9994	0.9994	0.9994	0.9995	0.9995	0.9995
3.3	0.9995	0.9995	0.9995	0.9996	0.9996	0.9996	0.9996	0.9996	0.9996	0.9997
3.4	0.9997	0.9997	0.9997	0.9997	0.9997	0.9997	0.9997	0.9997	0.9997	0.9998
3.5	0.9998	0.9998	0.9998	0.9998	0.9998	0.9998	0.9998	0.9998	0.9998	0.9998
3.6	0.9998	0.9998	0.9999	0.9999	0.9999	0.9999	0.9999	0.9999	0.9999	0.9999
3.7	0.9999	0.9999	0.9999	0.9999	0.9999	0.9999	0.9999	0.9999	0.9999	0.9999
3.8	0.9999	0.9999	0.9999	0.9999	0.9999	0.9999	0.9999	0.9999	0.9999	0.9999
3.9	1.0000	1.0000	1.0000	1.0000	1.0000	1.0000	1.0000	1.0000	1.0000	1.0000
4.0	1.0000	1.0000	1.0000	1.0000	1.0000	1.0000	1.0000	1.0000	1.0000	1.0000

Author Index

A
Adler, R.J., 576
Aitchison, J., 188
Aizerman, M., 732
Alexander, K., 545
Alon, N., 20
Aronszajn, N., 726
Ash, R., 1, 249
Athreya, K., 614, 667, 671, 694
Azuma, K., 483

B
Balakrishnan, N., 54
Barbour, A., 34
Barnard, G., 623
Barndorff-Nielsen, O., 583
Basu, D., 188, 208, 600, 604
Basu, S., 570
Beran, R., 545
Berlinet, A., 731, 733
Bernstein, S., 51, 52, 89
Berry, A., 308
Besag, J., 615, 623
Bhattacharya, R.N., 1, 71, 76, 82, 249, 308, 339, 402, 408, 409, 414, 423
Bickel, P.J., 230, 249, 282, 583, 603, 690, 693, 704, 714
Billingsley, P., 1, 421, 531
Blackwell, D., 188
Borell, C., 572
Bose, R., 208
Braverman, E., 732
Breiman, L., 1, 249, 402
Brémaud, P., 339, 366, 614, 652, 654, 655, 658
Brown, B.M., 498
Brown, L.D., 76, 416, 429, 583, 600, 601, 612
Brown, R., 401

Bucklew, J., 560
Burkholder, D.L., 481, 482

C
Cai, T., 76
Carlin, B., 614, 673
Carlstein, E., 702
Casella, G., 583, 600, 614, 704, 714
Chan, K., 671, 714
Cheney, W., 731, 734
Chernoff, H., 51, 52, 68–70, 89
Chervonenkis, A., 540, 541, 736
Chibisov, D., 534
Chow, Y.S., 249, 259, 264, 267, 276, 463
Chung, K.L., 1, 381, 393, 394, 463
Cirelson, B.S., 572
Clifford, P., 615, 623
Coles, S., 238
Cowles, M., 673
Cox, D., 157
Cramér, H., 249
Cressie, N., 570
Cristianini, N., 736
Csáki, E., 535
Csörgo, M., 421, 423, 425, 427, 527, 536
Csörgo, S., 421

D
DasGupta, A., 1, 3, 71, 76, 215, 238, 243, 249, 257, 311, 317, 323, 327, 328, 333, 337, 421, 429, 505, 527, 570, 695, 699, 704, 714, 723
Dasgupta, S., 208
David, H.A., 221, 228, 234, 236, 238, 323
Davis, B., 481, 482
de Haan, L., 323, 329, 331, 332, 334
Deheuvels, P., 527, 553

757

del Barrio, E., 527, 553
Dembo, A., 560, 570
Dempster, A., 705
den Hollander, F., 560
Devroye, L., 485–488, 560, 725
Diaconis, P., 67, 339, 505, 614, 615, 652, 657, 671
Dimakos, X.K., 615
Do, K.-A., 634
Dobrushin, R.L., 652, 657
Doksum, K., 249, 282, 583, 603, 704, 714
Donsker, M., 421, 422, 530
Doob, J.L., 463
Doss, H., 614, 667, 671
Dubhashi, D., 560
Dudley, R.M., 1, 505, 508, 527, 530, 531, 541, 578
Durrett, R., 402
Dvoretzky, A., 242
Dym, H., 390

E
Eaton, M., 208
Efron, B., 570, 571, 690
Eicker, F., 535
Einmahl, J., 535
Einmahl, U., 425
Embrechts, P., 238
Erdös, P., 421, 422
Esseen, C., 308
Ethier, S., 243
Everitt, B., 54

F
Falk, M., 238
Feller, W., 1, 62, 71, 76, 77, 249, 254, 258, 277, 285, 307, 308, 316, 317, 339, 375, 386, 389
Ferguson, T., 188, 249
Fernique, X., 576
Fill, J., 615, 657
Finch, S., 382
Fisher, R.A., 25, 212, 586, 602
Fishman, G.S., 614, 624
Freedman, D., 339, 402, 693
Fristedt, B., 463, 472, 493, 496
Fuchs, W., 381, 394

G
Galambos, J., 221, 238, 323, 328, 331
Gamerman, D., 614
Garren, S., 674

Gelman, A., 614, 674
Geman, D., 614
Geman, S., 614
Genz, A., 157
Geyer, C., 614
Ghosh, M., 208
Gibbs, A., 505, 508, 509, 516
Gilks, W., 614
Giné, E., 527, 538, 541, 542, 545, 548, 570, 694
Glauber, R., 646
Gnedenko, B.V., 328
Götze, F., 570
Gray, L., 463, 472, 493, 496
Green, P.J., 615
Groeneboom, P., 563
Gundy, R.F., 481, 482
Györfi, L., 560, 725

H
Haff, L.R., 208, 327, 429
Hall, P., 34, 71, 224, 231, 249, 308, 331, 421, 463, 497, 498, 560, 570, 623, 634, 690, 691, 694, 701, 704
Hastings, W., 614
Heyde, C., 249, 421, 463, 497, 498
Higdon, D., 615
Hinkley, D., 174
Hobert, J., 671
Hoeffding, W., 483, 570
Horowitz, J., 704
Hotelling, H., 210
Hüsler, J., 238

I
Ibragimov, I.A., 508
Isaacson, D., 339

J
Jaeschke, D., 535
Jing, B., 704
Johnson, N., 54
Jones, G., 671

K
Kac, M., 421, 422
Kagan, A., 69, 157
Kamat, A., 156
Karatzas, I., 414, 416, 417, 463
Karlin, S., 402, 411, 427, 437, 463

Kass, R., 519
Kemperman, J., 339
Kendall, M.G., 54, 62, 65
Kendall, W., 615
Kesten, H., 254, 257
Khare, K., 614, 671
Kiefer, J., 242
Kingman, J.F.C., 437, 439, 442, 451, 456, 457
Klüppelberg, C., 238
Knight, S., 673
Komlós, J., 421, 425, 426, 537
Körner, T., 417
Kosorok, M., 527
Kotz, S., 54
Krishnan, T., 705, 711, 714, 739
Künsch, H.R., 702

L

Lahiri, S.N., 690, 702–704
Laird, N., 705
Landau, H.J., 577
Lange, K., 714
Lawler, G., 402, 437
Le Cam, L., 34, 71, 509, 704
Leadbetter, M., 221
Ledolter, J., 714
Ledoux, M., 560
Lehmann, E.L., 238, 249, 583, 600, 699, 704
Leise, F., 505
Levine, R., 714
Light, W., 731
Lindgren, G., 221
Linnik, Y., 69, 157
Liu, J., 614, 657
Logan, B.F., 570
Lugosi, G., 560, 725

M

Madsen, R., 339
Mahalanobis, P., 208
Major, P., 421, 425, 426, 537
Mallows, C.L., 570
Martynov, G., 238
Mason, D.M., 535, 537, 570
Massart, P., 242
McDiarmid, C., 485, 487, 488, 560
McKean, H., 390
McLachlan, G., 705, 711, 714, 739
Mee, R., 157
Mengersen, K., 667, 669, 673

Metropolis, N., 614
Meyn, S., 339
Mikosch, T., 238
Millar, P., 545
Minh, H., 734
Morris, C., 612
Mörters, P., 402, 420
Murray, G.D., 712
Mykland, P., 674

N

Niyogi, P., 734
Norris, J., 339, 366

O

O'Reilly, N., 534
Olkin, I., 208
Oosterhoff, J., 563
Owen, D., 157

P

Paley, R.E., 20
Panconesi, A., 560
Parzen, E., 437
Patel, J., 157
Peres, Y., 402, 420
Perlman, M., 208
Petrov, V., 62, 249, 301, 308–311
Pitman, J., 1
Plackett, R., 157
Politis, D., 702, 704
Pollard, D., 527, 538, 548
Pólya, G., 382
Port, S., 239, 264, 265, 302, 303, 306, 315, 437
Propp, J., 615
Pyke, R., 421

R

Rachev, S.T., 505
Rao, C.R., 20, 62, 69, 157, 210, 505, 519, 520, 522
Rao, R.R., 71, 76, 82, 249, 308
Read, C., 157
Reiss, R., 221, 231, 234, 236, 238, 285, 323, 326, 328, 505, 516
Rényi, A., 375, 380, 389
Resnick, S., 221, 402
Révész, P., 254, 420, 421, 423, 425, 427, 527

Revuz, D., 402
Rice, S.O., 570
Richardson, S., 614
Ripley, B.D., 614
Robert, C., 614, 624, 673
Roberts, G., 614, 667
Romano, J., 702
Rootzén, H., 221
Rosenblatt, M., 721
Rosenbluth, A., 614
Rosenbluth, M., 614
Rosenthal, J.S., 615, 667, 671
Ross, S., 1, 614, 624
Roy, S., 208
Rozonoer, L., 732
Rubin, D., 614, 674, 705
Rubin, H., 337, 628
Rudin, W., 726, 730

S
Saloff-Coste, L., 505, 614, 652, 671
Sauer, N., 539, 540
Schmeiser, B., 624
Sen, P.K., 328, 329
Seneta, E., 339, 361
Serfling, R., 249, 311, 323, 326
Sethuraman, J., 614, 667, 671
Shao, J., 690, 695, 701
Shao, Q.M., 560, 570, 571
Shawe-Taylor, J., 736
Shepp, L., 570, 577
Shorack, G., 238, 527
Shreve, S., 414, 416, 417, 463
Singer, J., 328, 329
Singh, K., 693, 699, 701
Sinha, B., 208
Smith, A.F.M., 614
Smith, R.L., 674
Sparre-Andersen, E., 389
Spencer, J., 20
Spiegelhalter, D., 614
Spitzer, F., 375, 389, 390
Steele, J.M., 34
Stein, C., 67
Stern, H., 614
Stigler, S., 71
Stirzaker, D., 1, 339
Strassen, V., 425
Strawderman, W.E., 429
Stroock, D., 560, 614
Stuart, A., 54, 62, 65
Sturmfels, B., 615
Su, F., 505, 508, 509, 516

Sudakov, V.N., 572
Sundberg, R., 705

T
Talagrand, M., 572, 578
Tanner, M., 614, 714
Taylor, H.M., 402, 411, 427, 437, 463
Teicher, H., 249, 259, 264, 267, 276, 463
Teller, A., 614
Teller, E., 614
Thomas-Agnan, C., 731, 733
Thönnes, E., 615
Tibshirani, R., 690
Tierney, L., 614, 667, 668, 671
Titterington, D.M., 623
Tiwari, R., 188
Tong, Y., 138, 154, 201, 203, 204, 208, 210, 211, 215, 698
Tu, D., 690, 695, 701
Tusnady, G., 421, 425, 426, 537
Tweedie, R., 339, 667, 669

V
Vajda, I., 505
van de Geer, S., 527, 553
van der Vaart, A., 249, 310, 527, 546, 547, 557
van Zwet, W.R., 537
Vapnik, V., 540, 541, 736
Varadhan, S.R.S., 456, 560
Vos, P., 519

W
Wang, Q., 560
Wasserman, L., 67
Waymire, E., 1, 339, 402, 408, 409, 414, 423
Wei, G., 714
Wellner, J., 238, 310, 527, 538, 546, 547, 557
Wermuth, N., 157
White, A., 704
Whitt, W., 421
Widder, D., 53
Williams, D., 463
Wilson, B., 615
Wolf, M., 702
Wolfowitz, J., 242
Wong, W., 614
Wu, C.F.J., 711–713

Y
Yang, G., 704
Yao, Y., 734
Yor, M., 402
Yu, B., 674

Z
Zabell, S., 67
Zeitouni, O., 560
Zinn, J., 538, 694
Zolotarev, V.M., 505
Zygmund, A., 20

Subject Index

A
ABO allele, 709–711
Acceptance distribution, 664
Accept-reject method
  beta generation, 627–628
  described, 625
  generation, standard normal values, 626–627
  scheme efficiency, 628
Almost surely, 252, 253, 256, 259, 261–262, 267, 287, 288, 314, 410, 461, 470, 491–494, 496, 498, 503, 535, 537, 542, 555, 576, 577, 578, 615, 618, 632, 641, 694, 725
Ancillary
  definition, 601
  statistic, 602, 604, 605
Anderson's inequality, 214, 218
Annulus, Dirichlet problem, 417–418
Aperiodic state, 351
Approximation of moments
  first-order, 278, 279
  scalar function, 282
  second-order, 279, 280
  variance, 281
Arc sine law, 386
ARE. See Asymptotic relative efficiency
Arrival rate, 552, 553
  intensity function, 455
  Poisson process, 439, 441, 442, 446, 460–462
Arrival time
  definition, 438
  independent Poisson process, 445
  interarrival times, 438
Asymptotic independence, 337
Asymptotic relative efficiency (ARE)
  IQR-based estimation, 327
  median, 335
  variance ratio, 325
Asymptotics
  convergence, distribution
    CDF, 262
    CLT, 266
    Cramér–Wold theorem, 263–264
    Helly's theorem, 264
    LIL, 267
    multivariate CLT, 267
    Pólya's theorem, 265
    Portmanteau theorem, 265–266
  densities and Scheffé's theorem, 282–286
  laws, large numbers
    Borel-Cantelli Lemma, 254–256
    Glivenko-Cantelli theorem, 258–259
    strong, 256–257
    weak, 256–258
  moments, convergence (see Convergence of moments)
  notation and convergence, 250–254
  preservation, convergence
    continuous mapping, 260–261
    Delta theorem, 269–272
    multidimension, 260
    sample correlation, 261–262
    sample variance, 261
    Slutsky's theorem, 268–269
    transformations, 259
    variance stabilizing transformations, 272–274
Asymptotics, extremes and order statistics
  application, 325–326
  several, 326–327
  single, 323–325
  convergence, types theorem
    limit distributions, types, 332
    Mills ratio, 333
  distribution theory, 323

763

Asymptotics, extremes and order
        statistics (*cont.*)
    easy applicable limit theorem
        Gumbel distribution, 331
        Hermite polynomial, 329
    Fisher–Tippet family
        definition, 333
        theorem, 334–335
Avogadro's constant, 401
Azuma's inequality, 483–486

B

Ballot theorem, 390
Barker's algorithm, 643
Basu's theorem
    applications, probability
        convergence result, 606–607
        covariance calculation, 605
        expectation calculation, 606
        exponential distribution result, 605
        mean and variance independence,
            604–605
    exponential family, 602
    and Mahalanobis's D^2, 611
    and Neyman–Fisher factorization
        definition, 602–603
        factorization, 603–604
        general, 604
Bayes estimate, 147–152, 467–468, 493–494
Bayes theorem
    conditional densities, 141, 147, 149
    use, 147
Berry–Esseen bound
    definition, 76
    normal approximation, 76–79
Bessel's inequality, 742
Best linear predictor, 110–111, 154
Best predictor, 142, 154
Beta-Binomial distribution
    Gibbs sampler, 965–966
    simulation, 643–644
Beta density
    defintion, 59–60
    mean, 150
    mode, 89
    percentiles, 618
Bhattacharya affinity, 525
Bikelis local bound, 322
Binomial confidence interval
    normal approximation, 74
    score confidence, 76
    Wald confidence, 74–76

Binomial distribution
    hypergeometric distribution problems, 30
    negative, 28–29
Bivariate Cauchy, 195
Bivariate normal
    definition, 136
    distribution, 212
    five-parameter distribution, 138–139
    formula, 138
    joint distribution, 140
    mean and variance independence, 139–140
    missing coordinates, 707–709
    property, 202
    simulation, 136, 137, 200–201
Bivariate normal conditional distributions
    conditional expectation, 154
    Galton's observation, 155
Bivariate normal formulas, 138, 156
Bivariate Poisson, 120
    Gibbs chain, 683
    MGF, 121
Bivariate uniform, 125–126, 133
Block bootstrap, 701, 702, 703
Bochner's theorem, 306–307
Bonferroni bounds, 5
Bootstrap
    consistency
        correlation coefficient, 697–698
        Kolmogorov metric, 692–693
        Mallows-Wasserstein metric, 693
        sample variance, 696
        Skewness Coefficient, 697
        t-statistic, 698
    distribution, 690, 691, 694, 696, 738
    failure, 698–699
    higher-order accuracy
        CBB, 702
        MBB, 702
        optimal block length, 704
        second-order accuracy, 699–700
        smooth function, 703–704
        thumb rule, 700–701
    Monte Carlo, 694, 697, 737
    ordinary bootstrap distribution, 690–691
    resampling, 689–690
    variance estimate, 737
Borel–Cantelli lemma
    almost sure convergence, binomial
        proportion, 255–256
    pairwise independence, 254–255
Borell inequality, 215
Borel's paradox
    Jacobian transformation, 159

Subject Index

marginal and conditional density, 159–160
mean residual life, 160–161
Boundary crossing probabilities
  annulus, 417–418
  domain and harmonic, 417
  harmonic function, 416
  irregular boundary points, 416
  recurrence and transience, 418–419
Bounded in probability, 251–252
Box–Mueller transformation, 194
Bracket, 546
Bracketing number, 546
Branching process, 500
Brownian bridge
  Brownian motion, 403
  empirical process, iid random variables, 404
  Karhunen–Loeve expansion, 554
  maximum, 410
  standard, 405
Brownian motion
  covariance functions, 406
  d-dimensional, 405
  Dirichlet problem and boundary crossing probabilities
    annulus, 417–418
    domain and harmonic, 417
    harmonic function, 416
    irregular boundary points, 416
    recurrence and transience, 418–419
  distributional properties
    fractal nature, level sets, 415–416
    path properties and behavior, 412–414
    reflection principle, 410–411
  explicit construction
    Karhunen–Loéve expansion, 408
  Gaussian process, Markov
    continuous random variables, 407
    correlation function, 406
  invariance principle and statistics
    convergence, partial sum process, 423–424
    Donsker's, 424–425
    partial sums, 421
    Skorohod embedding theorem, 422, 423
    uniform metric, 422
  local time
    Lebesgue measure, 419
    one-dimensional, 420
    zero, 420–421
  negative drift and density, 427–428
  Ornstein–Uhlenbeck process
    convergence, 430–431
    covariance function, 429
    Gaussian process, 430
  random walks
    scaled, simulated plot, 403
    state visited, planar, 406
  stochastic process
    definition, 403
    Gaussian process, 404
    real-valued, 404
    standard Wiener process, 404–405
    strong invariance principle and KMT theorem, 425–427
    transition density and heat equation, 428–429

C

Campbell's theorem
  characteristic functional Π, 456
  Poisson random variables, 457
  shot effects, 456
  stable laws, 458
Canonical exponential family
  description, 590
  form and properties
    binomial distribution, 590
    closure, 594–596
    convexity, 590–591
    moment and generating function, 591–594
    one parameter, 589–590
  k-parameter, 597
Canonical metric
  definition, 575
Capture-recapture, 32–33
Cauchy
  density, 44
  distribution, 44, 48
Cauchy order statistics, 232
Cauchy random walk, 395
Cauchy–Schwarz inequality, 21, 109, 259, 516, 731
CDF. *See* Cumulative distribution function
Central limit theorem (CLT)
  approximation, 83
  binomial confidence interval, 74–76
  binomial probabilities., 72–73
  continuous distributions, 71
  de Moivre–Laplace, 72
  discrete distribution, 72
  empirical measures, 543–547
  error, 76–79
  iid case, 304
  martingale, 498
  multivariate, 267

Central limit theorem (CLT) *(cont.)*
  random walks, 73
  and WLLN, 305–306
Change of variable, 13
Change point problem, 650–651
Chapman–Kolmogorov equation, 256, 315, 318
  transition probabilities, 345–346
Characteristic function, 171, 263, 383–385, 394–396, 449, 456, 458, 462, 500, 533
  Bochner's theorem, 306–307
  CLT error
    Berry–Esseen theorem, 309–310
    CDF sequence, 308–309
    theorem, 310–311
  continuity theorems, 303–304
  Cramér–Wold theorem, 263
  Euler's formula, 293
  inequalities
    Bikelis nonuniform, 317
    Hoeffding's, 318
    Kolmogorov's maximal, 318
    partial sums moment and von Bahr–Esseen, 318
    Rosenthal, 318–319
  infinite divisibility and stable laws
    distributions, 317
    triangular arrays, 316
  inversion and uniqueness
    distribution determining property, 299–300
    Esseen's Lemma, 301
    lattice random variables, 300–301
    theorems, 298–299
  Lindeberg–Feller theorem, 311–315
  Polýa's criterion, 307–308
  proof
    CLT, 305
    Cramér–Wold theorem, 306
    WLLN, 305
  random sums, 320
  standard distributions
    binomial, normal and Poisson, 295–296
    exponential, double exponential, and Cauchy, 296
    n-dimensional unit ball, 296–297
  Taylor expansions, differentiability and moments
    CDF, 302
    Riemann–Lebesgue lemma, 302–303
Chebyshev polynomials, 742
Chebyshev's inequality
  bound, 52
  large deviation probabilities, 51

Chernoff's variance inequality
  equality holding, 68–69
  normal distribution, 68
Chibisov–O'Reilly theorem, 534–535
Chi square density
  degree of freedom, 45
  Gamma densities, 58
Chung–Fuchs theorem, 381, 394–396
Circular Block Bootstrap (CBB), 702
Classification rule, 724, 725, 734, 735, 736, 743, 744
Communicating classes
  equivalence relation, 350
  identification, 350–351
  irreducibility, 349
  period computation, 351
Completeness
  applications, probability
    covariance calculation, 605
    expectation calculation, 606
    exponential distribution result, 605
    mean and variance independence, 604
    weak convergence result, 606–607
  definition, 601
  Neyman–Fisher factorization and Basu's theorem
    definition, 602–603
Complete randomness
  homogeneous Poisson process, 551
  property, 446–448
Compound Poisson Process, 445, 448, 449, 459, 461
Concentration inequalities
  Azuma's, 483
  Burkholder, Davis and Gundy
    general square integrable martingale, 481–482
    martingale sequence, 480
  Cirel'son and Borell, 215
  generalized Hoeffding lemma, 484–485
  Hoeffding's lemma, 483–484
  maximal
    martingale, 477–478
    moment bounds, 479–480
    sharper bounds near zero, 479
    submartingale, 478
  McDiarmid and Devroye
    Kolmogorov–Smirnov statistic, 487–488
    martingale decomposition and two-point distribution, 486–487
  optional stopping theorem, 477
  upcrossing, 488–490

Subject Index

Conditional density
  Bayes theorem, 141
  best predictor, 142
  definition, 140–141
  two stage experiment, 143–144
Conditional distribution, 202–205
  binomial, 119
  definition, 100–101
  interarrival times, 460
  and marginal, 189–190
  and Markov property, 235–238
  Poisson, 104
  recurrence times, 399
Conditional expectation, 141, 150, 154
  Jensen's inequality, 494
  order statistics, 247
Conditional probability
  definition, 5
  prior to posterior belief, 8
Conditional variance, 143, 487, 497
  definition, 103, 141
Confidence band, continuous CDF, 242–243
Confidence interval
  binomial, 74–75
  central limit theorem, 61–62
  normal approximation theorem, 81
  Poisson distribution, 80
  sample size calculation., 66
Confidence interval, quantile, 246, 326
Conjugate priors, 151–152
Consistency bootstrap, 693–699
Consistency of kernel classification, 743
Continuity of Gaussian process
  logarithm tail, maxima and Landau–Shepp theorem, 577
  Wiener process, 576–577
Continuous mapping, 260–262, 268, 270, 275, 306, 327, 694
Convergence diagnostics
  Garren–Smith multiple chain Gibbs method, 675
  Gelman–Rubin multiple chain method, 674
  spectral and drift methods, 673–674
  Yu–Mykland single chain method, 674–675
Convergence in distribution
  CLT, 266
  Cramér–Wold theorem, 263
  Delta theorem, 269–272
  description, 262
  Helly's theorem, 264
  LIL, 267
  multivariate CLT, 267
  Pólya's theorem, 265

Portmanteau theorem, 265–266
Slutsky's theorem, 268–269
Convergence in mean, 253
Convergence in probability
  continuous mapping theorem, 270
  sure convergence, 252, 260
Convergence of Gibbs sampler
  discrete and continuous state spaces, 671
  drift method, 672–673
  failure, 670–671
  joint density, 672
Convergence of martingales
  L_1 and L_2
    basic convergence theorem, 493
    Bayes estimates, 493–494
    Pólya's urn, 493
  theorem
    Fatou's lemma and monotone, 491
Convergence of MCMC
  Dobrushin's inequality and
    Diaconis–Fill–Stroock bound, 657–659
  drift and minorization methods, 659–662
  geometric and uniform ergodic, 653
  separation and chi-square distance, 653
  SLEM, 651–652
  spectral bounds, 653–657
  stationary Markov chain, 651
  total variation distance, 652
Convergence of medians, 287
Convergence of moments
  approximation, 278–282
  distribution, 277–278
  uniform integrability
    conditions, 276–277
    dominated convergence theorem, 275–276
Convergence of types, 328, 332
Convex function theorem, 468, 495–496
Convolution, 715
  definition, 167–168
  density, 168
  double exponential, 195
  n-fold, 168–169
  random variable symmetrization, 170–172
  uniform and exponential, 194
Correlation, 137, 195, 201, 204, 271–272
  coefficient, 212–213, 697–698
  convergence, 261–262
  definition, 108
  exponential order statistics, 234
  inequality, 120
  order statistics, 244, 245

Correlation (*cont.*)
  Poisson process, 461
  properties, 108–109
Correlation coefficient distribution
  bivariate normal, 212–213
  bootstrapping, 697–698
Countable additivity
  definition, 2–3
Coupon collection, 288
Covariance, 544, 574, 735
  calculation, 107, 605
  definition, 107
  function, 412, 429
  inequality, 120, 215
  matrix, 207, 216
  multinomial, 610
  properties, 108–109
Covariance matrix, 136–137, 200, 203, 205, 207, 210, 216, 217, 267, 272, 280, 326, 530, 597, 611, 709
Cox–Wermuth approximation
  approximations test, 157–158
Cramér–Chernoff theorem
  Cauchy case, 564
  exponential tilting technique, 563
  inequality, 561
  large deviation, 562
  rate function
    Bernoulli case, 563–564
    definition, 560
    normal, 563
Cramér–Levy theorem, 293
Cramér–von Mises statistic, 532–533
Cramér–Wold theorem, 263, 266, 305, 306
Cubic lattice random walks
  distribution theory, 378–379
  Pólya's formula, 382–383
  recurrence and transience, 379–381
  recurrent state, 377
  three dimensions, 375
  two simulated, 376, 377
Cumulants
  definition, 25–26
  recursion relations, 26
Cumulative distribution function (CDF)
  application, 87–88, 90, 245
  continuous random variable, 36–37
  definition, 9
  empirical, 241–243
  exchangeable normals, 206–207
  and independence, 9–12
  joint, 177, 193
  jump function, 10
  Markov property, 236

moments and tail, 49–50
nonnegative integer-valued random variables, 16
PDF and median, 38
Pólya's theorem, 265
quantile transformation, 44
range, 226
standard normal density, 41–42, 62
Curse of dimensionality, 131–132, 184
Curved exponential family
  application, 612
  definition, 583, 608
  density, 607–608
  Poissons, random covariates, 608–609
  specific bivariate normal, 608

D

Delta theorem, continuous mapping theorem
  Cramér, 269–270
  random vectors, 269
  sample correlation, 271–272
  sample variance and standard deviation, 271
Density
  midrange, 245
  range, 226
Density function
  continuous random variable, 38
  description, 36
  location scale parameter, 39
  normal densities, 40
  standard normal density, 42
Detailed balance equation, 639–640
Diaconis–Fill–Stroock bound, 657–658
Differential metrics
  Fisher information
    direct differentiation, 521
    multivariate densities, 520
    Kullback–Leibler distance curvature, 519–520
  Rao's Geodesic distances, distributions
    geodesic curves, 522
    variance-stabilizing transformation, 522
Diffusion coefficient
  Brownian motion, 427, 428
Dirichlet cross moment, 196
Dirichlet distribution
  density, 189
  Jacobian density theorem, 188–189
  marginal and conditional, 189–190
  and normal, 190
  Poincaré's Lemma, 191
  subvector sum, 190

Subject Index

Dirichlet problem
  annulus, 417–418
  domain and harmonic, 417
  harmonic function, 416
  irregular boundary points, 416
  recurrence and transience, 418–419
Discrete random variable
  CDF and independence, 9–12
  definition, 8
  expectation and moments, 13–19
  inequalities, 19–22
  moment-generating functions, 22–26
Discrete uniform distribution, 2, 25, 277
Distribution
  binomial
    hypergeometric distribution problems, 30
  Cauchy, 44, 48
  CDF
    definition, 9
  Chebyshev's inequality, 20
  conditional, 203–205
  continuous, 71
  correlation coefficient, 212–213
  discrete, 72
  discrete uniform, 2, 25, 277
  lognormal, 65
  noncentral, 213–214
  Poisson, 80
  quadratic forms, Hotelling's T^2
    Fisher Cochran theorem, 211
  sampling, statistics
    correlation coefficient, 212–213
    Hotelling's T^2 and quadratic forms, 209–212
    Wishart, 207–208
    Wishart identities, 208–209
DKW inequality, 241–242, 541, 553
Dobrushin coefficient
  convergence of Metropolis chains, 668–669
  defined, 657–658
  Metropolis–Hastings, truncated geometric, 659
  two-state chain, 658–659
Dobrushin's inequality
  coefficient, 657–658
  Metropolis–Hastings, truncated geometric, 659
  stationary Markov chain, 658
  two-state chain, 658–659
Domain
  definition, 417
  Dirichlet problem, 416, 417

  irregular boundary point, 416
  maximal attraction, 332, 336
Dominated convergence theorem, 275, 285
Donsker class, 544
Donsker's theorem, 423, 531
Double exponential density, 39–40
Doubly stochastic
  bistochastic, 340
  one-step transition matrix, 347
  transition probability matrix, 342
Drift
  Bayes example, 672–673
  drift-diffusion equation, 429
  and minorization methods, 659–662
  negative, Brownian motion, 427–428
  time-dependent, 429
Drift-diffusion equation., 429

E
Edgeworth expansion, 82–83, 302, 622, 699, 700
Ehrenfest model, 342, 373
EM algorithm, 689, 704–744
  ABO allele, 709–711
  background plus signal model, 738
  censoring, 740
  Fisher scoring method, 705
  genetics problem, 739
  mixture problem, 739
  monotone ascent and convergence, 711
  Poisson, missing values, 706–707
  truncated geometric, 738
Empirical CDF
  confidence band, continuous, 242, 243
  definition, 527
  DKW inequality, 241–242
Empirical measure
  CLTs
    entropy bounds, 544–546
    notation and formulation, 543–544
  definition, 222, 528
Empirical process
  asymptotic properties
    approximation, 530–531
    Brownian bridge, 529–530
    invariance principle and statistical applications, 531–533
    multivariate central limit theorem, 530
    quantile process, 536
    strong approximations, 537
    weighted empirical process, 534–535
  CLTs, measures and applications
    compact convex subsets, 547

Empirical process (*cont.*)
    entropy bounds, 544–546
    invariance principles, 543
    monotone and Lipschitz functions, 547
    notation and formulation, 543–544
    maximal inequalities and symmetrization
        Gaussian processes, 547–548
    notation and definitions
        cadlag functions and Glivenko–Cantelli
            theorem, 529
    Poisson process
        distributional equality, 552
        exceedance probability, 553
        randomness property, 551
        Stirling's approximation, 553
    Vapnik–Chervonenkis (VC) theory
        dimensions and classes, 540–541
        Glivenko–Cantelli class, 538–539
        measurability conditions, 542
        shattering coefficient and Sauer's
            lemma, 539–540
        uniform convergence, relative
            frequencies, 542–543
Empirical risk, 736
Entropy bounds
    bracket and bracketing number, 546
    covering number, 545–546
    Kolmogorov–Smirnov type statistic, 545
    P-Donsker, 546
    VC-subgraph, 545
Equally likely
    concept, 3
    finite sample space, 3
    sample points, 3
Ergodic theorem, 366, 640, 641
Error function, 90
Error of CLT
    Berry–Esseen bound, 77
Error probability
    likelihood ratio test, 565, 580
    type I and II, 565
    Wald's SPRT, 476–477
Esseen's lemma, 301, 309
Estimation
    density estimator, 721
    histogram estimate, 683, 720–721, 737, 740
    optimal local bandwidth, 723
    plug-in estimate, 212, 630, 704, 720, 740
    series estimation, 720–721
Exceedance probability, 553
Exchangeable normal
    variables
        CDF, 206–207

        construction, 205–206
        definition, 205
Expectation
    continuous random variables, 45–46
    definitions, 46–47
    gamma function, 47
Experiment
    countable additivity, 2–3
    countably infinite set, 2
    definition, 2
    inclusion–exclusion formula, 5
Exponential density
    basic properties, 55
    definition, 38
    densities, 55
    mean and variance, 47
    standard double, 40–41
Exponential family
    canonical form and properties
        closure, 594–596
        convexity, 590–591
        definition, 589–590
        moments and moment generating
            function, 591–594
    curved (*see* Curved exponential family)
    multiparameter, 596–600
    one-parameter
        binomial distribution, 586
        definition, 585–586
        gamma distribution, 587
        irregular distribution, 588–589
        normal distribution, mean and variance, 584–586
        unusual gamma distribution, 587
        variables errors, 586–587
    sufficiency and completeness
        applications, probability, 604–607
        definition, 601–602
        description, 600
        Neyman–Fisher factorization and
            Basu's theorem, 602–604
Exponential kernel, 736, 740, 742, 744
Exponential tilting
    technique, 563
Extremes. *See also* Asymptotics, extremes and order statistics
    definition, 221
    discrete-time stochastic sequence, 574–575
    finite sample theory, 221–247

F

Factorial moment, 22
Factorization theorem, 602–604

Subject Index

Failure of EM, kernel, 712–713
Fatou's lemma, 491, 492
F-distribution, 174–175, 218
f-divergence
  finite partition property, 507, 517
Feature maps, 732–744
Filtered Poisson process, 443–444
Finite additivity, 3
Finite dimensional distributions, 575
Finite sample theory
  advanced distribution theory
    density function, 226
    density, standard normals, 229
    exponential order statistics, 227–228
    joint density, 225–226
    moment formulas, uniform case, 227
    normal order statistics, 228
  basic distribution theory
    empirical CDF, 222
    joint density, 222–223
    minimum, median and maximum variables, 225
    order statistics, 221
    quantile function, 222
    uniform order statistics, 223–224
  distribution, multinomial maximum
    equiprobable case, 243
    Poissonization technique, 243
  empirical CDF
    confidence band, continuous CDF, 242–243
    DKW inequality, 241–242
  existence of moments, 230–232
  quantile transformation theorem, 229–230
  records
    density, values and times, 240–241
    interarrival times, 238
  spacings
    description, 233
    exponential and Réyni's representation, 233–234
    uniform, 234–235
First passage, 354–359, 383–386, 409, 410, 427, 475
Fisher Cochran theorem, 211
Fisher information
  and differential metrics, 520–521
  matrix, 597–600, 610–611
Fisher information matrix
  application, 597–610
  definition, 597
  and differential, 520–521
  multiparameter exponential family, 597–600

Fisher linear, 735
Fisher's Iris data, 744
Fisher–Tippet family
  definition, 333
  theorem, 334–335
Formula
  change of variable, 13
  inclusion-exclusion, 4
  Tailsum, 16
Full conditionals, 645–646, 648–651, 671, 672
Function theorem
  convex, 468, 495–496
  implicit, 569

G

Gambler's ruin, 351–353, 370, 392, 464, 468–470, 475, 501
Gamma density
  distributions, 59
  exponential density, 56
  skewness, 82
Gamma function
  definition, 47
Garren–Smith multiple chain Gibbs method, 675
Gartner–Ellis theorem
  conditions, 580
  large deviation rate function, 567
  multivariate normal mean, 568–569
  non-iid setups, 567–568
Gaussian factorization, 176–177
Gaussian process
  definition, 404
  Markov property, 406–407
  stationary, 430
Gaussian radial kernel, 736
Gegenbauer polynomials, 741
Gelman–Rubin multiple chain method, 674
Generalized chi-square distance, 525, 652, 653
Generalized Hoeffding lemma, 484–485
Generalized negative binomial distribution, 28–29, 609
Generalized Wald identity, 475–476
Generalized Wasserstein distance, 525
Generating function
  central moment, 25–26
  Chernoff–Bernstein inequality, 51–52
  definition, 22
  distribution-determining property, 24–25
  inversion, 53
  Jensen's inequality, 52–53
  Poisson distribution, 23–24
  standard exponential, 51

Geometrically ergodic
  convergence of MCMC, 653
  Metropolis chain, 669–670
  spectral bounds, 654
Geometric distribution
  definition, 28
  exponential densities, 55
  lack of memory, 32
Gibbs sampler
  beta-binomial pair, 647–648
  change point problem, 650–651
  convergence, 670–673
  definition, 646
  Dirichlet distributions, 649–650
  full conditionals, 645–646
  Gaussian Bayes
    formal and improper priors, 648
  Markov chain generation problem, 645
  random scan, 646–647
  systematic scan, 647
Glivenko–Cantelli class, 539, 541
Glivenko–Cantelli theorem, 258–259, 336, 487, 529, 538, 541, 545, 555, 690
Gumbel density
  definition, 61
  standard, 61

H

Hájek–Rényi inequality, 396
Hardy–Weinberg law, 609
Harmonic
  definition, 417
  function, 416
Harris recurrence
  convergence property, 672
  drift and minorization methods, 660
  Metropolis chains, 667–668
Heat equation, 429, 435
Hellinger metric, 506–507, 514
Helly's theorem, 264
Helmert's transformation, 186–187
Hermite polynomials, 83, 320, 329, 741
High dimensional formulas, 191–192
Hilbert norm, 729
Hilbert space, 725–732, 734, 735, 743
Histogram estimate, 683, 720–721, 737, 740
Hoeffding's inequality, 318, 483–484
Holder continuity, 413
Holder's inequality, 21
Hotelling's T^2
  Fisher Cochran theorem, 211
  quadratic form, 210

Hypergeometric distribution
  definition, 29
  ingenious use, 32

I

IID. *See* Independent and identically distributed
Importance sampling
  Bayesian calculations, 629–630
  binomial Bayes problem, 631
  described, 619–620
  distribution, 629, 633–634
  Radon–Nikodym derivatives, 629
  theoretical properties, 632–633
Importance sampling distribution, 633–634
Inclusion-exclusion formula, 4, 5
Incomplete inner product space, 728
Independence
  definition, 5, 9
Independence of mean and variance, 139–140, 185–187, 211, 604–605
Independent and identically distributed (IID)
  observations, 140
  random variables, 20
Independent increments, 404–405, 409, 412, 427, 433, 449
Independent metropolis algorithm, 643
Indicator variable
  Bernoulli variable, 11
  binomial random variable, 34
  distribution, 11
  mathematical calculations, 10
Inequality
  Anderson's, 214
  Borell concentration, 215
  Cauchy–Schwarz, 21
  central moment, 25
  Chen, 215
  Chernoff–Bernstein inequality, 51–52
  Cirel'son, 215
  covariance, 215–216
  distribution, 19
  Jensen's inequality, 52–53
  monotonicity, 214–215, 593
  positive dependence, 215
  probability, 20–21
  Sidak, 215
  Slepian's I and II, 214
Infinite divisibility and stable laws
  description, 315
  distributions, 317
  triangular arrays, 316

Subject Index 773

Initial distribution, 369, 641, 651, 652, 655, 658, 668
   definition, 340
   Markov chain, 361
   weak ergodic theorem, 366
Inner product space, 726, 727, 728, 740–742
Inspection paradox, 444
Integrated Brownian motion, 432
Intensity, 557, 682
   function, 454, 455, 460
   piecewise linear, 454–456
Interarrival time, 238, 240, 241
   conditional distribution, 453
   Poisson process, 438
   transformed process, 453
Interquartile range (IQR)
   definition, 327
   limit distribution, 335
Invariance principle
   application, 434
   convergence, partial sum process, 423–424
   Donsker's, 424–425, 538
   partial sums, 421
   Skorohod embedding theorem, 422, 423
   and statistics
      Cramér–von Mises statistic, 532–533
      Kolmogorov–Smirnov statistic, 531–532
   strong
      KMT theorem, 425–427
      partial sum process, 530–531, 537
   uniform metric, 422
   weak, 536
Invariant measure, 663
Inverse Gamma density, 59
Inverse quadratic kernel, 736, 744
Inversion of mgf
   moment-generating function, 53
Inversion theorem
   CDF, 298–299
   failure, 307–308
   Plancherel's identity, 298
IQR. *See* Interquartile range
Irreducible, 569, 581, 640–643, 645, 647, 653–655, 660, 661, 663, 667, 671, 685
   definition, 349
   loop chains, 371
   regular chain, 363
Isolated points, 415, 416
Isotropic, 717, 742, 743
Iterated expectation, 105, 119, 146, 162, 172
   applications, 105–106

   formula, 105, 146, 167, 468
   higher-order, 107
Iterated variance, 119, 623
   applications, 105–106
   formula, 105

J

Jacobian formula
   CDF, 42
   technique, 45
Jacobi polynomials, 741, 742
Jensen's inequality, 52–53, 280, 468, 491, 494, 516
Joint cumulative distribution function, 97, 112, 124, 177, 193, 261, 263, 447
Joint density
   bivariate uniform, 125–126
   continuous random vector, 125
   defined, 123–124
   dimensionality curse, 131–132
   nonuniform, uniform marginals, 128–129
Joint moment-generating function (mgf), 112–114, 121, 156, 597
Joint probability mass function (pmf), 121, 148, 152, 585, 599, 608
   definition, 96, 112
   function expectation, 100

K

Karhunen–Loéve expansion, 408, 533, 554
Kernel classification rule, 724–731
Kernels and classification
   annhilator, 743
   definition, 715
   density, 719–723
   estimation
      density estimator, 721
      histograms, 720
      optimal local bandwidth, 723
      plug-in, 720
      series estimation, 720–721
   Fourier, 717
   kernel plots, 742
   product kernels, 743
   smoothing
      definition, 715–716
      density estimation, 718
   statistical classification
      exponential kernel, 736
      Fisher linear, 735
      Gaussian radial kernel, 736
      Hilbert norm, 729
      Hilbert space, 725–727

Kernels and classification (*cont.*)
    inner product space, 728
    linearly separable data, 735
    linear span, 726–727, 729, 742
    Mercer's theorem, 732–744
    operator norm, 729
    parallelogram law, 727
    polynomial kernel, 736
    positive definiteness, 732
    Pythagorean identity, 727
    reproducing kernel, 725
    Riesz representation theorem, 730
    rule, 724–725
    sigmoid kernel, 736
    support vector machines, 734–736
    Weierstrass kernels, 719
Kolmogorov metric, 506, 511, 692, 693, 695, 696
Kolmogorov–Smirnov statistic
    invariance principle, 531–532
    McDiarmid and Devroye inequalities, 487–488
    null hypothesis testing, 545
    weighted, 534–535
Komlós, Major, Tusnady theorem
    rate, 537
    strong invariance principle, 425–427
Kullback–Leibler distance
    definition, 507
    inequality, 711, 712
Kurtosis
    coefficient, 82
    defined, 18
    skewness, 82

L

Lack of memory
    exponential densities, 55
    exponential distribution, 55–56
    geometric distribution, 32
Laguerre polynomials, 741
Landau–Shepp theorem, 577
Laplacian, 417
Large deviation, 51, 463, 483
    continuous time
        extreme statistics, 574
        finite-dimensional distribution (FDD), 575
        Gaussian process, 576–577
        metric entropy, supremum, 577–579
        Cramér–Chernoff theorem, 560–564
        Cramér's theorem, general set, 566–567

        Gärtner–Ellis theorem and Markov chain, 567–570
    Lipschitz functions and Talagrand's inequality
        correlated normals, 573–574
        outer parallel body, 572–573
        Taylor expansion, 572
    multiple testing, 559
    rate function, properties
        error probabilities, likelihood ratio test, 565–566
        shape and smoothness, 564
    self-normalization, 560
    t-statistic
        definition, 570
        normal case, 571–572
        rate function, 570
Law of iterated logarithm (LIL)
    application, 433
    Brownian motion, 419
    use, 267
Legendre polynomials, 741
Level sets
    fractal nature, 415–416
Lévy inequality, 397, 400
Likelihood function
    complete data, 708, 710
    definition, 148, 565
    log, 712
    shape, 153
Likelihood ratios
    convergence, 490–491
    error probabilities, 565–566
    Gibbs sampler, 645
    martingale formation, 467
    sequential tests, statistics, 469
Lindeberg–Feller theorem
    failure condition, 315
    IID variables, linear combination, 314–315
    Lyapounov's theorem, 311–312
Linear functional, 729, 730, 743
Linear span, 726, 727, 729, 742
Local limit theorem, 82
Local time
    Brownian motion
        Lebesgue measure, 419
        one-dimensional, 420
        zero, 420–421
Lognormal density
    finite mgf, 65
    skewness, 65
Loop chains, 369, 371

Subject Index

Lower semi-continuous
    definition, 564
Lyapounov inequality, 21–22, 52–53

M

Mapping theorem, 261, 262, 275, 327, 422,
    424, 694
    continuous, 269–272
    intensity measure, 452, 453
    lower-dimensional projections, 452
Marginal density
    bivariate uniform, 125–126
    function, 125
Marginal probability mass function (pmf), 117
    definition, 98
    $E(X)$ calculation, 110
Margin of error, 66, 91
Markov chain, 381, 463, 466, 500, 567–570,
    581, 613–686
    Chapman–Kolmogorov equation
        transition probabilities, 345–346
    communicating classes
        equivalence relation, 350
        identification, 350–351
        irreducibility, 349
        period computation, 351
    gambler's ruin, 352–353
    long run evolution and stationary
        distribution
        asymmetric random walk, 365–366
        Ehrenfest Urn, 364–365
        finite Markov chain, 361
        one-step transition probability matrix,
            360
        weak ergodic theorem, 366
    recurrence, transience and first passage
        times
        definition, 354
        infinite expectation, 355–358
        simple random walk, 354–355
Markov chain large deviation, 567–570
Markov chain Monte Carlo (MCMC) methods
    algorithms, 638
    convergence and bounds, 651–662
    general spaces
        Gibbs sampler, convergence, 670–673
        independent Metropolis scheme, 665
        independent sampling, 664
        Markov transition kernel, 662–663
        Metropolis chains, 664
        Metropolis schemes, convergence,
            666–670
        nonconventional multivariate
            distribution, 666
        practical convergence diagnostics,
            673–675
        random walk Metropolis scheme, 664
        simulation, t-distribution, 665
        stationary Markov chain, 663–664
        transition density, 663
    Gibbs sampler, 645–651
    Metropolis algorithm
        Barker and independent, 643
        beta–binomial distribution, 643–644
        independent sampling, 642–643
        Metropolis–Hastings algorithm, 643
        proposal distribution, 645
        transition matrix, 642
        truncated geometric distribution,
            644–645
    Monte Carlo
        conventional, 614
        ordinary, 615–624
    principle, 614
    reversible Markov chains
        discrete state space, 640–641
        ergodic theorem, 641–642
        irreducible and aperiodic stationary
            distribution, 640
        stationary distribution, 639–640
    simulation
        described, 613
        textbook techniques, 614–615, 624–636
    target distribution, 613, 637–638
Markov process
    definition, 404
    Ornstein–Uhlenbeck process, 431
    two-dimensional Brownian motion, 433
Markov's inequality
    description, 68
Markov transition kernel, 662–663
Martingale
    applications
        adapted to sequence, 464–465
        Bayes estimates, 467–468
        convex function theorem, 468
        gambler's fortune, 463–464
        Jensen's inequality, 468
        likelihood ratios, 467
        matching problem, 465–466
        partial sums, 465
        Pólya urn scheme, 466
        sums of squares, 465
        supermartingale, 464
        Wright–Fisher Markov chain, 466–467

Martingale (*cont.*)
    central limit theorem, 497–498
    concentration inequalities, 477–490
    convergence, 490–494
    Kolmogorov's SLLN proof, 496
    optional stopping theorem, 468–477
    reverse, 494–496
Martingale central limit theorem
    CLT, 498
    conditions
        concentration, 497
        Martingale Lindeberg, 498
Martingale Lindeberg condition, 498
Martingale maximal inequality, 477–478
Matching problem, 84, 465–466
Maximal inequality
    Kolmogorov's, 318, 396
    martingale and concentration, 477–480
    and symmetrization, 547–551
Maximum likelihood estimate (MLE)
    closed-form formulas, 705
    definition, 152–153
    endpoint, uniform, 153–154
    exponential mean, 153
    normal mean and variance, 153
McDiarmid's inequality
    Kolmogorov–Smirnov statistic, 487–488
    martingale decomposition and two-point distribution, 486–487
Mean absolute deviation
    definition, 17
    and mode, 30
Mean measure, 450, 557
Mean residual life, 160–161
Median
    CDF to PDF, 38
    definition, 9
    distribution, 55
    exponential, 55
Mercer's theorem, 732–744
Metric entropy, 545, 577–579
Metric inequalities
    Cauchy–Schwarz, 516
    variation *vs.* Hellinger, 517–518
Metropolis–Hastings algorithm
    beta–binomial distribution, 644
    described, 643
    SLEM, 656–657
    truncated geometric, 659
Midpoint process, 462
Midrange, 245
Mills ratio, 160, 333
Minkowski's inequality, 21

Minorization methods
    Bernoulli experiment, 661–662
    coupling, 661
    drift condition, 660
    Harris recurrence, 660
    hierarchical Bayes linear models, 660–661
    nonreversible chains, 659–660
Mixed distribution, 88, 434
Mixtures, 39, 214, 300, 307, 611, 661, 721, 739
MLE. *See* Maximum likelihood estimate
Mode, 30–31, 89, 91, 232, 252, 282, 552, 627, 628, 714
Moment generating function (MGF)
    central moment, 25–26
    Chernoff–Bernstein inequality, 51–52
    definition, 22
    distribution-determining property, 24–25
    inversion, 53
    Jensen's inequality, 52–53
    Poisson distribution, 23–24
    standard exponential, 51
Moment generating function (mgf), 116, 121
Moment problem, 277–278
Moments
    continuous random variables, 45–46
    definitions, 46–47
    standard normal, 48–49
Monotone convergence theorem, 491, 492
Monte Carlo
    bootstrap, 691–692
    conventional, 614–615
    EM algorithm, 706
    Markov chain, 637–645
    ordinary
        Bayesian, 618–619
        computation, confidence intervals, 616
        P-values, 622–623
        Rao–Blackwellization, 623–624
Monte Carlo P-values
    computation and application, 622–623
    statistical hypothesis-testing problem, 622
Moving Block Bootstrap (MBB), 702
Multidimensional densities
    bivariate normal
        conditional distributions, 154–155
        five-parameter, density, 136, 138–139
        formulas, 138, 155–158
        mean and variance, 139–140
        normal marginals, 140
        singular, 137
        variance–covariance matrix, 136–137
    Borel's paradox, 158–161

Subject Index

conditional density and expectations, 140–147
posterior densities, likelihood functions and Bayes estimates, 147–152
Multinomial distribution, 243, 542, 599–600
  MGF, 116
  pmf, 114
  and Poisson process, 451–452
Multinomial expansion, 116
Multinomial maximum distribution
  maximum cell frequency, 243
Multiparameter exponential family
  assumption, 596–597
  definition, 596, 597
  Fisher information matrix definition, 597
  general multivariate normal distribution, 598–599
  multinomial distribution, 599–600
  two-parameter
    gamma, 598
    inverse gaussian distribution, 600
    normal distribution, 597–598
Multivariate Cauchy, 196
Multivariate CLT, 267
Multivariate Jacobian formula, 177–178
Multivariate normal
  conditional distributions
    bivariate normal case, 202
    independent, 203
    quadrant probability, 204–205
  definition and properties
    bivariate normal simulation, 200, 201
    characterization, 201–202
    density, 200
    linear transformations density, 200–201
  exchangeable normal variables, 205–207
  inequalities, 214–216
  noncentral distributions
    chi-square, 214
    t-distribution, 213

N
Natural parameter
  definition, 589, 599
  space, 589, 590, 596, 598
  and sufficient statistic, 600
Natural sufficient statistic
  definition, 585, 596
  one-parameter exponential family, 595–596
Nearest neighbor, 462, 554, 556
Negative
  binomial distribution, 28–29, 609
  drift, 427–428
Neighborhood recurrent, 394, 418, 419
Noncentral chi square distribution, 214, 218

Noncentral F distribution, 210
Noncentral t-distribution, 213, 218
Non-lattice, 701
Nonreversible chain, 659–660
Normal approximation
  binomial
    confidence interval, 74–76
Normal density
  CDF, 41–42
  definition, 40
Normalizing
  density function, 39–40
Normal order statistics, 228
Null recurrent
  definition, 356
  Markov chain, 373

O
Operator norm, 729
Optimal local bandwidth, 723
Optional stopping theorem
  applications
    error probabilities, Wald's SPRT, 476–477
    gambler's ruin, 475
    generalized Wald identity, 475–476
    hitting times, random walk, 475
    Wald identities, 474–475
  stopping times
    defined, 469
    sequential tests and Wald's SPRT, 469
Order statistics
  basic distribution theory, 221–222
  Cauchy, 232
  conditional distributions, 235–236
  density, 225
  description, 221
  existence of moments, 230–232
  exponential, 227–228
  joint density, 222–223
  moments, uniform, 227
  normal, 228–229
  quantile transformation, 229–230
  and range, 226–227
  spacings, 233–235
  uniform, 223–224
Orlicz norms, 556–557
Ornstein–Uhlenbeck process
  convergence, 430–431
  covariance function, 429
  Gaussian process, 430
Orthonormal bases, Parseval's identity, 728–729
Orthonormal system, 741

P

Packing number, 548
Parallelogram law, 727
Pareto density, 60
Partial correlation, 204
Partial sum process
    convergence, Brownian motion, 423–424
    interpolated, 425
    strong invariance principle, 537
Pattern problems
    discrete probability, 26
    recursion relation, 27
    variance formula, 27
Period
    burn-in, 675
    circular block bootstrap (CBB), 702
    computation, 351
    intensity function, 454, 455
Perron–Frobenius theorem, 361–363, 569, 654
Plancherel's identity, 298–299
Plug-in estimate, 212, 630, 704, 720, 740
Poincaré's lemma, 191
Poisson approximation
    applications, 35
    binomial random variable, 34
    confidence intervals, 80–82
Poisson distribution
    characteristic function, 295–296
    confidence interval, 80–81
    moment-generating function, 23–24
Poissonization, 116–118, 121, 243, 244
Poisson point process
    higher-dimensional
        intensity/mean measure, 450
        Mapping theorem, 452–453
        multinomial distribution, 451–452
        Nearest Event site, 451
    Nearest Neighbor, 462
    polar coordinates, 460
Poisson process, 551–553, 557, 636, 682
    Campbell's theorem and shot noise
        characteristic functional Π, 456–457
        shot effects, 456
        stable laws, 458
    1-D nonhomogeneous processes
        intensity and mean function, plots, 455
        mapping theorem, 453–454
        piecewise linear intensity, 454–456
    higher-dimensional Poisson point
        processes
        distance, nearest event site, 451
        mapping theorem, 452–453
        stationary/homogeneous, 450

Polar coordinates
    spherical calculations
        definition, 182–183
        dimensionality curse, 184
        joint density, 183–184
        spherically symmetric facts, 185
    two dimensions
        n uniforms product, 181–182
        polar transformation usefulness, 181
        transformation, 180–181
        use, 134–135
Polar transformation
    spherical calculations, 182–185
    use, 181
Pólya's criterion
    characteristic functions, 308
    inversion theorem failure, 307–308
    stable distributions, 307
Pólya's formula, 382–383
Pólya's theorem
    CDF, 265, 509
    return probability, 382–383
Pólya's urn, 466, 493
Polynomial kernel, 736
Portmanteau theorem
    definition, 265
Positive definiteness, 568, 717, 732
Positive recurrent, 356–357, 361, 363, 365, 651
Posterior density
    definition, 147–148
    exponential mean, 148–149
    normal mean, 151–152
    Poisson mean, 150
Posterior mean
    approximation, 651
    binomial, 149–150
Prior density
    definition, 148
    uses, 151
Probability metrics
    differential metrics, 519–522
    metric inequalities, 515–518
    properties
        coupling identity, 508
        f-divergences, 513–514
        Hellinger distance, 510–511
        joint and marginal distribution
            distances, 509–510
    standard probability, statistics
        f-divergences, 507
        Hellinger metric, 506–507
        Kolmogorov metric, 506
        Kullback–Leibler distance, 507

Subject Index

Lévy–Prokhorov metric, 507
 separation distance, 506
 total variation metric, 506
 Wasserstein metric, 506
Product kernels., 743
Product martingale, 499
Projection formula, 742
Prophet inequality
 application, 400
 Bickel, 397
 Doob–Klass, 397
Proportions, convergence of EM, 282, 711–713
Proposal distribution, 645, 664, 665
Pythagorean identity, 727

Q

Quadrant probability, 204
Quadratic exponential family, 612
Quadratic forms distribution
 Fisher Cochran theorem, 211
 Hotelling's T^2 statistic, 209–210
 independence, 211–212
 spectral decomposition theorem, 210
Quantile process
 normalized, 528–529
 restricted Glivenko–Cantelli property, 536
 uniform, 529
 weak invariance principle, 536
Quantile transformation
 accept–reject method, 625
 application, 242
 Cauchy distribution, 44
 closed-form formulas, 625
 definition, 44
 theorem, 229–230
Quartile ratio, 335

R

Random permutation, 636
Random scan Gibbs sampler, 646, 647, 672
Random walks
 arc sine law, 386
 Brownian motion, 402, 403
 Chung–Fuchs theorem, 394–396
 cubic lattice
  distribution theory, 378–379
  Pólya's formula, 382–383
  recurrence and transience, 379–381
  recurrent state, 377
  three dimensions, 375
  two simulated, 376, 377

Erdös–Kac generalization, 390
first passage time
 definition, 383
 distribution, return times, 383–386
inequalities
 Bickel, 397
 Hájek–Rényi, 396
 Kolmogorov's maximal, 396
 Lévy and Doob–Klass Prophet, 397
 Sparre–Andersen generalization, 389, 390
Wald's identity
 stopping time, 391, 392
Rao–Blackwellization
 Monte Carlo estimate, 623–624
 variance formula, 623
Rao's geodesic distance, 522
Rate function, 52
Ratio estimate, 630
Real analytic, 564–565
Records
 density, values and times, 240–241
 interarrival times, 238
Recurrence, 354–359, 379–381, 383, 385, 393, 396, 399, 418–419, 472, 660, 667, 668
Recurrent state, 355, 357–359, 370, 373, 377, 394
Recurrent value, 393
Reflection principle, 410–411, 432
Regular chain
 communicating classes, 349
 irreducible, 363
Regular variation, 332, 333
Reproducing kernel, 725–731, 733, 735, 743
Resampling, 543, 689, 690, 691, 692, 701
Restricted Glivenko–Cantelli property, 536
Return probability, 382–383
Return times
 distribution, 383
 scaled, asymptotic density, 385
Reverse martingale
 convergence theorem, 496
 defined, 494
 sample means, 494–495
 second convex function theorem, 495–496
Reverse submartingale, 494–496
Reversible chain, 373, 640, 645, 652, 685
Réyni's representation, 233–234
Riemann–Lebesgue lemma, 302, 303
Riesz representation theorem, 730, 743
RKHS, 743
Rosenthal inequality, 318–319, 322

S

Sample maximum, 224, 235, 274, 276, 287, 337, 603–604
  asymptotic distribution, 330
  Cauchy, mode, 232
  domain of attraction, 336
Sample points
  concept, 3
Sample range, 337
Sample space
  countable additivity, 2–3
  countably infinite set, 2
  definition, 2
  inclusion–exclusion formula, 5
Sauer's lemma, 539–540
Schauder functions, 408
Scheffés theorem, 284–285
Score confidence interval, 74, 76
Second Aresine law, 410
Second largest eigenvalue in modulus (SLEM)
  approximation, 686
  calculation, 685
  convergence, Gibbs sampler, 671
  vs. Dobrushin coefficients, 686
  Metropolis–Hastings algorithm, 656–657
  transition probability matrix, 651–652
Second order accuracy, 699–701
Separation distance, 506, 653
Sequential probability ratio test (SPRT), 469, 476–477
Sequential tests, 391, 469
Series estimate, 720–721
Shattering coefficient, 539, 540, 542
Shot noise, 456–458
Sidak inequality, 215
Sigmoid kernel, 736
Simple random walk, 354, 375, 393, 395, 397, 399, 474
  mathematical formulation, 345
  periodicity and, 369
  simulated, 376, 377
Simulation
  algorithms, common distributions, 634–636
  beta-binomial distribution, 643–644
  bivariate normal, 137, 201
  textbook techniques
    accept–reject method, 625–628
    quantile transformation, 624–625
  truncated geometric distribution, 644–645
Skewness
  defined, 18
  normal approximation, 82
Skorohod embedding, 422, 423

SLEM. *See* Second largest eigenvalue in modulus
Slepian's inequaliity, 214
SLLN. *See* Strong law of large numbers
Slutsky's theorem
  application, 269, 270
Spacings
  description, 233
  exponential and Réyni's representation, 233–234
  uniform, 234–235
Spatial Poisson process, 459
Spectral bounds
  decomposition, 655–656
  Perron–Frobenius theorem, 653–654
  SLEM, Metropolis–Hastings algorithm, 656–657
Spherically symmetric, 135, 180, 182–186, 191
SPRT. *See* Sequential probability ratio test
Stable laws
  index 1/2, 385–386
  infinite divisibility
    description, 315
    distributions, 317
    triangular arrays, 316
  and Poisson process, 458
Stable random walk, 395
Standard discrete distributions
  binomial, 28
  geometric, 28
  hypergeometric, 29
  negative binomial, 28–29
  Poisson, 29–34
Stationary, 152, 153, 163, 404, 406, 430, 431, 440, 449, 450, 466, 569, 575, 576, 614, 639, 640, 641, 645–648, 651–656, 658–664, 667, 672, 674, 684, 685, 701–703, 711, 713
Stationary distribution, 614, 639, 640, 646–648, 651, 653, 658, 659, 661, 663, 667, 672, 674, 684, 685
Statistical classification, kernels
  exponential kernel, 736
  Fisher linear, 735
  Gaussian radial kernel, 736
  Hilbert norm, 729
  inner product space, 728
  linear functional, 729, 730
  linearly separable data, 735
  linear span, 726–727, 729, 742
  Mercer's theorem, 732–744
  orthonormal bases, 728–729
  parallelogram law, 727

Subject Index

polynomial kernel, 736
Pythagorean identity, 727
Riesz representation theorem, 730
sigmoid kernel, 736
support vector machines, 734–736
Statistics and machine learning tools
    bootstrap
        consistency, 692–699
        dependent data, 701–704
        higher-order accuracy, 699–701
    EM algorithm
        ABO allele, 709–711
        modifications, 714
        monotone ascent and convergence, 711–713
        Poisson, missing values, 706–707
    kernels
        compact support, 718
        Fourier, 717
        Mercer's theorem, 722–734
        nonnegativity, 717
        rapid decay, 718
        smoothing, 715–717
Stein's lemma
    characterization, 70
    normal distribution, 66
    principal applications, 67–68
Stochastic matrix, 340, 654
Stochastic process
    continuous-time, 422
    d-dimensional, 418
    definition, 403
    standard Wiener process, 404
Stopped martingale, 472
Stopping time
    definition, 391, 478
    optional stopping theorem, 470–471
    sequential tests, 469
    Wald's SPRT, 469
Strong approximations, 530–531, 537
Strong ergodic theorem, 641
Strong invariance principle
    and KMT theorem, 425–427
    partial sum process, 530–531, 537
Strong law of large numbers (SLLN)
    application, 261
    Cramér-Chernoff theorem, 579
    Kolmogorov's, 258, 496, 615, 632
    Markov chain, 640
    proof, 496
    Zygmund–Marcinkiewicz, 693–695
Strong Markov property, 409, 411, 423
Submartingale
    convergence theorem, 491–492

convex function theorem, 468
nonnegative, 477–478
optional stopping theorem, 470–471
reverse, 494–496
upcrossing inequality, 489
Sufficiency
    Neyman–Fisher factorization and Basu's theorem
        definition, 602
        general, 604
Sum of uniforms
    distribution, 77
    normal density, 78, 79
Supermartingale, 464, 491, 494
Support vector machine, 734–736
Symmetrization, 170–172, 319, 547–551
Systematic scan Gibbs sampler
    bivariate distribution, 647–648
    change point problem, 650–651
    convergence property, 672
    defined, 646
    Dirichlet distribution, 650

T

Tailsum formula, 16–19, 86, 357
Target distribution
    convergence, MCMC, 651–652
    Metropolis chains, 664
Taylor expansion, characteristic function, 303
t confidence interval, 187–188, 290–291
t-distribution
    noncentral, 213, 218
    simulation, 665, 668
    student, 175–176, 187
Time reversal, 412
Total probability, 5, 6, 117
Total variation distance
    closed-form formula, 283
    definition, 282
    and densities, 282–283
    normals, 283–284
Total variation metric, 506, 509
Transformations
    arctanh, 274
    Box–Mueller, 194
    Helmert's, 186–187
    linear, 200–201
    multivariate Jacobian formula, 177–178
    n-dimensional polar, 191
    nonmonotone, 45
    polar coordinates, 180–185
    quantile, 44–45, 229–232, 624–628
    simple linear, 42–43
    variance stabilizing, 272–274, 291

Transience, 354–359, 379–381, 394, 399, 418–419
Transition density
  definition, 428, 662–663
  drift-diffusion equation, 429
  Gibbs chain, 672
Transition probability
  Markov chain, 341
  matrix, 340
  nonreversible chain, 684
  n-step, 346
  one-step, 344
Triangular arrays, 316
Triangular density
  definition, 39–40
  piecewise linear polynomial., 77

U
Uniform density
  basic properties, 54
Uniform empirical process, 528, 530, 531, 552, 553
Uniform integrability
  conditions, 276–277
  dominated convergence theorem, 275–276
Uniformly ergodic, 653, 669–670
Uniform metric, 422, 424, 425
Uniform order statistics, 223–225
  joint density, 235
  joint distribution, 230
  moments, 227
Uniform spacings, 234–235
Unimodal
  properties, 44
Unimodality, order statistics, 39, 246
Universal Donsker class, 544
Upcrossing, 488–490
Upcrossing inequality
  discrete time process, 488
  optional stopping theorem, 490
  stopping times, 488–489
  submartingale and decomposition, 489

V
Vapnik–Chervonenkis (VC)
  dimension, 539–541
  subgraph, 545, 546

Variance
  definition, 17, 46
  linear function, 68–69
  mean and, 19
Variance stabilizing transformations (VST)
  binomial case, 273–274
  confidence intervals, 272
  Fisher's, 274
  unusual, 274
Void probabilities, 450
VST. See Variance stabilizing transformations

W
Wald confidence interval, 74
Wald's identity
  application, 399
  first and second, 474–477
  stopping time, 391, 392
Wasserstein metric, 506, 693
Weak law of large numbers (WLLN), 20, 256, 257, 269, 270, 305
Weakly stationary, 404
Weibull density, 56
Weierstrass's theorem, 265–266, 268, 742
Weighted empirical process
  Chibisov–O'Reilly theorem, 534–535
  test statistics, 534
Wiener process
  continuous, 576, 577
  increments, 581
  standard, 404, 405
  tied down, 405
Wishart distribution, 207–208
Wishart identities, 208–209
WLLN. See Weak law of large numbers
Wright–Fisher Markov chain, 466–467

Y
Yu–Mykland single chain method, 674–675

Z
Zygmund–Marcinkiewicz SLLN, 693–694